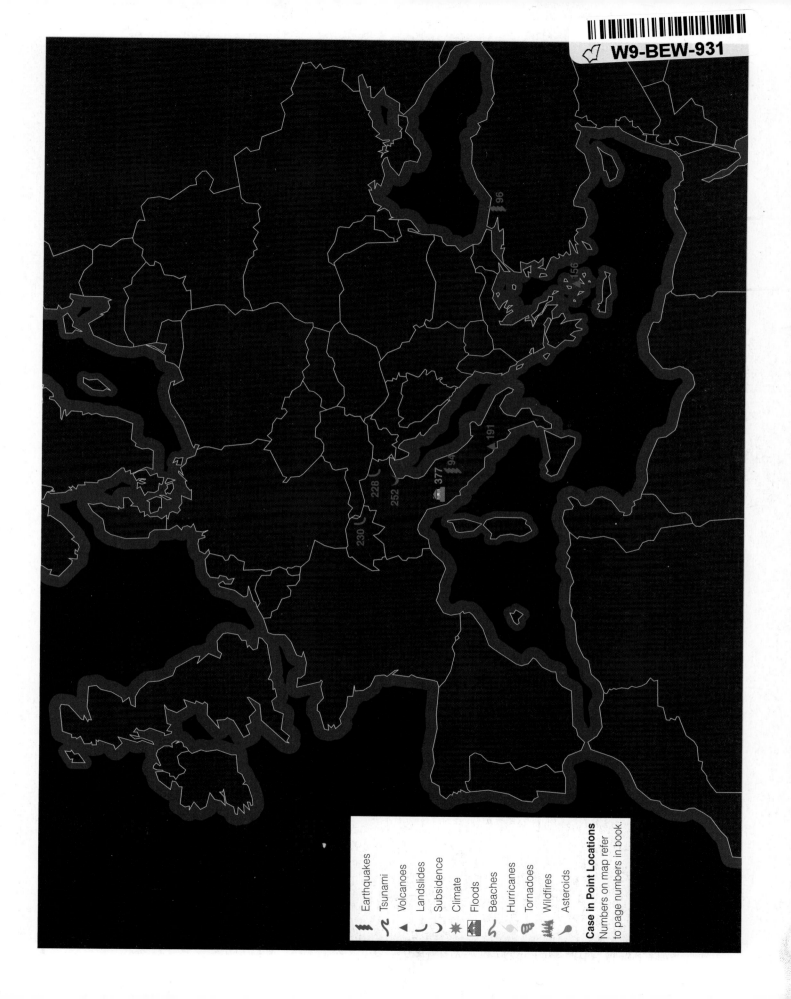

96

156

191

377

94

228

252

230

Earthquakes

Tsunami

Volcanoes

Landslides

Subsidence

Climate

Floods

Beaches

Hurricanes

Tornadoes

Wildfires

Asteroids

Case in Point Locations
Numbers on map refer
to page numbers in book.

Natural Hazards and Disasters

Third Edition

Natural Hazards and Disasters

Third Edition

Donald Hyndman
University of Montana

David Hyndman
Michigan State University

BROOKS/COLE
CENGAGE Learning™

Australia • Brazil • Japan • Korea • Mexico • Singapore • Spain • United Kingdom • United States

BROOKS/COLE
CENGAGE Learning™

Natural Hazards and Disasters: Third Edition
Donald Hyndman, David Hyndman

Publisher: Yolanda Cassio

Acquisitions Editor: Laura Pople

Developmental Editor: Rebecca Heider

Assistant Editor: Samantha Arvin

Editorial Assistant: Kristina Chiapella

Media Editor: Alexandria Brady

Marketing Manager: Nicole Mollica

Marketing Assistant: Kevin Carroll

Content Project Management: PreMediaGlobal

Design Director: Rog Hugel

Art Director: John Walker

Print Buyer: Karen Hunt

Rights Acquisitions Specialist, Text: Roberta Broyer

Rights Acquisitions Specialist, Images: Don Schlotman

Production Service: PreMediaGlobal

Text Designer: PreMediaGlobal

Photo Researcher: PreMediaGlobal

Cover Designer: Brian Salisbury

Cover image credt: Code Red (Getty Images)

Compositor: PreMediaGlobal

For product information and technology assistance, contact us at
Cengage Learning Customer & Sales Support, 1-800-354-9706.
For permission to use material from this text or product,
submit all requests online at **www.cengage.com/permissions.**
Further permissions questions can be e-mailed to
permissionrequest@cengage.com

Library of Congress Control Number: 2010923645

Student Edition:

ISBN-13: 978-0-538-73752-4

ISBN-10: 0-538-73752-2

Instructor's Edition:

ISBN-13: 978-0-538-73753-1

ISBN-10: 0-538-73753-0

Brooks/Cole
20 Davis Drive
Belmont, CA 94002-3098
USA

Cengage Learning is a leading provider of customized learning solutions with office locations around the globe, including Singapore, the United Kingdom, Australia, Mexico, Brazil, and Japan. Locate your local office at **www.cengage. com/global**.

Cengage Learning products are represented in Canada by Nelson Education, Ltd.

To learn more about Brooks-Cole, visit www.cengage.com/Brooks-Cole Purchase any of our products at your local college store or at our preferred online store **www.CengageBrain.com**.

Printed in Canada
2 3 4 5 14 13 12 11

To Shirley and Teresa
for their endless encouragement and patience

About the Authors

DONALD HYNDMAN is an emeritus professor in the department of geology at the University of Montana, where he has taught courses in natural hazards and disasters, regional geology, igneous and metamorphic petrology, volcanology, and advanced igneous petrology. He continues to lecture on natural hazards. Donald is co-originator and co-author of six books in the Roadside Geology series and one on the geology of the Pacific Northwest; he is also the author of a textbook on igneous and metamorphic petrology. His B.S. in geological engineering is from the University of British Columbia, and his Ph.D. in geology is from the University of California, Berkeley. He has received the Distinguished Teaching Award and the Distinguished Scholar Award, both given by the University of Montana. He is a Fellow of the Geological Society of America.

DAVID HYNDMAN is a professor in and the chairman of the department of geological sciences at Michigan State University, where he teaches courses in natural hazards and disasters, the dynamic Earth, and advanced hydrogeology. His B.S. in hydrology and water resources is from the University of Arizona, and his M.S. in applied earth sciences and Ph.D. in geological and environmental sciences are from Stanford University. David was selected as a Lilly Teaching Fellow and has received the Ronald Wilson Teaching Award. He was the 2002 Darcy Distinguished Lecturer for the National Groundwater Association and is a Fellow of the Geological Society of America.

Brief Contents

Appendix 1: Geological Time Scale, and Appendix 2: Mineral and Rock Characteristics Related to Hazards, can be found online at www.cengagebrain.com/shop/ISBN/0538737522.

Contents

Appendix 1: Geological Time Scale, and Appendix 2: Mineral and Rock Characteristics Related to Hazards, can be found online at www.cengagebrain.com/shop/ISBN/0538737522.

Preface

The further you are from the last disaster, the closer you are to the next.

Why We Wrote This Book

In teaching large introductory environmental and physical geology courses for many years—and, more recently, natural hazards courses—it has become clear to us that topics involving natural hazards are among the most interesting for students. Thus, we realize that employing this thematic focus can stimulate students to learn basic scientific concepts, to understand how science relates to their everyday lives, and to see how such knowledge can be used to help mitigate both physical and financial harm. For all of these reasons, natural hazards and disasters courses should achieve higher enrollments, have more interested students, and be more interesting and engaging than those taught in a traditional environmental or physical geology framework.

A common trend is to emphasize the hazards portions of physical and environmental geology texts while spending less time on subjects that do not engage the students. Students who previously had little interest in science can be awakened with a new curiosity about Earth and the processes that dramatically alter it. Science majors experience a heightened interest, with expanded and clarified understanding of natural processes. In response to years of student feedback and discussions with colleagues, we have reshaped our courses to focus on natural hazards.

Students who take a natural hazards course greatly improve their knowledge of the dynamic Earth processes that will affect them throughout their lives. They should be able to make educated choices about where to live and work. Perhaps some who take this course will become government officials or policy makers who can change some of the current culture that contributes to major losses from natural disasters.

Undergraduate college students, including nonscience majors, should find the writing clear and stimulating. Our emphasis is to provide them a basis for understanding important hazard-related processes and concepts. This book encourages students to grasp the fundamentals while still appreciating that most issues have complexities that are beyond the current state of scientific knowledge and involve societal aspects beyond the realm of science. Students not majoring in the geosciences may find motivation to continue studies in related areas and to share these experiences with others.

Natural hazards and disasters can be fascinating and even exciting for those who study them. Just don't be on the receiving end!

Living with Nature

Natural hazards, and the disasters that accompany many of them, are an ongoing societal problem. We continue to put ourselves in harm's way, through ignorance or a naïve belief that a looming hazard may affect others but not us. We choose to live in locations that are inherently unsafe.

The expectation that we can control nature through technological change stands in contrast to the fact that natural processes will ultimately prevail. We can choose to live *with* nature or we can try to fight it. Unfortunately, people who choose to live in hazardous locations tend to blame either "nature on the rampage" or others for permitting them to live there. People do not often make such poor choices willfully, but rather through their lack of awareness of natural processes. Even when they are aware of an extraordinary event that has affected someone else, they somehow believe "it won't happen to me." These themes are revisited throughout the book, as we relate principles to societal behavior and attitudes.

People often decide on their residence or business location based on a desire to live and work in scenic environments without understanding the hazards around them. Once they realize the risks, they often compound the hazards by attempting to modify the environment. Students who read this book should be able to avoid such decisions. Toward the end of the course, our students sometimes ask, "So where is a safe place to live?" We often reply that you can choose hazards that you are willing to deal with and live in a specific site or building that you know will minimize impact of that hazard.

It is our hope that by the time students have finished reading this textbook, they should have the basic knowledge to evaluate critically the risks they take and the decisions they make as voters, homeowners, and world citizens.

Our Approach

This text begins with an overview of the dynamic environment in which we live and the variability of natural processes, emphasizing the fact that most daily events are small and generally inconsequential. Larger events are less frequent, though most people understand that they can happen. Fortunately, giant events are infrequent;

regrettably, most people are not even aware that such events can happen. Our focus here is on Earth and atmospheric hazards that appear rapidly, often without significant warning.

The main natural hazards covered in the book are earthquakes and volcanic eruptions; extremes of weather, including hurricanes; and floods, landslides, tsunami, wildfires, and asteroid impacts. For each, we examine the nature of the hazard, the factors that influence it, the dangers associated with the hazard, and the methods of forecasting or predicting such events. Throughout the book, we emphasize interrelationships between hazards, such as the fact that building dams on rivers often leads to greater coastal erosion. Similarly, wildfires generally make slopes more susceptible to floods, landslides, and mudflows.

The book includes chapters on dangers generated internally, including earthquakes, tsunami, and volcanic eruptions. Society has little control over the occurrence of such events but can mitigate their impacts through a deeper understanding that can afford more enlightened choices. The landslides section addresses hazards influenced by a combination of in-ground factors and weather, a topic that forms the basis for many of the following chapters. A chapter on sinkholes, subsidence, and swelling soils addresses other destructive in-ground hazards that we can, to some extent, mitigate and that are often subtle yet highly destructive.

The following hazard topics depend on an understanding of the dynamic variations in climate and weather, so we begin with a chapter to provide that background and an overview of global warming. Chapters on streams and floods begin with the characteristics and behavior of streams and how human interaction affects both a stream and the people around it. Chapters follow on wave and beach processes, hurricanes and Nor'easters, thunderstorms and tornadoes, and wildfires. The final chapters discuss asteroid impacts and future concerns related to natural hazards. Appendixes present the geological time scale and a brief discussion of the nature of rocks and minerals, primarily as background for some of the physical hazards.

The book is up-to-date and clearly organized, with most of its content derived from current scientific literature and from our own personal experience. It is packed with relevant content on natural hazards, the processes that control them, and the means of avoiding catastrophes. Numerous excellent and informative color photographs, many of them our own, illustrate scientific concepts associated with natural hazards. The diagrams are clear, straightforward, and instructive.

Extensive illustrations and Case in Point examples bring reality to the discussion of principles and processes. These cases tie the process-based discussions to individual cases and integrate relationships between them. They emphasize the natural processes and human factors that affect disaster outcomes. Illustrative cases are interwoven with topics as they are presented. We attempt to provide balanced coverage of natural hazards across North America and the rest of the world. As our global examples illustrate, although the same fundamental processes lead to natural hazards around the world, the impact of natural disasters can be profoundly different depending on factors such as economic conditions, security, and disaster preparedness.

End-of-Chapter material also includes Critical View Photos with paired questions, a list of Key Points, Key Terms, Questions for Review, and Discussion Questions.

New to the Third Edition

With such a fast-changing and evolving subject as natural hazards, we have extensively revised and added to the content, with emphasis not only on recent events but on those that best illustrate important issues. We have endeavored to keep breaking material as up-to-date as possible, both with new Cases in Point and in changes in governmental policy that affect people and their hazardous environments. New hazard maps help the reader quickly determine the locations of important events, including those of Cases in Point.

In response to feedback from instructors, we have made changes throughout to strengthen key pedagogical features of the book:

- The Critical View feature has been revised so that photos are now larger.and each photo now includes a descriptive caption and questions specifically related to the photo. This important feature promotes both critical thinking and visual learning.

- The art program has been significantly revised to enhance visual learning. New figure titles show students the relevance of each piece at first glance. More labels have been added to photos to indicate key features. Photos are more often paired with diagrams to encourage students to relate real-world and scientific views. There are more multi-part figures, setting up constructive comparisons between related figures. Throughout the text, art and captions have been revised to ensure a close and clear connection between the text and art.

- Much improved shaded relief maps replace most earlier maps to better illustrate and permit recognition of environments of interest.

- End-of-chapter material now includes a new category of Discussion Questions. These questions will help students apply concepts from the chapter to difficult real-life situations such as zoning and development issues, the balance between personal freedoms and government protections, or the cost and benefits of hazard mitigation.

- Key Concepts and Review Questions have been expanded to include more key information from each chapter.

- Explanations of key processes have been closely edited to improve clarity.

In addition to these overall changes, some significant additions to individual chapters include the following:

- **Chapter 1, Natural Hazards and Disasters,** includes a new review of the percentages of fatalities from each of the major natural hazards, emphasizing that heat and drought, perhaps surprisingly, constitute by far the greatest numbers.

- **Chapter 2, Plate Tectonics and Physical Hazards,** has been reorganized to present the essentials of Plate Tectonics before reviewing the Earth features and processes that led to development of the theory itself. Students can better understand what led to this fundamental theory when they can picture its key aspects.

- **Chapters 3** and **4, Earthquakes and Their Causes** and **Earthquake Prediction and Mitigation,** now include a much amplified discussion of the fascinating prediction of the Haicheng, China, earthquake, including the key people involved and decisions by government agencies that led to warnings that saved thousands of lives. Chapter 3 begins with a detailed discussion of the tragic 2010 Haiti earthquake that killed more than 230,000. New Cases in Point include the 2009 L'Aquila, Italy, earthquake; and the 2008 Wenchuan earthquake that killed almost 90,000.

- **Chapter 5, Tsunami,** now includes coverage of the 2009 Samoa tsunami, the 2010 Chile tsunami, and the 2010 Samoa earthquake and tsunami. Potential tsunami from the Puerto Rico Trench and its effect on eastern U.S. coasts has been added, as well as new information on the pending Cascadia earthquake and tsunami and its effect on coastal communities. We have also added new discussion of towns on the Oregon coast, their vulnerability to tsunami, and what they are doing and can do about specific hazards.

- **Chapters 6** and **7, Volcanoes: Tectonic Environments and Eruptions** and **Volcanoes: Hazards and Mitigation,** includes the 2010 mid-Atlantic Rift eruption of a volcano in Iceland and its dramatic effect on aircraft flights in Europe. We have also added new discussion of likely warnings of a new eruption from Yellowstone Volcano.

- **Chapter 8, Landslides and Other Downslope Movements,** now includes amplified discussion of avalanches, the increasing numbers of people affected by them, and the reasons for the increase. We added coverage of a terrible 2009 landslide in Utah, where lives were claimed from a known hazard, and lessons that should be learned from the event.

- **Chapter 9, Sinkholes, Land Subsidence, and Swelling Soils,** includes discussion of some new sinkholes that affected homes and extended coverage of sinking ground and damage to buildings as Mexico City grapples with its water supply and waste-water problems. We added discussion of major moveable barriers to encroaching seas that threaten low-lying populous areas in an era of rising sea levels.

- **Chapter 10, Climate Change and Weather Related to Hazards,** is extensively updated. Major additions include additional information on the North Atlantic Oscillation, on drought and widespread bark-beetle infestations of trees, and on winter blizzards, along with a massive ice storm in the Midwest. Climate change is newly addressed with the carbon cycle, long-term issues for increasing levels of carbon dioxide and its consequences, along with possible means of minimizing that increase, cap-and-trade, carbon sequestration, wind turbines, and other non-fossil power sources.

- **Chapters 11** and **12, Streams and Flood Processes** and **Floods and Human Interactions,** are reorganized and now include extensive new coverage of the 2008 floods in the upper Mississippi Valley and a new 2009 flood event initiated by winter freezing and spring thaw on the Red River in North Dakota.

- **Chapter 13, Waves, Beaches, and Coastal Erosion,** has new, improved photos and artwork that amplify and clarify these important processes.

- **Chapter 14, Hurricanes and Nor'easters,** now includes an analysis of storm-surge outwash erosion; accounts of the 2009 Typhoon Morakot in Taiwan; analysis of the 2008 Typhoon Nargis that killed tens of thousands in Myanmar; the 2008 damage by Hurricane Gustav that wreaked additional destruction in New Orleans three years after Katrina; along with extensive and exclusive personal coverage of 2008 Hurricane Ike that devastated areas near Galveston, Texas. A new Case in Point addresses the hazards and future prospects for barrier islands of the North Carolina Outer Banks.

- **Chapter 15, Thunderstorms and Tornadoes,** includes updated coverage and new photos.

- **Chapter 16, Wildfires,** now includes the 2009 wildfires that blackened fire-prone coastal areas of southern California. A new Case in Point presents the 1991 Oakland–Berkeley Hills firestorm, how it developed, and lessons learned that can be applied to other areas of urban-wildland interface. The 2008 lightning-triggered wildfires that ravaged northern California provided a vivid reminder of that event, along with updates for the 2007 to 2009 wildfires in southern California.

- **Chapter 17, Impact of Asteroids and Comets,** is extensively updated and improved.

- **Chapter 18, The Future: Where Do We Go From Here?,** has been significantly amplified with discussion of persistent natural hazard and disaster problems.

- **Appendix 2, Rocks,** has new and improved photos of representative rocks.

Acknowledgments

We are grateful to a wide range of people for their assistance in preparing and gathering material for this book, far too many to list individually here. However, we particularly appreciate the help we received from the following:

- We especially wish to thank Rebecca Heider, Development Editor for Brooks/Cole, who not only expertly managed and organized all aspects of the second and third editions but suggested innumerable and important changes in the manuscript. In large measure, the enhancements are in response to her insight, perception, and skillful editing. Beth Kluckhohn, Senior Project Manager at PreMediaGlobal, and her crew skillfully and cheerfully organized the artwork and assembled everything into the final book.

- For editing and suggested additions: Dr. Dave Alt, University of Montana, emeritus; Ted Anderson; Tony Dunn (San Francisco State University); Shirley Hyndman; Teresa Hyndman; Dr. Duncan Sibley, Dr. Kaz Fujita, and Dr. Tom Vogel from Michigan State University; Dr. Kevin Vranes, University of Colorado, Center for Science and Technology Policy Research; Peter Adams, executive editor for Earth Sciences for Brooks/Cole during the second edition.

- For information and photos on specific sites: Dr. Brian Atwater, USGS; Dr. Rebecca Bendick, University of Montana; Karl Christians, Montana Department of Natural Resources and Conservation; Susan Cannon, USGS; Jack Cohen, Fire Sciences Laboratory, U.S. Forest Service; Dr. Joel Harper, University of Montana; Bretwood Hickman, University of Washington; Peter Bryn, Hydro.com (Storegga Slide); Dr. Dan Fornari, Woods Hole Oceanographic Institution, MA; Dr. Kaz Fujita, Michigan State University; Colin Hardy, Program Manager, Fire, Fuel, Smoke Science, U.S. Forest Service; Dr. Benjamin P. Horton, University of Pennsylvania, Philadelphia; Dr. Roy Hyndman, Pacific Geoscience Center, Saanichton, British Columbia; Bernt-Gunnar Johansson photo, Sweden; Sarah Johnson, Digital Globe; Walter Justus, Bureau of Reclamation, Boise, ID; Ulrich Kamp, Geography, University of Montana; Bob Keane, Supervisory Research Ecologist, U.S. Forest Service; Dr. M. Asif Khan, Director, National Center of Excellence in Geology, University of Peshawar, Pakistan; Karen Knudsen, executive director, Clark Fork Coalition; Dr. David Loope, University of Nebraska; Martin McDermott, McKinney Drilling Co.; Dr. Ian Macdonald, Texas A&M University, Corpus Christi; Andrew MacInnes, Plaquemines Parish, LA; Dr. Jamie MacMahan, Naval Postgraduate School, Monterey, CA; Andrew Moore, Kent State University; Jenny Newton, Fire Sciences Laboratory, U.S. Forest Service; Dr. Mark Orzech, Naval Postgraduate School, Monterey, CA; Jennifer Parker, Geography, University of Montana; Dr. Stanley Riggs, Institute for Coastal Science and Policy, East Carolina University; Dr. Steve Running, Numerical Terradynamic Simulation Group, University of Montana; Todd Shipman, Arizona Geological Survey; Dr. Duncan Sibley, Michigan State University; Robert B. Smith, University of Utah; Donald Ward, Travis Co., TX; Karen Ward, Terracon, Austin, TX; John M. Thompson; Dr. Ron Wakimoto, Forest Fire Science, University of Montana; Dr. Robert Webb, USGS, Tucson, AZ; Vallerie Webb, Geoeye.com, Thornton, CO; Ann Youberg, Arizona Geological Survey.

- For providing access to the excavations of the Minoan culture at Akrotiri, Santorini: Dr. Vlachopoulos, head archaeologist, Greece.

- For assisting our exploration of the restricted excavations at Pompeii, Italy: Pietro Giovanni Guzzo, the site's chief archaeologist.

- For logistical help: Roberto Caudullo, Catania, Italy; Brian Collins, University of Montana; and Keith Dodson, Brooks/Cole's earth sciences textbook editor at the time the first edition was published.

In addition, we appreciate chapter reviewers who suggested improvements that we made in the Third Edition including: Donald J. Stierman, University of Toledo; Ingrid Ukstins Peate, University of Iowa; and Shawn Willsey, College of Southern Idaho.

We are indebted to reviewers who helped focus our attention on issues and specifics that led to many improvements in the Second Edition: Eric M. Baer at Highline Community College; David M. Best, Northern Arizona University; M. Stanley Dart, University of Nebraska at Kearney; Richard W. Hurst, California State University, Los Angeles; Mary Leech, San Francisco State University; Tim Sickbert, Illinois State University; Christiane Stidham, Stony Brook University; Kent M. Syverson, University of Wisconsin–Eau Claire.

Donald Hyndman and David Hyndman
July 2010

Natural Hazards and Disasters

1

Flooding during
Hurricane Ike in 2008
undermined tall posts
supporting houses
on the barrier island
east of Galveston,
Texas, toppling
them into the surf.

Hyndman.

Those who cannot remember the past are condemned to repeat it.
—GEORGE SANTAYANA (SPANISH PHILOSOPHER), 1905

Living in Harm's Way

Why would people choose to put their lives and property at risk? Large numbers of people around the world live and work in notoriously dangerous places—near volcanoes, in floodplains, or on active fault lines. Some are ignorant of potential disasters, but others even rebuild homes destroyed in previous disasters. Sometimes the reasons are cultural or economic. Because volcanic ash degrades into richly productive soil, the areas around volcanoes make good farmland. Large floodplains attract people because they provide good agricultural soil, inexpensive land, and natural transportation corridors. Some people live in a hazardous area because of their job or because they find the place appealing. For understandable reasons, such people live in the wrong places. Hopefully they recognize the hazards and understand the processes involved so they can minimize their risk.

But people also crowd into dangerous areas for frivolous reasons. They build homes at the bases or tops of large cliffs for scenic views, not realizing that big sections can give way in

FIGURE 1-1 THE UNEXPECTED

This four-year-old house near Zion National Park in southern Utah was built near the base of a steep rocky slope capped by a sandstone cliff. Early one morning in October 2001, the owner awoke with a start as a giant boulder 4.5 meters (almost 15 feet) across crashed into his living room and bedroom, narrowly missing his head.

landslides or rockfalls (**FIGURE 1-1**). They long to live along edges of sea bluffs where they can enjoy ocean views, or they want to live on the beach to experience the ocean more intimately. Others build beside picturesque, soothing rivers. Far too many people build houses in the woods because they enjoy the seclusion and scenery of this natural setting.

Some experts concerned with natural catastrophes say these people have chosen to live in "idiot zones." But people don't usually reside in hazardous areas knowingly—they generally don't understand or recognize the hazards. However, they might as well choose to park their cars on a rarely-used railroad track. Trains don't come frequently, but the next one might come any minute.

Catastrophic natural hazards are much harder to avoid than passing freight trains; we may not recognize the signs of imminent catastrophes because these events are infrequent. So decades or centuries may pass between eruptions of a large volcano that most people forget it is active. Many people live so long on a valley floor without seeing a big flood that they forget it is a floodplain. The great disaster of a century ago is long forgotten, so folks move into the path of a calamity that will occur on some unknowable future date. The hazardous event may not arrive today or tomorrow, but it is just a matter of time.

Catastrophes in Nature

Everyday geologic processes, like erosion, have produced large effects over the course of Earth's vast history, carving out valleys or changing the shape of coastlines. While some processes operate slowly and gradually, infrequent catastrophic events have sudden and major impacts. For instance, streams that run clear most of the year are muddy during a few high water days or weeks, when they carry most of their annual load of sediment. That sediment reflects a short and intense erosion period.

FIGURE 1-2 A LOOMING CATASTROPHE

Orting Washington, with spectacular views of Mount Rainier, is built on a giant, ancient mudflow from the volcano. If mudflows happened in the past, they almost certainly will happen again.

FIGURE 1-3 A DISASTER TAKES A HIGH TOLL

The Haiti earthquake of January 12, 2010, killed more than 222,000, mostly in concrete and cinder block buildings with little or no reinforcing steel. Searchers dig for survivors.

Major floods occurring once every ten or twenty years do far more damage and move more material than all of the intervening floods put together. Soil moves slowly downslope by creep, but occasionally a huge part of a slope may slide. Mountains grow higher, sometimes slowly, but more commonly by sudden movements. During an earthquake, a mountain can abruptly rise several meters above an adjacent valley.

Some natural events involve disruption of a temporary *equilibrium*, or balance, between opposing influences. Unstable slopes, for example, may hang precariously for thousands of years, held there by friction along a slip surface until some small perturbation, such as water soaking in from a large rainstorm, sets them loose. Similarly, the opposite sides of a fault may stick until continuing stress finally tears them loose, triggering an earthquake. A bulge may form on a volcano as molten magma slowly rises into it; then it collapses as the volcano erupts. The behavior of these natural systems is somewhat analogous to a piece of fabric or plastic wrap that remains intact as it stretches until it suddenly tears.

People watching Earth processes move at their normal and unexciting pace rarely pause to imagine what might happen if that slow pace were suddenly punctuated by a major event. The fisherman enjoying a quiet afternoon trout fishing in a small stream can hardly imagine how a 100-year flood might transform the scene. Someone gazing at a serene, snow-covered mountain can hardly imagine it erupting in an explosive blast of hot ash (**FIGURE 1-2**) followed by destructive mudflows racing down its flanks. Large or even gigantic events are a part of nature. Such abrupt events produce large results that can be disastrous if they affect people.

Human Impact of Natural Disasters

When a natural process poses a threat to human life or property, we call it a **natural hazard**. Many geologic processes are potentially hazardous. For example, streams flood, as part of their natural process, and become a hazard to those living nearby. A hazard is a **natural disaster** when the event causes significant damage to life or property. A moderate flood that spills over a floodplain every few years does not often wreak havoc, but when a major flood strikes, it may lead to a disaster that kills or displaces many people. When a natural event kills or injures large numbers of people or causes extensive property damage, it is called a **catastrophe**.

The potential impact of a natural disaster is related not only to event size but also to its effect on the public. A natural event in a thinly populated area can hardly pose a major hazard. For example, the magnitude 7.6 earthquake that struck the southwest corner of New Zealand on July 15, 2009, was severe but posed little threat because it happened in a region with few people or buildings. However, the magnitude 7.6 Kashmir earthquake occurred in heavily populated valleys of the southern Himalayas and killed more than 80,000 people, and the much smaller January 12, 2010, magnitude 7.0 earthquake in Haiti killed more than 222,000 (**FIGURE 1-3**). The May 2, 2008, cyclone in Myanmar killed an estimated 138,000 in a mostly rural area. By contrast, super typhoon Choi-Wan, a monstrous category 5 storm that passed directly over the Northern Marianas Islands south of Japan on September 15, 2009, resulted in no deaths because few people live there. The eruption of Mount St. Helens in 1980 caused few fatalities and remarkably little property damage simply because the area surrounding the mountain is sparsely populated. On the other hand, a similar eruption of Vesuvius, on the outskirts of Naples, Italy, could kill hundreds of thousands of people and cause property damage beyond reckoning.

People often associate natural hazard deaths with dramatic events, such as large earthquakes, volcanic eruptions, floods, hurricanes, or tornadoes. However, some of the most dramatic natural hazards occur infrequently or in restricted areas, so they cause fewer deaths than more common and less dramatic hazards. **FIGURE 1-4** shows the approximate proportions of fatalities caused by typical natural hazards in the United States.

In the United States, heat and drought account for the largest numbers of deaths. In fact, there were more U.S. deaths from heat waves between 1997 and 2008 than from any other type of natural hazard. In addition to heat stress, summer heat wave fatalities can result from dehydration and other factors; the very young, the very old, and the poor are affected the most. The same populations are vulnerable during winter weather, the third most deadly hazard in the U.S. Winter deaths often involve hypothermia, but some surveys include, for example, auto accidents caused by icy roads.

Flooding is the second most deadly hazard in the U.S., accounting for 16 percent of fatalities between 1986 and 2008. Fatalities from flooding can result from hurricane-driven floods; some surveys place them in the hurricane category rather than floods.

The number of deaths from a given hazard can vary significantly from year to year due to rare, major events. For example, there were about 1,800 hurricane-related deaths in 2005 when Hurricane Katrina struck, compared with zero in other years. The rate of fatalities can also change over time as a result of safety measures or trends in leisure activities. Lightning deaths were once amongst the most common hazard-related causes of death, but associated casualties have declined by a factor of five from 1940–1959 compared with 1989–2009, due in part to satellite radar and better weather forecasting. In contrast, avalanche deaths have increased by a factor of five from 1952–1971 compared with 1989–2008, a change that seems to be associated with snowmobile use and skiing in mountain terrains.

Some natural hazards can cause serious physical damage to land or manmade structures, some are deadly for people, and others are destructive to both. The type of damage sustained as a result of a natural disaster also depends on the economic development of the area where it occurs. In developing countries, there are increasing numbers of deaths from natural disasters, whereas in developed countries, there are typically greater economic losses. Developing countries show dramatic increases in populations relegated to marginal and hazardous land on steep slopes and near rivers. They have less ability to evacuate as hazards loom. Developed countries show lower population growth, forewarning is more immediate, and people can easily move.

The average annual cost of natural hazards has increased dramatically over the last several decades (**FIGURE 1-5**). This is due in part to the increase in world population overall, but it is also a function of human migration to more hazardous areas. Overall losses have increased even faster than population growth. Population increases in urban and coastal settings result in more people occupying land that is subject to major natural events. In effect, people place themselves in the path of unusual, sometimes catastrophic events. Economic centers of society are increasingly concentrated in larger urban areas that tend to expand into regions previously considered undesirable, including those with greater exposure to natural hazards.

Predicting Catastrophe

A catastrophic natural event is unstoppable, so the best way to avoid it is to predict its occurrence and get out of the way. Unfortunately, for those who would predict the occurrence of a natural disaster on a particular date, the result is most often dejection. So far, there have been few well-documented cases of accurate prediction, and even the ones on record may have involved luck more than science. Use of the same techniques in similar circumstances has resulted in false alarms and failure to correctly predict disaster.

Many people have sought to find predictable cycles in natural events. Natural events that occur at predictable intervals are called **cyclic events**. However, even most recurrent events are generally not really cyclic; too many variables control their behavior. Even with cyclic events, overlapping cycles make resultant extremes noncyclic, which affects the predictability of an event. So far as

FIGURE 1-4 HAZARD-RELATED DEATHS

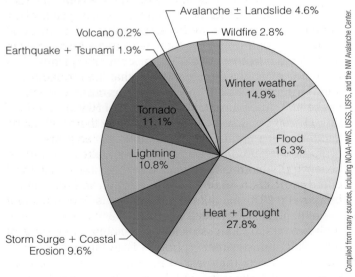

Compiled from many sources, including NOAA-NWS, USGS, USFS, and the NW Avalanche Center.

Approximate percentages of U.S. fatalities due to different groups of natural hazards from 1986 to 2008, when such data are readily available. For hazardous events that are rare or highly variable from year to year (earthquakes and tsunami, volcanic eruptions, and hurricanes), a 69-year record from 1940–2008 was used.

FIGURE 1-5 INCREASING COSTS OF NATURAL HAZARDS

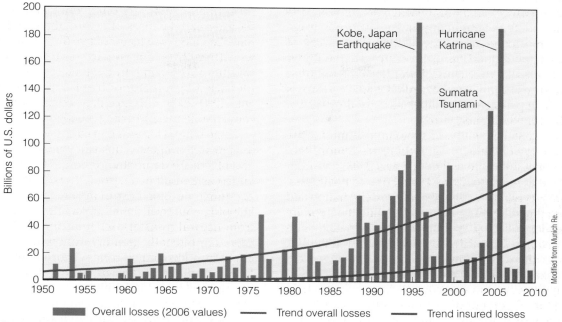

The cost of natural hazards is increasing worldwide, partly because world population doubled from 3 billion to 6 billion in only 40 years, from 1959 to 1999. By 2009 it reached 6.8 billion.

anyone can tell, most episodes, large and small, occur at seemingly random and essentially unpredictable intervals. The calendar does not predict them.

Nevertheless, scientists who make it their business to understand natural disasters can provide some guidance to people at risk. They cannot predict exactly when an event will occur. However, based on past experience, they can often **forecast** the occurrence of a hazardous event in a certain area within decades with an approximate percentage probability. They can forecast that there will be a large earthquake in the San Francisco Bay region over the next several decades, or that Mount Shasta will likely erupt sometime in the next few centuries. In many cases, their advice can greatly reduce the danger to lives and property.

Ask a stockbroker where the market is going, and you will probably hear that it will continue to do what it has done during recent weeks. Ask a scientist to forecast an event, and he or she will probably look to the geologically recent past and forecast more of the same; in other words *the past is the key to the future*. Most predictions of any kind are based on linear projections of past experience. However, we must be careful to look at a long enough sample of the past to see prospects for the future. Many people lose money in the stock market because *short-term* past experience is not always a good indicator of what will happen in the future.

Statistical predictions are simply a refinement of past recorded experiences. They are typically expressed as **recurrence intervals** that relate to the probability that a natural event of a particular size will happen–within a certain period of time. For example, the past history of a fault may indicate that it is likely to produce an earthquake of a certain size once every hundred years on average.

A recurrence interval is not, however, a fixed schedule for events. Recurrence intervals can tell us that a 50-year flood is likely to happen sometime in the next several decades but not that such floods occur at intervals of 50 years. Many people do not realize the inherent danger of an unusual occurrence, or they believe that they will not be affected in their lifetimes because such events occur infrequently. That inference often incorrectly assumes that the probability of another severe event is lower for a considerable length of time after a major event. In fact, even if a 50-year flood occurred last year, that does not indicate that there will not be another one this year or next or for the next ten years.

To understand why this is the case, take a minute to review probabilities. Flip a coin, and the chance that it will come up heads is 50 percent. Flip it again, and the chance is again 50 percent. If it comes up heads five times in a row, the next flip still has a 50 percent chance of coming up heads. So it goes with floods and many other kinds of apparently random natural events. The chance that someone's favorite fishing stream will stage a 50-year flood this year and every year is 1 in 50, regardless of what it may have done during the last few years.

As an example of both the usefulness and the limitations of recurrence intervals, consider the case of Tokyo. This enormous city is subject to devastating earthquakes that for more than 500 years came at intervals of close to 70 years.

The last major earthquake ravaged Tokyo in 1923, so everyone involved awaited 1993 with considerable consternation. The risk steadily increased during those years as both the population and the strain across the fault zone grew. More than 15 years later, no large earthquake has occurred. Obviously, the recurrence interval does not predict events at equal intervals, in spite of the 500-year Japanese historical record. Nonetheless, the knowledge that scientists have of the pattern of occurrences here helps them assess risk and prepare for the eventual earthquake. There was a magnitude 7.1 event 325 km south of Tokyo on August 9, 2009, and experts project that there is a 70 percent chance that a major quake will strike that region in the next 30 years.

To estimate the recurrence interval of a particular kind of natural event, we typically plot a graph of each event size versus the time interval between sequential individual events. Such plots often make curved lines that cannot be reliably extrapolated to larger events that might lurk in the future (**FIGURE 1-6**). Plotting the same data on a logarithmic scale often leads to a straight-line graph that can be extrapolated to values larger than those in the historical record. Whether the extrapolation produces a reliable result is another question.

The probability of the occurrence of an event is related to the magnitude of the event. We see huge numbers of small events, many fewer large events, and only a rare giant event (**By the Numbers 1-1:** Relationship between Frequency and Magnitude). The infrequent occurrence of giant events means it is hard to study them, but it is often rewarding to

> ## By the Numbers 1-1

Relationship between Frequency and Magnitude

$M \propto 1/f$

Magnitude (M) of an event is inversely proportional to frequency (f) of the type of event.

study small events because they may well be smaller-scale models of their uncommon larger counterparts that may occur in the future.

Many geologic features look the same regardless of their size, a quality that makes them **fractal**. A broadly generalized map of the United States might show the Mississippi River with no tributaries smaller than the Ohio and Missouri rivers. A more detailed map shows many smaller tributaries. An even more detailed map shows still more. The number of tributaries depends on the scale of the map, but the general branching pattern looks *similar across a wide range of scales* (**FIGURE 1-7**). Patterns apparent on a small scale quite commonly resemble patterns that exist on much larger scales that cannot be easily perceived. This means that small events may provide insight into huge ones that occurred in the distant past but are larger than any seen in historical time; we may find evidence of these big events if we search. The scale of some natural catastrophes that have affected the Earth, and will do so again, is almost too large to fathom. Examples include catastrophic failure of the flanks of oceanic volcanoes or the impact of large asteroids. For these, reality is more awesome than fiction. Yet each is so well documented in the geologic record that we need to be aware of the potential for such extreme events in the future.

It is impossible in our current state of knowledge to predict most natural events, even if we understand in a general way what controls them. The problem of avoiding

FIGURE 1-6 RECURRENCE INTERVAL

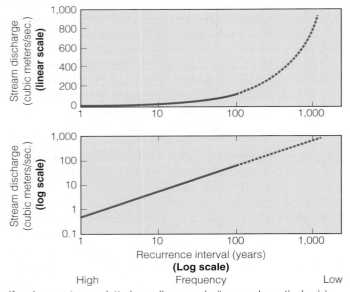

If major events are plotted on a linear scale (top graph, vertical axis), the results often fall along a curve that cannot be extrapolated to larger possible future events. If the same events are plotted on a logarithmic scale (bottom graph), the results often fall along a straight line that can be projected to possible larger events.

FIGURE 1-7 THE BRANCHING OF STREAMS IS FRACTAL

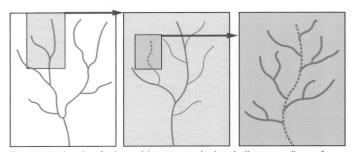

The general style of a branching stream looks similar regardless of scale—from a less-detailed map on the left to the most-detailed map on the right.

natural disasters is like the problem drivers face in avoiding collisions with trains. They can do nothing to prevent trains, so they must look and listen. We have no way of knowing how firm the natural restraints on a landslide, fault, or volcano may be. We also do not generally know what changes are occurring at depth. But we can be confident that the landslide or fault will eventually move or that the volcano will erupt. And we can reasonably understand what those events will involve when they finally happen.

Relationships among Events

Although randomness is a factor in forecasting disasters, not all natural events occur quite as randomly as floods or tosses of a coin. Some events are directly related to others—formed as a direct consequence of another event (**FIGURE 1-8**). For example, the slow movement of the huge outer layers of the Earth colliding or sliding past one another clearly explains the driving forces behind volcanic eruptions and earthquakes. Heavy or prolonged rainfall can cause a flood or a landslide. But are some events unrelated? Could any of the arrows in Figure 1-8 be reversed? Given all of the interlocking possibilities, the variability, and the uncertainties, we could call this figure a "chaos net" for natural hazards.

Past events can also create a contingency that influences future events. It is certainly true, for example, that sudden movement on a fault causes an earthquake. But the same movement also changes the stress on other parts of the fault and probably on other faults in the region, so the next earthquake will likely differ considerably from the last. Similar complex relationships arise with many other types of destructive natural events.

FIGURE 1-8 INTERACTIONS AMONG NATURAL HAZARDS

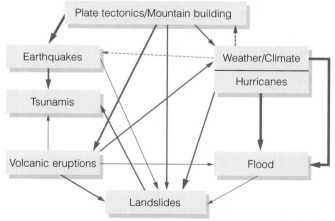

The bolder arrows in this flowchart indicate stronger influences. Can you think of others?

Some processes result in still more rapid changes—a *cascading* or *domino effect*. For example, global warming causes more rapid melting of Arctic sea ice. The resulting darker sea water absorbs more of the sun's energy than the white ice, and in turn, this causes even more sea ice melting. Similarly, global warming causes faster melting of the Greenland and Antarctic ice sheets. More meltwater pours through fractures to the base of the ice, where it lubricates movement, accelerating the flow of ice toward the ocean. This leads to more rapid crumbling of the toes of glaciers to form icebergs that melt in the ocean.

In other cases, an increase in one factor may actually lead to a decrease in a related result. Often as costs of a product or service go up, usage goes down. With increased costs of hydrocarbon fuels, people conserve more and thus burn less. A rapid increase in the price of gasoline in 2008 led people to drive less and to trade in large SUVs and trucks for smaller cars. In some places, commuter train, bus, and bicycle use increased dramatically. With the rising cost of electricity, people are switching to compact fluorescent bulbs and using less air conditioning. These changes had a noticeable effect on greenhouse gases and their effect on climate change (discussed in Chapter 10).

Sometimes major natural events are preceded by a series of smaller **precursor events**, which may warn of the impending disaster. Geologists studying the stirrings of Mount St. Helens, Washington, before its catastrophic eruption in 1980 monitored swarms of earthquakes and decided that most of these recorded the movements of rising magma as it squeezed upward, expanding the volcano. Precursor events alert scientists to the potential for larger events, but events that appear to be precursors are not always followed by a major event.

The relationships among events are not always clear. For example, an earthquake occurred at the instant Mount St. Helens exploded, and the expanding bulge over the rising magma collapsed in a huge landslide. Neither the landslide nor the earthquake caused the formation of molten magma, but did they trigger the final eruption? If so, which one triggered the other—the earthquake, the landslide, or the eruption? One or more of these possibilities could be true in different cases.

Events can also overlap to amplify an effect. Most natural disasters happen when a number of unrelated variables overlap in such a way that they reinforce each other to amplify an effect. If the high water of a hurricane storm surge happens to arrive at the coast during the daily high tide, the two reinforce each other to produce a much higher storm surge (**FIGURE 1-9**, p. 8). If this occurs on a section of coast that happens to have a large population, then the situation can become a major disaster. Such a coincidence caused the catastrophic hurricane that killed 8,000 people in Galveston, Texas, in 1900. Bad luck prevailed.

FIGURE 1-9 AMPLIFICATION OF OVERLAPPING EFFECTS

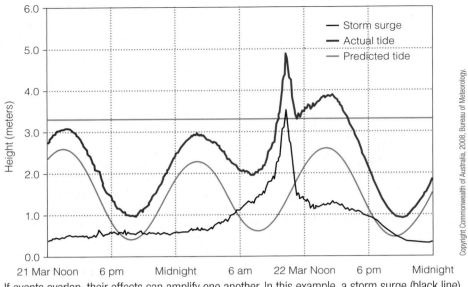

If events overlap, their effects can amplify one another. In this example, a storm surge (black line) can be especially high if it coincides with high tide (red line). The blue line shows the much higher tide that resulted when the tide overlapped with the storm surge.

Mitigating Hazards

Because natural disasters are not easily predicted, it falls to governments and individuals to assess their risk and prepare for and mitigate the effects of disasters. **Mitigation** refers to efforts to prepare for a disaster and reduce its damage. Mitigation can include engineering projects like levees, as well as government policy and public education. In each chapter of this book, we examine mitigation strategies related to specific disasters.

Land-Use Planning

One way to reduce losses from natural disasters is to find out where disasters are likely to occur and restrict development there, using **land-use planning**. Ideally, we should prevent development along major active faults by reserving that land for parks and natural areas. We should also limit housing and industrial development on floodplains to minimize flood damage and along the coast to reduce hurricane and coastal erosion losses. Limiting building near active volcanoes and the river valleys that drain them can curtail the hazards associated with eruptions.

It is hard, however, to impose land-use restrictions in many areas because such imposition tends to come too late. Many hazardous areas are already heavily populated, perhaps even saturated with inhabitants. Many people want to live as close as they can to a coast or a river and

resent being told that they cannot; they oppose attempts at land-use restrictions because they feel it infringes on their property rights. Almost any attempt to regulate land use in the public interest is likely to ignite intense political and legal opposition.

Developers, companies, and even governments often aggravate hazards by allowing—or even encouraging—people to move into hazardous areas. Many developers and private individuals view restrictive zoning as an infringement on their rights to do as they wish with their land. Developers, real estate agents, and some companies are reluctant to admit the existence of hazards that may affect a property for fear of lessening its value and scaring off potential clients (**FIGURE 1-10**, p. 9). Most local governments consider news of hazards bad for growth and business. They shun restrictive zoning or minimize possible dangers for fear of inhibiting improvements in their tax base. As in other venues, different groups have different objectives. Some are most concerned with economics, others with safety, still others with the environment.

Insurance

Some mitigation strategies help with recovery once a disaster occurs. Insurance is one way to lessen the financial impact of disasters after the fact. People buy property **insurance** to shield themselves from major losses they cannot afford. Insurance companies use a formula for risk to establish premium rates for policies. **Risk** is essentially a

FIGURE 1-10 RISKY DEVELOPMENT

David Hyndman.

Some developers seem unconcerned with the hazards that may affect the property they sell. High spring runoff floods this proposed development site in Missoula, Montana.

hazard considered in the light of its recurrence interval and expected costs (**By the Numbers 1-2**: Assessing Risk). The greater the hazard and the shorter its recurrence interval, the greater the risk.

In most cases, a company can estimate the cost of a hazard event to a useful degree of accuracy, but its recurrence interval is hardly better than an inspired guess. The history of experience with a given natural hazard in any area of North America is typically less than 200 years. Large events recur, on average, only every few decades or few hundred years or even more rarely. Estimating risk for these events becomes a perilous exercise likely to lose a company large amounts of money. In some cases, most notably floods, the hazard and its recurrence interval are both firmly enough established to support a rational estimate of risk. But the amount of risk and the potential cost to a company can be so large that a catastrophic event would put the company out of business. Such a case explains why private insurance companies are not eager to offer disaster policies.

The uncertainties of estimating risk make it impossible for private insurance companies to offer affordable policies to protect against many kinds of natural disasters. As a result, insurance is generally available for events that present relatively little risk, mainly those with more or less dependably

long recurrence intervals. The difficulty of obtaining policies from private insurers for certain types of natural hazards has inspired a variety of governmental programs. Earthquake insurance is available in areas such as Texas, where the likelihood of an earthquake is low. In California, where the risks and expected costs are much higher, insurance companies are required to provide earthquake coverage. As a result, companies now make insurance available through the California Earthquake Authority, a consortium of compaies. Similarly, most hurricane-prone southeastern states have mandated insurance pools that provide property insurance where individual private companies are unwilling to provide it.

Insurance for some natural hazards is simply not available. Landslides, most mudflows, and ground settling or swelling are too risky for companies, and each potential hazard area would have to be individually studied by a scientist or engineer who specialized in such a hazard. The large number of variables makes the risk too difficult to quantify; it is too expensive to estimate the different risks for the relatively small areas involved.

A critical question arises for people who lose their houses in landslides and are still paying on a mortgage. They may not only lose what they have already paid into the mortgage or home loan, but can be obligated to continue paying off a loan on a house that no longer exists. However, California, for example, has a law that generally prevents what are called "deficiency judgments" against such mortgage holders. This permits home owners to walk away from their destroyed homes, and the bank cannot go after them for the remainder of the loan. However, the situation is not always clear, because federal law may overrule state law. A federal agency, such as the Veterans Administration, which guarantees some mortgages, may pay a bank the balance of a loan and then go after the borrower for the remainder.

The Role of Government

The United States and Canadian governments are involved in many aspects of natural hazard mitigation. They conduct and sponsor research into the nature and behavior of many kinds of natural disasters. They attempt to find ways to predict hazardous events and mitigate the damage and loss of life they cause. Governmental programs are split among several agencies.

The U.S. Geological Survey (USGS) and Geological Survey of Canada (GSC) are heavily involved in earthquake and volcano research, as well as in studying and monitoring stream behavior and flow. The National Weather Service monitors rainfall and severe weather and uses this and the USGS data to try to predict storms and floods.

The Federal Emergency Management Agency (FEMA) was created in 1979, primarily to bring order to the chaos of relief efforts that seemed invariably to emerge after natural disasters. After the hugely destructive

▶ # By the Numbers 1-2

Assessing Risk

Insurance costs are actuarial: they are based on past experience. For insurance, a "hazard" is a condition that increases the severity or frequency of a loss.

Risk $\propto$ [probability of occurrence] $\times$ [cost of the probable loss from the event]

Midwestern floods of 1993, it has increasingly emphasized hazard reduction. Rather than pay victims to rebuild in their original unsafe locations, such as floodplains, the agency now focuses on relocating them. Passage of the Disaster Mitigation Act in 2000 signals greater emphasis on identifying and assessing risks before natural disasters strike and taking steps to minimize potential losses. The act funds programs for hazard mitigation and disaster relief through FEMA, the U.S. Forest Service, and the Bureau of Land Management.

To determine risk levels and estimate loss potential from earthquakes, federal agencies such as FEMA use a computer system called HAZUS (Hazard United States). It integrates a group of interdependent modules that include potential hazards, inventories of the hazards, direct damages, induced damages, direct economic and social losses, and indirect losses.

Unfortunately, some government policy can be counterproductive, especially when politics enter the equation. In some cases, disaster assistance continues to be provided without a large cost-sharing component from states and local organizations. Thus, local governments continue to lobby Congress for funds to pay for losses but lack incentive to do much about causes. FEMA is charged with rendering assistance following disasters; it continues to provide funds for victims of earthquakes, floods, hurricanes, and other hazards. It remains reactive to disasters, as it should be, but is only beginning to be proactive in eliminating the causes of future disasters. Congress continues to fund multimillion-dollar Army Corps of Engineers projects to build levees along rivers and replenish sand on beaches. The Small Business Administration disaster loan program continues to subsidize credit to finance rebuilding in hazardous locations. The federal tax code also subsidizes building in both safe and hazardous sites. Real estate developers benefit from tax deductions, and ownership costs, such as mortgage interest and property taxes, can be deducted from income. A part of uninsured "casualty losses" can still be deducted from a disaster victim's income taxes. Such policies do not discourage future damages from natural hazards.

The Role of Public Education

Much is now known about natural hazards and the negative impacts they have on people and their property. It would seem obvious that any logical person would avoid such potential damages or at least modify their behavior or their property to minimize such effects. However, most people are not knowledgeable about potential hazards, and human nature is not always rational. Until someone has a personal experience or knows someone who has had such an experience, most people subconsciously believe *It won't happen here* or *It won't happen to me*. Even knowledgeable scientists aware of the hazards, the odds of their occurrence, and the costs of an event do not always act appropriately. Compounding the problem is the lack of tools to reliably predict specific locations and timing of many natural hazards.

Unfortunately, a person who has not been adversely affected in a serious way is much less likely to take specific steps to reduce the consequences of a potential hazard. Migration of the population toward the Gulf and Atlantic coasts accelerated in the last half of the twentieth century and still continues. Most of those new residents, including developers and builders, are not very familiar with the power of coastal storms. Even where a hazard is apparent, people are slow to respond. Is it likely to happen? Will I have a major loss? Can I do anything to reduce the loss? How much time will it take, and how much will it cost? Who else has experienced such a hazard?

Several federal agencies have programs to foster public awareness and education. The Emergency Management Institute—in cooperation with FEMA, the National Oceanic and Atmospheric Administration (NOAA), USGS, and other agencies—provides courses and workshops to educate the public and governmental officials. Some state emergency management agencies, in partnership with FEMA and other federal entities, provide workshops, reports, and informational materials on specific natural hazards.

Given the hesitation of many local governments to publicize natural hazards in their jurisdictions, people need to educate themselves. Being aware of the types of hazards in certain regions allows people to find evidence for their past occurrence. It also prepares them to seek relevant literature and ask appropriate questions of knowledgeable authorities.

One of the best means of protecting ourselves from natural hazards is an ability to recognize landscapes and rocks and to understand the processes that shape them. Volcanoes not only shed lava flows but ash and mudflows that can be recognized in ancient deposits. Old landslides often leave lumpy landscapes, and sinkholes can leave closed, undrained depressions. Streams meander across flat floodplains, shifting their channels by eroding meander banks. Storm waves undercut sea cliffs and churn sand from beaches. Offshore barrier islands are eroded on their seaward sides, depositing sand landward; that moves the islands landward. Homes in the urban fringe are often burned by wildfires that creep along the ground in dry leaves and needles, even though surrounding trees survive. We study landscapes and the processes that shape them in the chapters that follow.

Some people are receptive to making changes in the face of potential hazards. Some are not. The distinction depends partly on knowledge, experience, and whether they feel vulnerable. A person whose house was badly damaged in the 1989 Loma Prieta, California, earthquake is likely to either move to a less earthquake-prone area or live in a house that is well braced for earthquake resistance. A similar person losing his home to a landslide is more likely to avoid living near a steep slope. The best window of opportunity for effective hazard reduction is immediately following a disaster of the same type. Studies show that this opportunity is short—generally, not more than two or three months.

Successful public education programs, such as some of those on earthquake hazards in parts of California presented by the USGS, have shown that information must come from multiple credible sources and be presented in nontechnical terms that spell out specific steps people can take. Broadcast messages can be helpful, but written material that people can refer to should accompany them. Discussion among potentially affected groups can help them understand hazards and act on the information. If people think the risk is plausible, they tend to seek additional reliable information to validate what they have heard. And the range of additional sources must be trustworthy to different groups of people. Some groups believe scientists; others favor structural engineers. Some seek out information online. Successful education programs must include specialists and should adapt material to the different interests of specific groups, such as homeowners, renters, and corporations. Overall, natural hazard education depends on tailoring a clear message to different audiences using nontechnical language. It must not only convey the nature of potential events but also show that certain relatively simple and inexpensive actions can substantially reduce potential losses.

Living with Nature

Catastrophic events are natural and expected, but the most common human reaction to a current or potential catastrophe is to try to stop ongoing damage by controlling nature. In our modern world, it is sometimes hard to believe that scientists and engineers cannot protect us from natural disasters by predicting them or building barriers to withstand them. But there are limits to scientific understanding and engineering capabilities. In fact, although scientists and engineers understand much about the natural world, they understand less than many people suppose.

Unfortunately, we cannot change natural system behaviors, because we cannot change natural laws. Most commonly, our attempts tend only to temporarily hinder a natural process while diverting its damaging energy to other locations. In other cases, our attempts cause energy to build up and produce more severe damage later.

If, through lack of forethought, you find yourself in a hazardous location, what can you do about it? You might build a river levee to protect your land. Or you might build a rock wall in the surf to stop sand from leaving your beach and undercutting the hill on which your house is built.

If you do any of these things, however, you merely transfer the problem elsewhere, to someone else, or to a later point in time. For example, if you build a levee to prevent a river from spreading over a floodplain and damaging your property, the flood level past the levee will be higher than it would have been without it. Constricting river flow with a levee also backs up floodwater, potentially causing flooding of an upstream neighbor's property. Deeper water also flows faster past your levee, so it may make for worse erosion of a downstream neighbor's riverbanks. As in the stock market, individual stocks go up and down. If you make money because you bought a stock when its price was low and sold it when its price was high, then you effectively bought it from someone else who lost money. In the stock market, over the short term, the best we can do, from a selfish point of view, is to shift disasters to our neighbors. The same is true when tampering with nature. We need to understand the consequences.

Individually and as a society, we must learn to live with nature, not try to control it. Mitigation efforts typically seek to avoid or eliminate a hazard through engineering. Such efforts require financing from governments, individuals, or groups likely to be affected. Less commonly, but more appropriately, mitigation requires changes in human behavior. Behavioral change is usually much less expensive and more permanent than the necessary engineering work. In recent years, governmental agencies have begun to learn this lesson, generally through their own mistakes. In a few places along the Missouri and Sacramento Rivers, for example, some levees are being reconstructed back from the riverbanks to permit water to spread out on floodplains during future floods.

Natural hazards exist worldwide. They depend on climate, topography, tectonic environment, and proximity to rivers and coasts. However, they are not constrained by national boundaries. The same natural hazards and processes that we see in the United States also operate, for example, in France, Argentina, New Zealand, and China. Although many of the examples you read about and photos you see in this book are from other regions, most are relevant to places much closer to home. We use good examples from other regions to amplify what can happen here. We use many of our own photos to help you recognize natural hazards and to emphasize that there is nothing unusual about what is shown in these pictures. You can easily learn to spot hazards wherever you are. The Critical View exercises at the end of many chapters provide practice for observing and analyzing hazards around you.

In reality, few places are completely free of all natural hazards. Given the constraints of health, education, and livelihood, we can minimize living in the most hazardous areas. We can avoid one type of hazard while tolerating a less ominous one. Above all, we can educate ourselves about natural hazards and their controls, how to recognize them, and how to anticipate increased chances of a disaster. Although prediction may not be realistic, we can forecast the likelihood of certain types of occurrence that may endanger our property or physical safety. This book provides the background you need to be knowledgeable about natural hazards.

Chapter Review

Key Points

Catastrophes in Nature

- Many natural processes that we see are slow and gradual, but occasional sudden or dramatic events can be hazardous to humans.

- Hazards are natural processes that pose a threat to people or their property.

- A large event becomes a disaster or catastrophe only when it affects people or their property. Large natural events have always occurred but do not become disasters until people place themselves in harm's way.

- More common and less dramatic hazards, such as heat, cold, and flooding, often have higher associated fatalities than rare but dramatic hazards, such as earthquakes and volcanoes. **FIGURE 1-4**.

- Developed countries lose large amounts of money in a major disaster; poor countries lose larger numbers of lives.

Predicting Catastrophe

- Events are often neither cyclic nor completely random.

- Although the precise date and time for a disaster cannot be predicted, understanding the natural processes that control them allows scientists to forecast the probability of a disaster striking a particular area.

- Statistical predictions or recurrence intervals are average expectations based on past experience.

- There are numerous small events, fewer larger events, and only rarely a giant event. We are familiar with the common small events but, because they come along so infrequently, we tend not to expect the giant events that can create major catastrophes. **FIGURE 1-6**, **By the Numbers 1-1**.

- Many natural features and processes are fractal—that is, they have similarities across a broad range of sizes. Large events tend to have characteristics that are similar to smaller events. **FIGURE 1-7**.

Relationships among Events

- Different types of natural hazards often interact with, or influence, one another. **FIGURE 1-8**.

- Natural processes can have a cascading, or domino effect, with one change triggering other, more rapid changes.

- Overlapping influences of multiple factors can lead to the extraordinarily large events that often become disasters. **FIGURE 1-9**.

Mitigating Hazards

- Mitigation involves efforts to avoid disasters rather than merely dealing with the resulting damages.

- Risk is proportional to the probability of occurrence and the cost from such an occurrence. **By the Numbers 1-2**.

- People need to be educated about natural processes and how to learn to live with and avoid the hazards around them.

Living with Nature

- Erecting a barrier to some hazard will typically transfer the hazard to another location or to a later point in time.

- Humans need to learn to live with some natural events rather than trying to control them.

Key Terms

catastrophe, p. 3

cyclic events, p. 4

forecast, p. 5

fractal, p. 6

insurance, p. 8

land-use planning, p. 8

mitigation, p. 8

natural disaster, p. 3

natural hazard, p. 3

precursor events, p. 7

recurrence intervals, p. 5

risk, p. 8

Questions for Review

1. What are some of the reasons people live in geologically dangerous areas?

2. Is the geologic landscape controlled by gradual and unrelenting processes or intermittent large events with little action in between? Provide an example to illustrate.

3. Some natural disasters happen when the equilibrium of a system is disrupted. What are some examples?

4. What factors influence how many fatalities result from a given disaster?

5. What are the three most deadly hazards in the United States?

6. What are the main reasons for the ever-increasing costs of catastrophic events?

7. Why are most natural events not perfectly cyclic, even though some processes that influence them are cyclic?

8. What is the difference between prediction and forecast?

9. Give an example of the domino effect in natural processes.

10. Give an example of a fractal system.

11. Describe the general relationship between the frequency and magnitude of an event.

12. If the recurrence interval for a stream flood has been established at 50 years and the stream flooded last year, what is the probability of the stream flooding again this year?

13. When an insurance company decides on the cost of an insurance policy for a natural hazard, what are the two main deciding factors?

14. When people or governmental agencies try to restrict or control the activities of nature, what is the general result?

Discussion Questions

1. If people should not live in especially dangerous areas, what beneficial uses could there be for those areas? What are some examples?

2. What responsibility does the government have to ensure that its citizens are safe from natural hazards? Conversely, what freedom should individuals have to choose where they want to live?

3. A small town suffering economic losses from the closure of a factory considers a plan to build a new housing development in an area where there is a record of infrequent flooding. Make a case for and against this development. In your case for the development, stipulate what measures need to be taken to minimize hazards.

4. Should people be permitted to build in hazardous sites? Should they expect government help in case of a disaster? Should they be required to pay for all costs incurred in a disaster?

5. Contrast the general nature of catastrophic losses in developed countries versus poor countries. Explain why this is the case.

2 Plate Tectonics and Physical Hazards

The Andes of Peru rise along the collision zone between the Pacific Ocean floor and continental South America.

Donald Hyndman.

The Big Picture

Why are mountain ranges commonly near coastlines? Why are some of these mountains volcanoes that erupt molten rocks? What causes giant tsunami waves, and why do most originate near mountainous coastlines? Why are most devastating earthquakes near those same coastlines? Why do the opposite sides of the Atlantic Ocean look like they would match? (**FIGURE 2-1**) Giant areas of the upper part of Earth move around, grind sideways and collide, or sink into the hot interior of the planet, where they cause melting of rocks and formation of volcanoes. Those collisions between plates squeeze up and maintain high mountain ranges, even as landslides and rivers erode them away. Those same collisions can generate giant tsunami waves. To understand where and when these hazards occur, we need to understand the forces that drive them. Without the movements of Earth's plates, there would be no high mountain ranges to cause rockfalls and other landslides or for rivers to flow down. Those same mountain ranges even have a big effect on weather and climate. All of these processes ultimately drive natural hazards.

FIGURE 2-1 MOUNTAIN TOPOGRAPHY

This shaded relief map shows the continents standing high. Mountain ranges in red tones concentrate at some continental margins. Light blue ridges in oceans are mountain ranges on the ocean floor.

Earth Structure

Scientists now know that at the center of the Earth is its core, surrounded by the thick mantle and covered by the much thinner crust (**FIGURE 2-2**). The distinction between the mantle and the crust is based on rock composition.

We also distinguish between two zones of the earth based on rock rigidity or strength. The stiff, rigid outer rind of the Earth is called the **lithosphere**, and the inner, hotter, more easily deformed part is called the **asthenosphere**. The lithosphere is made up of large blocks, called **tectonic plates**. Continental lithosphere includes silica-rich

FIGURE 2-2 EARTH STRUCTURE

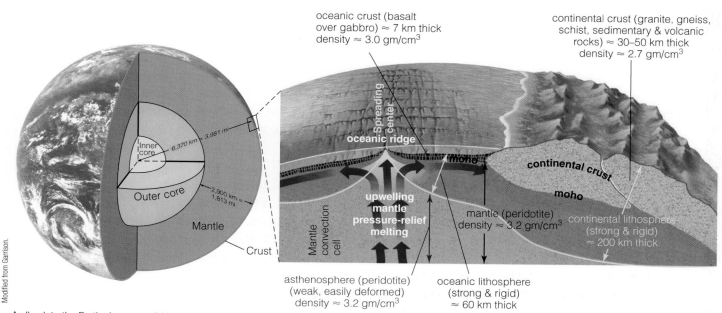

A slice into the Earth shows a solid inner core and a liquid outer core, both composed of nickel-iron. Peridotite in Earth's mantle makes up most of the volume of the Earth. Earth's crust, on which we live, is as thin as a line at this scale. The right-hand diagram shows an expanded view with details of Earth's lithosphere.

crust 30 to 50 kilometers thick, underlain by the upper part of the **mantle** (see Appendix 2 online for detailed rock compositions). Oceanic lithosphere is generally only about 60 kilometers thick; its top 7 kilometers are a low-silica basalt-composition crust. Continental crust is largely composed of high-silica-content minerals, which give it the lowest density (2.7 g/cm^3) of the major regions on Earth. Oceanic crust is denser (3 g/cm^3) because it contains more iron- and magnesium-rich minerals.

As shown in the right-hand diagram of FIGURE 2-2, the low-density continental crust is thicker and stands higher than the denser oceanic crust. The shaded-relief map of Earth's topography also clearly shows the continents standing high relative to the ocean basins because of isostacy, the thick mountain-ranges of Earth's crust sinking deeper into the Earth (**By the Numbers 2-1**: Height of a Floating Mass). The thin lithosphere of the ocean basins stands low; the continents with their thick lower-density lithosphere float high and sink deep into the asthenosphere.

The elevation difference between the continental and oceanic crusts is explained by the concept of **isostacy**, or buoyancy. A floating solid object displaces an amount of liquid of the same mass. Although Earth's mantle is not liquid, its high temperature (above 450°C or 810°F) permits it to flow slowly as if it were a viscous liquid. As a result, the proportion of a mass immersed in the liquid can be calculated from the density of the floating solid divided by the density of the liquid. Similarly, where the weight of an extremely large glacier is added to a continent, the crust and upper mantle slowly sink deeper into the mantle. That happened during the last ice age when thick ice covered most of Canada and the northern United States. As the ice melted, these areas gradually rose back toward their original heights.

Almost 150 years ago, measurements showed that gravitational attraction of the huge mountain mass of the Himalayas pulled plumb bobs of very precise surveying instruments toward the mountain range more than would be expected based on the height of the mountains above sea level. A scientist, George Airy, inferred that the mountains must be thicker than they appeared, not merely standing on the Earth's crust but extending deeper into it. Based on measurements of the density and velocity of earthquake waves through the crust, it now seems that many major mountain ranges do in fact have roots, and their crust is much thicker than adjacent older crust. As the mountains grew higher, their roots sank deeper into a fluid Earth, like a block of wood floating in water.

The temperature of the crust also affects its elevation. The crust of the mountainous Cordillera of western Canada and the northwestern United States is no thicker than the 40-km-thick continental crust to the east, and in some areas it is even thinner. Why then does it rise higher above sea level? Measurement of the temperature of the deep crust shows it to be hotter than old, cold continental crust to the east. In certain mountain ranges such as the Cordillera, the hot, more expanded, crust of the mountain range is less dense, so it floats higher than old, dense continental crust on the underlying asthenosphere (see FIGURE 2-2). Heat in the thin crust may have been provided from the hot underlying mantle asthenosphere that stands relatively close to the Earth's surface.

▶ By the Numbers 2-1

Height of a Floating Mass

Iceberg: 10% above water

90% of volume submerged

Water

Major mountain range: about 16% above average continent

Mountain root (crust)

Earth's mantle

The height to which a floating block of ice rises above water depends on the density of water compared with the density of ice. For example, when water freezes, it expands to become a lower density (ice density is 0.9 g/cm^3 relative to liquid water at 1.0 g/cm^3). Thus, 90 percent of an ice cube or iceberg will be underwater. Similarly, for many large mountain ranges, approximately 84 percent of a mountain range of continental rocks (2.7 g/cm^3) will submerge into the mantle (3.2 g/cm^3) as a deep mountain root.

Because we do not have direct observations of crustal thickness, scientists measure the gravitational attraction of Earth (which is greater over denser rocks) and analyze the velocity and timing of seismic waves as they radiate away from earthquake epicenters to provide indirect evidence of the density, velocity, and thickness of subsurface materials. The boundary between Earth's crust and mantle has been identified as a major difference in density that we call the or *Moho* (see FIGURE 2-2). It marks the base of the continental crust.

Deeper in the mantle, the next major change in material properties occurs at the boundary between the strong, rigid lithosphere above and the weak, deformable asthenosphere below. This boundary was first identified as a near-horizontal zone of lower velocity earthquake waves that move at several kilometers per second. The so-called *low-velocity zone* is concentrated at the top of the asthenosphere and may contain a small amount of molten basalt over a zone a few hundred kilometers thick. The cold, rigid lithosphere rides on that asthenosphere made weak by its higher temperatures and perhaps also by small melt contents. The lithosphere moves over the weak, deforming asthenosphere at a few centimeters per year.

Plate Movement

The lithosphere is not continuous like the rind on a melon. It is broken into a dozen or so large plates and about another dozen much smaller plates (**FIGURE 2-3**). Even though they are uneven in size and irregular in shape, the plates fit neatly together almost like a mosaic that covers the entire surface of the Earth. The plates do not correspond to continent versus ocean areas; most plates consist of a combination of the two. The South American Plate is about half below the Atlantic Ocean and half continent. Even the Pacific Plate, which is mostly ocean, includes a narrow slice of western California and part of New Zealand.

Earth's plates move up to 11 centimeters (4.2 inches) per year, as confirmed by satellite Global Positioning System (GPS) measurements. Many move in roughly an east-west direction, but some don't. **Plate tectonics** is the big picture theory that describes the movements of Earth's plates. We will present the evidence for plate tectonics at the end of this chapter.

Some plates separate, others collide, and still others slide under, or over, or past one another (**FIGURE 2-4**).

FIGURE 2-3 LITHOSPHERIC PLATES

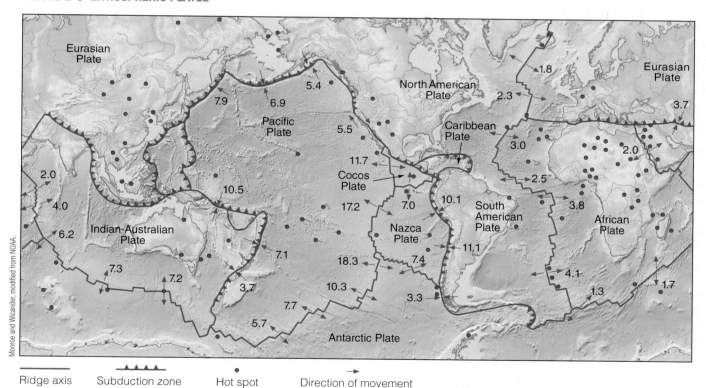

Ridge axis   Subduction zone   Hot spot   Direction of movement

Most large lithospheric plates consist of both continental and oceanic areas. Although the Pacific Plate is largely oceanic, it does include parts of California and New Zealand. General direction and velocities of plate movement (compared with hotspots that are inferred to be anchored in the deep mantle), in centimeters per year, are shown with red arrows.

FIGURE 2-4 PLATE BOUNDARIES

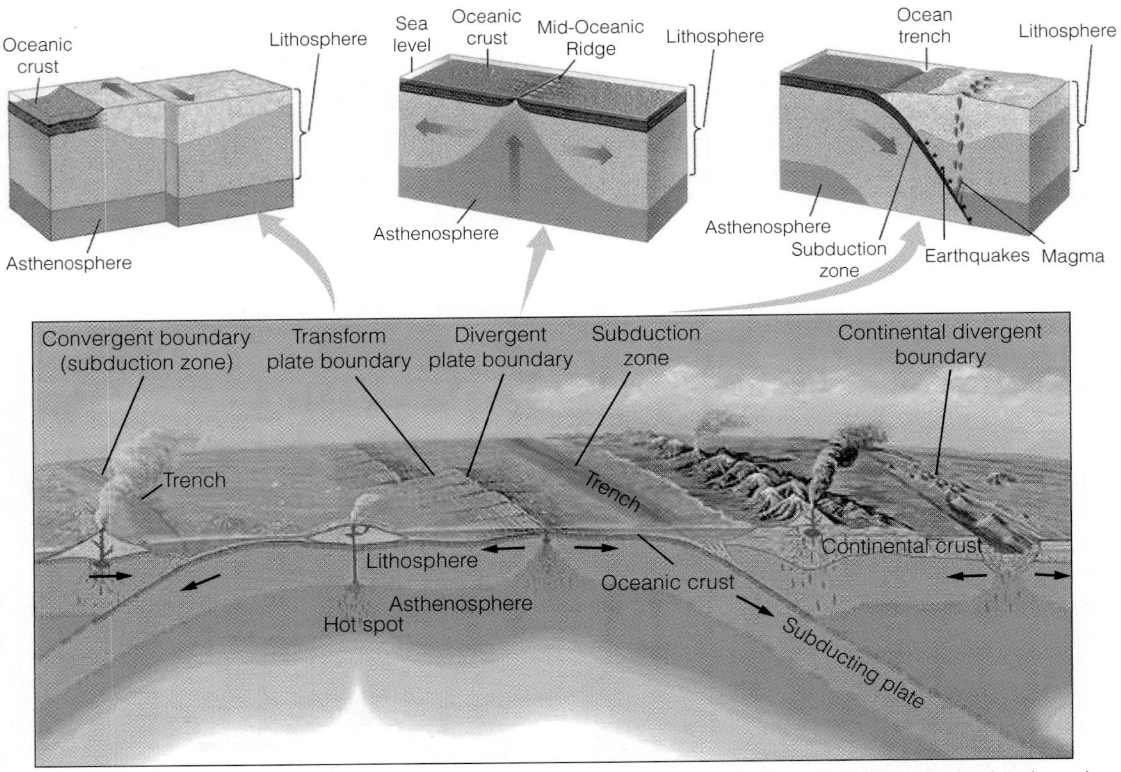

Modified from Garrison.

USGS.

This three-dimensional cutaway view shows a typical arrangement of the different types of lithospheric plate boundaries: transform, divergent, and convergent.

In some cases, their encounters are head on; in others, the collisions are more oblique. Plates move away from each other at **divergent boundaries**. Plates move toward each other at collision or **convergent boundaries**. In cases where one or both of the plates are oceanic lithosphere, the denser plate will slide down, or be *subducted*, into the asthenosphere, forming a **subduction zone**. When two continental plates collide, neither side is dense enough to be subducted deep into the mantle, so the two sides typically crumple into a thick mass of low-density continental material. This type of convergent boundary is where the largest mountain ranges on Earth, such as the Himalayas, are built. In the remaining category of plate interactions, two plates slide past each other at a **transform boundary**, such as the San Andreas Fault.

New oceanic crust wells up at the mid-oceanic ridges, spreads apart (a process called **seafloor spreading**), and finally sinks into the deep oceanic **trenches** (top of the subduction zone) along the edges of some continents (**FIGURE 2-5**). Plates continue to pull apart at the Mid-Atlantic Ridge, for example, making the ocean floor wider and moving North America and Europe farther apart. In the Pacific Ocean, the plates diverge at the East Pacific

Rise; their oldest edges sink in the deep ocean trenches near the western Pacific continental margins.

In some places, different types of plate edges intersect. Because transform boundaries are often at a large angle to spreading centers or subduction zones, two or even three such boundaries may intersect. For example, at the Mendocino *triple junction* just off the northern California coast, the Cascadia subduction zone at the Washington-Oregon coast joins both the San Andreas transform fault of California and the Mendocino transform fault that extends offshore (**FIGURE 2-6**). The north end of the same subduction zone joins both the Juan de Fuca spreading ridge and the Queen Charlotte transform fault at a triple junction just off the north end of Vancouver Island.

Hazards and Plate Boundaries

Most of Earth's earthquake and volcanic activity occurs along or near plate boundaries (**FIGURES 2-7** and **2-8**, p. 20). Most of the convergent boundaries between oceanic and continental plates form subduction zones along the Pacific coasts of North and South America, Asia, Indonesia, and New Zealand. Collisions between continents are best

FIGURE 2-5 SEAFLOOR SPREADING

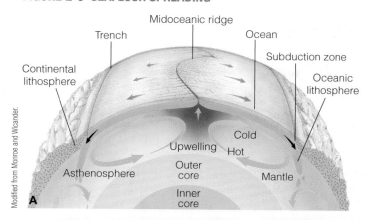

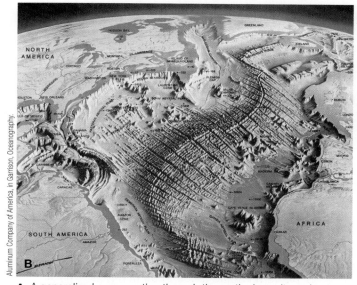

Modified from Monroe and Wicander.

Aluminum Company of America, in Garrison, Oceanography.

A. A generalized cross section through the earth shows its main concentric layers. The more rigid lithosphere moves slowly over the less rigid asthenosphere, which is thought to circulate slowly by convection. The lithosphere pulls apart at ridges and sinks at trenches. **B.** The spreading Mid-Atlantic Ridge, fracture zones, and transform faults are dramatically exhibited in this exaggerated topography of the ocean floor.

FIGURE 2-6 THREE TYPES OF PLATE BOUNDARIES

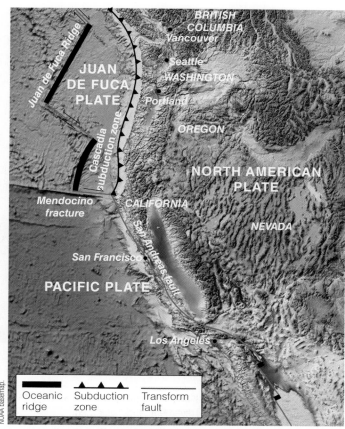

NOAA basemap.

Good examples of plate boundaries are located at the western edge of North America: the San Andreas Fault runs through the western edge of California; the Cascadia subduction zone parallels the coast off Oregon, Washington, and southern British Columbia; and the Juan de Fuca spreading ridge lies farther offshore. Spreading at the Juan de Fuca Ridge carries ocean floor of the Juan de Fuca Plate down the Cascadia subduction zone.

expressed in the high mountain belts extending across southern Europe and Asia. Most rapidly spreading divergent boundaries follow oceanic ridges. In some cases, slowly spreading continental boundaries, such as the East African Rift zone, pull continents apart. Each type of plate boundary has a distinct pattern of natural events associated with it.

Divergent Boundaries

Divergent boundaries, where plates pull apart by the sinking of heavy lithosphere at oceanic trenches, make a system of more-or-less connected oceanic ridges that wind through the ocean basins like the seams on a baseball.

Iceland is the only place where that ridge system rises above sea level; elsewhere, it is submerged to an average depth of a few thousand meters.

A broad valley doglegs from south to north across Iceland. The hills east of it are on the Eurasian Plate that extends all the way east to the Pacific Ocean; the hills to the west are on the North American Plate. Repeated surveys over several decades have shown that the valley is growing wider at a rate of several centimeters per year. The movement is the result of the North American and Eurasian Plates pulling away from each other, making the Atlantic Ocean grow wider at this same rate.

Iceland's long recorded history shows that a broad fissure opens in the floor of its central valley every 200 to 300 years. It erupts a large basalt lava flow that covers as much as several thousand square kilometers. The last fissure

FIGURE 2-7 EARTHQUAKES AT PLATE BOUNDARIES

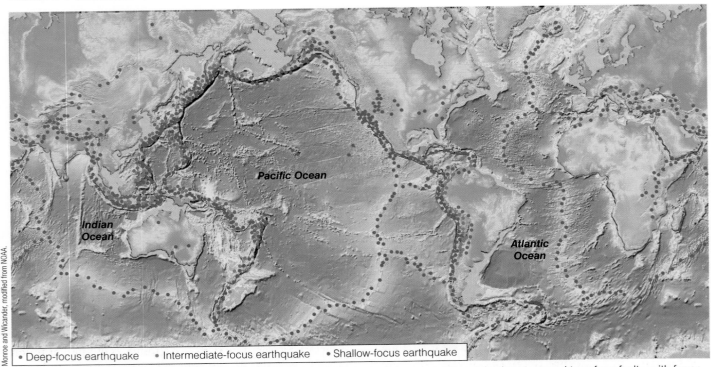

• Deep-focus earthquake   • Intermediate-focus earthquake   • Shallow-focus earthquake

Most earthquakes are concentrated along boundaries between major tectonic plates, especially subduction zones and transform faults, with fewer along spreading ridges.

FIGURE 2-8 VOLCANOES NEAR PLATE BOUNDARIES

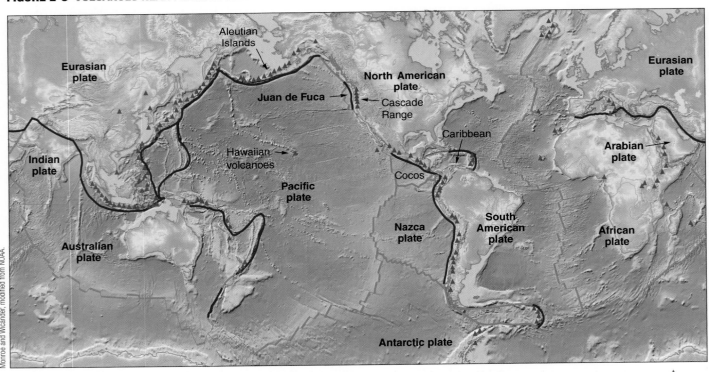

Divergent plate boundary   Transform plate boundary   Convergent boundary   Volcano

Most volcanic activity also occurs along plate tectonic boundaries. Eruptions tend to be concentrated along the continental side of subduction zones and along divergent boundaries, such as rifts and mid-oceanic ridges.

FIGURE 2-9 EVOLUTION OF A SPREADING RIDGE

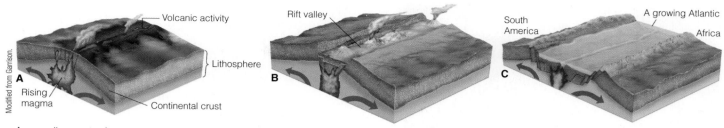

A spreading center forms as a continent is pulled apart to form new oceanic lithosphere. This process separated the supercontinent of Pangea into South America and Africa, thereby forming the Atlantic Ocean.

opened in 1821. Finally in April 2010, rifting under a glacier again erupted basalt magma. The hot magma melted the ice causing flooding and an immense ash cloud that spread over most of northern Europe, curtailing air traffic for days. Another such event could happen anytime. Fortunately, the sparse population of the region limits the potential for a great natural disaster.

It now seems clear that similar eruptions happen fairly regularly all along the oceanic ridge system. These spreading centers are the source of the basalt lava flows that cover the entire ocean floor, roughly two-thirds of Earth's surface, to an average depth of several kilometers. The molten basalt magma rises to the surface, where it comes in contact with water. It then rapidly cools to form pillow-shaped blobs of lava with an outer solid rind initially

encasing molten magma. As the plate moves away from the spreading center, it cools, shrinks, and thus increases in density. This explains why the hot spreading centers stand high on the subsea topography. New ocean floor continuously moves away from the oceanic ridges as the oceans grow wider by several centimeters every year (**FIGURE 2-9**). The only place where frequent earthquakes and volcanic eruptions along oceanic ridges pose a danger to people or property is in Iceland, where the oceanic ridge rises above sea level.

Spreading centers in the continents pull apart at much slower rates and do not generally form plate boundaries. The Rio Grande Rift of New Mexico and the Basin and Range of Nevada and Utah are active North American examples (**FIGURE 2-10**). The East African Rift zone that

FIGURE 2-10 CONTINENTAL SPREADING

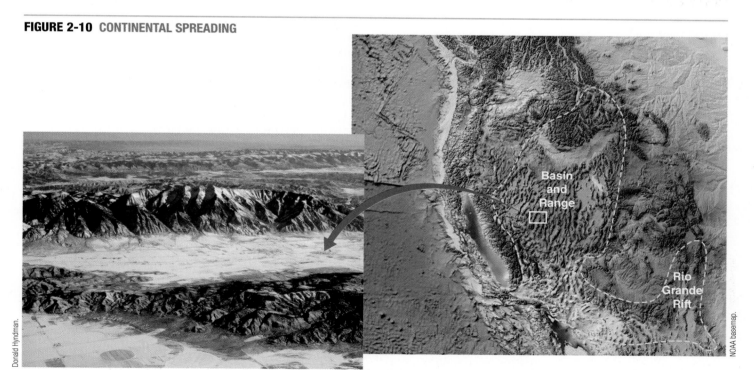

The Basin and Range terrain is found southwest of Salt Lake City, Utah. This broad area of spreading in the western United States is marked by prominent basins and mountain ranges. Centered in Nevada and western Utah, it gradually decreases in spreading rate to the north across the Snake River Plain, near its north end. Its western boundary includes the eastern edge of the Sierra Nevada Range, California, and its main eastern boundary is at the Wasatch Front in Utah. An eastern branch includes the Rio Grande Rift of central New Mexico.

FIGURE 2-11 BEGINNING OF AN OCEAN

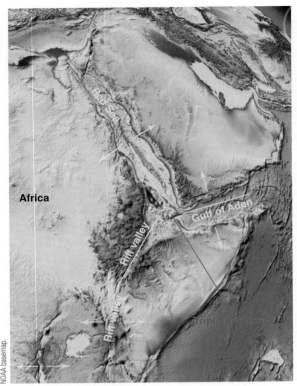

NOAA basemap.

The East African Rift Valley spreads the continent apart at rates 100 times slower than typical oceanic rift zones. This rift forms one arm of a triple junction, from which the Red Sea and the Gulf of Aden form somewhat more rapidly spreading rifts.

extends north-south through much of that continent (**FIGURE 2-11**) may be the early stage of a future ocean. A few earthquakes—sometimes large—and volcanic eruptions accompany the up-and-down, "normal fault" movements. Volcanic activity is varied, ranging from large rhyolite calderas in the Long Valley Caldera of the Basin and Range region of southeastern California and the Valles Caldera of the Rio Grande Rift of New Mexico, to small basaltic eruptions at the edges of the spreading center. The Red Sea Rift, at the northeastern edge of Africa, is a location where the rift forms a plate boundary. Continental rifts, such as the Rio Grande Rift, spread so slowly that they cannot split the continental plate to form new ocean floor.

Most of the magmas that erupt in continental **rift zones** are either ordinary rhyolite or basalt with little or no intermediate andesite (see Appendix 2 online). But some of the magmas, as in East Africa, are peculiar, with high sodium or potassium contents. Some of the rhyolite ash deposits in the Rio Grande Rift and in the Basin and

Range provide evidence of extremely large and violent eruptions of giant rhyolite volcano activity. But those events appear to be infrequent, and much of the region is sparsely populated, so they do not pose much of a volcanic hazard.

Convergent Boundaries

SUBDUCTION ZONES As the Earth generates new oceanic crust at boundaries where plates pull away from each other, it must destroy old oceanic crust somewhere else. It swallows this old crust in subduction zones (**FIGURE 2-12**). If not, our planet would be growing steadily larger at the same rate as new oceanic crust forms. That is clearly not the case.

The idea of two tectonic plates colliding is truly horrifying at first thought—the irresistible force finally meets the immovable object. But Earth solves its dilemma as one plate slides beneath the other and dives into the hot interior. Grinding rock against rock, the slippage zone sticks and occasionally slips, with an accompanying earthquake. The plate that sinks is the denser of the two, the one with oceanic crust on its outer surface. It absorbs heat as it sinks into the much hotter rock beneath.

Where an oceanic plate sinks in a subduction zone, a line or *arc* of picturesque volcanoes rises inland from the trench. The process begins at the oceanic spreading ridge, where fractures open in the ocean floor. Seawater penetrates the dense peridotite of the upper mantle, where the two react to make a greenish rock called serpentinite. That altered ocean floor eventually sinks through an oceanic trench and descends into the upper mantle, where the serpentinite heats up, breaks down, releases its water, and reverts back to peridotite. The water rises into the overlying mantle, which it partially melts to make basalt magma that rises toward the surface. If the basalt passes through continental crust, it can heat and melt some of those rocks to make rhyolite magma. The basalt and rhyolite may erupt separately or mix in any proportion to form andesite and related rocks, the common volcanic rocks in stratovolcanoes. The High Cascades volcanoes in the Pacific Northwest are a good example; they lie inland from an oceanic trench, the surface expression of the active subduction zone (**FIGURE 2-13**).

Recall that most mountain ranges stand high. They stand high because they are either hot volcanoes of the volcanic arc or part of the hot *backarc*, the area behind the arc, above the descending subduction slab (see FIGURE 2-12). The backarc environment stands high because it weakens, perhaps due to circulating hot water-bearing rocks of the asthenosphere that spread, expand, and rise. In some cases, an oceanic plate descends beneath another section of oceanic plate attached to a continent. The same melting

FIGURE 2-12 SUBDUCTION ZONE HAZARDS

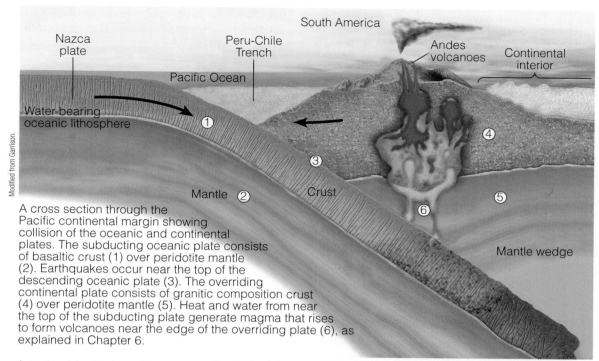

Modified from Garrison.

A cross section through the Pacific continental margin showing collision of the oceanic and continental plates. The subducting oceanic plate consists of basaltic crust (1) over peridotite mantle (2). Earthquakes occur near the top of the descending oceanic plate (3). The overriding continental plate consists of granitic composition crust (4) over peridotite mantle (5). Heat and water from near the top of the subducting plate generate magma that rises to form volcanoes near the edge of the overriding plate (6), as explained in Chapter 6.

A continental volcanic arc forms on a continent where the oceanic lithosphere descends beneath the continental margin. Earthquakes are generated in the subduction zone where the overriding lithosphere sticks against the descending lithosphere and then suddenly slips.

FIGURE 2-13 VOLCANOES NEAR SUBDUCTION ZONES

NOAA basemap.

John Pallister, USGS

The Cascade volcanic chain forms a prominent line of peaks parallel to the oceanic trench and 100 to 200 kilometers inland. Mount St. Helens (in foreground) and Mount Rainier (behind) are two of the picturesque active volcanoes that lie inland from the Cascadia subduction zone.

process described previously generates a line of basalt volcanoes because there is no overlying continental crust to melt and form rhyolite.

Volcanoes above a subducting slab present hazards to nearby inhabitants and their property. Deterring people from settling near these hazards can be difficult, because volcanoes are very scenic, and the volcanic rocks break down into rich soils that support and attract large populations. Volcanoes surrounded by people are prominent all around the Pacific basin and in Italy and Greece, where the African Plate collides with Europe.

The sinking slab of lithosphere also generates many earthquakes, both shallow and deep. Earth's largest earthquakes are generated along subduction zones; some of these cause major natural catastrophes. Somewhat smaller—but still dangerous—earthquakes occur in the overlying continental plate between the oceanic trench and the line of volcanoes.

Sudden slippage of the submerged edge of the continental plate over the oceanic plate during a major earthquake can cause rapid vertical movement of a lot of water, which creates a huge tsunami wave. The wave both washes onto the nearby shore and races out across the ocean to endanger other shorelines.

Collision of Continents

Where two continental plates collide, called a continental **collision zone**, the results can be catastrophic. Neither plate sinks, so high mountains, such as the Himalayas, are pushed up in fits and starts, accompanied by large earthquakes (**FIGURE 2-14**). During the continuing collision between India against Asia to form the Himalayas, and between the Arabian Plate and Asia to form the Caucasus range farther west, earthquakes regularly kill thousands of people. These earthquakes are distributed across a wide area because of the thick, stiff crust in these mountain ranges.

FIGURE 2-14 CONTINENTAL COLLISION ZONES

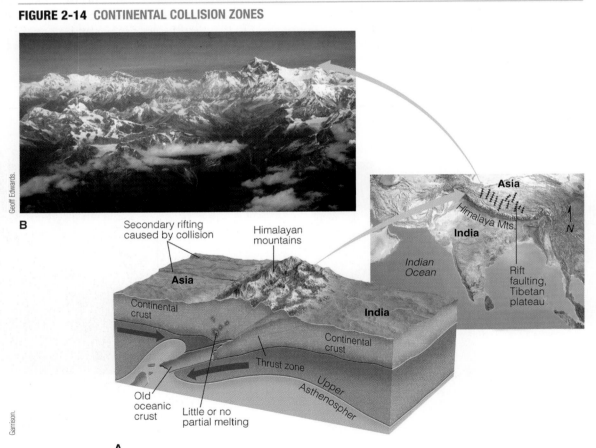

A. Collision of two continental plates generally occurs after subduction of oceanic crust. The older, colder, denser plate may continue to sink, or the two may merely crumple and thicken. Collision promotes thickening of the combined lithospheres and growth of high mountain ranges. **B.** The Himalayas, which are the highest mountains on any continent, were created by collision between the Indian and Eurasian Plates.

Transform Boundaries

In some places, plates simply slide past each other without pulling apart or colliding. Those are called transform plate boundaries or transform faults. Some of them offset the mid-oceanic ridges. Because the ridges are spreading zones, the plates move away from them. The section of the fault between the offset ends of the spreading ridge has significant relative movement (**FIGURE 2-15A**). Lateral movement between the ridge ends (areas of earthquakes shown by red stars) occurs in the opposite direction compared to beyond the ridges, where there is no relative movement across the same fault. Note also that the offset between the two ridge segments does not indicate the direction of relative movement on the transform fault.

Oceanic transform faults generate significant earthquakes without causing casualties because no one lives on the ocean floor. On continents it is a different story. The San Andreas Fault system in California (**FIGURE 2-15B**) is a well-known continental example. The North Anatolian Fault in Turkey is another that is even more deadly. The San Andreas Fault is the dominant member of a swarm of more-or-less parallel faults that move horizontally. Together, they have moved a large slice of western California, part of the Pacific Plate, north more than 350 kilometers so far.

Transform plate boundaries typically generate large numbers of earthquakes, a few of which are catastrophic. A sudden movement along the San Andreas Fault caused the devastating San Francisco earthquake of 1906, with its large toll of casualties and property damage. The San Andreas system of faults passes through the metropolitan areas south of San Francisco and just east of Los Angeles. Both areas are home to millions of people, who live at risk of major earthquakes that have the potential to cause enormous casualties and substantial property damage with little or no warning. Even moderate earthquakes in 1971 and 1994 near Los Angeles, in 1989 near San Francisco, and in 2003 near Paso Robles, between them, killed almost 200 people. The threat of such sudden havoc in a still larger event inspires much public concern and major scientific efforts to find ways to predict large earthquakes.

For reasons that remain mostly unclear, some transform plate boundaries are also associated with volcanic activity. Several large volcanic fields have erupted along the San Andreas system of faults during the last 16 million or so years. One of those, in the Clear Lake area north of San Francisco, erupted recently enough to suggest that it may still be capable of further eruptions.

FIGURE 2-15 TRANSFORM FAULT

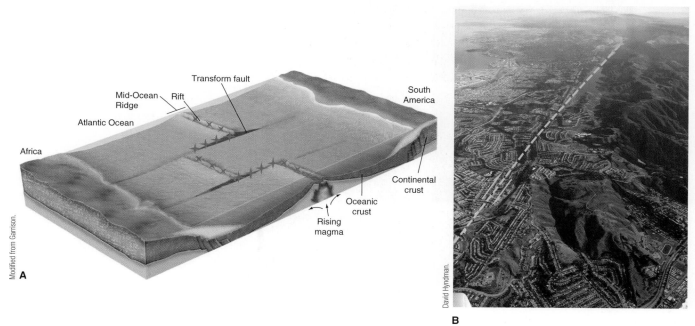

A. In this perspective view of an oceanic spreading center, earthquakes (stars) occur along spreading ridges and on transform faults offsetting the ridge. **B.** The San Andreas Fault, indicated with a yellow, dashed line, is an example of a continental transform fault. The heavily populated area, here viewed south from above San Francisco, California, straddles the San Andreas Fault, under San Andreas Lake and Crystal Springs Reservoir seen in the distance.

Hotspot Volcanoes

Despite being remote from any plate boundary, **hotspot volcanoes** provide a record of plate tectonic movements. Hotspots are the surface expressions of hot columns of partially molten rock anchored (at least relative to plate movements) in the deep mantle. Their origin is unclear, but many scientists infer that they arise from deep in the mantle, perhaps near the boundary between the core and the mantle. At a hotspot, plumes of abnormally hot but solid rock rising within Earth's mantle begin to melt as the rock pressure on them drops. Wherever peridotite of the asthenosphere partially melts, it releases basalt magma that fuels a volcano on the surface. If the hotspot is under the ocean floor, the basalt magma erupts as basalt lava. If the hot basalt magma rises under continental rocks, it partially melts those rocks to form rhyolite magma; that magma often produces violent eruptions of ash.

The melting temperature of basalt is more than 300°C above that of rhyolite, so a small amount of molten basalt can melt a large volume of rhyolite. The molten rhyolite rises in large volumes, which may erupt explosively through giant rhyolite calderas, such as those in Yellowstone National Park in Wyoming and Idaho, Long Valley Caldera in eastern California, and Taupo Caldera in New Zealand.

The rising column or plume of hot rock appears to remain nearly fixed in its place as one of Earth's plates moves over it, creating a track of volcanic activity. The movement of the plate over an oceanic hotspot is evident in a chain of volcanoes, where the oldest volcanoes are extinct, and possibly submerged, while newer, active volcanoes are created at the end of the chain. Mauna Loa and Kilauea, for example, erupt at the eastern end of the Hawaiian Islands, a chain of extinct volcanoes that become older westward toward Midway Island (**FIGURE 2-16**). Beyond Midway, the Hawaiian-Emperor chain doglegs to a more northerly course. It continues as a long series of defunct volcanoes that are now submerged. They form seamounts to the western end of the Aleutian Islands west of Alaska. So far as anyone knows, the hotspot track of dead volcanoes will continue to lengthen until eventually the volcanoes and the plate carrying them slide into a subduction zone and disappear.

Hotspot volcanoes leave a clear record of the direction and rate of movement of the lithospheric plates. Remnants of ancient hotspot volcanoes show the direction of movement

FIGURE 2-16 OCEANIC HOTSPOTS

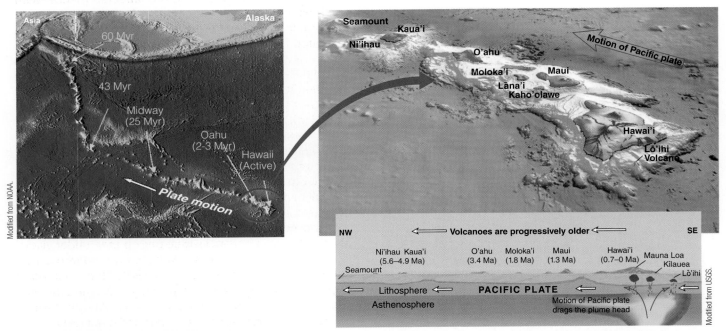

The relief map of the Hawaiian-Emperor chain of volcanoes clearly shows the movement of the crust over the hotspot that is currently below the Big Island of Hawaii, where there are active volcanoes. Two to three million years ago, the part of the Pacific Plate below Oahu was over the same hotspot. The approximate rate and direction of plate motion can be calculated using the common belief that the hotspot is nearly fixed in space through time. The distance between two locations of known ages divided by the time (age difference) indicates a rate of movement of about 9 cm per year. The lithospheric plate, moving across a stationary hotspot in the Earth's mantle (moving to the left in this diagram), leaves a track of old volcanoes. The active volcanoes are over the hotspot.

in the same way that a saw blade cuts in the opposite direction of movement of a board being cut. The ages of those old volcanoes provide the rate of movement of the lithospheric plate. The assumption, of course, is that the mantle containing the hotspot is not itself moving. Comparison of different hotspots suggests that this is generally valid compared with migration of the tectonic plates, but many researchers suggest that it is not absolutely so.

The Snake River Plain of southern Idaho is probably the best example of a continental hotspot track. Along this track is a series of extinct *resurgent calderas*, depressions where the erupting giant volcano collapsed. Those volcanoes began to erupt some 14 million years ago. They track generally east and northeast in southern Idaho, becoming progressively younger northeastward as the continent moves southwestward over the hotspot (**FIGURE 2-17**). They are a continental hotspot track that leads from its western end near the border between Idaho and Oregon to the Yellowstone resurgent caldera at its active northeastern end in northwestern Wyoming.

Hotspot tracks are one piece of evidence that Earth's plates are in motion. This theory of plate movements, colliding, spreading, and sliding past each other, and the mountain ranges that form by such movements, has been confirmed by repeated wide-ranging studies, tests, and many predictions, all of which confirm its validity. What was once a series of **hypotheses** or ideas which remained to be confirmed, has been so thoroughly examined and tested, that it has been elevated to the category of **theory**— that is, it is now considered to be *fact*. What prompted the original hypotheses and how did it finally lead to the present understanding?

Development of a Theory

When you look at a map of the world, you may notice that the continents of South America and Africa would fit nicely together like puzzle pieces. In fact, as early as 1596, Abraham Ortelius, a Dutch map maker, noted the similarity of the shapes of those coasts and suggested that Africa and South America were once connected and had since moved apart. In 1912, Alfred Wegener detailed the available evidence and proposed that the continents were originally part of one giant supercontinent that he called **Pangaea** (**FIGURE 2-18A**, p. 28). Wegener noted that the match between the shapes of the continents is especially good if we use the real edge of the continents, including the shallowly submerged continental shelves.

To test this initial hypothesis, Wegener searched for connections between other aspects of geology across the Atlantic Ocean: mountain ranges, rock formations and their ages, and fossil life forms. Continued work showed that ancient rocks, their fossils, and their mountain ranges also matched on the other side of the Atlantic (**FIGURE 2-18B**, p. 28). This analysis is similar to what you would use to put a jigsaw puzzle together; the pieces fit and the patterns match across the reconnected pieces. With confirmation of former connections, he hypothesized that the continents had moved apart; North and South America separated from Europe and Africa, widening the Atlantic Ocean in the process. He suggested that the continents drifted through the oceanic crust, forming mountains along their leading edges. This hypothesis, called **continental drift**, remained at the center of the debate about large-scale Earth movements into the 1960s.

As research has continued, other lines of evidence supported the continental drift hypothesis. Exposed surfaces of ancient rocks in the southern parts of Australia, South America, India, and Africa show grooves carved by immense areas of continental glaciers (**FIGURE 2-19**, p. 28). The grooves show that glaciers with embedded rocks at their bases may have moved from Antarctica into India, eastern South America, and Australia. The rocks were once buried under glacial ice, yet many of these areas now have warm to tropical climates. In addition, the remains of fossils that formed in warm climates are found in areas such as Antarctica and the present-day Arctic: coal with fossil impressions of tropical leaves, the distinctive fossil fern Glossopteris, and coral reefs.

Despite this evidence, many scientists rejected Wegener's whole hypothesis because they could show that his proposed mechanism was not physically possible. English geophysicist Harold Jeffreys argued that the ocean floor rocks were far too strong to permit the continents to plow through them. Others who were willing to consider different possibilities eventually came up with a mechanism that fit all of the available data.

FIGURE 2-17 CONTINENTAL HOTSPOTS

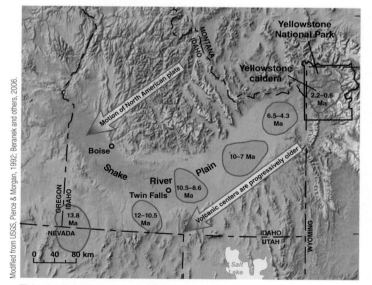

Modified from USGS, Pierce & Morgan, 1992; Beranek and others, 2006.

This shaded relief map of the Snake River Plain shows the outlines of ancient resurgent calderas leading northeast to the present-day Yellowstone caldera. Caldera ages are shown in millions of years before present.

FIGURE 2-18 CONTINENTS ONCE FIT TOGETHER

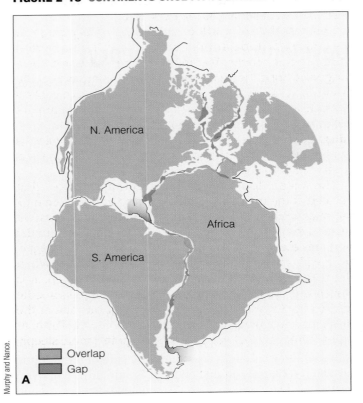

Murphy and Nance.

Overlap
Gap

A.

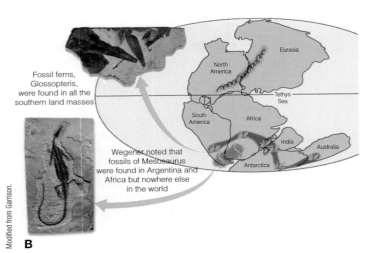

Modified from Garrison.

Fossil ferns, Glossopteris, were found in all the southern land masses

Wegener noted that fossils of Mesosaurus were found in Argentina and Africa but nowhere else in the world

Eurasia

North America

Tethys Sea

South America

Africa

India

Australia

Antarctica

B.

A. Before continental drift a few hundred million years ago, the continents were clustered together as a giant "supercontinent" that has been called Pangaea. The Atlantic Ocean had not yet opened. The pale blue fringes on the continents are continental shelves, which are part of the continents. The areas of overlap and gap (in red and darker blue) are small. **B.** Some distinctive fossils and mountain ranges lie in belts across the Atlantic and Indian oceans.

FIGURE 2-19 GLACIATION IN WARM AREAS

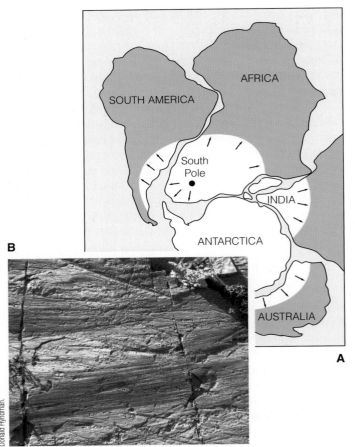

Donald Hyndman.

AFRICA

SOUTH AMERICA

South Pole

INDIA

ANTARCTICA

AUSTRALIA

B

A

A. Continental masses of the southern hemisphere appear to have been parts of a supercontinent 300 million years ago, from which a continental ice sheet centered on Antarctica spread outward to cover adjacent parts of South America, Africa, India, and Australia. After separation, the continents migrated to their current positions. **B.** The inset photo shows glacial grooves like those found in the glaciated areas of those continents.

The first step in understanding how the continents were separating was to learn more about the topography of the ocean floor, what it looked like, and how old it was. Oceanographers from Woods Hole Oceanographic Institute in Massachusetts, who were measuring depths from all over the Atlantic Ocean in the late 1940s and 1950s, found an immense mountain range down the center of the ocean, extending for its full length—a **mid-oceanic ridge** (**FIGURE 2-20**). Later, scientists recognized that most earthquakes in the Atlantic Ocean were concentrated in that central ridge.

Although the anti-continental drift group dominated the scientific literature for years, in 1960 Harry Hess of Princeton University conjectured that the ocean floors acted as giant conveyor belts carrying the continents. Hess calculated the spreading rate to be approximately 2.5 centimeters (1 inch)

FIGURE 2-20 OCEAN-BOTTOM TOPOGRAPHY

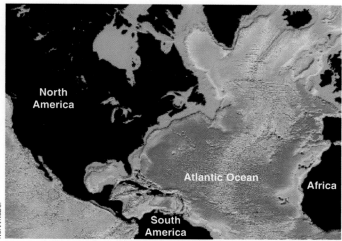

NOAA/NGDC.

In this seafloor topographic map for the Atlantic Ocean, shallow depths at the oceanic ridges are shown in orange to yellow, deeper water off the ridge crests are in green, and deep ocean is blue. Shallow continental shelves are shown in red.

per year across the Mid-Atlantic Ridge. If that calculation was correct, the whole Atlantic Ocean floor would have been created in about 180 million years.

Confirmation of seafloor spreading finally came in the mid-1960s through work on the magnetic properties of ocean floor rocks. We are all aware that Earth has a **magnetic field** because a magnetized compass needle points toward the north magnetic pole. Slow convection currents in the Earth's molten nickel-iron outer core are believed to generate that magnetic field (**FIGURE 2-21**). Because of changes in those currents, this field reverses its north-south orientation every 10,000 to several million years (every 600,000 years on average).

The ocean floor consists of basalt, a dark lava that erupted at the mid-oceanic ridge and solidified from molten magma. Iron atoms crystallizing in the magma orient themselves like tiny compass needles, pointing toward the north magnetic pole. As a result, the rock is slightly magnetized with an orientation like the compass needle. When the magnetic field reverses, that reversed magnetism is frozen into rocks when they solidify. A compass needle at the equator remains nearly horizontal but one at the north magnetic pole points directly down into the Earth. At other latitudes in between, the needle points more steeply

FIGURE 2-21 EARTH'S MAGNETIC FIELD

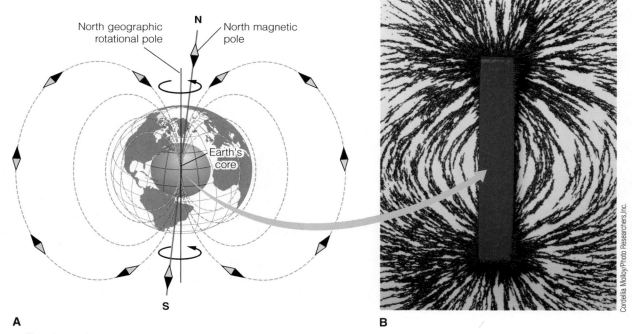

Cordellia Molloy/Photo Researchers, Inc.

A. The shape of Earth's magnetic field suggests the presence of a huge bar magnet in the Earth's core. But instead of a magnet, Earth's rotation is thought to cause currents in the liquid outer core. Those currents create a magnetic field in a similar way in which power plants generate electricity when steam or falling water rotates an electrical conductor in a magnetic field.
B. Metal filings align with the magnetic field lines from this bar magnet.

downward as it approaches the poles. Thus we can tell the latitude at which the rock formed when it solidified by the inclination of its magnetism.

British oceanographers Frederick Vine and Drummond Matthews, studying the magnetic properties of ocean-floor rocks in the early 1960s, discovered a striped pattern parallel to the mid-oceanic ridge (**FIGURE 2-22**). Some of the stripes were strongly magnetic; adjacent stripes were weakly magnetic. They realized that the magnetism was stronger where the rocks solidified while Earth's magnetism was oriented parallel to the present-day north magnetic pole. Where the rock magnetism was pointing toward the south magnetic pole, the recorded magnetism was weak—it was partly canceled by the present-day magnetic field. Because it reversed from time to time, Earth's

magnetic field imposed a pattern of magnetic stripes as the basalt solidified at the ridge. As the ridge spread apart, ocean floor formed under alternating periods of north-versus south-oriented magnetism to create the matching striped pattern on opposite sides of the ridge.

These magnetic anomalies provide the relative ages of the ocean floor; their mapped widths match across the ridge, and the rocks are assumed to get progressively older as they move away from mid-oceanic ridges. Determination of the true ages of ocean-floor rocks eventually came from drilling in the deep-sea floor by research ships of the Joint Oceanographic Institute for Deep Earth Sampling (JOIDES), funded by the National Science Foundation. The ages of basalts and sediments dredged and drilled from the ocean floor showed that those near the Mid-Atlantic Ridge were young (up to 1 million years old) and had only a thin coating of sediment. Both results contradicted the prevailing notion that the ocean floor was extremely old. In contrast, rocks from deep parts of the ocean floor far from the ridge were consistently much older (up to 180 million years) (**FIGURE 2-23**).

All of this evidence supports the modern theory of plate tectonics, the big picture of Earth's plate movements. We now know that the world's landmasses once formed one giant supercontinent, called Pangaea, 225 million years ago. As the seafloor spread, Pangaea began to break up, and the plates slowly moved the continents into their current positions (**FIGURE 2-24**).

As it turns out, Wegener's hypothesis that the continents moved apart was confirmed by the data, although his assumption that they plowed through the ocean was not. The evolution of this theory is a good example of how the scientific method works.

FIGURE 2-22 MAGNETIC RECORD OF OCEAN-FLOOR SPREADING

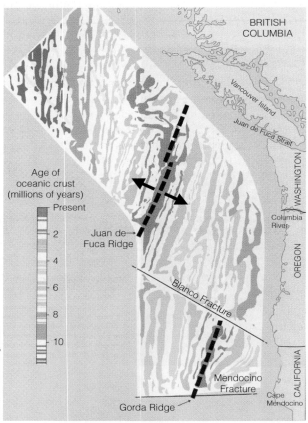

The magnetic polarity, or orientation, across the Juan de Fuca Ridge in the Pacific Ocean shows a symmetrical pattern, as shown in this regional survey (a similar nature of stripes exists along all spreading centers). Basalt lava erupting today records the current northward-oriented magnetism right at the ridge; basalt lavas that erupted less than 1 million years ago recorded the reversed, southward-oriented magnetic field at that time. The south-pointing magnetism in those rocks is largely canceled out by the present-day north-pointing magnetic field, so the ocean floor shows alternating strong (north-pointing) and weak (south-pointing) magnetism in the rocks.

FIGURE 2-23 AGES OF OCEAN FLOOR

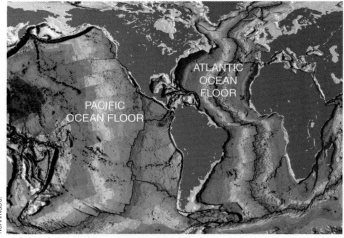

Ocean-floor ages are determined by their magnetic patterns. Red colors at the oceanic spreading ridges grade to yellow at 48 million years ago, to green 68 million years ago, and to dark blue some 155 million years ago.

FIGURE 2-24 CONTINENTS SPREAD APART

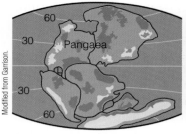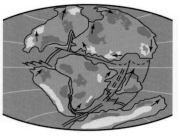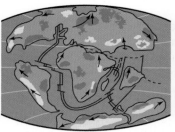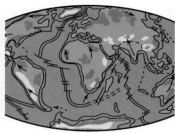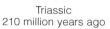

Triassic 210 million years ago	Late Jurassic 135 million years ago	Late Cretaceous 70 million years ago	Present

Mid-ocean ridges
Oceanic Trench

The supercontinent Pangaea broke up into individual continents starting approximately 225 million years ago.

The **scientific method** is based on logical analysis of data to solve problems. Scientists make observations and develop tentative explanations—that is, hypotheses—for their observations. A hypothesis should always be testable, because science evolves through continual testing with new observations and experimental analysis. Alternate hypotheses should be developed to test other potential explanations for observed behavior. If observations are inconsistent with a hypothesis, it can either be rejected or revised. If a hypothesis continues to be supported by all available data over a long period of time, and if it can be used to predict other aspects of behavior, it becomes a theory.

After a century of testing, Wegener's initial hypothesis of continental drift was modified to be the foundation for the modern theory of plate tectonics. Plate tectonics is supported by a large mass of data collected over the last century. Modern data continue to support the concept that plates move, substantiate the mechanism of new oceanic plate generation at the mid-oceanic ridges, and support the concept of plate destruction at oceanic trenches. This theory is a fundamental foundation for the geosciences and important for understanding why and where we have a variety of major geologic hazards, such as earthquakes and volcanic eruptions.

Chapter Review

Key Points

Earth Structure and Plates

■ The concept of isostacy explains why the lower-density continental rocks stand higher than the higher density ocean-floor rocks and sink deeper into the underlying mantle. This behavior is analogous to ice (lower density) floating higher in water (higher density). **FIGURE 2-1** and **By the Numbers 2-1**.

■ A dozen or so nearly rigid lithospheric plates make up the outer 60 to 200 kilometers of the Earth. They slowly slide past, collide with, or spread apart from each other. **FIGURES 2-5** and **2-6**.

Hazards and Plate Boundaries

■ Much of the tectonic action, in the form of earthquakes and volcanic eruptions, occurs near the boundaries between the lithospheric plates. **FIGURES 2-9** and **2-10**.

- Where plates diverge from each other, new lithosphere forms. If the plates are continental material, a continental rift zone forms. As this process continues, a new ocean basin can develop, and the spreading continues from a mid-oceanic ridge, where basaltic magma pushes to the surface. **FIGURES 2-9** and **2-11**.

- Subduction zones, where ocean floors slide beneath continents or beneath other slabs of oceanic crust, are areas of major earthquakes and volcanic eruptions. These eruptions form volcanoes on the overriding plates. **FIGURES 2-12** and **2-13**.

- Continent–continent collision zones, where two continental plates collide, are regions with major earthquakes and the tallest mountain ranges on Earth. **FIGURE 2-14**.

- Transform faults involve two lithospheric plates sliding laterally past one another. Where these faults cross continents, such as along the San Andreas Fault through California, they cause major earthquakes. **FIGURE 2-15**.

- Hotspots form chains of volcanoes within individual plates rather than near plate boundaries. Because lithosphere is moving over hotspots fixed in the Earth's underlying asthenosphere, hotspots grow as a trailing track of progressively older extinct volcanoes. **FIGURES 2-16** and **2-17**.

Development of a Theory

- Continental drift was proposed by matching shapes of the continental margins on both sides of the Atlantic Ocean, as well as the rock types, deformation styles, fossil life forms, and glacial patterns. **FIGURES 2-18** and **2-19**.

- Continental drift evolved into the modern theory of plate tectonics based on new scientific data, including the existence of a large ridge running the length of many deep oceans, matching alternating magnetic stripes in rock on opposite sides of the oceanic spreading ridges, and age dates from oceanic rocks that confirmed a progressive sequence from very young rocks near the rifts to older oceanic rocks toward the continents. **FIGURES 2-20** to **2-24**.

- The scientific method involves developing tentative hypotheses that are tested by new observations and experiments, which can lead to confirmation or rejection.

Key Terms

asthenosphere, p. 15
collision zone, p. 24
continental drift, p. 27
convergent boundaries, p. 18
divergent boundaries, p. 18
hotspot volcanoes, p. 26

hypotheses, p. 27
isostacy, p. 16
lithosphere, p. 15
magnetic field, p. 29
mantle, p. 16
mid-oceanic ridge, p. 28

Mohoroviçic Discontinuity, p. 17
Pangaea, p. 27
plate tectonics, p. 17
rift zones, p. 22
scientific method, p. 31

seafloor spreading, p. 18
subduction zone, p. 18
tectonic plates, p. 15
theory, p. 27
transform boundary, p. 18
trenches, p. 18

Questions for Review

1. Distinguish among Earth's crust, lithosphere, asthenosphere, and mantle.

2. What does oceanic lithosphere consist of and how thick is it?

3. What are the main types of lithospheric plate boundaries – described in terms of relative motions? Provide a real example of each (by name or location).

4. Why does oceanic lithosphere almost always sink beneath continental lithosphere at convergent zones?

5. Along which type(s) of lithospheric plate boundary are large earthquakes common? Why?

6. Along which type(s) of lithospheric plate boundary are large volcanoes most common? Provide an example.

7. What direction is the Pacific Plate currently moving, based on FIGURE 2-16? How fast is this plate moving?

8. Before people understood plate tectonics, what evidence led some scientists to believe in continental drift?

9. If the coastlines across the Atlantic Ocean are spreading apart, why isn't the Atlantic Ocean deepest in its center?

10. What evidence confirmed seafloor spreading?

11. Why are high volcanoes such as the Cascades found on the continents and in a row parallel to the continental margin?

Discussion Questions

1. Discuss how the modern theory of plate tectonics developed in the context of the scientific method.

2. Discuss the height of a mountain range compared with the thickness of the crust or lithosphere below the mountain, and relate this to the percentage of an iceberg above the water line.

3. Explain the role of density of Earth materials with respect to Earth's features such as mountains and mid-ocean ridges. For example, why is the top of basaltic crust below sea level while the surface of granitic crust is generally above sea level.

4. The Basin and Range region of Nevada and Utah is a continental spreading zone. Because it is pulling apart, why isn't its elevation low, rather than as high as it is? Why isn't it an ocean?

3 Earthquakes and Their Causes

The Haiti earthquake on January 12, 2010, destroyed the capital city of Port-au-Prince and surrounding areas, killing more than 230,000 people.

Candice Villareal, U.S. Navy.

Deadly Effects with Poor Quality Construction

At 4:53 in the afternoon of January 12, 2010, the sprawling city of Port-au-Prince, capital of Haiti, shook violently in a magnitude 7.0 earthquake that killed an estimated 250,000 people out of a population of 1.2 million. It was on the east-west-trending Enriquillo-Plaintain Garden Fault that last produced a large quake in 1860. In 1751, a big earthquake struck the island; a month later, another destroyed Port-au-Prince. The left-lateral strike-slip fault accommodates just under half of the motion between the eastward-moving Caribbean Plate on the south and the North American Plate. The two sides of the fault move past one another at about 7 mm per year. In 1946, a magnitude 8.1 earthquake in the Dominican Republic, east of Haiti, caused a tsunami that killed 1,790 people. The new break was only 6 kilometers below the surface so the shaking was not weakened much by distance. About 1.85 million people experienced violent shaking at Mercalli Intensity IX. Survivors described the earthquake arriving with the sound of a freight train or jet engine, immediately followed by a sudden jolt and continuous violent shaking for about 35 seconds. Clouds of dust billowed up from collapsing buildings and the sky turned gray with the dust. The power failed so the city was mostly dark after the sun went down and communications were almost non-existent.

Haiti has virtually no construction standards. Its rampant corruption and poverty, the worst in the western hemisphere, led to shoddy construction that was unsafe even before the earthquake. Thousands of people were crushed in collapsing buildings. Walls were commonly poor quality, composed of handmade and light-weight cinder blocks with minimal cement and little or no reinforcing steel. Thin, poor-quality concrete floors often had only small amounts of light-weight reinforcing bar. People caught in collapsing buildings described floors giving way beneath them and ceilings and walls coming down around them. Survivors told of being trapped in pitch dark, feeling completely helpless, surrounded by concrete, with broken bones, and sometimes their legs or arms held or crushed under debris with no possible escape. People on sidewalks were killed by falling parts of buildings. The central cathedral, the best hotel in the city, and the local United Nations Development Program headquarters collapsed, and the Presidential Palace was severely damaged. Severely injured people sat in the street, pleading for medical help, but most hospitals had either collapsed or were badly damaged and many doctors and nurses had been killed. At least 1.5 million residents were left homeless and damages reached between 7.2 and 13.2 billion dollars. Garment factories collapsed, killing many workers and ending the source of employment for thousands. One shirt factory owner managed to restart part of a factory that suffered minimal damage and put hundreds back to work within a few days; they earn only about $7 a day but that is more than most. A large part of the population lives on only two dollars per day; they have few possessions and no financial resources. The earthquake contrasts strikingly with the February 27, 2010, Chile earthquake, Magnitude 8.8 (about 500 times stronger), that killed fewer than one thousand people. In Chile, construction standards are enforced and the country is much more affluent.

When the dust settled the city was in chaos. Buildings everywhere were destroyed; bodies lay in the street or protruded from under debris. People immediately began searching for relatives, coworkers, and neighbors. Many children were separated from their parents or left orphans. With relief workers focused on search and rescue, and most streets blocked by debris, dozens of bodies remained stacked in piles. With no reasonable burial sites, the marginally functional government removed tens of thousands of bodies with bulldozers and trucks, dumping them in mass burial sites in swamps at the edge of the city. More than 150,000 ended up there, mostly unidentified.

UN Development Program.

The Haiti earthquake collapsed numerous buildings, mostly of concrete and cinder block construction with little or no reinforcing steel. Rescue teams searched accessible spaces under collapsed buildings.

When people gave up any hope the authorities would come to take the decaying corpses of relatives, some burned the bodies in impromptu funeral pyres using wood craters, scrap wood, and old tires.

Aftershocks continued for days, including two of magnitude 5.9 to 6.0. People feared the aftershocks would collapse already damaged buildings. They camped out on streets in the open, or under tarps, scraps of cardboard, plywood, or corrugated sheet metal. In late January and early February, aftershocks and heavy rains collapsed a school and killed three children. In separate events, four people were trapped when a building collapsed and an aftershock collapsed more of a damaged supermarket. Days later, even with more searchers, fewer and fewer were found alive; after more than a week, teams began giving up. However, one 84-year-old woman was pulled, severely dehydrated and in shock, but alive, from her flattened house 10 days after the quake. One man survived 11 days buried in a collapsed hotel grocery store.

Aid came slowly. Aid groups already on the ground, from UN countries, charities, and religious organizations, provided initial searches as best they could, but they had also lost many of their members in the earthquake. Water lines failed, leading to contaminated water, and food distribution channels were disrupted. Port-au-Prince's seaport, its main conduit for food imports, was destroyed in the earthquake. Its single-runway international airport was damaged, and the main highway—of poor quality to begin with—was further hindered by damage and debris. Many cars and trucks were damaged or destroyed. The airport was quickly repaired but its single runway, limited aircraft parking, and limited supply of aircraft fuel made it an aid bottleneck. Except for medical help to the injured, clean water is most important. People can survive much longer without food than without water. Aid flights brought in bottled water, and the desalination plant of a U.S. aircraft carrier standing off the coast soon provided more than 750,000 liters of fresh water per day.

By the end of January, 700,000 were living in scattered, chaotic, unsanitary tent camps. Some tent camps had only one portable toilet for 2,000 people! Most were forced to use a nearby gutter. Earthquake relief teams were also leveling ground outside the city to move half of those in squalid tent settlements to safer areas—where tents could be spread out on drier ground, latrines could be constructed, and clean water and food more easily provided.

Although the main rainy season begins in May, early heavy rains in mid-February flooded low areas and turned streets and the ground under tent camps to mud. Many remained without even plastic sheets for cover. Poor sanitation and widespread contaminated water led to concern about diarrhea, typhoid, cholera, and malaria, which would increase in the rainy season and with higher summer temperatures. Malaria and dengue fever are endemic in Haiti. Some hospitals reported that half of the children being treated had malaria.

Two weeks after the quake, with little clean water, food, or shelter, about 200,000 people abandoned the city, heading to the countryside or other towns, often to stay with relatives. They crammed onto buses, ferries, or even walked. Later, with no work, many of them would return to Port-au-Prince, hoping to at least cook beans or fried dough to sell on the street. Some were employed by aid groups as hand laborers for debris clearing. Cleanup and reconstruction, hopefully with better-built buildings, will take months or more likely years, even with massive external aid. Although Haiti imports a large percentage of its food, its agricultural sector largely escaped damage. Unfortunately much of the imported food is subsidized rice from the United States that undercuts and discourages local rice production.

Faults and Earthquakes

To understand why earthquakes happen, remember that the plates of Earth's crust move, new crust forms, and old crust sinks into subduction zones. It is these movements that give rise to earthquakes, which form along **faults**, or ruptures, in the Earth's crust. Faults are simply fractures in the crust along which rocks on one side of the break move past those on the other. Faults are measured according to the amount of displacement along the fractures. Over several million years, for example, the rocks west of the San Andreas Fault of California have moved at least 450 kilometers north of where they started. Thousands of other faults have moved much less than 1 kilometer in the same amount of time.

Some faults produce earthquakes when they move; others produce almost none. Some faults have not moved for such a long time that we consider them inactive; others are clearly still active and potentially capable of causing earthquakes. Active faults are rare in regions such as the American Midwest and central Canada, where the continental crust has been stable for hundreds of millions of years. Such stable regions contain many faults that geologists have yet to recognize. Some of these first announce their presence when they cause an earthquake; others are marked by the line of a fresh break near the base of a mountainside (**FIGURE 3-1**). Earthquakes are common in the mountainous western parts of North America, where the rocks are deformed into complex patterns of faults and folds.

Faults can be classified according to the way the rocks on either side of the fault move in relation to each other (**FIGURE 3-2**, p. 38). **Normal faults** move on a steeply inclined surface. Rocks above the fault surface slip down and over the rocks beneath the fault. Normal faults move when Earth's crust pulls apart, during crustal extension. **Reverse faults** move rocks on the upper side of a fault up and over those below. **Thrust faults** are similar to reverse faults, but the fault surface is more gently inclined (see **FIGURE 3-2a**). Reverse and thrust faults move when Earth's crust is pushed together, during crustal compression. **Strike-slip faults** move horizontally as rocks on one side of a fault slip laterally past those on the other side. If rocks on the far side of a fault move to the right, as in FIGURE 3-2b, it is a right-lateral fault. If they moved in the opposite direction, it would be a left-lateral fault.

The orientations of rock layers and faults are described in terms of *strike* and *dip*. *Strike* is the orientation of a horizontal line on a rock surface, illustrated as the line of water standing against the rock surface. *Dip* is the inclination angle (perpendicular to the strike direction) down from horizontal to the rock surface (**FIGURE 3-3**, p. 38).

Causes of Earthquakes

At the time of the great San Francisco earthquake of 1906 (see **Case in Point**: Devastating Fire Caused by an Earthquake—San Francisco, California, 1906, p. 97), the cause of earthquakes was a complete mystery. The governor of California at the time appointed a commission to find the cause of earthquakes. The director of this commission, Andrew C. Lawson, was a distinguished geologist and one of the most colorful personalities in the history of California. Lawson and his students at the University of California (UC)–Berkeley had already recognized the San Andreas Fault and mapped large parts of it, but

---

FIGURE 3-1 FAULT SCARPS

A. This fault scarp near West Yellowstone, Montana, formed during the 1959 Hebgen Lake, Montana, earthquake. Such lines indicate an active fault. **B.** An aerial view of an active, basin-and-range normal-fault scarp in southwestern Montana. The mountains in the upper-left rose as the valley in the lower-right dropped.

FIGURE 3-2 FAULT MOVEMENTS

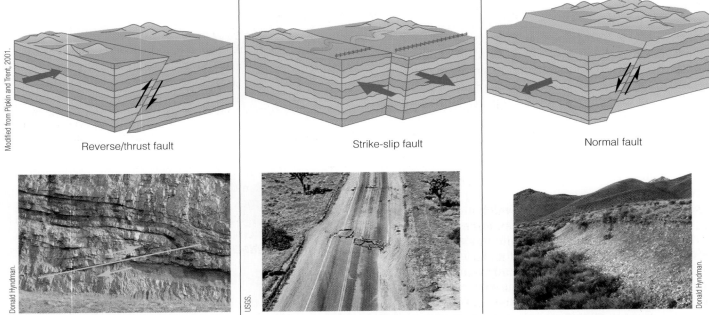

Modified from Pipkin and Trent, 2001.

Reverse/thrust fault

Strike-slip fault

Normal fault

Donald Hyndman.

USGS.

Donald Hyndman.

A. A small thrust fault (see arrows) east of Vail, CO. **B.** A strike-slip fault offset the road after the Landers earthquake in California.
C. A normal fault near Challis, Idaho, moved in the 1983 earthquake.

FIGURE 3-3 STRIKE AND DIP

Donald Hyndman.

dip

Strike direction

Strike is the direction of a horizontal line (standing water intersects the layer). Dip is the slope of the layer measured down from horizontal.

until the 1906 event they had no idea that it could cause earthquakes. During their investigation, members of the commission found numerous places where roads, fences, and other structures had broken during the 1906 earthquake just where they crossed the San Andreas Fault. In every case, the side west of the fault had moved north as much as 7 meters (23 feet). That led to the theory of how fault movement causes earthquakes.

The earthquake commission hypothesized that as Earth's crust moved, the rocks on opposite sides of the fault had bent, or deformed, instead of slipping, over many years. As the rocks on opposite sides of the fault bent, they accumulated energy. When the stuck segment of the fault finally slipped, the bent rocks straightened with a sudden snap, releasing energy in the form of an earthquake (**FIGURE 3-4**). Imagine pulling a bow taut, bending it out of its normal shape, and then releasing it. It would snap back to its original shape with a sudden release of energy capable of sending an arrow flying. This explanation for earthquakes, called the **elastic rebound theory**, has since been confirmed by rigorous testing.

We now know enough about the behavior of rocks in response to stress to explain why faults either stick or slip. We think of rocks as brittle solids, but rocks are elastic, like a spring, and can bend when a force is applied. We use the term **stress** to refer to the forces imposed on a rock and **strain** to refer to the change in shape of the rock in response to the imposed stress. The larger the stress applied, the greater the strain.

Rocks bend, or *deform*, in broadly consistent ways in response to stress. Typical rocks will deform *elastically* under low stress, which means that they revert to their former shape when the force is relieved. At higher stress,

FIGURE 3-4 ELASTIC REBOUND THEORY

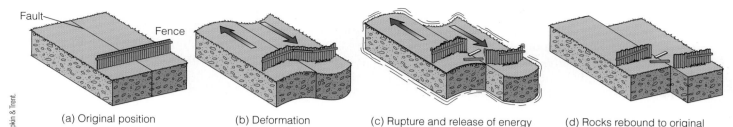

(a) Original position (b) Deformation (c) Rupture and release of energy (d) Rocks rebound to original undeformed shape

Rocks near a fault are slowly bent elastically (b) until the fault breaks during an earthquake (c), when the rocks on each side slip past each other, relieving the stress.

Pipkin & Trent.

these rocks will deform *plastically*, which means they permanently change shape or flow when forces are applied. Deformation experiments show that most rocks near Earth's surface, where they are cold and not under much pressure from overlying rocks, deform elastically when affected by small forces. Under other conditions, such as deep in the Earth where they are hot and under high pressure imposed by the overlying load of rocks, it is much more likely that rocks will deform plastically.

Rocks can bend, but they also break if stretched too far. In response to smaller stresses, rocks may merely bend, while in response to large stresses, they fracture or break. As stress levels increase, rocks ultimately succumb to *brittle failure*, causing fault slippage or an earthquake (**FIGURE 3-5** and **By the Numbers 3-1**: Compression and Rock Shear). Under these conditions, a fault may begin to fail, with smaller

FIGURE 3-5 STRESS AND STRAIN

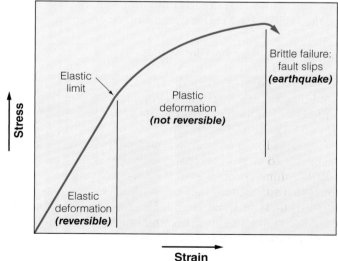

With increasing stress, a rock deforms elastically, then plastically, before ultimately failing or breaking in an earthquake. A completely brittle rock fails at its elastic limit.

▶ **By the Numbers 3-1**

Compression and Rock Shear

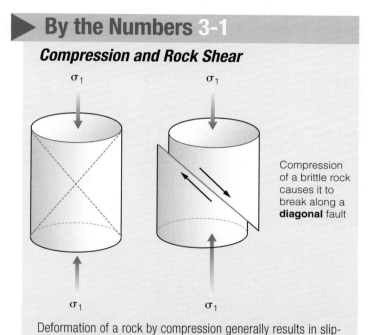

Compression of a brittle rock causes it to break along a **diagonal** fault

Deformation of a rock by compression generally results in slippage diagonal to the direction of compression.

Experimental study of compression of a cylinder of rock from the top and bottom similarly breaks the rock on diagonal shear planes. In FIGURE 3-2c, for example, the Earth's gravity is pulling straight down, but the rock breaks along a dipping fault.

Shear is generally on one plane only, as shown by the red line. σ_1 is the maximum principal stress.

slips, called **foreshocks**, preceding the main earthquake. It then continues to adjust with small slips called **aftershocks** after the event.

Along a fault, differential plate motions apply stresses continuously. Because those plate motions do not stop, elastic deformation progresses to plastic deformation within meters to kilometers of the fault, and the fault finally ruptures in an earthquake. When stress on a section of a fault releases as slippage during a large earthquake, some of that stress is often transferred to increasing stress on a part of the fault beyond the slip zone or to adjacent faults. That makes those adjacent areas more prone to slip than they were before.

The size of an earthquake is related to the amount of movement on a fault. The displacement, or **offset**, is the distance of movement across the fault, and the **surface rupture length** is the total length of the break (**FIGURE 3-6**). The largest earthquake expected for a particular fault generally depends on the *total fault length*, or the longest segment of the fault that typically ruptures. A 1,000-kilometer-long (620-mile) subduction zone, such as that off the coast of Oregon and Washington, or a 1,200-kilometer-long transform fault, such as the San Andreas Fault of California, if broken along its full length in one motion, would generate a giant earthquake. Although the full length of such faults does occasionally break in a single earthquake, shorter segments commonly break at different times. Normal faults, such as along the Wasatch Front of Utah, generally break in shorter segments, so somewhat smaller earthquakes are to be expected in such areas (compare FIGURE 4-13).

This relationship between fault-segment length and earthquake size puts a theoretical limit on the size of an earthquake at a given fault. A short fault only a few kilometers long can have many small earthquakes but not an especially large one because the whole fault is not long enough to break a large area of rock. The 1,000-kilometer-long San Andreas Fault, however, could conceivably break its whole length in one shot, causing a giant earthquake and catastrophic damage in this heavily populated region. Although the full length of such faults does occasionally break in a single earthquake, shorter segments commonly break at different times. Scientists have so far observed earthquakes breaking only as much as half the length of the San Andreas Fault. The type of fault also has an effect on earthquake size. *Normal* faults, such as the Wasatch Front of Utah, generally break in shorter segments, so somewhat smaller earthquakes are to be expected in such areas (compare FIGURE 4-13).

Sometimes a fault moves almost continuously, rather than suddenly snapping. A segment of the central part of the San Andreas Fault south of Hollister slips slowly and nearly continuously without causing significant earthquakes. In this zone, strain in the fault is released by **creep** and thus does not accumulate to cause large earthquakes. Why that segment of the fault slips without causing major earthquakes, whereas other segments stick until the rocks break during a tremor, is not entirely clear. Presumably, the rocks at depth are especially weak, such as might be the case with shale or serpentine; or perhaps water penetrates the fault zone to great depth, making it weak. The Hayward Fault also creeps but can produce large earthquakes (**FIGURE 3-7**). The continuously creeping section

FIGURE 3-6 OFFSET

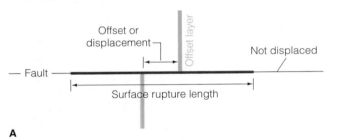

A. This diagram shows displacement and surface rupture length on a fault. Beyond the ends of the rupture, the fault does not break or offset. **B.** A fence near Point Reyes, north of San Francisco, was offset approximately 2.6 meters in the 1906 San Francisco earthquake on the San Andreas Fault.

FIGURE 3-7 A CURB CREEPS

This curb in Hayward, California, has been offset by creep along the Hayward Fault. The right photo shows further offset of the same curb two years later.

of the fault almost halfway between San Francisco and Los Angeles seems to be a zone of soft rocks that would be unable to build up a significant stress. Perhaps that means that no more than half of the length of the fault is likely to break in one sudden movement. Slip of only half the length of the fault could still generate a catastrophic earthquake.

Tectonic Environments of Faults

Because earthquakes are triggered by the motion of the Earth's crust, it follows that earthquakes are associated with plate boundaries. The sense of motion during a future earthquake is dictated by the relative motion across a plate boundary—strike-slip faults move along transform boundaries, thrust faults are typically associated with subduction zones and continent–continent collision boundaries, and normal faults move in spreading zones.

This section discusses examples of faults in these major tectonic environments, as well as fault systems isolated from plate boundaries. Chapter 4 will explore the human impact of earthquake activity in some of these earthquake zones.

Transform Faults

The most important example of a transform fault in the United States is the San Andreas Fault, a zone that slices through a 1,200-kilometer length of western California, from just south of the Mexican border to Cape Mendocino in northern California (**FIGURE 3-8**). The trace of the San Andreas Fault appears from the air and on topographic maps as lines of narrow valleys, some of which hold long lakes and marshes, that have eroded because rocks along the fault are crushed by its movements (see FIGURE 2-15b and 3-9b).

The San Andreas Fault is a continental transform fault in which the main sliding boundary marks the relative motion between the Pacific Plate, which moves northwest, and the North American Plate, which moves slightly south of west. As shown in **FIGURE 3-9**, p. 42, the total motion of the westernmost slice of California moves more than the slices closer to the continental interior. Areas where there is the greatest *difference* between those arrow lengths, which represent movement rates across a fault, have the greatest likelihood of new fault slippage or an earthquake.

The west side of the northwest-trending fault moves northwestward at an average rate of 3.5 centimeters per year, or 3.5 meters every 100 years, relative to the east side. The rupture length for a magnitude 7 earthquake for a

FIGURE 3-8 A GIANT CRACK IN THE EARTH

Robert E. Wallace, USGS, 1990.

The San Andreas Fault and other major faults nearby appear as a series of straight valleys slicing through the Coast Ranges in this shaded-relief map of California. In the view from the air, streams jog abruptly (yellow arrows) where they cross the San Andreas Fault in the Carrizo Plain north of Los Angeles. The 1857 Fort Tejon earthquake caused 9.5 meters of this movement.

3.5-meter offset would be 50 kilometers; release of all the strain accumulated in 100 years would require a series of such earthquakes along the length of the fault. However, earthquakes frequently occur in clusters separated by periods of relative seismic inactivity.

The San Andreas Fault stretches from the San Francisco Bay area southward toward Los Angeles. The northward drag of the Pacific Plate against the continent is slowly crushing the Los Angeles basin northward at roughly 7 millimeters per year a small part of the overall plate movement. The sedimentary formations buckle into folds and break along thrust faults, both of which shorten the basin as they move

FIGURE 3-9 SLICES THROUGH THE CITY

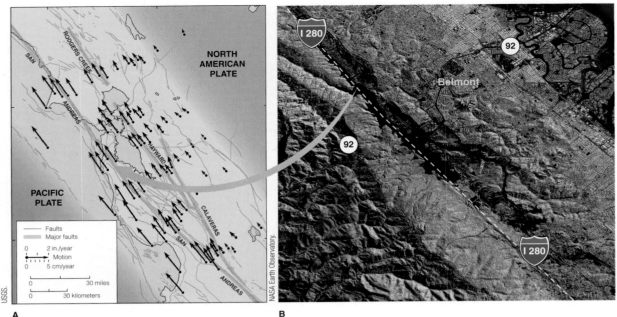

A. The San Andreas Fault system is a wide zone that includes nearly the entire San Francisco Bay area. Based on Global Positioning System measurements, the black arrows are proportional to the rates of ground movement relative to the stable continental interior. Energy builds up when there is a differential movement, as can be seen across each of the major faults shown in orange. **B.** The San Andreas Fault follows a prominent straight valley trending northeast through Crystal Springs Reservoir, southeast of San Francisco. The approximate location of the fault is shown with a dashed white line.

FIGURE 3-10 THE NEXT BIG ONE?

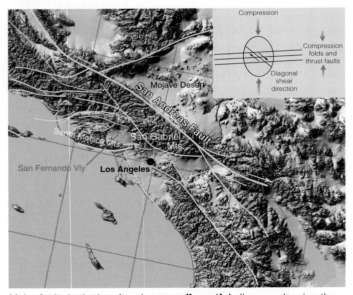

Major faults in the Los Angeles area. **(Inset)** A diagram showing the stresses that cause movement on blind thrust faults in the area. Note that the San Andreas Fault is parallel to the diagonal shear direction, and the blind thrusts are oriented perpendicular to the compression direction.

(**FIGURES 3-10** and **3-11**). The thrust faults are called **blind thrusts** because they do not break the surface. A tight fold at the surface marks the shallow end of the fault (see FIGURE 3-2a). Similarly, an anticline (up-fold) is shown in FIGURE 3-11 at the point of the left-most yellow arrow. Blind thrusts are dangerous because many of them remain unknown until they cause an earthquake (**Case in Point:** A Major Earthquake on a Blind Thrust Fault—Northridge Earthquake, California, 1994).

A total of seventeen earthquakes greater than magnitude 4.8 shook the Los Angeles region between 1920 and 1994. Especially dangerous faults include the Sierra Madre–Cucamonga Fault system that follows 100 kilometers of the northern edge of the San Fernando and San Gabriel Valleys. The blind thrust along the westernmost 19 kilometers of the fault moved to cause the 1971 San Fernando Valley earthquake. The fault systems beneath downtown Los Angeles include thrust faults that dip down to the northeast. The Santa Monica Mountains fault zone near downtown Los Angeles extends west along the Malibu coast for 90 kilometers. It includes blind thrust faults that do not break the surface and strike-slip faults that do. The Oak Ridge Fault system north of the Malibu coast generated the 1994 Northridge earthquake. The Palos Verdes thrust fault, along the coast south of downtown Los Angeles, slips approximately 3 millimeters per year; this progressive slip

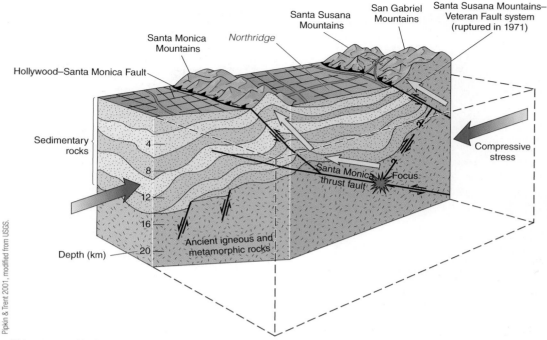

FIGURE 3-11 BLIND THRUSTS

Labels on figure:

Santa Monica Mountains

Northridge

Santa Susana Mountains

San Gabriel Mountains

Santa Susana Mountains–Veteran Fault system (ruptured in 1971)

Hollywood–Santa Monica Fault

Sedimentary rocks

Compressive stress

Santa Monica thrust fault

Focus

Ancient igneous and metamorphic rocks

Depth (km)

4 8 12 16 20

Pipkin & Trent 2001, modified from USGS.

This cutaway block diagram shows the blind thrust fault movement that caused the 1994 Northridge earthquake. Yellow arrows indicate the relative direction of movement.

releases accumulating strain that might someday cause an earthquake on an adjacent fault segment.

The San Andreas Fault appears to have accumulated a total displacement of 235 kilometers in the approximately 16 million years since it began to move. The fault has been stuck along its big bend, south of Parkfield, since the Fort Tejon earthquake of 1857. More recent earthquakes near the southern San Andreas Fault are associated with blind thrusts, which means that some of the crustal movement is being taken up in folding and thrust faults near the main fault instead of in slippage along the main fault itself (FIGURE 3-11). Note also that the response to overall horizontal compression is diagonal dipping faults, as shown earlier in **By the Numbers 3-1**.

Subduction Zones

Subduction zones are another tectonic environment in which earthquakes occur, including the largest earthquake on record, a magnitude 9.5, which struck the coast of Chile in 1960 and another of magnitude 8.8 in February 2010. In 1868, a magnitude 9 event in Peru (now in northern Chile) killed several thousand. In 2001, a magnitude 8.4 earthquake on the same subduction zone may have increased stress on nearby parts of the boundary. It was followed on August 15, 2007, by a magnitude 8.0 event that struck the coast of Peru and killed more than 510 people, many from collapse of their adobe-brick homes.

The most important example of subduction-zone faults in the United States is in the Pacific Northwest. We know from several lines of evidence that an active subduction zone lies off the coast for the 1,200 kilometers between Cape Mendocino in northern California and southern British Columbia (**FIGURE 3-12**, p. 44). The magnetic stripes parallel to the Juan de Fuca Ridge show that the plate on the east is moving to the southeast; in contrast, the Yellowstone hotspot track shows that the North American Plate is moving to the southwest. The collision between ocean floor and continent is along the Cascadia subduction zone. In addition, the line of active Cascade volcanoes about 100 kilometers inland indicates an active subduction zone at depth. We also know that subduction zones often generate giant earthquakes and that such sudden shifts of the ocean floor can generate huge ocean waves, called tsunamis. A comparable zone in Sumatra generated a giant earthquake and tsunami in December 2004 that killed about 220,000 people (see Chapter 5 for details). On March 6, 2007, a magnitude 6.3 earthquake struck southeast of the 2004 event, killing 70 people. Then on September 12, 2007, a magnitude 8.4 earthquake farther southeast on the same zone, near Penang, Sumatra, caused considerable damage and a 3-meter tsunami. These events prompted concern that the post-2004 events could be precursors to a still larger earthquake. Much farther southeast along the same plate boundary in May 2006, a magnitude 6.3 earthquake in Java killed more than 6,000 people and left about 650,000 homeless. Then

FIGURE 3-12 GIANT QUAKE IN THE NORTHWEST

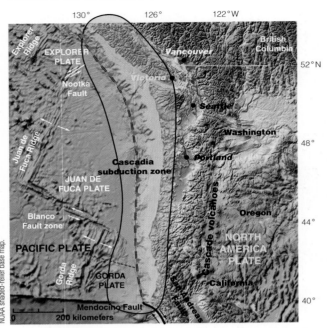

The Cascadia oceanic trench dominates the Pacific continental margin off Washington, Oregon, and southern British Columbia. The December 2004 magnitude 9.15 earthquake rupture (superimposed here with yellow shading) is comparable to the size of the potential Cascadia subduction zone slip.

Table 3-1	Largest World Earthquakes Since 1700	
EARTHQUAKE	**DATE**	**MAGNITUDE**
Chile	May 22, 1960	9.5
Anchorage, Alaska	Mar. 28, 1964	9.2
Northern Sumatra	Dec. 26, 2004	9.15
Kamchatka	Nov. 4, 1952	9.0
Cascadia	Jan. 26, 1700	9.0
Arica, Chile	Aug.13, 1868	9.0
Ecuador	Jan. 13, 1906	8.8
Maule, Chile	Feb. 27, 2010	8.8

in September 2009, a magnitude 7.0 earthquake struck the same offshore boundary south of Jakarta, in Java. It caused considerable damage and widespread panic.

The largest recorded earthquakes since A.D. 1700 have all occurred in subduction zones (**Table 3-1**).

Slabs of oceanic lithosphere sinking through an oceanic trench at subduction zone boundaries typically generate earthquakes from as deep as several hundred kilometers, so the apparent absence of those deep earthquakes in the Pacific Northwest has worried geologists for years. Several lines of evidence now show that major earthquakes do indeed happen but at such long intervals that none have struck within the 300 or so years of recorded Northwest history.

A giant prehistoric earthquake. Radiocarbon dating of the peat and buried trees in the Pacific Northwest helped place the last major earthquake in the region within a decade or two of the year 1700. In a separate analysis, careful counting of tree rings from killed and damaged trees indicates that the event happened shortly after the growing season of 1699.

In a clever piece of sleuthing, geologists of the Geological Survey of Japan found old records with an account of a great wave 2 meters high that washed onto the coast of Japan at midnight on January 27, 1700. No historical record tells of an earthquake at about that time on other Pacific

margin subduction zones, in Japan, Kamchatka, Alaska, or South America. That leaves the Northwest coast as the only plausible source. Correcting for the day change at the International Date Line and the time for a wave to cross the Pacific Ocean, the earthquake would have occurred on January 26, 1700, at approximately 9 p.m.

Coastal Indians in the Pacific Northwest have oral traditions that tell of giant waves that swept away villages on a cold winter night. Archaeologists have now found flooded and buried Indian villages strewn with debris. These many lines of data help confirm the timing of the last giant earthquake in this area.

Those analyses indicate similar dates at most, though not all, sites along the coast between Cape Mendocino and southern British Columbia. That probably means that the fault generally broke simultaneously along this entire 1,200-kilometer length of coast, an extremely long rupture that would likely correspond to an earthquake of about magnitude 9, similar to that of the December 2004 Sumatra earthquake (see FIGURE 3-12). Such an enormous earthquake offshore would generate a wave large enough to cause considerable damage all around the Pacific Ocean. Shaking in such a major earthquake, with accelerations of at least 1 **g**, would make it difficult to stand. Strong motion would continue for several minutes. Tsunami waves could arrive at the coast within 15 minutes of the earthquake, leaving little time to evacuate.

A giant future earthquake. Radiocarbon dating of the peat and buried trees covered by beach-derived sand in the Pacific Northwest bays helped establish a record of major earthquakes in that region that could be a precedent for future events. The oceanic plate sinking through the trench off the Northwest coast is now stuck against the overriding continental plate. The continental plate is bulging up, as shown by precise surveys (**FIGURE 3-13**). The locked zone is 50 to 120 kilometers off the coasts of Oregon, Washington, and southern British Columbia. Just inland, the margin is now rising at a rate between 1 and 4 millimeters per year and shortening horizontally by as much as 3 centimeters per year.

FIGURE 3-13 FLEXING THE CONTINENTAL EDGE

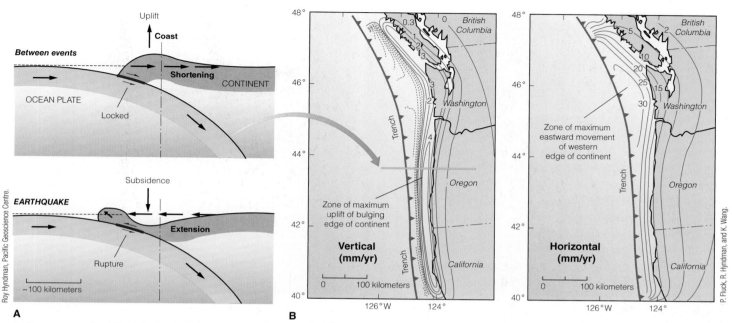

A. Denser oceanic plate sinks in a subduction zone. As strain accumulates, a bulge rises above the sinking plate while an area landward sinks. Those vertical displacements reverse when the fault slips to cause an earthquake. **B.** The subduction zone is locked east of the oceanic trench in the area near the coast.

Radiocarbon dating of leaves, twigs, and other organic matter in the buried soils at Willapa Bay, Washington, indicates seven deep earthquakes in the past 3,500 years, an average of one per 500 years. Elsewhere along the coast, the records show that twelve have occurred in the last 7,000 years, dating from the eruption of Mount Mazama in Oregon which deposited ash layers on the seafloor at an average interval of 580 years. The intervals between them range from 300 to 900 years. The last one was a little more than 300 years ago, so the next could come at any time.

We now know that giant earthquakes in the southern part of the subduction zone occur at intervals as short as 250 years—less than the time since the A.D. 1700 earthquake! It has also become apparent that earthquakes on major faults can trigger earthquakes on adjacent faults. Even more ominous, the latest research from Oregon State University and the USGS, reported in 2010, indicates there is an 80 percent chance of a giant subduction-zone earthquake along the southern part of the fault off southern Oregon and Northern California in the next 50 years! A major earthquake on the northern San Andreas could possibly trigger one on the southern Cascadia subduction zone—and vice versa, a disturbing thought.

Earthquakes above the subduction zone. In contrast, most of the earthquakes in the Puget Sound area of northwestern Washington State do not involve slip on the collision boundary at the oceanic trench offshore. Instead they accompany movement at shallow depth on faults that trend west or northwest and straddle Puget Sound (**FIGURE 3-14A**, p. 46). Every three or four years, the Puget Sound area feels the jolt of a moderate to large earthquake with a Richter magnitude of 5 to 7.

The Seattle Fault is the best known, and perhaps most dangerous, inland fault in the region. It trends east through the southern end of downtown Seattle, almost through the interchange between Interstate 5 and Interstate 90 (see **FIGURE 3-14B**, p. 46). Seventy kilometers of the fault are mapped; the part that reaches the surface dips steeply down to the south. Studies show that the rocks south of the fault rose 15.6 meters in a large earthquake about A.D. 900–930. That movement generated large tsunami waves in the water in Puget Sound and caused landslides into Lake Washington at the eastern edge of downtown Seattle (also discussed in Chapter 5). In 2001, movement on a related fault not far to the south during the Nisqually earthquake caused more than $2.45 billion* in property damage (**Case in Point:** Damage Mitigated by Depth of Focus—Nisqually Earthquake, Washington, 2001, p. 60).

---

* For the sake of comparison, 2010 dollars are used in discussion of earthquake damages.

FIGURE 3-14 EARTHQUAKE UNDER A CITY

This map of northwestern Washington shows the Seattle Fault and related major recent active fault zones in the Seattle area. The Seattle Fault runs east-west through the interchange of I-90 (foreground) and I-5 (middle right) at the southern edge of Seattle.

Continental Spreading Zones

In western North America, the best-known area of continental extension and associated normal faults is the Basin and Range of Nevada, Utah, and adjacent areas (**FIGURE 3-15**; see also Chapter 2, FIGURE 2-10). This broad region is laced with numerous north-trending faults that separate raised mountain ranges from dropped valleys.

The Wasatch Front, the eastern face of the Wasatch Range of central Utah, is a high fault scarp that faces west across the Salt Lake basin and defines the eastern margin of the Basin and Range (**FIGURE 3-16**). It is the eastern counterpart to the Sierra Nevada front of California. The Wasatch Front overlooks the deserts of Utah in the same way that the Sierra Nevada overlooks those of Nevada. Many small earthquakes shake the Wasatch Front, but none of any consequence have been felt since Brigham Young's party founded Salt Lake City in 1847.

One way to interpret the modest size of many deposits of stream sand and gravel at the mouths of canyons at the base of the Wasatch Front is to suggest that the fault movement has dropped the valley relative to the Wasatch Range during the geologically recent past, probably within tens of thousands of years. That would roughly correspond to the time in which the Sierra Nevada last rose. In fact, both faults remain active as their ranges rise. The active fault zone extends from central Utah, north to southeastern Idaho. The central section near Weber, Salt Lake City, Provo, and Nephi is the most active, but even the end segments are capable of causing magnitude 6.9 earthquakes.

Intraplate Earthquakes

Earthquakes occasionally strike without warning in places that are remote from any plate boundary and lack any recent record of seismic activity. These intraplate, or within-continent, earthquakes can be devastating, especially because most local people are unaware of their threat. Some of these isolated earthquakes are enormous, easily capable of causing a major natural catastrophe. Although many geologists have offered tentative explanations for these earthquakes, their causes remain generally obscure.

The intraplate earthquakes that struck southeastern Missouri in 1811 and 1812 were among the most severe to hit North America during its period of recorded history. The three great earthquakes that struck near New Madrid, Missouri, in December 1811, January 1812, and February 1812 were felt throughout the eastern United States, toppling chimneys in Ohio, Alabama, and Louisiana and causing church bells to ring in Boston.

Although there has not been another large earthquake in the region since then, the area is seismically active enough that people as far away as St. Louis and Memphis, the nearest big cities, occasionally hear the ground rumble as their dishes and windows rattle (**FIGURE 3-17**, p. 48). A repetition of an earthquake in this magnitude range would cause enormous loss of life and major property damage in Memphis, St. Louis, Louisville, Little Rock, and many smaller cities that have older masonry buildings. Few of the buildings in such cities are designed or built to resist significant earthquakes.

FIGURE 3-15 A SPREADING CONTINENT

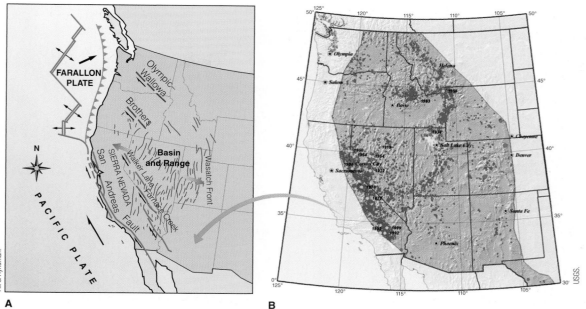

A

B

A. The north–south faults of the Basin and Range of Nevada, western Utah, and adjacent areas occupy a spreading zone accompanying the northwestward drag of the Pacific Ocean floor. That drag also causes shear to form the San Andreas Fault. The western margin of the Basin and Range is marked by the precipitous eastern edge of the Sierra Nevada; the eastern margin is the equally precipitous Wasatch Front at Salt Lake City. **B.** Most of the earthquake activity of the Basin and Range is concentrated along the east face of the Sierra Nevada and the Wasatch Front.

Another isolated earthquake, which struck Charleston, South Carolina, in 1886, caused many casualties and heavy property damage. The Charleston event, near the eastern coast of the United States, was along what has been called a "trailing continental margin." This is not a current plate margin but the margin between the North American continent and the Atlantic Ocean basin; it was originally near a plate margin when the Atlantic Ocean floor began to spread more than 150 million years ago.

FIGURE 3-16 MOUNTAINS FACE THE BASIN

A

B

A. The Wasatch Front, on the east, and **B.** the Sierra Nevada Front, on the west, mark the tectonically active and earthquake-prone edges of the Basin and Range.

FIGURE 3-17 RESTLESS MIDCONTINENT

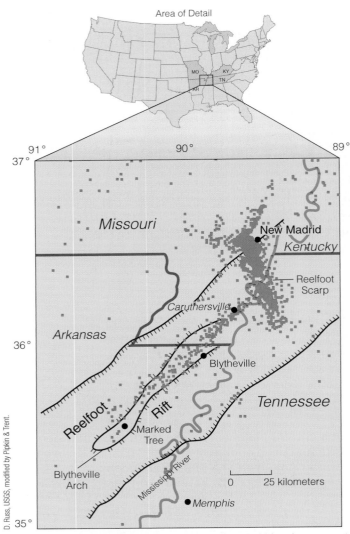

Recent microearthquake epicenters in the New Madrid region appear to outline three fault zones responsible for the earthquakes of 1811 and 1812. Two northeast-trending lateral-slip faults are offset by a short fault that pushed the southwestern side up over the northeastern side.

FIGURE 3-18 EAST COAST FAULT SYSTEM

The structural brick walls of this house at 157 Tradd Street in Charleston collapsed during the 1886 earthquake. This map of the East Coast fault system between South Carolina and Virginia shows how the fault zone lies close to the buried boundary between the continental crust of the Piedmont and the Atlantic oceanic crust. The Piedmont is rising.

An earthquake struck 20 kilometers northwest of Charleston on August 31, 1886, sending hundreds of people into the streets and toppling several chimneys in Lancaster, Ohio, 800 kilometers away. It shook plaster from the walls on the fourth floor of a building in Chicago, 1,200 kilometers away; 14,000 chimneys fell, and many buildings were destroyed on both solid and soft ground. One hundred people were killed, most of them in areas where the soil liquefied.

The Charleston earthquake and many others may be the result of movements along segments of the East Coast fault system, a swarm of aligned segments near the modern East Coast, that trend generally northeast (**FIGURE 3-18**). They were first recognized between South Carolina and

Virginia but may extend much farther. The fault zone is near the buried boundary between the continental crust of the Piedmont and Atlantic oceanic crust; the coastal plain is dropping. Renewed movement on faults may be associated with the early stages of opening of the Atlantic Ocean more than 150 million years ago.

Earthquake hazards may be significant in at least some parts of the eastern United States, where past quakes have left

FIGURE 3-19 SHAKING AND DAMAGE BY REGION

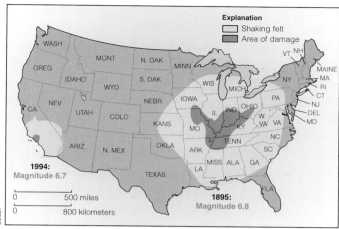

A comparison of similar magnitude earthquakes shows that the damage would be much greater for an earthquake in the Midwest than in the mountainous West.

FIGURE 3-20 EPICENTER AND FOCUS

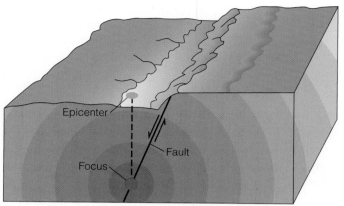

The epicenter of an earthquake is the point on the Earth's surface directly above the focus where the earthquake originated.

FIGURE 3-21 DIFFERENT EARTHQUAKE WAVES

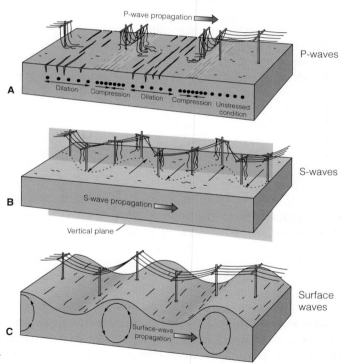

A. Compressional P-wave propagation (see front face of diagram): waves of compression alternating with extension move through the rock. **B.** Shear (S) wave propagation: S waves travel in a wiggling motion perpendicular to the direction of wave travel (only the horizontal direction is shown here). **C.** In the rolling motion of surface waves, individual particles at the Earth's surface move in a circular motion (opposite the direction of travel) both in a vertical plane and a horizontal plane: see front face of diagram.

little memory or lasting concern. Although large earthquakes are more frequent in the West, the few large ones that have occurred in eastern North America have been much more damaging because the Earth's crust in eastern North America transmits earthquake waves more efficiently, with less loss of energy, than that in the West, which is hotter and more broken along faults. That explains why the area of significant damage for an earthquake of a given size is greater in the East than in the West (**FIGURE 3-19**). Good land-use planning and building codes for new structures cost little and may someday avert enormous loss of life and property damage in a city that does not now suspect it is living dangerously.

Earthquake Waves

When a fault slips, the released energy travels outward in **seismic waves** from the place where the fault first slipped, called the **focus**, or hypocenter, of the earthquake. The **epicenter** is the point on the map directly above the focus (**FIGURE 3-20**). The behavior of earthquake waves explains both how we experience earthquakes and the types of damage they cause.

Types of Earthquake Waves

Observant people have noticed for centuries that many earthquakes arrive as a distinct series of shakings that feel different. The different types of shaking are a result of the different types of earthquake waves (**FIGURE 3-21**). The first event is the arrival of **P waves**, the primary or compressional waves, which come as a sudden jolt. People indoors might wonder for a moment whether a truck just hit the house. P waves consist of a train of compressions and expansions. P waves travel roughly 5 to 6 kilometers per second in the less-dense continental crust and 8 kilometers per second in the dense, less-compressible rocks of the upper mantle. People sometimes hear the low rumbling of the P waves of an earthquake. Sound waves are also compressional and closely comparable to P waves, but they travel through the air at only 0.34 kilometer per second.

After the P waves comes a brief interval of quiet while the cat heads under the bed and plaster dust sifts down from cracks in the ceiling. Then come the **S waves** (secondary, or shear, waves), moving with a wiggling motion—like that of a rhythmically shaking rope—and making it hard to stand. Chimneys may snap off and fall through floors to the basement. Streets and sidewalks twist and turn. Buildings jarred by the earlier P waves distort and may collapse. S waves are slower than P waves, traveling at speeds of 3.5 kilometers per second in the crust and 4.5 kilometers per second in the upper mantle. Their wiggling motions make them more destructive than P waves. The P and S waves are called **body waves** because they travel through the body of the Earth. These shear waves do not travel through liquids.

After the body waves, the **surface waves** arrive—a long series of rolling motions. Surface waves travel along Earth's surface and fade downward. Surface waves include Love and Rayleigh waves, which move in perpendicular planes. Love waves move from side to side, and Rayleigh waves move up and down in a motion that somewhat resembles ocean swells.

Surface waves generally involve the greatest ground motion, so they cause a large proportion of all earthquake damage. Surface waves find buildings of all kinds loosened and weakened by the previous body waves, vulnerable to a final blow. Inertia tends to keep people and loose furniture in place as ground motion yanks the building back and forth beneath them. Shattering windows spray glass shrapnel as plaster falls from the ceiling. If the building is weak or the ground loose, it may collapse. Although there are more complex, internal refractions of waves as they pass between different Earth layers, those complications do not much affect the damage that earthquakes inflict because the direct waves are significantly stronger.

The differences people feel during this series of earthquake waves can be explained by the different characteristics of those waves. To describe the vibrations of earthquake waves, we use a variety of terms (**FIGURE 3-22**). The time for one complete cycle between successive wave peaks to pass is the **period**; the distance between wave crests is the **wavelength**; and the amount of positive or negative wave motion is the **amplitude**. The number of peaks per second is the **frequency** in cycles per second, or Hertz (Hz).

When you bend a stick until it breaks, you hear the snap and feel the vibration in your hands. When the Earth breaks along a fault, it vibrates back and forth with the frequency of a low rumble, although the frequencies of earthquake waves

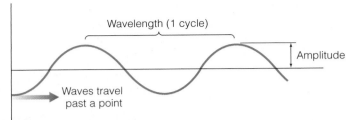

FIGURE 3-22 WAVE FORMS

The definition of wavelength and amplitude for earthquake waves.

are generally too low to be heard with the human ear. P and S waves generally cause vibrations in the frequency range between 1 and 30 cycles per second (1–30 Hz). Surface waves generally cause vibrations at much lower frequencies, which dissipate less rapidly than those associated with body waves. That is why they commonly damage tall buildings at distances as great as 100 kilometers from an epicenter.

Seismographs

A **seismograph** records the shaking of earthquake waves on a record called a **seismogram**. When recording seismographs finally came into use during the early part of the twentieth century, it became possible to see those different shaking motions as a series of distinctive oscillations that arrive in a predictable sequence (**FIGURE 3-23**). Imagine the seismograph as an extremely sensitive mechanical ear clamped firmly to the ground, constantly listening for noises from the depths. It is essentially the geologist's stethoscope.

We normally stand firmly planted on solid ground to watch things move, but how do we stand still and watch Earth move? Seismographs consist of a heavy weight suspended from a rigid column that is firmly anchored to the ground. The whole system moves with the earthquake motion, except the suspended weight, which stays relatively still due to its inertia. In seismographs designed to measure horizontal motion, the weight is suspended from

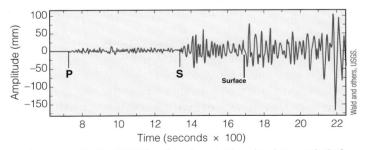

FIGURE 3-23 AN EARLY RECORD OF SHAKING

A seismogram for the 1906 San Francisco earthquake shows arrival of the main seismic waves in sequence—P, S, and surface waves.

FIGURE 3-24 RECORDING THE SHAKES

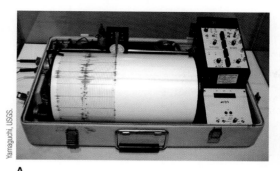

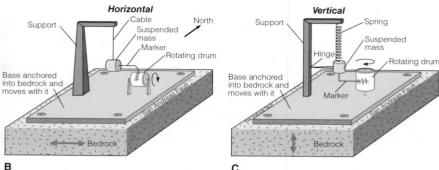

A. Although many seismograph stations now record earthquake waves digitally, a recording drum seismograph, like this one, is especially useful for visualizing the nature of earthquake waves: their amplitude, wavelength, and frequency of vibration. Different seismographs are used to measure, for example, **B.** horizontal versus **C.** vertical motion and small versus large earthquakes.

a wire, whereas in those designed for vertical motion, it is suspended from a weak spring (**FIGURE 3-24**).

The first seismograph used in the United States was at UC Berkeley in 1887. It used a pen attached to the suspended weight to make a record on a sheet of paper that was attached to the moving ground. Most modern seismographs work on the same basic principle but detect and record ground motion electronically.

Seismograms can help scientists understand more about how a fault slipped, as well as where it slipped and how much. Faults with different orientations and directions of movement generate various patterns of motion. Specialized seismographs are designed to measure those various directions of earthquake vibrations—north to south, east to west, and vertical motions. Knowing the directions of ground motion makes it possible to infer the direction of fault movement from the seismograph records.

Locating Earthquakes

The time interval between the arrivals of P and S waves recorded by a seismograph can also help scientists locate the epicenter of an earthquake. Imagine the P and S waves as two cars that start at the same place at the same time, one going 100 kilometers per hour, the other 90 kilometers per hour. The faster car gets farther and farther ahead with time. An observer who knows the speeds of the two cars could determine how far they are from their starting point simply by timing the interval between their passage. In exactly the same way, because we know the velocity of the waves, the time interval between the arrivals of the P and S waves reveals the approximate distance between the seismograph and the place where the earthquake struck (**FIGURE 3-25**, p. 52). This calculation is explained in greater detail in **By the Numbers 3-2**: Earthquake-Wave Velocities.

The arrival times of P and S waves at a single seismograph indicates how far from the seismograph an earthquake

originated, but it does not indicate in which direction the earthquake occurred. This means the earthquake could have happened anywhere on the perimeter of a circle drawn with the seismograph at its center and the distance to the earthquake as its radius. In order to better pinpoint the location of the earthquake, this same type of data is needed from at least three different seismograph stations. The three circles will intersect at only one location, and that is where the earthquake struck (see FIGURE 3-25b). In fact, because earthquake waves travel at slightly different velocities through different rocks on their way to a seismograph, their apparent distances are slightly different, and the circles intersect in a small triangle of error.

In practice, seismograph stations communicate the basic data to a central clearing house that locates the earthquake, evaluates its magnitude, and issues a bulletin to report when and where it happened. The bulletin is often the first news of the event. That is why we so often find the news media reporting an earthquake before any information arrives from the scene of the earthquake itself.

Earthquake Size and Characteristics

A question that comes to mind when people feel an earthquake or see the wild scribbling of a seismograph recording its ground motion is, "How big is it?" This question can be answered by describing its perceived effects—its intensity, or by measuring the amount of energy released—its magnitude.

Earthquake Intensity

After the great Lisbon earthquake of 1755, the archbishop of Portugal sent a letter to every parish priest in the country asking each to report the type and severity of damage in his

FIGURE 3-25 LOCATING EARTHQUAKES

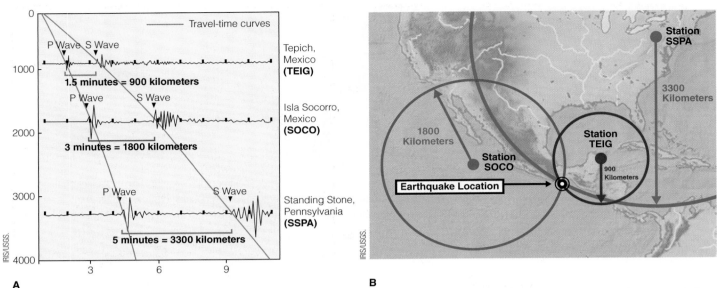

A. The difference in arrival time between the P and S waves reveals the distance from a seismograph to an earthquake. This plot shows records from seismographs at different distances from a single earthquake. **B.** Circles of distance to the earthquake drawn from at least three seismograph stations locate the earthquake on a map, in this case in the Mexico trench.

▶ By the Numbers 3-2

Earthquake-Wave Velocities

Surface wave arrival times increase linearly with distance from an earthquake because the waves travel with nearly constant velocity in shallow rocks. In contrast, P waves travel at 5–6 km/sec in the continental crust but about 8 km/sec in the more dense rocks of the mantle. S waves travel at about 3.5 km/sec in the crust but about 4.5 km/sec in the mantle. Because P and S waves travel faster deeper in the Earth, waves at greater depths can reach a seismograph faster along those curving paths (see Figure 3-25a).

Table 3-2	Mercalli Intensity Scale
MERCALL INTENSITY (APPROX.) AT EPICENTER	**EFFECT ON PEOPLE AND BUILDINGS**
I–II	Not felt by most people.
III	Felt indoors by some people.
IV–V	Felt by most people; dishes rattle, some break.
VI–VII	Felt by all; many windows and some masonry cracks or falls.
VIII–IX	People frightened; most chimneys fall; major damage to poorly built structures.
X– XI	People panic; most masonry structures and bridges destroyed.
XII	Nearly total damage to masonry structures; major damage to bridges, dams; rails bent.
>XII	Near total destruction; people see ground surface move in waves; objects thrown into air.

parish. Then the archbishop had the replies assembled into a map that clearly displayed the pattern of damage in the country. Jesuit priests have been prominent in the study of earthquakes ever since.

Italian scientist Giuseppe Mercalli formalized the system of reporting in 1902 with his development of the Mercalli Intensity Scale. It is based on how strongly people feel the shaking and the severity of the damage it causes. The original Mercalli Scale was later modified to adapt it to construction practices in the United States.

The **Modified Mercalli Intensity** Scale is still in use. The U.S. Geological Survey sends questionnaires to people it considers qualified observers who live in an earthquake area and then assembles the returns into an earthquake intensity map, on which the higher Roman numerals record greater intensities (**Table 3-2**).

Mercalli Intensity Scale maps reflect both the subjective observations of people who felt the earthquake and an

FIGURE 3-26 SHAKEMAP

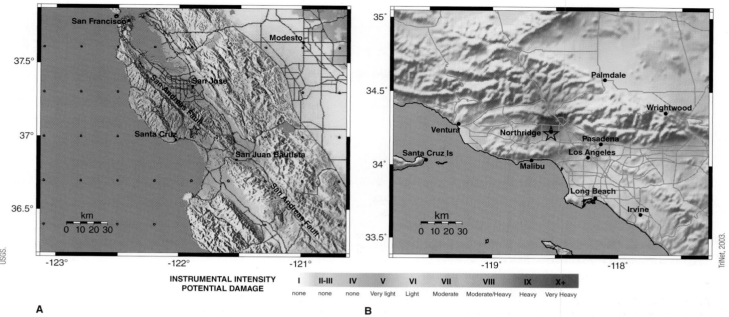

INSTRUMENTAL INTENSITY
POTENTIAL DAMAGE

I	II-III	IV	V	VI	VII	VIII	IX	X+
none	none	none	Very light	Light	Moderate	Moderate/Heavy	Heavy	Very Heavy

A **B**

A. This "Shakemap" of earthquake intensities of the 1989 Loma Prieta earthquake near Santa Cruz, California, shows a Mercalli Intensity VIII at the epicenter northeast of Santa Cruz. **B.** This TriNet ShakeMap shows the distribution of shaking during the 1994 Northridge earthquake. These maps were created after the earthquakes because the system was not available at the time. The system is now operational in California.

objective description of the level of damage. They typically show the strongest intensities in areas near the epicenter and areas where ground conditions cause the strongest shaking. In the case of the map generated for the Loma Prieta earthquake, shown in **FIGURE 3-26A**, the zones of greater intensity are, as expected, elongated parallel to the San Andreas Fault. The greater intensities shown along San Francisco Bay can be explained by the fact that its loose, wet muds amplified the shaking. The Northridge earthquake in 1994 caused significant damage across a broad area Northwest of Los Angeles (**FIGURE 3-26B**). Such maps are especially useful in land use planning because they predict the pattern of shaking in future earthquakes along the same fault.

The maps shown in FIGURE 3-26 are examples of recently developed computer-generated maps of ground motion called **ShakeMaps**, which show the distribution of maximum acceleration and maximum ground velocity for many potential earthquakes; they can be used to infer the likely level of damage. Such real-time maps can help send emergency-response teams quickly to areas that have likely suffered the greatest damage. Near San Francisco, the zone of greatest shaking is parallel to the San Andreas Fault because earthquakes occur along the fault and the soft muds of San Francisco Bay amplify the shaking there. Near Los Angeles, recent earthquakes form a broad patch localized around blind thrust faults that are not parallel to the San Andreas Fault.

Earthquake Magnitude

Suppose you were standing on the shore of a lake on a perfectly still evening admiring the flawless reflection of a mountain on the opposite shore. Then a ripple arrives, momentarily marring the reflection. Did a minnow jump nearby, or did a deranged elephant take a flying leap into the lake from the distant opposite shore? Nothing in the ripple as you would see it could answer that question. You also need to know how far it traveled and spread before you saw it, because the size of the wave decreases with distance. Useful as it is, the Mercalli Intensity Scale does not help answer some of those basic questions. That is the problem that Charles Richter of the California Institute of Technology addressed when he first devised a new earthquake magnitude scale in 1935.

Richter developed an empirical scale, called the **Richter Magnitude** Scale, based on the maximum amplitude of earthquake waves measured on a seismograph of a specific type, the Wood-Anderson seismograph. Although wave amplitude decreases with distance, Richter designed the magnitude scale as though the seismograph were 100 kilometers from the epicenter.

Seismograms vary greatly in amplitude for earthquakes of different sizes. To make that variation more manageable, Richter chose to use a logarithmic scale to compare earthquakes of different sizes. At a given distance from an

FIGURE 3-27 THE RICHTER SCALE

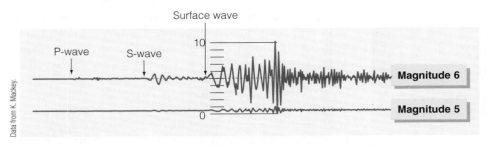

An earthquake of magnitude 6 registers with 10 times the amplitude as an earthquake of magnitude 5 from the same location and on the same seismograph. That difference is an increase in 1 on the Richter scale. The horizontal axis on these seismograms is time, and the vertical axis is the ground motion recorded.

earthquake, an amplitude 10 times as great on a seismograph indicates a magnitude difference of 1.0—an earthquake of magnitude 6 sends the seismograph needle swinging 10 times as high as one of magnitude 5 (**FIGURE 3-27**).

The Richter Magnitude Scale is simple in principle and easy to use, but actual practice leads to all sorts of complications, which have inspired many modifications. Seismographs, like buildings and people, sense shaking at different frequencies. Tall buildings, for example, sway back and forth more slowly than short ones—they have longer periods of oscillation. P waves, S waves, and surface waves have different amplitudes and different periods. With this variability in earthquake waves, Richter chose to use as the standard for his **local magnitude**, M_L, waves with periods, or back-and-forth sway times, of 0.1 to 3.0 seconds. The Richter magnitude is now known as M_L.

Seismologists, the scientists who study earthquakes, use different magnitude scales, specifically based on the amplitudes of surface waves, P waves, or S waves, or the moment magnitude, M_W (see the following). Larger earthquakes have longer periods, so different seismographs are used to measure short- and long-period wave motion.

Distant earthquakes travel through Earth's interior at higher velocities and frequencies. To work with distant earthquakes, Beno Gutenberg and Charles Richter developed two more-specific magnitude scales in 1954. M_S, the surface-wave magnitude, is calculated in a similar manner to that described for M_L. The number quoted in the news media is generally the surface-wave magnitude, as it is in this book, unless specified otherwise. Surface waves with a period of 20 seconds or so generally provide the largest amplitudes on seismograms. Special seismographs record earthquake waves with such long periods. M_B, the body-wave magnitude, is measured from the amplitudes of P waves.

To estimate the magnitude of an earthquake, we need the amplitude (from the S wave or surface wave). Because the amplitude of shaking decreases with distance, we also need the distance to the epicenter (from the P minus S time). These calculations can be made using a graphical method, the earthquake nomograph, on which a straight line is plotted between the P – S time and the S-wave amplitude (**FIGURE 3-28**). This line

intersects the central line at the approximate magnitude of the earthquake.

For earthquakes with M_L above 6.5, the strongest earthquake oscillations, which have a lower frequency, may lie below the frequency range of the seismograph. This may cause saturation of earthquake records, which occurs when the ground below the seismograph is still going in one direction while the seismograph pendulum, which

FIGURE 3-28 ESTIMATING EARTHQUAKE MAGNITUDE

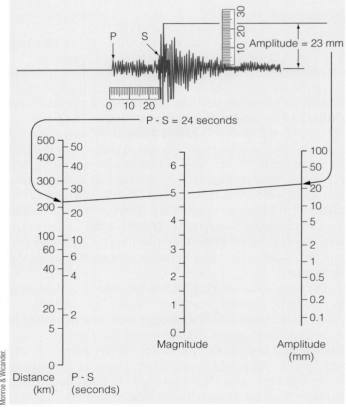

A Nomograph chart uses the distance from the earthquake (P – S time in seconds) and the S-wave amplitude (in mm) to estimate the earthquake magnitude.

swings at a higher frequency, has begun to swing back the other way. Then the seismograph does not record the maximum amplitude. So the Richter magnitude becomes progressively less accurate above M_L 6.5, and a different scale, such as that of M_W, becomes more appropriate.

An earthquake of magnitude 6 indicates ground motion or seismograph swing 10 times as large as that for an earthquake of magnitude 5, but the amount of energy released in the earthquakes differs by a factor of about 32 (**By the Numbers 3-3:** Energy of Different Earthquakes). Below M_L 6 or 6.5, the various measures of magnitude differ little; but above that, the differences increase with magnitude. For

larger earthquakes, the energy released is a better indicator of earthquake magnitude than ground motion.

Moment magnitude, M_W is essentially a measure of the total energy expended during an earthquake. It is determined from long-period waves taken from broadband seismic records that are controlled by three major factors that affect the energy expended in breaking the rocks. Calculation of M_W depends on the seismic moment, which is determined from the shear strength of the displaced rocks multiplied by both the surface area of earthquake rupture and the average slip distance on the fault. The largest of these variables, and the one most easily measured, is the offset or slip distance.

Small offsets of a fault release small amounts of energy and generate small earthquakes. If the length of fault and the area of crustal rocks broken is large, then it will cause a large earthquake. Because the relationships are consistent, it is possible to estimate the magnitude of an ancient earthquake from the total surface rupture length (**FIGURE 3-29**). For typical rupture thicknesses, a fault offset of 1 meter would generate an earthquake of approximately M_W 6.5, whereas a fault offset of 13 meters would generate an earthquake of approximately M_W 9. If you find a fault with a measurable offset that occurred in a single earthquake, then you can infer the approximate magnitude of the earthquake it caused.

In 1954, Gutenberg and Richter worked out the relationship between frequency of occurrence of a certain size of earthquake and its magnitude. Recall from Chapter 1 that there are many small events, fewer large ones, and only

FIGURE 3-29 HOW BIG WAS IT?

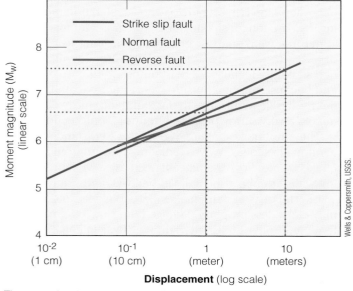

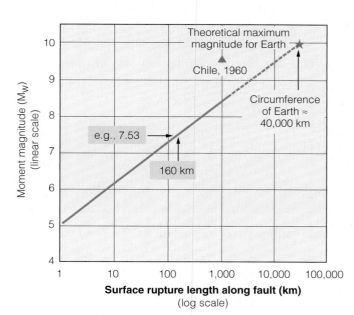

These graphs show the relationships between the magnitude of the earthquake and the maximum fault offset (during earthquakes on all types of faults), the surface rupture length.

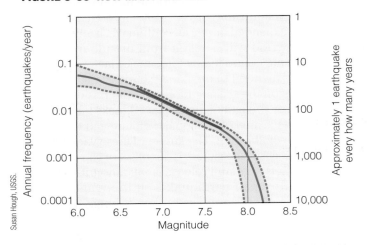

FIGURE 3-30 HOW MANY ARE BIG?

Susan Hough, USGS.

This graph plots the Gutenberg-Richter frequency–magnitude relationship for the San Francisco Bay region. The logarithm of the annual frequency of earthquakes plotted against their Richter magnitude generally plots as a straight line, or nearly so (red line). The curved line (green) provides the best fit to the data, which mostly plot between the dashed blue lines. Different faults would plot as somewhat different lines.

rarely a giant event. Quantitatively, that translates as a *power law*. Plotted on a graph of earthquake frequency versus magnitude, the power law can be plotted as a log scale: 10^1 or 10 to the power of 1 is 10; 10^2 or 10 to the second power is $10 \times 10 = 100$; $10^3 = 10 \times 10 \times 10 = 1,000$; and so on.

The Gutenberg-Richter frequency–magnitude relationship tells us that if we plot all known earthquakes of a certain size against their frequency of occurrence (on a logarithmic axis), we get a more or less straight line that we can extrapolate to events larger than those on record (**FIGURE 3-30**). Small earthquakes are far more numerous than large earthquakes, and giant earthquakes are extremely rare, which is why we have not had many in the historical record.

Most of the total energy release for a fault occurs in the few largest earthquakes (**FIGURE 3-31**). Each whole-number increase in magnitude corresponds to an increase in energy release of approximately 32 times. Thus, 32 magnitude 6 earthquakes would be necessary to equal the total energy release of 1 magnitude 7 earthquake. And more than 1,000 earthquakes of magnitude 6 would release energy equal to a single earthquake of magnitude 8 ($32 \times 32 = 1,024$).

Ground Motion and Failure During Earthquakes

How much and how long the ground shakes during an earthquake is related to how much and where the fault moves. **Table 3-3** summarizes the relationship between earthquake magnitude and ground motion. Local conditions can also amplify shaking and increase damage.

Ground Acceleration and Shaking Time

Sometimes it helps to think of ground motion during an earthquake as a matter of acceleration, that is, the strength of the shaking. **Acceleration** is normally designated as some proportion of the acceleration of gravity (g); 1 g is the acceleration felt by a freely falling body, such as what you feel when you step off a diving board. Most earthquake accelerations are less than 1 g; a few are more. A famous photograph taken after the San Francisco earthquake of 1906 shows a statue of the eminent nineteenth-century scientist Louis Agassiz stuck headfirst in a courtyard on the campus of Stanford University, its feet in the air (**FIGURE 3-32**). Perhaps the statue was tossed off its pedestal at a moment when the vertical ground acceleration was greater than 1 g. It is commonplace after a strong earthquake to find boulders tossed a meter or more.

The duration of strong shaking in an earthquake depends on the size of the earthquake. The time that the ground moves in one direction during an earthquake, before the oscillation moves back in the other direction, is similar to the time of initial fault slip in one direction. The total duration of motion is longer because the ground oscillates back and forth.

An increase in magnitude above 6 does not cause much stronger shaking; rather, it increases the area and total time of shaking. Earthquakes of magnitude 5 generally last only 2 to 5 seconds; those of magnitude 7 from 20 to 30 seconds; and those of magnitude 8 almost 50 seconds (see Table 3-3). A magnitude 6 earthquake, shaking only 10–15 seconds, provides only a short time to evacuate. A magnitude 7 earthquake provides more time, but evacuation is harder to do, with accelerations approaching 1 g. The longer shaking lasts, the more damage occurs; structures weakened or cracked in the first few seconds of an earthquake are commonly destroyed with continued shaking. Because there is almost no time to evacuate, and because running outside can result in death by falling debris, it is generally best to duck under a sturdy desk or lie next to a very sturdy piece of furniture for protection.

The amount of shaking also relates to distance from an earthquake's focus. Waves radiating outward from an earthquake source show a significant decrease in violence of shaking with distance, especially in bedrock and firmly packed soil. For this reason, earthquakes that occur deep underground may cause less property damage than smaller earthquakes that occur near the Earth's surface (**Case in Point:** Damage Mitigated by Depth of Focus—Nisqually Earthquake, Washington, 2001, p. 60). The focus for most earthquakes is generally at depths shallower than 100 kilometers, because rocks at greater depths behave plastically and slip continuously.

Shaking severity is also affected by the type of material waves are traveling through. For example, upon reaching an area of loose, water-saturated soils, such as old

FIGURE 3-31 HOW BIG ARE EARTHQUAKES?

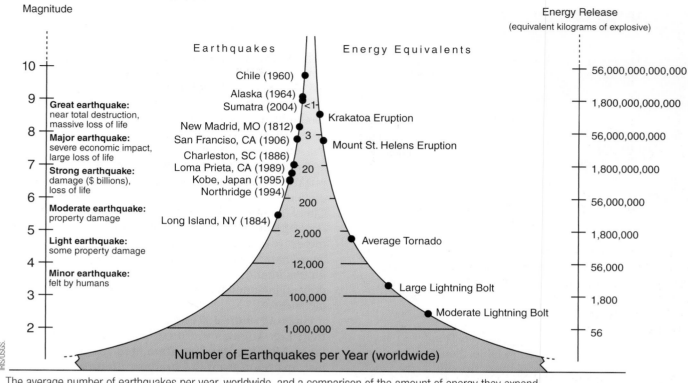

The average number of earthquakes per year, worldwide, and a comparison of the amount of energy they expend.

Table 3-3	Characteristics of Earthquakes of Different Magnitudes					
RICHTER MAGNITUDE (M_L)	MAXIMUM ACCELERATION (APPROX. % OF g)[a]	MAXIMUM VELOCITY OF BACK-AND-FORTH SHAKING[a]	APPROXIMATE TIME OF SHAKING NEAR SOURCE (sec)	DISPLACEMENT DISTANCE, OR OFFSET	SURFACE RUPTURE LENGTH (km)
<2	0.1–0.2				
~3	<1.4	<1	0–2		
~4	1.4–9	1–8	0–2		
~5	9–34	8–31	2–5	~1 cm	1
~6	34–124	31–116	10–15	60–140 cm	~8
~7	>124	>116	20–30	~2 m	50–80
~8	>124	>150	~50	~4 m	200–400
~9+	>124		>80	>13 m	>1,200
M_W = 9.7–9.8					Circumference of Earth

[a]$g = 9.8 \ m/sec^2$

lakebed clay or artificial fill at the edge of bays, earthquake waves are strongly amplified to accelerations many times greater than nearby waves in bedrock (**Case in Point:** Amplified Shaking over Loose Sediment—Mexico City Earthquake, 1985, p. 63). The violence of shaking depends on the frequency of the earthquake waves compared with the frequency of the ground oscillation. The lower-frequency oscillations of surface waves often correspond to the natural oscillation frequency of loose, water-saturated ground.

FIGURE 3-32 UPENDED

After the 1906 San Francisco earthquake, Louis Agassiz's head was firmly planted in the sidewalk at Stanford University.

Secondary Ground Effects

Earthquakes often trigger landslides (see Chapter 8). If you place a pile of loose, dry sand on a table, then sharply whack the side of the table, some of the sand will immediately slide down the pile. In nature, if sand or soil in the ground is saturated with water, the quick back-and-forth acceleration from a quake has a pumping effect on the water between the grains. Water is forced into spaces between the grains with each pulse of an earthquake. This sudden increase in water pressure in the pore spaces can effectively push the grains apart and permit the mass to slide downslope.

Earthquakes can also cause **liquefaction**—in which soils that ordinarily seem perfectly stable become almost liquid when shaken and then solidify again when the shaking stops. Many soft sediment deposits consist of extremely loose sand or silt grains with water-filled pore spaces between them. An earthquake can shake these deposits down to a much tighter grain packing, expelling water from the pore spaces. The escaping water carries sediment along as it rapidly flows to the surface, creating sand boils and mud volcanoes that are typically a few meters across and several centimeters high.

Liquefaction can cause significant damage to buildings and roads on soft sediment. During the 1964 Alaska earthquake, the ground in Anchorage, 100 kilometers from the epicenter, began to shake and continued for 72 seconds. Clays liquefied in the Turnagain Heights district, causing bluffs up to 22 meters high to collapse along 2.8 kilometers of coastline. The swiftly flowing liquid clay carried away many modern frame houses that were as much as 300 meters inland. The $3.8 billion in property damage (2010 dollars) included roads, bridges, railroad tracks, and harbor facilities.

Shaking of the 1971 San Fernando Valley earthquake near Los Angeles induced liquefaction of the upper face of the Van Norman Dam, nearly causing its failure just upstream from the homes of tens of thousands of people. On June 16, 1964, a magnitude 7.4 earthquake in Niigata, Japan, caused liquefaction and settling of approximately one-third of the city. Strongly constructed apartment buildings in some areas remained intact but toppled over on their sides (**FIGURE 3-33A**). A magnitude 7.4 earthquake in Izmit, Turkey, had a similar effect (**FIGURE 3-33B**).

FIGURE 3-33 LIQUEFACTION

A. These strong buildings in Niigata, Japan, remained intact during the 1964 earthquake but fell over when the sediments below them liquefied.
B. Liquefaction of the foundation under the left side of this building during the August 17, 1999, earthquake in Izmit, Turkey, caused settling of the left section and destruction of the middle section.

Wasatch Fault

Donald Hyndman.

Broad, low mounds in a nearly flat area downslope from the Wasatch Front west of Farmington, near Ogden, Utah, mark an area of liquefaction of soft clays laid down in Glacial Lake Bonneville, the ancestor of Great Salt Lake.

The deep fill of soft sediments and high groundwater levels in the Salt Lake Valley, Utah, have a significant likelihood of liquefaction during earthquakes. Together, these factors may also amplify ground motion more than 10 times. The ground is saturated with water only a few meters below the surface in Salt Lake City. Liquefaction of wet clays would cause loss of bearing capacity and downslope flow (**FIGURE 3-34**).

In the next chapter, we use the principles and related discussions from Chapter 3 to consider the possibilities of earthquake forecasts and prediction, and we discuss how to avoid or minimize damages caused by earthquakes. A few prominent examples illustrate results of some of those effects.

Cases in Point

A Major Earthquake on a Blind Thrust Fault
Northridge Earthquake, California, 1994 ▶

On January 17, 1994, at 4:31 a.m., a magnitude=6.7 earthquake struck Northridge, California; its epicenter was 20 kilometers southwest of that of the San Fernando Valley earthquake of similar size in 1971. Both accompanied movement on faults in the San Fernando Valley north of Los Angeles. The earthquake was caused when a thrust fault slipped at a depth of 10 kilometers. The offset reached to within 5 kilometers of the surface but did not break it, so the fault is a blind thrust (see FIGURE 3-11). The fault, unknown before the earthquake, is now known as the Pico thrust fault. The fault movement raised Northridge 20 centimeters (8 inches); the Santa Susana Mountains north of the San Fernando Valley rose 40 centimeters.

Ground acceleration reached almost 1 **g** in many areas and approached 2 **g** at one site in Tarzana, and ground velocity locally exceeded 1 meter per second. A few people felt the ground motion as much as 300 to 400 kilometers away from the epicenter. Local topography and bedrock structure may have amplified ground shaking. Anomalously strong shaking occurred on some ridge tops and at the bedrock boundaries of local basins.

Damage inflicted during the Northridge earthquake reached Mercalli Intensity IX near the epicenter. Intensity V effects were noted as far as 120 kilometers to the north, 180 kilometers to the west, and 200 kilometers to the southeast.

Severity of damage depended on the distance from the epicenter, type of ground under the foundation, and structural design. Walls fell from older buildings constructed from structural brick. Many buildings collapsed because they had weak first floors, most commonly apartments above garages. Concrete around steel reinforcing bars shattered in overpasses and parking garages, permitting the reinforcing steel to buckle and then collapse. Such failures could have been prevented if the concrete had been tightly

Collapsed structural brick wall

FEMA.

▶ *Older structural brick buildings were heavily damaged in the 1994 Northridge earthquake.*

wrapped with steel. Even a carefully designed and nearly new parking garage constructed from flexible materials thought able to withstand a strong earthquake failed. Many welds broke where they attached horizontal beams to vertical

(continued)

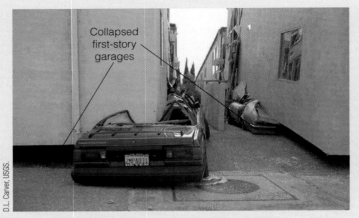

Collapsed first-story garages

D.L. Carver, USGS.

▶ *Many first-floor garages under apartment buildings were not well braced, so the buildings collapsed on the cars below.*

USGS.

▶ *A nearly new Northridge stadium parking structure collapsed during the 1994 earthquake in spite of being built from flexible materials designed to withstand strong earthquakes.*

columns, permitting the beams to fall. Mobile homes were jarred off their foundations, and broken gas lines ignited fires.

The Northridge earthquake caused $58.5 billion in property damage (2010 dollars). Some 3,000 buildings were closed to all reentry; another 7,000 buildings were closed to occupation. Sixty-one people were killed. In fact, the toll would probably have been in the thousands had the earthquake happened later in the day, when more people would have been up and about. Most buildings and parking garages that collapsed were essentially unoccu-

pied. Freeway collapse occurred at seven locations, 170 bridges were damaged, and a few unlucky people were on the road.

Many California faults are capable of causing large earthquakes, tens of times larger than those in the San Fernando Valley in 1971 and Northridge in 1994. Future earthquakes could cause stronger shaking that would last longer and extend over broader areas than those previous California earthquakes. The consequences in densely populated areas would be tragic.

FEMA.

▶ *This driver had the misfortune of driving off a collapsed freeway in the early morning darkness during the Northridge earthquake.*

Damage Mitigated by Depth of Focus
Nisqually Earthquake, Washington, 2001 ▶

On February 28, 2001, the Nisqually earthquake struck the Seattle-Tacoma area with a Richter magnitude of 6.8; it was the largest earthquake since the magnitude 7.1 Olympia event of April 1949 and the magnitude 6.5 Seattle event of 1965. The Nisqually earthquake inflicted widespread minor damage: windows broke, cornices fell, and objects rattled off shelves. The shaking continued for 30 to 45 seconds. Property damage amounted to $2.45 billion (2010 dollars), mostly to old buildings on reclaimed

and poorly compacted tidelands. One person died, and more than 250 were injured. Although the Seattle area was not severely damaged, scientists there are particularly concerned with the safety of the Alaska Way double-deck elevated highway around Seattle's waterfront. Geotechnical studies of the damage and the level of shaking suggest that if the earthquake had lasted 10 seconds longer, footings on bay mud near the south end of Alaska Way would have failed, and the highway would have collapsed. In

2009, plans were developed to replace that highway with a new surface road and tunnel. Construction began in 2010.

So what caused the Nisqually earthquake? It is unclear, but one possibility is bending of the ocean floor as it begins its descent under the continental margin. Another possibility is that parts of the

(continued)

subducting plate descend at different rates or different dips or inclinations than adjacent areas to the north or south, and the stress on the plate caused it to break. Alternatively, a northward drag of the ocean floor could have deformed the continental margin.

Although the magnitude of the Nisqually earthquake matched that of the Northridge quake that struck Los Angeles in 1993, it inflicted much less damage and many fewer casualties due to several factors. First, the fault movement that caused the Nisqually earthquake occurred at a depth of 50 kilometers, so no building on the Earth's surface was any closer to the focus. Second, deep earthquakes, though commonly strong, do not generate much surface wave motion, which is the most damaging. Third, recent bracing of buildings and bridges to better withstand earthquakes helped immensely.

▶ *A support beam (right photo) on the Alaska Way viaduct in Seattle was damaged (see arrows) in the 2001 Nisqually earthquake. The viaduct is in danger of collapse should there be another large earthquake on the Seattle Fault.*

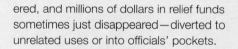

Collapse of Poorly Constructed Buildings
Kashmir Earthquake, Pakistan, 2005 ▶

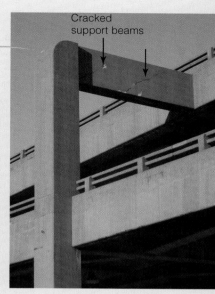

At 8:51 a.m., on October 8, 2005, in Pakistani Kashmir, the ground shifted and continued shaking violently when a magnitude 7.6 earthquake struck the western Himalayas 105 kilometers north-northeast of Islamabad, Pakistan. Many homes, stores, and schools collapsed, crushing occupants. Huge boulders and landslides crashed down from steep mountainsides onto more houses. Landslides and rockfalls closed many highways and mountain roads for days—in some places, months—cutting off access to injured and buried people. Other roads were open for only one or two hours per day because of heavy rains and the danger of continuing slides. Tens of thousands migrated to shelters at lower elevations, but many refused to leave their homes, in part because they feared others would occupy them and take their few belongings.

In late November, temperatures dropped below freezing, and snow fell; hundreds of thousands of people who remained in villages in remote mountain valleys were without tents, warm clothes, blankets, and sufficient heat; 2,000 died from the cold.

People burned furniture to keep warm. Food saved for the winter months was inaccessible, buried in their collapsed houses. Corrugated iron sheets were needed to keep snow from collapsing light tents.

Helicopters brought food, medicine, and other relief supplies but were hindered by weather and the steep terrain that offered no flat areas to land. Many boxes of supplies dropped on slopes merely slid down and disappeared into rivers, sometimes a thousand meters below. Trucks carried supplies via reopened roads, and mule trains sufficed when only makeshift trails across landslides were available. Hospitals treated many people for flu, pneumonia, hypothermia, measles, and tetanus after lack of treatment for open wounds and broken bones.

Some $580 million for earthquake relief was pledged by dozens of countries, including the United States and Canada, but only $15.8 million was immediately available, much of it in the form of goods and services provided directly by a country, rather than money that could cover numerous needs. Unfortunately, much of the pledged aid was never deliv-

ered, and millions of dollars in relief funds sometimes just disappeared—diverted to unrelated uses or into officials' pockets.

In all, about 87,000 people died and tens of thousands more were injured, mostly from collapse of heavy masonry structures, including a ten-story building in Islamabad, and from landslides triggered by the earthquake. It was by far the most deadly earthquake on record in India, Pakistan, and surrounding areas. About 3.5 million people lost their homes. Buildings with weak walls supporting heavy concrete floors collapsed, crushing their occupants. A few well-built structures survived near the earthquake source, but the largest numbers of casualties were in poorly built schools, hospitals, and police stations. In many cases, these showed evidence of shoddy construction practices or low-quality materials. Some buildings on river floodplains were constructed from rounded river rocks, poorly cemented together with

(continued)

Rebecca Bendick.

▶ *Heavy concrete roofs collapsed on people in Muzaffarabad, Pakistan, when poor-quality masonry walls and support posts crumbled.*

Asif Hassan/AFP/Getty Images.

▶ *Victim lies buried in the rubble of the ten-story Margalla Towers in Islamabad.*

little or no reinforcing steel. Total damages were estimated at $5 billion, mostly uninsured. A 1935 earthquake along the same zone had killed 30,000 people.

The enormous hazard for people in the region resulted from a combination of shaking intensity, the enormous population, and poor-quality building materials and construction in the mountainous westernmost Himalayan foothills. More than 80 percent of the concrete frame and masonry structures in some large cities were destroyed. Hundreds of aftershocks, some greater than magnitude 6, continued to shake the area for months. They hindered rescue efforts by causing further building collapses and triggering more slides.

The earthquake happened at a depth of 28 kilometers, making that people's minimum distance from the shock. The quake occurred along the northwest-trending Muzaffarabad Fault, part of the main continent–continent collision boundary between the colder, more rigid Indian Plate on the south and the hotter, softer southern edge of the Eurasian Plate on the north. The Indian Plate descends at a gentle angle beneath Asia in an enormous collision zone that deforms and thickens the earth's crust to form the Himalayas. Continental convergence is about 2 centimeters per year, causing the Himalayas to rise at about 1 centimeter per

year. The ongoing deformation is accompanied by geologically frequent earthquakes, many of which are very large and devastating. The October earthquake involved 3 to 4 meters of slip on a 70- to 80-kilometer-long fault but may have released only one-tenth of the elastic energy accumulated in that area during the time since the last great earthquake, a magnitude 8.0 event in September 1555. It partly filled a seismic gap in the line of previous earthquakes, overlapping the western end of the 1885 break, an area overdue for a large earthquake.

Ulrich Camp.

▶ *Rockslides, shaken loose, crushed and buried homes, sometimes leaving only their corrugated metal roofs exposed.*

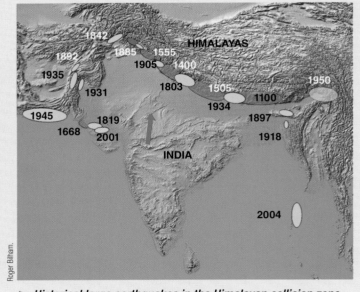

Roger Bilham.

▶ *Historical large earthquakes in the Himalayan collision zone between the Indian and Eurasian Plates. The 2005 Kashmir earthquake is labeled in red.*

Amplified Shaking over Loose Sediment
Mexico City Earthquake, 1985 ▶

On September 19, 1985, the oceanic lithosphere sinking through the trench off the west coast of Mexico suddenly slipped in an area 350 kilometers from Mexico City. The resulting earthquake struck with a magnitude of 7.9. Slippage along almost 200 kilometers of the fault filled an earthquake gap (compare Figure 4-6) that had worried researchers for more than a decade. Large aftershocks two and five days after the main earthquake registered magnitudes of 7.5 and 7.3, respectively.

Accelerations of 16 percent of **g** near the coast decreased with distance to 4 percent of **g** in Mexico City. Most of the older parts of Mexico City are built on clays and sands deposited in the bed of old Lake Texcoco. These materials amplified the shaking to cause severe damage to 500 buildings despite their distance from the epicenter. Meanwhile, the buildings of the National University of Mexico stood undamaged on the bedrock hills around the old lakebed.

Many of the damaged buildings were 6 to 16 stories tall, a height that vibrates in resonance with the 1- to 2-second frequencies of the ground shaking. Shorter buildings were generally not structurally damaged; nor was the 37-story Latin American Tower, which was built in the 1950s. Its vibration period of 3.7 seconds was not in resonance with the ground vibration.

Many of the damaged buildings were poorly designed. Some had weak first stories that collapsed. Adjacent buildings of different heights swayed at different frequencies, banging into one another until one or both collapsed. Tall buildings with setbacks, to smaller widths at higher floors, commonly failed at the setback, again because of different shaking frequencies above and below. Total property damage was $8 billion (2010 dollars). Roughly 9,000 people died in collapsing buildings in Mexico City, and 30,000 were injured. Many of the identified problems resulted from the gap between building codes and actual construction, as has been the case in Mexico City and many other earthquake-prone locations.

Numerous windows in line along each floor

M. Celebi, USGS.

▶ *This fifteen-story reinforced concrete building collapsed during the 1985 Mexico City earthquake. The problem may have been too many windows in every floor, or poor-quality construction, or the building shaking at the same frequency as ground movement in the earthquake.*

Critical View

Test your observational skills by interpretation of the following scenes relevant to earthquakes. Discuss the issues in detail—what happened and why?

A This central Idaho irrigation ditch was offset during an earthquake in the 1980s (see arrows).

G. Reagor, USGS.

1. What type of fault movement is illustrated?
2. What tectonic region is this in?
3. What regional movement causes this type of fault?

C This curb is in the San Francisco Bay area.

Kious & Tilling, USGS.

1. What type of fault movement is illustrated?
2. What direction of slip is illustrated? Explain why.
3. What specific fault or fault zone is this likely to be?

B This scarp across a glacial ridge at the eastern base of the Sierra Nevada Range of eastern California shows that the right side dropped compared with the left side of the scarp.

Glacial ridge

Glacial ridge

C.D. Miller, USGS.

1. What type of fault movement is illustrated?
2. What tectonic region is this in?
3. What regional movement causes this type of fault?

D This view in southwestern Montana shows three ridges with their ends lopped off to form triangles. (One of the ridges is dotted and its end is outlined in red).

Donald Hyndman.

1. What type of fault movement would have lopped off the ridge ends?
2. Does that same movement explain why the mountain range stands high? Why?

E This southern California view from above shows recent movement on a fault (between arrows).

USGS.

1. What type of fault movement is illustrated?
2. What direction of slip is illustrated? Explain why.
3. Assuming the tire tracks are from a car, approximately how much offset occurred during the earthquake?
4. How would you determine the size of earthquake that caused the offset?

F Industry on San Francisco Bay mud.

NOAA/NGDC and University of Colorado.

1. This industrial area is at the edge of San Francisco Bay, near the airport. What fault near here might cause an earthquake?
2. In case of a large earthquake, what processes would affect the buildings?
3. Compared with a location a little closer to the fault, would damages here be greater or less? Why?

G This pile of sand with a crater in its top is about 1 meter across. It shows signs of water having spilled out of the crater and flowed down the pile's sides.

J.R. Tinsley, USGS.

1. What would cause the sand pile to form?
2. Where did the sand come from?
3. Why did the sand move? Explain how the process works.

H This fault (between arrows) just south of San Francisco, California, shows movement that bent the layers.

Hyndman.

1. Which way did the fault move (right side went which way)?
2. What type of fault movement is this called?
3. What direction of force (direction of push) wound be required to cause this movement?

Chapter Review

Key Points

Faults and Earthquakes

- Faults move in both vertical and lateral motions controlled by earth stresses in different orientations. **FIGURE 3-2**.

- Earthquakes are caused when stresses in the earth deform or strain rocks until they finally snap. Small strains may be **elastic**, where the stress is relieved and the rocks return to their original shape; **plastic**, where the shape change is permanent; or **brittle**, where the rocks break in an earthquake. **FIGURES 3-4** to **3-5**.

Tectonic Environments of Faults

- Earthquakes typically occur along plate boundaries. Strike-slip faults move along transform boundaries; thrust faults are typically associated with subduction zones and continent–continent collision boundaries; and normal faults move in spreading zones.

- Intraplate earthquakes, as in the case of the New Madrid, Missouri, events of 1811–1812, though less frequent, can also be quite large.

- Large earthquakes along the eastern fringe of North America are less frequent but can be significant and highly damaging. **FIGURE 3-18**.

Earthquake Waves

- Earthquakes are felt as a series of waves: first the compressional P waves, then the higher-amplitude shear-motion S waves, and finally the slower surface waves. **FIGURE 3-21**.

- Earthquakes are measured using a seismograph, which is essentially a suspended weight that remains relatively motionless as the Earth moves under it. **FIGURE 3-24**.

- We can estimate the distance to an earthquake using the time between arrival of the S and P waves to a seismograph and then locate the earthquake by plotting the distances from at least three seismographs in different locations. **FIGURE 3-25**.

Earthquake Size and Characteristics

- The size of an earthquake, as estimated from the degree of damage at various distances from the epicenter, is recorded as the Mercalli Intensity; the strength of an earthquake, recorded as the Richter magnitude, can be determined from its amplitude on a specific type of seismograph. An increase of one Richter magnitude corresponds to a 10-fold increase in ground motion and about a 32-fold increase in energy. **FIGURE 3-27**, By the Numbers 3-3.

- Small, frequent earthquakes are caused by short fault offsets and rupture lengths; larger earthquakes are less frequent, with longer offsets and rupture lengths; and giant earthquakes are infrequent and caused by extremely long fault offsets and rupture lengths. **FIGURES 3-29** and **3-30**.

Ground Motion and Failure During Earthquakes

- The rigidity of the ground has a large effect on damage to buildings. Soft sediments with water-filled pores shake more violently than solid rocks. The ground may fail by liquefaction, clay flake collapse, or landsliding. **FIGURES 3-33** and **3-34**.

Key Terms

acceleration, p. 56

aftershocks, p. 39

amplitude, p. 50

blind thrusts, p. 41

body waves, p. 50

creep, p. 40

elastic rebound theory, p. 38

epicenter, p. 49

faults, p. 37

focus, p. 49

foreshocks, p. 39

frequency, p. 50

liquefaction, p. 58

local magnitude, p. 54

Modified Mercalli Intensity, p. 52

moment magnitude, p. 55

normal faults, p. 37

offset, p. 40

period, p. 50

P waves, p. 49

reverse faults, p. 37

Richter Magnitude, p. 53

seismic waves, p. 49

seismogram, p. 50

seismograph, p. 50

ShakeMap, p. 53

strain, p. 38

stress, p. 38

strike-slip faults, p. 37

surface rupture length, p. 40

surface waves, p. 50

S waves, p. 50

thrust faults, p. 37

wavelength, p. 50

Questions for Review

1. Explain how earthquakes occur according to the elastic rebound theory. Draw sketches to support your explanation.

2. Give examples of four significant and active earthquake zones in North America. Tell what type of fault characterizes each zone and what kind of earthquake activity is typical for each zone.

3. Why does the ground near the coast drop dramatically during a major subduction-zone earthquake? Draw a diagram to illustrate.

4. In what sequence do different types of earthquake waves occur?

5. Which type or types of earthquake waves move through the mantle of the Earth?

6. Which type of earthquake waves shake with the largest amplitudes (largest range of motion)?

7. What is the approximate highest frequency of vibration (of back-and-forth shaking) in earthquakes?

8. What is the difference between the focus of an earthquake and its epicenter?

9. How is the distance to an earthquake epicenter determined?

10. What does the Richter Magnitude Scale measure?

11. On a seismogram, how much higher is ground motion from a magnitude 7 earthquake than a magnitude 6 earthquake?

12. Name three factors that help determine the strength of shaking during an earthquake. What combination of factors would produce the most intense shaking?

13. Explain the role water can play in ground failure during an earthquake.

Earthquake Discussion Questions are at the end of Chapter 4.

4 Earthquake Prediction and Mitigation

The Tangshan earthquake of July 27, 1976, leveled the city and killed at least 250,000 people. The damage in many areas was nearly complete.

Chinastock.

Predicted Earthquake Arrives on Schedule

Small earthquakes were a familiar occurrence in the area near Haicheng, China, but became somewhat more frequent in late 1974. Early on February 4, 1975, officials near Haicheng warned people of an imminent large earthquake and told them to remain outside to avoid being injured or killed if their houses collapsed. Already frightened by frequent small earthquakes, most people complied, in spite of the winter cold. The warnings and evacuation orders were not based on rigorous scientific studies nor handed down from the central government. They arose through an unusual mix of evidence and circumstance.

There was some background for these extraordinary interpretations and predictions. Although earthquakes are not uncommon, there had been few large tremors in northeastern China in three centuries. Then in 1966, 1967, 1969, there were four earthquakes of magnitudes 6.3 to 7.4 between 400 and 800 km southeast of Haicheng. In response, the Chinese government increased monitoring of earthquakes and anomalies that could relate to them. A regional earthquake office published a pictorial brochure for the public that explained earthquake concepts including magnitude, intensity,

Earthquake

foreshocks, deformation, and electrical currents of the Earth's crust. They explained how people could make their own observations that might aid in earthquake prediction—anomalous changes in the soil or ground, well-water levels or color, and animal behavior.

In June, 1974, a scientific conference on earthquakes in North China focused on determining earthquake potential for the next one to two years. Most scientists attending felt that earthquakes of magnitude 5 to 6 were likely in the region within one to two years. The Chinese Academy of Sciences report from the conference added that they should "prevent such assumption from causing panic and disturbance to the masses, interrupting production and people's living." Based on careful studies of 2.7 mm ground elevation changes in the area of the nearby Jinzhou Fault, a seismologist in the provincial Earthquake Office estimated a magnitude 6 earthquake on one of the nearby faults would likely occur in the first half of 1975, perhaps even in January or February.

A swarm of earthquakes in December prompted concern among earthquake researchers until they realized the filling of a major reservoir had triggered them. However, in the last week of December, almost to mid-January, reports surfaced of anomalous water-level declines in wells, water turning muddy and bitter, and strange animal behavior—horses neighing and cows mooing for no obvious reasons, noisy chickens flying into trees, and snakes and frogs coming out of ground hibernation and freezing to death.

Earthquake offices were formed in all communes of a nearby county; rescue teams, food, and winter clothes were stockpiled. Another series of earthquakes occurred in the third week in January, then activity spiked to high levels on February 1 through 4. On the evening of February 3, a surge in earthquake activity, both in numbers and size, caused alarm in the provincial Earthquake Office 25 km southwest of Haicheng. Over the next 12 hours, Mr. Zhu of that office wrote three Earthquake Information reports, and then relayed his concerns that there would be a large earthquake within the next week or so to Mr. Hua, Vice Chairman of the "Revolutionary Committee" of the province. Sensing the urgency, Mr. Hua ordered a small group of officials to organize an emergency meeting of 12 government officials in Haicheng. They ordered cancellation of all public entertainment and sports activities, business activities, and production work. At 10:30 a.m. on February 4, the group distributed the earthquake information by organized telephone groups; they emphasized a general warning that earthquake magnitudes were increasing and advised all nearby regions to be on high alert. The greatest concern was expressed by a Mr. Cao of the county earthquake office; although not a scientist, he had been delegated to form the county earthquake office a few months before and he enthusiastically learned as much as possible about earthquakes. On February 4, he repeatedly insisted that a large earthquake would occur that day. At the same time, reports of slight damages caused by small foreshock earthquakes began to trickle in. The local commanding general then distributed directives including: "Those who have unsafe houses should sleep elsewhere," those responsible for "guarding of factories, mining structures, reservoirs, bridges . . . and high-voltage power lines" should be on high alert and "report [any] urgent situation." By afternoon, the foreshocks had caused substantial damage to chimneys and overhanging roofs.

The written report on the situation was distributed to the provincial government at 2:00 pm. Because of that bulletin and the damages reported in it, and without waiting for higher-level instructions, some communes apparently made their own decisions for people to stay outside. After 1:00 p.m., foreshock activity abruptly decreased both in size and frequency. Educational materials had noted that in 1966, foreshocks quieted shortly before the main earthquake, so this alarmed many people. Most people chose to sleep outside on both February 3 and 4; many who decided to remain inside wore winter clothes to bed in case an earthquake should force them outside but some of them died when their houses collapsed. However, the distribution of warnings and citizen compliance was very uneven. The difficulty of dealing with the cold was emphasized by the hundreds of people who died from hypothermia after the earthquake.

At 7:36 p.m. on February 4, the magnitude 7.3 earthquake struck about 20 km southwest of Haicheng. Ninety percent of the city buildings were destroyed or severely damaged. Relatively modern two- and three-story buildings built with bricks and unreinforced concrete were most susceptible to collapse, including office buildings, factories, guesthouses, and schools. It was fortunate that most of these buildings were not occupied because the earthquake struck after work. Despite this severe destruction, there were only 2,014 deaths and 4,292 individuals with severe injuries out of one million people. Of those, 372 died from freezing or carbon monoxide poisoning, and 341 died in fires. Were it not for the prediction, the earthquake would have likely killed hundreds of thousands of people.

The Haicheng prediction is the most notable successful prediction on record. It was based primarily on earthquake foreshocks but also on general long-term knowledge and experience with earthquakes, educational efforts directed to the public, organization of data-gathering by the public, and conscientious work, intuitive judgment, and just plain good luck by amateur scientists. Some of the past experience had been recorded in ancient Chinese documents. Fatalities were relatively low in some areas with wood-frame house construction that is resistant to ground shaking, even though those people were not warned to stay outside.

In a related example of earthquake prediction, there was a forecast early in 1976 suggesting that there would be a magnitude 5 to 7 earthquake in the nearby Tangshan area by July or August of that year. On July 26, a scientist with the earthquake bureau noted large changes in the electrical properties of the ground that suggested expanded ground cracking and issued a warning of an impending quake with a magnitude greater than 7.3. Steam coming from Tangshan's water company became irregular. However, because of the strained political climate at the time, local officials were reluctant to convey this information (release of prediction information was then a criminal offense) to Tangshan residents before a sudden right-lateral shift of the northeast-trending fault leveled the city at 3:43 a.m. on July 28, 1976 (**FIGURE 4-1**). Unfortunately the magnitude 7.8 earthquake did not provide foreshocks. That earthquake, some 400 kilometers to the southwest of Haicheng, killed 242,769 and injured almost 165,000 out of a population of 1 million.

Tectonically, both the Haicheng and Tangshan earthquakes formed in response to east-west compressive stress from Pacific Plate subduction near Japan and continental collision of India with Asia that formed the Himalayas (see also: eastward spreading of the Himalayas in the Case in Point on the 2008 Wenchuan earthquake).

FIGURE 4-1 LIGHTLY REINFORCED CONCRETE COLLAPSES

Poorly-built masonry buildings collapsed. Many buildings were lightly reinforced concrete frames with brick-wall infill.

Predicting Earthquakes

Charles Richter, the developer of the magnitude scale, once remarked, "Only fools, charlatans, and liars predict earthquakes." In fact, psychics, astrologers, crackpots, assorted cranks, and miscellaneous prophets regularly predict earthquakes, but never with any success. A prediction of a new big earthquake between St. Louis and Memphis on the New Madrid fault, which suffered three major quakes in 1811–1812, caused considerable concern and mobilization of rescue resources in 1990 but there was no earthquake. The self-proclaimed earthquake expert who made the prediction actually had no background in earthquake studies.

Not even scientists can effectively predict the date and time when an earthquake will strike, although they do understand which regions are likely to experience them (**Case in Point:** A Strong Earthquake Jolts the Collision Zone between Africa and Europe L'Aquila, Italy, April 6, 2009, p. 95). It is perfectly reasonable to say that earthquakes are far more likely to strike California than North Dakota. However, it is still just as impossible to know exactly when earthquakes will strike California as it is to know in July the dates of next winter's blizzards in North Dakota.

One main objective of earthquake research has been to provide people with some reasonably reliable warning of impending events. Scientists have been able to identify some short-term earthquake indications, and there is a possibility of developing a short-term prediction system for them. However, until predictions are more reliable, they raise a host of complex ethical and political issues for governments. For now, most efforts are going into development of early warning systems for areas away from epicenters, to provide warning after an earthquake starts but before shaking reaches that area (see "Early Warning Systems", p. 73).

Earthquake Precursors

One main avenue of research in earthquake forecasting has been to explore phenomena that warn of an imminent earthquake. Although few of those efforts have proved useful to date, that may be changing.

Some researchers hope that tracking the movement of a fault will allow them to anticipate when the fault will break. Careful surveys monitoring deformation along the San Andreas Fault are now made by Global Positioning Systems (GPS) at an accuracy of 1 to 2 millimeters horizontal distance and about 1.5 centimeters vertical. Straight lines surveyed across an active fault gradually become bent before the next earthquake and fault slip (**FIGURE 4-2**). A total of 250 permanent GPS stations, for example, have been set up in the Los Angeles Basin and surrounding areas to monitor ongoing ground movements.

Tectonic plates appear to move at constant speeds, and it seems reasonable to suppose that rock strength in a major fault zone could be nearly constant through time. If so, stresses might accumulate enough to break a stuck fault loose at a predictable time. Unfortunately, it is fairly common to find that a stress applied at a constant rate does not produce results at a constant rate. Broadly speaking, this is because the natural situation is far more complicated than our mental image. More specifically, every time a fault slips, for example, it juxtaposes different rocks of different strengths, thus changing the terms of the problem of predicting when it will slip again. A large fault movement will change the stress patterns along other faults, which dramatically changes their chances of future movement.

Some hope persists that swarms of minor earthquakes, or **foreshocks**, may warn of major earthquakes if they announce the onset of fault slippage. Foreshocks precede one-third to one-half of all earthquakes. However, large

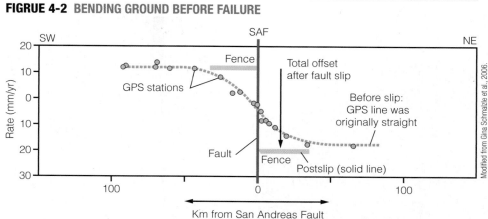

FIGRUE 4-2 BENDING GROUND BEFORE FAILURE

Global positioning system (GPS) stations placed in a straight line across the San Andreas Fault become curved as the Earth's crust on one side bends elastically with respect to the crust on the other side. Carrizo Plain, north of Los Angeles.

earthquakes often occur with no foreshocks. Even when foreshocks do precede large earthquakes, they are rarely distinguishable from the ongoing background of common small earthquakes.

Recently, it became possible to determine precisely the focus of large numbers of minute earthquakes, or *micro-earthquakes*. This can reveal the presence of a previously unknown fault system, as it did in the New Madrid, Missouri, area. The changes with time in the detailed distribution of tiny crackling movements along a fault may ultimately lead to some kinds of predictions. Unfortunately, no one knows until after the fact whether the preceding small earthquake was a foreshock or the main event.

A change in levels of the radioactive gas radon is another possible earthquake precursor. Radon is one step that uranium atoms pass through in their long decay chain, which finally ends with the formation of lead. Radon is one of the rare gases. It forms no chemical compounds and remains in rocks only because it is mechanically trapped within them. Uranium is a widely distributed minor element, so most rocks contain small amounts of radon, which escapes along fractures. Radon is responsible for nearly all of the background radiation in well water and long-term exposure is a hazard in many parts of the world.

Geologists in the Soviet Union reported during the 1970s that they had observed a strange set of symptoms in the days immediately before an earthquake. These included a rise of a few centimeters in the surface elevation above the fault, a drop in the elevation of the water table, and a rise in background radioactivity in wells. Evidently, the rocks along the fault were beginning to break. The water table dropped because water was draining into the new fractures, and radon was escaping through the fractures. It all seemed to make sense as a set of precursor events.

A change in ground elevation, such as that observed in the Soviet Union, is another possible indication of an impending earthquake. In 1975, geologists of the U.S. Geological Survey (USGS) noticed that the surface elevation had risen in the area around Palmdale, California, near Los Angeles. In addition, the groundwater level was dropping while the level of background radiation was rising. Furthermore, Palmdale is on a part of the fault that had not shown a recent large earthquake. Many members of the geologic community began to hold their breath waiting for a major earthquake, but there has been no major earthquake, at least not yet.

High zones of fluid pressure in a fault zone may also localize fault movement. Detailed study of the Parkfield area of the San Andreas Fault approximately half way between San Francisco and Los Angeles suggests that fluid pressure may be an important factor there. In fact, there have been cases where the injection of water underground inadvertently triggered earthquakes. In the early 1960s, the U.S. Army pumped waste fluids into an ancient inactive fault zone in Colorado, which apparently triggered a series of earthquakes in an area not noted for them. When the Army stopped pumping in fluids, the earthquake frequency dropped from twenty to fewer than five per month. In Basel, Switzerland, thousands small earthquakes, including two larger than magnitude 3, occurred from December 2006 through January 2007 following injection of water into deep hot rocks for geothermal power production (**FIGURE 4-3**). Following the events, a local prosecutor started an investigation to determine whether the company injecting the water should pay damages.

From these correlations between fluid pressure and earthquakes, geologists infer that the addition of fluid increases pressure between mineral grains until the rocks break in an earthquake. The situation is analogous to the way water addition can trigger landslide movement (see Chapter 8). Some people have suggested deliberately injecting water into a fault to cause movement and initiate modest-size earthquakes. This might relieve stress on the fault and preempt the occurrence of a giant earthquake. Unfortunately, there are consequences for public safety and

FIGURE 4-3 EARTHQUAKES CORRELATE WITH INJECTION OF FLUIDS

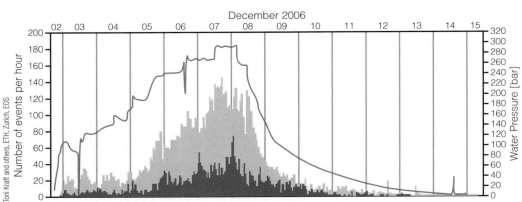

Injection of fluids for geothermal power production triggered a series of earthquakes at Basel, Switzerland, in December 2006. Red is large earthquakes; gray is small earthquakes (left scale). Blue line is water pressure injected (right scale; bar = atmospheric pressure). As water was pumped in, earthquake frequency increased.

liability that would require careful consideration before taking such an action.

Early Warning Systems

It is reasonable to hope that a large earthquake 100 kilometers or more from vulnerable structures could be detected in time to provide a useful warning. Earthquake waves traveling 4 kilometers per second from the focus of a large earthquake would require 25 seconds to reach points 100 kilometers away. Therefore, an early warning center close to an epicenter might provide as much as 25 seconds' warning in advance of an earthquake. This could be just enough time for the automatic shut-down of critical facilities, such as power plants, pumping stations, and trains. It might even be enough time for people to flee the ground floor of some buildings.

A consortium of federal, state, university, and private groups began building such a system, called TriNet, in southern California in 1997. By 2002, 600 seismometers were installed to monitor earthquakes throughout the region. Within minutes of a tremor, information from these seismometers is used to produce a ShakeMap of the ground-shaking severity in the area of the earthquake. By 2009, similar capability, the Advanced National Seismic System (ANSS), was installed nationally in areas of moderate to high seismic risk.

Although an early warning system provides a little advance warning of earthquake shaking, for most of us it will not provide enough time to evacuate from a building. Important strategies are outlined in sections below on Structural Damage, Retrofitting Buildings, and Earthquake Preparedness.

Prediction Consequences

Although short-term forecasts or early warnings would provide a good opportunity to save lives and property, they also raise complex political issues. Most earthquake predictions are false alarms, which can be expensive and disruptive. Even though the successful prediction of the Haicheng earthquake, described at the beginning of this chapter, saved many lives, the Chinese government clamped down on unofficial predictions in the late 1990s after 30 false alarms severely curtailed industry and business operations.

Imagine yourself the governor of a state, perhaps California. Now imagine that your state geological survey has just given you a perfectly believable prediction that an earthquake of magnitude 7.3 will strike a large city next Tuesday afternoon. What would you do?

If you were to appear on television to announce the prediction, you would run the risk of causing a hysterical public reaction that might result in major physical and economic damage before the date of the predicted earthquake. Imagine the consequences if you broadcast the warning and nothing happened. If you do not announce the prediction, you could stand accused of withholding vital information from the public. The problem is a political and ethical minefield.

However, the potential benefits in saved lives and minimized economic damage make short-term predictions

a desirable goal. If, for example, warning of an impending earthquake could be given several minutes before a large event, people could evacuate buildings, bridges and tunnels could be closed, trains stopped, critical facilities prepared, and emergency personnel mobilized.

With the current state of knowledge, earthquakes seem as inherently unpredictable in the short term as the oscillations of the stock market. Until prediction methods become sounder, governments will have to weigh the consequences of a false prediction carefully before taking action.

Earthquake Probability

Although scientists cannot make specific predictions about when and where an earthquake will occur, they can make reasonably reliable forecasts about where and how frequently events of different sizes *are likely* to occur. Being aware of the tectonic environments that control earthquakes helps scientists quantify risks for different regions and predict generally what earthquakes in those regions would be like. Establishing a record of past events helps determine the likely frequency of future events.

Forecasting Where Faults Will Move

Understanding the plate tectonic movements that control earthquakes allows scientists to identify the probable locations for future earthquakes. Remember from Chapter 3 that the faulting that leads to earthquakes commonly occurs on plate boundaries, and different types of boundaries lead to different types of faults. By establishing a pattern of movement on a particular fault, scientists get a better idea of where the next earthquake is likely to strike.

Through **paleoseismology**, the study of prehistoric fault movements, seismologists can establish fault movement trends that predate written records. Some basins and ranges in western North America show distinctive white stripes along the lower flanks of the ranges (see FIGURE 3-1a). Many of these stripes mark the fault scars where the basin dropped and the range rose during a past earthquake. They are most obvious in the arid Southwest, where vegetation does not obscure fault scarps.

Trenches dug across segments of an active fault generally expose the fault, along with sediments deposited across it (**FIGURE 4-4**, p. 74). Earthquakes displace sediment layers, so older sedimentary layers are more offset because they have experienced multiple fault movements. The amount of offset during fault movement is generally proportional to earthquake magnitude (**FIGURES 3-29, 4-5**, p. 74). This relationship means that trenches also provide a record of the magnitudes of prehistoric earthquakes. Evidence of past earthquakes can also be found in features that indicate soil liquefaction, sand boils (see Chapter 8), and sediment compaction.

Major faults are segmented by smaller faults, called *cross faults*, that run perpendicular to the main fault (see FIGURE 4-5a). By studying the activity along different

FIGURE 4-4 EARTHQUAKE HISTORY EXPOSED IN A TRENCH WALL

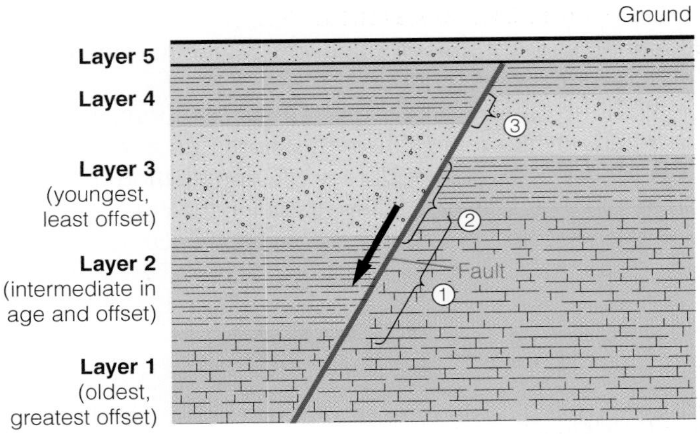

Ground

Layer 5

Layer 4

Layer 3
(youngest,
least offset)

③

Layer 2
(intermediate in
age and offset)

②

Fault

①

Layer 1
(oldest,
greatest offset)

This schematic diagram of a trench dug across an earthquake fault shows the relative displacement of layers offset by movements at different times. Offset (top of brick pattern) during first earthquake (1); offset (top of gray layer) during second earthquake (2), offset (top of orange layer) during last earthquake. Note that more sediment is deposited, in each stage, on the down-dropped side of the fault, and the first offset (1) gets progressively greater during each following offset.

segments of a fault, geologists can understand which segments are likely to move next and how big the resulting earthquake might be. Cross faults limit the size of an earthquake that can occur on the main fault because typically only one segment of the fault breaks at a time. Recall from Chapter 3 that earthquake magnitude depends on the surface rupture length (see FIGURE 3-29b).

The pattern of earthquakes along fault segments also provides clues into future earthquake activity. Segments of a major fault that have less earthquake activity than neigh-

boring segments represent a **seismic gap**. Experience shows that these seismic gaps are far more likely to be the locations of large earthquakes than the more active segments of the same fault. The Loma Prieta earthquake of 1989 filled a seismic gap in the San Andreas Fault that geologists had identified as an area statistically likely to produce an earthquake within a few decades (**FIGURE 4-6** and **Case in Point:** Earthquake Fills a Seismic Gap—Loma Prieta Earthquake, California, 1989, p. 92). The length of a seismic gap provides an indication of the maximum possible length of a future break. Because the length of fault rupture during an earthquake is proportional to the magnitude of the accompanying earthquake, the maximum likely length of the rupture of a fault provides an indication of the magnitude of an earthquake in that segment.

Earthquakes are known to migrate along certain major faults, which can also help scientists anticipate future movements. One of the best examples of **migrating earthquakes** is along the North Anatolian Fault in Turkey, which trends east over a distance of 900 kilometers. It closely resembles the San Andreas Fault in California and is comparable in length and slip rate. Earthquakes along the North Anatolian Fault are caused as the Arabian and African plates jam northward against Eurasia at a rate of 1.8 to 2.5 centimeters per year. Between 1939 and 1999, earthquakes moved sequentially westward along the North Anatolian Fault (**FIGURE 4-7**), including the 1999 earthquake that devastated Izmit. If that pattern of earthquakes continues, the next one should be a little farther west (**Case in Point:** One in a Series of Migrating Earthquakes—Izmit Earthquake, Turkey, 1999, p. 96). Is Istanbul the next major disaster?

FORECASTING WHEN FAULTS WILL MOVE Although scientists cannot predict the date on which an earthquake will occur, they can make forecasts about the likelihood of

FIGURE 4-5 WHEN A FAULT SHIFTS

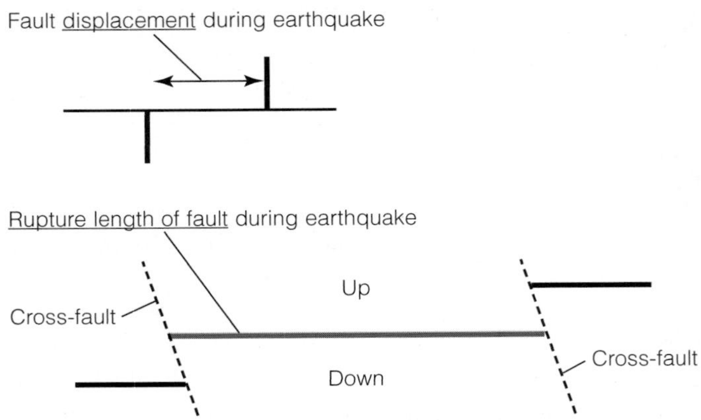

Fault <u>displacement</u> during earthquake

<u>Rupture length of fault</u> during earthquake

Cross-fault

Up

Cross-fault

Down

A

B

A. Both the displacement along a fault during an earthquake and the rupture length of the fault are proportional to the fault magnitude (large offsets or displacements are caused by larger earthquakes). Cross faults limit the length of the main fault. **B.** Strike-slip fault offset of a highway near Landers, California, in 1992.

FIGURE 4-6 SEISMIC GAP ON THE SAN ANDREAS

A seismic gap (two purple rectangles) in the historic earthquake pattern along the San Andreas Fault south of San Francisco was filled in 1989 by the Loma Prieta earthquake (small black circle) and aftershocks (blue dots).

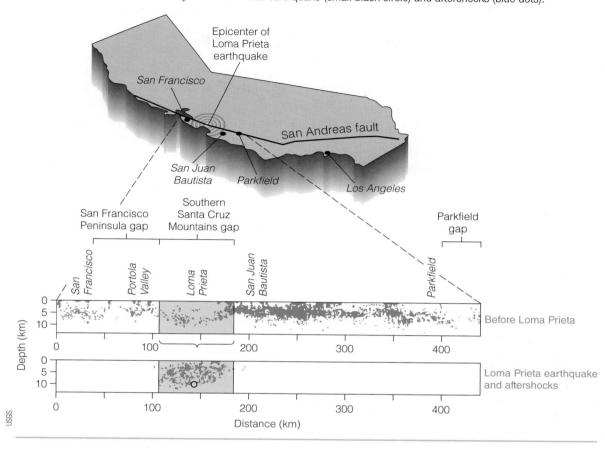

FIGURE 4-7 MIGRATION OF EARTHQUAKES ALONG A FAULT

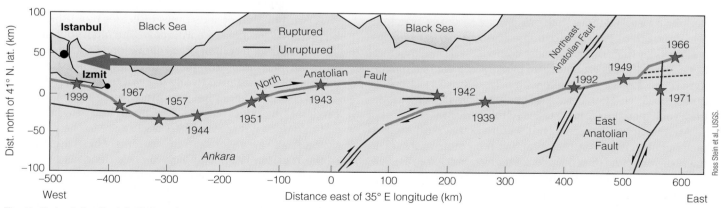

The North Anatolian Fault in Turkey shows earthquake migration with time. The red arrow in the top map shows earthquake migration from 1939 to 1999. Is Istanbul next?

FIGURE 4-8 TOKYO LEVELED IN 1923

A magnitude 8.2 earthquake in 1923 destroyed Tokyo from an epicenter 90 kilometers away.

an earthquake striking during a given timeframe, for example, over a period of decades. Making forecasts about earthquake frequency generally requires knowledge of past earthquakes along a fault. That makes it possible to estimate the **recurrence interval** for earthquakes of various sizes. Once scientists have established a recurrence interval for a certain area, they can statistically estimate the probability that an earthquake greater than a given magnitude will strike within a specified period in the future.

A few faults move at somewhat regular intervals, although this kind of behavior is so uncommon, and the time series so unreliable, that it has never been a useful method of earthquake prediction. One of the most reliable cases of equal-interval earthquakes was on the San Andreas Fault, near Parkfield, north of Los Angeles. Until recently, a series of six moderate-size earthquakes occurred about every 22 years on average. After that, there was a 38-year gap to the most recent earthquake on September 28, 2004.

For more than 500 years, especially strong earthquakes devastated Tokyo, Japan, at intervals of roughly 70 years, most recently in 1923, when 105,000 people died (**FIGURE 4-8**). Many people, especially those in the insurance business, watched the approach of 1993 with extreme concern. No earthquake has yet occurred, but it may just come later than expected. USGS seismologists estimate a 70 percent chance of a magnitude 6.7 or larger earthquake within 75 kilometers of Tokyo before 2035. The Japanese government estimates that losses from such a quake would exceed a trillion dollars.

Another recently recognized pattern is that earthquakes may occur in groups. A major fault can be quite active over many decades and then lie quiet for many more before it again becomes active. In North America, our written history may be a bit short to demonstrate such patterns clearly, but in Turkey the historical record is much longer. Beginning in A.D. 967, the North Anatolian Fault moved to cause earthquakes at intervals of 7 to 40 years, followed by no activity for 204 years. The fault saw more earthquakes from 1254 to 1784, then another long hiatus. Then again in 1939, activity picked up at intervals ranging from 1 to 21 years and continues to the present day.

The coast of northern California, Oregon, Washington, and southern British Columbia has a record of strong earthquakes but is currently experiencing a long gap in such earthquakes (**FIGURE 4-9**). The last major earthquake here was just over 300 years ago. As with some other subduction zones, the expected magnitude 8 to 9 earthquake would make up for that long interval.

Scientists look for clues to establish dates for earthquakes that occurred before recorded history. Radiocarbon dating of organic matter, most commonly charcoal, in the sedimentary layers broken along a fault can reveal the maximum dates of fault movements. Radiocarbon dates on offset sediments exposed in a trench across the Reelfoot Fault scarp in Tennessee reveal that the New Madrid fault system has moved three times within the last 2,400 years, which, in combination with other data, suggests that the recurrence interval for earthquakes in the New Madrid area may be 500 to 1,000 years, although the uncertainty is large. One limitation of radiocarbon dating is that it can be used to date events that happened only in the last 40,000 years.

A combination of such clues indicates that Utah should be prepared for a major earthquake. The intermountain seismic zone through central Utah, Yellowstone Park, and western Montana feels more earthquakes than one might expect given the significant distance to any plate boundary (see FIGURE 3-15b). No movement has occurred on the main fault near Brigham City, north of Salt Lake City, for at least 1,300 years. But that does not offer much reassurance to the people of the Salt Lake area. Nor does the extremely fresh appearance of the Wasatch Front. The lack of large deposits of eroded material at the base of the Wasatch Front suggests that the range is rising rapidly, presumably with accompanying earthquakes. The people of the Salt Lake area should consider themselves at high risk for major earthquakes.

An obvious fault scarp shows evidence of a large earthquake in Little Cottonwood Canyon in the southeastern part of Salt Lake City (**FIGURE 4-10A**). Radiocarbon dates on scraps of charcoal show that the scarp rose

FIGURE 4-9 BIG EARTHQUAKES CONCENTRATE NEAR THE COAST

Earthquake epicenter maps for Canada and the United states from 1899 to 1990. A seismic gap remains along the west coast. Larger symbols represent larger earthquakes. Shallow epicenters are purple, green are deeper, and yellow are deepest, primarily along the Aleutian subduction zone of southwestern Alaska.

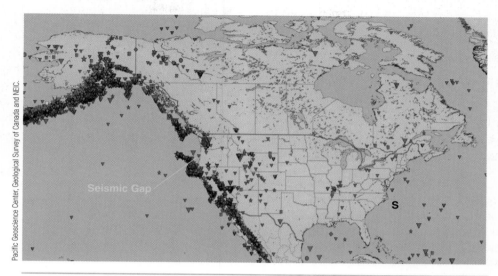

Pacific Geoscience Center, Geological Survey of Canada and NEIC.

approximately 9,000 years ago. More radiocarbon dates on offset sedimentary layers exposed in trenches cut across the fault reveal evidence of movements 5,300, 4,000, 2,500, and 1,300 years ago. The average recurrence interval appears to be about 1,350 years, with an average offset of 2 meters per event. It is not reassuring to see that developers continue to encourage building homes next to the fault (**FIGURE 4-10B**).

Movement on the Wasatch scarp appears to occur in segments (**FIGURE 4-11**, p. 78). Within a 6,000 year record, seventeen fault offsets exposed in trenches suggest an earthquake of magnitude greater than 6.5 every 350 years on average. Maximum segment lengths and offsets found in some trenches suggest earthquake magnitudes up to 7.5. Despite these warning signs, development continues near, and even within, the fault zone (see FIGURE 4-10b).

FIGURE 4-10 FAULT SCARPS AND DEVELOPMENT

Donald Hyndman.

Craig Nelson, USGS.

A. Wasatch front fault scarp cuts a 19,000-year-old glacial moraine from Cottonwood Canyon near Salt Lake City. The scarp (between arrows) formed during a large earthquake since that time. **B.** In spite of real hazards and the educational efforts of the Utah Geological Survey, development continues, even along the scarp itself. Fault scarps marked by arrows.

FIGURE 4-11 EARTHQUAKE ACTIVITY ON A SEGMENTED FAULT

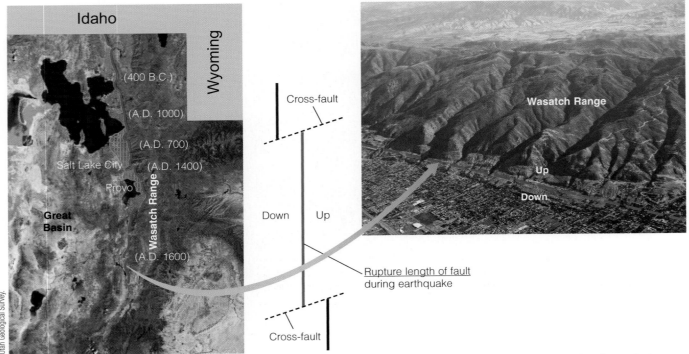

The historical record of earthquake activity along the north-south Wasatch Fault near Salt Lake City suggests an earthquake of magnitude greater than 6.5 every 350 years on average. The fault is broken into segments by east-west faults (not shown on map). Individual fault segments typically break one at a time; longer segments can generate larger earthquakes.

Populations at Risk

"We only get a few small quakes here, nothing to worry about. They have some pretty big ones around San Francisco and Los Angeles, but we don't get them." We heard this remark recently in Santa Rosa, some 70 kilometers (43.5 miles) north of San Francisco. Unfortunately, the speaker lacked some critical information about the probable location and magnitude of future earthquakes. Although the San Andreas Fault lies 40 kilometers to the west, a major earthquake on that fault could still devastate Santa Rosa, as the 1906 San Francisco earthquake did. In addition, Santa Rosa lies on the Hayward-Rodgers Creek Fault, a major strand of the overall San Andreas system that is currently considered more dangerous than the main San Andreas Fault in the San Francisco Bay area. Finally, the speaker is apparently unaware that even though they occasionally feel small earthquakes in an earthquake-prone area, larger ones are likely, and a very large earthquake is quite possible.

Seismologists use what they have learned about probabilities of where and when an earthquake will strike to estimate the risk for a given area on a **risk map** (**FIGURES 4-12** and **4-13**). These maps are based on past activity of both frequency and magnitude, and they are invaluable when choosing sites for major structures—such as dams, power plants, public buildings, and bridges—and important for insurance purposes.

Notice on the map in FIGURE 4-13 that one of the highest-hazard areas in the United States is the heavily populated coast of California. California gets far more than its share of North American earthquakes because of its location along the boundary between the Pacific and North American tectonic plates. That part of the plate boundary is marked by the San Andreas Fault, one of Earth's largest and most active transform faults. This well-researched earthquake zone provides a good example of how forecasting allows scientists to estimate the probability of earthquakes and their likely levels of damage.

Roughly one-third of California's future earthquake damage will probably result from a few large events in Los Angeles County. A large proportion of the remainder will most likely occur in the San Francisco Bay area. Before the 1989 Loma Prieta event, earthquakes in the United States cost an average of $230 million per year. Costs escalate as more people move into more dangerous areas and as property values rise. Some authorities now expect future losses to average more than $4.4 billion per year, with 75 percent of that in California. Given these high stakes, any taxpayer in these areas should be interested in the earthquake hazard.

FIGURE 4-12 WORLD EARTHQUAKE HAZARD

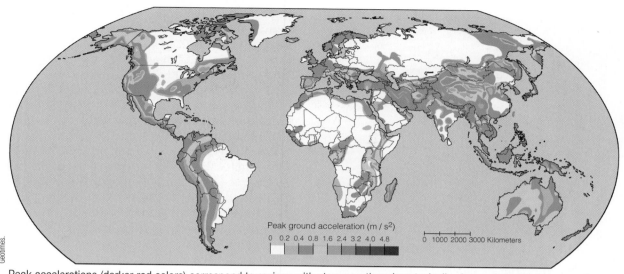

Peak accelerations (darker red colors) correspond to regions with strong earthquakes, typically along subduction zones such as those along the west coasts of North and South America and continental collision zones such as in southern Europe and Asia.

FIGURE 4-13 UNITED STATES AND CANADA EARTHQUAKE HAZARD

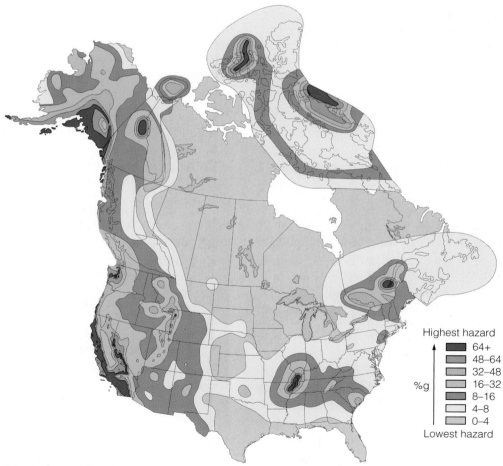

Seismic zones of Canada and the United States show the greatest risk along the west coast, where subduction zones and transform faults dominate the plate boundary.

FIGURE 4-14 CHANCE OF THE NEXT BIG CALIFORNIA EARTHQUAKE

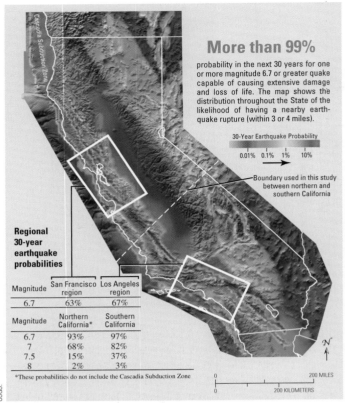

More than 99%

probability in the next 30 years for one or more magnitude 6.7 or greater quake capable of causing extensive damage and loss of life. The map shows the distribution throughout the State of the likelihood of having a nearby earthquake rupture (within 3 or 4 miles).

30-Year Earthquake Probability

0.01% 0.1% 1% 10%

Boundary used in this study between northern and southern California

Regional 30-year earthquake probabilities

Magnitude	San Francisco region	Los Angeles region
6.7	63%	67%

Magnitude	Northern California*	Southern California
6.7	93%	97%
7	68%	82%
7.5	15%	37%
8	2%	3%

*These probabilities do not include the Cascadia Subduction Zone

0   200 MILES

0   200 KILOMETERS

USGS.

California is almost certain (>99 percent probability) to have a major earthquake in the next 30 years. The Los Angeles and San Francisco regions (see white boxes and table in lower left), respectively, have a 67 and 63 percent chance of suffering a major earthquake in that time.

In fact, the USGS predicts that for California there is more than a 99 percent chance of an earthquake with magnitude 6.7 or larger striking in the next 30 years. In addition they project a 46 percent chance of a much larger magnitude 7.5 event will occur somewhere in the state in the next 30 years (**FIGURE 4-14**).

The San Francisco Bay Area

The San Andreas Fault system is a wide zone that includes nearly the entire San Francisco Bay area. This fault began to move along most of its length approximately 16 million years ago and has likely inflicted thousands of earthquakes on the San Francisco Bay area during this period. The most significant in modern history, the devastating 1906 San Francisco earthquake, reduced that city to ruins (**Case in Point:** Devastating Fire Caused by an Earthquake—San Francisco, California, 1906, p. 97). The people who live in the Bay area dread the next catastrophic earthquake, the "Big One." It will happen just as surely as the sun will rise tomorrow, but no one knows when it will occur or how big it will be.

The Gutenberg-Richter frequency-versus-magnitude relationship suggests that any segment of the San Andreas Fault 100 kilometers long should release energy equivalent to an earthquake of magnitude 6 on average every 8 years, one of magnitude 7 on average every 60 years, and one of magnitude 8 on average about every 700 years (see FIGURE 3-31). If this is correct, then several fault segments are overdue. Energy builds up across the fault until an earthquake occurs, so the chance of a major event grows as the time since the last large earthquake increases.

Researchers have calculated different overall slip rates on the fault using different types of data. Using a relatively low slip rate of 2 centimeters per year, strain should accumulate on stuck faults at a rate that would cause an earthquake of moment magnitude (M_w) 7.5 every 120 to 170 years. Alternatively, there could be six earthquakes of magnitude 6.6 in the same period. Because the actual number has been far fewer, this again suggests that the area is overdue.

Although the San Andreas is the main fault, others in that fault system also pose risk to populations in the area. The Hayward Fault splays north from the San Andreas and runs along the base of the hills near the east side of San Francisco Bay through a continuous series of cities, including Hayward, Oakland, Berkeley, and Richmond. The Hayward Fault has not caused a major earthquake since 1868, but many geologists believe it is one of the most dangerous faults in California and may be ready to move (**FIGURE 4-15**). A moderate earthquake (magnitude 5.6) struck the Hayward fault northeast of San Jose on October 30, 2007, without doing much damage, although it did serve as a reminder of the hazard. An offset of as much as 3 meters on the Hayward Fault would probably cause a magnitude 7 earthquake that would shake for 20 to 25 seconds. The USGS estimates that several thousand people might perish and at least 57,000 buildings would be damaged, eleven times as many as in the 1989 Loma Prieta earthquake. The magnitude 6.8 earthquake that hit Kobe, Japan, in January 1995 killed more than 6,000 people in an equally modern city built on similar ground.

The Rodgers Creek Fault continues the Hayward Fault trend from the north end of San Francisco Bay north through Santa Rosa. Along with the San Andreas Fault, these two, and their southward extension, the Calaveras Fault, pose the greatest hazards in the region, partly because they are likely to cause large earthquakes and partly because they traverse large metropolitan areas with rapidly growing populations. The risk is not only in the low-lying areas of San Francisco Bay muds, or in the heavily built-up cities along the fault trace, but in the precipitous hills above the fault. Those hills are blanketed with expensive homes, including those at the crest of the range along Skyline Drive, where the outer edges of many are propped on spindly-looking stilts.

In the 70 years before the 1906 San Francisco event, earthquakes of magnitude 6 to 7 occurred every 10 to 15 years.

FIGURE 4-15 SAN FRANCISCO AREA EARTHQUAKES

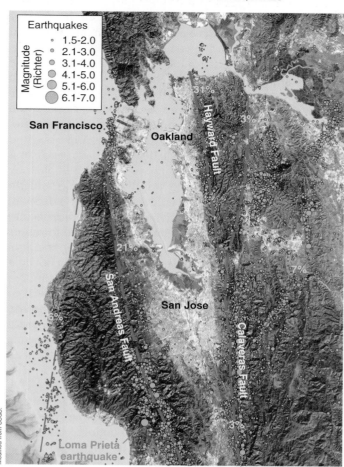

Modified from USGS.

The San Francisco Bay area faults show a probability of a magnitude 6.7 or larger earthquake on the main faults before 2036. The Loma Prieta epicenter is shown as a larger orange dot. The 1906 earthquake broke the fault from near the Loma Prieta event to far north of this map.

No earthquakes greater than magnitude 6 struck the San Francisco Bay area between 1911 and 1979. Four earthquakes of magnitudes 6 to 7, including the 1989 Loma Prieta event, struck the region between 1979 and 1989. We may be in the midst of another cluster of strong earthquakes. A recent assessment by the USGS of the earthquake probabilities in the Bay Area in the next 30 years suggests a 21 percent probability of a magnitude 6.7 or larger earthquake on the San Andreas Fault, 31 percent on the Hayward-Rodgers Creek Fault, and 7 percent on the Calaveras Fault. The total probability of at least one magnitude 6.7 earthquake in the next 30 years somewhere in the San Francisco Bay area is 63 percent (see FIGURE 4-15). The same size earthquake caused severe damage in the Los Angeles area in 1994.

Just as the risk of a large earthquake increases as more time passes without one, so does the potential damage such an earthquake would cause. People build in areas

FIGURE 4-16 DANGEROUS HOUSING

USGS.

A satellite view south along the San Andreas Fault zone in Daly City and Pacifica, just south of San Francisco, shows where housing developments straddle the fault, as located by U.S. Geological Survey mapping. The San Andreas Fault apparently runs through the corner of this school just south of San Francisco.

close to or even on top of faults, right on potential future epicenters. In some areas immediately south of San Francisco, developers have filled fault-induced depressions and built subdivisions right across the trace of the fault (**FIGURE 4-16**). Within San Francisco itself, zoning prevents building in some areas near the fault. All those people and all that development guarantee that the next Big One will be far more devastating than the last major earthquake. A November 2005 study by Allstate Insurance Company indicated that if San Francisco had the same size quake as in 1906, it could cost $400 billion to rebuild the city. For comparison, the state's entire budget in 2004 was $164 billion. This means that it could cost every man, woman, and child in California more than $11,000 to rebuild San Francisco.

The Los Angeles Area

The San Andreas Fault lies 50 kilometers northeast of Los Angeles. A large earthquake on that part of the fault would undoubtedly cause significant loss of life and property damage, but the prospect of lesser earthquakes on any of the many related faults that cross the Los Angeles basin is a matter of more immediate concern because of their proximity to heavily populated areas (see FIGURE 3-10).

Several moderate earthquakes have shaken Los Angeles and the Transverse Ranges area during the last 150 years. Such a history guarantees that we can expect more. Moderate earthquakes similar to the magnitude 6.7 Northridge event of 1994 are likely to shake the Los Angeles basin with an average recurrence interval of less than 10 years.

Given the overall slip rate in the region, the number of observed moderate earthquakes is fewer than these estimates would lead us to expect. A larger earthquake of M_w 7.5, requiring a rupture length of roughly 160 kilometers, should occur on average once every 300 or so years. Far too few moderate earthquakes have occurred in the Los Angeles area to relieve the observed amount of strain built up across the San Andreas Fault between 1857 and 2007. Releasing the accumulated strain would have required seventeen moderate earthquakes, but only two (1971 and 1994) have been greater than magnitude 6.7. The reason for the deficiency is unclear. Elsewhere in the world, clusters of destructive earthquakes have occurred at intervals of a few decades. Perhaps several faults that were mechanically linked ruptured at roughly the same time to generate a large earthquake. Such combination ruptures appear to have occurred in the Los Angeles area in the past.

Seismic activity has been notably absent along the segment of the San Andreas Fault south of the Fort Tejon slippage of 1857, almost to the Mexican border. Fault segments with such seismic gaps are far more likely to experience an earthquake than fault segments that have recently moved.

Trenches dug across the San Andreas Fault northeast of Los Angeles exposed sand boils that record sediment liquefaction during nine large earthquakes during the 1,300 years before the Fort Tejon earthquake of 1857. They struck at intervals of between 55 and 275 years, the average interval between these events being 160 years. Based on this history, the area appears due for a major earthquake.

A magnitude 7.4 earthquake would cause the ground to shake over a much larger area and for a longer duration than the Northridge earthquake did in 1994, and the greater area and longer shaking would result in far more casualties and property damage. Studies of the fault in 2006 showed that it was sufficiently stressed to break with a magnitude 7 earthquake but could, in the future, see a magnitude 8 (the Northridge and San Fernando Valley earthquakes were both magnitude 6.7). The nearby, lesser-known San Jacinto Fault, which runs northwest through San Bernardino and Riverside, is being stressed about twice as fast as formerly thought. It is capable of generating magnitude 7 earthquakes.

Minimizing Earthquake Damage

Throughout history, earthquakes have had devastating effects, destroying cities and decimating their populations (Table 4-1). Given the high probability of a destructive earthquake in a major urban area, what is the potential damage from such an earthquake, and how can this damage be mitigated?

The primary cause of deaths and damage in an earthquake is the collapse of buildings and other structures, which we explore in greater detail below. But the shaking during an earthquake can trigger other damage as well. In the aftermath of the earthquake, fire becomes a serious hazard. The fires sparked in the 1906 San Francisco earthquake caused more damage than the initial shaking (see **Case in Point:** San Francisco, California, 1906, p. 97). During an earthquake, electric wires fall to an accompaniment of great sparks that are likely to start fires. Release of gas or other petroleum products in port areas, such as Los Angeles, could spark a firestorm. Buckled streets, heaps of fallen rubble, and broken water mains would hamper firefighters' efforts.

Without immediate outside help, the days after a major earthquake commonly bring epidemics caused by impure water, because broken sewer mains leak contaminants into broken water mains. Decomposing bodies buried in rubble also contribute to the spread of disease. It is not uncommon for many people to die from diseases following a major earthquake, especially in the poorest countries and warm climates. Meanwhile, fires continue to burn and expenses mount.

Scientists know a number of ways to mitigate earthquake damage, but these mitigation efforts are expensive and hard to enforce. The costs should be weighed against the large number of lives that could be saved if these dollars were spent on improved medical care or mitigation of other hazards.

Structural Damage and Retrofitting

Earthquakes don't kill people—falling buildings and highway structures do. No one suffers much injury from the few seconds of shaking during an earthquake. The main danger is overhead.

Load-bearing masonry walls of any kind are likely to shake apart and collapse during an earthquake, dropping heavy roofs on people indoors. Some bridge decks and floors of parking garages are not strongly anchored at their ends because they must expand and contract with changes in temperature. A strong earthquake can shake them off their supports (**FIGURE 4-17A**, p. 84). Many external walls are loosely attached to building frameworks. They may break free during an earthquake and collapse onto the street below (**FIGURE 4-17B**, p. 84).

Reinforced concrete often breaks in large earthquakes, leaving the formerly enclosed reinforcing steel free to buckle and fail (**FIGURE 4-18**, p. 84). Prominent recent examples include the collapse of elevated freeways during the 1971 San Fernando Valley earthquake, the 1989 Loma Prieta earthquake, the 1994 Northridge earthquake, and the 1995 Kobe, Japan earthquake. Reinforced concrete fares much better if it is wrapped in steel to prevent crumbling. Much of the strengthening of freeway overpasses in California during the 1990s involved fitting steel sheaths around reinforced concrete columns. New construction often involves wrapping steel rods around the vertical reinforcing bars in concrete supports.

Table 4-1		Some of the Most Catastrophic Earthquakes in Terms of Casualties*	
EARTHQUAKE	DATE	MAGNITUDE (M_S)	CASUALTIES
Haiti, Port-au-Prince	Jan. 12, 2010, 4:53 p.m.	7.0	>222,000—poor building quality.
China, Wenchuan	May 12, 2008, 2:28 p.m.	7.9	87,587—collapse of recent, poorly built schools and apts.
Kashmir	Oct. 8, 2005, 8:50 a.m.	7.6	80,361—mostly in collapse of schools and apartment buildings
Sumatra	Dec. 26, 2004, 7:59 a.m.	9.15	230,000—including from the tsunami caused by the quake
Iran, Bam	Dec. 26, 2003, 5:26 a.m.	6.7	~26,000—mostly in buildings of mud and brick
India, Bhuj	Jan. 26, 2001	7.7	~30,000—mostly in buildings of mud and brick
Iran	June 20, 1990	7.7	~50,000—landslides and adobe and unreinforced masonry
Armenia	Dec. 7, 1988	7.0	25,000—mostly in precast, poorly constructed concrete bldgs.
China, Tangshan	July 27, 1976	7.6	250,000—mostly in collapsed adobe houses
Peru	May 31, 1970	7.8	66,000—in rock slide that destroyed Yungay
India, Quetta	May 31, 1935	7.5	60,000
Japan, Kwanto	Sept. 1, 1923	8.2	143,000—incl. deaths in great Tokyo fire started by the quake
China, Kansu	Dec. 16, 1920	8.5	180,000
Italy, Sicily, Messina	Dec. 28, 1908	7.5	120,000
Ecuador and Colombia	Aug. 16, 1868	?	70,000 total: 40,000 in Ecuador, 30,000 in Colombia
Ecuador, Quito	Feb. 4, 1797		40,000
Italy, Calabria	Feb. 4, 1783		50,000
Portugal, Lisbon	Nov. 1, 1755		70,000—including deaths in tsunami caused by the quake
India, Calcutta	Oct. 11, 1737		300,000
Italy, Catania	Jan. 11, 1693		60,000
Caucasia, Shemakha	Nov. 1667		80,000
China, Shaanxi, Shensi	Jan. 23, 1556		830,000—mostly in collapse of homes dug into silt
Portugal, Lisbon	Jan. 26, 1531		30,000

*Information gathered from various sources. M_s = surface-wave magnitude.

Weak floors at any level of a building are a major problem during an earthquake. Upper floors move back and forth during shaking, leading to the collapse of either individual floors or the whole building (**FIGURE 4-19**, p. 85). Garages and storefronts commonly lack the strength to resist major lateral movements, as do commercial buildings with too many unbraced windows on any floor. The addition of diagonal beams can provide the support needed to resist lateral movement (**FIGURE 4-20**, p. 85). Heavy reinforced concrete floors and corner supports with unreinforced fill-in walls of brick are very common throughout the developing world. Because the bricks have no strength in an earthquake, they provide no lateral strength; such structures frequently collapse.

Glass is too rigid to fare well during large earthquakes. Shattering glass is one of the most common causes of injuries in earthquakes. Broken glass rained down on the street from the windows of a large department store in downtown San Francisco during the Loma Prieta earthquake of 1989. Safety glass is now required in the ground floor windows of commercial buildings but not in the upper floors. Glass systems in modern high-rise buildings are designed to accommodate routine sway from wind, and they also perform fairly well during earthquakes. Safety glass similar to that used in cars helps, as does polyester film bonded to the glass.

The walls of many houses built before 1935 merely rest on their concrete foundations. The only thing holding the house in place is its weight. Large earthquakes often shake older houses off their foundations and destroy them (**FIGURE 4-21A**, p. 86). Older buildings can be particularly dangerous during an earthquake. Many structures 100 or more years old have wooden floor beams loosely resting in notches in brick or stone walls (**FIGURE 4-21B**). Earthquake shaking may pull these beams out of the walls and cause the building to collapse. Weak foundations can be strengthened with diagonal bracing or sheets of plywood nailed to the wall studs (**FIGURE 4-22**, p. 86).

Overhanging parapets are common on old brick and stone buildings. Earthquakes commonly crack them off where the external wall joins the roofline (**FIGURE 4-23**, p. 86). Even if the building remains standing, falling portions of walls may crush people on the street or in cars.

FIGURE 4-17 JOINS AND BEAMS CAN FAIL

Steel plates welded on floor beam and ledge

Precast floor beam

Support beam

Mehmet Celebi USGS.

A

A. The precast floor of the Northridge Fashion Center's three-year-old parking garage, which was supported on the ledge of a supporting beam, failed during the 1994 Northridge, California, earthquake. **B.** The exterior façade loosely hung on a building framework in the community of Reseda collapsed in the 1994 Northridge earthquake. Directional shaking detached one wall while leaving the perpendicular wall intact.

Furniture shaking inside may have hammered wall

James W. Dewey USGS.

B

FIGURE 4-18 CRUMBLING CONCRETE COLUMNS

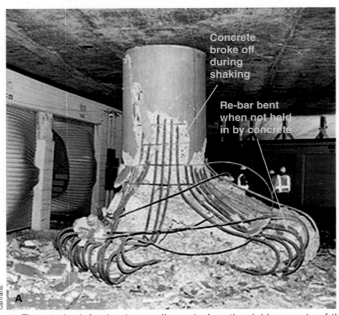

Concrete broke off during shaking

Re-bar bent when not held in by concrete

CalTrans.

A

Remains of steel reinforcing bars and heavy concrete column

James Dewey USGS.

B

A. The steel reinforcing bars collapsed when the rigid concrete of this freeway support column shattered during the 1994 Northridge earthquake. **B.** Segments of the Interstate 5 / California Highway 14 interchange collapsed while under construction in the 1971 San Fernando Valley earthquake. They were not retrofitted but rebuilt with the same specifications and collapsed again, as shown here in the 1994 Northridge earthquake.

FIGURE 4-19 WEAK FLOORS

A. A weak ground floor without diagonal braces is a recipe for lateral failure. **B.** Unbraced ground-floor garages of this apartment building in Reseda, California, collapsed on cars in the 1994 Northridge earthquake. **C.** Too many windows on a single floor can lead to collapse, as was the case in the second floor of this building in the Kobe, Japan, earthquake.

FIGURE 4-20 DIAGONAL STRENGTH

This U.S. Geological Survey building in Menlo Park, California, near the San Andreas Fault, has prominent diagonal braces to withstand earthquake shaking. Constructed with a precast concrete floor and roof slabs hoisted onto a steel frame, the diagonal braces were added later.

Fortunately, buildings can be built to withstand a severe earthquake well enough to minimize the risk to people inside. Houses framed with wood generally have enough flexibility to bend without shattering. A house may bounce off its foundation and be wrecked during an earthquake, but it is unlikely to collapse and kill anyone. Commercial buildings with frames made of steel beams welded or bolted together are also generally flexible enough to resist collapse. Most building codes now require builders to bolt the framing to the foundation—an excellent example of an inexpensive change that can make an enormous improvement in the ability of a house to withstand an earthquake (FIGURE 4-22).

Modifying existing buildings to minimize damage during strong earthquake motion, or **retrofitting**, can provide additional protection. Large earthquakes have a low probability of occurrence, but they may happen during the average life of a building, which is considered to be 50 to 150 years. Retrofitting existing buildings to survive these large but rare events is extremely expensive; it is much less expensive to construct new buildings to a higher standard. This leaves scientists and policymakers unsure of the best

FIGURE 4-21 WEAK FLOOR CONNECTIONS

A. Even single-story houses can have problems. This wood-frame house in Watsonville, California, southeast of Santa Cruz, was shaken off its foundation during the 1989 Loma Prieta earthquake. **B.** Floor joists rest loosely in this structural brick wall in a century-old building in Missoula, Montana. Strong shaking could pull the floor joists free from the walls, permitting collapse of the building.

FIGURE 4-22 STRENGTHENING HOUSES

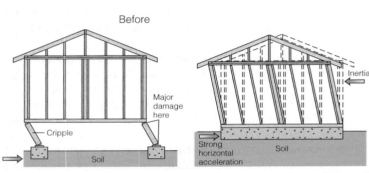

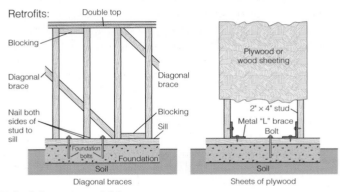

Houses can be easily strengthened with diagonal braces or sheets of plywood (pair of diagrams on right).

strategy when developing codes for new or retrofitted buildings.

Taller buildings experience a particular range of earthquake damage caused by the sway of the building. Even buildings without weak floors may collapse if their upper floors move in one direction as the ground snaps back in the other. Research in response to the 1994 Northridge earthquake included numerical modeling of the effect of an M_w 7 earthquake on a 20 story building framed with I-beams and using the latest earthquake codes for the Los Angeles area. Over a ground area of 1,000 square kilometers (386 square miles), the model predicted that the first story would move 35 centimeters (13.8 inches) off-center, three times what was then considered severe. At the top of the building, the predicted lateral displacement would reach 3.5 meters (11.5 feet) within 7 seconds, followed by continued swaying of 1.5 meters with a period of 2.5 seconds. The most damaging movement in these models was forward just when the back motion of the ground strikes the building, creating a whiplash effect.

Another simulation showed cracks opening at the welds between vertical columns carrying only 25 percent of their load capacity. In some locations, an exterior column that broke during back motion could no longer prevent lateral motion in the first aboveground story, and the building collapsed. Even when a building doesn't collapse, its oscillation may cause significant damage. Taller buildings sway more slowly than their shorter neighbors. Adjacent tall and short buildings tend to bang against each other. This commonly breaks the taller building at about the level of the top of its shorter neighbor (**FIGURE 4-24**). Buildings should be either firmly attached or stand far enough apart that they do not bash one another during an earthquake.

The amount of sway for buildings is related to the frequency of earthquake waves. Earthquakes shake the ground with frequencies of 0.1 to 30 oscillations per second. Small earthquakes generate a larger proportion of higher-frequency vibrations. If the natural oscillation frequency of a building is similar to that of the ground, it may resonate in

FIGURE 4-23 STRUCTURAL BRICK AND OVERHANGS

Brick parapet

Brick parapet failed

Bricks crushed cars

Edgar V. Leyendecker USGS.

The fourth-story wall and overhanging brick parapet of an unreinforced building in San Francisco collapsed onto the street during the 1989 Loma Prieta earthquake. It crushed five people in their cars.

the same way as a child on a swing pulls in resonance with the oscillations of the swing. In both cases, the resonance greatly amplifies the motion. In general, soft mud shakes at a low frequency whereas solid rock shakes at a high frequency. Tall, heavy structures, like some raised freeways built across mud-filled bays, frequently collapse unless their supports are deeply anchored and designed to minimize low-frequency shaking (**FIGURE 4-25**, p. 88). Buildings and other structures need to be designed to avoid matching their natural vibration frequency with that of the shaking ground beneath.

Tall buildings are most vulnerable to lower-frequency vibrations. To understand why, you can simulate the swaying of a tall building by dangling a weight at the end of a string and moving the hand holding the string. For example, if you move your hand back and forth 30 centimeters

(1 foot) each second, the weight at the end of a string 30 centimeters long will cause a large swing back and forth. This swing is in resonance with the oscillation period of the pendulum (**By the Numbers 4-1:** Movement of a Pendulum). On the other hand, if you move your hand back and forth three times per second, the weight will hardly move. Similarly, at three times per second, the weight will hardly move if the string is much longer than 30 centimeters. A tall building sways back and forth more slowly than a short one, so low-frequency earthquake vibrations are more likely to damage tall buildings (**By the Numbers 4-2:** Frequency of Building Vibration). Short buildings up to several stories high vibrate at high frequencies and do not sway much.

The ground also vibrates at different frequencies depending on the sediment type. Soft sediment vibrates at low frequency; it makes a dull thud when hit with a hammer. In contrast, bedrock vibrates at high frequency, so it rings when hit with a hammer. Buildings sustain the most damage when they oscillate at a frequency similar to that of the ground. Because short, rigid buildings have a high frequency, they survive earthquakes better when built on soft sediment (without taking into account secondary ground effects such as liquefaction and landslides). A tall, flexible building, which sways at a low frequency, often survives an earthquake well on high-frequency bedrock.

Guidelines for new construction as well as retrofitting of older buildings can reduce damage to buildings during an earthquake. The Federal Emergency Management Agency (FEMA) responded in 2001 with new guidelines for steel construction in areas prone to earthquakes. Most high buildings built in the last 30 years have welded steel frames designed to resist earthquake motions. In recent years, seismic engineers have tried to minimize the shaking, and therefore the damage, by isolating buildings from the shaking ground in a procedure called **base isolation** (**FIGURE 4-26**).

FIGURE 4-24 UNSYNCHRONIZED SWAY

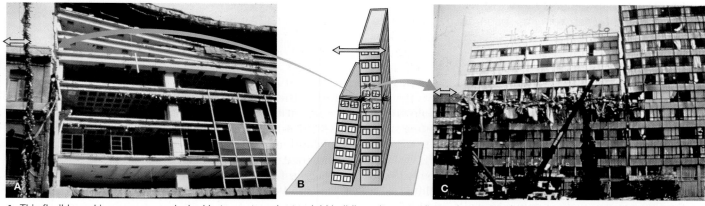

Christopher Arnold, Building Systems Development.

A

B

C

A. This flexible parking garage was locked between two shorter rigid buildings. Its upper floors, free to sway, collapsed in the 1985 Mexico City earthquake. **B.** Adjacent buildings of different heights will sway at different frequencies, so they collide during earthquakes. **C.** During the 1985 Mexico City earthquake, the building on the left hammered its taller neighbor, collapsing one story.

FIGURE 4-25 WEAK FREEWAY UNDERPINNINGS

Heavy concrete support was not anchored deeply enough in the soft bay muds

Brooks/Cole, Cengage Learning.

Broken ends of roadway

Monroe and Wicander.

A. A major elevated freeway in the low-lying waterfront area of Kobe, Japan, fell over on its side during the 1995 earthquake.
B. Even elevated freeways that were held together with bolts and welds did not fare well in the Kobe earthquake. This bus managed to stop just before crashing down onto the dropped continuation of the roadway (lower left).

They place the building on thick rubber pads, which act like a car's springs and shock absorbers that isolate us from many bumps in the road.

Base isolation pads are generally installed during initial construction; however, historic buildings can be retrofitted with them. In 1989, the historic City and County Building in Salt Lake City, five floors of stone masonry construction, was detached from its foundation, reinforced, and fitted with base isolation pads. That building is now designed for an earthquake of Richter magnitude 5 or 6. Seismic risk maps indicate a 10 percent probability of such an earthquake within 50 years. In a still larger project in the mid-2000s, the Utah State Capitol building in Salt Lake City was also jacked up and placed on rubberized pads about 50 centimeters thick (FIGURE 4-26b).

Earthquake Preparedness

Preparing homes for an earthquake and knowing what to do when an earthquake occurs can reduce damage and save lives. Evaluating structural weaknesses in a home and retrofitting are the first steps (**FIGURE 4-27**, p. 90). In general, walls of all kinds should be well anchored to floors and the foundation. Other recommendations include bolting bookcases and water heaters to walls and securing chimneys and vents with brackets. Earthquakes commonly break gas mains and electric wires, which start fires that

> ## By the Numbers 4-1
>
> ### *Movement of a Pendulum*
>
> The back-and-forth oscillation of a pendulum depends only on the length of the swinging object. Specifically, the period (*P*) of the pendulum, or total time for a back-and-forth movement, is equal to the square root of the pendulum length (*L*):
>
> $$P \approx \sqrt{L}$$
>
> Note also that the earthquake wave frequency multiplied by its wavelength equals the wave velocity. For example, a wave frequency of 2 cycles/sec multiplied by a wavelength of 3 km equals a velocity of 6 km/sec. That is comparable to the velocity of typical P waves in the continental crust.

firefighters cannot readily combat if water mains are also broken (**FIGURE 4-28**, p. 90). Matches or candles are likely to ignite gas in the air. It helps if water and gas mains are flexible and if stoves, refrigerators, and television sets are well anchored to floors or walls.

If you live in an area with significant earthquake risk, you should consider purchasing earthquake insurance. Although most well-built wood-frame houses will not collapse during an earthquake, damage may make the house uninhabitable and worthless. Because a house

Frequency of Building Vibration

Buildings of different heights sway at different frequencies, like inverted pendulums, as shown in the following figure.

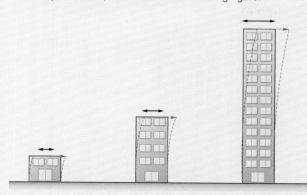

SHORT BUILDING	MID-HEIGHT BUILDING	TALL BUILDING
• Rigid 1- to 2-story building oscillates at 5–10 Hz.*	• 5- to 10-story building oscillates at 0.5–3.0 Hz.	• Flexible 20-story building oscillates at ~0.2 Hz.
• Shakes back and forth rapidly (high frequency).	• Shakes back and forth less rapidly (intermediate frequency).	• Sways back and forth slowly (low frequency).
• Thus, period is 1/5 to 1/10 = 0.2–0.1 sec.	• Thus, period is 1/0.5 to 1/3 = 2.0–0.3 sec.	• Thus, period is 1/0 or 5 sec.

*Hz = Hertz = cycles of back-and-forth motion per second.

FIGURE 4-26 SHOCK ABSORBERS FOR BUILDINGS

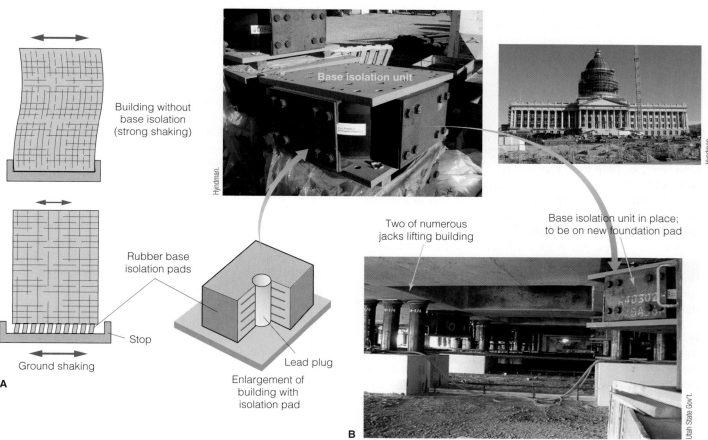

Building without base isolation (strong shaking)

Rubber base isolation pads

Stop

Ground shaking

A

Enlargement of building with isolation pad

Lead plug

Base isolation unit

Two of numerous jacks lifting building

Base isolation unit in place; to be on new foundation pad

B

A. Base isolation pads permit a building to shake less than the violent shaking of the ground. They often consist of laminations of hard rubber and steel in a stack about 50 centimeters high. A lead plug in the middle helps dampen vibrations. Base isolation unit ready for installation below Utah state capitol building. **B.** In a large project retrofit, base isolation pads are emplaced under jacked-up Utah State Capitol building (inset) in 2006.

FIGURE 4-27 EARTHQUAKE-PROOFING A HOUSE

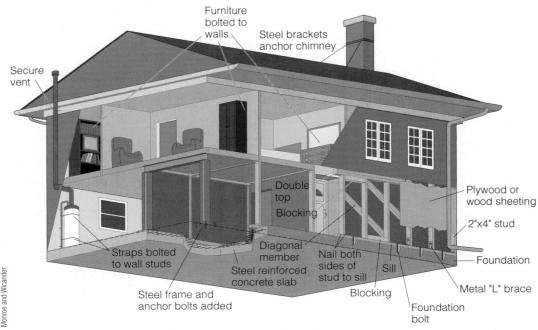

Diagram of some useful means of strengthening a house for earthquakes.

Monroe and Wicander.

FIGURE 4-28 WEAK UTILITY LINES

USGS.

The Northridge earthquake severed gas lines and caused fires.

is typically the largest investment for most people, total loss of its value can be financially devastating. In earthquake-prone areas, it may be worth buying earthquake insurance, even though it can be extremely expensive, because basic homeowners insurance does not cover such damage. However, earthquake insurance can cover only the cost of replacing the house, not the land, which can be a big difference, as in the expensive real estate of western California.

Plan ahead of time what you will do when an earthquake occurs. Most earthquakes last less than a minute. People who find themselves inside a building during an earthquake are well advised to stay there because the earthquake will probably end before they can get out. Remember from TABLE 3-3 that for earthquakes with magnitudes less than 6, the shaking time is so short that by the time you realize what is happening, there is little time to move to safety. For greater magnitudes, there may be enough time, but the accelerations are so high that it is hard to stay on your feet. If you do attempt to leave a building, avoid elevators. People outdoors and away from buildings are much safer than those indoors because no roof is overhead to collapse on them. A car parked next to a building provides little safety from falling debris (**FIGURE 4-29**). Glass and other debris falling from nearby buildings might make it advisable to run for open ground.

To increase your chances of surviving the collapse of a building during an earthquake, the latest guidelines suggest exiting and moving away from the building if there is time. Earthquake safety experts say to "duck, cover, and hold," which means to get under a sturdy desk or table to protect yourself from falling light fixtures, book shelves, and other objects. In some older buildings and in many foreign countries where floors consist of concrete poured in place, lying next to a very sturdy object could protect a person from a heavy collapse. Structural brick and prestressed concrete floors seem especially dangerous.

Land Use Planning and Building Codes

Governments also have a role in preparing for earthquakes and mitigating their damage. Land use planning and building codes are the best defenses against deaths, injuries, and property damage in earthquakes.

These cars were crushed by falling bricks during the moderate-sized March 4, 2001, earthquake in Olympia, Washington.

Building codes in areas of likely earthquake damage should require structures that are framed in wood, steel, or appropriately reinforced concrete. They should forbid masonry walls made of brick, concrete blocks, stone, or mud that support roofs. The Uniform Building Code provides a seismic zonation map of the United States that indicates the level of construction standards required to provide safety for people inside a building (compare FIGURE 4-13).

Enforcement of building codes is one reason the largest earthquakes do not necessarily kill the most people. Typically, large earthquakes in developed countries with modern construction codes—and strong enforcement of those codes—cause significant damage but fewer deaths. In poor countries with substandard construction or little enforcement of existing construction codes, such earthquakes result in high death tolls (**Chapter 3 Opener:** Deadly effects with poor quality construction—Haiti, 2010, p. 34; Chapter 3 **Case in Point:** Collapse of Poorly Constructed Buildings—Kashmir Earthquake, Pakistan, 2005, p. 61; Chapter 4 **Case in Point:** Collapse of Poorly Constructed Buildings that did not follow Building Codes, Wenchuan (Sichuan), China, Earthquake, May 12, 2008, p. 98). Contrast these death tolls (tens of thousands) with 486 killed in the February 2010 Magnitude 8.8 earthquake, the 7th or 8th largest on record in Chile, a country with very strong earthquake building codes.

Most high death tolls from earthquakes come from countries notable for poor-quality building construction or unsuitable building sites (see TABLE 4-1). The high death toll from the January 2001 earthquake in San Salvador was the result of both a huge landslide in a prosperous part of the capital and poorly built adobe houses in the poor areas (**FIGURE 4-30**). The tens of thousands of deaths in the January 2001 earthquake in Bhuj, India, derived mainly from the collapse of houses that were poorly built with heavy materials.

A cursory examination of TABLE 4-1 also shows that death tolls from earthquakes are not significantly declining with time. More than 200 years ago, and even in the past 50 years, many tens of thousands of people, and occasionally hundreds of thousands, died in major earthquakes. Unfortunately, there are now far more people living in crowded conditions and often in poorly constructed buildings (**FIGURE 4-31**, p. 92). Developed countries tend to be better off in this respect but can still experience devastating loss of life in an earthquake.

Zoning should strictly limit development in areas along active faults, on ground prone to landslides, or on soft

FIGURE 4-30 ADOBE-BRICK CONSTRUCTION

A   **B**

Construction in poor areas of developing countries often involves soft mud bricks, or "adobe," that have been merely sun-dried rather than kiln-fired. Even a small earthquake jeopardizes buildings constructed from such materials. **A.** Cairo, Egypt. **B.** near Cuzco, Peru.

FIGURE 4-31 HIDDEN FLAWS BEHIND PLASTER

Textured plaster designed to look like stone blocks

Actual structure built from cemented rocks and soft bricks

A

Hyndman.

Local rounded rocks with little mortar

B

Donald Hyndman.

A. What appears to be a nice, strong stone wall may be merely textured plaster over weak masonry. South of Madrid, Spain. **B.** Houses that were damaged in old parts of Athens, Greece, in a magnitude 5.9 earthquake in September 1999, were shoddily constructed from local rocks weakly cemented together. The collapsing houses killed 143 people. Would you stay in a hotel with this type of construction? Could you tell the type of construction if its walls were covered with plaster or stucco? Red "X" placed after earthquake to designate "no re-entry."

mud or fill. Parks and golf courses are better uses for such areas. If people did not live near faults, their sudden shifts would not create problems, but cities and towns grew in those areas for reasons that had nothing to do with Earth's movements. Now with millions of people residing in haz-ardous environments, societies are beginning to realize that we have a worsening problem, both for individuals and civilization as a whole. In order to deal with the hazards, we need to know more about what creates each danger.

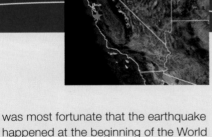

Cases in Point

Earthquake Fills a Seismic Gap
Loma Prieta Earthquake, California, 1989 ▶

On October 17, 1989, Game Three of the World Series between the San Francisco Giants and the Oakland Athletics was about to begin. The teams were warming up, and the crowd was settling into its seats in Candlestick Park on the southern edge of San Francisco. At 5:04 p.m., the sudden shock of an earthquake jarred everyone to a stop. That was the P wave arriving. Ten seconds later, the shaking suddenly intensified enough to knock a few people off their feet. That was the S wave arriving. The ten-second interval between the P and S waves indicated that the earthquake was some 80 kilometers away. Then the light towers began swaying, and the entire country experienced

the Loma Prieta earthquake on television. By morning, it was clear that the San Andreas Fault had moved near Loma Prieta in the Santa Cruz Mountains 80 kilometers southeast of the stadium with a magnitude of 6.9.

Meanwhile, cars swerved and traffic stopped on a two-and-a-quarter-mile double-decked stretch of Interstate 880 across the bay in Oakland. At first, some drivers thought they had flat tires. Initial excitement turned to terror for those on the lower deck, when chunks of concrete popped out of the support columns as the upper deck collapsed onto their vehicles. It seemed a miracle to rescuers that no more than 42 motorists died. Overall, it

was most fortunate that the earthquake happened at the beginning of the World Series in San Francisco. Most people in the San Francisco Bay area who were not at the game were home watching it on television. Freeways that would normally have been filled with rush-hour traffic were nearly empty. If not for this, many more people would have been killed.

Soft mud along the edge of San Francisco Bay amplified the ground motion under the freeway by a factor of 5 to 8, despite the 90-kilometer distance from the epicenter. It was especially unfortunate that the sediments reverberated with the vibration frequency of the elevated freeway

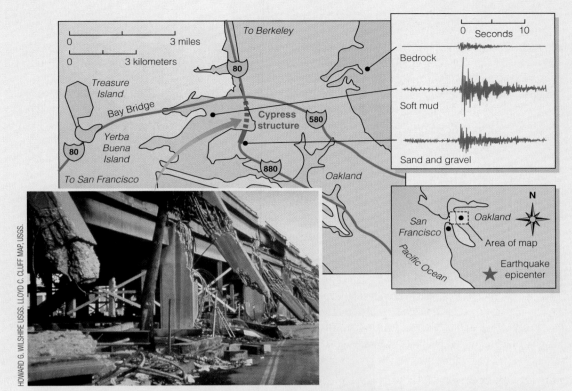

▶ Severe shaking of the Interstate 880 double-deck freeway in Oakland sheared off heavily reinforced concrete supports, and much of the upper deck collapsed onto the lower deck. Note that the heavy column in mid-photo, for example, failed at the level of the lower deck, where the two parts were joined during construction. The Cypress section of Interstate 880 collapsed during the 1989 Loma Prieta earthquake. Seismographs show that the shaking was much stronger on mud than on bedrock.

▶ The 1989 Loma Prieta earthquake severely wrecked the unreinforced masonry of the Pacific Garden Mall in Santa Cruz.

▶ This apartment building, constructed on the artificial fill of the Marina district of San Francisco, collapsed during the Loma Prieta earthquake in 1989. The car parked next to the building did not fare well, but people inside the building and above the first floor probably survived. Note that the second-floor balcony behind the car is now at street level.

(2 to 4 lateral cycles per second) because that greatly amplified the damage.

Santa Cruz, less than 16 kilometers west of the epicenter, was badly damaged, as were neighboring towns. Landslides closed highways in the Coast Ranges north of Santa Cruz. Buildings in the Marina district were especially vulnerable because much of the construction on the edge of San Francisco Bay stands on fill that amplified ground motion. The water-saturated sediments of the landfill turned to a mushy fluid, causing settling that broke gas lines and set fires. It also broke water mains, forcing firefighters to string hoses to pump seawater from the bay.

Though the Loma Prieta earthquake was not by any means the "Big One" the people of the Bay Area have come

(continued)

to dread, collapsing structures did kill 62 people and injured 3,757. Some 12,000 people were displaced from their homes. The earthquake destroyed 963 homes and inflicted $6 billion in property damage ($10.5 billion in 2010 dollars). Damage to the San Francisco–Oakland Bay Bridge closed it for a month. The collapsed I-880 freeway slowed traffic along the East Bay freeway for years until a single-level thoroughfare was completed to replace the fallen multilevel structure.

▶ *Violent shaking of the San Francisco–Oakland Bay Bridge sheared off the array of heavy bolts securing one section of the westbound upper roadway and dropped it onto the lower roadway. The inset view is westward from near the east end of the bridge. Note the skid marks on the roadway in lower right. The western half of the bridge, a suspension design, was not damaged.*

The fault slippage that caused the Loma Prieta earthquake began at a depth of 18 kilometers, where rocks west of the fault moved 1.9 meters (6.2 feet) north and 1.3 meters up. The offset did not break the surface or even the upper 6 kilometers of the crust.

The USGS had convened the Working Group on California Earthquake Probabilities. It had already identified the fault segment that caused the Loma Prieta earthquake as one likely to pro-

duce an earthquake of magnitude 6.5 or greater in the 30 years after 1988. As with many other earthquakes, this one in the Santa Cruz Mountains segment of the fault filled a seismic gap, an area along the fault that had not seen recent earthquakes. Seismographs did not detect any precursor events that might have warned of an imminent earthquake.

Edgar V. Leyendecker USGS.

A Strong Earthquake Jolts the Collision Zone between Africa and Europe
L'Aquila, Italy, April 6, 2009 ▶

In mid-January, 2009, small earthquakes began rattling the central Apennines range of mountains that runs down the spine of Italy, a zone above subduction to the northeast and spreading of the sea to the southwest (see FIGURE 7-27). Earthquakes are not uncommon in Italy but seldom strong or deadly. Seven large earthquakes have damaged the region at irregular intervals in the past 700 years.

An earthquake of moment magnitude 6.3 struck near L'Aquila, Italy, about 120 km northeast of Rome, at 3:32 am on April 6. Residents reported that shaking lasted about 20 seconds. The earthquake,

at a depth of 10 km was caused by extensional movement on a NW-SE normal fault. A technician had predicted that a quake would occur, but he was off by a week and 60 km to the northwest. Was this a valid prediction?

L'Aquila is a medieval city of 70,000 people, with buildings dating back to the 1,200s. It lies in an old lake bed that would amplify the ground shaking. 297 people died, about 1,500 were injured; one nearby village lost 38 people out of 350 residents. 28,000 were left homeless. Tents were put up to house 12,000 of the people displaced. Many

buildings collapsed and thousands were damaged, including churches and palaces of historical significance. Parked cars were crushed with rubble that filled narrow streets. Aftershocks, including one of magnitude 5.3, followed the main event, shaking down more rubble on rescuers in hardhats. Like most medieval buildings these were largely of unreinforced masonry construction—stone, often with ancient wood timbers.

Whether the claimed forecast was valid is an open question. As noted early

(continued)

in Chapter 4, Soviet geologists noted in the 1970s an increase in radioactivity from radon in water wells a few days before an earthquake. Another study, reported in 1981, showed mild correlation of radon changes in groundwater with earthquakes in Iceland. Of 57 comparisons for earthquakes of magnitude 1 to 4.3 (all about 1,000 times weaker than the L'Aquila event), 48 showed no correlation. However, 9 showed radon changes before an earthquake and 7 with high water-flow rates showed false alarms. Although the radon correlation is not well documented, its changes, along with the frequent earthquakes in January through March could provide cause for concern. There is insufficient data at present to predict a specific date for an earthquake based on these or other anomalous events (compare the Haicheng, China, earthquake prediction at the opener of this chapter).

If Giuliani's evacuation of Sulmona had been heeded, many evacuees would probably have been housed in the larger town of L'Aquila – where the earthquake actually caused most of the damage. A case can probably be made for explaining the symptoms, and possible precautions, to the endangered public in a straightforward, honest way and letting people decide individually what action they wish to take. Perhaps the potential threat could be expressed as a percent chance of an event of a certain size within a certain time period—much as is done for storms by the National Weather Service.

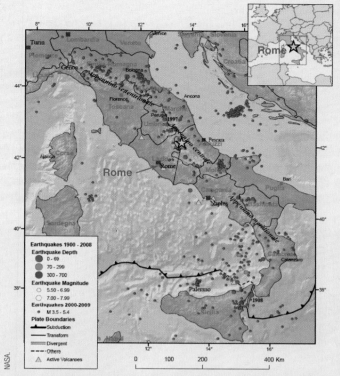

▶ *Distribution of large earthquakes in Italy since 1900. 2009 Aquila event is shown as a yellow star (USGS). Shuttle radar image of central Italy.*

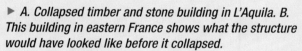

▶ *A. Collapsed timber and stone building in L'Aquila. B. This building in eastern France shows what the structure would have looked like before it collapsed.*

A

B

One in a Series of Migrating Earthquakes
Izmit Earthquake, Turkey, 1999 ▶

The North Anatolian Fault slipped on August 17, 1999, in the area 100 kilometers east of Istanbul, causing a magnitude 7.4 earthquake. The surface displacement was horizontal, between 1.5 and 2.7 meters along a fault break of 140 kilometers. More than 30,000 people died. Property damage reached $9.1 billion (in 2010 dollars).

This fault has caused eleven major earthquakes larger than magnitude 6.7 in the last century. A series of six large earthquakes progressed steadily westward between 1939 and 1957. Between 1939 and 1944, the fault ruptured along an incredible 600 kilometers of continuous length (see FIGURE 4-8). Meanwhile, three

other earthquakes occurred near both ends of the fault, beyond the sequence of six. The earthquake of August 17, 1999, filled a seismic gap. Less than three months later, a 43-kilometer section immediately east of the previous

▶ **A.** *This building, which was unfinished at the time of the Izmit earthquake, had its upper floors hanging out over the street. Its heavy superstructure rested on concrete posts with weak links to the foundation and the concrete floors. The floors and foundation also lacked diagonal braces.* **B.** *This modern, multistory apartment building in Izmit pancaked over to the right. It is unlikely that anyone would have survived this type of collapse.*

▶ *A group of older apartment buildings in Izmit lies in ruins after the poorly braced lower floors of most of them collapsed. Poor-quality construction contributed. Diagonal braces and shear walls would have prevented lateral shift.*

▶ *Some buildings sank or tilted because of both liquefaction of the ground and heavy construction, featuring upper floors overhanging streets.*

(continued)

movement slipped, causing a magnitude 7.1 earthquake that killed 850 people in Düzce. On May 1, 2003, a magnitude 6.4 earthquake struck just south of the east end of the North Anatolian Fault. If current trends continue, Istanbul, a short distance farther west and only 20 kilometers north of the North Anatolian Fault, is next. For a very old city of 12.6 million inhabitants, such an event would be a catastrophe.

Although building codes in Turkey match those of the United States with respect to earthquake safety, enforcement is poor and much of the construction is shoddy. Because buildings are taxed on the area of the street-level floor, developers favor the erection of buildings with second stories that extend over the street. Many of these buildings collapsed during the earthquakes, and liquefaction caused buildings to tilt or fall over. In addition, uppermost floors are sometimes added illegally during election years when builders hope that politicians and inspectors will overlook the work.

Devastating Fire Caused by an Earthquake
San Francisco, California, 1906 ▶

The San Francisco earthquake came at 5:12 a.m., before dawn on April 18, 1906. It began with a foreshock that rudely awakened nearly everyone. Then, 20 to 25 seconds later, the main shock struck with a magnitude of approximately 7.8. One survivor recalled a rumble and roar like old cannons. Others heard roaring sounds or dull booms. People on the street recalled that the ground rose and fell in waves as the earthquake approached.

Strong shaking lasted 45 to 60 seconds—it seemed like it would never stop. Thousands of chimneys snapped off and fell through houses into their basements. Many brick or stone buildings collapsed into heaps of rubble. New skyscrapers with steel frames along Market Street survived with little damage, as did most wood-frame buildings. Building design and materials played a big role in their survival. Unfortunately, many of the wood-frame buildings on filled areas along the edges of San Francisco Bay collapsed. Others sank as the mud under them compacted. Aftershocks destroyed more buildings weakened in the main event. Buildings on the bedrock hills survived relatively well.

Cooking fires and broken gas and electrical lines sparked fires in many of the wood buildings. Dozens of fires ignited within a half hour, then coalesced into two major fires that spread across much of the city, one north and another south of Market Street. Broken water mains hampered the fire department to the point that it resorted to dynamite to cut fire lines. Three days of fire destroyed Chinatown, the skyscrapers along Market Street that survived the earthquake, and the wharves along the edge of San Francisco Bay. More than 28,000 buildings were destroyed, 10.6 square kilometers (4 square miles) of the city.

North of San Francisco, the earthquake damaged buildings almost to the Oregon border; to the south, damage reached a third of the way to Los Angeles. The death toll was generally quoted at 700, mostly in San Francisco, and epidemics that followed killed many more. However, the mortality count appears to have been understated, apparently because of an anticipated negative impact on the local economy. In fact, probably more than 3,000 people died from the effects of the earthquake and its aftermath. Damage was $578 million from the earthquake and $8.4 billion from the fire (2010 dollars). Some 225,000 people, out of a population of 400,000, lost their

▶ **The Hibernia Bank building in San Francisco was destroyed by the 1906 earthquake.**

W. C. Mendenhall USGS.

Courtesy Bancroft Library, University of California, Berkeley.

▶ **This view of San Francisco looking down Market Street shows the devastation following the 1906 earthquake and fire.**

BANC PIC.

▶ **People and horses were killed by falling bricks and walls. Even streets between buildings were not safe.**

Doorway was once at street level

Donald Hyndman.

homes. Santa Rosa, Healdsburg, and San Jose were severely damaged.

The surface offset extended 430 kilometers from the area east of Santa Cruz to Cape Mendocino. The maximum horizontal displacement was 2.6 meters near Point Reyes (see broken fence in FIGURE 3-4), 50 kilometers north of San Francisco. Neither the concept of earthquake magnitude nor seismographs suitable for measuring magnitudes existed in 1906. The size of the area damaged to Mercalli intensity VII or higher suggests a moment magnitude of approximately 7.8.

▶ *Houses in the Marina District of San Francisco sank into the artificial fill at the edge of San Francisco Bay during the 1906 earthquake. They are now below street level.*

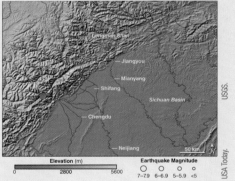

California Historical Society.

▶ *Broken water mains prevented use of fire hydrants.*

Collapse of Poorly Constructed Buildings that did not follow Building Codes
Wenchuan (Sichuan), China, Earthquake, May 12, 2008. ▶

At 2:28 p.m. on May 12, 2008, sudden movement along almost 300 km of the northeast-trending Longmenshan thrust belt moved the eastern high Tibetan Plateau, northeast and upward against the Sichuan Basin on its east. The upper plate, on the west, shifted as much as 4.9 meters east and 6.5 meters up, a total offset of more than 8 meters. The resulting magnitude 7.9 earthquake began 19 km below the surface and 75 km west-northwest of Chengdu, a city in southwestern China with a population of about 10 million.

A sudden jolt and violent back and forth shaking threw people off their feet. It felt like free-fall except that they fell sideways rather than down. People were both stunned and unable to run because the floor under them moved rapidly, first one way and then the other. Plaster fell from ceilings and walls buckled. As buildings rocked back and forth for as long as two minutes, much of the rigid masonry construction cracked and ultimately collapsed. The rigidity of the terrain, shallow depth of the earthquake focus, and proximity of such a

large population amplified damages and deaths.

Cities and rural villages lie in narrow, steep-sided valleys where people live in multistory concrete and brick apartment complexes built since the 1950s or in small one- and two-story homes built of brick or concrete blocks. On older homes a sparse framework of timber beams and wood slats supported heavy tile roofs.

Most children were in school when the earthquake leveled most of the buildings near the epicenter, killing thousands of students and teachers. Thirteen hundred died in one Middle School built in 1998; at least 1,000 were buried when another school, 7 stories high and 160 km from the epicenter, collapsed into a 2-meter-high pile of rubble. Examination of collapsed buildings quickly showed that inferior building materials had been used in many of them. Most of the schools had large rooms with insufficient support; weak floors and walls with large window areas and no diagonal bracing were typical. Parents soon protested angrily about building standards for schools and shoddy construction fostered by corrupt

USGS.

USA Today.

▶ *The Wenchuan earthquake struck at the eastern edge of the Himalayas.*

and incompetent officials. Because China's population-control policy permits only one child per family, almost all of the families lost their only child. In their old age most Chinese depend on their children for support because pensions are almost nonexistent.

Most buildings did not meet seismic design standards because China did not have adequate seismic design codes until after the disastrous Tangshan earthquake in 1976.

(continued)

Many newer buildings in the area were designed to strict earthquake standards but enforcement of the standards was lax, especially in smaller towns. Migration of large numbers of people to rapidly growing urban centers led to hurriedly built shoddy construction and little government oversight. Multistory masonry buildings used insufficient mortar and too few steel reinforcing bars.

Engineering reports indicated that school construction problems in rural areas were aggravated by poor funding and exploitation by government and school officials. Officials in Beijing and Sichuan agreed that building codes in rural areas were weak and stated that they were investigating collapses and drafting changes to construction standards. Those whose homes were destroyed were promised permanent housing within three years. However, they pressed parents to sign documents, in exchange for money and agreements they would not hold protests. Officials ordered the media to stop reporting on school collapses, riot police broke up protests, and some protesting parents were detained and threatened.

Highways and rail lines into the area were buried by landslides or had collapsed, so rescue teams were delayed. Bridges, power lines, and cell-phone towers fell, blocking communication, especially to rural towns in rugged valleys. Immediately following the earthquake, rescue teams rushed to the larger towns where the largest numbers of people would need help, often bypassing small communities where people urgently needed help. They lacked shelter, warm clothes, blankets, food, and clean water; cold spring rains added to their misery. Rains brought down more landslides. Even where damaged buildings could provide some shelter, hundreds of aftershocks threatened to collapse them so most survivors remained outside. Without external help, the only rescue came from local people who had no heavy equipment to move the rubble that covered roads and to lift heavy concrete slabs. The first outside help reached the epicenter on May 14 when troops on foot trudged through mud with heavy packs of tents, generators, and medicine and 100 paratroopers brought supplies. Extremely rugged terrain in Tibetan villages, at elevations of 4,000 meters and above, prevented the landing of helicopters.

Although for the first 4 days, the government rejected foreign help, it soon appealed for help from medical and rescue teams and for tents. By May 16, 135,000 Chinese troops, along with rescuers from South Korea, Japan, Taiwan, Singapore, and Russia, were involved. The United States provided supplies by cargo aircraft and furnished high-resolution satellite imagery of the stricken area.

Some remote areas lacked food and potable water, even for doctors and nurses. Medical needs were overwhelming; nurses guarded boxes of medical supplies. Even some hospitals moved patients outside because their buildings were unstable. Many shops, on the ground floors of buildings, collapsed because the full width of the storefront had no supports or diagonal braces. Magnitude 6 to 6.1 aftershocks occurred on July 23 and August 1, and on August 6, almost three months after the main earthquake, another struck southwest Sichuan. Buildings swayed and people ran into the streets in panic. On August 30, a magnitude 5.7 earthquake struck 500 km to the south at a depth of 10 km along a related fault zone, killing at least 40 and destroying another 258,000 homes.

Although many people were rescued from the rubble initially, by eight days after the earthquake the numbers dwindled to only one or two. Without food, they can survive for up to 60 days but without water few can last for more than 5 to 7 days. A few survived by drinking their own urine; doctors say this is safe because it is naturally antiseptic. The only help for some people buried in collapsed buildings came from family members or co-workers. In many cases they could not reach victims because concrete slabs were too heavy to move by hand or blocks were linked by steel reinforcing bars. Former residents, who had left to work in industrial cities to the east, returned to search for relatives. They asked friends and searched nearby shelters, in many cases without success; a lot of survivors were taken to larger shelters many miles away.

Three months later, the total confirmed dead reached more than 89,000, with thousands still missing. More than 5 million remained homeless. On May 25 and 27, hundreds of thousands of houses collapsed in three aftershocks. Although the buildings had survived the initial earthquake, they had been cracked and weakened.

A few weeks after the earthquake, experts rushed to assess buildings in

Weak ground-floor garages and stores collapsed.

Weak ground-floor garages collapsed on cars

A

B

(continued)

Kelin Wang, Pacific Geoscience Center, Canada.

Hollow concrete panels and no reinforcing bar

Collapsed ground-level stores

▶ *Masonry construction, with little or no reinforcing bar, crumbled easily. Stores and apartments occupying ground floors and garages had no diagonal bracing. Being "weak floors" they collapsed during the earthquake. Wenchuan, China.*

the earthquake zone to determine how to strengthen them or whether they should be demolished. At the same time, men were straightening rusty tangles of old reinforcing bars that they planned to resell to families needing to rebuild their homes. Engineers say that such re-used rebar is not as strong as new material but it sells for about $100 less per metric ton.

Huge landslides shaken loose by the earthquake and aftershocks blocked 35 rivers, causing the water to back up and flood narrow mountain valleys, including those with tent camps housing earthquake survivors. Heavy thunderstorms accelerated the problem. The water level rose several meters per day at Tangjiashan Lake, which formed when a large river

was dammed by a huge landslide; water reached a depth of 725 meters two weeks after the earthquake. Engineers, using heavy machinery, and fuel airlifted onto the landslide dam, dug a channel to drain the water; relief officials ordered evacuation of towns of more than a million people that would be swept away if the dam failed.

Hundreds of newspaper, radio, and television reporters traveled to the disaster area and spoke with foreign reporters, both in defiance of government orders. However, in contrast to other officials, Chinese Premier Wen Jiabao, who was trained as a geoscientist, immediately flew to the area to encourage rescue and relief operations; his prompt and aggressive action quickly led to his being viewed as a national hero. Quick rescue response from the government contrasted with the closed-door response to the Cyclone Nargis disaster 10 days earlier in Myanmar and to the painfully slow response to Hurricane Katrina in 2005, where the U.S. government turned away international and out-of-state help and volunteers. In addition to formal international organizations, relief came from unofficial and impromptu groups and individual volunteers. Internet bulletin boards, blogs, emails, and text messages led to rapid and widespread dispersal of aid. Total costs of reconstruction from the earthquake were estimated as $124 billion (2010 dollars).

Tibet Earthquake, April 14, 2010 ▶

At 7:49 am (local time) on Wednesday April 14, 2010, a magnitude 6.9 earthquake struck southern Qinghai province of western China, at a depth of 17 kilometers. At least 2,183 people were killed in Yushu Prefecture in the eastern interior of the high Tibetan Plateau. Eighty-five percent of the houses in Jiegu, the largest town in the area, were destroyed. The school day had not yet begun and many students were able to escape from their dorms before complete collapse. The much larger and more-disastrous Wenchuan (Sichuan) earthquake of May 2008 was at the eastern edge of the plateau about 650 km to the southeast. Both

earthquakes reflect the ongoing collision of the Indian Plate with the Asian Plate, building the Himalayas that spread to the east and north. Left-lateral strike-slip subsurface movement of up to 1.7 meters was on the southeast-trending Yushu fault but motion did not break the surface. Historic earthquakes on the same fault zone include magnitudes 7.0 in 1904, 7.6 in 1973, and 6.9 in 1981.

The sparsely populated area, at an elevation of almost 4,000 meters (13,000 feet), is mountainous and poor with herders who raise sheep, yaks and horses. About 97 percent are ethnic Tibetans. Relief supplies were quickly flown to the nearest

airport but were slowed by the 860-km, 12-hour drive on the heavily damaged highway from the Xining, the largest city and regional airport. China's top leaders visited the area and promised new homes and schools, and to restore a treasured 800-year-old Buddhist monastery. Rescuers from lower elevations and relief efforts were hindered by aftershocks and working at the high-altitude in snow and temperatures well below freezing. In contrast to the Wenchuan earthquake, in which proportionally more schools collapsed than other buildings, the majority of buildings in Yushu were built from wood and mud, and most collapsed.

Critical View

A This building in western China was destroyed during the 2009 Wenchuan earthquake.

Brick walls

Collapsed wall

Cancrete floor slab and concrete posts

USGS.

1. What tectonic environment controlled the location of that earthquake?
2. What aspects of this building construction would lead to damage during an earthquake? Why?
3. During a major earthquake, what aspects of the building are likely to pose a hazard to people inside the building?
4. What hazards would affect people outside the building and why?

B These buildings in Machu Picchu Village, Peru, lies east of the South American subduction zone. Most are constructed from cinder blocks cemented together, and upper floors sometimes overhang the ground floor. Plaster covers some outer walls to cover the rough construction.

Cinder-block construction

Plaster covering cinder blocks

Overhanging second floor

Concrete floor slab

Reinforcing bars cemented into cinder blocks

Donald Hyndman.

1. What aspects of this building construction would lead to building damage during an earthquake?
2. During a major earthquake, what aspects of the building are likely to pose a hazard to people inside the building?
3. What hazards would affect people outside the building and why?

C This building is in downtown Portland, Oregon, about 200 kilometers from the Cascadia subduction zone. Note the heavy-duty diagonal steel I-beams. Note also the display windows on the ground floor.

Hyndman.

1. Explain what role these I-beams play in stabilizing the building during an earthquake.
2. Does it appear that the I-beams were added later, after completion of the building? Why?
3. During a major earthquake, what aspects of the building are likely to pose an indoor hazard?
4. What hazards would affect people outside the building and why?

D This modern overpass in downtown Philadelphia, Pennsylvania, has floor-supporting beams resting on heavy cross-supports.

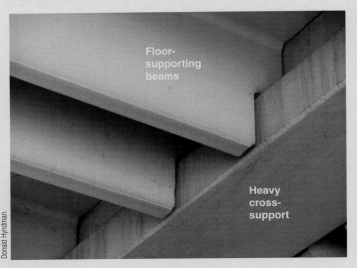

Floor-supporting beams

Heavy cross-support

Donald Hyndman.

1. Is there any earthquake danger in this city and, if so, why?
2. Would this type of construction be more hazardous in St. Louis, Missouri, than here? Why?
3. In case of a moderate-size earthquake, describe what might happen to the structure and why.

E This building in Talca, Chile, was destroyed in the 27 February, 2010, magnitude 8.8 earthquake.

Donald Hyndman.

1. What tectonic environment would have caused this earthquake?
2. Describe the apparent construction materials, whether was any structural strengthening, and if so, what is was.
3. Is this material inherently strong or weak? What makes it strong or weak?
4. What structural features make the structure especially weak or especially strong?
5. Before the earthquake, what could have been done to strengthen the structure to minimize this type of damage?

Chapter Review

Key Points

Predicting Earthquakes

- Precursors that suggest an imminent earthquake include foreshocks, changes in ground elevation, radon gas emissions, and changes in groundwater levels.

- Early warning systems are triggered when a fault moves, so they provide less than a minute to prepare for the arrival of an earthquake.

- The life-saving potential of short-term earthquake forecasts and early warnings must be weighed against the political and economic costs of false alarms.

Earthquake Probability

- Earthquake forecasts, which are now fairly reliable, specify the probability of an earthquake in a magnitude range within a region over a long time period, such as a few decades.

- Paleoseismology, the study of prehistoric fault movements, helps us understand future fault behavior. Past fault movements are visible in offset sedimentary layers exposed in trenches and fault scars. **FIGURE 4-4**.

- Examining trends in past fault movements helps scientists determine where future earthquakes are likely to occur. Seismic gaps, where movement has not occurred on a fault segment, are likely areas for future fault movements. Earthquakes can also migrate along a fault over time.

- Establishing a recurrence interval for earthquakes helps scientists determine when earthquakes are likely to occur. Some faults slip at regular intervals. Others may be very active and then experience a gap in earthquake activity.

- The magnitude of future earthquakes can be estimated based on the lengths of fault segments that are likely to break.

Populations at Risk

- The San Andreas transform fault, running much of the length of coastal California, is the dominant earthquake fault in North America. Its most dangerous areas are the large population centers around San Francisco Bay and Los Angeles. Blind thrusts occur off the main fault.

- The probability of a major earthquake on the San Andreas Fault can be assessed from the relationship between frequency and magnitude or strain accumulated by the overall slip rate on the fault for the San Francisco Bay area or the Los Angeles area. **FIGURE 4-14**.

Minimizing Earthquake Damage

- Earthquake damages include collapsed buildings and other structures, as well as fire and, over the longer term, disease.

- People generally die in earthquakes because things fall on them, not because of the shaking itself. Therefore, weak, rigid buildings and highway overpasses are most dangerous. Flexible, well-built wood-frame houses may be damaged but are not likely to collapse on people.

- Weak floors are likely to collapse, as are floors without lateral bracing. Walls not anchored to floors and roofs can separate and fall. **FIGURES 4-19** and **4-22**.

- Where the frequency of vibration or back-and-forth oscillation is the same in the building as in the ground under it, the shaking is strongly amplified and the building is more likely to fall. By the Numbers 4-2.

- Earthquake damage can be mitigated through building codes, retrofitting, and land-use planning, as well as educating the public about earthquake preparedness.

Key Terms

base isolation, p. 87

foreshocks, p. 71

migrating earthquakes, p. 74

paleoseismology, p. 73

recurrence interval, p. 76

retrofitting, p. 85

risk map, p. 78

seismic gap, p. 74

Questions for Review

1. To what extent can earthquakes be predicted?

2. List several of the precursors that indicate an earthquake may be coming.

3. What can a trench dug across an active fault show about past fault movement? Use a sketch to illustrate your answer.

4. What is a seismic gap, and what is its significance in determining future fault activity?

5. What information indicates the probable magnitude of future earthquakes along a specific fault segment?

6. An earthquake on the North Anatolian Fault in Turkey caused more than 30,000 deaths in 1999. What North American fault is it similar to and in what way?

7. Why does the North Anatolian Fault kill many more people than its North American counterpart?

8. What kinds of structural materials make walls dangerously weak during an earthquake?

9. What type of wall strengthening is commonly used to prevent a building from lateral collapse during an earthquake?

10. Why do the floor or deck beams of parking garages and bridges sometimes fail and fall during an earthquake?

11. When a tall building stands next to a short building, why is the tall building often damaged during an earthquake? Where on the tall building does the damage occur?

12. What feature is sometimes used to prevent a building from shaking too much during an earthquake?

13. Name three things that would prepare a home for an earthquake.

14. What should you do if you are in a building, in the event of an earthquake?

Discussion Questions

1. When water was injected into deep hot rocks to generate geothermal power in Switzerland it caused many small earthquakes—who should be held responsible (see discussion related to FIGURE 4-3)? The geothermal power company, the drilling company, the governmental agency that provided the drilling permit, or someone else and if so, who?

2. If injection of fluids into a fault zone can trigger earthquakes, why not trigger some small earthquakes instead of waiting for "The Big One?" Consider numbers of earthquakes of a given size, consequences for public safety, what might go wrong, and who should be liable.

3. The section on Prediction Consequences discusses the pros and cons of making predictions. If you were the mayor of Los Angeles and U.S. Geological Survey earthquake experts advised you an earthquake of magnitude 7.5 will strike the city the day after tomorrow, what would you do? What are the consequences of notifying the public? What would you tell them to do? What are the consequences of withholding that information?

4. Many older buildings were not designed for safety in earthquakes. If you were a policymaker for an earthquake-prone area such as California, would you design laws that require expensive retrofitting or require that buildings be demolished and rebuilt to a safer standard? Why? Why would some groups argue against such a law?

Tsunami

5

A massive tsunami up to the eaves of houses, surges into Khao Lak, Thailand, carrying sand, debris, and struggling people.

John M. Thompson.

Swept Away

Eight a.m. December 26, 2004. Two friends had just sat down for a leisurely breakfast on the ground floor of a two-story lodge in Phuket, a beach community in southern Thailand. Suddenly the ground shook, cups and plates rattled, and they looked quizzically at one another. That felt like an earthquake! It didn't do much damage and they went back to their menus. A little less than two hours later, they heard people yelling and screaming from the beach nearby (**FIGURE 5-1**). A giant wave suddenly surged in from the beach, carrying sand, driftwood, plastic chairs, and struggling people. They both ran to the second story, but the fast-rising water swept them off as they reached the top of the stairs. Both were strong swimmers but could do little in the raging flood except try to avoid being struck by floating debris and to swim laterally to avoid being rammed into buildings, trees, and bobbing cars. One was carried more than a kilometer inland; both survived but one was badly battered. More than 5,000 people died along the coast of Thailand that morning, a total of more than 230,000 around the Indian Ocean (**Case in Point:** Lack of Warning and Education Costs Lives—Sumatra Tsunami, 2004, p. 123).

FIGURE 5-1 A TSUNAMI FLOODS ASHORE

A. The force of the tsunami waves destroyed almost everything in some low-lying coastal areas of Sumatra. **B.** Startled people near the beach react to tsunami striking in Koh Raya, Thailand.

Tsunami Generation

Tsunami, the Japanese name for *harbor wave*, are so named because the waves rise highest where they are focused into bays or harbors. Although tsunami are sometimes called tidal waves, this term is misleading because tsunami are not related to tides.

Tsunami are most commonly generated by earthquakes, but they can also be caused by other mechanisms that cause sudden displacement of large volumes of water. These include volcanic eruptions, landslides or rockfalls, volcano flank collapses, and asteroid impacts.

Earthquake-Generated Tsunami

Most tsunami are generated during shallow-focus underwater earthquakes associated with the sudden rise or fall of the seafloor, which displaces a large volume of water.

Earthquake-generated tsunami occur most commonly by displacement of the ocean bottom on a reverse- or thrust-fault movement on a subduction-zone fault (and occasionally on a normal fault). Strike-slip earthquakes seldom generate tsunami because they do not displace much water.

In a subduction zone, recall that oceanic lithosphere typically slides under continental lithosphere. Although the plates move at a nearly constant rate, the boundary between two plates sticks for many years. Where these plates stick, the continental edge is pulled downward and toward the continent as the subducting plate moves under it. This causes the overlying plate to flex upward in a **coastal bulge**, in the same way that a piece of paper bulges upward if you pull its far edge toward you (**FIGURE 5-2**; also see Chapter 3, FIGURE 3-13). When the stuck zone, which commonly stretches a considerable length parallel to the coast, finally ruptures in an earthquake, the edge of the continent snaps up and

FIGURE 5-2 TSUNAMI FORMATION

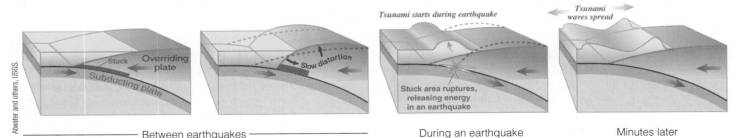

A subduction-zone earthquake snaps the leading edge of a continent up and forward, displacing a huge volume of water to produce a tsunami.

oceanward. This suddenly displaces a huge volume of water, creating tsunami waves that move in both directions (towards and away from the continent) from the location where they are generated (**FIGURE 5-3**).

The height of a tsunami wave depends on the magnitude of the shallow-focus earthquake, area of the rupture zone, rate and volume displaced, sense of ocean floor motion, and depth of water above the rupture. Vertical movement on a fault, as during a subduction-zone earthquake, causes a large displacement of water, whereas horizontal movement on a strike-slip fault does not.

The size of an earthquake-generated tsunami in the open ocean is limited to the maximum displacement or offset on a fault. Based on the relationship between displacement and fault magnitude (see FIGURE 3-29), an earthquake of moment magnitude (M_w) 8 from vertical displacement on a normal fault or a thrust (especially subduction) fault—the most likely types to displace significant water—could have a vertical offset of 15 meters. Because the tsunami wave height approximates the vertical displacement on a fault, the maximum wave height from an earthquake is about 15 meters, which would increase as waves are pushed into shallow water and bays. The most vulnerable parts of the United States and Canada are Hawaii and the Pacific coast (California, Oregon, Washington, British Columbia, and Alaska). On average, a major tsunami forms somewhere around the Pacific Ocean roughly once a decade; and once every 20 years, a 30-meter-high wave hits. Subduction-zone earthquakes off Japan, Kamchatka, the Aleutian Islands / Gulf of Alaska, Mexico, Peru, and Chile are the most frequent culprits. The subduction zone off the coast of Washington and Oregon is like a tightly drawn bow waiting to be released (see Chapter 3, FIGURE 3-13).

Even in southern California, south of the Cascadia subduction zone, a nearby earthquake poses a potential problem. A tsunami from an earthquake on the Santa Catalina Fault offshore from Los Angeles would reach the community of Marina Del Rey, just north of the Los Angeles Airport, in only *eight minutes*. Given the large population and near sea-level terrain, the results could be tragic.

Tsunami Generated by Volcanic Eruptions

Tsunami are also caused by volcanic processes that displace large volumes of water. Water is also driven upward or outward by fast-moving flows of hot volcanic ash or submarine volcanic explosions into a large body of water. Volcanoes can also collapse in a giant landslide, as addressed below. More than one of these mechanisms can occur at an individual volcano. Tsunami generated by volcanic eruptions are occasionally catastrophic, but they are poorly understood; their maximum size is unknown. We do not know enough about the mechanism of water displacement from an underwater eruption to do much more than speculate.

In July 2003, Montserrat Island's Soufrière Hills volcano collapsed and spilled volcanic material into the ocean; it generated tsunami that ran up as high as 21 meters on nearby islands. One of the most infamous and catastrophic events involving a volcano-generated tsunami was at Krakatau in 1883. On August 27, the mountain exploded in an enormous eruption, the climax of activity that had been going on for several months. Thirty-five minutes later, a series of waves as high as 30 meters (almost 100 feet) flattened the coastline of the Sunda Strait between Java and Sumatra, including its palm trees and houses. Only a few who happened to be looking out to sea saw the incoming wave in time to race upslope to safety. More than 35,000 people died. Studies of the distribution of pyroclastic flow deposits and seafloor materials in the Sunda Straits between Krakatau and the islands of Java and Sumatra suggest that seawater seeping into the volcano interacted with the molten magma to generate huge underwater explosions and upward displacement of a large volume of seawater.

In an earlier event, the cone of the eastern Mediterranean island volcano of Santorini collapsed into its erupting magma chamber between 1630 and 1550 B.C. The collapse displaced a series of huge tsunami waves that washed ashore in Crete to the south, southwestern Turkey to the east, and Israel, at the east end of the Mediterranean. Some researchers correlate this event to the legend of Atlantis, the city that disappeared under the sea.

FIGURE 5-3 INITIATION OF A TSUNAMI

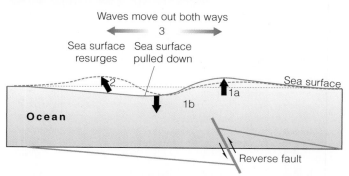

The sequence of events that creates a tsunami generated by a subsea reverse- or thrust-fault in the ocean floor are: (1a) Seafloor snaps up, pushing water with it; (1b) sea surface drops to form a trough; (2) displaced water resurges to form wave crest; and (3) gravity restores water level to its equilibrium position, sending waves out in both directions.

Tsunami from Fast-Moving Landslides or Rockfalls

When major fast-moving rockfalls or landslides enter the ocean, they can displace immense amounts of water and generate tsunami. You might expect that the height of the tsunami depends primarily on the volume of the mass that displaces water. However, a more important parameter is the height of fall. A striking example is the 1958 tsunami in Lituya Bay, Alaska, which was generated when a nearby earthquake detached a large section of cliff into a coastal fjord (**Case in Point:** Immense Local Tsunami from a Landslide—Lituya Bay, Alaska, 1958, p. 126). Although it killed only two people, this is the highest tsunami in the historical record. A similar event in southern Chile on April 21, 2007, was also triggered by a magnitude 6.2 earthquake. A large landslide plunged into a narrow fjord and caused 7.6-meter-high waves that swept away ten people at a beach and destroyed some boats.

Submarine landslides, those that occur underwater, can also generate tsunami. A giant submarine landslide 8,100 years ago, the Storegga slide offshore from Norway, caused a tsunami 11 meters high that ran up on the coasts of eastern Scotland and Norway (**FIGURE 5-4**). The slide moved 800 km out into the deep ocean floor and affected 95,000 square km, an area larger than Scotland and nearly the size of Virginia. The slide, at the end of the last ice age, may have been triggered by an earthquake that destabilized frozen methane–ice layers (discussed in Chapter 10) in the continental shelf sediments.

Submarine landslides from the outer continental shelf off the eastern United States and Canada can also generate tsunami. On November 18, 1929, tsunami from the magnitude 7.2 (M_w) Grand Banks earthquake killed 27 people on the southern coast of Newfoundland. The epicenter 250 km south of Newfoundland and 610 km east of Halifax, Nova Scotia, triggered a submarine landslide from the edge of the continental shelf that broke 12 submarine telegraph cables on the ocean floor. A series of three waves, each 3 to 8 meters high, arrived at the coast at 105 km per hour (65 mph), amplifying to run up as high as 13 m at the head of narrow bays. The waves were recorded in Newfoundland, the east coast of Canada, and the United States as far south as Charleston, South Carolina, almost 6 hours later. They were even recorded across the Atlantic Ocean in Portugal. Similar submarine slides all along the edge of the Atlantic continental shelf are documented in the ocean-floor record. The recurrence interval for a magnitude 7 earthquake off the New England coast is 600 to 3,000 years. Any of those could cause catastrophic slope failure and a major tsunami. Recently discovered fractures along a 40-kilometer stretch of the continental shelf 160 kilometers off Virginia and North Carolina suggest the possibility of a future undersea landslide. Such a slide could generate a tsunami like the one that occurred 18,000 years ago just south of those fractures.

A major subduction-zone earthquake in the Caribbean—for example, north of Puerto Rico—may well trigger a large tsunami that could inundate low-lying areas of the Gulf Coast and East Coast states, even at a considerable distance from the epicenter (**FIGURE 5-5**). Tsunami-deposited sand layers have also been discovered at several sites on islands west of Norway. A smaller, more recent subsea slide in 1998 generated a tsunami that killed 2,200 people in Papua New Guinea.

FIGURE 5-4 CONTINENTAL SHELF COLLAPSE

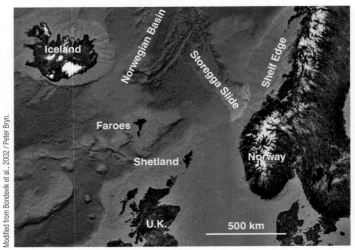

Modified from Bondevik et al., 2002 / Peter Bryn.

The Storegga Slide collapsed the continental shelf off Norway, causing tsunami that inundated nearby coasts.

FIGURE 5-5 FUTURE TSUNAMI IN PUERTO RICO

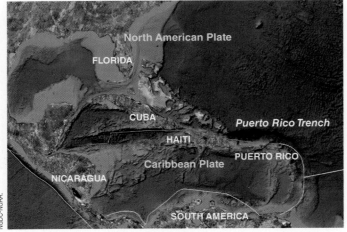

NGDC-NOAA.

The east-west Puerto Rico trench could generate a major subduction-zone earthquake and tsunami that would inundate much of the east coast of the United States and Canada.

Tsunami can also occur in large, deep lakes. On the west side of Lake Tahoe, on the California-Nevada border, a broad shelf of lake sediments collapsed some time after the melting of glaciers from the last ice age (**FIGURE 5-6**). It caused a catastrophic landslide and tsunami. Giant blocks as much as 0.5- to 1-kilometer-across dropped 0.5 km and moved 10 to 15 km across the bottom of the lake. Although the north half of the sediment shelf, including much of Tahoe City, did not collapse, it may yet do so in the future. Study of lake-floor sediments suggests magnitude 7 earthquakes every 2,000 to 3,000 years; such earthquakes could trigger slumps below the lake surface.

Coral limestone boulders up to 9 meters across in Tonga, in the southwest Pacific Ocean, appear to have been swept ashore less than seven thousand years ago by tsunami generated by huge submarine landslides from adjacent submarine volcanoes. The boulders are 10–20 m above sea level and 100 to 400 m inland.

FIGURE 5-6 LAKE TAHOE LANDSLIDES AND TSUNAMI

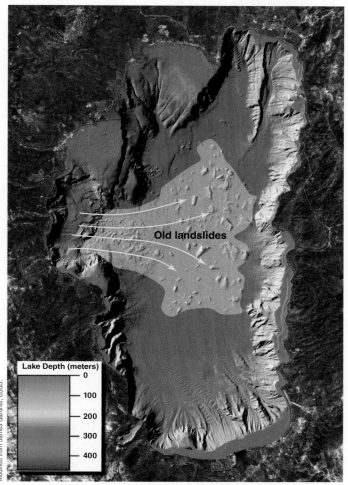

Modified from James Gardner, USGS.

A detailed survey of the bottom of Lake Tahoe shows clear topography of old large landslides, which could cause tsunami in the lake.

Tsunami from Volcano Flank Collapse

The flanks of many major oceanic volcanoes, including those of the Hawaiian Islands in the Pacific Ocean and the Canary Islands in the Atlantic Ocean, apparently collapse on occasion and slide into the ocean, suddenly displacing thousands of cubic kilometers of water. Resultant tsunami can be hundreds of meters high.

Volcanoes, such as those that make up the Hawaiian Islands, grow from the seafloor for 200,000 to 300,000 years before breaking sea level, then build a reasonably solid sloping dome above sea level, called a lava shield, for a similar time. Mega-landsliding occurs near the end of the shield-building stage, when the growth rate is fastest, heavy rock load on top is greatest, and slopes are steepest and thus least stable. The lower part of each volcano, below sea level, consists largely of loose volcanic rubble formed when the erupting lava chilled in seawater and broke into fragments. It has little mechanical strength.

The three broad ridges that radiate outward from the top of a volcano spread slightly under their own enormous weight, producing rift zones along their crests. The volcano eventually breaks into three enormous segments that look on a map like a pie cut into three slices of approximately equal size. One or more of the three volcano segments may begin to move slowly seaward.

The rifts between segments provide easy passage for molten magma rising to the surface. Those rifts that become the sites of most of the eruptions also form weak vertical zones in the volcano. There is a long history of one or more volcano segments breaking loose to slide into the ocean, sometimes slowly but sometimes catastrophically. Studies of the ocean floor using side-scanning radar around the Hawaiian Islands reveal 68 giant **debris avalanche** deposits, each more than 20 km long (**FIGURE 5-7**, p. 110). Some extend as far as 230 km from their source and contain several thousand cubic km of volcanic debris. At least some of those deposits are the remains of debris avalanches that raised giant tsunami waves, which washed high onto the shores of the Hawaiian Islands as one of the enormous pie segments of an active volcano plunged into the ocean.

Similar situations are now known to exist on volcanoes of the Lesser Antilles in the Caribbean, Mount Etna in the Mediterranean Sea, and the Marquesas Islands in the Pacific Ocean. Specific sites on the north end of a volcano on Dominica in the Caribbean could fail and produce tsunami that would inundate tourist beaches and a population of 30,000 people on the island of Guadeloupe. A flank collapse of Reunion Island in the Indian Ocean could cause tsunami inundation and destruction of many coastal areas, including the dense sea-level populations of Bangladesh. Collapse of a flank of the Canary Islands, off the northwestern coast of Africa, could generate a giant tsunami that would cross the Atlantic Ocean to obliterate coastal cities on the eastern coast of North America and, perhaps, those in western Europe (see details, pg. 122).

FIGURE 5-7 HAWAII FLANKS COLLAPSE

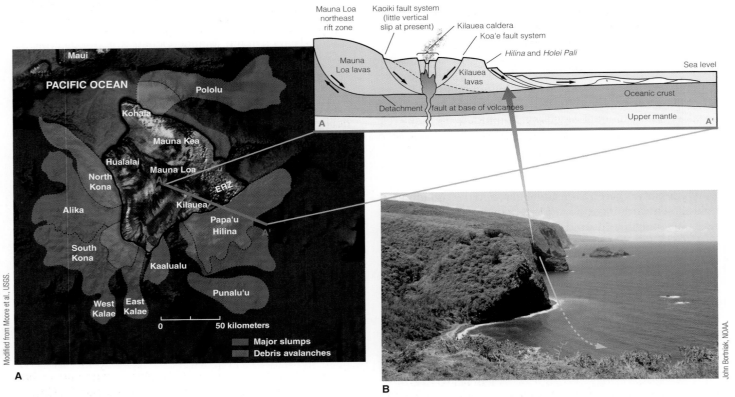

A. This map of the island of Hawaii shows the major slumps and debris avalanches formed by collapse of the island's flanks. The northwest-southeast cross section of Kilauea volcano shows the probable failure surfaces that lead to collapse of the volcano's flanks. Cross-section location is line A–A' on map. **B.** Giant cliffs, or pali, amputate the lower slopes of the big island of Hawaii.

Although earthquake-generated tsunami are more common, there is a limit to the size of those waves. It seems likely that a catastrophic tsunami, many times larger than any in the historical record, is likely to come from the flank collapse of an oceanic volcano. Our geologic record of such events is clear enough to show that they have happened, and they will again.

Tsunami from Asteroid Impact

The impact of a large asteroid into the ocean would generate huge tsunami radiating outward from the impact site—much as happens with any other tsunami (see related discussion in Chapter 17, Asteroid Impacts). The average frequency of such events is low, but a 1-km asteroid falling into a 5-km-deep ocean might generate a 3-km-deep cavity. Cavity walls would collapse at speeds up to supersonic, sending a plume high into the atmosphere. Initial kilometer-high waves would crest, break, and interfere with one another. Waves with widely varying frequencies would radiate outward. The behavior of such complex waves is not well understood, but they are thought to decrease fairly rapidly in size away from the impact site. The different wave frequencies would, however, interfere and locally pile up on one another to cause immense run-ups at the shore.

The chance of a 1-km asteroid colliding with Earth is not great—only about one per million years, so such a hazard is not likely in our lifetime. However, the chance of a catastrophic tsunami from the flank collapse of an oceanic island such as one of the Hawaiian or Canary islands is perhaps ten times as great, or one per 100,000 years.

Tsunami Movement

A wave can be described by its wavelength, height, and period (**FIGURE 5-8**). Tsunami wave heights in the open ocean are small. Tsunami generated by ocean earthquakes are often no more than a meter or two high far out in the ocean, with a maximum of about 15 meters near an earthquake epicenter. The average **wavelength** of a tsunami is 360 kilometers, so the slopes on the wave flanks are extremely gentle. The time between waves, or the **period**, can be half an hour. For example, for a wave with a 30-minute period, a ship would go from wave trough to crest and back to trough in 30 minutes. A ship at sea would not even notice such a gentle wave.

FIGURE 5-8 CHARACTERISTICS OF WAVES

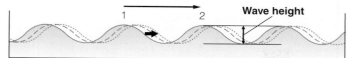

Wavelength (distance from peak to peak)
Period (time for passage of one wave)

Wavelength is the distance between two wave crests, while *period* is the time between the passage of two successive wave crests (for example, between waves 1 and 2). *Wave height* is measured from the bottom of the trough to the top of the crest.

Water particles in waves travel in a circular motion that fades downward. At depths of less than approximately half their wavelengths, this circular motion reaches the seafloor and waves are said to *feel bottom*. As a wave drags on the bottom, it slows and the waves become much shorter, perhaps one-sixth of their former wavelength. Because wave volume remains the same, its height must rise dramatically, perhaps to six times its open-ocean height (**FIGURE 5-9A**). For example, a 3-meter-high wave in the open ocean could rise in shallow water to 18 meters!

The mechanisms that drive tsunami waves are different from those that drive typical waves, which are driven by wind. Wind waves have relatively short wavelengths so they feel bottom only near shore, then lean forward and break on the shore (discussed in Chapter 13). Tsunami waves, however, have extremely long wavelengths, so they drag on the bottom everywhere in the ocean. Because a tsunami feels bottom far from shore, it gains amplitude to become a virtual wall of water; as it rides over the continental shelf and approaches land, it continues to flow forward like a flash flood (**FIGURE 5-9B**).

This same relationship between wavelength and depth affects the speed of a wave. In the open ocean, tsunami can travel as fast as 870 km per hour, but as they reach shallower water, they slow because their circular motions at depth drag even more strongly on the ocean bottom. Thus, water depth is related to wave velocity (**FIGURE 5-10**, p. 112 and **By the Numbers 5-1:** Velocity of Tsunami Waves). On the continental shelf, a 1-meter open-ocean tsunami wave may slow to 150 to 300 kilometers per hour (90–180 miles per hour). Clearly this is too fast for escape after you see it coming , unless there is high ground nearby.

Tsunami on Shore

In some cases, tsunami may appear much like ordinary breaking waves at the coast, except that their velocities are much greater and they are much larger. Some come in as a high breaking wave that destroys everything in its path. Others advance as a rapid rise of sea level, a swiftly flowing, churning, and rising "river" without much of a wave.

Even tsunami that rise without a breaking wave are extremely dangerous because they advance much faster than a person can run. Even a strong swimmer caught in the swift current as the wave retreats will be swept inland or out to sea, with minimal chance of survival (**FIGURE 5-11**, p. 112). Loose debris picked up as the waves advance act as battering rams that impact both structures and people. Coastal regions can be annihilated (**FIGURE 5-12**, p. 113).

Coastal Effects

As tsunami waves approach the shore, the mouths of rivers and coastal bays funnel the waves and dramatically raise their height. If they arrive at high tide, their height is further

FIGURE 5-9 TSUNAMI AMPLITUDE

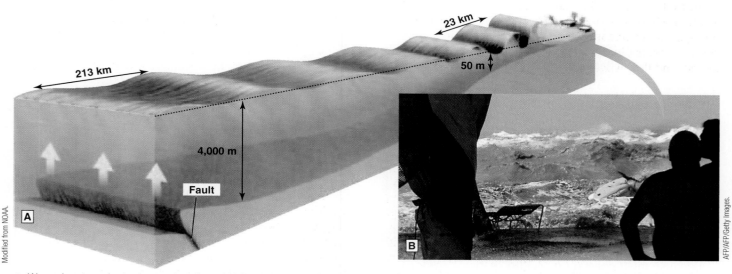

A. Waves become shorter in wavelength and higher in amplitude before rushing onshore. **B.** Tsunami wave-front full of sand and debris arrives at Panang, Malaysia.

FIGURE 5-10 WAVE VELOCITY AND WATER DEPTH

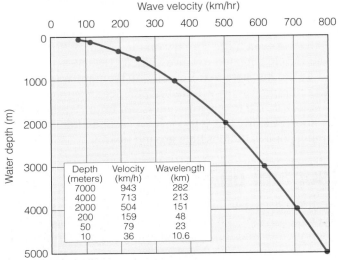

Depth (meters)	Velocity (km/h)	Wavelength (km)
7000	943	282
4000	713	213
2000	504	151
200	159	48
50	79	23
10	36	10.6

Waves are faster in deeper water and slower in shallower water.

FIGURE 5-11 AFTER THE INCOMING WAVE

Tsunami wave pours back offshore Chennai (Madras), India, pulling loose debris torn up by the incoming wave of the 2004 Sumatra tsunami.

By the Numbers 5-1

Velocity of Tsunami Waves

The velocity of tsunami waves depends on the water depth and gravity

$$C = \sqrt{gD}$$

Where

C = velocity in meters per second

D = depth in meters

g = gravitational acceleration (9.8 m/sec²)

Thus, $C = 3.13\sqrt{D}$

For example, if D = 4,600 meters (deep ocean):

$C = \sqrt{4,600}$ meters = 3.13 x 67.8 meters per second,

or 763 kilometers per hour (the speed of some jet aircraft!)

If D = 100 meters (near shore):

$C = 3.13\sqrt{100} = 3.13$ x 10 = 31.3 meters per second,

or 112.7 kilometers per hour (the speed of freeway traffic).

amplified. Sloshing back and forth from one side of a bay to the other, waves can interfere with one another, combining to form still higher waves. Tsunami waves also curve progressively to face toward the shore as they drag bottom; moving around one end of an island, they can merge with another part of the wave as it curves around the opposite side of the island. This creates a wave larger than either of the parts.

Because most coastal towns and seaports are located in bays, enhanced damage results from these waves. Even though the 1960 tsunami emanated from Chile, far to the

southeast, and Hilo Bay faces northeast, refraction of the waves around the island left the head of the bay vulnerable to waves four meters above sea level (**Case in Point:** An Ocean-Wide Tsunami from a Giant Earthquake—Chile Tsunami, 1960, p. 128). On December 12, 1992, a magnitude (M_S) 7.5 earthquake in Indonesia generated a tsunami in the Flores Sea. The southern coast of the small island of Babi, situated opposite the direction from which the waves came, was hit by 26-meter-high waves, twice as high as the northern coast. In this case, the waves reaching the northern coast split and refracted around the circular island, interfering and combining with one another on the opposite coast. More than 1,000 people died.

The coastal geography of other regions can protect them from tsunami. Many low-lying Pacific and Caribbean islands are surrounded by offshore coral reefs that drop steeply into deep water. Thus, tsunami waves are forced to break on the reef, providing some protection for the islands themselves.

Run-Up

When a wave reaches shore, we talk about the characteristics of its **run-up**, or the height that a wave reaches as it rushes onshore. Run-up is greater than incoming-wave height; it varies depending on distance from a fault rupture and whether the wave strikes the open coast or a bay. For the largest earthquakes, such as the 1964 subduction event in Alaska and the 1960 earthquake in Chile, run-up heights were generally 5 to 10 meters above normal tide level. Local run-up reached as high as 30 meters in Chile.

Water levels can change rapidly, as much as several meters in a few minutes. A wave will typically run up onto shore in a direction perpendicular to the orientation of the wave crest; the wave then drains back offshore straight downslope. Driftwood, trees, and the remains of

FIGURE 5-12 TSUNAMI DESTRUCTION

A. The December 2004 tsunami destroyed all the near-shore buildings in this community on the east coast of Sri Lanka. **B.** It destroyed almost all of the buildings in Banda Aceh, Sumatra. The road here has been cleared to provide access.

boats, houses, and cars commonly mark the upper limit of tsunami run-up. Even long after the fact, evidence of past tsunami can be seen in a **trimline**, or the line across a mountainside where tall trees upslope are bordered by distinctly shorter trees downslope (see Case in Point: Immense Local Tsunami from a Landslide, p. 126).

Period

The mound of water suddenly appearing at the sea surface in response to a major event generates a series of waves that may cross the entire Pacific Ocean. Because

the initial mound of water oscillates up and down a few times before fading away, it generates a series of waves, just like a stone thrown into a pond. Thus the arrival of a giant wave is followed by others that are often larger than the first. As the initial wave slows, those following catch up and thus arrive more frequently. At a velocity of 760 km-per-hour and a wavelength of 200 km offshore, a wave would pass any point in the ocean or arrive onshore every 15 minutes.

What seems like a calm sea or a sea in retreat can be the trough before the next wave (**FIGURE 5-13**). Survivors of tsunami often report an initial withdrawal of

FIGURE 5-13 THE WAVES FROM SPACE

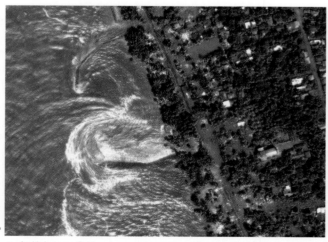

A. At Kalutara, Sri Lanka, on December 26, 2004, the first wave of the tsunami surrounds all of the houses in this satellite view. Here it begins to drain back to the ocean. **B.** A view of a larger area shows the broad offshore beach exposed after the first wave drains back offshore. The red dotted line is the normal beach edge.

FIGURE 5-14 FROM SHOCK TO PANIC

People on the beach at Krabi, Thailand, were stunned to see a giant wave breaking on the horizon and headed their way. Within a few seconds the tsunami was on them.

the sea with a hissing or roaring noise. In many cases, curious people drown when they explore the shoreline as the sea recedes before the first big wave or subsequent waves (**FIGURE 5-14**). In 1946, people in Hilo, Hawaii, assuming the danger had past, went out to see the wide, exposed beach littered with stranded boats and sea creatures. There they were caught in the second, larger wave.

As a wave recedes into the trough before the next wave, its current and load of debris flowing back offshore are almost as fast and dangerous to people as the initial tsunami. Because the time between tsunami waves is often more than a half hour, the wave trough is well offshore; people and wreckage are carried out to sea.

Tsunami waves may continue for several hours, and the first wave is commonly not the highest (**FIGURE 5-15**). In

FIGURE 5-15 TSUNAMI TIDE GAUGE

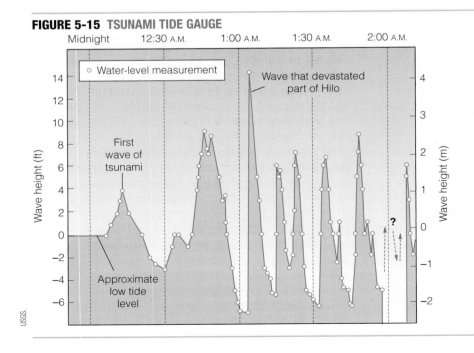

This tide gauge record shows the tsunami waves in Hilo, Hawaii, May 23, 1960, following the Chilean earthquake. In this case, the first wave was relatively low, followed by successively higher waves that rose to more than 4 meters above the low tide that preceded the tsunami. After the first couple of waves, wavelength and frequency increased.

harbors, tsunami have wave periods of 10 to 35 minutes, and the series of large waves may continue arriving for up to six hours. In Hawaii in 1960, in spite of several hours of tsunami warning, people went down to the beach to watch the spectacular wave, only to be overwhelmed by it.

Tsunami Hazard Mitigation

Tsunami hazards can be mitigated by land-use zoning that limits building to elevations above those that would potentially be flooded by a tsunami. In Hilo, Hawaii, the waterfront area at the head of the bay where the worst damage occurred from disastrous tsunami in both 1946 and 1960 was converted into a park to minimize future damage.

If lower elevations are to be developed, potential tsunami impact should be taken into account during planning. Coastal developments that orient streets and buildings perpendicular to waves tend to survive better than those aligned parallel to the shore. This layout allows waves to penetrate farther and dissipate as they flow through open streets, limiting debris impact. Structures should also be designed to resist erosion and scour. Landscaping with vegetation capable of resisting wave erosion and scour can help. Trees can slow waves while permitting water to flow between them, but they need to be well rooted or they can themselves become projectiles. A large ditch or reinforced concrete wall placed in front of houses can help reduce the impact of the first wave and may provide a little extra evacuation time. Both options are locally used in Japan (**FIGURE 5-16**).

Tsunami Warnings

Depending on their distance from an earthquake's epicenter, tsunami are most likely to appear within a few minutes to several hours after an earthquake that involves major vertical

FIGURE 5-16 TSUNAMI BARRIER

Tsunami protection wall in Japan.

motion of the seafloor. Tsunami warning systems have now been perfected for *far-field* tsunami, those far from the source that generated them, although not all regions have invested in such systems (**Case in Point:** Sumatra Tsunami, 2004, p. 123).

A tsunami warning network around the Pacific Ocean monitors large earthquakes and ocean waves and signals the possibility of tsunami generation and arrival time to 26 participating countries. A world network of seismographs locates the epicenters of major earthquakes, and the topography of the Pacific Ocean floor is so well known that the travel time for a tsunami to reach a coastal location can be accurately calculated (**FIGURE 5-17**). In addition, environmental satellites take readings from tidal sensors along the coasts, and ocean

FIGURE 5-17 TRAVEL TIMES FOR DIFFERENT EARTHQUAKE ZONES

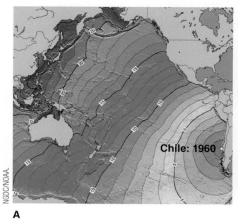

A

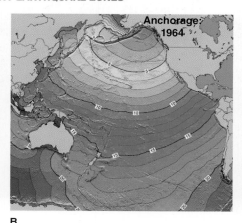

B

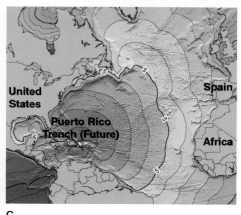

C

Estimated tsunami travel times across the Pacific and Atlantic oceans from the **A.** Chile, 1960; **B.** Alaska, 1964; and **C.** possible future Puerto Rico Trench subduction-zone earthquakes. Concentric arcs are travel-time estimates in hours after each earthquake. From the Alaska earthquake, for example, the first tsunami wave reached Hawaii in approximately six and a half hours. It would reach the north island of Japan after seven hours.

bottom sensors detect ocean surface heights as waves radiate outward across the Pacific Ocean (**FIGURE 5-18**). This information now permits prediction of tsunami arrival times within five minutes, at any coastal location around the Pacific Ocean. Pacific tsunami warning centers are located in Honolulu, Hawaii, and Palmer, Alaska. Some low-lying areas, such as parts of Hawaii, are equipped with sirens mounted on high poles to warn people in dangerous coastal areas.

The Pacific Tsunami Warning System has two levels: a **tsunami watch** and a **tsunami warning**. A watch is issued when an earthquake of magnitude 7 or greater is detected somewhere around the Pacific Ocean. If a significant tsunami is identified from the buoy system, the watch is upgraded to a warning, and civil defense officials order evacuation of low-lying areas that are in jeopardy.

A new warning system for detection of tsunami waves is likely to come from satellites. Because wavelengths of

FIGURE 5-18 BUOY WARNING SYSTEM

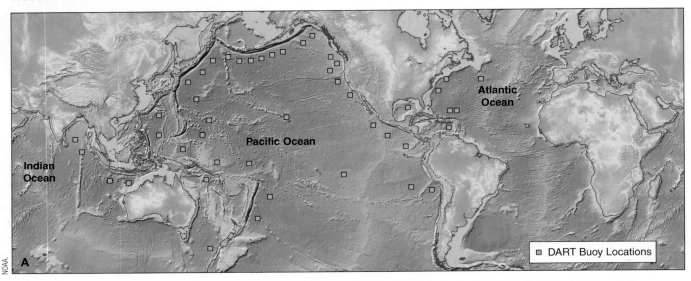

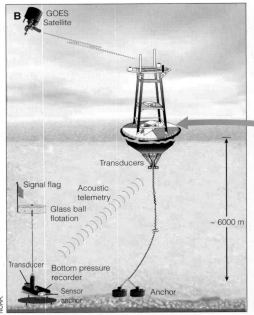

A. Map of the 2007 Deep-Ocean Assessment and Reporting of Tsunami (DART) buoy system, which has been greatly expanded since the catastrophic December 2004 tsunami. **B.** A pressure sensor on the ocean floor detects changes in wave height because a higher wave puts more water and therefore more pressure above the sensor. The pressure sensor transmits a signal to a buoy floating at the surface and to the warning center via satellite.

tsunami are very long and the speed of tsunami waves is so much faster, they differ from other disturbances on the ocean surface. Those differences can be detected by satellites. The main limitation is providing real-time data to scientists so that warnings can be issued in time to warn vulnerable population centers.

Although tsunami warning systems can be quite effective for far-field tsunami, they rarely provide enough warning of tsunami generated by a nearby earthquake. The Pacific Tsunami Warning System was put to the test on September 29, 2009, at 6:48 in the morning, when a major earthquake struck the westward-dipping Tonga subduction zone near Samoa, halfway between Hawaii and Australia (**FIGURE 5-19**). Shaking lasted two to three minutes. At a depth of 18 kilometers and in deep water it generated a tsunami that inundated Samoa, American Samoa, and other nearby islands.

Within ten minutes after the shaking stopped, huge waves swept onto low-lying coasts of the mountainous islands and amplified into bays. Some reached more than 1.5 kilometers inland. Waves rose as high as 3.14 meters at Pago Pago, in American Samoa, 1.4 meters in Samoa, and almost 0.5 meters in Rarotonga. Coastal houses and tourist resorts were flattened or swept off their foundations and demolished; cars were lifted and smashed into buildings, and some, along with people, were swept out to sea (**FIGURE 5-20**). Many people drowned or were crushed by debris floating in the surging water. Some could not swim or were trapped under water. 149 people were killed in Samoa, 34 in American Samoa, about 200 kilometers from the epicenter, and 9 on Tonga.

The Pacific Tsunami Warning Center in Hawaii sent out an alert when the earthquake struck, but the local population had only about ten minutes to respond before the arrival of the first wave. Many did not realize how urgent the

FIGURE 5-20 SAMOA TSUNAMI

A. The second wave into the harbor at Pago Pago carried cars and the shattered remnants of houses. **B.** Survivors search for belongings in the remnants of their homes.

situation was. Reportedly, some radio stations did not interrupt their music to provide any warning. Although Samoa has a text-message system for warning residents, mobile-phone service is scattered in more remote areas, and some people's phones were not turned on. With such a brief window of time for evacuation, warning dissemination measures taken by local authorities, as well as the education of the local population about how to respond to those warnings, is key to saving lives.

Surviving a Tsunami

As was the case in Samoa in 2009, a nearby earthquake allows almost no time for official tsunami warning. Whenever coastal residents feel an earthquake, they should move immediately to higher ground. Tsunami warning signs in coastal Oregon suggest fleeing to higher ground if you feel an earthquake (**FIGURE 5-21**, p. 118). After a nearby subduction-zone tremor, there is little time before the first wave arrives, possibly only 10 to 20 minutes. You need to get to high ground or well inland immediately. Roads heading directly inland are escape routes, but blocked roads and traffic jams are likely. Climb a nearby slope as far as possible.

FIGURE 5-19 SAMOA EARTHQUAKE

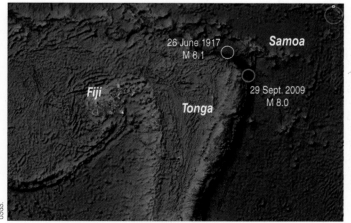

The 2009 earthquake and tsunami struck in the Tonga trench, northeast of New Zealand.

FIGURE 5-21 TSUNAMI ZONE SIGNS

This sign warns of potential tsunami along the Oregon coast.

Even if there are no nearby hills, quickly moving inland can still help, because the energy, height, and speed of a tsunami dissipates on land. Moving to an upper floor of a well-reinforced building away from the beach may also help. When determining how far to evacuate, keep in mind that a wave reaching shore may either break on the beach or rush far onshore with a steep front. Large tsunami can reach more than a kilometer inland in low-lying areas.

In addition to drowning in an incoming wave, tsunami dangers include death or injury from being thrown against solid objects or hit by debris, severe abrasions from dragging along the ground at high speeds, and being carried out to sea when a wave recedes. Hazardous fragments can include boards and other remains of houses, trees, and cars. Even when a wave slows to, say, 55 km-per-hour as it drags on shallow bottom, it is much too fast to outrun.

Even if you don't live in a coastal area at risk for tsunami, you might be vacationing in such an area when a tsunami hits, as many tourists in Thailand were during the 2004 Sumatra tsunami. Keep in mind these key points for survival:

- Protect yourself during the earthquake—take cover from falling objects until it ends.

- Even without warning, an unexpected rise or fall in sea level may signal an approaching tsunami. Get to high ground or well inland immediately. Climb a nearby slope as far as possible, certainly higher than 30 meters.

- Never go to the shore to watch a tsunami. Tsunami move extremely fast, and traffic jams in both directions are likely to require abandoning your vehicle where you least want to do so.

- Do not return to the shore after the first wave. Although the sea may pull back offshore for a kilometer or more following that first wave, other, even higher, waves often arrive for several hours, with long intervals in between. Wait until officials provide an all-clear signal before you return.

- Stay tuned to your radio or television.

Future Giant Tsunami

The largest tsunami that are likely to impact people are caused by giant earthquakes in subduction zones and, less frequently, flank collapse of an island shield volcano. We describe three cases of known major hazards that almost certainly will affect North America in the future:

1. Giant tsunami following a huge subduction-zone earthquake in the Pacific Northwest
2. Catastrophic flank collapse of a shield volcano on the big island of Hawaii
3. Catastrophic flank collapse of a shield volcano on the Canary Islands in the Atlantic Ocean

Pacific Northwest Tsunami: Historical Record of Giant Tsunami

The lack of recent earthquakes along the Pacific Northwest coast from northern California through Oregon and Washington to southern British Columbia is more of a concern than a comfort. As discussed in Chapter 4, the last major earthquake in the area was in January 1700, but there have been many other giant earthquakes at intervals of a few hundred years. Although the subduction zone was known to exist, and almost all such zones are marked by major earthquakes, evidence of major earthquakes here remained elusive.

Finally, in the 1980s, Brian Atwater of the U.S. Geological Survey (USGS) found the geologic record of giant earthquakes in marshes at the heads of coastal inlets. Evidence of giant tsunami generated by the earthquakes included a consistent and distinctive sequence of sedimentary layers. A bed of peat, consisting of partially decayed marsh plants that grew just above sea level, lies at the base of the sequence (**FIGURE 5-22A**). Above the peat lies a layer of sand notably lacking the sort of internal layering contained in most sand deposits. Above the sand is a layer of mud that contains the remains of saltwater plants. That sequence tells a simple story that begins with peat accumulating in a salt marsh barely above sea level. It appears that a large earthquake caused huge tsunami that rushed up on shore and into tidal inlets, carrying sand swept in from the continental shelf. The sand covered the old peat soils in low-lying ground inland from the bays as the salt marsh suddenly dropped as much as 2 meters below sea

FIGURE 5-22 THE GHOST FOREST

John Clague.

Donald Hyndman.

Tsunami sand

Peat

Tsunami

Sea Level

Peat marsh soil

Before earthquake

Peat marsh soil

Just after earthquake

Tidal mud

Stump

Tsunami sand

Peat marsh soil

Centuries after earthquake

C

A. Tsunami sand from a mega-thrust earthquake deposited in 1700 over dark brown peat in a British Columbia coastal marsh. The scale is in tenths of a meter. **B.** This ancient Sitka Spruce forest in the bay at Neskowin, Oregon, was felled by a giant tsunami following the huge subduction-zone earthquake of January 1700. Stumps of the giant trees punctuate low tide at this beach 25 kilometers north of Lincoln City. The forest—with trees as old as 2,000 years—was suddenly dropped into the surf during a mega-thrust earthquake and then felled by the huge tsunami that followed. **C.** Simplified sketch showing tsunami sand deposited immediately after a subduction earthquake when a tidal marsh suddenly drops below sea level.

level. Then the mud, with seaweed (later fossilized), accumulated on the sand (**FIGURE 5-22B**). The sequence of peat, sand, and mud is repeated over and over. In some cases, forests were drowned by the invading salt water or trees were snapped off by a huge wave (**FIGURE 5-22C**). Huge tsunami-flattened forests in low-lying coastal inlets are found all down the Pacific coast from British Columbia to southern Oregon. These stumps are now at and below sea level because the coastal bulge dropped during the 1700 earthquake.

Sand sheets were deposited at elevations up to 18 meters above sea level. Tsunami of this size expose coastal communities to extreme danger. For an earthquake that breaks the entire 1,100-kilometer length of the subduction zone, mathematical models suggest that Victoria's harbor, at the southern end of Vancouver Island, could see 4-meter-high waves 80 minutes after the earthquake. The harbor area of Seattle could see 1-meter-high waves three hours after the earthquake and 2-meter-high waves after six hours. Tsunami from an earthquake on the Seattle Fault pose a greater danger for Seattle. A magnitude 7.6 event could generate a wave of up to 6 meters in the Seattle harbor area. Calculations suggest that Portland, Oregon, would likely not be at significant tsunami risk from such a subduction-zone earthquake because it is well up the shallow water of the Columbia River; waves are expected to largely dissipate before reaching it.

Communities on the open coast or smaller coastal bays, however, are in real peril. An earthquake near the coast could generate a tsunami wave that could sweep ashore in less than 20 minutes, leaving too little time for warning and evacuation of danger areas. The small coastal community of Seaside, Oregon, west of Portland, for example, swells with 40,000 summer residents and visitors, many of whom are unaware of the hazard or what to do when it comes. Feeling an earthquake along the coast, people should immediately move inland to higher ground. The first indication along the coast of Oregon, Washington, or British Columbia may be an unexpected rise or fall of sea level. If the compressional bulge where the continental plate overrides the subduction zone is under water offshore (for example, off the coast of Oregon; see FIGURE 3-13), the bulge will drop, pulling water down with it. Thus the first sign at the coast will be a drop in water level as the water pulls back offshore. If the compressional bulge is onshore (for example, in Washington), the bulge will drop but the toe of the plate, under water, thrusts up. The first sign of a tsunami at the Washington coast may be a rapid rise in water level. Later waves are likely to be larger than this first wave. Commonly the initial wave withdraws because the bulge is offshore underwater.

Based on past events, estimates provide about one chance in ten for such a giant earthquake and tsunami in the next fifty years. That chance may seem small but, as with the weather, the event may come at any time. Recent research at Oregon State University shows 3–4 major tsunami per 1000 years between 4600 and 2800 years ago, a gap of about 1000–1200 years—then 4 during the next 1000 years, followed by another gap of 670–750 years. Finally in A.D. 1700, a giant earthquake and tsunami. Perhaps most disturbing is that although the average recurrence time in the northern part of the fault is 525 years, the average in the southern part is only 278 years. That area may be overdue for the next earthquake and tsunami.

The sudden drop of the coastal area will raise sea level. Thus, the tsunami will rush ashore to higher levels than would otherwise be expected. Given the length of the subduction zone, comparable in size to the one that caused the catastrophic 2004 earthquake and tsunami in Sumatra (see FIGURE 3-12), the next event is likely to be disastrous. Calculations suggest that tsunami up to 16 meters high may invade some coastal bays. The record of the last event indicates that waves rose as high as 20 meters where they funneled into some inlets. Computer models estimate that the heights of those waves will be approximately 10 meters high offshore. These heights would be amplified by a factor of 2 to 3 in some bays and inlets. Port Alberni, at the head of a long inlet on the west coast of Vancouver Island, British Columbia, in the 1964 Alaska earthquake, for example, saw the waves amplified by a factor of 3 compared with the open ocean. Inland, where the wave sloshed up against a slope, the run up reached 25–30 meters (80–100 feet). Although approximate, similar numbers are obtained from studies of onshore damage. The general pattern of ground movements described above occurred in the Alaska earthquake of 1964 (**Case in Point:** Subduction-Zone Earthquake Generates a Major Tsunami—Anchorage, Alaska, 1964, p. 129).

Coastal communities in the earthquake zone are endangered. Cannon Beach and Seaside, both in northern Oregon, are well aware of the problem and are beginning to address solutions (**FIGURE 5-23**).

Kilauea, Hawaii: Potentially Catastrophic Volcano Flank Collapse

Hawaiian geologists wondered for years about blocks of coral and other shoreline materials strewn across the lower slopes of the islands more than 6 kilometers inland and to elevations up to 400 meters above sea level. The flanks of some of the islands have lost much or all of their soil up to a similar elevation. It now seems clear that both the displaced coral and the scrubbed slopes are evidence of monstrous waves that washed up on the flanks of the islands as one of the enormous pie segments of an active volcano plunged into the ocean. The head scarps of such collapsed segments become gigantic coastal cliffs, some more than 2,000 meters high (see FIGURE 5-7c).

Tsunami formed by island-flank collapse are documented from deposits as high as tens of meters above sea level. On Molokai, tsunami left cemented fragments of limestone reef and basalt 70 meters above the sea; on Lanai, they left blocks of coral up to 326 meters above sea level. An eventual repetition of those events seems inevitable, and much of the population of the Hawaiian Islands lives below these levels.

The landslides appear to occur during major eruptive cycles and have a recurrence interval of roughly 100,000

FIGURE 5-23 TSUNAMI EVACUATION MAP

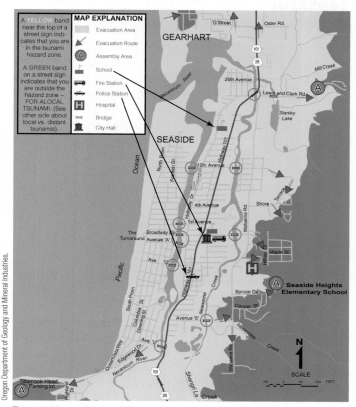

Tsunami hazard map of Seaside, Oregon, showing the area of possible inundation in yellow. Bridges (circled) are likely to fail in an M=8.5+ earthquake. Evacuation time is too short for schools, and emergency response of firefighters and police is endangered because their headquarters are in the hazard zone. The barrier bar, west of the north-south channel, is especially endangered because of the need to evacuate a large number of people when bridges may not survive.

FIGURE 5-24 KILAUEA SLIDING SEAWARD

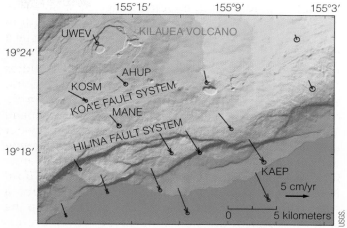

The south flank of Kilauea volcano is slowly slumping seaward. The arrows indicate directions and rates of movement as measured by Global Positioning System.

years. Head scarps of the slides are the giant pali, or cliffs, that mark one or more sides of each of the Hawaiian Islands. Despite the existence of such evidence, the frequency of these horrifying events remains unclear. If we can judge from the age of coral fragments washed onto the flanks of several islands, the most recent slide detached a large part of the island of Hawaii 105,000 years ago. That slide, on the island of Lanai, raised tsunami waves to the above-mentioned 326 meters above sea level. Mauna Loa, the gigantic volcano on the big island of Hawaii, the youngest and largest of the Hawaiian Islands, has collapsed repeatedly to the west. Two of these collapses were slumps and two were debris avalanches. Most collapses were submarine, though the head scarp of the North Kona slump grazes the west coast of Hawaii.

Kilauea, the youngest and most active volcano in Hawaii, is now slowly slumping. The potential for future collapse of its flank is emphasized in the Hilina scarps. A mass 100 kilometers wide and 80 kilometers long is moving seaward at 10 to 15 centimeters per year, sometimes suddenly (**FIGURE 5-24**). On November 19, 1975, the southern flank of the volcano moved more than 7 meters seaward and dropped more than 3 meters during a magnitude 7.2 earthquake. The resulting relatively small tsunami drowned two people nearby, destroyed coastal houses, and sank boats in Hilo Bay on the northeastern side of the island. What this portends for further movement is not clear. Will the flank of the volcano continue to drop incrementally at unpredictable intervals or could it fail catastrophically?

If this huge slump suddenly collapses into the sea, perhaps triggered by a large earthquake or major injection of magma, it could generate tsunami more than 100 meters high. Many low-lying coastal communities in Hawaii would be obliterated with little or no warning. However, to put these numbers in perspective, if 100,000 people were killed in such an event every 100,000 years, the average would be one person per year. Although unimaginably catastrophic when it does happen, there are certainly greater dangers, on average, in one person's lifetime.

The danger is not limited to Hawaii. If the flank of Kilauea, now moving seaward, should fail catastrophically, it could generate a tsunami large enough to devastate coastal populations all around the Pacific Ocean. Those in Hawaii would have little warning. The Pacific coast of the Americas would get several hours. It remains to be seen how many people could be warned and how many of those would heed the warning. The transportation networks of major urban centers, such as San Francisco and Los Angeles, could not accommodate enough traffic

FIGURE 5-25 A TSUNAMI ON THE ATLANTIC COAST?

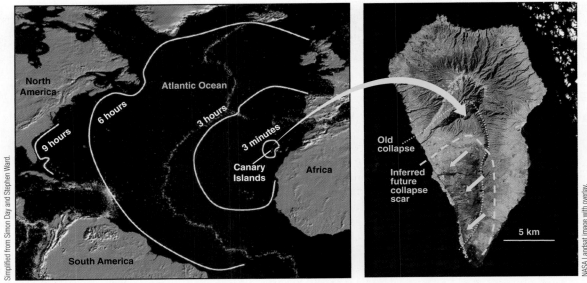

A. A large landslide from La Palma, Canary Islands, could generate immense tsunami waves that would fan out into the Atlantic Ocean. Computer simulations suggest that huge waves would reach the east coast of North America in six to seven hours. **B.** La Palma, Canary Islands, showing a major old collapse scar and a possible future collapse site inferred by W. J. McGuire and others.

to permit evacuation. We hope that the next event will not be anytime soon, but we have no way of knowing.

Canary Islands: Potential Catastrophe in Coastal Cities Across the Atlantic

Like other large basaltic island volcanoes, Tenerife—in the Canary Islands off the northwestern coast of Africa—shows evidence of repeated collapse of its volcano flanks. Tenerife reaches an elevation of 3,718 meters, almost as high as Mauna Loa. It is flanked by large-volume submarine debris deposits that have left broad valleys on the volcano flanks. Lava filling these valleys is as thick as 590 meters and overlies volcanic rubble along an inferred detachment surface that dips seaward at about 9 degrees.

Collapse may have been initiated by subsidence of the 11- to 14-kilometer-wide summit caldera into its active magma chamber. The most recent caldera and island flank collapse was 170,000 years ago. That event carried a large debris avalanche from the northwestern coast of Tenerife onto the ocean floor. It carried 1,000 cubic kilometers of debris, some of which moved 100 kilometers offshore. The much larger El Golfo debris avalanche detached 15,000 years ago from the northwestern flank of El Hierro Island. It carried 400 cubic kilometers of debris as much as 600 kilometers offshore.

An average interval of 100,000 years between collapse events on the Canary Islands may be long, and some scientists argue that it is unlikely to happen, but the consequences of such an event would be catastrophic. And the interval is merely an average. The next collapse could come at any time, and the giant tsunami caused by such a collapse would not only catastrophically inundate heavily populated coastal areas around the northern Atlantic Ocean but also reach coastal Portugal in only two hours, England in little more than three hours, and the east coasts of Canada and the United States in six to nine hours (**FIGURE 5-25**). Because large populations live in low-lying coastal cities and on unprotected barrier islands along the coast, millions would be at risk. Predicted wave heights in Florida, for example, would reach 20 to 25 meters, more than the height of a five- to seven-story building! Even if warning were to reach endangered areas as much as six hours before arrival of the first wave, we know from experience with hurricanes that evacuation would likely take much longer that that. Imagine hundreds of thousands of people trying to evacuate without a well-thought-out plan and in traffic that is heavy under normal circumstances. What about congestion on the single two-lane bridges that link most barrier islands to the mainland? How many would ignore the warning, not realizing the level of danger? The death toll could be staggering.

Cases in Point

Lack of Warning and Education Costs Lives
Sumatra Tsunami, December 2004, March 2005, and July 2006 ▶

The Indian Plate is moving northeast at 6 centimeters per year relative to the Burma Plate. In the 10 years preceding this event, there were 40 events larger than magnitude 5.5 in the area, but none generated tsunami. In the last 200 or so years, several other earthquakes larger than magnitude 8 have generated moderate-sized tsunami that have killed as many as a few hundred people. Paleoseismic studies show that giant events occur in the region on an average of once every 230 years.

The subduction boundary had been locked for hundreds of years, causing the overriding Burma Plate to slowly bulge like a bent stick; it finally slipped to cause a magnitude 9.15 earthquake on December 26, 2004. Given the size of the earthquake, offset on the thrust plane was some 15 meters, with the seafloor rising several meters. The subduction zone broke suddenly, extending north over approximately 1,200 kilometers of its length, shaking violently for as long as eight minutes.

For people nearby in Sumatra, Indonesia, the back-and-forth distance of shaking, with accelerations greater than that of a falling elevator, made it impossible to stand or run while poorly reinforced buildings collapsed around them. The first reports in northern Sumatra indicated that the earthquake severely damaged bridges and knocked out electric power and telephone service. Buildings were heavily damaged. People ran into the streets in panic. Smaller earthquakes quickly followed farther north along the subduction zone.

The sudden rise of the ocean floor generated a huge wave that moved away from the earthquake source at speeds of more than 700 kilometers per hour; it reached nearby shores within 15 minutes. A short time later, tsunami waves 5 meters high struck northern-most Sumatra, wiping out 25 square kilometers of the provincial capital of Banda Aceh. Locally, the wave swept inland as far as 8 kilometers; it had a 24-meter-high run-up on one hill almost a kilometer inland.

Most people in this steeply mountainous country live in low-lying coastal areas, around river mouths, for example; they had little or no warning of the incoming wave. Most people were preoccupied with the earthquake, and few were aware of even the possibility of tsunami. For some who did not happen to be looking out to sea, the first indication was apparently a roaring sound similar to fast-approaching locomotives. Elsewhere there was no sound as the sea rose.

In less than two hours, the first of several tsunami waves crashed into western Thailand, the east coast of Sri Lanka, and shortly thereafter the east coast of India; seven hours later, it reached Somalia on the east coast of Africa. In Sri Lanka, a coastal train carrying 1,000 passengers was washed off the tracks into a local swamp; more than 800 bodies were recovered. By January 13, more than 230,000 people were dead and tens of thousands more remained missing. Even with many hours between the earthquake and the first waves, hundreds of people died in Somalia on the northeastern African coast.

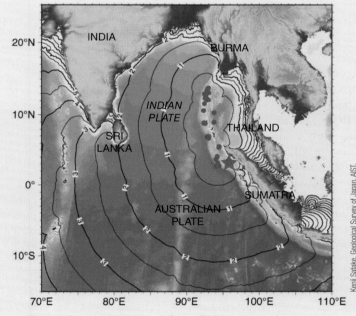

Kenji Satake, Geological Survey of Japan, AIST.

▶ *Tsunami wave-front travel times (in hours) are shown emanating from the rupture zone, which spans from the earthquake epicenter (red star) through the area of aftershocks (red dots).*

(continued)

At least 31,000 died in Sri Lanka; 10,750 in India; and 5,400 in Thailand. In Indonesia, at least 168,000 were dead or missing. In Banda Aceh alone, 30,000 bodies may remain in the area in which no buildings were left standing. Relief organizations were overwhelmed by the unprecedented scale of the disaster encompassing eleven countries. Five million people in the region lost their homes; hundreds of thousands of survivors huddled in makeshift shelters. This tsunami was the most devastating natural disaster of its kind on record.

Although the massive earthquake was recorded worldwide, people along the affected coasts were not notified of the possibility of a major tsunami. Unfortunately, the Indian Ocean did not have a tsunami warning network. The Pacific Tsunami Warning Center in Hawaii alerted member countries around the Pacific and tried to contact some countries around the Indian Ocean about potential tsunami that might have been generated. Because tsunami in the Indian Ocean are infrequent, no notification framework was in place to rapidly disseminate the information between or within countries. By the summer of 2006, a warning system was finally up and working.

An official in West Sumatra recorded the earthquake and spent more than an hour unsuccessfully trying to contact his national center in Jakarta. An official in Jakarta later sent email notices to other agencies but did not call them. A seismologist in Australia sent a warning to the national emergency system and to Australia's embassies overseas but not to foreign governments because of concern for breaking diplomatic rules. Officials in Thailand had up to an hour's notice but apparently failed to disseminate the warning. Among the public, few people had any knowledge of tsunami or that earthquakes could produce them. Ironically, the country's chief meteorologist, now retired, had warned in the summer of 1998 that the country was due for a tsunami. Fearing a disaster for the tourist economy, government officials had labeled him crazy and dangerous. He is now considered a local hero.

Although scientists have expressed concern about the lack of a warning system in the Indian Ocean, most officials in Thailand and Malaysia viewed tsunami as a Pacific Ocean problem, and the tens of millions of dollars it would cost to set up a network left it a low priority in a region with limited finances. However, by 2009 they had placed ocean-floor sensors in the Indian Ocean for detection of any future tsunami.

Even if there had been an Indian Ocean warning system in place, it would not have been able to save most of the lives in the most devastated region of Sumatra because the time between the earthquake and the wave arrival was short. Compounding the problem was the time delay in determining the size of the earthquake. The location of the quake was determined quickly and automatically from the arrival times of seismic waves from several locations. The magnitude was apparently large, initially estimated by Indonesian authorities at 6.6, a size that would not generate a significant tsunami. However, because the magnitude of giant earthquakes is determined by the amplitude of the surface waves, and such large earthquakes have lower-frequency waves that move more slowly, it often takes more than an hour to determine the magnitude. By that time, it would have been too late for a local warning because waves had already battered Sumatra; however, it could have saved many lives in more distant locations, such as Sri Lanka and India.

As with all hazards, public education could have done much to save lives. There was a lack of knowledge, even among officials, that a large earthquake could generate large tsunami. On the other hand, a ten-year-old girl who had recently learned about tsunami in school saw the sea recede before the first wave and yelled to those around her to run uphill. A dock worker on a remote Indian island had seen a television special on tsunami, felt the earthquake, and ran to warn those in a nearby community that giant waves were coming. Together, these two saved more

▶ **A.** *The northern part of Banda Aceh, Sumatra, on June 23, 2004, before the tsunami.* **B.** *The same area on December 28, 2004, after the tsunami. Note that virtually all of the buildings were swept off the heavily populated island. The heavy rock riprap along the north coast of the island before the tsunami remains only in scattered patches afterward. A large part of the island south of the riprap has disappeared, as has part of the southern edge, between the two bridges, where closely packed buildings were built on piers in the bay.*

(continued)

▶ **A.** *The coastal area of Meulaboh, Indonesia, before the tsunami.* **B.** *After the tsunami, nothing remains of most houses. Distinctive buildings in the upper right part of each image can be matched.*

than 1,500 people. Knowledge of hazard processes can save lives.

Could there be another earthquake and tsunami along the same subduction zone any time soon? The catastrophic 2004 earthquake relieved stress on that part of the fault but shifted some of that stress to other nearby faults. That additional stress effectively triggered earthquakes on those faults sooner than otherwise. On March 28, 2005, a Mw 8.7 earthquake struck the zone just to the southeast.

It was under less than 1 kilometer of water and generated only a small tsunami; 1,313 people died.

On July 17, 2006, a magnitude Mw 7.7 earthquake struck the same zone along the south coast of Java, about 2,000 kilometers to the southeast. Waves reached 5–7 meters in some areas. Although alerts were provided from Japan and Hawaii tsunami warning centers, authorities did not evacuate residents until too late—730 died in Java. Indonesian authorities acknowledged but did not forward warnings

▶ **Wreckage from houses and boats piled up inland, along with the bodies of those caught in the tsunami.**

to the public because they didn't want to cause "unnecessary alarm." They finally sent text messages 7 minutes before the tsunami hit, but it was too late for local communities. Again, on September 12, 2007, a magnitude 8.5 earthquake and tsunami struck southern Sumatra, killing 25 people. Although inhabitants were concerned that it would generate a giant tsunami, the largest swells were 3 meters high, apparently because the waves moved out to sea rather than toward shore and because the earthquake occurred under shallow water.

▶ **Low-lying coastal areas in northern Indonesia were completely destroyed, leaving only parts of some house foundations, a huge cement barge, and its tugboat.**

Immense Local Tsunami from a Landslide
Lituya Bay, Alaska, 1958 ▶

One of the most spectacular tsunami resulting from a land-based rockfall occurred in Lituya Bay, Alaska, in 1958. Lituya Bay, a deep fjord west of Juneau, Alaska, and at the western edge of Glacier Bay National Park, was the site of one of the highest tsunami run-ups ever recorded. On July 9, 1958, 60 million cubic meters of rock and glacial ice, loosened by a nearby magnitude 7.5 earthquake on the Fairweather Fault, fell into the head of Lituya Bay. The displaced water created a wave 150 meters high—the height of a 50-story building. It surged to an incredible 524 meters over a nearby ridge and removed forest cover up to an average elevation of 33 meters and up to 152 meters over large areas. This was a huge wave compared with common tsunami, which may be 10 to 15 meters high; it swept through Lituya Bay at between 150 and 210 kilometers per hour.

Although three fishing boats, with crews of two each, were in the bay at the time, only those on one boat died, when their boat was swept into a rocky cliff. On another boat, on the south side of the island in the center of the bay, Howard Ulrich and his seven-year-old son hung on as their boat was carried high over a submerged peninsula into another part of the bay. They were actually able to motor out of the bay the next day. On the third boat, Mr. and Mrs. William Swanson, anchored on the north side of the bay, were awakened as the breaking wave lifted their boat bow first and snapped the anchor chain. The boat was carried at a height of 25 meters above the tops of the highest trees, over the bay-mouth bar, and out to the open sea. Their boat sank, but they climbed onto a deserted skiff and were rescued by another fishing boat two hours later.

Examination of the forested shorelines of Lituya Bay, above the trimline of the 1958 event, shows two, much higher trimlines produced by earlier tsunami. Scientists from the USGS

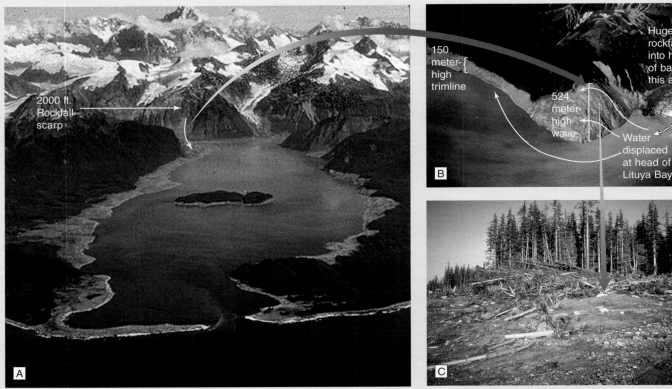

▶ **A.** *A huge rockfall into the head of Lituya Bay, Alaska, generated a giant tsunami wave that stripped the forest and soil from a ridge. This view to the northeast shows the broad areas of forest that the tsunami swept from the fringes of the bay. The scarp left by the rockfall is visible at the head of the bay (arrow).* **B.** *The rockfall crashed into the head of the bay, displacing water up to 524 meters, over a ridge and back into the bay.* **C.** *Detailed view of the forest damage at the crest of that ridge.*

x

(continued)

1874 wave

1936 wave

1958 wave

Lituya Bay

▶ *This view of Lituya Bay shows trimlines from two previous tsunami that were even larger than the 1958 event. Do narrow, steep-sided inlets elsewhere show similar healed trimlines? Could they provide hazard information for future rockfall tsunami?*

examined trees at the level of these higher trimlines and found severe damage caused by the earlier events. Counting tree rings that had grown since then, they determined that the earlier tsunami had occurred in 1936, 1874, and 1853 or '54.

Glacier Bay, 50 kilometers east of Lituya Bay, is in a similar spectacular environment. Could it be the site of the next big landslide-generated tsunami? It is a deep fjord bounded by precipitous rock cliffs and glaciers. It lies between two major active strike-slip faults, the Fairweather Fault and the Denali Fault, each 50 or 60 kilometers away, and each capable of earthquakes of magnitudes greater than 7. Glacier Bay is a prominent destination for cruise ships touring from Seattle or Vancouver to Alaska, so a tsunami generated by a large landslide into the bay is a concern. Study by USGS geologists suggests that an unstable rockslide mass on the flank of a tributary inlet to Glacier Bay would generate waves with more than 100-meter run-ups near the source and tens of meters within the inlet.

In the deepwater channel of the western arm of Glacier Bay, the wave amplitude would decrease with distance into the bay. Ships near the mouth of the tributary inlet could encounter a 10-meter-high wave only 4 minutes after the slide hit the water, then 20-meter-high waves after 20 minutes. The waves would likely strike the cruise ships broadside (left photo). If the ships kept to this central channel, the largest waves would likely be approximately 4 meters high. The response of a ship to waves near the mouth of the tributary inlet would depend on the wave heights, the wave frequencies relative to the ship's rocking frequency, and the height of the lowest open areas on the ship. Because cruise operators are now aware of this risk, they avoid the dangerous near-inlet waters.

Glacier Bay

Wave heights from cliff collapse:

~ 4 m

~ 20 m

Side channel

main Glacier Bay

▶ *View is from the apex of the slide with the tidal inlet in the foreground; two cruise ships in Glacier Bay are visible in mid-photo. The tidal inlet landslide mass next to Glacier Bay, Alaska, includes the rock face from the new higher scarp to below water.*

An Ocean-Wide Tsunami from a Giant Earthquake
Chile Tsunami, 1960 ▶

The largest earthquake in the historical record (moment magnitude [M_w], 9.5) happened on the subduction zone along the coast of Chile on May 22, 1960. Fifteen minutes after the quake, the sea rose rapidly by 4.5 meters. Fifty-two minutes later, an 8-meter-high second tsunami arrived at 200 kilometers per hour, crushing boats and coastal buildings. A third, slower wave was 10.7 meters high. More than 2,000 people died.

In Maullín, Chile, the tsunami washed away houses on low ground or carried them off their foundations. Some of those houses were moved more than a kilometer inland; others were demolished or washed out to sea. Many people wisely ran for higher ground. Some who returned for valuables were not so lucky. One group survived by climbing to the loft of a barn; several others climbed trees. One person in a tree watched water rise to his waist. A farmer who watched his house on a river floodplain collapse later found 10 centimeters of sand covering his fields. Forests on low ground dropped abruptly below sea level, permitting saltwater to flow in and kill the trees.

Fifteen hours later, as predicted, the tsunami reached Hawaii. Coastal warning sirens sounded at 8:30 p.m. When the 9 p.m. news from Tahiti reported that waves there were only 1 meter high, many Hawaiians relaxed. Few people in Hawaii realized that well-developed reefs protect the Tahitian islands. Warned hours earlier by radio and sirens, a third of the people in low areas of Hilo evacuated; others did not because previous warnings had involved only small tsunami that had caused little damage. The first wave just after midnight was little more than 1 meter high.

Many people thought the danger was over and returned to Hilo. At 1:04 a.m., a low rumbling sound, like that of a distant train, became louder and louder, followed by crashing and crunching as buildings collapsed in the 4-meter-high, nearly vertical wall of the largest wave; 282 people were badly injured and 61 died, all in Hilo, including sightseers who went to the shore to see the tsunami. The waves destroyed water mains, sewage systems, homes, and businesses. Most of the deaths were avoidable; people heard the warnings but misinterpreted the severity of the hazard.

Hilo is the most vulnerable location on the Hawaiian Islands. Although it has a particularly good harbor, Hilo Bay also focuses the damage. Even tsunami waves that approach from the southeast refract in the shallower waters around the island to focus their maximum height and energy in the bay. The bay also has an unfortunate shape that, as a first tsunami wave drains back offshore, reinforces the incoming second wave that arrives about a half hour later.

Nine hours after the tsunami hit Hawaii, it reached the island of Honshu, Japan. Its wave height decreased only a little; 185 people died, 122 of them on Honshu. Following a few unusual waves

Brian Atwater, USGS.

▶ *A tsunami wave swept material inland from the beach near Maullín, Chile, in 1960. It left a sand layer over a soil horizon on a farmer's field.*

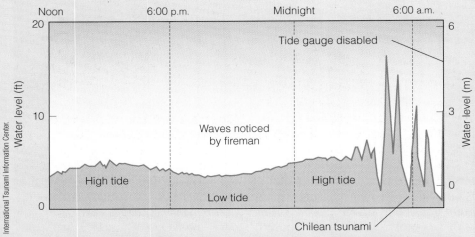

International Tsunami Information Center.

▶ *These tide gauges record the tsunami waves following the 1960 Chilean earthquake, at Onagawa, Japan, May 23–24. It recorded a dramatic drop in sea level, which, along with rapid rises in sea levels, serves as a warning sign for incoming tsunami.*

(continued)

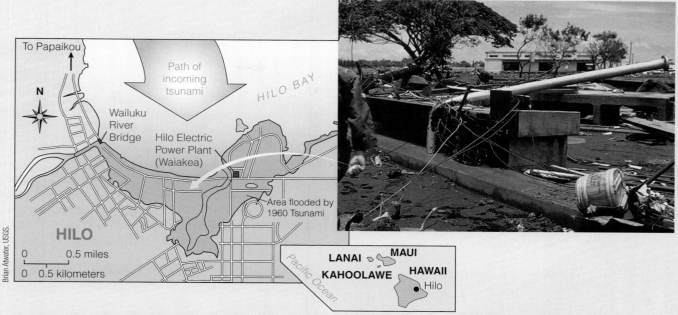

▶ This low-lying waterfront area of downtown Hilo, at the head of Hilo Bay, was destroyed by the Chilean tsunami of 1960.

up to 1 meter high, the first sign of the tsunami was sea retreat accompanying a rapid 1.5-meter drop in sea level (see FIGURE 5-15). Then the first tsunami wave arrived more than 4 meters above the previous low. It was that withdrawal followed immediately by a rapid rise that caught people off guard and drowned many of them. Five waves over six hours culminated in a huge wave more than 5 meters high that disabled the tide gauge and preventing further record of the tsunami. Note that the highest wave was far from the first, and the waves can be an hour or more apart.

Subduction-Zone Earthquake Generates a Major Tsunami
Anchorage, Alaska, 1964 ▶

The giant magnitude 8.6 (moment magnitude [M_W], 9.2) Anchorage, Alaska, earthquake on March 27, 1964, showed large vertical effects of slip on the subduction zone. The slab sinking beneath the Aleutian oceanic trench slipped over a 1,000-kilometer length and across a 300-kilometer-wide zone. A 500 X 150 kilometer strip of extremely shallow seafloor rose 10 meters above sea level and moved 19.5 meters seaward. Another inland belt, fully comparable in length and more than 100 kilometers wide, sank as much as 2.3 meters. Low-lying coastal areas dropped below sea level. Twenty-seven years later (in 1991), the same areas had slowly rebounded, rising above sea level again.

The sudden change in seafloor elevation displaced the overlying water into giant tsunami waves that washed ashore on the Kenai Peninsula within 19 minutes and onto Kodiak Island in 34 minutes. The maximum tsunami wave run-up occurred where it funneled into Valdez Inlet, just west of the Kenai Peninsula, where the earthquake caused a submarine landslide. Of the 131 people killed in the earthquake, 122 drowned in the 61-meter-high waves that funneled into and devastated waterfront areas in Valdez and Seward. Smaller wave heights extended all the way to Crescent City, northern California, destroying much of the waterfront area.

Eight minutes after the Anchorage earthquake, an alarm sounded at the Honolulu Observatory. The location and magnitude of the earthquake were determined from seismograms within an hour. The California State Disaster Office received warnings of a possible "tidal wave" two hours after the earthquake. The county sheriff at Crescent City received the warning one and a half hours

(continued)

1964

Brian Atwater, USGS.

1991

▶ **A.** *Immediately after the Alaska earthquake of 1964, the coastal area flooded when the coastal bulge collapsed.* **B.** *This is the same area 27 years after the bulge again began to rise.*

USGS.

▶ *Following the March 1964 Prince William Sound, Alaska, tsunami, cars in a full parking lot were thrown about at the head of a bay like so many toys.*

USGS.

▶ *This truck, 10 meters above sea level, was wrapped around a tree.*

later and notified people in low-lying coastal areas to evacuate. An hour after that, a 1.5-meter-high wave reached Crescent City, amplified by the shallowing water near shore and narrowing of the harbor. According to the files of the Del Norte Historical Society, the curator of Battery Point Lighthouse, on a small rocky island just offshore, recalled that it was a clear moonlit night, and the waves were clearly visible as they deluged the town. The first wave carried giant logs, trees, and other debris, demolishing buildings

and cars. This debris-laden wave receded as quickly as it arrived, leaving battered cars, houses, logs, and boats. The sea receded to a kilometer offshore.

After a second wave entered the harbor, some people returned to clean up. A third and larger wave then washed inland more than 500 meters, drowning five people; it knocked out power and ignited a fire before the sea receded even farther. The fourth and largest wave was 6.3 meters high; it killed ten people who went back down to check their houses.

That wave submerged the damaged Citizen's Dock and lifted a big, loaded lumber barge, setting it down and crushing the dock. Fuel from ruptured tanks at the Texaco bulk plant spread to the fire and ignited. One after another the tanks exploded. Pieces of everything imaginable drained back offshore with the outgoing wave. The fifth wave was somewhat smaller. In all, the tsunami destroyed 56 blocks of the town.

Critical View

Test your observational skills in interpretation of the following scenes relevant to tsunami. Discuss the issues in detail—what happened and why?

A This damage is in the coastal area of Samoa after the recent earthquake and tsunami.

Casey Deshong, FEMA.

1. Is this damage likely from the earthquake or the tsunami? Why?
2. If you were caught in the incoming tsunami, what would cause injuries or death?
3. Would you be better off in a vehicle or outside it? Why?
4. What would be a safer choice if you have five minutes of warning?

C This damage is in the coastal area of Sri Lanka after the Dec., 2004 earthquake.

USGS.

1. Where did the loose sand and the scattered bricks come from and how did they get here?
2. What would severely injure or kill you in this event?
3. If you had only 2 or 3 minutes as the wave approached, what would be your safest plan of action?

B This damage is in the coastal area of Samoa after the recent tsunami.

Gordon Yamasaki, USGS.

1. Comparing this photo with the one in "A," might you be better off in a vehicle or out of it and explain why?
2. If you can't get far enough upslope or away from the shoreline, what are other options?

D This damage is in the coastal area of Sri Lanka after the Dec., 2004 earthquake.

Bruce Jaffe, USGS.

1. Many of these homes appear to be of sturdy brick construction. Why did many of them crumble with the incoming waves?
2. Given the severe damage especially in the foreground in which direction is the beach?
3. If you think most of the houses in the area are not strongly built, the only roads are parallel to the beach, and there are no nearby hills, what is your safest plan of action?

Chapter Review

Key Points

Tsunami Generation

- Tsunami are caused by any large, rapid displacement of water, including earthquake offsets or volcanic eruptions underwater; landslides; and asteroid impacts into water.

- In coastal subduction zones, the leading edge of the continental plate slowly bulges upward before suddenly dropping during an earthquake. The displacement of water caused by this rupture is a common source of tsunami. **FIGURES 5-2 and 5-3**.

- The height of an earthquake-generated tsunami is related to the magnitude of the earthquake, which in turn depends on the length of the rupture and the vertical extent of offset under water.

- Volcanic processes—such as an underwater eruption, lava flow into the sea, or volcano flank collapse—can all trigger tsunami.

- Landslides into the ocean as well as landslides that occur underwater can trigger tsunami. The height of the waves produced depends in large part on the height of the fall.

- The impact of a large asteroid into the ocean, while rare, would displace a huge amount of water and generate a massive tsunami.

Tsunami Movement

- Tsunami have such long wavelengths that they always drag on the ocean bottom. As they drag on bottom in shallow water, they slow, and their wave height increases.

- Wave velocity depends on water depth. Tsunami waves in the deep ocean are low and far apart but move at velocities of several hundreds of kilometers per hour. They slow and build much higher in shallow water near the coast, especially in coastal bays. **FIGURES 5-9 and 5-10**.

Tsunami on Shore

- The effects of tsunami are amplified by coastal bays, river mouths, and tides.

- Tsunami run-up is the height the water reaches when it sweeps up on shore.

- Tsunami reach shore as a series of waves, often tens of minutes apart, and may continue for hours. The third and subsequent waves are often the largest. **FIGURE 5-15**.

Tsunami Hazard Mitigation

- Tsunami damage can be mitigated by land-use planning and appropriate considerations for development. Structures and vegetation can also reduce tsunami impact.

- Warning systems are able to predict tsunami travel time fairly accurately, even to coastlines far from an earthquake epicenter.

- Coastal residents should be aware of danger signals for tsunami, including earthquakes and a rapid rise or fall in sea level. Once these signs are noticed, inhabitants should run upslope or drive directly inland. The safest areas are a kilometer or more inland and several tens of meters above sea level.

Future Giant Tsunami

- Tsunami from major Pacific Coast subduction earthquakes occur every few hundred years and reach shore within 20 minutes of the quake, destroying coastal communities, particularly those in bays and inlets.

- The record of subduction-zone tsunami is based on sand sheets over felled forests and marsh vegetation pushed into coastal bays. **FIGURE 5-22**.

- A volcano flank collapse that suddenly moves an enormous amount of water, such as in the Canary Islands, could generate giant tsunami that would be catastrophic for much of the east coast of North America, especially low-lying coastal communities. **FIGURE 5-25**.

Key Terms

coastal bulge, p. 106

debris avalanche, p. 109

period, p. 110

run-up, p. 112

submarine landslides, p. 108

trimline, p. 113

tsunami, p. 106

tsunami warning, p. 116

tsunami watch, p. 116

wavelength, p. 110

Questions for Review

1. What are three of the main causes of tsunami?

2. Of the three main types of fault movements—strike-slip faults, normal faults, and thrust faults—which can and which cannot cause tsunami? Why?

3. How high are the largest earthquake-caused tsunami waves in the open ocean compared to when they reach shore?

4. How fast do tsunami waves tend to move in the deep ocean compared to when they reach shore?

5. Do tsunami speed up or slow down as they approach the coast? Why?

6. How does the height of a tsunami wave change as it reaches shore? Provide a sketch to illustrate your explanation.

7. Why should you stay away from the beach even after a tsunami wave has retreated? How long should you stay away?

8. Why is even the side of an island facing away from the source earthquake not safe from a tsunami?

9. Name three ways to mitigate damages from tsunami.

10. In what situations would a tsunami warning system not be effective for evacuating a population at risk?

11. Name three signs of an imminent tsunami.

12. If you are in a coastal area, what should you do to increase your chances of surviving a tsunami?

13. For a subduction-zone earthquake off the coast of Oregon or Washington, how long would it take for a tsunami wave to first reach the coast?

14. What specific evidence is there for multiple tsunami events having struck coastal bays of Washington and Oregon?

15. Because the Atlantic coast experiences fewer large earthquakes, what other specific event could generate a large tsunami wave that would strike the Atlantic coast of North America?

Discussion Questions

1. Given the past record of major earthquakes and tsunami along the Pacific coasts of Northern California, Oregon, Washington, and British Columbia, what should be done about people currently living in especially hazardous areas such as:

   a. Those living in narrow bays along the coast—for example, Seaside, Oregon and Newport, Oregon

   b. Those living on barrier bars along the coast—for example, bay-mouth bars in southwestern Washington State (Westport, Long Beach)?

   Consider possibilities such as moving endangered houses or even whole towns, building major barriers to slow the advance of a tsunami, building safe evacuation structures, building safe evacuation routes away from the coast, or doing nothing.

2. Given the dangers noted in question 1, what should be done to prevent more people from moving into such hazardous areas?

3. Because both permanent residents and visitors occupy such hazardous areas and thousands of people form inland cities visit the same coastal areas, who should pay for the costs of mitigation or removal?

6 Volcanoes: Tectonic Environments and Eruptions

From 2004 to 2006, renewed activity in Mt. St. Helen's crater showed deep magma slowly pushing up a dome of rock that had solidified in the vent since the climactic eruption of 1980. The volcano remains active.

Willie Scott, USGS.

Cascade Range Volcanoes Are Active

Until the 1970s, most of the Cascades volcanoes were thought by almost everyone to be extinct or at least dormant. The rule of thumb was that if a volcano had significant glacial erosion that had not been erased by later eruptions, it probably had not erupted since the last ice age some 10,000 years ago. The inference was that because it had not erupted in such a long time, it was unlikely to erupt again. It seems that our human timeframe colored our view of what a volcanic timeframe should be. Mount Lassen, in northern California, staged a series of eruptions that started in 1914 and intermittently continued with declining vigor until 1921, but most geologists regarded it as an unusual event. This eruption did not persuade them that the High Cascades were an active volcanic chain. With the new understanding of plate tectonics and the fact that active subduction zones are associated with earthquakes and overlying active volcanoes, geologists began to wonder whether in fact the Cascades volcanoes might really be active. If a subduction zone is active, based on either measured convergence of two plates or major earthquakes,

Volcanoes

the chain of volcanoes above it is almost certainly active. As the term is now used by volcanologists, an active volcano is one likely to erupt again. Evidence for such activity is generally provided by documentation of its past eruptions and their average frequency. Non-specialists or the news media may refer to a volcano as dormant because it has not erupted for hundreds or thousands of years, but some volcanoes have even longer periods between eruptions. We now know that several Cascade volcanoes are probably active. Glacier Peak, Mt. Rainier, Mt. St. Helens, Mt. Adams, Mt. Hood, Three Sisters, Mt. Shasta, and Mt. Lassen, have all erupted in the last 200 years. Mt. Baker and Crater Lake erupted somewhat earlier.

Introduction to Volcanoes: Generation of Magmas

A **volcano** is typically a cone-shaped hill or mountain formed at a vent from which molten rock, called **magma**, or magmatic gases reach the Earth's surface and erupt. Once the magma reaches the surface, it is called **lava**. Because most of the Earth beneath the surface is not molten, volcanoes erupt in a limited number of geologic settings where magma is generated at depth and can rise to erupt at the surface.

To understand the generation of magmas, it is important to appreciate the basic difference between states of matter: solid, liquid, and gas, which are related to the movement of their particles. Molecules in a solid are tightly bound together in a rigid shape, so they hardly move. Those in a liquid are loosely held together by flexible bonds that permit them to move fluidly. Molecules in a gas are far apart, free to move and fill available space. With very few exceptions, solids are most dense, liquids are less dense, and gases are least dense (**FIGURE 6-1**).

As temperature and pressure change, substances undergo a change from one state to another. For example, as you compress a gas, it first becomes a liquid and then a solid. The reverse is also true: as you decrease the pressure on a hot solid rock, it may expand and melt into a liquid and then expand even more to vaporize into a gas. These principles govern the

transition from rock to magma and the release of gases from magma. A very hot rock may melt if pressure decreases. A gas dissolved in a liquid may expand and separate with decreasing pressure. The **melting temperature**, which controls when a rock becomes magma, depends on pressure and the availability of water. A hot rock deep within the Earth may melt if temperature rises, pressure falls, or water is added; addition of water shifts the melting curve to lower temperature (**By the Numbers 6-1:** Melting Temperature of a Rock).

Masses of magma rise through the crust because they are less dense than the surrounding rocks, much as an iceberg rises into the air because ice has lower density than the water surrounding it (see Chapter 2, **By the Numbers 2-1,** p. 16). Magmas may even rise into rocks of lower density; as long as the column of magma is less dense than the surrounding rock, the magma will float in it. Magma may rise through cracks, sometimes breaking off and incorporating pieces of the adjacent rocks. **Magma chambers** are large masses of molten magma that rise through Earth's crust, often erupting at the surface to build a volcano. The behavior of earthquake waves, minute irregularities in Earth's gravitational attraction, or a slight rise or tilt of the ground surface may indicate an expanding magma chamber. Erosion eventually exposes and dissects old volcanoes and their magma chambers, which are now crystallized into masses of solid **igneous rock** solidified from magma.

Magma Properties and Volcanic Behavior

No two volcanoes are quite alike; nor are any two eruptions from the same volcano. What happens during an eruption depends mainly on how fluid the magma is (its viscosity), the quantity of water vapor and other volcanic gases it contains (its volatiles), and the type and amount of magma (its volume) that erupts.

VISCOSITY The nature of a volcanic eruption depends in part on the composition of the magma, including its content of water and other gases, and its temperature, both of which affect its viscosity. **Viscosity** refers to how fluid magma is; high-viscosity magmas are thick and pasty. The viscosity of a magma depends on its chemical composition—the internal

FIGURE 6-1 SPREADING MOLECULES

 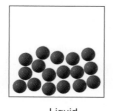 

Solid   Liquid   Gas

Molecules in a solid are tightly packed, while molecules in a liquid are held together loosely, and molecules in a gas can spread out to fill a container.

Melting Temperature of a Rock

The melting temperature of a rock depends on its depth within the Earth and the amount of available water. As shown in the following illustration (see green arrows), a hot rock deep within the Earth (e.g., a rock above a subduction zone at "A") may melt as a result of an increase in temperature, a decrease in pressure, or the addition of water that shifts the melting curve to lower temperatures.

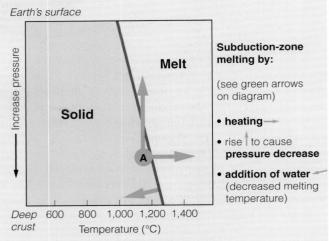

Newly formed magmas rising toward the surface may not always reach it. The magma may crystallize by cooling or loss of water. A water-rich magma must lose water and begin to crystallize as it rises to lower pressures.

Examples: Most basalt magma (almost dry) forms at high temperatures deep in the mantle. At a mid-oceanic ridge, hot mantle peridotite rises, and the new lower pressure causes it to melt to form basalt magma.

Deep in a subduction zone, water boils off the subducted slab and lowers the melting temperature of overlying hot mantle peridotite. The newly formed basalt magma rises into the continental crust, heats it, and partly melts it with the little available water to form granite (rhyolite) magma. Both basalt and rhyolite magmas rise to erupt in stratovolcanoes, the large, steep-sided volcanic cones like those of the Cascades

arrangement of its atoms and molecules. By far the most abundant atoms in a magma are oxygen and, to a lesser extent, silicon and aluminum. Almost every silicon atom in its natural state is surrounded by four oxygen atoms arranged in the shape of a tetrahedron (**FIGURE 6-2**). All silicate rocks and minerals, including the common volcanic rocks, consist of an array of submicroscopic silica tetrahedra that generally are linked to atoms of aluminum, iron, magnesium, calcium, potassium, sodium, and other elements. The chemical bonds between silicon and oxygen atoms are too strong to bend or break easily, so silicate structures are rigid, like an assemblage of Tinkertoy parts.

Differences in viscosity among the major magma types (basalt, andesite, and rhyolite) are due mainly to different percentages of silica. In general, the higher the percentage of silica, the more viscous the magma. As shown in **Table 6-1**, the composition spectrum ranges from low-viscosity (fluid) basalt magmas, which have around 50 percent silica, to high-viscosity rhyolite magmas, which have around 70 percent silica. Andesite magmas fall in between these two extremes, with around 60 percent silica. (See Appendix 2 online.)

At the more fluid end of the spectrum, **basalt** is black or brownish black, and its magma is about as fluid as cold molasses. Dark magmas have fewer silica tetrahedra, and these either are not linked directly to others or share two oxygen atoms to form long chains of tetrahedra. The tetrahedra chains link to others with weaker bonds provided by charged atoms of calcium, magnesium, or iron; they can wiggle about in magma like worms in a can, which explains why dark magmas are so fluid.

Rhyolite comes in white or pale shades of gray, yellow, pink, green, and lavender. Rhyolite magma is extremely viscous, causing it to flow stiffly, if at all. Flows of rhyolite glass, obsidian, are rare; the few that exist are so viscous that thick flows move and solidify to form flow fronts 50 to 100 meters high with 45-degree slopes. Pale magmas, such as rhyolite, contain more silicon atoms to attach to oxygen atoms. Most of the silica tetrahedra share oxygen atoms to build exceptionally rigid frameworks of tetrahedra linked in all directions. The other atoms, including aluminum, sodium, and potassium, do not provide much freedom of movement, so pale magmas are extremely viscous.

The viscosities of **andesite** magmas are intermediate, falling between low-viscosity basalt magmas and high-viscosity rhyolite magmas. Andesite magma is the most abundant type, erupting from subduction-zone volcanoes like those in the Cascades. These magmas contain enough silica to be quite viscous, and as a result they flow very slowly, if at all; they solidify to form thick flows. Because of its high viscosity, andesite lava solidifies on steep slopes, resulting in steep-sided volcanoes. The ash and broken-andesite rubble tumble down the same slopes and come to rest at angles near 30 degrees. Gas content of the dark andesite magmas is generally quite low, and these magmas generally erupt in sluggish flows that are considerably thicker than basalt flows, or they blow out chunks of solidified lava or cinders. Dacite, with compositions between andesite and rhyolite, tend to erupt more vigorously to make clouds of ash and fields of broken rubble.

Temperature also plays a role in viscosity, because as magma cools, more bonds link between atoms and molecules and the magma becomes more and more viscous. At temperatures reaching 1,100° to 1,200°C, red-hot basalt lava generally pours down stream valleys or spreads across flat ground like pancake batter across a griddle, making relatively thin flows that commonly cover dozens of square kilometers (**FIGURE 6-3**, p. 42). In contrast, the more viscous rhyolite magmas erupt at temperatures between 800° and 900°C, a dull red heat.

FIGURE 6-2 MOLECULES UNSTICK

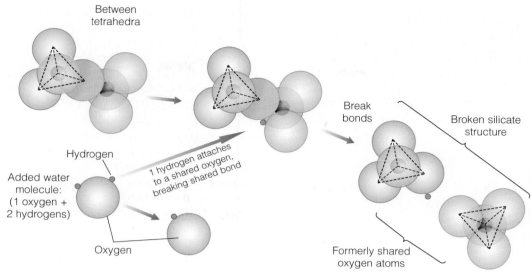

Four oxygen atoms (blue) surround each silicon atom (gray) to make a silica tetrahedron (shape outlined by dashed lines). Other silicon atoms share oxygen atoms (blue-green atom shared) to form mineral molecules. A water molecule reacts with a silicate structure and its shared oxygen atoms, breaking some of the strong bonds.

Table 6-1	Characteristics of the common magmas or lavas*		
MAGMA	**BASALT GRADATIONAL TO →**	**ANDESITE TO DACITE GRADATIONAL TO →**	**RHYOLITE**
Viscosity	Fluid (low viscosity)	Medium viscosity	Thick and pasty (high viscosity)
Color	Black	Dark shades of gray, green, and other intermediate colors	Pale colors, including white, pink, yellow
Composition	Magnesium, iron, and calcium-rich	Gradational to →	Potassium and silica-rich
Silica content	≈ 50%	≈ 60%	≈ 70%
Temperature	1,100° to 1,200°C (yellow to red hot)	Gradational to →	800° and 900°C (dull red)
Volatiles	Little water content unless magma encounters groundwater. May contain significant carbon dioxide.	Gradational to →	Generally more water
Erupted material**	Mostly lava flows	Lava flows, ash, broken fragments (rubble)	Mostly ash

*All of these rock types are gradational. ** Discussed later in the chapter*

VOLATILES The **volatiles** of a volcano refer to the dissolved gases it contains. Almost all volcanoes emit gases, often quietly but sometimes violently. The gases rise either from seemingly random locations high on the volcano or at an eruptive vent, sometimes during an eruption, sometimes long after. Water vapor is the most abundant volcanic gas and the most important in governing what happens when magma erupts. Under volcanic conditions, water in magmas exists mostly as vapor—not familiar teakettle steam—that can be hot enough to glow and instantly ignite anything flammable in its path. Carbon dioxide is normally second to water in abundance among the volcanic gases, but it has much less influence on the explosive nature of an eruption. It is relatively more abundant in basalt magmas than in those closer to rhyolite in composition.

Water content is critical because this is mainly what drives volcanoes to violence. Water at the high temperature

FIGURE 6-3 RED-HOT LAVA

A. Lava erupting as a bright-red curtain from a Hawaiian rift zone. **B.** Lava flowing downslope as a fiery stream.

of magma expands to form steam at low pressures near the Earth's surface. Pressure holds in any dissolved water, so the amount of water a magma can contain decreases dramatically as it rises into levels where the rock pressure is lower. Rhyolite magma with 2 percent water at a depth of 20 kilometers could hold only half that much at a depth of 2 or 3 kilometers. By the time the magma reaches the low pressure of Earth's surface, it can dissolve virtually no water or other gases; any gases that were dissolved at depth must separate from the magma and bubble out (**FIGURE 6-4**). It is those separating gases that drive explosive volcanic eruptions. Even if magma stops rising, steam can separate and build up pressure to drive an eruption. Most of the new crystals growing in the magma contain little or no water. As they grow, they displace the water or steam into an ever-smaller volume of the magma. This may eventually drive an eruption.

Magmas that contain little water erupt quietly as lavas. Basalt magmas, for example, are fluid enough to let their small amounts of water and carbon dioxide escape without causing much commotion. Those that contain large amounts of dissolved water, however, are likely to explode unless they are fluid enough to let the steam fizz quietly away. Rhyolite magmas contain 0 to 10 percent water by weight and much more than that by volume. When rhyolite magma reaches the surface nearly devoid of water, it erupts quietly as a dome that may grow to the size of a small mountain. When it arrives at the surface with a heavy charge of water, rhyolite magma explodes into clouds of steam full of foamy pumice and white rhyolite ash.

Because basaltic magmas feeding a volcano above a subduction zone stream up from deep in the Earth's mantle, they may rise into a water-rich rhyolite magma chamber. The extremely hot basalt heats the rhyolite, causing its water to boil into steam. Rapid expulsion of this steam near the surface causes explosive eruption of the rhyolitic magma. Dark inclusions of basalt sometimes found in pale rhyolite help document this process. Such magma interactions may be a frequent trigger for explosive eruptions; the process adds to the complications of predicting the style and timing of eruptions.

VOLUME Viscosity and volatiles of a magma are the properties that determine the nature of an eruption, and volume is the property that determines its size or magnitude. The volume of magma expelled in a single eruption has a significant bearing on the degree of hazard. It affects the size of a lava flow, as well as the volume, areal extent, and time span of an ash eruption. Recognition of various types of volcanoes from their size and the slopes of their flanks permits us to interpret their behavior, even from a distance.

For andesite to rhyolite gas-driven explosive eruptions, the volume of fragmental material erupted depends on the volume of magma reaching Earth's surface, the viscosity of the magma, and the proportion of dissolved gas it contains. Because the dissolved gases tend to rise toward the top of the magma chamber and the expanding gases drive the eruption, the upper gas-rich part of the magma (the upper third or so of the magma chamber) erupts explosively. The remaining magma stays underground, or a little may erupt as a lava flow late in the eruption process. Because these magmas are highly viscous, it may take decades to centuries for more gas to collect in the upper part of the magma chamber to again drive an eruption.

The composition and total volume of magma rising under a subduction-zone volcano depends on the subduction rate; the temperature and water content of the descending slab; the composition, temperature, and water

FIGURE 6-4 PRESSURE RELEASE

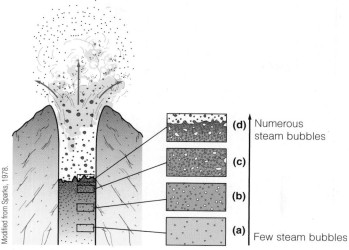

Modified from Sparks, 1978.

(**d**) Numerous steam bubbles

(**c**)

(**b**)

(**a**) Few steam bubbles

(**a**) and (**b**) represent steam separating from a magma in open bubbles. (**c**) and (**d**) show that the bubbles grow and the magma begins to froth, expand, and rise. That pushes magma upward, and its pressure decreases, launching a chain reaction of increasingly rapid bubble formation that leads to an explosive eruption.

content of the overlying crustal rocks; the ease with which the magma can rise through the crust; and other intangible factors. For an individual volcano, most of these factors are unlikely to change much over time. Thus, the behavior of an individual volcano is likely to be more or less predictable.

Tectonic Environments of Volcanoes

Plate tectonic environments, where Earth's tectonic plates spread apart or converge (see FIGURE 2-4), are common locations of changes in temperature, pressure, or water content. Most of the world's volcanoes are along these plate boundaries. Spreading zones, where plates pull apart, and subduction zones, where one plate pushes under another, are the most common locations for eruptions. Hotspots produce large volcanoes even though they are not on plate boundaries. Few volcanoes occur in continent–continent collision zones, such as in the Himalayas. Transform boundaries (faults that slide laterally past one another) rarely show volcanic eruptions.

The nature of an eruption is influenced by the tectonic environment in which it forms. Relatively peaceful eruptions are characteristic of spreading zones, while violent eruptions are characteristic of subduction zones.

Spreading Zones

Ocean-floor rifts or spreading zones, such as in the Pacific and Atlantic Oceans, erupt basaltic lavas that spread out on the adjacent ocean floor. These don't provide much hazard to humans except in Iceland, where the Mid-Atlantic Ridge extends above sea level. There the ridge crest makes a broad valley that follows a dogleg course across the island, its opposite walls moving an average of a few centimeters farther away from each other per year. Every few hundred years, a long **fissure** or crack opens in the floor of the valley and erupts a large basalt flow. Until 2010 the most recent such occasion was in 1783, when the Laki fissure erupted 12 cubic kilometers of basalt lava that spread 88 kilometers down a gentle slope. Given the events of the last 1,000 years of written Icelandic record it is not surprising that the spreading zone erupted again, producing basalt lava, along with ash that drifted over much of Europe. In that case water to drive the ash eruption came from the overlying glacier melted by the basalt magma.

In Iceland, as everywhere along the oceanic ridge system, peridotite, the hot dark rock of Earth's mantle, rises at depth to fill the gap between lithospheric plates that separate at an oceanic ridge. As the dry mantle rock slowly rises, the lessening load of overlying rock subjects it to progressively lower pressures that permit it to partly melt to make basalt magma (see **By the Numbers 6-1**). The lower density of the magma lets it rise to erupt at the ridge. Its low viscosity (high fluidity) permits it to spread out as a thin basalt lava flow. Similar flows erupt in the crest of the oceanic ridge system along its entire length, to build the basalts that make up the entire ocean floor.

Watch wax dribbling down a candle and think about molten basalt lava erupting on the ocean floor. The cooling dribble of wax soon acquires a thin skin of nearly solid wax. Then the molten wax within bursts through that skin and dribbles down a new path. As that happens again and again, the original dribble of wax forms a network of dribbles. Basalt lava erupting into water behaves the same way. A chilled skin of solid basalt forms on the outside of the flow; then the molten basalt within bursts out and pours off in a new direction. The result is a pile of basalt cylinders, about the size of small barrels, called **pillow basalt**. Exposed in a cliff or road cut, they look like a pile of oversized pillows in dull shades of greenish to brownish black (**FIGURE 6-5**, p. 144).

On continents, the largest flood-basalt flows are simply giant lava flows with volumes of several hundred cubic kilometers, more than 100 times those of ordinary basalt flows. Individual flows may cover tens of thousands of square kilometers to depths of 30 to 40 meters. Some in the Pacific Northwest of Washington, Oregon, and western Idaho cover areas almost the size of Maine.

Continental rifts or spreading zones, such as the Basin and Range of Nevada and vicinity, the Rio Grande Rift of New Mexico, and the East African Rift Zone, move apart at much slower rates than those on the ocean floor. The mantle rocks under the Basin and Range and Rio Grande Rift rise more slowly, so their decrease in pressure is slower,

FIGURE 6-5 PILLOW BASALTS

David Hyndman.

A thick pile of pillow basalts from uplifted ocean floor in Oman.

and they melt at lower rates and produce much less magma. Much of the magma erupted through the continents in the last million years or so rose along rift-zone faults. Some of it formed cone-shaped piles of basalt cinders a few tens or hundreds of meters high, called **cinder cones**, along with associated lava flows. These are not particularly dangerous. Fewer but much larger and more violent eruptions spread incinerating flows of rhyolite over large areas; those would have been catastrophic for any nearby populations.

Subduction Zones

Subduction zones, the locations where oceanic plates slide under either oceanic or continental plates, are widespread and spawn most active volcanoes (see FIGURE 2-10). Volcanoes formed in this environment are the most spectacular and most hazardous on Earth. Here, cold ocean-floor lithosphere (basaltic crust and hydrated upper mantle) collides with and descends beneath warmer, lower-density ocean-floor or continental rocks. Cold ocean-floor rocks formed tens of millions of years ago at oceanic rift zones; ocean water percolating deep in cracks warms up at depth, circulates widely, and over millions of years combines with the minerals in the crustal basalt and mantle peridotite. By the time these altered rocks descend into the oceanic trench, they contain enough water to make a difference in the way magmas are produced.

The descending oceanic plate or slab slowly heats up and, at depths of 100 km or so, begins to boil off some of the water. The water rises into hot mantle rocks under the continental crust (see FIGURE 2-12). In that environment, the melting boundary for the hot mantle rocks moves to lower temperatures because of the added water.

Whenever mantle peridotite melts, it forms basalt magma; this new lower-density magma rises into the overlying crustal rocks, most commonly continental crust. Basalt magma is hot enough to melt granitic composition rocks of the continental crust to form rhyolite magma. Thus the basalt and rhyolite magmas intermingle. Either or both can rise to Earth's surface to erupt in a volcano. If the two magmas mingle and mix, they form magma of intermediate composition—andesite—that can also rise to erupt in the same volcano. Thus, subduction-zone volcanoes, including those in the Cascade Range of western North America, lie on the continental side of oceanic trenches and are characterized by basalt, andesite, and rhyolite compositions. Proportions of those rocks depend on a variety of factors, including the composition and thickness of continental crust. Magmas are commonly generated near a depth of 100 kilometers on the inclined subducted slab. Volcanoes above that location will be farther inland if the inclination of the slab is gentler, and closer to the trench if the inclination is steeper.

Because magma in the subduction-zone environment is generated due to the release of water from the subducted slab, the magmas formed generally contain some water. That water contributes to the eruption behavior of the magmas. Instead of flowing out quietly as lavas, water in the magma near the surface separates into gas bubbles that expand rapidly under the lower pressure. The expanding bubbles blast the magma into fragments in a violent eruption.

Hotspots

Hotspot volcanoes are far fewer in number (see FIGURE 2-3) but generally produce large volumes of magma. Instead of appearing at plate boundaries, they grow within tectonic plates at what appear to be random locations. Because an active hotspot volcano lies at the end of a series of older inactive volcanoes, the source of the magma must be in the underlying relatively stationary asthenosphere, rather than in the moving lithospheric plate, as discussed in Chapter 2. Melting of mantle peridotite under an ocean basin produces dry basalt magma that rises through only peridotite and basalt-composition rocks, so it continues to the surface to erupt as basalt lava. Basalt magma derived from deep, dry mantle peridotite is not inherently explosive because it contains no water. Thus a hotspot volcano in an ocean basin, such as in Hawaii, typically erupts basalt lavas that flow out quietly without blasting into fragments. If the magma picks up some groundwater en route, the water can vaporize and blast the magma into fragments during eruption.

Although few hotspot volcanoes appear on the continents, Yellowstone is a prominent example. As with oceanic hotspot volcanoes, the basalt magma source must be in the mantle, but this magma rises through thick continental crust. Little of the basalt magma makes it all the way to the surface, but its intense heat melts the silica-rich continental crust that has a lower melting temperature. That new magma rises to erupt at the surface as light-colored rhyolite lava; its dissolved water drives explosive eruptions.

Volcanic Eruptions and Products

Characteristics of volcanoes depend on the same factors that control the nature of eruptions. The magma volume, viscosity, and volatile content control the size of the volcano, the steepness of its slopes, and its eruption products. Depending on its style of eruption, a volcano can produce lava, pyroclastic materials (air-fall ash, pumice, and pyroclastic flow deposits), and lahars (volcanic mudflows) (**FIGURE 6-6**). Table 6-2 gives an overview of volcanic materials.

FIGURE 6-6 LAVAS, ASH, AND MUD

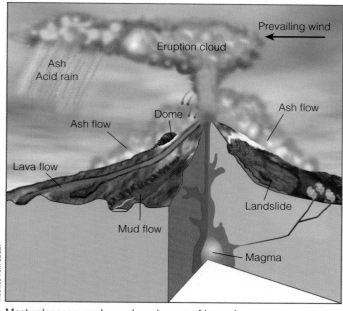

Most volcanoes produce a broad range of hazards.

Nonexplosive Eruptions: Lava Flows

Most basaltic magmas do not explode, because they are fluid and contain modest amounts of gas, mainly water and carbon dioxide, which bubbles harmlessly out of the melt. Instead, basaltic magma tends to spill out of a volcano and flow down its sides in the form of a **lava flow**. Lava flows fed by a hidden magma chamber may spill out from a central **crater** (a depression created as ash blasts out) or pour from a spreading crack or rift on the flank of the cone.

Where fluid basalt lava flows dominate eruptions, the flows are sometimes called **Hawaiian-type lava**, after one of the places that best characterizes them. Some of those basalt flows have a smooth, ropy, or billowy surface generally called **pahoehoe**, a Hawaiian term pronounced *pah-hoy-hoy* (**FIGURE 6-7A**, p. 146). *Pahoehoe* surfaces develop on lavas rich in steam and other volcanic gases. The ropy form develops when fluid lava drags a thin cooling skin into small wrinkles or folds. These flows commonly develop open passages as molten lava escapes from under a solidified crust several centimeters to a meter or two thick. This crust is important because it makes naturally insulated pipes through which lava can flow for several kilometers without much cooling. The molten lava can then break out downslope and again flow on the surface. People walking across recently erupted but solidified lava flows risk breaking through a thin crust and falling into the molten lava. Because of their low viscosity, basalt flows spread out easily and solidify on gentle slopes.

Basaltic lavas that are charged with less steam and other gases, and that have partially crystallized, develop an extremely rubbly and clinkery surface called **aa**, another Hawaiian term, pronounced *ah-ah*. An *aa* surface (**FIGURE 6-7B**, p. 146), with its sharp and rough rocks, is easily capable of ruining a nice pair of boots during a short walk. The rubble tumbles down the slowly advancing front of the flow, which runs over it the way a bulldozer lays down

Table 6-2		Generalized Products of Volcanoes	
	LAVA	**PYROCLASTIC MATERIAL (ASH, PUMICE, OTHER FRAGMENTS)**	**LAHARS**
DEFINITION	Molten magma that flows out and onto Earth's surface	Fragments and shreds of solidified magma blown out of a volcano. May be deposited by a pyroclastic flow or by air-fall ash	Volcanic ash and other fragments transported downslope with water (mudflows)
GENERAL CHARACTERISTICS	Molten magma that solidifies as coherent sheets or broken jumbles of volcanic rock	Fragments range from less than 2 mm ash to tens of cm. across. Larger pieces may be broken from older volcanic rocks on the sides of the vent.	Angular to rounded; unsorted particles from mud to boulders

FIGURE 6-7 LAVA TYPES

A Pahoehoe

B Aa

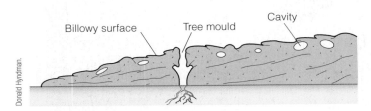

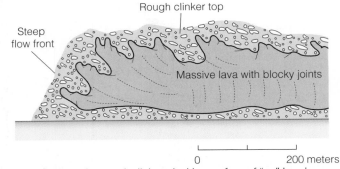

A. Smooth-topped pahoehoe lava in cross section (bottom) and flowing on Kilauea (top). **B.** A ragged, clinkery-looking surface of "aa" lava in cross section (bottom) and from Mauna Loa Volcano, Waikoloa, Hawaii (top).

and runs over its tread. *Aa* lavas flow more slowly and form thicker flows than *pahoehoe*.

Explosive Eruptions: Pyroclastic Materials

Explosive eruptions produce fragments of solidified magma known generally as **pyroclastic material**. The more viscous the magma and the greater its gas content, the more likely it is to explode. Steam bubbles separate with difficulty from rhyolitic or dacitic magma as it approaches the surface; the froth expands to form a foam of glassy bubbles called pumice. The bubbles expand as the magma continues to rise until they burst into ash, which consists mostly of curving shards of glass, formerly the walls of bubbles resembling soap bubbles. When the pumice bursts into ash, the steam escapes, and the whole mass explodes into a cloud of volcanic gases and suspended ash (**FIGURE 6-8**).

Finer particles of pyroclastic material—volcanic ash—may rain down from an eruption cloud or be blown downwind as air-fall ash. Heavier parts of the eruption cloud may collapse onto the volcano flank as a **pyroclastic flow**. Pyroclastic flows may also spill downslope directly from the crater rim. Pyroclastic flows sometimes travel distances of 20 kilometers—in a few cases, much farther,

destroying everything in their path (**FIGURE 6-9**). Continuing expansion and expulsion of steam, and heating and expansion of air trapped below the ash cloud maintain the internal turbulence and high speeds of pyroclastic flows. If ash and fragment deposits on a volcano's flank soak up sufficient water from rain or melting snow, the mixture may pour downslope as a **lahar**, or mudflow (**Case in Point:** Deadly Lahar—Mount Pinatubo, Philippines, 1991, p. 157).

On March 18, 2007, partial collapse of Mt. Ruapehu's crater rim, on the north Island of New Zealand, released a large lahar. Seepage through the crater rim following several days of heavy rain caused its collapse, releasing about 1.3 million cubic meters of water flowing at 2,500 cubic meters per second in a valley 7 kilometers downstream. Because the area is almost uninhabited, there was little damage.

The size of an ash eruption depends on many factors, including the amount of magma, the magma viscosity, and its water-vapor content. The **Volcanic Explosivity Index** (VEI) crudely quantifies eruption size, volume, and violence. Steam explosions eject particles of magma at velocities that depend upon the amount and pressure of the steam and the narrowness of the vent. Those factors dictate the rate at which the steam can expand and escape.

FIGURE 6-8 SAINT HELENS BLOWS

A. This explosive eruption of Mount St. Helens on July 22, 1980, pushed an ash cloud to a height of 14 kilometers, dwarfing the mountain visible in the lower right. **B.** A fragment of volcanic ash from the main Mount St. Helens eruption in 1980, magnified about 3,000 times.

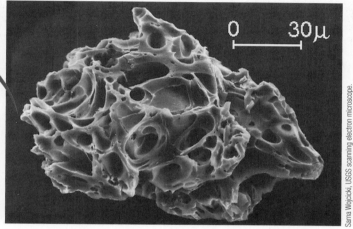

James Valance, USGS.

Vent

Volcano

A

Sarna Wojcicki, USGS scanning electron microscope.

B

FIGURE 6-9 HOT ASH CHARS TREES

Donald Hyndman.

Charred logs

Charcoal logs are all that remain of the forest that was overwhelmed by the Taupo ash-flow eruption in New Zealand some 2,000 years ago. The pyroclastic flow must have been hot enough to burn the logs.

Escaping gas slows as it expands and cools. Erupting steam commonly moves between 400 and 700 kilometers per hour but may exceed the speed of sound in air—1,200 kilometers per hour. Steam explosions may throw large blocks and blobs of magma, called *bombs*, as far as 5 to 10 kilometers. Smaller particles drift farther downwind.

Styles of Explosive Eruptions

Styles of explosive eruptions range from frequent and mild to infrequent and violent, as classified by the VEI (**Table 6-3**, p. 148). Many are typified by Italian volcanoes that provide their names.

PHREATIC AND PHREATOMAGMATIC ERUPTIONS **Phreatic eruptions** are violent steam-driven explosions generated by vaporization of shallow water in the ground, a nonvolcanic lake, a crater lake in a volcano, or shallow sea. The result is often a maar, a broad, bowl-shaped crater encircled by a low rim that commonly rises only slightly above the surrounding terrain. Steam dominates; magma is not erupted. If magma incorporates groundwater, it causes a phreatomagmatic eruption. In either case, such water-rich eruptions can be

VEI	VOLUME OF EJECTA (KM³)	ERUPTION COLUMN HEIGHT (KM)	ERUPTION STYLE	DURATION OF CONTINUOUS BLAST (HRS)	ERUPTION FREQUENCY (APPROXIMATE)	EXAMPLE ERUPTION
						Volcanic Explosivity Index (VEI) for Different Styles of Individual Eruptions
0	<0.0001	<0.1 (100 m)	Hawaiian	<1		Kilauea, Hawaii
1	0.0001–0.001	0.1–1	Hawaiian Strombolian	<1	100 per year	Kilauea, Hawaii Stromboli, 1996
2	0.001–0.01	1–5	Strombolian Vulcanian	<1	15 per year	Unzen, Japan, 1994
3	0.01–0.1	3–15	Vulcanian	1–6	2–3 per year	Nevado del Ruiz, Columbia, 1985
4	0.1–1	10–25		6–12	1 / 2 years	Iceland, 2010 El Chichon, 1982 Papua New Guinea, 1994
5	1–10	>25	Plinian	>12	1 / 10 years	Mount St. Helens, 1980
6	10–100	>25	Plinian	>12	1 / 40 years	Krakatau, 1883 Pinatubo, 1991 Thera (Santorini), 1600 B.C.
7	100–1,000	>25	Plinian	>12	1 / 200 years	Tambora, Indonesia, 1815
8	>1,000	>25	Yellowstone	Off scale	1 / 2000 to 1 / 1,000,000 years	Yellowstone, 600,000 B.C. Long Valley, California, 730,000 B.C. Taupo, New Zealand, 186 B.C.

Increasing volume → Increasing eruption height → Increasing duration → Decreasing frequency

especially dangerous; the erupting column of steam and ash can collapse to form a base surge that sweeps rapidly outward. The surge carries hot sand and rock fragments that can sandblast and uproot trees and overwhelm and kill people.

STROMBOLIAN ERUPTIONS Stromboli, off the west coast of Italy, is fed by magma that interacts with groundwater or seawater. Rapidly expanding steam bubbles in the magma blow it into cinders and bomb-size blocks that fall around the vent and tumble down steep slopes to form a cinder cone. This is a **Strombolian eruption** (see FIGURE 6-17). The fluid nature of the magma is associated with generally mild eruptions.

VULCANIAN ERUPTIONS Vulcano, an island off the north coast of Sicily, is fed by highly viscous, andesitic magmas that are rich in gas. Dark eruption clouds blow out blocks

of volcanic rock, along with ash, pea-size *lapilli*, and bombs. With a **Vulcanian eruption**, ash falls may dominate, but pyroclastic flows and lateral-blast eruptions can develop with—or follow—the ash fall.

PELÉAN ERUPTIONS Mount Pelée in Martinique, the West Indies, erupted in 1902 to obliterate the town of St. Pierre and its 28,000 inhabitants (see Chapter 7, **Case in Point:** Pyroclastic Flows Can Be Deadly—Mount Pelée, Martinique, West Indies, p. 194). Its rhyolite, dacite, or andesite eruptions can be violent, especially early in the eruption, when high columns of ash may be ejected. **Peléan eruptions** are characterized by giant ash columns that collapse to form incandescent pyroclastic flows. Most distinctive in the 1902 eruption, however, was the growth of Pelée's huge, steep-sided spine of solidified magma that the magma pushed up from the vent. At intervals, the sides of the expanding dome collapsed to form searing hot block and pyroclastic flows. Occasionally, the magma below slowly pushed a viscous lava spine outward from the surface of the dome, sometimes as much as 300 meters high.

PLINIAN ERUPTIONS The great eruptions of Vesuvius in a.d. 79 and Mount St. Helens in 1980 produced powerful continuous blasts of gas that carried huge volumes of pumice high into the atmosphere. Even larger than Peléan eruptions, **Plinian eruptions** can be truly catastrophic for any nearby population. Silica-rich ash falls accompany incinerating pyroclastic flows, including those consisting mostly of pumice flows. Ejection of a large volume of magma often causes collapse of the magma chamber to form a **caldera**, as in the case of Mount Mazama (Crater Lake), Oregon, 7,700 years ago; Santorini, Greece, in approximately 1620 B.C. (**Case in Point:** Long Periods Between Collapse—Caldera Eruptions—Santorini, Greece, p. 160); Krakatau in Indonesia in 1883; and Tambora in Indonesia in 1815 (**FIGURE 6-10**).

Small eruptions typically form craters at the **eruptive vent**; a depression or crater forms where the vent is excavated by the violent eruption. The distinction between crater and caldera is not so much size but the mechanism of depression formation. A cinder cone blows out vent material to form a crater. A giant Plinian eruption, such as the one that formed Crater Lake, Oregon, 7,700 years ago, collapses into the emptying magma chamber to form a caldera. To be precise, Crater Lake should be called Caldera Lake. Similarly, the eruption of the Bishop tuff and collapse of the Long Valley caldera about 760,000 years ago was a giant Plinian eruption.

Types of Volcanoes

Based on the appearance of a volcano, primarily its size and the slopes of its flanks, you can infer the magma composition that produced it and its volatile content; both reflect the eruptive style and the types of associated hazards and risks (**Table 6-4**). In this section, we describe the main differences between volcano types, using examples to illustrate the range of behaviors and hazards.

Shield Volcanoes

Persistent basalt eruptions of very fluid basalt lavas within a small area eventually build a gently sloping pile of thin flows. More than a century ago, geologists decided to call those piles **shield volcanoes** because of a fancied resemblance to the shape of an ancient Roman shield (**FIGURE 6-11**, p. 150). The flows are characterized by their low viscosity, low volatile content, broad and gently sloping sides, and large to giant volumes.

Most of the lava erupted from shield volcanoes does not flow from the peak but rather from the crests of three broad, equally spaced ridges that radiate outward from the volcano summit (**FIGURE 6-12**, p. 151). Gravity pulling down on the flanks of each ridge causes a slight spreading, which causes rifting of the ridge crest. Most of the lava rises within a rift, spreads out, and flows down the flanks of the ridge (**FIGURE 6-13**, p. 151). The shield volcano may have a caldera in the summit where the surface rocks sank into the magma chamber below. Many have cinder cones or lava cones along the crests of the three main **eruptive rifts**, sometimes flowing downslope to

FIGURE 6-10 DEADLY SEARING ASH

A. Pyroclastic flows from Mount Tambora in Indonesia killed more than 10,000 people in a giant eruption in 1815. **B.** Artifacts excavated from ash from the 1815 eruption. People and their communities were buried in the eruption.

Table 6-4			General Characteristics of Common Volcanoes				
VOLCANO TYPE	TECTONIC ENVIRONMENT	VISCOSITY	SLOPE OF FLANKS	VOLATILES	VOLUME (SIZE)	ERUPTION STYLE
Shield volcano (basalt), Figs. 6-11 and 12	Oceanic hotspots, some volcanic chains	Low	Gentle	Low	Large to giant	Quiet lavas
Cinder cone (basalt), Figs. 6-11, 17, 18, 19	Flanks of shield volcanoes, continental rifts	Low	Steep	Moderate	Small	Explosive but not very dangerous
Stratovolcano (andesite), Fig. 6-10, 11, 20	Above active subduction zones	Moderate	Moderate	Moderate to high	Moderate	Violent, dangerous pyroclastic flows and air-fall ash
Lava dome (rhyolite)	Above active subduction zones	High	Steep	Low to moderate	Small to moderate	Dangerous dome collapse and pyroclastic flows
Continental resurgent caldera (rhyolite), Fig. 6-22	Continental hotspots, some continental rift zones	High	Very gentle	Moderate	Giant	Violent, dangerous pyroclastic flows and ash

FIGURE 6-11 VOLCANOES IN ALL SIZES

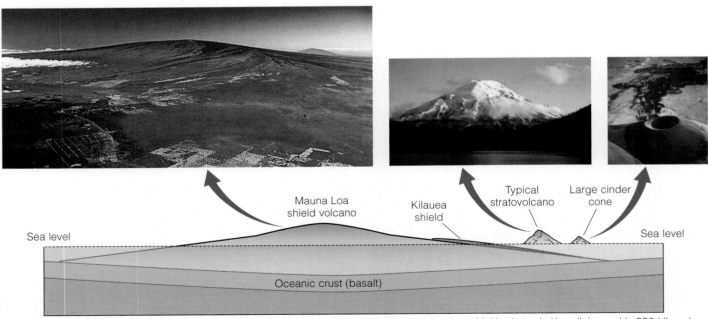

The different types of volcanoes have dramatically different sizes (or volumes). Mauna Loa, the giant shield volcano in Hawaii, is roughly 220 kilometers in diameter and rises some 9,450 meters above the seafloor; the 4,169 meters above sea level is less than half its total height. The gentle slopes of Mauna Loa are typical of giant shield volcanoes. A typical subduction-zone volcano such as Mt. Rainier, on a base near sea level is only 2,000 to 3,000 meters high. Cinder cones are still smaller.

FIGURE 6-12 A VOLCANO SAGS AND SPLITS

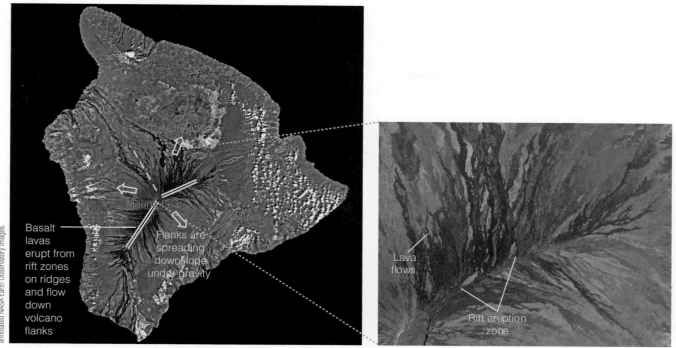

Basalt lavas erupt from rift zones on ridges and flow down volcano flanks

Flanks are spreading downslope under gravity

Lava flows

Rift eruption zone

annotated NASA Earth Observatory images.

Mauna Loa, the dominant volcano on the Big Island of Hawaii, erupts along three spreading rift zones. The sagging flanks (three arrows) of the crudely three-sided volcano maintain spreading at the three rifts, permitting basalt lavas to erupt and flow, like dark fingers, downslope from the rifts.

FIGURE 6-13 FREQUENT ERUPTIONS

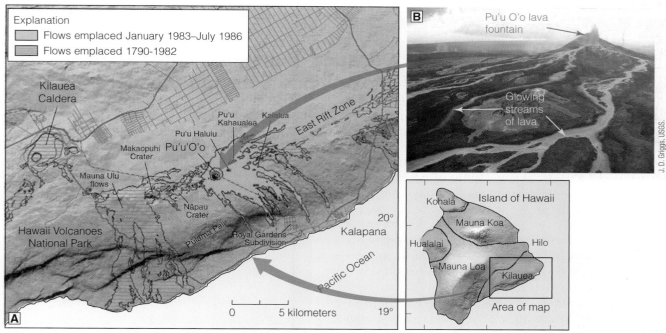

Explanation
- Flows emplaced January 1983–July 1986
- Flows emplaced 1790-1982

Kilauea Caldera

Pu'u Kahaualea

Kalalua

East Rift Zone

Pu'u Halulu

Makaopuhi Crater

Pu'u'O'o

Mauna Ulu flows

Nāpau Crater

Hawaii Volcanoes National Park

Pūlama Pali

Royal Gardens Subdivision

20°

Kalapana

Pacific Ocean

0   5 kilometers   19°

B Pu'u O'o lava fountain

Glowing streams of lava

J. D. Griggs, USGS.

Island of Hawaii

Kohala

Mauna Koa

Hualalai

Hilo

Mauna Loa

Kilauea

Area of map

Helikar et al., 2003, USGS.

This map of Kilauea shows the summit caldera, the southwestern and eastern rift zones that erupt most of the lava flows, and major fault zones on which the south side is sliding toward the ocean. Note that almost all eruptions occur along rifts along the main ridges, with lava flows spilling down their flanks. The flows in 1986 destroyed subdivisions of Royal Gardens and Kalapana. Eruptions of Pu'u O'o crater, Kilauea volcano, Hawaii, on the East Rift Zone send basaltic lava flows downslope.

populated areas. The largest volcano on Earth, Mauna Loa in Hawaii, rises 9,450 meters (31,000 feet) above its base on the floor of the Pacific Ocean, 4,270 meters (14,000 feet) above sea level. Somewhat smaller basaltic shield volcanoes include Newberry Volcano in central Oregon and Medicine Lake Volcano in northernmost California, both just east of the Cascades. Small shield volcanoes may cover as little as 10 to 15 square kilometers and rise as little as 100 meters above the surrounding countryside.

MAUNA LOA AND KILAUEA: BASALT GIANTS OVER AN OCEANIC HOTSPOT

Mauna Loa, Hawaii's largest volcano, has erupted 33 times since 1843. Recent eruptions include those in 1926, 1940–1942, 1949–1950, and 1975. Seven of its flows, including one in 1984, came within 6.5 kilometers of Hilo. Mauna Loa and Kilauea exhibit the mildest of eruptions, with VEIs of 0 to 1 (see Table 6.3).

Oceanic hotspot volcanoes go through three stages of activity. The first is a long series of eruptions below sea level that build the broad base of the volcano, a great heap of volcanic rubble with little mechanical strength. In the second or main stage, eruptions produce basalt lava flows that build the main, visible mass of the volcano. The late stage of activity comes as the volcano moves off the hotspot. Eruptions become smaller and less frequent. Before major eruptions, volcanoes slowly swell at a few centimeters per year with the pressure of inflating magma chambers.

Kilauea is the successor to Mauna Loa, the new volcano at the active end of the hotspot track. It is much younger than Mauna Loa, still just a small fraction of its size, and produces much smaller eruptions. Kilauea looks like a big ledge on Mauna Loa's southeastern flank. Kilauea is now in its main growth stage. Approximately 95 percent of the part of Kilauea above sea level has grown within the last 1,500 years. It has erupted more than 60 times since 1832. Some of its more recent eruptions happened in 1955, 1961–1974, 1977, and almost continuously from 1983 to 2010 as the main activity centered on Pu'u O'o on the East Rift. Most erupted material is lava, but cinder cones blast out bubbly chunks of basalt that tumble downslope.

Even though Kilauea is still young, its edifice has already split into pie-slice segments that have begun to spread—only two, because one side is buttressed against Mauna Loa. Most of its eruptions are along the **rift zones** that radiate from the summit and separate the segments. The southern flank of Kilauea broke along a scarp more than 500 meters high, Hilina Pali, and has begun slowly sliding toward the ocean.

Like many volcanoes, Kilauea announces an impending eruption with swarms of small earthquakes that originate at shallow depths, along with harmonic tremors that record magma movement. Sensitive tiltmeters may detect an inflation of the volcano summit as magma pressure increases. In some cases, these symptoms raise false alarms.

Kilauea's eruptions typically produce only basalt. Although lava flows cross and burn roads (**FIGURE 6-14**) and can torch buildings, they rarely injure or kill anyone. But hot gases, presumably water vapor, sulfur dioxide, and carbon dioxide, did drive a surge in 1790 that killed 80 warriors from King Keoua's army as they crossed a high flank of Kilauea. In February, 2007, 30 scientists and naturalists exploring underground tunnels in a Canary Islands' shield volcano off northwestern Africa got lost in the maze of caverns; 6 suffocated, probably from breathing carbon dioxide, and 6 were hospitalized. In March, 2008, a gas vent on the wall of Halema`uma`u Crater, the lava lake near the summit of Kilauea Volcano, became active and exploded, ejecting steam and ash (**FIGURE 6-15**). Hazardous sulfur dioxide gas in "vog" (volcanic gas and fog) required evacuation of nearby communities. Lava was seen sloshing in the vent in the floor of Halema`uma`u Crater in December, 2009.

MOUNT ETNA, SICILY

The huge mass of Mount Etna broods over Catania and other cities in eastern Sicily. It is 3,315 meters high, the largest continental volcano on Earth. Except for Stromboli, it is also the most active volcano in Europe, typically erupting basalt lava flows and cinders (**FIGURE 6-16**). Etna stands on an unstable base of soft muds deposited from seawater onto oceanic crust.

Flank eruptions from rifts that divide the volcano into three large pie slices have built three radial ridges similar to those of the Hawaiian volcanoes. Escaping gases blow cinders out of vents in these ridges, forming basalt cinder cones that also quietly produce lava flows that burst from the bases of the cones. People easily avoid the flows, but their houses cannot; ruined houses litter the flanks of the volcano (FIGURE 6-16a).

FIGURE 6-14 HOT BASALT MAGMA BURNS A ROAD

USGS.

Creeping pahoehoe lava from Kilauea flows over Kalapana road in October, 2009. The black, smooth-topped basalt on the surface of the flow covers the red-hot molten basalt.

FIGURE 6-15 A NEW VENT ON KILAUEA

On June 30, 2009, in the crusted-over lava lake in the floor of Halema`uma`u Crater on Kilauea, a new vent erupts gray ash and steam. White steam contains little or no ash.

Etna's almost continuous—but frequently erratic—eruptions feature occasional violent episodes. On July 22, 1998, for example, one of the craters produced an eruption column 10 kilometers high that dumped ash over a wide area to the east. It has also staged at least 24 sub-Plinian eruptions in the past 13,000 years. Plinian eruptions of basalt volcanoes are rare but hazardous, especially with towns creeping ever higher on Etna's flanks. A Plinian erup-tion in 122 B.C. wreaked havoc on the Roman town of Catania 25 kilometers south of Etna. The main Plinian deposit was 20 centimeters of coarse ash that started fires and collapsed roofs. As the eruption rate dropped, magma apparently boiled ground-water into steam that carried clouds of ash high into the air.

The low viscosity and low water content of basaltic mag-mas normally prevent Plinian eruptions that are dominated by voluminous ash. Etna's may develop when the magma

FIGURE 6-16 MOUNT ETNA

This is all that remains of a house that was in the path of a 1983 Mount Etna lava flow north of Nicolosi, Sicily. Mount Etna erupted a prominent plume of dark ash on October 30, 2002, while fires were ignited by lava pouring down its north flank. Light-colored plumes are gas emissions from a line of vents along a rift extending out from the summit. Roads and towns on the flanks of the vol-cano are faintly visible in the upper right and across the bottom of the photo.

rises so rapidly that gases cannot slowly escape as they normally do. Instead, bubbles of steam explode within the rapidly decompressing basalt magma.

Cinder Cones

Cinder cones are also basalt, but they are characterized by their small size, low viscosity, steep sides, and moderate volatile content (**FIGURE 6-17**). They erupt where rising basaltic magma encounters near-surface groundwater, and escaping steam coughs cinders of bubbly molten lava out of a vent or summit crater (**FIGURES 6-18, 6-19**). These cinders fall around the vent and build a loose, steep-sided pile.

FIGURE 6-17 CINDER CONE VOLCANO

Sunset Crater, near Flagstaff, Arizona, a typical cinder cone, shows a smooth-sided cone capped by a central crater.

The smallest shreds drift downwind in a black cloud and fall as basalt ash. When the water in the ground dries up, a single basalt lava flow pushes out from the base.

Along extensional faults, such as the Great Rift at Craters of the Moon in southern Idaho, basalt also rises to feed lava flows. Where it encounters water, it flashes into steam to build cinder cones.

Most cinder cones erupt over only one short period, a few months to a few years. They typically build a pile of cinders 100 to 200 meters high. When the supply of steam is exhausted, a basalt lava flow generally erupts from the base of the cone. The lava flow emerges from the base in the same way that water poured on top of a pile of gravel emerges from the bottom of the pile. The next eruption in the area will probably build a new cinder cone where water is present rather than reactivate an old one. Many volcanic areas, such as Haleakala on Maui in Hawaii, Lassen in northern California, and Newberry in Oregon, are liberally peppered with cinder cones, many visible within a single view.

Cinder cones provide an exciting nighttime fireworks display as glowing "bombs" and cinders arc through the air and then roll, still glowing, down flanks of the cone (FIGURE 6-18). The upper parts of lava flows look like glowing red streams running downslope; the lower parts darken as they cool.

Most people are sensible enough to evade the rain of glowing cinders from a growing cinder cone and agile enough to avoid cremation in its lava flows. As a result, these eruptions endanger only property.

The dark basalt ash that drifts downwind from erupting cinder cones eventually weathers into fertile soil that supports abundant and nutritious crops if water is available. Such areas generally support large agricultural communities.

FIGURE 6-18 CINDER CONE ERUPTION

 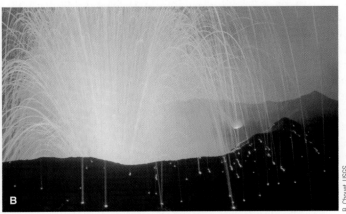

A. Red-hot cinders blasted out of the vent fall and tumble down the cone of Paricutín Volcano in Mexico. **B.** A close view of glowing cinders blown out from Stromboli's crater.

FIGURE 6-19 AN OLD PILE OF CINDERS

Donald Hyndman.

Cross-section exposure of a cinder cone showing layers parallel to slope and scattered large volcanic bombs. Northern California. Volcanic bombs in the cinder cone were twisted and tapered as the molten magma flew through the air. The two large bombs, broken open and filling most of the inset view, show chilled rims enclosing gas bubbles that could not escape through the rims. The red color forms later as volcanic gases percolate up through the porous pile. The pocket knife in the lower left is 8 centimeters long.

Stratovolcanoes

When most people imagine a volcano, they envision a **stratovolcano**—a large, steep-sided cone over a subduction zone (**FIGURE 6-20**). Stratovolcanoes, also called composite volcanoes, are characterized by their moderate volume and size, moderate viscosity and slope, and moderate-to-high volatile content. Big stratovolcanoes, such as those of the High Cascades, are dominated by andesitic compositions, but darker varieties range to basaltic andesite; lighter varieties range to dacite to rhyolite. The names stratovolcano and composite volcano both convey the fact that they typically consist of layers of lava flows, fragmental debris, and ash. The slopes of the flanks of stratovolcanoes depend on two factors:

1. Their magma has moderate viscosity, so lavas are not especially fluid; they flow only on moderately steep slopes before they cool and solidify.

2. Escape of dissolved volatiles through viscous magma typically causes large eruptions of ash and broken

rubble that concentrate near vents and tumble to form slopes of about 30 degrees; this builds the cone higher near the vent.

Lava flows give a stratovolcano's cone enough mechanical strength to hold an internal column of magma. This enables the volcano to erupt repeatedly from a summit crater and permits it to grow into a tall cone.

Examples of stratovolcanoes include the High Cascades of the western United States and Canada, and Mount Vesuvius in Italy (see the subsection in Chapter 7 on Vesuvius and Its Neighbors, p. 181) and Mount Fuji in Japan. Most stratovolcanoes grow in long chains above a slab of ocean floor that subducts into the interior of the Earth. The volcanic chain, generally between 75 and 200 kilometers inland, trends parallel to the oceanic trench that is swallowing the ocean floor.

Many large stratovolcanoes begin growing with a large basalt shield and then build a tall andesite cone on that base. Finally, they may erupt dacite or rhyolite, which may end in a massive explosion that destroys their andesite cone. In some of those cases, renewed activity builds a new andesite cone in the ruins of the old one. The eruptive behavior and intervals between eruptions vary widely. Mount St. Helens has erupted at least 14 times in the last 4,000 years. Mount Lassen, in northern California, remained almost dormant for 27,000 years before a moderate eruption about 1,000 years ago and another in 1914. An absence of activity for tens of thousands of years does not indicate that a volcano is no longer active.

Lava Domes

Lava domes are rhyolitic volcanoes characterized by their small-to-moderate size, high magma viscosity, steep flanks, and low-to-moderate volatile content. Rhyolite to dacite magmas sometimes erupt with little steam. They emerge

FIGURE 6-20 STRATOVOLCANO

Alaska Volcano Observatory, USGS.

Mount Griggs in the Aleutian Range of Alaska is a stratovolcano.

slowly and quietly expand over months or years, like a giant spring mushroom, to make a small mountain. As the eruption proceeds, the lava on the outside of the dome solidifies while the still-molten, but extremely-viscous, lava within continues to rise. The solid rock on the outside cracks off in pieces and tumbles down the side of the growing dome, making steep talus slopes of angular rubble. The typical result is a steep mountain so cloaked in sliding rubble that solid rock is exposed only on the summit.

A single lava dome may erupt only once, though it may be replaced by another dome as magma below continues to rise. Collapse of a big bulge growing on the flank of Mount St. Helens in 1980 released pressure on the underlying magma. That was the final trigger for its catastrophic eruption. Sometimes magma that solidifies in the throat of a stratovolcano is slowly extruded as magma below continues to rise; that extruded magma may boil out and pour down the volcano flanks as a pyroclastic flow. Many pyroclastic flows develop in the collapse of an expanding volcanic dome (**FIGURE 6-21**). When a dome collapses to form a pyroclastic flow, it is extremely dangerous for anything in its path.

Giant Continental Calderas

Giant **continental calderas** are rhyolite volcanoes characterized by high viscosity and high volatile content but gently sloping flanks due to a predominant ash content spread over large areas. Few of the millions of people who visit Yellowstone National Park realize they are on one of the world's largest volcanoes. The Yellowstone Volcano is a typical giant rhyolite caldera volcano (**Case in Point:** Future Eruptions of a Giant Caldera Volcano—Yellowstone Volcano, Wyoming, p. 161).

These giant volcanoes erupt rhyolite, typically in enormous volume, most of it explosively. Their pyroclastic flows cover tens of thousands of square kilometers, and sheets of airborne ash cover millions of square kilometers. The volumes of magma involved are typically in the range of hundreds to more than a thousand cubic kilometers.

As an emptying magma chamber withdraws support from the ground above, the surface collapses to open a broad caldera (**FIGURE 6-22**). Many large eruptions from dacite magma chambers open calderas up to 25 kilometers across. Eruptions from giant rhyolite magma chambers commonly open calderas 50 or more kilometers across. These eruptions generally fill the sinking caldera with enough rhyolite ash to ensure that only a subdued depression remains in the landscape.

As magma continues to rise beneath the filled caldera, it slowly raises a **resurgent dome** in its surface. The dome may become as large as a small mountain, and it may or may not develop into a new eruption, perhaps after hundreds of thousands of years. Some ancient rhyolite calderas still display an obvious resurgent dome. The resurgent dome of Long Valley Caldera is still rising, 0.75 meters in the past 33 years; bulging between 1982 and 1999 was likely caused by magma injection. Whether that foretells a new eruption is presently unknown.

Some giant rhyolite calderas erupt several times at intervals of hundreds of thousands of years. It is fortunate that their eruptions are so infrequent, because their pyroclastic flows would likely incinerate everything in valleys for 100 or more kilometers from the caldera. They would also almost certainly inject enough ash into the upper atmosphere to cause drastic climate change over a large region for years afterward.

FIGURE 6-21 SOURCES OF SEARING ASH

Continuous eruption with continuous or intermittent column collapse (e.g., Mount St. Helens, 1980, after initial blast)

Magma rises into vent with resulting collapse of the ash cloud.

Ash flow
Collapse of dome with or without gas explosion (e.g., Mt. Pelee, 1902; Unzen volcano, Japan, 1993)

Landslide   Magma eruption
Landslide of bulge releases pressure on magma, initiates eruption (e.g., Mount St. Helens, 1980)

Modified from Williams, 1932; Cas & Wright, 1988.

These four sketches show common mechanisms that generate pyroclastic flows.

FIGURE 6-22 COLLAPSE INTO A MAGMA CHAMBER

Cyrus Read, AVO, USGS.

S.R. Brantly, USGS.

Modified from Smith and Bailey, USGS.

Later lava dome within caldera

Approximate outline of caldera

B

C

A

Eruption of rhyolitic ash flows from ring fracture: partial evacuation of magma chamber

Caldera collapse along ring fracture zone
Pyroclastic flow deposits partly fill the caldera

Resurgent doming

Magma

Magma

Magma

A. The ground above an erupting rhyolite magma chamber subsides to make a caldera during the eruption of a giant pyroclastic flow; it then domes, or resurges, again as new magma refills the magma chamber. **B.** A small (2.5 km diameter) caldera in Kaguyak volcano on Katmai National Park, Alaska. A prominent lava dome rose in the caldera after collapse. **C.** The 32-kilometer-diameter Long Valley Caldera of southeastern California, as seen from rim to rim. Its tree-covered resurgent dome is behind the tree on the right.

Cases in Point

Deadly Lahar
Mount Pinatubo, Philippines, 1991 ▶

Mount Pinatubo is an andesitic volcano 1,745 meters high on the Philippine island of Luzon, some 90 kilometers northwest of Manila. On April 2, 1991, steam explosions suddenly piled ash on Pinatubo's upper slopes. This surprised and frightened the people who lived on the flanks of the mountain, the rice farmers on the plains below, and the 300,000 people

25 kilometers from the crater in Angeles City. Pinatubo had not erupted in more than 400 years, and few people were aware that it was an active volcano (in the sense that it has erupted repeatedly over thousands of years and will undoubtedly do so again).

Philippine volcanologists, with the help of scientists from the U.S.

Geological Survey, installed portable seismographs on and around the mountain to monitor its earthquake activity. Geologic mapping soon showed that an eruption 600 years ago had spread hot pyroclastic flows across densely populated areas south and east of the summit and over the site of Clark Air Force Base. Those pyroclastic flows reached

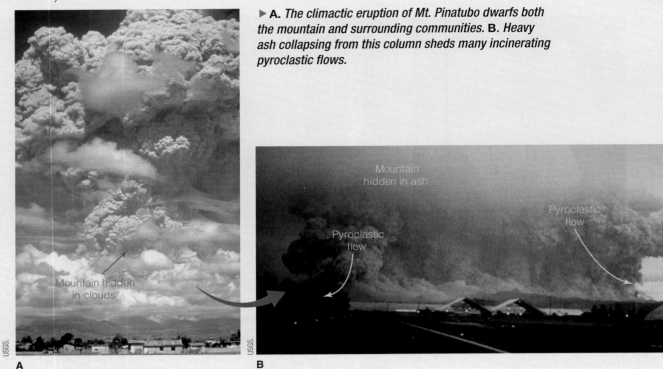

▶ **A.** *The climactic eruption of Mt. Pinatubo dwarfs both the mountain and surrounding communities.* **B.** *Heavy ash collapsing from this column sheds many incinerating pyroclastic flows.*

▶ *The satellite view, below, of Mount Pinatubo shows the distribution of mudflow deposits a few months after the eruption. The extent of one-centimeter-thick ash is shown in yellow dashes. The map of Mount Pinatubo shows the centimeter-depths of ash laid down between June 12 and 15, 1991, and the mudflow deposits two months after the eruption.*

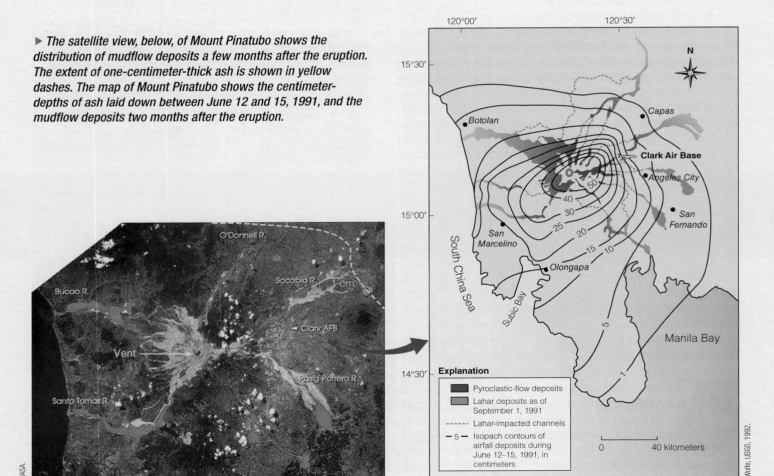

(continued)

John Major, USGS.

▶ *This village was buried in a mudflow after the eruption of Mount Pinatubo in 1991. Air-fall ash still mantles the roof on the right.*

T. J. Casadevall, USGS.

▶ *An air view over the Abacan River shows a bridge collapsed on August 12 by mudflows in Angeles City, Philippines, near Clark Air Force Base. In the lower left, people cross the river on temporary footbridges.*

20 kilometers east of the crater. Deep deposits of ash and widespread mudflows reached much farther.

At first, the blasts of steam and ash seemed fairly harmless to the volcanologists at the site because they contained only fragments of old altered rock, no freshly solidified ash from new magma. But then they saw the frequency of earthquakes increase and their origins migrate from deep below the north side of the volcano to shallow levels near the summit. By June 5, the numbers of small earthquakes and volumes of sulfur dioxide emissions had increased dramatically. Occasional pyroclastic flows swept down valleys. The volcanologists thought a major eruption could happen within two weeks. That led them to recommend evacuation of the area within 10 kilometers of the summit. The volcanologists worked closely with public officials, carefully explaining the looming dangers. Then the officials went to great lengths to educate the public.

A viscous lava dome began growing, and many small earthquakes and harmonic tremors suggested that magma was

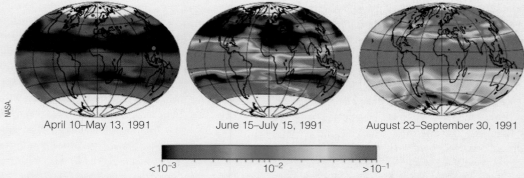

NASA.

| April 10–May 13, 1991 | June 15–July 15, 1991 | August 23–September 30, 1991 |

$<10^{-3}$ 10^{-2} $>10^{-1}$

▶ *Dust and SO_2 generated by the Pinatubo eruption encircled the Earth, scattering incoming sunlight and reducing global temperature. Pinatubo is at the red dot in left view.*

(continued)

moving at depth. Continuous ash eruptions accompanied expansion of the lava dome. On June 12, authorities evacuated people to a radius of 30 kilometers from the crater.

The climactic event, a classic Plinian eruption (VEI: 6), finally began early on June 12 with a blast and a huge plume of steam and ash. The eruption became continuous by early afternoon. It climaxed on June 15 when the eruption cloud towered to a height of 35 to 40 kilometers. Pyroclastic flows reached 16 kilometers from the old summit.

Ash was as much as 30 centimeters thick at a distance of 40 kilometers from the volcano. Then, in an unfortunate twist of fate, the climactic Plinian eruption coincided with passage of Typhoon Yunya,

which brought intense rains. Heavy loads of wet air-fall ash collapsed many roofs, and lahars rushed down nearby valleys. By June 16, another 200,000 people fled the mud. The rains continued to trigger mudflows every few days. By December 1991, almost every bridge within 30 kilometers had been destroyed.

Pinatubo erupted a total of 4 to 5 cubic kilometers of magma, leaving a crater 2 kilometers wide at its summit. Twenty million tons of sulfur dioxide gas combined with water in the atmosphere to make minute droplets of sulfuric acid. They hung in the air, encircling the earth and reflecting 2 to 4 percent of incoming ultraviolet radiation. Mean temperatures dropped as much as 1°C in parts of the northern hemisphere. Spectacular sunsets

with broad streaks of green continued for another two years.

In general, the efforts to predict and mitigate the hazards of a large eruption were an outstanding success. The volcano provided ample warning, the geologists correctly anticipated most of the major volcanic events, the local officials efficiently managed evacuations, and the local people cooperated.

In spite of a major program to educate everyone and evacuate 58,000 people, some 350 people died, mostly when heavy wet ash collapsed buildings. Another 932 died later from disease. The large death toll notwithstanding, there is no doubt that the timely warnings and broad evacuations saved tens of thousands of lives.

Long Periods Between Collapse—Caldera Eruptions
Santorini, Greece ▶

Santorini is an island volcano in the eastern Mediterranean Sea, south of mainland Greece. It now appears as a ring of islands, high places along the rim of an otherwise submerged caldera 6 kilometers across. Santorini is one of the most spectacular caldera volcanoes on Earth, similar in origin and size to Crater Lake in Oregon.

Santorini has staged twelve major explosive eruptions during the last 360,000 years, with a relatively long recurrence interval averaging one every 30,000 years. After each caldera collapse, a new andesite volcano grew within the old caldera until it sank into a new caldera during the next catastrophic eruption. Caldera eruptions happened 180,000; 70,000; 21,000; and 3,600 years ago. The intervals between eruptions become progressively shorter with time, much less than half the preceding interval.

In approximately 1620 B.C., a series of catastrophic Plinian eruptions of rhyolite ash and pumice erupted about 54.5 cubic kilometers of magma, evacuated the huge magma chamber, and culminated in the most recent caldera collapse. The volume was about the size of the 1815 eruption

of Tambora volcano in Indonesia, the largest eruption in historic time. The remnant flanks of the volcano slope gently outward from the much steeper cliffs that face into the caldera. Thera, the main town, is on the rim and its adjacent steep cliffs. Many of the homes and tourist hotels have rooms excavated from the caldera wall into the deep ash that fell in 1620 B.C.

Events during the initial stages of collapse included eruption of many meters of white rhyolite pumice. Its white lower part grades upward to dark gray andesite ash, presumably because the eruption was tapping progressively deeper levels of a differentiated magma chamber. A similarly graded pyroclastic flow exists around Crater Lake, Oregon (see Chapter 7, FIGURE 7-32).

Many of the rhyolite ash deposits contain inclusions of chilled basalt magma and compositionally banded pumice, indicating that basaltic magma injected the rhyolite magma chamber from below. That would cause rapid boiling and drive intense lateral blasts (see Chapter 7, FIGURE 7-10b) and a major Plinian eruption.

At the time of the eruption in 1620 B.C., the wealthy town of Akrotiri, on the lower

southern flank of the volcano, had paved streets, underground sewers lined with stone, beautiful wall paintings, decorated ceramics, and attractive jewelry. The people raised sheep and pigs, farmed using surprisingly modern techniques, made barley bread and wine, gathered honey, and imported olives and nuts.

When the volcano erupted, seawater pouring into the collapsing caldera probably caused tremendous steam explosions. Some estimates suggest that the eruption raised a plume of ash 36 kilometers into the atmosphere. It may have lasted for weeks. Most of the population of Akrotiri must have been sufficiently frightened to evacuate with their possessions before the main eruptions. Those who remained were buried under several meters of deposits dropped from the Plinian column, and another 56 meters of pumice and ash erupted during caldera collapse. Earthquakes,

(continued)

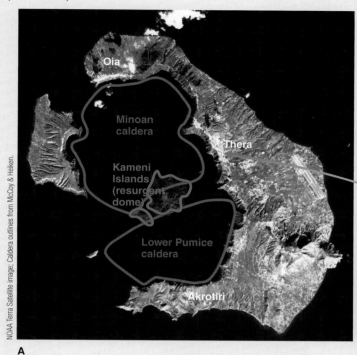

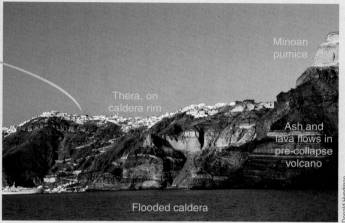

▶ **A.** *Santorini's main islands surround the caldera that collapsed during the great eruption of 1620 B.C. The eruption ended the Minoan civilization.* **B.** *The inside wall of Santorini caldera exposes white Minoan pumice at the top right. White houses along the lower caldera rim are built mostly into the same pumice erupted in 1620 B.C.*

A

B

ash falls, mudflows, and debris flows completed the destruction of buildings. One popular theory suggests that this eruption was the source of the Atlantis legend, in which earthquakes and floods accompanied the sudden disappearance of an island empire.

Renewed volcanic activity began in 197 B.C. and continued sporadically until 1950. Resurgent domes of andesite and dacite rose in the floor of the caldera. Earthquakes took a further toll in 1570 and 1672. In 1956, an earthquake of magnitude 7.8 wrecked Thera and raised a giant

tsunami wave that rose along the shores to heights between 25 and 40 meters. Hazard mitigation for Santorini since then involves restrictions on building on steep slopes of loose pumice in the caldera wall and monitoring of minor earthquakes and volcanic gases.

Future Eruptions of a Giant Caldera Volcano
Yellowstone Volcano, Wyoming ▶

A rhyolite giant over a continental hotspot, Yellowstone is a typical large resurgent caldera with a nearly flat summit and gently sloping sides, and it is one of the largest continental volcanoes on Earth, sometimes termed a supervolcano. The Yellowstone Volcano we visit today is the relic of three monstrous eruptions that occurred 2 million, 1.3 million, and 642,000 years ago, so the recurrence interval is, crudely, 700,000 years. Smaller eruptions occurred between the large ones. Each of the three left calderas approximately 50 kilometers across. Because the caldera collapsed into the magma

chamber below, that magma body must have been at least 50 kilometers across.

Resurgent caldera eruptions are by far the largest and presumably most destructive of all types of volcanic eruptions, often called supervolcano eruptions. No eruption even remotely comparable to the most recent eruption of the Yellowstone resurgent caldera has happened in historic time, perhaps not since the appearance of modern human beings. The first of those great eruptions produced 2,500 times the volume of magma erupted by Mount St. Helens in 1980.

Much of the ash discharged in gigantic rhyolite pyroclastic flows that reached more than 100 kilometers down adjacent valleys, filling most of them to the brim. They were hot enough when they finally settled to weld themselves into solid sheets of rock tens of meters thick. Airborne ash drifted east and south on the wind and settled on the High Plains. Ash from the last eruption covered the

High Plains as far east as Kansas and at least as far south as the Mexican border. A large proportion of the North American wheat crop grows on soil developed in Yellowstone rhyolite ash.

If such an eruption were to happen today, it would be one of the most cataclysmic natural disasters in human history. The destruction or disruption of transportation, communication, and energy systems in the western and central United States would be major. Estimates suggest deposition of 1 meter of ash over western Wyoming and southeastern Idaho and 0.3 meters (1 foot) over eastern Wyoming, northwestern Colorado, and most of Utah. About 0.03 meters (1 inch) would cover the western states, from San Francisco to Phoenix to much of Nebraska and Kansas. Those impacts would hardly matter relative to the serious consequences that would almost certainly follow from enormous volumes of rhyolite ash injected into the upper atmosphere. The repercussions would probably resemble those that followed the 1815 eruption of Mount Tambora in Indonesia but on a vastly larger scale. All that ash would probably block enough of the sun's radiation to cause much colder climates within a few short weeks, leading to a worldwide agricultural disaster and probably famine.

Will Yellowstone erupt again? Absolutely —but when is less clear. All of the steam and boiling water in Yellowstone National Park is clear evidence of tremendous amounts of heat at shallow depth. It is more ominous that a broad fringe of dead trees surrounds most of the thermal areas. This can mean either that the thermal areas are growing larger, overwhelming trees that were thriving just a few years ago, or that there is greater circulation of carbon dioxide gas up to the roots of the trees. Temperature measurements taken since the 1870s show that the thermal areas are also getting hotter.

Detailed studies of seismograph records show that the shear waves of earthquakes that pass beneath the Yellowstone resurgent caldera arrive at the seismograph later and weaker than they would in other areas. They clearly show that molten magma exists from a depth

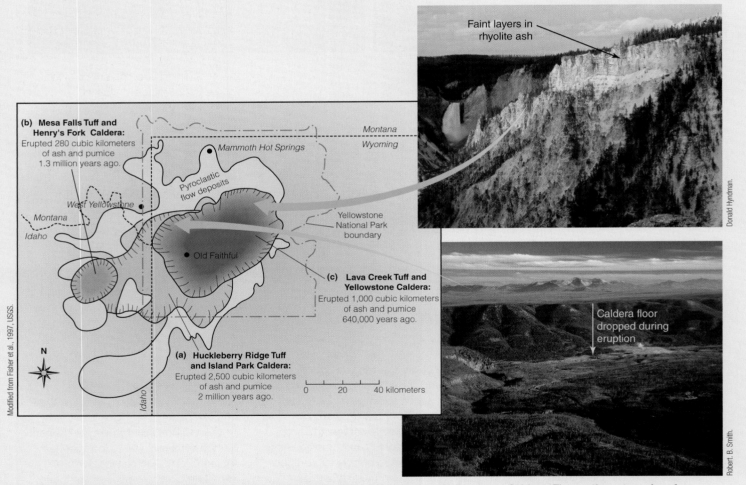

▶ *The Lava Creek ash erupted 642,000 years ago during collapse of the immense Yellowstone Caldera. The northwestern rim of Yellowstone Caldera formed during that collapse. Yellowstone Canyon has eroded down through 1,200 feet of the rhyolite ash that erupted and filled the Yellowstone Caldera.*

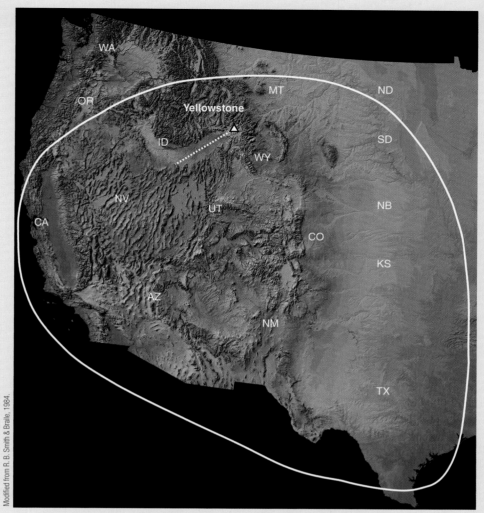

▶ *The Lava Creek ash from the Yellowstone Caldera, at the northeastern end of the Snake River Plain, covered most of the central plains of the United States. The hotspot track is marked by the white dotted line.*

of 8 to 16 kilometers or so under the caldera. The swarms of mini earthquakes that frequently rattle small areas in the resurgent caldera include harmonic tremors, probably an indication that the magma is moving. Two resurgent bulges in the floor of the caldera rise and fall over time, presumably because magma is rising below them. Between 1923 and 1984, it rose about 101 cm, then fell 20 cm until 1995. Between 2004 and 2007, precise GPS studies show a more-rapid 21-centimeter rise of the caldera floor.

These ups and downs are typical symptoms of a volcano getting ready to erupt. However, no one knows whether they also portend eruptions of such supervolcanoes.

What other warnings might we get? Because there has been no other eruption of this type or this size in historic time, we really don't know. However, smaller volcanoes generally show distinctive "harmonic" earthquakes from moving magma, earthquakes that become more frequent and closer to the surface—increasing surface temperature as the magma heats the

rocks above—and broad-scale bulging of the Earth's surface as the magma pushes up. This bulging stretches the crust near the surface, causing fractures likely to release water vapor, carbon dioxide, and gases richer in sulfur. These and other precursors are reviewed in Chapter 7, in the section on Eruption Warnings. As gases are released from the magma, pressure on it drops so more gases are released. The runaway reaction to these changes accelerates so that the gases dissolved in the rhyolite magma cannot diffuse fast enough. They expand within the magma violently so the magma itself froths out explosively. Because the rhyolite magma is so viscous, none of these changes are likely to happen very fast. Unfortunately, we cannot predict how fast they will happen. The enormous size of the volcano adds another unknown dimension. As the magma rises from below, the surface rocks settle into the space they leave (FIGURE 6-22).

An eruption equaling any of those of the past 2 million years would rise to almost 30 kilometers, nearly three times the flight height of commercial jet aircaft. It would expel sulfur that would mix with water vapor to form sulfate aerosols in quantities that reflect sunlight and dwarf the effects of much smaller volcanoes that dramatically cooled Earth's atmosphere for more than a year and caused widespread famine. Toba, a supervolcano that erupted about 74,000 years ago on northern Sumatra, annihilated much of the human population on Earth, through such a "volcanic winter," in this case one lasting 6 to 10 years. The eruption was about three times the size of the Yellowstone eruption 642,000 years ago.

A similar-size future eruption of Yellowstone would likely cause a volcanic winter that would last for years, causing extensive crop failures and probably worldwide famine, resulting in millions of deaths. Studies indicate that rainfall worldwide would drop by 50 percent because of decreased evaporation from the ocean.

Chapter Review

Key Points

Introduction to Volcanoes: Generation of Magmas

- A hot rock deep within the Earth may melt by increased temperature, decreased pressure, or addition of water. **By the Numbers 6-1**.

- The violence of a volcanic eruption depends on the magma's viscosity, volatiles, and volume.

- The viscosity of a magma is largely controlled by the silica content, with high-silica magmas (rhyolite) having higher viscosity than low-silica magmas (basalt).

- A volcanic eruption is likely to be more explosive for magmas with higher viscosity and larger quantities of volatiles, especially water.

Tectonic Environments of Volcanoes

- The tectonic environment dictates the volcano distribution, type, composition, and behavior.

- Most hazardous volcanoes are near subduction zones, and most of the remainder occur at spreading centers.

- Volcanoes that are not near plate boundaries are generally situated over hotspots.

Volcanic Eruptions and Products

- Basaltic magma commonly produces nonexplosive eruptions, spilling out in the form of lava. Types of lava include ropy *pahoehoe* and rubbly *aa*.

- Explosive eruptions produce pyroclastic material, solidified magma in the form of ash. Ash may rain down or be carried by the wind, or it may flow down a volcano flank in the form of a pyroclastic flow. Hot ash may combine with rain or melting snow to produce a lahar, or mudflow.

- The size of an explosive eruption depends on the amount of magma, the magma viscosity, and its water-vapor content.

Types of Volcanoes

- Shield volcanoes are characterized by gently sloping sides and are typically segmented by eruptive rifts.

- Stratovolcanoes have the classic volcano shape and moderate to high volatile content.

- Cinder cones are characterized by their small size and steep sides. Erupting cinder cones produce glowing fragments of cinders that rarely cause serious injury.

- As a lava dome rises and expands, lava fragments tumble down its sides while molten lava continues to rise within the dome. If the dome collapses, it may release a pyroclastic flow that can be extremely dangerous.

- Continental calderas are formed when the roof over a giant magma chamber collapses. Their infrequent eruptions produce huge volumes of pyroclastic material and are extremely destructive.

Key Terms

aa, p. 141
andesite, p. 136
basalt, p. 136
caldera, p. 145
cinder cone, p. 140
continental caldera, p. 152

crater, p. 141
eruptive rift, p. 145
eruptive vent, p. 145
fissure, p. 139
Hawaiian-type lava, p. 141
igneous rock, p. 135

lahar, p. 142
lava, p. 135
lava dome, p. 151
lava flow, p. 141
magma, p. 135
magma chamber, p. 135

melting temperature, p. 135
pahoehoe, p. 141
Peléan eruption, p. 143
pillow basalt, p. 139
Plinian eruption, p. 145
pyroclastic material, p. 142

Questions for Review

1. What can happen to heat, pressure, and water content to melt rock and create magma?

2. What factors control the violence or style of an eruption?

3. What properties of basalt magma control its eruptive behavior?

4. What properties of rhyolite magma control its eruptive behavior?

5. What drives an explosive eruption?

6. How does pahoehoe lava differ from aa lava?

7. On a huge shield volcano, such as Mauna Loa, what is the main type of eruptive site? Where on the volcano is (or are) such a site (or sites)?

8. Yellowstone Park has two huge calderas, each more than 20 kilometers across. How do such calderas form?

9. How is a caldera different from a crater?

10. Why do shield volcanoes have such a different shape than stratovolcanoes?

11. What is the driving force behind the explosive activity of a cinder cone? Where does it come from?

12. How does dacite or rhyolite magma form in a line of arc volcanoes, such as the Cascades?

13. Why do the Hawaiian Islands form a chain of volcanoes?

14. On what types of plate boundaries are volcanoes typically found? Explain how these tectonic environments give rise to volcanoes.

Discussion Questions

1. If Yellowstone volcano were to show signs of a pending eruption and you were placed in charge of evacuation, what would be the most important hazards and what would you do?

2. The Three Sisters cluster of steep-sided andesitic volcanoes, just west of Bend, Oregon, showed signs of bulging a few years ago. If that bulging were to accelerate, along with other signs of impending activity and you were placed in charge of evacuation, what would be the most important hazards and what would you do?

3. Long Valley Caldera in southeastern California, a few years ago, showed activity generally associated with rising magma—increasing slow rise of its resurgent dome, earthquake activity, and carbon dioxide concentrations that killed areas of trees. Should the hazard zones nearby be zoned to restrict development of new subdivisions? Why?

7 Volcanoes: Hazards and Mitigation

As Mount St. Helens erupts, the growing bulge above the magma chamber collapses in a landslide, permitting its gases to expand explosively. A plane by chance passes close overhead—note the wing in the upper right corner of the photo.

Keith and Dorothy Stoffel.

Mount St. Helens Erupts

May 18, 1980, was a brilliant late spring day throughout the Pacific Northwest. Early that morning, a few people were camping or logging on forest land in the restricted zone north and west of Mount St. Helens in southwestern Washington. Some had sneaked around official safety barriers erected in response to two months of minor eruptions. Some of the interlopers wanted to watch St. Helens from nearby. They hoped to see a major eruption. A bulge high on the north flank of the volcano had been growing for weeks, showing that a large mass of magma was rising below the mountain (**Case in Point:** Volcanic Precursors—Mount St. Helens Eruption, Washington, 1980, p. 186). At 8:32 a.m., Dr. David Johnston, the lone volcanologist from the U.S. Geological Survey (USGS) who was stationed on a high ridge facing the volcano at the time, radioed the Cascades Volcano Observatory in Vancouver, Washington: "Vancouver, Vancouver—this is it!" He died in the eruption. The

Volcanoes

night before, he had reluctantly agreed to replace someone else at that post. Ironically, he had been especially concerned about the hazards of an impending eruption, repeatedly referring to St. Helens as "a dynamite keg with the fuse lit."

P. and C. Hickson were 17 kilometers northeast of the crater when they saw the north side of the volcano suddenly look "fuzzy." It began to slide, the lower part faster; a dense black cloud blossomed from the summit and the north flank seemed to explode. Others nearby saw the horizontal blast and a shock wave racing ahead of the cloud. It looked like photos of nuclear explosions they had seen.

J. Downing, on the flank of Mount Rainier 75 kilometers to the north, saw two distinct "flows" 300 to 600 meters thick—they hugged the ground, disappeared into valleys, and then "hopped" over ridges. These were pyroclastic flows, masses of steam dark with suspended ash that made them so dense they stayed on the ground.

C. McNerney was 13 kilometers northwest of the crater, within the area doomed to imminent devastation, when he watched the north side of the volcano collapse. The leading wall of the black cloud climbed over a ridge, and a hot wind began to blow from the volcano. Two minutes after the eruption began, he started driving west at 120 kilometers per hour (kph), but the black cloud gained on him, so he sped up to 140 kph. The base of the black pyroclastic flow advanced "like avalanches of black chalk dust," one after another.

G. and K. Baker were 17 kilometers northwest of the crater, also within the area destined for devastation. They saw a "big, black, inky waterfall" a few miles up the valley and began driving west on Highway 50 at 160 kph. Even so, the black cloud almost reached them within four minutes. The cloud looked like it might be boiling oil with bubbles 2 meters in diameter. It was the same pyroclastic flow that others northwest of the volcano were fleeing.

B. Cole and three others were logging 20 kilometers northwest of the crater when the pyroclastic flow reached them. A "horrible crashing, crunching, grinding sound" came from the east. The air around them became totally dark and intensely hot. Cole and his companions gasped to breathe. Their mouths and throats burned, and they were knocked down along with the trees. Everything was covered by a foot of gray ash. The heat burned large parts of their bodies but not their clothing. Cole's three companions later died.

D. and L. Davis and A. Brooks were 19 kilometers north of the crater. They watched the black, boiling cloud of the pyroclastic flows bear down on them. Its leading edge snatched trees out of the ground and tossed them in the air. Then the black cloud swallowed them, and everything went pitch black and burning hot. A physician later said their burns were similar to those caused by a microwave.

Keith and Dorothy Stoffel were ready to drive home after attending a geoscience conference in Yakima, southeast of St. Helens, when they decided to hire a pilot to fly them around for a quick look at the mountain. As the plane rounded the north flank of the volcano, they saw that the snow had melted off the bulge overnight. Then they watched the bulge detach in a great landslide while an enormous cloud of steam, black with ash, spouted behind the slide and blossomed to fill the sky (see chapter opener photo). The pilot banked out of range and headed for Spokane.

Volcanic Hazards

In Chapter 6, we discussed the geologic processes that produce volcanoes and the products of eruptions. Here we describe how volcanoes impact people and what measures can be taken to mitigate volcanic hazards. Millions around the world live in risk of a volcanic hazard. A history of volcanic activity gives an area rich and fertile soil, ideal for agriculture. And landscapes of soaring, majestic mountain peaks make desirable spots for homes with sweeping vistas. As a result, a number of densely populated urban areas around the world exist in the shadow of active volcanoes. The main volcanic hazard areas of the United States, are shown in **FIGURE 7-1**.

Although few volcanic eruptions kill more than a few hundred people, those few can produce massive casualties. Deaths from eruptions depend heavily on the numbers of people living in close proximity to a volcano and on the eruption product. An assessment by the USGS of deadly volcanic events worldwide showed that pyroclastic flows killed about 70,800 in 19 major eruptions since 1631, forty-one percent of them at Mt. Pelée in 1902. Lahars killed 51,300 at 11 volcanoes in the same period, forty-four percent of them at Nevado del Ruiz in 1985. Ash falls killed relatively few, including 300 at Mt. Pinatubo in 1991. Tsunami initiated by eruptions killed about 50,900, sixty-two percent of them at Krakatau in 1883. In some cases, the high casualties were a result of misjudgments or politically motivated decisions by authorities.

FIGURE 7-1 VOLCANIC HAZARDS IN THE UNITED STATES

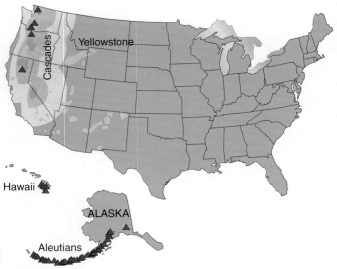

Red triangles are volcanoes. Dark orange is high hazard; lighter orange is lower hazard. Dark gray area has higher ash-fall hazard; light-gray area has lower ash-fall hazard.

Lava Flows

An erupting volcano can send molten rock, or lava, flowing or spewing forth. Even well downslope from vents, where the lava exterior is no longer red, it is still hot enough to ignite wooden structures (**FIGURE 7-2**). Even if an object does not burn, it is still overwhelmed and often buried by the flow. Where basalt lava flows surround a green tree, its heat boils moisture in the trunk and chills the lava next to the tree. The woody tissue either chars or later rots away, leaving a cast of the tree trunk. Excellent examples of such casts can be found on Kilauea Volcano and at Lava Cast Forest, south of Bend, Oregon.

Although lava flows are incredibly destructive to anything in their path, they move so slowly and cover such a small area that they generally pose little threat to human life. Basaltic lava flows commonly advance at speeds from

FIGURE 7-2 HOT LAVA IGNITES FIRES

Dark, but still hot, pahoehoe basalt lava from Kilauea torches the Waha'ula Visitor Center on June 22, 1989.

FIGURE 7-3 PYROCLASTIC FLOWS

C. Newhall, USGS.

Alberto Garcia/CORBIS.

A.

B.

A. Several pyroclastic flows pour down the slope of Mayon volcano in the Philippines during its 1984 eruption. **B.** A roiling pyroclastic flow from Mt. Pinatubo bears down on a fleeing vehicle.

1 meter per second near the vent to less than one-tenth that as the lava cools. Most travel up to 25 kilometers from the vent, but some may reach more than 50. Andesitic lava flows are more viscous and travel shorter distances. In a few cases, however, lava can flow rapidly. The especially fluid alkali-rich lava flow from Vesuvius killed 3,000 people in 1631. Even more extreme was lava that erupted from Vesuvius in 1805; it traveled from the crater to the base of the mountain in four minutes, twice the speed of an Olympic sprinter. Although basaltic magma typically forms lava flows, it may occasionally contain enough gas to blast out a Plinian eruption of hot fragments. Mount Etna in southern Italy occasionally erupts that way.

Pyroclastic Flows and Surges

A **pyroclastic flow** is a mixture of hot volcanic ash and steam that pours downslope because it is too dense to rise (**FIGURE 7-3**). A pyroclastic flow can also be referred to as an ash flow, *nuée ardente,* glowing avalanche, or ignimbrite deposit. Many flows develop when rapidly erupting steam carries a large volume of ash in a column that rises high above the volcano. When the rush of steam slows, part of the column collapses, and the cloud of ash pours down the flank of the volcano. The main flow hugs the ground, its less dense part billowing above as loose ash is stirred into the turbulent air. Fast-moving flows tend to hug valley bottoms, but their high velocity can carry them over intervening hills and ridges.

Glowing hot pyroclastic flows can race down the flank of a volcano at speeds from 50 to more than 200 kilometers per hour, incinerating everything flammable in their path, including forests and people (**FIGURE 7-4**; **Case in Point:** Pyroclastic Flows Can Be Deadly—Mount Pelée, Martinique, West Indies, p. 194; and **Case in Point:** The

Catastrophic Nature of Pyroclastic Flows—Mount Vesuvius, Italy, p. 195). People engulfed in a pyroclastic flow face certain death unless they are near its outer fringes, preferably in a building or vehicle. Even locations many kilometers from the base of a volcano may not be safe.

In some cases, a high-speed ash-rich shock wave called a **surge** may race across the ground ahead of a pyroclastic flow. They commonly originate as lateral blasts of ash and steam in the first stage of an ash-flow eruption. Because of their larger proportion of steam, they are generally less dense than standard pyroclastic flows. Some surges hug

FIGURE 7-4 NO MATCH FOR A PYROCLASTIC FLOW

USGS.

This car was singed, abraded, and crumpled by a pyroclastic flow from Mount St. Helens, 11 kilometers north of the crater.

FIGURE 7-5 SURGE DOWNS A FOREST

The directed blast of the May 1980 eruption of Mount St. Helens stripped and flattened trees on the windward side of a hill. Where the trees were sheltered from the volcano on the far side, they remained standing, but their tops were snapped off.

FIGURE 7-6 HOW FAR CAN IT BLAST?

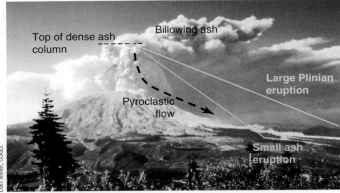

An energy line sloping from the top of the dense ash column of an eruption can estimate the height of hills that might be overridden by a pyroclastic flow from a stratovolcano eruption. The energy line has a lower slope from larger, more ash-rich eruptions, so their pyroclastic flows reach farther from the vent. The energy-line slopes are variable, depending on factors, such as the density of ash in the gas-thrust zone.

the ground and travel at speeds up to 600 kilometers per hour and may cover hundreds of square kilometers. They flatten forests and kill almost every living thing they meet, by heat, abrasion, and impact (**FIGURE 7-5**). Surges commonly pick rocks off the ground and carry them along, leaving in their wake deposits of ash mixed with rocks. Surges leave dunes of ash with cross beds much like those in sand dunes (see FIGURE 7-20b).

Both steam and ash in the hottest pyroclastic flows show a dull red heat, between 800° and 850°C—hot enough to glow in the dark. The ash and pumice particles in those flows are still extremely hot and plastic. As they come to rest and begin to cool, they may fuse into a solid mass of hard rocks. The hardened flows form sheets of *welded ash* that may cover hundreds of square kilometers, all deposited during a single eruption, in a matter of hours. Not only does the ash flow kill people caught in its path but also most animals and plants. Within a few months, some animals move in from other areas and new plants begin to grow.

The travel distance of a pyroclastic flow is related to the density and the quantity of the pyroclastic materials produced in a particular eruption. Heavy ash falling from the dense part of an erupting ash column accelerates down a volcano flank due to gravity and picks up speed that can carry it over nearby hills. We can approximate which hills the pyroclastic flow will go over by imagining an *energy line*, from the top of the dense ash column outward toward the ground (**FIGURE 7-6**).

For a small ash eruption, the energy-line slope may be as steep as 30 degrees. For the largest Plinian eruptions, the slope may be as flat as 12 degrees, so the danger zone extends much farther from the volcano. However, the actual hazard distance may be much greater if the ash is driven by a lateral blast, such as the one that felled the forest at Mount St. Helens in 1980. Because pyroclastic flows pour downslope, the most dangerous place to be at the time of an eruption is in the bottom of a valley. The best place to be when a pyroclastic flow forms is somewhere far away.

You might imagine yourself safe from a pyroclastic flow if you were watching from the other side of a body of water, but you could be wrong! Pyroclastic flows can actually cross rivers and bays or lakes. The lower part of the flow is the main mass. Less-dense ash billows into the air above the descending flow (**FIGURE 7-7**). Although pyroclastic flows are too dense to rise into the air, only part of the main flow may have particles large and dense enough to sink into water. That part may incorporate loose sediment in its path or trigger underwater sliding. The less-dense ash cloud above it may skim across bodies of water, even some that are tens of kilometers across. During the eruption of Vesuvius in A.D. 79, a glowing pyroclastic flow crossed some 30 kilometers of the Bay of Naples. Similarly, 2,000 people died in hot ash flows that crossed more than 40 kilometers of open sea in southeastern Sumatra during the eruption of Krakatau in 1883. In 1902, pyroclastic flows raced down the slopes of Mount Pelée in Martinique and continued offshore to capsize and burn boats anchored in the harbor. A body of water is clearly not an effective safety barrier from an erupting volcano.

FIGURE 7-7 PYROCLASTIC FLOW CROSSING WATER

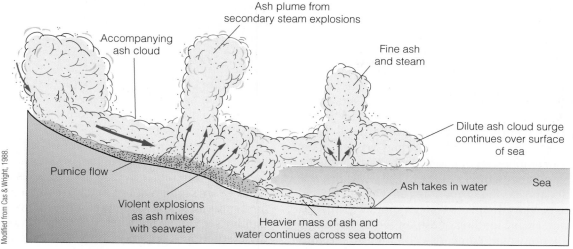

Ash plume from
secondary steam explosions

Accompanying
ash cloud

Fine ash
and steam

Dilute ash cloud surge
continues over surface
of sea

Sea

Ash takes in water

Pumice flow

Violent explosions
as ash mixes
with seawater

Heavier mass of ash and
water continues across sea bottom

Modified from Cas & Wright, 1988.

One explanation for the movement of hot pyroclastic flows over water is that some of the dense part of a pumice flow may sink as the still glowing lighter part races over the water surface.

Ash and Pumice Falls

Volcanic ash is composed of bits of pumice less than 2 millimeters across, light enough to drift some distance on the wind (see FIGURE 6-8). Ash erupts into a column that can rise 6 to 20 kilometers in the air. On occasion, a volcano may only blow off steam that rises in a nearly white column. On the other hand, if new magma has reached the surface, signaling a new eruption, the rising eruptive column will be dark with new ash (**FIGURE 7-8**). The ash particles are suspended in a cloud of steam that condenses into water droplets as it expands and cools. Much of the water coats the ash particles, which fall like snow downwind of the vent. The largest, heavier particles fall closer to the vent, where they form dangerous projectiles. Finer particles are carried downwind; especially fine ash is carried high into the atmosphere and may spread around the world.

Some ash may linger for several years in the upper atmosphere, where it blocks radiation from the sun. In one notorious case mentioned in Chapter 6, a giant eruption of Tambora Volcano in Indonesia in April 1815 blew an immense amount of ash into the upper atmosphere. For several years the ash remained suspended above the altitudes of weather, where sulfur dioxide and ash from the eruption effectively reflected the incoming solar energy, resulting in cooler temperatures around the planet. The following year was known as "the year without a summer." Crops failed in eastern Canada, New England, Britain, and elsewhere, and tens of thousands of people died of starvation. In another example of the devastating effects of volcanic ashfalls, the June 1783 flood-basalt eruption of Laki Craters in Iceland lasted for eight months, releasing about

FIGURE 7-8 DARK ASH IS NEW MAGMA

Ash cloud

Water vapor
condenses to
form a cloud

Dark lava
flows

Eruptive
vent

Pyroclastic
flow

NASA Astronaut.

Dark clouds of new volcanic ash blast out in the early stages of the June 12, 2009, eruption from Sarychev Peak in the Kurile Islands off eastern Asia. The white ring and cap atop the dark ash probably formed by rapid expansion, cooling, and condensation of water vapor in the eruption cloud.

122 megatons of sulfur dioxide and lesser amounts of other gases. It lowered northern hemisphere temperatures as much as 1°C, caused killing frosts in May and June 1816, shortened the growing season, and contributed to widespread famine. The disruptive 2010 ash erupted from the same rift zone.

Although individual pieces of ash are small and light, in the great quantities sometimes produced during eruptions, ash can be extremely destructive. Twenty centimeters of ash is enough to collapse most roofs; less than that is required in warm regions where roofs are not designed to bear snow loads. Wet ash is much heavier than dry ash. Rain falling during an eruption is common because ash eruptions generate their own **volcanic weather**. The rising hot ash draws in, heats, and lifts the surrounding air, which expands and cools, causing its dissolved water vapor to condense and fall as rain. Heavy loading of ash on roofs, especially if saturated by rainwater, can cause collapses that injure or kill people (**FIGURE 7-9**). The hazard can be especially severe in regions where people flock to their places of worship when an eruption begins.

Heavy ash falls pose several threats to humans, including the inhalation of fine particles. At the time of an eruption, people should move out from under a dense plume of ash by heading perpendicular to the wind direction. Even outside a heavy ash fall area, ash in the air or on the ground can cause serious health problems. Eye and throat irritation is common. Those with existing respiratory illnesses or damage—including bronchitis, emphysema, asthma, or those who are heavy smokers, face the greatest risk. Although silicosis has been a concern, the Centers for Disease Control does not believe that short-term exposure to volcanic ash presents significant risk to healthy people. Those with long-term or concentrated exposure should wear approved high-efficiency dust masks. If an approved mask is not available, people can breathe through a cloth moistened with water. In fine dust areas, goggles or eyeglasses are better than contact lenses.

Ash fall also poses problems for transportation, potentially complicating evacuations. Ash fall from an eruption, combined with rain, can lead to dangerous driving conditions. Even a millimeter of fine ash on roads can obscure lane markings and road shoulders. Ash causes slippery conditions, especially when it is wet. Rainfall at the time of an eruption is common; both steering and braking are impaired. Heavy ash falling or being stirred up by vehicles hinders visibility, especially at night. This can make evacuation difficult or impossible. Headlights cannot penetrate falling ash; roads and familiar landmarks disappear; ash on wiper blades can scratch windshields and further obscure visibility. Rear-end collisions are common due to very poor visibility.

Ash in the air can also damage vehicles. It can clog radiators and be drawn through air filters into engines, causing them to seize up. Tiny fragments of rock grains

FIGURE 7-9 ASH IS HEAVY!

A. Ash from Mount Pinatubo in the Philippines collapsed many roofs. **B.** Heavy ash from Rabaul Caldera, in September 1994, collapsed the roof on the left and thickly coated the roof on the right.

A

B

and frothy glass can abrade the surfaces of moving engines, transmissions, and brakes. These particles clog air filters, leading to engine overheating and even failure. Except for conditions with only very light ash, vehicles should be driven only when absolutely necessary; oil and air filters may need changing every 100 to 1,000 kilometers.

ASH AND AIRCRAFT Ash can also cause serious problems for aircraft because jet engines can freeze up and stall when they enter an ash cloud. Heavy air traffic from North America and Europe to eastern Asia—about 20,000 people per day and tens of thousands of flights per year—is of special concern because of the active subduction-zone volcanoes at the north edge of the Pacific tectonic plate.

Planes accidentally strayed into fairly dense clouds of volcanic ash more than 100 times between 1980 and 2004; 7 of those cases resulted in loss of engine power and near crashes. In June 1982, a Boeing 747 with 263 passengers and crew on a flight from Malaysia to Australia flew into an ash cloud erupting from a volcano in Java. All four engines failed. The plane fell from 11,470 meters to 4,030 meters before the crew was able to restart the engines. On December 14, 1989, a new Boeing 747-400, heading from the Netherlands to Japan with 245 passengers and crew, flew into an ash cloud from Redoubt volcano, 160 kilometers southwest of Anchorage, Alaska. The plane lost power in all engines and dropped from 7,500 meters to 3,500 meters in twelve minutes before pilots were able to restart the engines. Generators tripped, causing shutdown of airspeed indicators and other cockpit instruments not powered by batteries. The plane managed to reach Anchorage but repairs cost $80 million. In this and other cases, ash entered jet intakes, melted, and coated fuel injectors and turbine vanes. It abraded all forward-facing aircraft surfaces, including cockpit windows, making it difficult for pilots to see the runway during landing.

On March 22, 2009, Mount Redoubt again erupted, ejecting an ash plume 20,000 meters into the air. On March 29, ash rained down over Anchorage and its busy international airport that sees about 1,500 planes fly into, out of, or over it every day. The airport was forced to shut down that day.

Monitoring systems installed near some active volcanoes now warn of potential danger, although most Aleutian volcanoes are still unmonitored. Nine Volcanic Ash Advisory Centers gather and convey eruption information to air traffic control centers and aircraft. However, the warning time can be short; ash from a new eruption can reach flying altitudes within five minutes, as was the case with Mount St. Helens in 1980. Continuous satellite monitoring and pilot reports help, but clouds often obscure the view. Airplanes still fly into clouds of volcanic ash because cockpit radar cannot detect the fine particles. Warning signs for pilots include volcanic dust in the cockpit and cabin, acrid or sulfurous odors, heavy electrical static discharge around the windshield, a bright white flow in the engine exhaust, and engine surge or flameout, all in less than a minute. Pilots flying unexpectedly into a volcanic ash plume are instructed to slow the engines to lower their operating temperatures below the melting point of the ash, and then fly back out of the plume.

In mid-April, 2010, an eruption along the northern Mid-Atlantic Ridge in Iceland caused a massive disruption in European air travel. Basalt lava under Eyjafjallajökull glacier in southern Iceland began erupting on March 20 outside the edge of the glacier, but then expanded on April 14, along a two-kilometer-long north-south series of vents. It began melting ice under the glacier and rapidly chilled the magma. Lava fountains reacted with surface water and drove expanding steam eruptions that blasted fine ash into the air. By April 16, one-quarter of the ice in the active crater had melted; when all of the ice melted, steam abated, and continuing eruption was mostly basalt lava. However, authorities remained concerned because in the past, another, even larger eruption along the same zone repeated the process. The volcano had been dormant for almost 200 years; its last eruption, in December 1821, lasted intermittently for a year.

The ash cloud, following the jet stream, spread east and southeast across Europe at about 40 kph, rising to altitudes of 6 to 9 kilometers (20,000 to 30,000 feet). It severely disrupted air traffic for six days, not so much because of visibility, but because ash pulled into the hot turbines of jet engines causes abrasion and can fuse to the engine parts, shutting them down. Without power, a jet aircraft flying at an altitude of 7 to 12 kilometers can drop 5 or more kilometers within a few minutes and potentially crash.

The ash plume spread over almost all of Europe, as far south as southern Italy (**FIGURE 7-10.** p. 174). Of Europe's 28,000 flights scheduled for Friday, April 16, about 16,000 were grounded, stranding close to 2,000,000 passengers. It cost airlines more than $200 million per day in lost revenue.

Volcanic Mudflows

Volcanic **mudflows** form when ash combines with water, primarily on stratovolcanoes, and pours down their flanks at high speeds with a consistency similar to very wet concrete. Racing down valleys with the velocity of floodwater, mudflows inundate and fill lower valleys where people live (**FIGURES 7-11**, p. 174 and **7-12**, p. 175). As the mud moves downslope, it gathers rocks of all sizes and carries them along, accelerating as it goes. Many mudflows carry rocks the size of cars. People are often killed by boulders in fast-moving flows.

Mudflows can be triggered by the eruption of a volcano covered in ice or snow that is rapidly melted or mobilized by hot ash (**Case in Point:** Even a Small Eruption Can Trigger a Major Debris Avalanche—Nevado del Ruiz, Colombia, 1985, p. 192). A hot mudflow is called a **lahar**,

FIGURE 7-10 ASH FROM THE ICELAND ERUPTION SPREAD OVER MOST OF EUROPE.

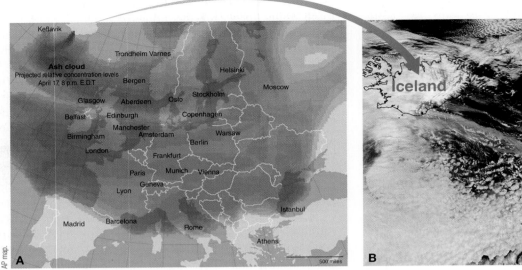

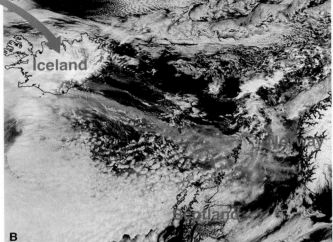

A. April 17 ash distribution. **B.** The dense plume spread from Iceland east and south.

though other volcanologists use the term for any volcanic mud or debris flow. Big blocks of rhyolitic magma buried within a lahar may blow steam for months before they finally cool.

Weather can also be a factor, because rain on loose ash washes it downslope as a heavy slurry. Even on an otherwise clear day, an eruption cloud may create thunderstorms and heavy rain due to the rapidly rising heat plume. Water pouring down the slope of a stratovolcano may develop into a mudflow whether or not the volcano is erupting. Thus mudflows can range from icy to boiling temperatures.

Stratovolcanoes pose an especially great mudflow risk because they are inherently unstable piles of lava and rubble. Many stratovolcanoes emit steam that quietly drifts out of rocks near the summit. In most such cases, the steam is simply meltwater from snow or rain on the high slopes that boils as it sinks into the hot rocks below. Although it probably does not foreshadow an eruption, it does tell us that the interior of the volcano is extremely hot. The rocks in there are stewing in the hot water and steam that fills the fractures.

In one example, Mount Rainier, in Washington State, poses serious mudflow risks for the rapidly developing

FIGURE 7-11 MUDFLOWS FILL A VALLEY

A. Mud lines high on tree trunks show the depth of the Toutle River mudflow on May 18, 1980. The person in yellow on the right is almost six feet tall. **B.** Lower in the Toutle River valley, mud swamped houses sit near the river.

FIGURE 7-12 STRONG HOUSE BURIED

Mudflow from Mount Unzen in Japan completely buried the first story of many houses.

communities of the Seattle-Tacoma area (**FIGURE 7-13**). Like many large stratovolcanoes, Mount Rainier consists of weak material. Hot volcanic gases and groundwater have degraded the edges of fractures in its rocks, and much of its andesite ash and rubble weather into soft clay that becomes even softer when wet. Enormous snow and ice deposits on the higher slopes of the mountain make it even less stable. An eruption, a season of unusually rapid snowmelt, or shaking by a major subduction-zone earthquake could easily mobilize immense volumes of ash and rubble into enormous mudflows.

The historical record of this area provides an indication of future hazards. Every 500 years, on average, large mudflows reach as far as 100 kilometers from Mount Rainier

to cover large parts of the Puget Sound lowland (**FIGURE 7-14**, p. 176). The Electron mudflow ran 48 kilometers down the Puyallup River valley 500 years ago, then spread onto the Puget Sound lowland, including the area where Orting now stands (see FIGURES 1-2 and 7-13). In places, the flow is 30 meters thick, 60 kilometers from its source on the volcano (**FIGURE 7-15**, p. 176). Smaller flows of mud and debris surge down valleys from Rainier on occasion, including one on August 14, 2001, after a week of hot weather accelerated melting of ice on the mountain. Still larger events, glacial outburst floods from subglacial meltwater have rushed down valleys from the mountain. More than 35 such floods occurred in the last century; many of these destroyed bridges and roads.

The Osceola mudflow poured 73 kilometers down the White River 5,000 years ago, burying the broad valley floor beneath as much as 20 meters of mud. The towns of Enumclaw and Buckley, home to tens of thousands of people, stand on that deposit. The Paradise mudflow filled the upper Nisqually River valley south and west of the mountain sometime between 5,800 and 6,600 years ago. Future years will certainly see more mudflows follow those old routes, covering everything on them. Every new development along those mudflow paths increases the hazard posed by Mount Rainier. Unfortunately, the flat valley floors that drain from Mount Rainier are attracting large housing developments to accommodate the rapidly growing population of the Seattle-Tacoma area. Today, tens of thousands of people now live on the surfaces of geologically recent mudflows.

The best way to avoid mudflows is to avoid the bottoms of stream valleys that drain away from volcanoes. Unfortunately, most towns develop in valley bottoms; these are the very places that mudflows follow as they move downslope. Trying to outrun a mudflow almost always

FIGURE 7-13 DEVELOPMENT AT RISK OF MUDFLOWS

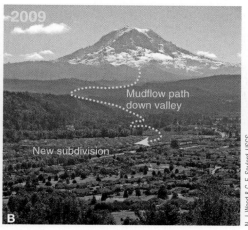

A. Mt. Rainier is heavily cloaked with snow and ice. Debris avalanches and mudflows in downstream valleys are the greatest threat. **B.** The town of Orting, pictured here, is one of several communities built on old mudflows. The bottom of the valley—an area with significant mudflow hazard—was extensively developed between 1995 and 2009.

FIGURE 7-14 HISTORICAL MUDFLOWS

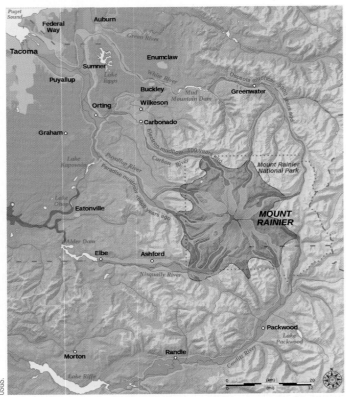

The Osceola mudflow inundated areas now occupied by hundreds of thousands of people. The Electron mudflow filled the floor of the valley that Orting now occupies.

FIGURE 7-15 EVIDENCE OF A HUGE MUDFLOW

This large Douglas Fir stump, buried by the 5-meter-thick Electron mudflow, was excavated at a new subdivision being built on the surface of the mudflow near Orting, Washington. Mt. Rainier, the source of the mudflow 60 kilometers away, is in the background.

results in death. Climbing well up the valley side or even running back into the forest, away from the channel, is a much better idea. Broad floodplains are especially dangerous because the valley sides may be too far away. Indonesia, a country with many volcanoes, has built safety hillocks several meters high where people can hopefully climb above the mudflows. (Additional details on debris flow and mudflow processes are included in Chapter 11.)

Poisonous Gases

Even if no eruption occurs, gases emitted by volcanoes can pose a hazard to people, animals, and trees. At depths of more than a few kilometers, gases dissolved in the magma are under tremendous pressure. As the magma rises, the pressure drops, which permits gases to *exsolve*, or come out of solution. Thus, an increasing volume of escaping gases commonly precedes—and may warn of—an impending eruption. These gases react with sunlight, moisture, and oxygen in the air to produce aerosols—tiny particles and droplets. The volcanic gases and aerosols create an acidic volcanic smog, or **vog**, which can pose a threat to life and health (**FIGURE 7-16**).

Although carbon dioxide (CO_2) is a familiar gas in the air we breathe, it is deadly at high concentrations. Expelled from some volcanoes it is heavier than air and can pour downslope, where it concentrates in depressions. Because it is colorless and odorless, it can suffocate people and other animals without warning.

A tragic case in 1986 in Cameroon, West Africa, involved CO_2 that bubbled out of Lake Nyos, a volcanic crater lake some 200 meters deep. The gas had seeped into and dissolved in the deeper waters of the lake over many years. On the evening of August 21, 1986, the CO_2-saturated deep waters of the lake rumbled loudly and belched an immense volume of the gas. A rainy season landslide into the lake,

FIGURE 7-16 DANGEROUS GASES

Bluish vog drifts downwind from Kupaianaha Lava Lake in Hawaii. The lake depression is about 100 meters in diameter.

FIGURE 7-17 DEADLY GASES LURK BELOW

Carbon dioxide in Lake Nyos, a water-filled volcanic crater in Cameroon, suddenly bubbled out and flowed downslope over a waterfall just beyond the right side of the photo. It suffocated both cattle and humans.

pushing some of the deep water upward, may have prompted overturn of the lake waters. A drop in pressure on the dissolved CO_2 would have rapidly released it from solution.

Heavier than air, the gas swept rapidly down through several small villages as a stream 50 meters thick and 16 kilometers long. More than 1,700 people, some 3,000 cattle, and countless small animals died of asphyxiation (**FIGURE 7-17**). An air mixture with ten percent CO_2 can kill people; in this case, it was nearly 100 percent.

There is risk of an even more catastrophic event at Lake Kivu, in the East African Rift Zone, at the eastern border of the Congo. Much larger than Lake Nyos, Kivu is 48 by 89 kilometers (or 2700 km^2); it contains 350 times as much CO_2 and methane below a depth of about 300 meters, posing a potentially fatal threat to more than 2 million people living near the lake.

In another example, carbon dioxide seeps from within the Long Valley Caldera north of Bishop, California. Evidence of the CO_2 concentration includes dying trees in several areas around Mammoth Mountain on the rim of the caldera (**FIGURE 7-18**). On March 11, 1990, Fred Richter, a forest ranger, found refuge from a blizzard in an old cabin surrounded by snowdrifts. Entering the cabin through a hatch, he almost suffocated before fighting his way back up the ladder into fresh air. The denser-than-air CO_2 had collected and concentrated in the nearly sealed cabin. Although people walking through an area are probably not in danger, someone camping in a depression out of the wind or entering a confined space below ground could be asphyxiated. Even the trees, located near caldera faults, often succumb to the CO_2 concentrating around their roots.

In addition to carbon dioxide, another dangerous volcanic gas is sulfur dioxide, a noxious gas with a sharply acrid smell and choking effect. Some eruptions produce large quantities of sulfur dioxide (**FIGURE 7-19**, p. 178). Sulfur dioxide is harmful to animals and extremely poisonous to plants,

even in small concentrations. It reacts with oxygen in the atmosphere to make sulfur trioxide, which reacts with water vapor to make minute droplets of sulfuric acid. They hang suspended in the atmosphere for months, partially blocking the incoming sunlight and thus cooling the climate.

FIGURE 7-18 DEADLY CARBON DIOXIDE

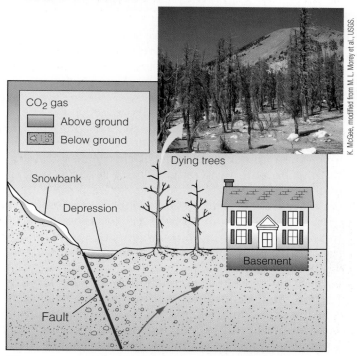

CO_2 gas rises along faults from molten magma at depth and can collect in depressions or confined areas, like the basement of a house. This gas killed these trees on the south side of Mammoth Mountain, California, shown in September 1996.

FIGURE 7-19 MONITORING VOLCANO EMISSIONS

fumarole vent

A. Volcanologists sample gases on Mount St. Helens. They are not wearing masks because they are upwind of the gases. **B.** Volcanic gases in a fissure on Kilauea, Hawaii, deposit bright yellow coatings of sulfur.

Some volcanoes also expel poisonous hydrogen sulfide, various chlorine compounds, fluorine, and small quantities of other gases. Most of these emissions have little to do with events during an eruption, though fluorine breaks bonds and can make magmas less viscous. Hydrogen sulfide settles into depressions and can quickly kill people or animals. Fluorine, a neurotoxin, expelled during the immense several-months-long Laki flood basalt eruption of June 1783 in Iceland, appears to have killed about 10,000 people (more than 20 percent of the population) and many more in Europe. Analysis of their malformed bones suggests that fluorine contaminated food supplies and drinking water. Fluorine compounds in the volcanic haze precipitated on grass contributed to the death of about 75 percent of their sheep, cattle, and horses.

Like other kinds of smog, vog droplets are small enough to be retained in lungs, where they degrade function and compromise immune systems. Especially endangered are children and those with asthma or other respiratory problems. Knowledge of other effects—especially over the long term—is limited, but it is clear that we should avoid breathing volcanic smog if possible, if necessary by staying indoors. If unexpectedly caught in a thick cloud of volcanic smog, breathing through a damp handkerchief may help.

Vog can also produce acid rain. In Hawaii, residents may unintentionally expose themselves to toxic metals when they collect rainwater from roofs for both washing and drinking. The acid rain caused by vog can leach metals from the roof that are then retained in drinking water. This means that people downwind of some volcanic vents unintentionally ingest significant amounts of metals, sometimes including lead, which can cause brain damage and other defects.

Predicting Volcanic Eruptions

As with most hazards, the best protection from volcanic hazards is to predict an occurrence well in advance in order to evacuate a population or take other precautions. If asked to anticipate future events, scientists commonly use historical records to assess the long-term prospects for volcanic activity in a certain location. They also rely on short-term indications to warn of impending eruptions.

Examining Ancient Eruptions

It is not possible in the current state of knowledge to predict exactly when a volcano will erupt—or, in many cases, whether it will erupt at all. But geologists have learned much about the history of many volcanoes. That knowledge makes it possible to assess the likely future behavior of a volcano and to mitigate the hazards it poses to life and property.

A volcano's historical record can help geoscientists understand patterns of recurrence in eruptions. The number of eruptions in the last 100 or even 1,000 years may be documented in populated areas with long historical records, such as in Italy or Japan. Elsewhere, like in the western United States, the historical record may cover only the past century or so. These records mean little if a volcano erupts once or twice in 10,000 years. In such cases, information is provided through **paleovolcanology**, which involves interpreting deposits from prehistoric eruptions and reconstructing a record using age dates on plant material charred in past eruptions or dates on the volcanic rocks themselves (see FIGURE 6-9). The archives are not in the written record but in the rocks.

In the eruptive products, we may recognize lavas, pyroclastic flows, ash-fall deposits, and mudflows, as well as the magnitude and lateral extent of the events. Ash after deposition forms a rock called **tuff**. Ash-fall material becomes ash-fall tuff; pyroclastic-flow material becomes pyroclastic-flow tuff (also called ash-flow tuff). Deposits of ash-fall tuff can be distinguished from pyroclastic-flow

FIGURE 7-20 IDENTIFYING VOLCANIC DEPOSITS

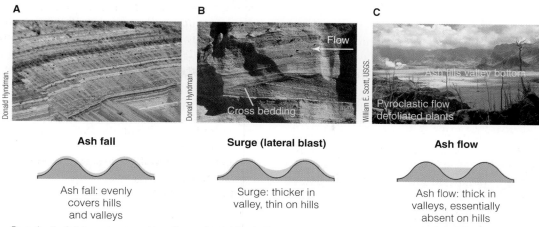

A — Donald Hyndman.

B — Donald Hyndman. Flow. Cross bedding.

C — William E. Scott, USGS. Ash fills valley bottom. Pyroclastic flow defoliated plants.

Ash fall

Ash fall: evenly covers hills and valleys

Surge (lateral blast)

Surge: thicker in valley, thin on hills

Ash flow

Ash flow: thick in valleys, essentially absent on hills

Pyroclastic fall layers, surge deposits, and pyroclastic flows can be distinguished by their distribution over hills. **A.** Thin layers of air-fall ash and pumice from the Cappadocia region of central Turkey. **B.** Cross-bedded surge deposit from the 1600 B.C. eruption of Santorini, Greece. **C.** Pyroclastic-flow deposit from the 1991 Mount Pinatubo eruption filled the valley bottom of Marella River. The foreground vegetation was stripped by hot ash in the upper part of the same pyroclastic flow. Ash coating surfaces in the foreground settled either from the billowy upper part of the pyroclastic flow or from the ash fall.

tuff based on characteristics and deposition over the landscape (**FIGURE 7-20**). The most obvious distinction is that ash-fall tuff is generally distinctly layered, whereas pyroclastic-flow tuff is unlayered, at least near the vent. Either may contain lumps of pumice, the frothy rock that is typically light enough to float on water. The distribution of deposits over hills and valleys is also an indication of what type of flow occurred. Ash fall deposits are spread evenly over hills and valleys because it falls from above, like snow. Surge deposits are thicker in valleys and thin on hills, whereas pyroclastic flow deposits are thick in valleys and virtually nonexistent on hills. Surge deposits also often show cross-bedding. By evaluating the type and extent of volcanic products in an area, geologists can estimate the magnitude, sequence, and timing of past eruptions. Past behavior may indicate what the volcano will do in the future.

However, the record in the rocks is open to careful interpretation. The exposed rocks on a volcano rarely provide a complete record of its previous eruptions because of erosion between events. Even where deposits are exposed, does each ash or pumice fall, pyroclastic flow deposit, or lava flow record a separate eruption? Or do they record episodes of a single eruptive sequence that may have lasted a matter of a few days or weeks? Such questions, if not resolved, make a shaky basis for statistical assessment of long-term recurrence intervals or hazards.

After scientists have determined how a volcano has behaved in the past, they are still faced with a key question:

is this volcano still active? Such a simple question begs for a simple answer. Recall that several decades ago, the answer for a Cascades volcano was that it was active if eruptions had covered evidence of glacial activity from the last ice age about 10,000 years ago. We now know that some volcanoes lie dormant for much longer periods before erupting again. Mount Lassen in northern California paused for some 27,000 years before again erupting in 1915. In the eastern Mediterranean Sea, Santorini rests for an average of 30,000 years between major eruptions, though the intervals are becoming shorter (see Chapter 6, **Case in Point:** Long Periods Between Collapse—Caldera Eruptions, Santorini, Greece, p. 156). And Yellowstone, the gigantic volcano in northwestern Wyoming, erupts only every 600,000 to 700,000 years (see Chapter 6, **Case in Point:** Future Eruptions of a Giant Caldera Volcano—Yellowstone Volcano, Wyoming, p. 157). Because the last of its massive eruptions occurred 640,000 years ago and seismic studies also show that molten magma lies just a few kilometers beneath the surface, the volcano is being closely monitored. Careful surveying within the Yellowstone Caldera shows resurgent domes that periodically rise and fall over decades or centuries, apparently in conjunction with magma movements underground. In 2002, geysers that had not erupted for a long time became quite active. Clearly, there is no simple answer for whether a volcano is active. Even if a volcano has not erupted in tens of thousands of years, scientists must be on the lookout for short-term indications of an impending eruption.

Eruption Warnings: Volcanic Precursors

Forecasting volcanic behavior for the long term is one thing. It is quite a different matter to predict what a volcano may do in the next few days or weeks. Accurate eruption predictions are especially critical in areas where a large population lives close to the base of a stratovolcano. Mount Vesuvius, near Naples, Italy, is such a case.

Seismograph records of volcanic earthquakes have been used to infer magma movement underground and to project the likelihood of an eruption. *Harmonic tremors* are the low-frequency rolling ground movement that precedes many eruptions. USGS volcanologists recorded a series of minor earthquakes originating beneath Kilauea Volcano in Hawaii in 1959. Over a period of two months, the earthquakes became more frequent as they rose from a depth of 60 kilometers, finally reaching the surface as the volcano erupted. Similar earthquakes have been recorded below many volcanoes. These earthquakes rapidly increase in their frequency and magnitude and, in some cases, migrate toward the summit before an eruption. That happened in the New Hebrides Islands during the 1950s and 1960s and at Mount Pinatubo in 1991 (see Chapter 6, **Case in Point:** Deadly Lahar—Mount Pinatubo, Philippines, 1991, p. 157). But earthquakes are not always a reliable indicator; the frequency and magnitude of earthquakes did not change much at Mount St. Helens in 1980 during the

two months between the first activity and the climactic eruption of May 18.

Changes in the surface temperatures of volcanoes and the steam they erupt can be another indication of an impending eruption. Telescopes fitted with thermometers instead of ordinary optical eyepieces can observe temperatures of distant objects with reasonable accuracy. But their usefulness in predicting eruptions is limited because it is sometimes hard to know whether observed temperature changes are the result of volcanic activity or extraneous factors, such as rainfall cooling rocks.

Small changes in summit elevations and slope steepness associated with eruptions have been observed at some Japanese volcanoes and at Kilauea Volcano in Hawaii. Tiltmeters, instruments that measure changes in the slope of a volcano, were first installed at Kilauea in the 1920s. Initial devices were simple levels made with water tubes 25 meters long that could detect tilts as gentle as 1 millimeter in a kilometer. More precise modern instruments use lasers to measure changes in both elevation and distance. Volcanologists also now commonly employ satellite-based Global Positioning System (GPS) devices to accurately measure changes in position. They show that volcano summits swell as magma rises into them and then deflate as they erupt (**FIGURE 7-21**).

A change in the gases emitted from a volcano is also associated with eruptions. As magma rises toward the

FIGURE 7-21 VOLCANIC PRECURSORS

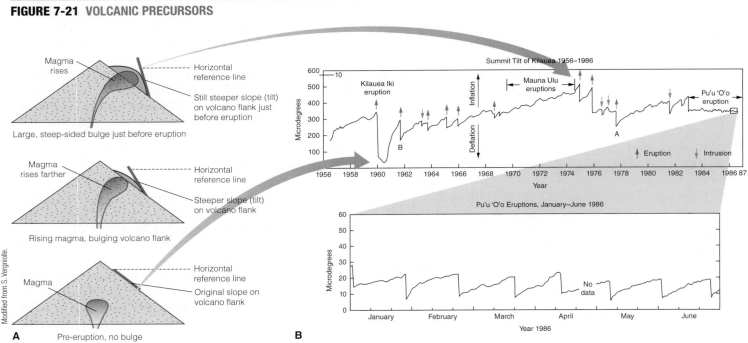

A. When rising magma gets close enough to the surface, it sometimes pushes overlying rocks to create a bulge in the flank of a volcano. **B.** A tiltmeter measures the changing increase in slope on the volcano flank as the bulge grows.

surface, steam and other gases are released, and in some volcanoes, there are abrupt increases in sulfur or the ratio of sulfur to chlorine. Geologists collect and analyze fumarole gases to watch for these ominous signs (see FIGURE 7-19).

Although none of these precursors is a sure sign of an eruption, they provide evidence of an increasing threat posed by a volcano. However, even when scientists feel reasonably sure that an eruption is coming, a prediction is useful only to the extent that authorities act on it. The USGS was reasonably successful in predicting eruption of Mount St. Helens as a major event in 1980 (**Case in Point:** Mount St. Helens Eruption, Washington, 1980, p. 186). But many people deeply resented the effects of the prediction upon their personal freedoms. The local loggers and timber companies fought closure of the nearby forests because they feared loss of income. The civil authorities relented and permitted their continued access. Loggers died, and logging equipment and millions of trees were destroyed in the pyroclastic flows and mudflows that accompanied the eruption of May 18. It is unfortunate civil authorities allowed them into the area and that the loggers who pressed for such access were not aware of the extent of the geologic hazards and the consequences of ignoring them. Local sightseers were equally ignorant of the real dangers, and, as discussed at the beginning of this chapter, some paid a heavy price.

Mitigation of Damage

Because it is not always possible to predict volcanoes, various strategies for mitigating volcano damage have been tried, with varying degrees of success.

Controlling Lava Flows

Attempts to slow or divert lava flows have brought only partial success. Perhaps the most effective approach is to cool and solidify the flow front with copious amounts of water delivered from fire hoses. A large basalt lava flow advancing on the town and harbor of Heimaey, Iceland, was cooled, slowed, and partly diverted in this way in 1973. Flows erupting from Mauna Loa were bombed in 1935 and 1942 to break the solid levee of cooling lava along the edges, thus diverting it into another path. These attempts were not successful enough, however, to inspire adoption as a standard procedure. Emergency diversion barriers were erected in nine days in 1960 to divert a flow erupting from Mauna Loa near Kapoho in westernmost Hawaii. These were partly successful in diverting the flow away from beach property and a lighthouse. Bulldozing levees of broken basalt lava into the path of a basalt lava flow proved unsuccessful, presumably because the broken rock of the levee was less dense than the flowing lava.

Warning of Mudflows

Fast-moving mudflows kill thousands of people living in valleys on the lower slopes of volcanoes. Automatic detection systems could provide enough warning for many people to evacuate from valley floors to higher ground. Warning systems high in valleys on the flanks of active volcanoes would provide warning when temperature sensors detected the passage of hot mudflows. New Zealand, for example, used them for many years on ski slopes on the flank of Ruapehu, an active volcano. Two vibration sensors have been installed high in each major valley on Rainier; they detect the frequency of mudflow ground vibration (between 0.6 and 20 Hz) and exclude the lower frequencies of earthquake tremors and volcanic eruptions.

As with all warning systems, however, identifying an imminent threat is only the first step toward saving lives. The travel time for a large mudflow from Mt. Rainier to reach the 4,000 plus residents of Orting, Washington, may be less than an hour. By the time a flow reaches the detector downslope, the event size is analyzed, likely false alarms eliminated, and the alarm sounded, residents might have only 30 minutes to flee. Successful evacuation will depend on a clear and timely warning, people's understanding of the hazard, and immediate response.

Populations at Risk

Although much has been learned about volcano behavior and warning of an impending eruption often can be provided, a false alarm could destroy the credibility of all involved. The social and economic consequences of a needless evacuation of hundreds of thousands of people would be devastating, if it were even possible. The region of Italy surrounding Mount Vesuvius and the Cascades Range of northwestern North America provide examples of how scientists assess risk based on historical records of volcanic activity as well as volcanic precursors. They also raise important questions about disaster planning and preparedness in these regions.

Vesuvius and Its Neighbors

Large numbers of people living in close proximity to active volcanoes in Italy provide glimpses into increasing volcanic hazards both there and in North America. People do not differ much from one place to another. Their behavior during the eruption of Vesuvius in A.D. 79 was probably similar to the way people now living in towns near an erupting volcano react.

How could 700,000 people near Vesuvius leave on roads and rail lines that are taxed beyond their capacity on normal days? And where could so many people find convenient refuge on short notice?

FIGURE 7-22 AN OMINOUS FUTURE?

Basemap: modified from NASA image; inset map: NASA; photo: Donald Hyndman.

A chain of active volcanoes, including Vesuvius and Etna, runs down the western side of Italy. Subduction zones are shown in green lines with pointers sloping down dip. Roads and houses crowd onto the lower flanks of Mount Vesuvius. Some new houses are built in a valley right next to historic lava flows.

The northeastern part of the African Plate descends under southern Europe along a short collision boundary that raises the Alps and drives a chain of volcanoes in Italy and the eastern Mediterranean region. These volcanoes have devastated populations and caused enormous property damage for thousands of years. Some have even completely destroyed ancient civilizations. This is one of the most dangerous volcanic environments on Earth.

MOUNT VESUVIUS The modern city of Naples, with a population of 3.7 million, nestles between two volcanoes—Mount Vesuvius, just 13 kilometers to the east (**FIGURE 7-22**), and Campi Flegrei, centered 13 kilometers to the west. Vesuvius has a long history of major eruptions (**Table 7-1**), including one around 3580 B.C. and another around 1800 B.C., in which

Table 7-1	Notable Eruptions of Mount Vesuvius
DATE	**EVENT**
3580 B.C.	Pyroclastic flows cover current area of Naples.
1800 B.C.	Pyroclastic flows kill many people in Pompeii and cover much of the area of present-day Naples.
A.D. 79	Destruction of Pompeii and Herculaneum kills about 4,000.
1631	Heavy pyroclastic flows and ash falls kill 4,000.
1906	More than 500 are killed.
1944	Lava flows partly destroy two communities.

FIGURE 7-23 THE PAST IS A SIGN

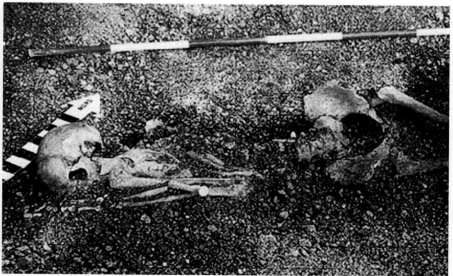

Guiseppe Mastroiorenzo and PNAS, National Academy of Sciences, USA.

A. People ran from an earlier, even larger eruption of Vesuvius 3,800 years ago, leaving these footprints preserved in ash. **B.** Some skeletons have been uncovered.

pyroclastic flows covered much of the present area of Naples and killed many people. In A.D. 79, during the reign of Nero, Vesuvius erupted catastrophically, burying Pompeii and Herculaneum and killing one fifth of the 20,000 people living there (**Case in Point:** Mount Vesuvius, Italy, p. 191).

Throughout its history, the area around Vesuvius has gone through cycles of volcanic destruction and rebuilding. The recent lack of significant activity is again leading to complacency as development continues in this high-risk area (**FIGURE 7-24**). Today, 700,000 people are living on borrowed time within the potentially fatal zone of the next major eruption of Vesuvius.

FIGURE 7-24 VESUVIUS LOOMS OVER NAPLES

Hyndman.

Naples and Naples Bay lie in the shadow of Mount Vesuvius (seen in the background).

So far as anyone knows, Vesuvius could erupt anytime. If past experience is any guide, some 0.5 km^3 of magma may have accumulated beneath Vesuvius after more than 60 years of inactivity. Seismic and tiltmeter studies suggest that molten magma now exists 5 to 10 kilometers below the crater of Vesuvius. Tiltmeters detected a half-meter of expansion from 1982 to 1984, with an increase in numbers of minor earthquakes. However, in approximately half of such cases, no eruption follows. This large uncertainty leaves volcanologists wondering how to advise civil authorities.

Experts suggest that the long pause since the last major eruption means that the next may be the largest since 1631. Some volcanologists think the next Vesuvius eruption will probably resemble those of 1906 and 1944, with lava flows and heavy ash falls that could collapse as many as 20 percent of the roofs in Naples and stop all traffic and activity. Pyroclastic flows racing downslope at more than 100 kilometers per hour with temperatures of 1,000°C would reach populated areas between five and seven minutes after the eruption column collapsed. Modern Pompeii and a dozen other towns, each with thousands of people, ring the base of Vesuvius hardly more than 6 kilometers from the crater. Instead of several thousand deaths, the toll of a new eruption resembling the one in A.D. 79 could kill several hundred thousand people.

In the current state of knowledge, disaster prevention depends on early sensing of volcanic warnings such as swarms of small earthquakes or compositions of emitted gases. A special commission formulated an eruption contingency plan in 1996, which assumed that a warning could be issued at least two weeks before an eruption and 600,000 people could be evacuated within a week. Authorities now argue that a complete evacuation would take about three days. Those who have driven in Naples or used trains and buses almost anywhere in Italy might consider such a goal hopelessly optimistic. Vocal critics argue that it is impossible to predict an eruption more than a few hours or days in advance. They suggest that even with the best available monitoring, an evacuation alarm would almost certainly come too late. The alternative of major urban planning in a region that already contains hundreds of thousands of people would require truly ruthless resettlement on a scale almost beyond imagining! If you were in charge, what would you do?

CAMPI FLEGREI Campi Flegrei is an unquestionably active resurgent caldera within the western suburbs of Naples. Two million people live on the floor of the caldera, with 400,000 residing within the most active part, making Campi Flegrei one of the most hazardous volcanic areas in the world. Giant rhyolitic pyroclastic flow eruptions, which spread searing pumice to the southeast, were hot enough for the fragments to flatten and fuse, forming a solid mass. These events left a caldera 16 kilometers in diameter 39,000 years ago and another caldera 21 kilometers in diameter, at the same site, 15,000 years ago. Intense explosive activity along the faults that define the margin of the caldera has happened repeatedly during the past 12,000 years. Eruptions came at intervals of approximately 50 to 70 years from 12,000 to 9,500 years ago; 8,600 to 8,200 years ago; and 4,800 to 3,800 years ago. That is an extremely high level of activity for any volcano, especially a giant rhyolitic caldera volcano.

As with the Yellowstone volcano, a pattern of swelling and subsiding magma under this area points to continued volcanic activity. Pozzuoli, a city of 83,000 people on the western outskirts of Naples, near the center of Campi Flegrei, has a history of rising and sinking three times in the last 1,360 years. The Temple of Jupiter Serapis, a Roman marketplace within Pozzuoli dating from the second century B.C., has spent some of its history below sea level. The lower parts of its columns show borings of mussels that live in seawater (**FIGURE 7-25**), indicating that the market sank about 12 meters after its construction and ended up below sea level. During the fifth century A.D., and right before the 1538 eruption of Monte Nuovo within the caldera, the pillars showed a total of 7 meters relative uplift, as magma rising under the caldera pushed it back above sea level. The marketplace generally subsided after 1538, but from 1969 to 1972 and from 1982 to 1984 it rose a total of 3.5 meters. In 1983 and '84, it rose at more than 1 meter per year! Rapid subsidence from 1985 to 2004 was followed by rapid 2-centimeter uplift in two months of late 2006. If the rise of this marketplace indicates that another eruption is coming, what impact would it have on this region today?

FIGURE 7-25 A MAGMA CHAMBER REFILLS

In the Roman market Temple of Jupiter Serapis in Pozzuoli, gray pitted parts of columns show borings from marine mollusks from when part of the columns were underwater.

Past eruptions dumped ash to a depth of a meter or more over much of Naples. If such an eruption were to happen now, it would kill many thousands of people, perhaps hundreds of thousands. Recall that 1 meter of volcanic ash is heavy enough to collapse almost any roof, especially if rain falls on it—and volcanic eruptions typically generate their own weather. The potential death toll and property damage from a resurgent caldera eruption in Campi Flegrei is frightening. It is hard to imagine any workable evacuation plan.

The Cascades of Western North America

The stratovolcanoes making up the Cascades Range lie over an active subduction zone and inland from an oceanic trench, although the trench, being filled with sediment, was not at first apparent. The volcanoes erupt along a nearly continuous line inland from the subduction zone (**FIGURE 7-26**). In an effort to assess volcanic hazards, the USGS embarked on a vigorous campaign of research in the High Cascades to determine when each volcano last erupted, how frequently, and with what types of activity. Those investigations revealed that the chain is far more active than most geologists had supposed; the Mount St. Helens eruption in 1980 confirmed those findings. Future large eruptions in the High Cascades should not surprise anyone.

Although the volcanic processes and products in the Cascades are similar to those in Italy, each region has unique hazards associated with it. The risk to human life

FIGURE 7-26 ACTIVE CASCADES VOLCANOES

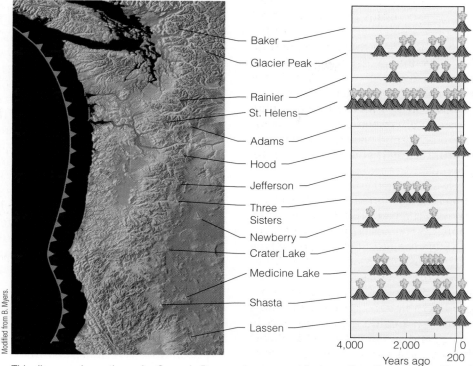

This diagram shows the major Cascade Range volcanoes and their eruptions through time. The chain of active volcanoes lies above the active subduction zone, shown as a green line with pointers sloping down dip.

and property in the Cascades is limited by the fact that although there are some large cities near volcanoes, they are sufficiently distant to provide some time for warning and minimize the numbers of people who need to be evacuated. With a few exceptions, the designation of volcanic areas as national parks, national monuments, national forests, or wilderness areas restricts settlement around many of the large and active volcanoes. One factor that compounds hazard in the Cascades region, however, is that these volcanoes lie in the belt of the westerly winds and close to the western coast of North America. Those winds carry heavy moisture from the Pacific Ocean, dump it on the mountains as rain and snow, and thus increase mudflow hazards. We discuss a few of the more dangerous Cascades volcanoes below.

MOUNT RAINIER Mount Rainier is the largest, highest, and most spectacular volcano in the main line of the High Cascades, rising from sea level to 4,393 meters and dominating the skyline of a large part of western Washington (**FIGURE 7-27**). Mount Rainier is arguably the most dangerous of the High Cascades volcanoes simply because of the large population close to its base. Rainier's status as a national park prevents most kinds of development that might attract even more people to crowd in too close.

Although Rainier last erupted in the 1840s, studies show eleven eruptions in the 10,000 years or so since the last ice age; nothing suggests that it is extinct. A substantial eruption could dump heavy ash falls on cities built on old ash

FIGURE 7-27 RAINIER HOVERS

Mount Rainier looms over Tacoma, Washington. Lahars from its flanks have reached all the way to Puget Sound, including the areas of Seattle and Tacoma.

falls downwind, most likely to the east and northeast. Several smaller towns lie within range of large rockfalls or pyroclastic flows. Rainier's greatest threat probably lies in the paths of mudflows likely to pour down the broad valleys to the north and west at speeds of more than 100 kilometers per hour. Such mudflows could develop without a volcanic eruption.

MOUNT ST. HELENS In 1980, Mount St. Helens was well known to be the most active volcano in the Cascade Range and the most likely to erupt. Its smooth, symmetrical shape showed that it must have erupted recently and filled deep valleys that mountain glaciers had carved in its flanks during the last ice age. When Mount St. Helens erupted on May 18, 1980 (**FIGURE 7-28**), it was observed as closely and studied as exhaustively as any volcanic event to that time. St. Helens had been quiet since 1857, when it produced andesite lava flows, and 1843 when it pushed up a dacite dome (a composition roughly between rhyolite and andesite). Pyroclastic flows from earlier eruptions

reached at least 20 kilometers down the valleys around the volcano, and mudflows traveled at least 75 kilometers.

Radiocarbon dates on wood charred and buried during previous eruptions were available before 1980 to show that most of the modern volcanic cone grew during the last 2,500 years. That pace of activity in the geologically recent past had already persuaded most geologists to consider St. Helens the Cascades volcano most likely to erupt.

The USGS issued hazard forecasts more than two years before St. Helens erupted in 1980. The forecasts proved generally accurate, although they understated the probable size and violence of the upcoming eruption. The agency informed the public of the dangers that existed and tried to dispel imaginary ones. It did not attempt to predict specific eruption times, though many people believed it could and should have done so. The USGS provided information but did not dictate which areas should be closed because such public policy decisions were beyond its authority. The whole episode provided a learning experience in both prediction and public policy for volcanic eruptions.

MOUNT HOOD Mount Hood, the spectacular volcano 75 kilometers east of Portland, Oregon (**FIGURE 7-29**), has not erupted since the 1790s. That is not a significant hiatus for a large andesite volcano. Radiocarbon dates on charcoal preserved in volcanic deposits reveal two major eruptions during the last 2,000 years. Volcanic domes near Crater Rock have repeatedly collapsed over the lifetime of the volcano to shed numerous hot pyroclastic flows down the southwest flank. Some reached as far as 11 kilometers from the peak.

There is little doubt that Mount Hood will produce more eruptions, probably like those of the past 2,000 years, but it is not currently possible to predict when. Close study of deposits that reveal the characteristics of the most recent eruptions provides the best guide to what might happen during the next one.

FIGURE 7-28 THE BIG BLAST

In the climactic eruption of Mount St. Helens in May 1980, hot ash-laden gas boiled out of the vent, expanding into the cooler surrounding air. The darker ash-rich upper and outer parts of the cloud are beginning to cool and descend.

FIGURE 7-29 MOUNT HOOD WATCHES OVER PORTLAND

Although Portland, Oregon, lies to its west, occasional storms have winds from the east that could dump large amounts of ash on the city.

FIGURE 7-30 SIGNS OF DEBRIS FLOW

Debris flows at the east base of Mount Hood have raced north toward the town of Hood River on the Columbia River, reaching it in just over an hour. The larger boulders are carried along near the top of the flow.

Another eruption would probably send pyroclastic flows as far as 10 kilometers down the south and west flanks of the cone. Mount Hood does not generally erupt large amounts of ash in clouds that drift downwind. Water from melting snow and ice would lift ash and rubble from the surface of the volcano and send mudflows racing down Sandy River and its tributaries (**FIGURE 7-30**), toward the eastern edge of Portland and probably north toward the town of Hood River. They would reach nearby towns to the west on the Sandy River in 30 minutes and the larger towns of Sandy and Troutdale in 2 hours and 3.5 hours respectively.

In all likelihood, Mount Hood will provide some warning of its next eruption in the form of swarms of small earthquakes and clouds of steam dark with ash. Unfortunately, its frequent small earthquakes decrease the usefulness of seismic activity in predicting an eruption. The slopes of the volcano are almost uninhabited, so property damage there would be limited. However, mud and debris flows pouring down river valleys toward Portland or Hood River would inflict immense property damage. If warning did not come several days before an eruption, mudflows could kill large numbers of people.

THREE SISTERS Among the lesser known of the Cascades volcanoes are the Three Sisters, a spectacular trio of volcanic cones just west of Bend, Oregon. Although any of the Cascades volcanoes could reawaken, the Three Sisters were not high on most people's list of the most likely mountains to erupt anytime soon. All that changed in March 2001 when USGS scientists using satellite radar interferometry discovered a low but broad bulge that had been rising since early 1997. By spring 2002, the 16-kilometer-wide bulge had risen a total of about 16 centimeters—3.3 centimeters per year. The crest of the bulge is 5 kilometers west of South Sister. Its shape suggests a magma chamber 5 or 6 kilometers below

the surface. A large eruption on the east flank of South Sister could be catastrophic for Bend, with a population of 60,000 people. Recent earthquakes add to the concern about renewed volcanic activity.

MOUNT MAZAMA (CRATER LAKE) Mount Mazama was an enormous andesite volcano that hovered over a large area of southwestern Oregon until around 7,700 years ago, when it destroyed itself in a gigantic rhyolitic eruption. Crater Lake now floods the caldera that opened where Mazama once stood (**FIGURE 7-31**). The geologic record of the Mazama eruption has provided much of our present understanding of such events.

The rim of Crater Lake is the stump of Mazama. It is easy to reconstruct an approximation of Mazama by projecting the outer slopes of the rim upward at the usual angle for large andesite volcanoes, then lopping something off the top for a crater. The result is a volcano fully on the scale of Mount Shasta. Enormous amounts of rhyolite pumice and ash surround Crater Lake for tens of kilometers. Rhyolite ash at least 40 centimeters thick that spread over much of the Pacific Northwest and northern Rocky Mountains provides evidence of an extremely large and violent eruption. Radiocarbon dates on the remains of plants charred in that eruption indicate that it happened 7,700 years ago. By most estimates, the Mazama eruption produced at least 35 times as much magma as the main 1980 St. Helens eruption. Mazama dumped at least 40 centimeters of volcanic ash across large areas of the northern Rocky Mountains. Today, an eruption that size would be a major natural disaster.

Deep deposits of the culminating pyroclastic flow from the Mazama eruption are pale and grade upward into much darker rock (**FIGURE 7-32**, p. 188). Most geologists interpret the gradation in color as evidence that the eruption tapped a magma mass that had separated into two components: the upper part richer in silica than the deeper part. The pale upper magma erupted first to make the lower part of the pyroclastic deposit, and the darker ash erupted last as the deeper part of the magma chamber emptied.

FIGURE 7-31 IT SHOULD BE NAMED CALDERA LAKE

Crater Lake is the caldera depression that formed when Mount Mazama erupted 7,700 years ago. The event was many times the size of the 1980 eruption of Mount St. Helens.

FIGURE 7-32 THICK ASH DEPOSITS

A

B

A. A thick off-white pyroclastic flow layer from Mazama grades up to a darker cap, evidence that the eruption tapped progressively deeper into a differentiated magma chamber. **B.** Pumice exposed in this quarry east of Crater Lake was deposited from the eruption of Mazama 7,700 years ago. Pumice lumps vary in size. Exposure is approximately 2 meters high.

For many years, geologists generally agreed that Mazama had blown itself to smithereens in an enormous steam explosion that opened a gaping crater 10 kilometers across. Then rain and groundwater filled the caldera to make Crater Lake. But more detailed study revealed an absence of old and altered volcanic rock in the debris blanket around Crater Lake. All the debris is freshly erupted volcanic ash and pumice. That discovery led geologists to conclude that Mazama's peak simply sank into the emptying magma chamber beneath the volcano. The basin that holds Crater Lake is a caldera formed by collapse, not a crater blasted out by explosion. Evidently, the eruption that destroyed Mazama's peak was not the end of volcanic activity on the site. A new version of Mazama may yet rise from the ruins of the old one. One thousand years ago, a smaller volcano grew in the floor of the caldera until it barely emerged above the surface of Crater Lake to become Wizard Island.

Crater Lake has become a picture of peace and serenity. It does not present any great hazard because the region supports few people and has enough roads to enable easy evacuation. However, we too easily forget that it was the site of an extraordinarily large and violent volcanic eruption.

MOUNT SHASTA French naval captain Jean-Francois de la Perouse and his crew cruised north along the coast of California in 1786 shortly after failing to see the Golden Gate and the great bay behind it, probably because of coastal fog. But they did report seeing an erupting volcano some distance inland at the latitude of Mount Shasta. That was indeed Shasta in its most recent eruption, the last of eleven in the past 3,400 years.

Shasta, at 4,318 meters, is the second highest of the High Cascades volcanoes (**FIGURE 7-33**). It looms over a large area of northern California, a constant hovering presence in the lives of thousands of people.

Some geologists note that Shasta is approximately the same size as Mount Mazama was before it destroyed itself and formed Crater Lake. And it may be significant that Shasta has produced small quantities of rhyolite during its most recent eruptions instead of the standard andesite that makes up most of the volcano. Finally, seismograph records of earthquake waves that pass beneath Shasta show weak S waves or none at all. This suggests that the earthquake waves pass through a liquid on their way to the seismographs, presumably molten magma in a shallow chamber. Some geologists interpret these observations as evidence that Shasta is getting ready to erupt a monstrous amount of rhyolite ash or lava. Of course, no one can tell when or even whether such an eruption will happen.

Nevertheless, Shasta is a major hazard just as it stands. It is essentially an enormous pile of loose andesite ash and rubble stacked at a steep angle to a height of 3,000 meters with only a few lava flows holding it together. It is likely to shed large masses of rubble with little or no provocation. Shasta did drop much of its north side in a giant rock avalanche 300,000 years ago (see FIGURE 8-13). The debris reached 51 kilometers northwest from the peak and covered at least 350 square kilometers of Shasta Valley. Later eruptions repaired the scar on the volcano. The towns of Mount Shasta City and Weed stand on older rock avalanche deposits directly downslope from valleys on Shasta. They also lie within easy range of large pyroclastic flows.

North America's highest volcano, Citalaltépetl, similar to Shasta and situated at the east end of the Trans-Mexican

FIGURE 7-33 SNOW-CAPPED PAIR

Mount Shasta punctuates the skyline of northern California. Shastina is the younger, less-eroded cone in the foreground of both photos.

Volcano Belt, has seen massive flank collapse twice in its history and may do so again. Its young summit cone is highly fractured and much has been altered to clays from long contact with acidic volcanic gases. Landslide volumes of 0.5 km³ have run out to distances of 20 km in less than 12 minutes. Collapse would endanger as many as 1 million people!

MOUNT LASSEN Visitors to Mount Lassen today see a quiet mountain that provides little indication of a potential eruption. Lassen is a grossly oversized lava dome of dacite that descends from a long history of volcanic violence. It rose within the wreckage of Tehama, a large andesitic volcano that sank into a caldera during a cataclysmic eruption 350,000 years ago. Brokeoff Volcano is a remnant of one of its flanks. Tehama Volcano had earlier risen in the caldera of an even larger volcano that similarly destroyed itself 450,000 years ago. Perhaps we should not be surprised to see more eruptions in this vicinity. Age dates show that six lava domes—called Chaos Crags—rose near Lassen 1,100 years ago. That was some 27,000 years after its previous large eruption. Clearly the time since its last eruption, regardless of how long, does not indicate that the volcano will not again erupt.

Most of the people who lived in northern California in 1914 thought Mount Lassen was thoroughly dead. Then on May 30, with no apparent warning, it suddenly produced an enormous cloud of steam with loud booming noises. Newspaper accounts tell of raw panic in the streets of Redding and Sacramento and places in between. Several lesser eruptions then overflowed the summit crater with extremely viscous pale dacite magma that broke off as blocks that tumbled downslope. On May 19, 1915, the north slope collapsed, sending a mass of hot blocks racing down over a 10-square-kilometer area at the base of the mountain. Eruptions climaxed on May 22 with a dark mushroom cloud of steam and ash that reached a height of 8 kilometers. Melting snow sent an enormous hot mudflow down the east slope that filled 50 kilometers of the Hat Creek valley overnight. Some of the blocks of hot lava that were buried in the mud spouted steam for months. Lassen continued to produce sporadic clouds of steam until 1921.

The area near Lassen is thinly populated, so any future eruptions will probably cause minimal loss of life or property. Redding, Red Bluff, and other towns to the west are 50 to 75 kilometers downslope and against the prevailing wind for the most likely hazard: falling ash. Mudflows could possibly reach that far. Susanville, a similar distance to the east, could be heavily impacted by a major ash eruption.

A Look Ahead

Despite our knowledge of volcanic hazards and our ability to monitor volcanic activity, the opportunities for volcanic catastrophe to people and property are greater today than ever before. Growing populations and the fertility of volcanic soils encourage people to settle close to active volcanoes. The chance of dying in an eruption is small enough that most people ignore the hazard.

However, the dangers are real. Many times in the past, mudflows and pyroclastic flows rushed down valleys that drain the flanks of Cascades volcanoes and onto the broader expanses beyond. A 1987 analysis of specific volcanic hazards and their distribution around the High Cascades volcanoes showed them to be far more dangerous than previously expected.

Most large volcanic eruptions are both dangerous and destructive. Some are genuine catastrophes that kill tens of thousands of people and destroy billions of dollars worth of property. Although the aggregate toll of death and devastation from volcanoes pales in comparison to those of earthquakes, landslides, and floods, the potential remains for a truly catastrophic volcanic event.

Cases in Point

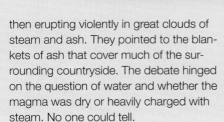

Volcanic Precursors
Mount St. Helens Eruption, Washington, 1980 ▶

The USGS had recently mapped and determined the age of the young eruptive deposits and concluded that indeed Mount St. Helens could erupt again at any time. Then, in March 1980, it showed renewed signs of life—swarms of small earthquakes and blasts of steam and ash—culminating in the climactic eruption in mid-May. Preliminary activity started suddenly in the afternoon of March 23, 1980, when seismographs began to detect swarms of small earthquakes followed by larger ones. Eruption prediction became an issue, and several research organizations rapidly installed arrays of portable seismographs around the volcano. Then, just after noon on March 27, St. Helens produced a large cloud of white steam with a loud boom. The steam rose 600 meters above the volcano while a new crater began to open in its summit. Observers watched a fissure 1,500 meters long open high across the north flank of the volcano, while lesser fractures opened and closed.

Those first clouds of steam were neither hot enough nor dark enough with suspended ash to convince geologists that they were the first phase of a genuine eruption. Many volcanoes blow off great clouds of steam in late spring and early summer as melting snow sends water percolating down into the hot rocks within. St. Helens again blew off clouds of steam about once per hour for seventeen hours starting in the early hours of March 28. Those blasts opened a second crater at the western edge of the first one. Flames began to light the craters at night on March 29; their pale blue color suggested burning methane. The ejected clouds of steam became hotter and much darker with ash. Intense study of the seismic records showed a longer-period wave-like pattern that came to be known as harmonic tremors. These began on April 1 and continued intermittently and with growing strength for twelve days.

Although the frequency of steam eruptions decreased to one per day between April 12 and April 22, they were becoming richer in ash. The situation was beginning to seem serious.

By the end of April, the two craters had combined into a single oval crater 300 to 500 meters across. Seismographs recorded more than 30 earthquakes with magnitudes greater than 3 every day. The arrays of seismographs revealed that the movements were originating below the north flank of the volcano at depths that became shallower day by day. Microscopic study of the newly erupted ash showed that the eruptions were still producing only fragments of old volcanic rocks altered by steam within the volcano. There was no new erupted magma.

A bulge slowly grew on the volcano's north flank in early May; it eventually expanded 106 meters outward from the original slope and about 1,067 meters across. A large mass of magma was clearly rising into the north flank of the volcano. Meanwhile, the clouds of steam began to drop ash that was clearly derived from new magma rising from below. It had the color and composition of dacite.

At this point, it seemed abundantly clear that St. Helens was ready to erupt, but the type of eruption remained open to vigorous debate. Many geologists expected a dome eruption in which the extremely viscous magma would rise quietly through the north flank of the volcano for months or a few years. They pointed to the bulge that swelled in the flank of the volcano. Other geologists reminded everyone that dacite magma is perfectly capable of absorbing water at depth and then erupting violently in great clouds of steam and ash. They pointed to the blankets of ash that cover much of the surrounding countryside. The debate hinged on the question of water and whether the magma was dry or heavily charged with steam. No one could tell.

Those who expected a dome eruption contended that if the mass of magma had sufficient steam to drive an explosive eruption, then it should be venting enough steam to melt the snow off the bulge growing on the volcano's flank—yet the bulge was still white. And many agreed that its growth, at a rate of 1 meter per day, would soon make it steep enough to pose an imminent landslide threat.

At 8:32 a.m. on May 18, a magnitude 5.1 earthquake triggered a massive landslide high on the north flank of the mountain.

▶ *The huge bulge on the flank of Mount St. Helens grew a few weeks before the climactic eruption in May 1980. The umbrella shades the surveying instrument.*

(continued)

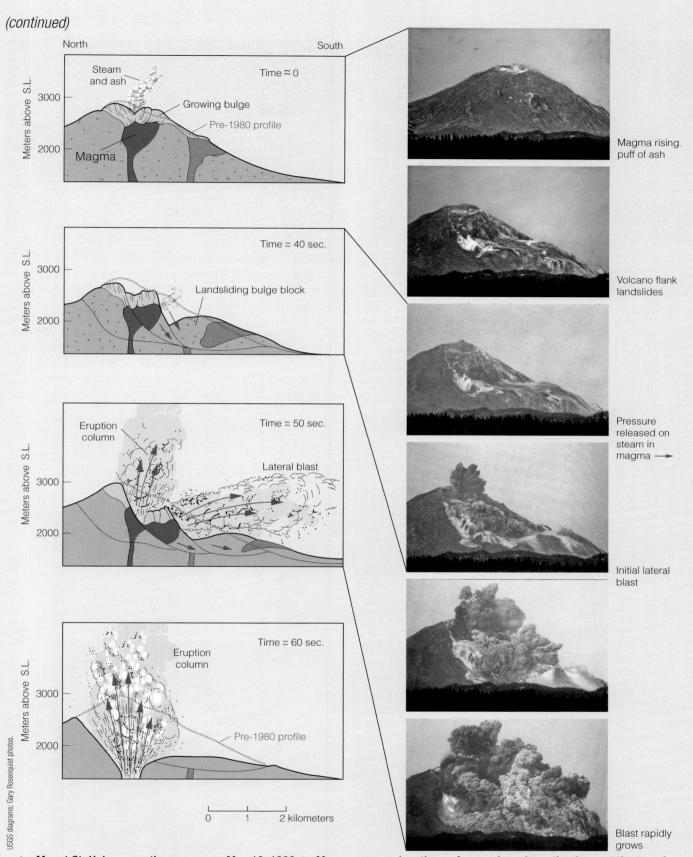

North   South

Steam and ash
Growing bulge
Pre-1980 profile
Magma
Time ≈ 0

Landsliding bulge block
Time ≈ 40 sec.

Eruption column
Lateral blast
Time ≈ 50 sec.

Eruption column
Pre-1980 profile
Time ≈ 60 sec.

0   1   2 kilometers

USGS diagrams; Gary Rosenquist photos.

Magma rising, puff of ash

Volcano flank landslides

Pressure released on steam in magma ⟶

Initial lateral blast

Blast rapidly grows

▶ *Mount St. Helens eruptive sequence, May 18, 1980:* **A.** *Magma approaches the surface and weakens the dome at the top of the magma chamber.* **B.** *The dome collapses in a giant landslide.* **C.** *Pressure release on magma permits gases to bubble out rapidly. The lateral blast blows out to north.* **D.** *A full Plinian-type eruption ensues. S. L. = sea level.*

(continued)

▶ *A map of Mount St. Helens after the 1980 eruption shows the crater open to the north, with pyroclastic flow, lateral blast, and mudflow—volcanic mud- or debris-flow—deposits.*

▶ *This forest was flattened by the lateral blast from the 1980 Mount St. Helens eruption. People circled in the lower right show the size of these large trees.*

That relieved the steam pressure in the magma beneath, an effect similar to popping the cork out of a bottle of champagne. The mass of steaming magma expanded into pumice as it felt the pressure relief, then exploded into a dark cloud of steam heavily laden with ash. A lateral blast shot north, as the eruption cloud soared to 24 kilometers above the volcano. (See sequence of diagrams and photos, p. 191). The event was far larger than the USGS volcanologists expected. The immense vertical column of ash continued to blow out of the crater for much of the day.

The lateral blast cloud, somewhere around 350°C, moved north at about 150 meters per second to reach a distance of 13 kilometers. Hot pyroclastic flows (>700°C) moved at least 36 meters per second to reach as far as 8 kilometers north of the crater. The lateral blast of steam and ash leveled the forest in an area 20 kilometers to the north and 30 kilometers from east to west. Trees were stripped of bark, uprooted, snapped off, and charred. Most on the ground pointed away from the direction of the blast. People in the blast zone who were out in the open were killed or severely burned.

The bulge on the north flank of the volcano collapsed to become a huge debris avalanche and mudflow with a volume of 2.8 cubic kilometers. Much of the debris followed the Toutle River all the way

downstream to the Columbia River, where it blocked navigation until a deep channel was dredged weeks later.

For the next nine hours—steam and carbon dioxide blowing from the crater

supported a column of ash more than 20 kilometers high: a Plinian-style eruption. Parts of the column of vigorously erupting ash collapsed at times to drop pyroclastic flows down the north flank of the volcano.

▶ *Mount St. Helens before and after the cataclysmic May 18, 1980, eruption that destroyed the peak.*

(continued)

The collapsing bulge and debris avalanche displaced the water from Spirit Lake at the base of the volcano and swept over the ridge to the north. It raced down the north fork of the Toutle River at speeds of 110 to 240 kilometers per hour. For 22 kilometers down the valley, it left a hummocky deposit averaging 45 meters thick.

The new crater left by the landslides and eruption lopped 400 meters off the top of the original peak; it was 1.5 by 3 kilometers across, oval, and open to the north. The total volume of newly erupted magma (before it spread out as ash) was approximately equivalent to 0.2 of a cubic kilometer of solid rock. Although the eruption of St. Helens in 1980 was modest by volcanic standards. By contrast, Mount Mazama (the remains of which now surround Crater Lake in Oregon) erupted 35 times that volume.

News of the May 18 eruption spread slowly. Many people in eastern Washington had no inkling that anything had happened until the eruption cloud suddenly appeared overhead and began to dump ash on their Sunday afternoon picnics. The eruption cloud moved directly east, dropping significant amounts of ash as far as 800 kilometers downwind in western Montana, with some continuing southeast to Colorado. The ash fall hampered transportation, utility systems, and outdoor activities until a heavy rain cleared the air four days later.

Few people in eastern Washington and the northern Rocky Mountains had experienced a volcanic eruption. Many found the ash fall frightening. As far east as Missoula, Montana, the late afternoon western sky turned a ghastly greenish black before fine light gray ash began falling like persistent snow. Under an otherwise blue sky with no wind, the finer particles hung suspended in the air for several days. People stayed indoors with the windows closed, and many wore masks for brief excursions to work or for groceries.

Water from rapidly melting snow, Spirit Lake, and the Toutle River created mudflows on May 18 that poured down the river's north fork at 16 to 40 kilometers per hour. They flushed thousands of cut logs downstream, destroyed 27 of 31 bridges, and deposited sediment in navigation channels, including the Columbia River shipping channel near Portland, Oregon. Mudflows diverted streams and raised valley floors as much as 3 meters and channel beds as much as 5 meters. Airborne ash blocked Interstate 90 and other highways. Airports more than 400 kilometers downwind were closed for a week while crews cleared 6 to 10 centimeters of ash.

▶ **A.** *Huge mounds were left by the May 18 debris avalanche and mudflow that raced down the Toutle River.* **B.** *Lines from the Toutle River mudflow were left high on trees after the flow continued to drain downvalley.*

Lyn Topinka, USGS.

Bouldery mudflow piles

A

Height of mudflow

Lyn Topinka, USGS.

B

Pyroclastic Flows Can Be Deadly
Mount Pelée, Martinique, West Indies ▶

Mount Pelée is a typical andesitic volcano, rising approximately 1,400 meters high near the north end of the island of Martinique, one of the larger islands in the West Indies. The town of St. Pierre is picturesquely sited along the curving head of a small bay with the symmetrical cone of Mount Pelée looming 10 kilometers to the north.

In 1900 St. Pierre was a beautiful small city of 28,000 people, a shipping and manufacturing center for sugar and rum. It also served as a port for a fishing fleet and large numbers of tourists from Europe and North America. The people of St. Pierre were quite aware that Pelée was a volcano—historical records contain brief accounts of minor eruptions in 1762 and 1851—but there was no hint of a major eruption or dangerous activity.

Swarms of small earthquakes began shaking Pelée in early spring 1902. As magma rose into its summit, Pelée began to pour occasional glowing andesite pyroclastic flows down its southwest flank in the general direction of St. Pierre but without reaching that far. The increasing level of volcanic activity soon convinced many people who lived on the flanks of Pelée that they were in grave danger. St. Pierre was still nearly unscathed, so many refugees stopped there; others fled farther south along the coast to Fort de France, the capital of Martinique.

As concern mounted, the French colonial governor convened an advisory committee of five prominent citizens who were knowledgeable in various aspects of public affairs and science. Only two of them considered the situation even mildly dangerous.

At noon on May 5, a hot mudflow boiled out of the shallow lake that flooded the crater and down the Rivière Blanche to destroy and bury the sugar factory at its mouth at the northern edge of St. Pierre. It killed 40 people and left an expanse of black mud studded with boulders that was more than 50 meters deep and 1,500 meters across the mouth of the river.

Thursday, May 8, was Ascension Day, a holiday in Martinique. The sky was blue except for the plume of ash and gases drifting west from the summit crater of Pelée. Soon the sky darkened to a dirty gray as the plume of ash spread across it. People swarmed the streets of St. Pierre. At 8 a.m., men on the deck of a ship 16 kilometers offshore and coming in to St. Pierre saw a dark cloud gush from the crater of Pelée and glow internally. They heard an accompaniment of sharp booming noises that sounded like heavy artillery. As the eruption cloud rushed down the mountain, its less dense part rose into the air as a plume of steam dark with ash.

The denser part formed a pyroclastic flow, a cloud of steam and ash hot enough to glow in the dark, which poured down the southwest flank of the volcano directly toward St. Pierre at a speed later estimated at 190 kilometers per hour. It engulfed the town within two minutes, arriving with enough force to flatten walls standing perpendicular to the flow direction.

People who had been on the street when the fast-moving glowing pyroclastic flow arrived were cremated where they fell, leaving little piles of bones and ashes on the sidewalk. People who were inside when the flow struck may have survived its blow, but the suffocating gases and intense heat of the glowing steam instantly ignited everything flammable in its path. Thousands of fires soon merged into a monstrous firestorm that flared high as it destroyed what remained of St. Pierre and sucked the oxygen out of the air.

The pyroclastic flow then rushed out across the bay, making great

1902 ash-flow path

▶ **A.** *Mount Pelée looms over the modern town of St. Pierre, Martinique. The path of the ash flow (pyroclastic flow) is shown in red.* **B.** *Pyroclastic flow from Pelée entered the sea north of St. Pierre, 1902, the first record of an active pyroclastic flow.* **C.** *The pyroclastic flow from Mount Pelée flattened walls and completely destroyed St. Pierre in 1902.*

roaring noises. This part of the cloud was too heavy to rise into the air, its denser part flowing down under the water. As one ship came into port, the people on board saw seventeen or eighteen large ships in the harbor unmanned and in flames, along with a large number of smaller boats. The ship's crew saw people dashing about the dock area just before the fatal cloud swept over them and the entire city burst into flames.

The exact death toll in St. Pierre is unknown, but the best estimates place the figure close to 29,000. That would include the citizens of St. Pierre who did not evacuate, the refugees from the flanks of the volcano, and those who died on boats in the harbor. Only two of the people in the direct path of the glowing cloud survived, both badly burned in the ash and steam. One was a prisoner locked in

a dungeon awaiting trial. The authorities dropped the charges against the jailed survivor because all of the records of his case had been destroyed in the eruption. He went on to become a minor celebrity on the carnival circuit as he toured Europe and North America displaying his dreadful scars.

People who visited St. Pierre soon after the tragedy were amazed at the extremely modest amounts of volcanic ash; little drifts lay here and there, but nothing even remotely resembling a blanket. The pyroclastic flow had been the major agent of death and destruction. A French geologist, Alfred Lacroix, visited St. Pierre a few weeks after the catastrophe to try to reconstruct and explain the events of the morning of May 8. Lacroix recognized that the lavas likely to

explode were those that were too viscous to rid themselves of a large content of steam and other volcanic gases. Instead of bubbling freely from the magma, the water vapor expands within the lava until it explodes, shattering the magma into a cloud of ash.

Nothing could have been done to mitigate the property damage, which included the value of the entire town plus the boats in the harbor, but any reasonable attempt to evacuate St. Pierre would have greatly reduced the death toll. This did not happen because the available advice suggested that the danger was not great enough to justify an evacuation order. In 1902, no one understood the destructive potential of clouds of volcanic gases loaded with enough ash to keep them on the ground.

The Catastrophic Nature of Pyroclastic Flows
Mount Vesuvius, Italy ▶

During the reign of Nero, residents of the bustling Roman towns of Pompeii and Herculaneum may not have thought much of potential danger from Mount Vesuvius, the volcano looming over them. They had forgotten during its 700 years of quiet that Vesuvius was capable of erupting. But in the summer of A.D. 79, they were reminded of the volcano by earthquakes, along with reports of rising ground levels. On the

morning of August 24, a series of steam explosions dropped a few centimeters of coarse ash close to the volcano. Pliny the Younger, whose account of the event has been preserved, sat across the bay, watching the eruption unfold. Meanwhile, his uncle, Pliny the Elder, who was commander of the Roman fleet in the Bay of Naples, dashed about in a ship equipped with oars, trying to establish some sort of order.

Vesuvius went into full eruption shortly after noon as ash and pumice rose in a scalding column of steam. High-altitude winds carried the ash plume directly over Pompeii, situated 8 kilometers south and east of the summit. Approximately 12 to

▶ *Vesuvius looms over the excavated ruins of Pompeii.*

▶ *These are casts of the cavities left by bodies killed in the ash, as they were found in Pompeii on top of the air-fall pumice.*

(continued)

15 centimeters of white pumice fell per hour until evening, a total of 1.3 meters. Roofs must have begun to collapse under the weight of the ash after the first few hours.

The volatile content of the magma decreased, or the vent widened, or both, after twelve hours of continuous eruption. The proportion of ash in the erupting column of steam increased to reach a height of 33 kilometers. The increasing ash content finally made the rising column of steam so dense that it collapsed onto the flanks of the volcano. It became a series of pyroclastic flows that killed everyone in their paths, including many who had escaped west to the shore of the Bay of Naples.

Many of the survivors walked around on the pumice during a pause in the activity. Then another big surge of scalding steam and ash swept over them and buried what remained of Pompeii, killing another 2,000 people. Archaeologists found many of those last victims inside their houses. They were lying on the earlier deposits of ash and pumice, some holding cloths over their faces. Perhaps they had not left their homes because it surely seemed safer than the frightening scene outside. Ultimately, those who remained were buried under 2.5 meters of ash and pumice.

In the early hours of August 25, a series of pyroclastic flows overwhelmed Herculaneum. They covered the 6 kilometers from the summit of Vesuvius in four minutes. Archaeologists found only six skeletons in the town of 5,000 and hundreds of others in the boathouses along the bay where people had sought shelter. They had died

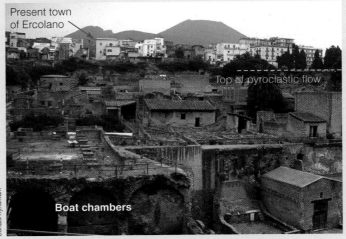

Present town of Ercolano

Top of pyroclastic flow

Boat chambers

Donald Hyndman.

▶ *Boat chambers at Herculaneum can be seen in the foreground. The current town of Ercolano in the middle ground rests on top of the ash that buried Herculaneum. Vesuvius looms in the background.*

instantly when they inhaled the hot ash and steam.

The pyroclastic-flow deposits are 20 meters deep at Herculaneum. They toppled walls of some buildings and buried nearly all of the others. The people who built the current town of Ercolano on top of these flows did not realize that its ancestor was entombed below. A farmer digging a well discovered part of Herculaneum in the early 1700s.

At breakfast time on August 25, a culminating sixth surge of ash and pumice swept across one town, 15 kilometers south of Vesuvius, where it dumped 2 centimeters of ash. Pliny the Younger wrote about the scene, across the Bay of Naples, 28 kilometers west of the crater. He fled with others as the dense black

cloud of hot ash and steam raced toward them across the surface of the water. The dark cloud followed them, hugging the ground—behind it, "fire." His must have been a narrow escape.

Rapid expansion of the eruption column quickly cooled the hot steam, condensing its water vapor on particles of ash, which fell as muddy rain onto the loose debris on the flanks of Vesuvius. Much of it poured rapidly downslope as mudflows that killed more people. As usual with Plinian eruptions, no lava flows appeared. The total volume of magma erupted was 3.6 cubic kilometers within less than 24 hours, approximately eighteen times the volume of magma that erupted from Mount St. Helens in 1980.

Even a Small Eruption Can Trigger a Major Debris Avalanche
Nevado del Ruiz, Colombia, 1985 ▶

Nevado del Ruiz, a 5,389-meter-high stratovolcano in the South American Andes of western Colombia, has been known to be active since the arrival of Spanish colonists in the 1500s. The greatest dangers are not so much from the direct products of eruptions but from the volcanic mud and debris flows that pour down valleys and far

beyond the steep slopes of the volcano. In 1595 and 1845, eruptions triggered mudflows that killed 636 and more than 1,000 people, respectively, in valley bottoms where they lived. Little happened in the next 140 years, so people forgot about the dangers and rebuilt on the site of the earlier disaster; they built the modern town of

Armero on the surface of the old mudflows 74 kilometers from the crater.

In October of 1985, a report on volcanic hazards around the volcano was completed by local and foreign

(continued)

R. Janda, USGS.

A

J. Marso, USGS.

B

▶ *Nevado del Ruiz, Colombia, and the mudflow that destroyed the town of Armero. A. Houses on the slopes above the valley bottom were safe. B. Mudflows cover the former location of Armero on the nearly flat area farther downslope.*

geologists, including some from the USGS. It indicated a high probability of mudflows from a moderate eruption and great danger for Armero and other towns; however, certain local officials dismissed the report as "too alarming." Then a month later, on November 13, the volcano erupted. At 3:06 p.m., an initial blast rained ash and pumice on Armero. The mayor and a local priest calmed people by saying the eruption was minor, which it was. The Red Cross ordered an evacuation at 7:00 p.m. but called it off when the ash stopped falling.

The next eruption came at 9:08 p.m.; it rained hot ash and debris on the snow-covered peak but a storm prevented people from seeing it. Pyroclastic flows melted snow and ice that mixed with the ash and volcanic debris, creating a mudflow that raced down-valley, picking up more stream water and loose material and increasing its volume by as much as four times. It moved downslope at about 60 kilometers per hour. At 11:35 p.m., after people had gone to bed, a flood of cold water from a lake just upstream poured through Armero. Fifteen minutes later, a roaring mass of hot mud, boulders, trees, and debris 2 to 5 meters deep swept through the town, crushing build-ings and sweeping away people. It killed more than 21,000 of the 28,700 people living there and buried them under as much as 50 meters of steaming debris.

The tragedy of Armero underscores the combinational danger of even a relatively small eruption on a volcano capped by a large area of snow and ice; mechanical mixing of hot rock and debris with the snow and ice; entrainment of additional water and sediment in a mudflow to increase its volume; and catastrophic impact from a mudflow confined to a valley that maintains its high velocity even 100 kilometers downstream.

Critical View

A Redoubt Volcano, Alaska, ash deposits.

C. Neil, USGS.

1. What type of eruption process would deposit each of the two main layers in this photo? What characteristics indicate the processes involved?

2. Where would you expect to find the thickest and thinnest parts of each layer?

3. Compare the areal extent of each type of layer.

B Hawaii lava flows.

Donald Hyndman.

1. What are these lava flows called and what kind of rock forms them?

2. What is the contrasting type of lava flow from the same source, which forms the thicker flows, and which moves downslope faster?

C Rio Grande Valley, New Mexico

NASA Int'l. Space Station.

1. In this astronaut view of the top of a giant volcano, a very large depressed area is flanked by stream valleys draining downslope. What is such a depressed area called and how does it form?

2. What is the minimum diameter of the underlying magma chamber?

3. Is this a pre- or post-eruption view?

4. Is such a volcano likely to erupt again? Why?

D Hawaii

USGS.

1. Why do the people in this picture seem so relaxed about its presence? Is it not endangering them? Why?

2. What is burning and why is it burning?

3. The slope on the horizon to the right of the brown hut is almost as flat as the surface of the ocean on the left. What built that slope and why is it almost flat?

4. What rock material forms when this lava pours into the ocean? Considering your answer, what would be the nature of the basalt more than 100 meters below this road?

E Foot of Mt. Pinatubo, Philippines

The inset shows a house before the 1991 eruption.

1. What material on the lower slopes of the volcano, makes up the lower half of the main photo and where did it come from?

2. Did this material likely form before, during, shortly after, or months or years after the eruption? Explain the possibilities and how each would develop?

G Near foot of Mt. St. Helens, early 1980s

This battered car is surrounded by dead trees that have no branches and are mostly lying on the ground.

1. What specific process or processes would have produced this kind of damage?

2. What is the light gray material on the car and on the log and specifically how did it get here?

3. Would the car driver have survived this event and if not, specifically what would have killed him or her?

F Near foot of Mt. St. Helens, early 1980s,

1. What material makes up the lumpy hills in the foreground?

2. Where did it come from and how did it get here?

3. Why are these small hills so lumpy whereas farther down valley (in distance) the valley floor is flat?

H Near Mammoth Mountain, southeastern California

1. What would have killed this large patch of dead trees?

2. What was the ultimate source of the material that killed the trees?

3. Could this same material kill animals and humans in the same area?

4. Have you ever come in contact with this same deadly material? When? Under what circumstances?

Chapter Review

Key Points

Volcanic Hazards

- The number of deaths in a volcanic eruption is highly dependent on the population near the erupting volcano as well as the eruption product.

- Lava flows are destructive to property, because they ignite and overwhelm everything in their path. They do not pose a significant threat to human lives because they are slow moving. (**FIGURE 7-2**)

- Pyroclastic flows consist of ash and steam that rush down volcano flanks. They may be preceded by an ash-rich shock wave called a surge. They can also cross bodies of water. The distance a flow will travel is related to the height of the main eruption column and the density of the ash in it. (**FIGURES 7-6 and 7-7**)

- Ash falls pose a threat of collapsing roofs if the deposits are thick enough, and even a small amount of ash can disrupt our main forms of transport. Planes are especially susceptible if they fly through an ash cloud. Suspended ash can also block solar radiation and lead to crop failure. (**FIGURE 7-9**)

- Mudflows can be triggered when hot ash combines with snow or rain. Hot mudflows are called lahars. Stratovolcanoes are particularly susceptible to mudflows because of their unstable makeup. Fast-moving mudflows often pick up boulders and are a major threat to communities downslope from volcanoes. (**FIGURES 7-11 to 7-15**)

- Volcanoes emit poisonous gases, called vog, including carbon dioxide, sulfur dioxide and hydrogen sulfide. High concentrations of vog can kill people and animals. Vog can also produce acid rain that can contaminate drinking water. (**FIGURES 7-16 to 7-19**)

Predicting Volcanic Eruptions

- Predicting a volcanic eruption well in advance is not currently possible, but scientists can provide long-term forecasts and, once activity begins, often warn of an impending eruption.

- Understanding the hazard presented by a particular volcano depends on analysis of the type, size, rock composition, and frequency of past eruptions.

- A volcano that has not erupted in thousands of years may still be active—intervals between eruptions can be very long.

- Precursors to eruptions include harmonic earthquake tremors that migrate toward the earth's surface, changes in the level or tilt of the ground surface, and changes in erupted gases (**FIGURE 7-21**).

Mitigation of Damage

- Lava flows cannot generally be stopped or redirected without extraordinary measures.

- Warning of ongoing mudflows can be given with modern technology, but the time before arrival can be quite short.

Populations at Risk

- Areas with significant volcano hazards include Italy and the Cascades of western North America.

Key Terms

lahar, p. 169
mudflows, p. 169
paleovolcanology, p. 174

pyroclastic flow, p. 165
surge, p. 165
tuff, p. 174

vog, p. 172
volcanic ash, p. 167
volcanic weather, p. 168

Questions for Review

1. What products of a volcano can kill large numbers of people long after an eruption has ceased? How?

2. Which is more dangerous, a lava flow or a pyroclastic flow? Why?

3. How can you tell whether a plume rising from a stirring volcano contains new magma that may soon erupt?

4. What product of a volcanic eruption causes a widespread drop in temperature and possible crop failures.

5. Why does it often rain during a volcanic eruption?

6. What commonly triggers mudflows?

7. Why is erupting volcanic ash dangerous to jet aircraft?

8. If you visit Mount St. Helens, Washington, you will see thousands of trees lying on the ground, all parallel to one another. Explain how they got that way.

9. If a pyroclastic flow approaches you from across a kilometer-wide lake, are you likely to be safe? Explain why or why not.

10. What characteristics of an old ash-fall tuff will permit you to distinguish it from an old pyroclastic-flow tuff?

11. Which erupting volcanic hazard kills more people than anything else? What accounts for this danger?

12. Scientists once believed that if a volcano had not erupted in the last 10,000 to 12,000 years, it was extinct. Give at least one example that shows this is not correct.

13. What evidence do scientists use to decide whether a volcano may be getting ready to erupt?

14. What causes a big bulge to slowly grow on the flank of an active Cascades volcano?

Discussion Questions

1. Mount Vesuvius, near Naples, Italy is considered to be an active volcano that is a hazard to nearby cities. If you were in charge of civil defence in Naples and the chief volcanologist in charge of monitoring Vesuvius advised you that Vesuvius had a 60 percent chance of a major eruption within two weeks, what would you do? Consider evacuation of large numbers of people, how you would do it, and when you would do it.

2. Discuss why volcanic eruptions are unpredictable in the short term? What factors likely contribute to difficulties in such predictions?

3. If you were designing a study to evaluate the risks of an eruption of Mount Rainier, what type of data would you collect and how would you propose to use this data to minimize risk to those living in the region?

8 Landslides and Other Downslope Movements

Homes in Laguna beach, southern California, collapsed in a 2005 landslide after the slope became water saturated from El Niño winter storms.

Donald Hyndman.

Unstable Hills

On October 2, 1978, a large slab landslide destroyed 24 homes in the steep, forested hills above the town of Laguna Beach, south of Los Angeles, California. Many houses were built on those slopes in the 1950s. The site had not been recognized as an old landslide although the bedrock sloped 30 to 35 degrees, parallel to the ground surface.

The 15 to 20 meter thick slide moved as an intact slab along a slide surface parallel to the slope. It travelled on a surface six meters above and parallel to an ancient slide surface. El Niño weather systems (see Chapter 10) during the winter of 1977–78 brought heavy rains that slowly soaked the ground. Months later this caused high water pressure within the tiny pore spaces between soil grains above a thin layer of slippery clay. That pore pressure helped buoy up the overlying mass of soil, permitting it to slide.

No earthquakes were recorded around that time. Leaking pipes were apparently not a factor in the slide because the composition of imported domestic water differs from that of the groundwater,

and that composition was not detected in the groundwater. Surprisingly, the next year the slope was regraded and the homes were rebuilt. Then, twenty-seven years later, on June 1, 2005, a mere 60 meters east of the 1978 slide, a new slide destroyed 19 homes. Again water was the main trigger; just as before, high water pressure in the sediments followed heavy winter rains that saturated the ground a few months earlier.

Forces on a Slope

As with many of the hazards discussed in previous chapters, downslope ground movement is a natural part of landscape evolution. Mountains rise up and crumble down. Rocks and other materials are constantly moving downslope. Gravity pulls a rock on a slope vertically downward. It may move slowly in a gradual process called "creep," but unless the underlying soil or rock prevents it, the rock can slide or roll down the hill. When a large volume of material moves downslope quickly, it can be catastrophic.

The ability of a slope to resist sliding depends on the total driving force pulling it down versus the resisting force holding it up. The **driving force** consists primarily of the force of gravity working on the weight of the material, while the **resisting force** consists of the strength of the material and the friction holding it in place. Factors such as slope steepness, material weight, and moisture content all play roles that determine when a slope will fail.

Slope and Load

The relationship between **slope angle**, or steepness of a slope, and the **load**, or weight of material on the slope, is a key factor in slope failure. The steeper the slope, the greater the driving force, and the greater the likelihood that the slope will fail. The **angle of repose** is the steepest angle at which any loose material is stable. Different materials stand at different angles of repose, depending mainly on the angularity and size of their grains and their moisture content. Rounded, dry sand grains on a pile will stand no steeper than approximately 30 degrees (measured down from horizontal). Adding more grains to the pile will cause some of them to roll down or patches of them to slide until the slope angle flattens to its angle of repose. In contrast, wet sand will stand almost vertical (**FIGURE 8-1**).

The downslope driving force is determined by the relationship between the slope steepness and the load, or weight, of materials on it. The load imposed by a rock is a factor in both the driving and resisting forces on a slope. The load can be separated into two components, the

FIGURE 8-1 ANGLE OF REPOSE

Donald Hyndman.

Dry sand slides down a slope until it reaches the angle of repose. Wet sand (dark pile on left) can stand nearly vertical.

driving force that pulls down parallel to the slope, and the resisting force that holds it back. On a steeper incline, a greater proportion of the sediment or rock load is directed parallel to the slope (**FIGURE 8-2**, p. 200),which increases the driving force and raises the chances of slope failure (**By the Numbers 8-1**: Slope Failure). Increasing the load on a slope will also increase the likelihood of slope failure.

Frictional Resistance and Cohesion

The driving force pulling materials down a slope is countered by the resisting forces of frictional resistance and cohesion. **Frictional resistance** depends on the slope angle (α) and the load (L) of the body. The area of contact between a mass and the underlying slope does not affect frictional resistance, which means that both small and large masses of the same material will slide at the same slope. The mass will slide, or the slope will fail, when the force (F) exceeds the frictional resistance (f) (FIGURE 8-2).

FIGURE 8-2 STEEPER SLOPES SLIDE

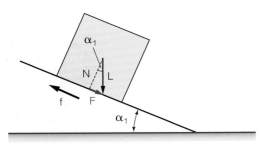

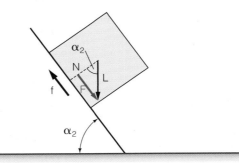

α = Slope angle
L = Load or weight
N = Force perpendicular to the slope
F = Force parallel to the slope
f = Friction holding the mass from sliding

For a moderate slope of 30°:
Force parallel to the slope =
Load × sin 30° = Load × 0.5
(e.g., 100 kg × 0.5 = 50 kg).

For a steep slope of 60°:
Force parallel to the slope =
Load × sin 60° = Load × 0.87
(e.g., 100 kg × 0.87 = 87 kg).

The driving force (F) parallel to the slope is proportional to slope angle.

These two diagrams show the forces on a mass resting on a slope. A steeper slope has a larger force parallel to the slope (red arrow) and is therefore more likely to slide. Note that for the gentler slope, the friction force (f) is larger than the force pulling parallel to the slope (F), whereas for the steeper slope the opposite is true. Clearly the mass on the steeper slope is more likely to slide.

▶ By the Numbers 8-1

Slope Failure

Mass will slide if the:

Frictional resistance + Cohesion < Driving Force

$$f + C < F \text{ or } (N - p) \times \tan \alpha + C < F$$

resisting force < driving force

where:

N = force perpendicular to the slope

$(N - p) \times \tan \alpha$ is "the force against the slope minus the pore pressure of water" times "the tangent of the slope angle"

p = pore pressure

α = slope angle

C = cohesion: soil cohesion includes the soil strength plus the root strength.

Especially important is the addition of water to the slope. When the ground is saturated with water, there is a buoyancy effect that decreases the force of the mass against the slope. Cohesion between grains is like weak glue that can be overcome if the sliding force is large enough. Clays that get wet also lose cohesion.

Anything that reduces the friction on the slope will increase the likelihood of slope failure.

Cohesion (C) is an important force holding soil grains together. It is generally provided by the **surface tension** of water between loose grains or cement between grains. **Cohesion** results from the static charge attraction between minute clay particles, the surface tension attraction of water between grains, or the strong chemical bonds of a cementing material. Most particles have tiny static charges on their outer surfaces. If you walk across a carpet on a cold day, you may feel the shock when you touch a doorknob and the negative-charged electrons you picked up from the carpet jump to the doorknob. Billions of extremely fine particles of clay (<30 microns, or <0.03 mm) have sufficient charges to hold tightly together.

For somewhat larger grain sizes, such as sand, the thin films of water between grains hold them together through cohesion. A child playing with sand quickly learns that a little water makes sand stick together so it can be molded into sand castles. That little bit of water is sufficient to wet the surfaces of sand grains but not fill the spaces between them (**FIGURE 8-3**). For the same reason, two wet boards are hard to separate, as is a piece of glass on a wet countertop. The surface tension attraction of water in a narrow space pulls the glass to the countertop.

In summary, a landslide occurs when a driving force is large enough to overcome both the force from a resisting mass and the friction along the surface, as well as the cohesion strength of materials. As the upper surface of a slide rotates back toward a hill, the driving mass decreases and the resisting mass increases until the forces are again in balance and the slide slows or stops. As with many geological processes, slopes maintain a dynamic equilibrium to keep those forces in balance. Over time, slopes adjust

FIGURE 8-3 WATERY GRAINS

Loose grains with water (blue) filling spaces between them. If water does not completely fill the pores, surface tension holds the grains together.

to near-equilibrium values controlled by the local environment; that is, they reflect the slope material, climate, and thus the water content of the soil.

Slope Material

The material that makes up a slope—as well as its topography and moisture content—also play a role in slides. Loose material above the bedrock is inherently weak. So are loose

sedimentary deposits not yet cemented into solid rocks or soft sedimentary materials, such as clay and shale. These materials are the most likely to slide.

MOISTURE CONTENT As with loose grains in a child's sand castle, the amount of water between grains determines what effect the water will have on the strength of a slope. A small amount of water provides cohesion and helps hold grains together. Too much water (that is, an increase in water pressure) eliminates cohesion and pushes the grains apart.

The spaces between grains are called *pore spaces*. Loose soils have from 10 to 45 percent pore space. When the pore spaces are saturated with water, we can imagine them as continuous vertical columns of water within the soil. The water pressure at the base of each column is determined by the height of the water above it and therefore its load. More water in the slope raises the level of water in the soil, called the *water table*, and increases the pressure in the pore spaces at depth (**FIGURE 8-4A**). The weight of water above a point in the ground provides water pressure that pushes mineral grains apart. That further weakens the soil and makes it more likely to slide. In some cases, you can tell the level of water in the ground, (the water table) by the fact that it is wet up to a certain height (**FIGURE 8-4B**). When the ground is fully saturated, water will ooze out of the surface, signaling an increased likelihood of slope failure (**FIGURE 8-5**, p. 202).

Internal Surfaces

Most solid rocks, such as granite, basalt, and limestone, are inherently strong and unlikely to slide. However, rocks commonly contain more or less planar internal surfaces of weakness that may be tilted at any angle. These include layers in sedimentary rocks, fractures in any kind of rock, contacts between rocks of different strengths,

FIGURE 8-4 WATER IN THE GROUND

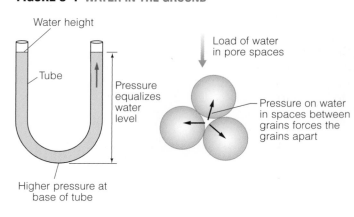

A. Water pressure at depth is equal to the load or weight of the overlying water. **B.** Water (dark) seeps out of the soil below the sharply defined saturation level exposed in a road cut in Glacier National Park, Montana.

FIGURE 8-5 SOGGY GROUND

Saturated and ready to slide

Water oozes out of a water-saturated slope east of Mendocino, northern California.

such as between soils and underlying bedrock, faults, and slip surfaces of old landslides. If any of these zones of weakness happen to lie nearly parallel to a slope, they are likely to become a slip surface. Such rocks are especially prone to slide if their zones of weakness are angled downslope.

Internal surfaces that dip at gentler angles than the slope of a hill, called **daylighted beds**, may intersect the lower slope. These daylighted beds, which are often exposed at their lower ends by a road cut or stream (**FIGURE 8-6**), make ideal zones for slippage. Because there is no resisting load holding them back, only the friction between the layers, along with cohesion, can keep the mass from sliding. The rock above the slip surface does not have to push any rock out of the way to start sliding.

Clays and Clay Behavior

Some soils or rock materials contain clays that absorb water and expand, thereby weakening the rock and even lifting it. Feldspars, the most abundant minerals in rocks, are basically grains of aluminum silicates, which also contain calcium, sodium, or potassium in various proportions. Chemical weathering of all minerals consists primarily of their reaction with water. As they weather, feldspars lose most or all of their calcium, sodium, and potassium, while their aluminum, silicon, and oxygen reorganize into clays with submicroscopic sheets of aluminum or silicon atoms that are each surrounded by four oxygen atoms. Two important clay minerals, kaolinite and smectite, have structures that can lead to landslides (these structures are on the scale of individual molecules and are too small to be seen with a microscope; see more details in Appendix 2 online).

Kaolinite flakes have no overall charge, but they do have weak positive charges on one surface (the top) and weak negative charges on the other. The weak positive and negative charges attract, holding the layers together. Individual kaolinite flakes do not absorb water and do not expand when wet. They form by weathering in warm, wet environments. However, their overall structure is soft and weak, soaks up water, and thus contributes to landslides.

Smectite flakes have an open structure between their molecular layers, which when filled with water, causes the clay to expand dramatically; these are swelling soils. Because water has virtually no strength, almost any load will cause layers to slide easily over other layers. Smectite forms readily by weathering of volcanic ash, so soils with old volcanic ash tend to be extremely slippery and prone to landslides when they are wet (**Case in Point:** Slippery Smectite Deposits Create Conditions for Landslide—Forest City Bridge, South Dakota, p. 225).

FIGURE 8-6 PRECARIOUS SLOPES

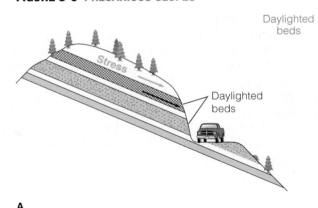

Stress

Daylighted beds

Daylighted beds

A

Daylighted beds

B

A. Rock layers dipping about parallel to a slope, with their edges coming to the surface, are said to "daylight." **B.** Steeply dipping sheets of granite daylight over Highway 3 in southern British Columbia.

Water-saturated muds in marine bays, estuaries, and old saline lakebeds are called **quick clays** because they are especially prone to collapse and flow when disturbed. Silt and clay grains are so fine that water cannot move through the tiny pore spaces quickly enough to escape. When flakes of clay are deposited in salty water, static negative charges on their tiny grains hold them apart; the flakes remain in random orientations so the mass has a total pore space of 50 percent or more. This "house of cards" with water and sea salts in between is unstable. Salt dissolved in water separates into positively charged sodium ions and negatively charged chlorine ions. Because tiny clay flakes have negative charges, the positive charges in the water hold the combination together. If this loose arrangement is disturbed by an earthquake or a heavy load above, the flakes can collapse, permitting the water to escape and "float" the flakes (**FIGURE 8-7**). The deposit liquefies and flows almost like water during the minutes that the clay flakes are moving into their new orientation. Then it again becomes solidly stable and will not liquefy again. The puddles of water that appear during the event filled the much larger volume of pore space that existed before the collapse.

Quick clays are common along northern coasts, including those of Canada, Alaska, and northern Europe. They make perilous foundations for almost any building. When tectonic movements raise a marine or salty lake sediment above the level of the surface water body, freshwater generally washes the salt from between the flakes of clay, leaving the mass even less stable. Water seeping through the deposit eventually removes positive sodium ions that attach to the negative-charged clay flakes. Then the deposit becomes a stack of randomly oriented clay flakes with nothing holding them rigidly in position. That is an unstable situation. One day, the deposit liquefies as the clay flakes collapse into their usual parallel orientation, like sheets of paper scattered across the floor. This commonly happens without warning, even with little or no triggering event.

FIGURE 8-8 FLOWING GROUND

The Lemieux flow in a horizontal terrace of the Leda Clay of the St. Lawrence River valley near Ottawa, Ontario, settled and flowed into the adjacent river on June 20, 1993.

Vibrations caused by an earthquake, pile driver, or heavy equipment can cause failure of quick clays. Even loading a surface can be enough to send a mass downslope as a muddy liquid. In Rissa, Norway, in 1978, a farmer piled soil at the edge of a lake, thereby adding a small load. It triggered a quick-clay slide covering 33 hectares of farmland. The widespread Leda Clay in the St. Lawrence River valley of Ontario and Quebec is another sensitive marine clay. On June 20, 1993, a nearly flat 2.8 million cubic meter clay terrace settled and slowly flowed down a gentle slope into the South Nations River near the town of Lemieux (**FIGURE 8-8**).

Causes of Landslides

Recall from the beginning of this chapter that slope equilibrium involves balancing the relationship between slope angle and load. Changes in slope imposed by external factors—such as undercutting a slope by a stream or road building; loading of the upper part of a slope by construction; adding water by various means; or removing vegetation to permit more access of water—can destabilize this equilibrium and promote sliding. Moist conditions, instability of slope material, or earthquakes can also cause landslides.

Oversteepening and Overloading

The balance between the forces acting on a slope can be upset by making the slope steeper or by increasing the load on the slope (recall FIGURE 8-2). The load can be

FIGURE 8-7 COLLAPSE OF QUICK CLAYS

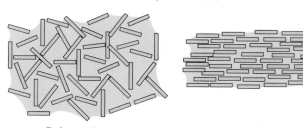

Before collapse,
"a house of cards"
with water in between. (EA)

After collapse,
much less pore space.
The water is displaced.

A. Clay grains deposited in random orientation have especially large pore spaces—a "house of cards" arrangement. **B.** After collapse, the compacted clays take up much less space, so water in the pore spaces must escape.

FIGURE 8-9 RECIPE FOR A LANDSLIDE

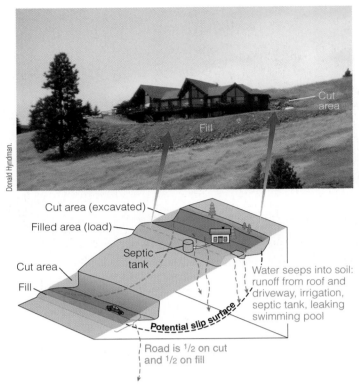

Adding weight or load (fill), slope steepening (cut and fill), building with loose material (fill), and adding water all contribute to landslide risk.

increased by adding material or other load at the top, or removing material at the toe. Load is sometimes added in the form of buildings and soil fill or naturally by rain or snow. Slope angle is increased when fill is added above or when slopes are undercut below. Because steeper slopes are less stable, oversteepening of a slope increases the likelihood of slope failure.

Oversteepening can occur through natural means, such as erosion at the base of a slope. Or it can be human-induced through excavation or adding fill. Often the problem is a combination of such factors. One notorious example is Highway 1 along the coast of central California south of Monterey. Excavated through steep slopes, this highway is frequently damaged and sometimes closed due to landslides. The largest recent landslide events in the area coincided with large El Niño events in 1982–83 and 1997–98, when big waves, strong coastal erosion, and heavy rains hammered the coast. Wave action at the base of the coastal cliffs and rains soaked the fractured rocks to add internal pore pressure, providing ideal conditions for landsliding.

Aside from convenience of location, people build at the tops of steep slopes to take in magnificent views of the sea, lakes, or rivers. Unfortunately, these slopes are steep for a reason; they are being oversteepened by processes that remove material at their bases. Waves or rivers undercut cliffs, and roads undercut the slope bases. Some of these slopes are soft materials, able to stand for long periods until people dig into them to provide flat areas for houses or yards. In doing so, they undercut the slope above and overload the slope below them, making the ground more prone to sliding (**FIGURE 8-9**). This type of development often includes septic systems or swimming pools, which increase soil moisture content and can further contribute to landslide risk.

On some particularly steep slopes, such as the face of a terrace steepened by wave action, the toe of a slide may provide little or no resisting mass. In a recent example along the wave-eroded coast north of Newport, Oregon, expensive homes at the top of 20-meter-high cliffs lost more than 6 meters of property on November 6, 2006. Landslides threatened to take the homes as well. Unusually heavy rains triggered the slides, closing roads in northwestern Oregon; most rivers in western Washington and Oregon reached flood stage. The huge waves and heavy warm rains are a common result of an El Niño event.

In some such cases, a slump may move rapidly, with tragic results. In one such case, part of the face of a steep tree- and brush-covered bluff collapsed suddenly, crushing a home in which a teacher and his young family were sleeping, instantly killing them (**FIGURE 8-10**). The home was part of a single row of houses on the narrow strip of beach on Puget Sound near Seattle. The teacher's parents had expressed concern about the home's location but had received assurances that it had been there for 70 years and there had never been a problem.

Adding Water

Adding water to a slope makes it more likely to slide, because water increases the load on the slope and reduces its strength. Periods of heavy or prolonged rainfall tend to saturate soil, increasing pore water pressure, thereby pushing apart grains and causing slides. Other things being equal, slopes in wet climates are generally more prone to landsliding.

Human actions also add water to slopes and increase landslide potential. Leaking water or sewer pipes or cracked swimming pools feed water into the soil, as do septic drainage fields and irrigation canals (**Case in Point:** Water Leaking from a Canal Saturates a Steep Slope and Triggers a Deadly Landslide, Logan, Utah, 2009, p. 226). Prolonged watering of lawns or excessive crop irrigation will raise the water table and make a slope less stable.

Water infiltration triggered deadly landslides in La Conchita, California (**FIGURE 8-11**). The near-vertical face of the 100-meter-high terrace above the town is made of soft, weak, and porous sediments that are not cemented into rock. An irrigated avocado orchard at the top of the terrace and heavy winter rains soaked it with water. Open cracks visible on the slope in the summer of 1994 were feeding water into the sediments making up the bluff, and water seeping out along

FIGURE 8-10 DANGEROUS HILLSIDES

A. A section of coastal cliff at the edge of Puget Sound, near Seattle, collapsed, crushing and burying a home on the narrow strip of beach. **B.** Magnolia Bluffs, a suburb of Seattle, with spectacular views over Puget Sound, is continually plagued by landslides that threaten and destroy homes. Slides are often promoted by wave erosion at the cliff base.

FIGURE 8-11 REPEATED DISASTER

The community of La Conchita, California, huddles at the base of a soft-sediment terrace that gave way when soaked with too much water. **A.** The 1995 landslide buried houses at the edge of the town but no one died. **B.** In 2005, a second collapse of the earlier landslide mass destroyed more houses to the right, killing ten people. In both **A.** and **B.**, the irrigated avocado field is visible above the slide, and an ancient slide scarp is visible just below the terrace, to the left of the slide. **C.** Note the pronounced tilt of the mid-floor window frames on the right, indicating that the upper floor pushed to the right. Brush behind the house grew over the slide deposit in the three years after the slide moved.

a horizontal surface suggested an old slip plane that could easily be reactivated. A primitive road excavated across the face of the steep slope made it weaker and likely permitted the infiltration of even more water. On January 10, 2005, a huge debris flow of material swept rapidly downslope and buried fifteen houses under ten meters of lumpy wet sand and clay, killing ten people. Even after the second fatal tragedy, many residents chose to remain, including one man whose home was destroyed and brother was killed.

Filling a reservoir behind a dam may also raise the water table enough to cause slides around the edges of the reservoir (**FIGURE 8-12**, and **Case in Point:** Landslide—Forest City Bridge, South Dakota, 1958, p. 225). In some cases, the rising water fills fractures in surrounding sedimentary layers sloping toward the reservoir, causing massive sliding into the reservoir (**Case in Point:** A Coherent Translational Slide Triggered by Filling a Reservoir—The Vaiont Landslide, Italy, 1963, p. 228).

Overlapping Causes

Most causes and triggers of landslides, such as steep slopes, undercutting, overloading, excess water, and earthquakes are independent of one another. The weather does not depend on removal of vegetation, changes in the slope angle, or added load. However, it is when multiple factors

FIGURE 8-12 RISING RESERVOIRS

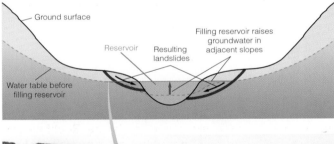

A. Filling a reservoir behind a dam raises groundwater in the adjacent slopes, often leading to sliding into the reservoir. **B.** Filling Roosevelt Lake, at Grand Coulee Dam, Washington, initiated landslides as the water rose.

FIGURE 8-13 REALLY BIG HUMMOCKS

The hummocks that cover most of this photo were deposited in a massive landslide from Mount Shasta between 300,000 and 380,000 years ago.

combine that the greatest catastrophes occur. In one such unfortunate coincidence, the devastating mudflows that accompanied the 1991 eruption of Mount Pinatubo in the Philippines developed in large part because a major tropical typhoon hit the area just in time for the eruption.

Consider a possible scenario for Mount Rainier, the highest and largest volcano in the Cascades. It has not erupted in 2,500 years but is still thought to be active. Its snow-clad flanks are steep but weak; they are heavily fractured and have altered to clays. Winters are wet in the Pacific Northwest, with especially heavy snow packs at high elevations. If the next giant megathrust earthquake (since the last one in January, 1700) were also to happen in winter, strong shaking lasting three minutes or longer could cause a large part of the mountain's flank to collapse, potentially similar to the gigantic flank collapse of Mount Shasta 300,000 years ago (**FIGURE 8-13**). If Mount Rainier collapsed toward communities to the northwest, the consequences would be tragic. What about the possibility that water-bearing magma could be resting within the volcano at the time of collapse? A sudden decrease in pressure by flank collapse could trigger rapid expansion of magmatic gases and eruption, as in the case of Mount St. Helens in 1980. None of these possibilities is implausible. The results could be truly disastrous.

Types of Downslope Movement

Landslides and other downslope movements are generally classified on the basis of material type, movement type, and rate of movement. Materials are classified into categories of blocks of solid bedrock; debris of various sizes mostly coarser than 2 millimeters; and earth or soil mostly finer than 2 millimeters. Water plays a major role in many of these, especially those involving smaller-size particles.

Rates of downslope movements are highly variable, even for individual mechanisms of movement. Rates depend on

Table 8-1 — Typical Velocities for Various Types of Downslope Movement

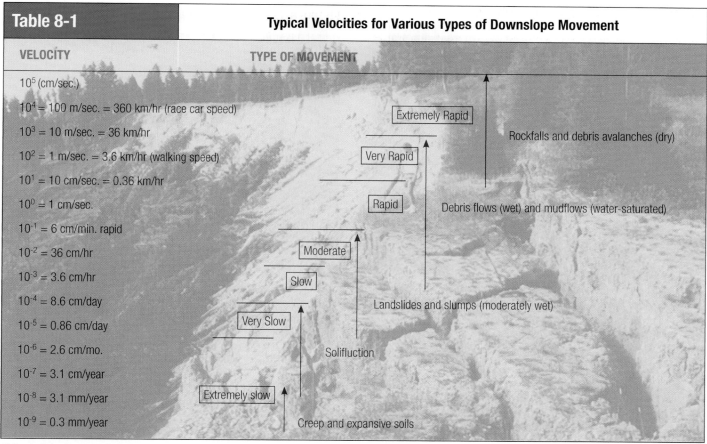

VELOCITY	TYPE OF MOVEMENT
10^5 (cm/sec.)	
$10^4 = 100$ m/sec. = 360 km/hr (race car speed)	Extremely Rapid — Rockfalls and debris avalanches (dry)
$10^3 = 10$ m/sec. = 36 km/hr	Very Rapid
$10^2 = 1$ m/sec. = 3.6 km/hr (walking speed)	Rapid — Debris flows (wet) and mudflows (water-saturated)
$10^1 = 10$ cm/sec. = 0.36 km/hr	
$10^0 = 1$ cm/sec.	
$10^{-1} = 6$ cm/min. rapid	Moderate
$10^{-2} = 36$ cm/hr	Slow — Landslides and slumps (moderately wet)
$10^{-3} = 3.6$ cm/hr	
$10^{-4} = 8.6$ cm/day	Very Slow
$10^{-5} = 0.86$ cm/day	Solifluction
$10^{-6} = 2.6$ cm/mo.	
$10^{-7} = 3.1$ cm/year	
$10^{-8} = 3.1$ mm/year	Extremely slow
$10^{-9} = 0.3$ mm/year	Creep and expansive soils

© Brooks/Cole, Cengage Learning

many factors, including slope steepness, grain size, water content, thickness of the moving mass, clay mineral type, and amount of clay. **Table 8-1** provides approximate movement rates.

Styles of movement include falls from cliffs, topples, slides, lateral spreads, and flows. Note that any of them can involve rock, debris, or soil. A continuous range of characteristics exists between most of these types of materials and styles of movement, and one type often transforms into another as it moves downslope. Moving masses are typically described by names that are a combination of the type of moving material and the style of movement. Among the most common of these are rockfalls, rock slides (talus), debris slides, debris flows, earth flows, mudflows, and snow avalanches.

Rockfalls

Rockfalls develop in steep, mountainous regions marked by cliffs with nearly vertical fractures or other zones of weakness. Large masses of rock separate from a steep slope or cliff, sometimes pried loose by freezing water, to fall, break into smaller fragments, and sweep down the slopes below. Factors that favor rockfalls include cliffs or steep slopes of at least 40 degrees. Rocks most likely to cause rockfalls are those that break easily into fragments: granite, metamorphic rocks, and sandstone. Rockfalls can be triggered by strong earthquakes, passing trains, slope undercutting, and blasting during mining (**Case in Point: A Rockfall Triggered by Blasting—Frank Slide, Alberta, 1903**, p. 229).

These rocks often collect in **talus** slopes, the fan-shaped piles of rock fragments banked up against the base of a cliff (**FIGURE 8-14**, p. 208). Individual rocks, especially large boulders, may bounce or roll well away from the base of the slope. These can be particularly dangerous because of their size, speed, and distance they travel. There is a growing number of cases in which a giant boulder has demolished a house (see FIGURE 1-1).

Rockfall hazards are widespread wherever rocky cliffs cap steep slopes. Denver is on the edge of the High Plains, but communities just to the west are in the foothills of Colorado, where steeply tilted layers of sandstone make ridges and cliffs. Tertiary volcanic rocks cap some bluffs, such as North Table Mountain, which sits just north of Golden. Big boulders that tumbled from its caprock cliffs

FIGURE 8-14 ANGLE OF REPOSE

Donald Hyndman.

These sloping accumulations—talus slopes—of volcanic-rock blocks formed as rocks fell from the cliffs and rolled downslope, coming to rest at their angles of repose along the Salmon River, Idaho.

FIGURE 8-15 ROCKFALL HAZARD

Donald Hyndman.

Some rockfall problems arise where a strong layer, such as sandstone, overlies a weak layer, like shale or clay, as in this southwestern Utah photo. The largest rocks tend to roll well out from the base of the cliff. Is this a safe place to live?

litter the steep, grassy slopes below. Houses now cover the lower parts of the slopes. The county zoned undeveloped land immediately north of it as a no-build area in response to a rockfall hazard map published by the U.S. Geological Survey (USGS). A developer, determined to build a subdivision in the area, challenged this designation less than ten years later and staged a rock-rolling demonstration to prove his claim that the area was safe. Unfortunately for him, several boulders rolled into the area of his proposed subdivision, one bouncing high enough to take out a power-line tower at the mountain's base. His application was denied.

Even in areas of relatively subdued topography and horizontal sedimentary layers, cliffs or high road cuts can be prone to rockfalls. Strong layers of sandstone or thick beds of limestone often break on vertical fractures. When intervening soft layers, such as shale, weather back to leave an overhang, the strong-but-fractured layers can break off and fall. The base of a steep slope capped by vertical cliffs that have shed big boulders in the past is no place for houses (**FIGURE 8-15**). Rockfalls can be hazardous not only to houses and people but also to highways. Roads often follow the base of precipitous cliffs or are built by excavating high road cuts that destabilize original slopes. For example, huge rockfalls repeatedly occur at the Debeque Canyon landslide along Interstate 70 in Glenwood Canyon, western Colorado. The highway in this area is crowded between a massive cliff of sandstone with shale interlayers and the river.

Colorado landslides are most common in areas of Jurassic- and Cretaceous-age marine shales interlayered with stronger sedimentary rocks. Such formations range from

FIGURE 8-16 HARD AND SOFT

Donald Hyndman.

Weak shales and hard, fractured sandstones of the Cretaceous-age Dakota Formation pose a rockfall hazard above I-70, east of Vail, Colorado. Note car at right edge of photo.

about 140 meters for the Morrison formation to 1800 or 2100 meters for the Mancos and Pierre formations. These muddy units, especially the Pierre, with its volcanic-ash-derived swelling clays, are especially weak when wet. Cliffs on some mountainsides or roadcuts are held at near-vertical angles by inter-layered sandstones, but the shales erode back into the cliff, leaving the sandstone layers protruding. Fractures across the sandstone layers permit boulders to split off and tumble into canyons or onto highways (**FIGURE 8-16**). A relatively small rockfall on November 10, 2003, closed Washington Highway 20 in the northern Cascades for an extended period of time (**FIGURE 8-17**).

FIGURE 8-17 ROCKS HIT THE ROAD

Washington State Department of Transportation.

Rockfalls can be hazardous to highways. Granite has spaced fractures that often create large blocks. This rockslide came down across a state highway in the northern Cascades on November 10, 2003.

FIGURE 8-18 POTENTIAL ENERGY TO KINETIC ENERGY

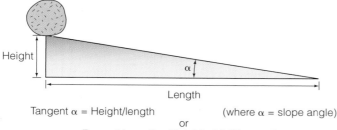

Tangent α = Height/length (where α = slope angle)

or

Runout Length = Height of fall/tangent α

The runout distance traveled by a boulder or debris avalanche depends principally on the height of the mass before it falls, which determines its potential energy. The values are variable, but an approximation is provided by the formula in bold.

ROCKFALL RUNOUT The distance a rockfall will travel, called its **runout**, is generally related to the height from which a rock falls as well as its mass—that is, to its potential energy before it accelerates downslope (**FIGURE 8-18** and **By the Numbers 8-2:** Potential Energy of a Rock on a Slope). Higher, more massive rockfalls travel farther because they have greater potential energy. The same principle guides a ski jumper or a child on a sled: they start higher on a hill to reach farther out on the flat below. Much smaller rockfalls from lesser heights run out much shorter distances. This principle can be demonstrated by a slope of boulders in which the largest rocks are concentrated at the base of the field (**FIGURE 8-19**, p. 210). The biggest boulders that fall from a cliff generally reach the base of the slope for two reasons: first, large blocks high on a cliff have more potential energy than smaller rocks and thus roll faster and farther; second, larger blocks roll easily on the slope's smaller material and

> ## ▶ By the Numbers 8-2
>
> ### *Potential Energy of a Rock on a Slope*
>
> Potential energy of a given rock on a slope depends on its mass and its height on the slope.
>
> #### Potential Energy = m × g × h
>
> where:
>
> m = mass (kilograms)
>
> g = gravitational acceleration (meters per second per second)
>
> h = height (meters)
>
> When the rock falls, its potential energy becomes kinetic energy—the energy of movement. Its highest velocity is near the bottom of its fall, so its kinetic energy is greatest there.
>
> #### Kinetic Energy = ½ m × v²
>
> where:
>
> m = mass (kilograms)
>
> v = velocity (meters per second)
>
> Thus, a larger rock higher on a cliff has more potential energy, will generally accelerate to a higher velocity, and will travel farther out from the base of the slope.

keep going. Little rocks tend to get caught between any larger boulders on a slope.

A formula for the relationship between height and runout can predict the distance of runout for a given rockfall. For small rockfalls of 500,000 cubic meters or less (roughly 80 × 80 × 80 meters), the slope angle α is about 33 degrees. Thus, a mass falling 100 meters would run out about 165 meters. For larger rockfalls, such as that in Elm, Switzerland, in 1887, with a volume of 10,000,000 cubic meters, α is about 17 degrees. That size mass falling 100 meters would run out 325 meters. For an especially large rockfall with a volume of 1,000,000,000 cubic meters (1 kilometer on a side), α is 5 degrees. The mass falling 100 meters would run out over a kilometer. The relationship is similar to the energy line for volcanic ash flows.

Fast-moving rockfalls can run out for distances greater than the height of an original slide scarp. This happens through conversion of the potential energy of a material's original height into the energy of movement. Very large rockfalls can run out horizontally as much as five to twenty times their vertical fall. Thus, a mountainside 1,000 meters high may collapse to run out as far as 20 kilometers. In midwinter 1965, the massive Hope landslide, east of Vancouver, British Columbia, buried Highway 3 when it swept across a valley and continued well up the opposite slope (**FIGURE 8-20**, p. 210).

Debris Avalanches

Rockfalls in which a material breaks into numerous small fragments that flow at high velocity as a coherent stream are called **debris avalanches**. Many of the largest and most

FIGURE 8-19 NO PLACE FOR A CAMPGROUND!

Largest boulders
at bottom

A big talus slope, shed from the cliff above, shows a pronounced concentration of huge boulders at its base. Canyon campground, south of Livingston, Montana.

FIGURE 8-20 A SLIDE BURIES A HIGHWAY

Headscarp

Slide mass

The massive Hope Slide, 18 kilometers east of Hope, British Columbia, buried Highway 3 on January 9, 1965, with no warning and no apparent triggering mechanism. In 2009, when this photo was taken, the extent of the slide is still apparent.

destructive landslides in recorded history started as ordinary rockfalls that developed into debris avalanches. A classic case at Elm, Switzerland, more than a century ago shows how a large rockfall can transform into a fast-moving catastrophic debris avalanche (**Case in Point:** A High-Velocity Rock Avalanche Buoyed by Air—Elm, Switzerland, p. 230).

Debris avalanches can flow downslope at speeds of 100 to 300 kilometers per hour. Some contain boulders as big as houses. A catastrophic debris avalanche struck Yungay, Peru, on May 31, 1970. A magnitude 7.7 earthquake 130 kilometers away, along the subduction zone offshore, triggered the slide on Mount Nevados Huascarán, the highest mountain in Peru. It began with a loud boom and a cloud of dust as 50 to 100 million cubic meters of granite, glacial debris, and ice fell 400 to 900 meters from a vertical cliff to the surface of a glacier and raced down the valley. It disintegrated and then picked up water from ice, streams, irrigation ditches, and soil, traveling downslope at speeds of 270 kilometers per hour (**FIGURE 8-21**).

Halfway down the slope, much of the debris, including huge boulders, hit a ridge of sediment and rocks deposited by a glacier and launched into the air, raining down on houses, people, and animals. Boulders weighing several tons flew as far as four kilometers. Trees blew down for at least half a kilometer beyond the area of boulder impact. Mud splattered more than one kilometer farther, in a blast of wind

FIGURE 8-21 SLIDE BURIES A TOWN

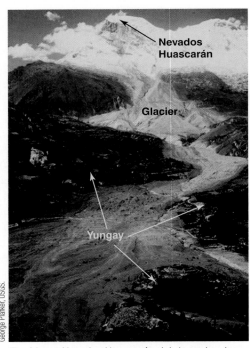

Nevados
Huascarán

Glacier

Yungay

The Mount Nevados Huascarán debris avalanche in 1970 fell from the peak at the top of the photo and raced down the valley to bury the town of Yungay, which occupied the lower half of the photo.

strong enough to knock people off their feet and shred bare skin. Survivors recalled that the strong wind arrived first, followed by flying rocks, then a huge wave of wet debris with a "rolling confused motion." The 1970 avalanche buried the entire city of Yungay, near the end of the flow, in 30 meters of bouldery mud, killing 18,000 people. A mudflow on January 10, 1962, had killed 4,000 people in the same area.

Eyewitness accounts indicate that the total time from the earthquake to the flow's arrival at Yungay (14 kilometers downslope) was approximately three minutes! The average velocity must have been 270 kilometers per hour, but the initial velocity on the mountain must have been faster than 750 kph to fling rocks as far as four kilometers. The victims never had a chance; they could not have seen this one coming. Equally precipitous cliffs of the avalanche scar now mark the peak, and broad fresh cracks parallel the cliffs in ice on the peak. The hazard of further collapse remains. People should avoid building on fans or in valleys that show evidence of previous debris deposits, especially after a recent large event like the 1962 flow.

A mass of rock that falls but does not disintegrate will not run far. The mechanism that permits high speeds and long runouts of debris avalanches has been a matter of considerable debate that has still not been resolved. Some authorities argue that rockfalls ride on a cushion of compressed air trapped beneath them. Air entrained between rock fragments lubricates the mass. But major debris avalanches, such as the one at Elm, Switzerland, scoured deep furrows and excavated the ground beneath them, so some parts could not have ridden on a cushion of air.

Many characteristics of rockfall deposits coupled with witness descriptions suggest that rockfalls flow as a fluid composed of rock fragments suspended in air. The mechanism, called **fluidization**, may work if air cannot readily escape the small spaces between rock fragments during the extremely brief period of movement. With tiny spaces between small particles, rockfalls composed of small grains should be more subject to fluidization than those composed of coarser particles. The air briefly supports the fragments and lubricates their flow.

Rotational Slides and Slumps

One of the most common landslide types is a **rotational slide**, or slump (**FIGURE 8-22**). Homogeneous, cohesive, soft materials, those that lack a planar surface that guides

FIGURE 8-22 ROTATIONAL LANDSLIDE

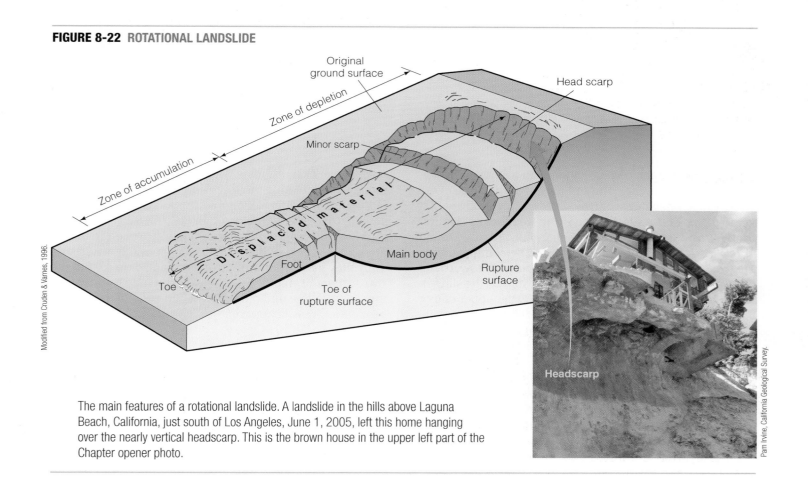

Modified from Cruden & Varnes, 1996.

Pam Irvine, California Geological Survey.

The main features of a rotational landslide. A landslide in the hills above Laguna Beach, California, just south of Los Angeles, June 1, 2005, left this home hanging over the nearly vertical headscarp. This is the brown house in the upper left part of the Chapter opener photo.

FIGURE 8-23 ANATOMY OF A ROTATIONAL SLIDE

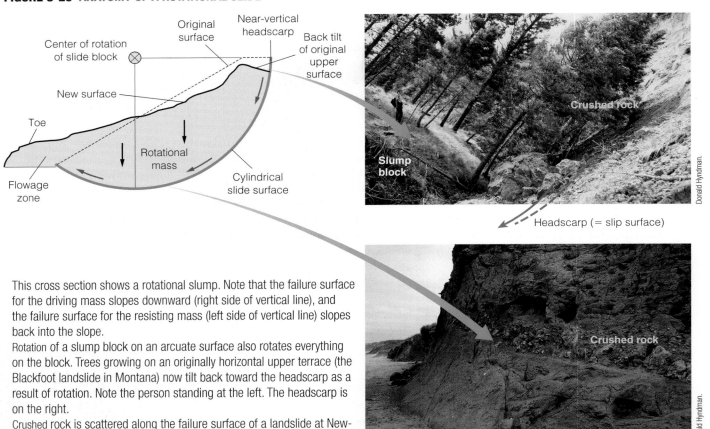

This cross section shows a rotational slump. Note that the failure surface for the driving mass slopes downward (right side of vertical line), and the failure surface for the resisting mass (left side of vertical line) slopes back into the slope.

Rotation of a slump block on an arcuate surface also rotates everything on the block. Trees growing on an originally horizontal upper terrace (the Blackfoot landslide in Montana) now tilt back toward the headscarp as a result of rotation. Note the person standing at the left. The headscarp is on the right.

Crushed rock is scattered along the failure surface of a landslide at Newport, Oregon.

landslide movement, commonly slide on a curving slip surface concave to the sky. The surface curves because at the top of the moving mass, gravity pulls it straight down; that vertical part of the slip surface is the **headscarp**. Farther downslope, the mass is also pushing outward, toward the open air where less load pushes down. The combination of the two forces rotates more and more outward toward the slope (**FIGURES 8-23** and **8-24A**). The curvature of the slip surface rotates the slide mass as it moves, so the upper end of the slide block tilts backward into the original slope while it moves. The lower part of the mass moves outward from the slope, leading finally to the lowest end, the **toe**. Additional slump surfaces may also develop within the rotating block. Below the original rotating surface, the excess material above the original slope may collapse as an incoherent flowing mass at the toe of the landslide. Examples of rotational slides include many in coastal southern California, such as the 2005 Laguna Beach landslide. The lower part of the slip surface commonly dips back into the slope. That provides some resistance so movement ultimately stops.

Engineers can estimate whether a rotational slump will move by calculating the forces on a slope. Because a rotational slump generally moves on a curving cylindrical surface (the green circular arc in FIGURE 8-23), engineers find the *center of rotation* by projecting perpendicular to any exposed part of that surface—such as the headscarp. They may also drill holes through a landslide to find the slip surface at depth. The slip surface may appear as a thin zone of smeared-out soil or a thin zone of thoroughly broken rock, as is apparent in FIGURE 8-23 (lower right photo).

Having determined the shape of the moving mass, engineers calculate the total of the forces pulling the slide downslope (the driving mass) and compare that with the total forces holding it back (the resisting mass). If the driving mass is large enough to overcome the force from the resisting mass, the friction along the potential slip surface, and cohesion, then the slide will move.

In some cases, backward rotation of the block leaves closed depressions at the top that may collect even more

FIGURE 8-24 ROTATIONAL VERSUS TRANSLATIONAL SLIDES

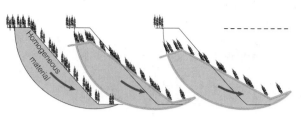

Deep, homogeneous material moves as a relatively coherent mass.
Tree roots don't penetrate deep enough to hinder movement.

A. Rotational landslide

Material slides downslope over top of stronger surface (loose material such as soil over stronger material such as bedrock).
Some tree roots may penetrate deep enough to hinder movement.

B. Translational landslide

A **rotational slide** penetrates deep into the ground; the pre-slide surface and growing trees tilt backward as movement progresses.
A **translational slide** is shallow and parallel to the ground surface; the pre-slide surface may break up, especially at the slide toe. Growing trees remain vertical on the slope but jumble erratic orientations at the toe.

---

water or snow that can soak into the slide. In addition to load, this water increases pore pressure and may facilitate further movement.

Translational Slides

Translational slides move on preexisting weak surfaces that lie more or less parallel to a slope. These may be planes along inherently weak layers, such as shale, old fault or slide surfaces, or fractures. Some involve soil sliding off underlying bedrock. Compared to a rotational slide, a translational slide is shallow, which is demonstrated by the fact that trees slip down the surface and remain vertical rather than rotating with the sliding surface (**FIGURE 8-24B**).

Translational slides are especially dangerous because they often move faster and farther than rotational slides. The range of internal behavior is large. Some move as coherent masses, while others break up in transit to become debris slides. The Vaiont slide of northern Italy is a good example of a coherent translational slide (**Case in Point:** Translational Slide—The Vaiont Landslide, Italy, 1963, p. 228).

Smaller examples that have destroyed homes—sometimes because of clearing for subdivisions or farming—include the destructive Aldercrest slide of southwestern Washington State and many small slides around Pittsburgh, Pennsylvania. In the Finger Lakes area of New York, the Tully Valley landslide in glacial lake clays is an example of an incoherent translational slide initiated by a long period of heavy rainfall and rapidly melting snow: on April 27, 1993, a large area of gently sloping farmland collapsed and flowed downslope (**FIGURE 8-25**). The gently sloping bench of soft glacial lake clays, downslope from steeper forest land, was capped by bouldery dirt left

FIGURE 8-25 SLIDING OF A GENTLE SLOPE!

The Tully Valley landslide affected gently sloping farmland in western New York.

by melting glaciers more than 10,000 years ago. The slide moved as a rapid slump and earthflow. It was not a unique event; a similar slide more than 200 years old lies immediately north of the new one, and many recent but smaller slides can be found in the region. The slope of the affected farmland was only 9 to 12 degrees, but shear strength of the clays was reduced by the high pore-water pressure.

In another example, between 9 and 10 a.m. on February 17, 2006, a steep, unstable slope collapsed following ten days of heavy rain (totaling 63 centimeters) that saturated mountainsides on the island of Leyte in the Philippines. It buried the village of Guinsaugon, including all 246 children and 7 teachers in an elementary school; 1,350 people were missing and presumed buried under rocks and mud as deep as 30 meters. The wet, dense mud left little chance for survival. Survivors blamed not only the rain but illegal logging on the slope above the 375-home village.

FIGURE 8-26 THE GROUND DROPS

Slump scarp

Slump scarp

USGS.

Subsidence and seaward spreading of the marine terrace in Anchorage, Alaska, during the 1964 earthquake, left most wood-frame buildings intact but severed or tilted some of them. This dropped trough formed at the head of the L Street landslide.

LATERAL-SPREADING SLIDES A variant of the translational slide type is sometimes called a lateral-spreading slide. One such event occurred on the marine terrace in Anchorage during the 1964 earthquake (**FIGURE 8-26**). If loose, water-rich sands or quick clays are present at shallow depth, then liquefaction or collapse may send the mass moving downslope. Parts of the moving mass may sink, leaving some blocks standing higher than others. Those in quick clay (described previously) can be fast and destructive when the randomly oriented clay flakes collapse and glide on thin, water-rich zones. Liquefaction, as grains settle into a closer packing arrangement with a lower porosity, expels a fluid mix of sand and water—a quicksand. Liquefaction can occur on a flat surface, where it does not cause a slide but can still collapse buildings.

Soil Creep

Soil creep, the slow downslope movement of surface soils and weak rock, involves near-surface movement and is not especially dangerous. It tilts fences, power poles, and walls. Rates of movement decrease at greater depth because most driving processes operate close to the surface. Alternating soil expansion and shrinkage from several processes, including wetting and drying or freezing and thawing, causes creep to accelerate. When the soil expands, it moves out perpendicular to the slope; when it shrinks, it moves more nearly straight down under the pull of gravity.

The net change is a slight movement downslope. When burrowing animals tunnel into the soil, their cavities eventually collapse. Plant roots expand the soil and then later rot, resulting in the collapse of their cavities. Even trampling by animals and people is significant over a long period of time (**FIGURE 8-27**). Trees that stand straight but with their bases curving back into a slope, so-called pistol-butt trees, have trunks that initially grow upward but become tilted downslope by creep. Even bedrock layers can bend downslope.

FIGURE 8-27 CREEPING GROUND

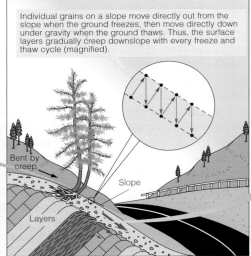

Individual grains on a slope move directly out from the slope when the ground freezes, then move directly down under gravity when the ground thaws. Thus, the surface layers gradually creep downslope with every freeze and thaw cycle (magnified).

Bent by creep

Slope

Layers

Donald Hyndman.

NOAA.

These trees along California's Highway 1 west of Leggett have been bent by soil creep, the slow downslope movement of the upper layers of soil or soft rocks. They originally grew upright but were tilted outward by creep. Continued upward growth produced the bending. Creep of sedimentary layers near the ground surface bends their upturned ends downslope (to the right, in this photo).

Solifluction is another type of near-surface downslope movement that occurs in extremely cold Arctic or alpine areas where water-saturated ground freezes to great depth. It is common downslope from snowdrifts. When near-surface layers thaw, the water cannot drain downward because the soil below is still frozen. The soggy near-surface layers slowly ooze downslope.

Snow Avalanches

In cold climates, **snow avalanches** can be deadly examples of downslope ground movements (**FIGURE 8-28**). An avalanche as little as 30 centimeters deep (1 foot) moves tons of snow downslope at high speed, sweeping victims off their feet and burying them. During the winter of 2007–08, 36 people died in avalanches. In the winter 2009-2010 avalanche season 36 people died in U.S. avalanches, 17 of them on snowmobiles. Conditions for avalanche formation depend upon slope steepness, weather, temperature, slope orientation (north or south), wind speed and direction, vegetation, and conditions within the snowpack. Although snow avalanches do sometimes occur spontaneously, they are often triggered by human actions. A skier crossing a slope can add enough load to trigger a snowpack failure. The weight and vibration of a heavy snowmobile is even more likely to trigger failure.

Steep slopes of 30 to 45 degrees are most prone to avalanching, but avalanches can occur on much gentler slopes given the right weather conditions. The Rutschblock test is commonly used to examine slope stability. A large block of snow more than a meter across is cut from the slope, and if the block collapses when a skier or snowboarder stands on it, the slopes are unsafe. If it doesn't collapse at first, jumping up and down may do the trick. Then it's time to go home!

Particular weather circumstances cause unstable layers to form within a snowpack, leading to extreme avalanche danger. New snow is one risk factor for avalanches. A heavy snowstorm adds a load of loose snow to a slope, just the conditions favored by many skiers. Within 24 hours after a snowfall, the snow is least stable. Ten centimeters of new snow rarely produces dangerous conditions, but more than 30 centimeters is hazardous.

Temperature changes lasting hours or days can also cause unstable layers to form in the upper part of a snowpack. Higher temperatures during a spring day can melt grain surfaces enough to fill any pore spaces with water. Water trickles down through the snowpack and collects at its base. If the ground below is not frozen, that water percolates down to raise the water table. If the ground is frozen, the water may run downslope, over the ground surface but under the snow, leading to slope failure as an avalanche. If water seeps down to an internal boundary between layers of open-textured dry snow against tightly packed or frozen snow, the internal boundary between

FIGURE 8-28 ICE CRYSTALS

A powder snow avalanche races downslope.

the layers provides a zone of weakness that can also lead to sliding.

Unstable layers are also formed when hoar frost on the surface is buried by new snow. Loss of ground heat by radiation at night under a clear sky can cause formation of *hoar frost*—loose, flaky ice crystals—on a snow surface (**FIGURE 8-29**, p. 216). Buried under later snow, hoar frost leaves an extremely weak layer on which an overlying snowpack can slide. Recognition of such loose, flaky layers can help keep you out of trouble in the backcountry.

Wet weather in fall and early winter can leave mountains coated with wet, heavy snow that can freeze as nights get colder. Newer, drier snow that falls on the solid base may leave a weak boundary that can be prone to sliding (**FIGURE 8-30**, p. 216). Winter storms loaded with moisture from the Pacific Ocean often follow warm, moist air that settles a snowpack, making it denser. An arriving storm brings colder air and new snow that collects to form oversteepened—and often overhanging—cornices high on the leeward sides of mountains. Such unstable masses of snow may break loose if triggered by animals, skiers, or even strong wind gusts.

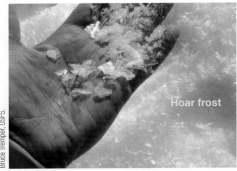

Bruce Tremper, USFS.

Hoar frost

Loose, flaky ice crystals can grow at night under very dry conditions. Their layers lead to very dangerous conditions in a snowpack.

FIGURE 8-30 BREAKING LOOSE

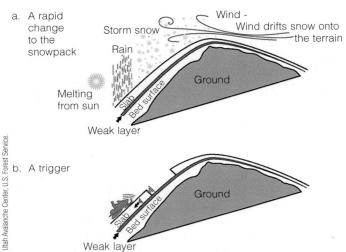

a. A rapid change to the snowpack

Wind -
Wind drifts snow onto the terrain
Storm snow
Rain
Ground
Melting from sun
Slab
Bed surface
Weak layer

b. A trigger

Ground
Slab
Bed surface
Weak layer

Utah Avalanche Center, U.S. Forest Service.

Ingredients for a snow avalanche include a weak layer in the snowpack; a rapid change in the snow from sun melting, rain, or heavy snowfall; and a trigger, such as a snowmobile or skier.

In November of 2008, an early snowpack in the Pacific Northwest states and southern British Columbia was soaked by a mild rain that refroze to form a continuous sheet of ice. Light, fluffy snow fell for the next few weeks; it was extremely cold and dry, so the grains did not stick together nor freeze to the underlying ice. High winds blew some of the light snow over ridges, building thick deposits in downwind areas and gullies as heavier wind-packed snow. In the last few days of December, high temperatures packed the upper parts of the snowpack into a heavy, dense slab that loaded the underlying layer of still-cold,

fluffy snow. That heavy slab was not well anchored to the ground, leading to severe avalanche conditions. Additional snowfall in some areas provided too much of a load on the underlying snow, and throughout January, avalanches swept downslope on the icy base. Elsewhere the slab stayed put until unsuspecting snowmobilers, snowboarders, and skiers added their weight, causing it to let loose.

Recognition of previous dangerous weather conditions or excavation of a section through a snowpack can sometimes reveal hazardous unstable layers. Those who monitor snow-packed slopes to warn against avalanche danger dig trenches in snow to find such weak internal zones.

The orientation of a slope is also a factor in assessing avalanche risk. Winds in the westerlies belt of the northwestern United States and western Canada blow upslope from the west, depositing snow on the leeward eastern sides of ridges. Those downwind sides are often steeper because the snow is deeper, and snow and freezing conditions endure longer in the spring. Because these avalanche-prone slopes collect more snow, many ski areas locate there. North-facing slopes that remain in shadow all day in the winter are dangerous because they don't warm enough during the day to cause the localized melting and refreezing that helps solidify a snowpack.

Landscape features on a slope can also signal heightened risk for snow avalanches. Among the most dangerous places on a slope are under a cornice or in a "bowl," just the places that attract adventuresome skiers, snowboarders, and snowmobile riders. Cracks or breakaway zones tend to form where the snowpack thins as it goes over a rock outcropping or convex slope, at the base of a cliff, immediately downslope from a tree, or under an overhanging cornice (FIGURE 8-31). A fresh avalanche on an adjacent slope facing the same direction and with a similar slope is a very dangerous sign.

The breakaway for an avalanche triggered by a skier or snowmobile rider can appear as a tension crack (compression of a low-density layer at depth, causing localized bending and crack formation). Load can be added to a slope by the weight of a person, a heavy snowfall, or even wind force against the surface of a snowpack. Wind-blown snow in a cornice is tightly packed and heavy; it can easily break off and avalanche downslope. Standing on top of such a ridge is especially dangerous because it is almost impossible to see where the snow overlies solid ground rather than an unstable overhang. A snow ridge's highest point can sit well over an overhang.

Gullies, or the chutes left by past avalanches, are other high-risk avalanche areas. Snow blows off higher surfaces and collects in such hollows. Avalanching down steep mountainside gullies, snow tears out trees and brush, leaving an avalanche chute or "greenslide"

FIGURE 8-31 HIGHLINING BY SNOWMOBILES

A. Avalanches triggered by snowmobile "highlining" at Baker Peak (tracks visible at left of left slide).
B. An alpine skier's paradise on Mt. Baker, Washington. A smooth, uncut slope below a cornice is a recipe for a disaster. One skier died here in 2006.

(**FIGURE 8-32**). A forested hillside with no trees or distinctly smaller trees in a fall-line gulley often signifies an avalanche chute. The greatest danger for mountain skiers or snowmobile riders occurs when their weight triggers an avalanche while crossing a ridge-crest cornice or avalanche chute. These chutes can also lead to sites of dangerous debris slides.

On an active glacier, an ice-fall avalanche can occur at a steep drop. Huge chunks of ice break off steep faces and fall onto slopes below. Such an avalanche can be triggered by movement in the glacier; changes in temperature; external vibrations, such as earthquakes; or disturbance by a hiker or ice climber. They can happen at any temperature or time of day. Intermittent cracking or explosive sounds in the ice indicate instability.

SURVIVING AN AVALANCHE The first rule in avalanche safety is to avoid travelling alone in avalanche country. If you must cross a dangerous slope, do so one at a time, testing the edge of the open slope carefully first. Friends should pay close attention to each skier's location in case of a problem. Even if the first few people safely traverse a slope, the next one may trigger a huge avalanche. The safest areas are on high patches of rocks or in areas of large trees on the upper edges of a slope. All backcountry skiers should carry a shovel, probe, and an avalanche beacon that emits a tracking signal for rescuers.

If you are caught in an avalanche and you are not near the surface as the snow slows, take a deep breath to expand your chest so you have room to breathe later

FIGURE 8-32 AVALANCHE CHUTE

A late-season dirty avalanche over an earlier clean avalanche just west of Stevens Pass, Washington, U.S. 2. The avalanche crosses a railroad snow shed near the base. After the snow melts, avalanche chutes remain as treeless gashes through the forest.

and punch out an air space around your face. Conserve energy by not panicking and yell only when rescuers are almost on top of you; snow rapidly dampens sound waves.

If you see someone caught in an avalanche, watch carefully to determine where you last saw them, most likely in the debris pile of the runout zone. Look for ski poles or clothing to determine the path of the victim. People traveling in potential avalanche terrain should carry lightweight but strong portable shovels in case someone is caught. Collapsible probes that join to form a pole at least 3 meters long can be used to locate a buried person. The quickest way to find a buried victim is through avalanche transceivers carried by each member of a ski party and always set to "transmit." Others in the party then can hone in on a buried transceiver. Deaths sometimes occur by blunt trauma when a victim is rammed into a tree or rock, but more people die by suffocation under the snow. Time is critical because if the person is dug out within 15 minutes, they have a 90 percent chance of surviving. In the next 15 minutes, the chance drops to 50 percent, and after another 90 minutes, there is virtually no chance.

As with all hazards, the best safety measure against snow avalanches is to avoid being caught in one. Evaluate weather and mountain conditions critically and take appropriate precautions. Those experienced in winter backcountry travel are generally better equipped and more aware of the hazards, but even they are likely to be caught in avalanches. Familiarity with snow conditions and steep snow terrain, especially in an area that one has used before, can lead to complacency and comfort with a situation. In one recent case, a group of highly experienced, strong skiers intended to ski a slope they had skied before. The most experienced person of the group skied out onto the slope, trying to trigger a small slab avalanche below that would make the slope safer for them to ski. Unfortunately, a hard slab avalanche more than a meter thick released suddenly above him; it rapidly swept him downslope into a narrow chute and over a cliff. He was knocked off his feet and his mouth was jammed full of snow; he lost a ski, repeatedly banged his head and body against ice and rocks but was lucky to remain conscious. As the slide slowed he managed to "swim" to the surface. He and his friends were amazed he didn't die. Reflecting on the accident later, he emphasized his errors. He was lulled by a false sense of security being with a group of strong skiers on a slope they had skied before. As a result, he didn't critically evaluate the snow conditions and left himself no escape route if the slope failed. He had always felt that if he got caught in an avalanche, he would just point his skis straight downslope and ski out of trouble; he no longer believes this.

Hazards Related to Landslides

Landslides are closely related to several other hazards (see FIGURE 1-8). They can be triggered by storms and flooding or by earthquakes. When a landslide blocks a waterway and then collapses after water backs up, it leads to flooding.

Earthquakes

Many eyewitness accounts tell of great clouds of dust rising from hillsides during and after earthquakes. In most cases, they rise from slides an earthquake has just shaken loose. If a slope is at all unstable, an earthquake is likely to send it downslope. Even without water, sudden shaking may trigger failure.

Earthquakes below magnitude 4 trigger few landslides. Progressively larger earthquakes trigger more and more landslides, especially closer to an earthquake's epicenter. Larger earthquakes may also start rockfalls. The 1959 magnitude 7.3 West Yellowstone earthquake, for example, triggered the massive Madison landslide and rockfall that collapsed from a 340-meter-high cliff above the Madison River channel, west of West Yellowstone, Montana. At 11:37 p.m., the fast-moving slide mass spread out in both directions as it reached the valley floor. It buried 23 people in a campground in the valley bottom. The toe of the slide in the Madison Valley is 1.5 kilometers wide, twice as wide as the slide scar on the south wall of the canyon. A blast of air from under the falling slide mass swept away two people and tumbled one automobile. The bedrock on the mountainside, a huge mass of weathered schist in which the layering was nearly parallel to the south canyon wall, was an obvious rockfall hazard. A strong mass of dolomite marble anchoring the base of the mountain broke during the earthquake, leaving the weak schist and gneiss above it unsupported. The mass crossed the valley floor and moved 200 meters up the opposite slope. The earthquake also triggered smaller slides farther away that took out the main highway (**FIGURE 8-33**).

Of all the different downslope movements an earthquake may trigger, debris avalanches and rapid soil flows make up less than one percent. However, because they move at high speeds on slopes as gentle as a few degrees, they are even more deadly than rockfalls, often killing more people than the earthquake that triggers them. In some cases, they bury towns several kilometers from their starting points. Slow-moving soil and rock slumps and lateral spreads rarely kill many people, but they do collapse buildings.

Earthquakes also often cause the failure of inherently unstable slopes (**FIGURE 8-34**). Among the most susceptible

FIGURE 8-33 EARTHQUAKES TRIGGER LANDSLIDES

The 1959 Yellowstone earthquake triggered many landslides in the vicinity. This one took out the highway and this house.

FIGURE 8-34 LANDSLIDE OVER ICE

A major earthquake on the Denali Fault in 2003 sent the side of this mountain down on top of the Black Rapids Glacier in the Alaska Range.

are recently raised marine terraces composed of soft, wet clays and associated sediments. A prominent case involved Anchorage, Alaska, in the 1964 magnitude 9.2 (M_W) earthquake. Much of Anchorage is built on a flat to gently sloping terrace as much as 22 meters above sea level. Shaking lasted 72 seconds, causing liquefaction of clays in the terrace and collapse (see FIGURE 8-26). Wood-frame houses and other buildings survived moderately well except where they happened to straddle a slump scarp.

Many accounts of earthquakes include reports of sand spouting from the ground or surfacing in big sandboils. This activity is evidence of **liquefaction**, as discussed in Chapter 3. When some earthquake waves pass through soil saturated with water, the sudden shock jostles the grains, causing them to settle into a more closely packed arrangement with less pore space (**FIGURE 8-35**). Because this leaves more water than the remaining space can accommodate, water must escape. During this event, the soil grains are largely suspended in water rather than pressing tightly against one another. The soil settles and is free to flow down any available slope.

You can easily demonstrate the process by filling a glass of water with fine sand. Then repeatedly tap the side of the glass with something hard. The sand will progressively settle as the grains rearrange themselves; the displaced water rises to cover the mass of sand grains. A loosely packed sand with 45 percent porosity might collapse to about 30 percent porosity. Buildings on liquefied soils can tilt and may even fall over when their underpinnings settle or spread (**FIGURES 8-36** and **8-37**, p. 220).

FIGURE 8-35 LOOSELY STACKED GRAINS

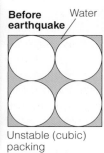

Before earthquake

Water

Unstable (cubic) packing

After earthquake

Excess pore water in liquefied soil carries load of overburden.

Water

Stable (hexagonal) packing

A. Loosely packed grains provide large pore spaces for water. B. If the grains collapse to a tighter arrangement, much of the water must be squeezed out to cause liquefaction.

FIGURE 8-36 FAILING GROUND

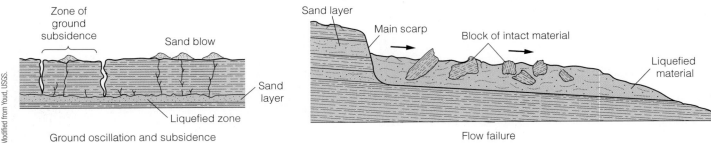

Ground oscillation and subsidence

Flow failure

Two typical types of ground failure occur during liquefaction. As water is expelled from between loose grains, the sediment structure settles, along with anything built above it.

FIGURE 8-37 THE GROUND SETTLES

A. Tilting of a house as the soil below it liquefies. **B.** These houses on loose fill at the edge of San Francisco Bay did not collapse during the 1906 earthquake but settled as the fill liquefied. **C.** Sunken doorways are still visible in the Marina district of San Francisco.

Failure of Landslide Dams

Any moderately fast-moving landslide can block a river or stream to create a dam (**FIGURE 8-38**). When a landslide dam fails, flooding can be catastrophic. Examples of landslide dams include those listed in **Table 8-2**. Often, water will back up to form a temporary lake before the dam eventually fails. Of those dams that failed, roughly a quarter eroded through in less than a day; half failed within ten days; and some lasted for a long time. What controls this behavior, and what are the downstream dangers? The time before a dam fails and the size of a resulting flood depends on:

- The size, height, and geometry of a dam.
- The material making up a dam.
- The rate of stream flow and how fast a lake rises.
- The use of engineering controls, such as the excavation of artificial breaches, artificial spillways, or tunnels.

Mudflows, debris flows, and earth flows create many natural dams that block rivers quickly, are not high, are composed of noncohesive material, and breach soon

FIGURE 8-38 MOUNTAINS FLOW DOWNSLOPE

The 1983 Thistle landslide at Thistle, southeast of Salt Lake City, Utah, began flowing down valley in response to rising groundwater levels from heavy spring rains during the melting of a deep snowpack. Within a few weeks, the slide dammed the Spanish Fork river and took out U.S. Highway 6 and a major railroad line. Water behind the slide dam submerged the town of Thistle. With costs exceeding $400 million, it was the most expensive single landslide event in U.S. history.

Table 8-2 — Prominent Landslide Dams Showing the type of Landslide and Whether the Dam Failed

LANDSLIDE DAM	TYPE (AND MATERIAL)	YEAR	RIVER BLOCKED	DAM HEIGHT (m)	DAM LENGTH (m)	DAM WIDTH (m)	LAKE LENGTH (km)	DAM FAILED?
Madison, Montana	Rock slide (broken rock)	1959	Madison	60–70	500	1,600	10	No
Mayunmarca, Peru	Rock slide (broken rock)	1974	Mantaro	170	1,000	3,800	31	Yes
Gros Ventre, Wyoming	Slide (soil)	1925	Gros Ventre	70	900	2,400	6.5	Yes
Slumgullion, Colorado (background)	Earth flow (altered volcanic rock, smectite clay, soil)	1300	Gunnison	40	500	1,700	3	No
Polallie Creek, Oregon	Debris flow (gravel, boulders)	1980	Hood, East Fork	11	—	230	—	Yes
Thistle, Utah (FIGURE 8-38)	Earth slide	1983	Spanish Fork	60	200	600	5	No
Usoy, Tajikistan	Landslide (broken rock)	1911	Murgab	500–700	1,000	1,000	60	Partial, may yet fail

Costa and Schuster, 1988 photo, USGS.

after formation. Other kinds of flows are often long-lived. Most landslide dams fail because the water behind them overflows and erodes a spillway that drains the lake. That may not happen if the dam consists of large rocks and is so permeable that the lake drains by seepage instead of through an overflow spillway. A small landslide dam is likely to fail soon after formation if it blocks a large stream. Permeable, easily eroded sediment is vulnerable to piping, seepage, and undermining that can lead to dam failure.

To minimize the chance of a catastrophic flood downstream if a dam fails, officials commonly try to stabilize landslide dams by constructing a channelized spillway across or around them. The U.S. Army Corps of Engineers handled the Madison rockslide this way. At the Thistle slide, the Corps used pipe and tunnel outlets.

The height and volume of impounded water, and therefore its potential energy, control the maximum height of a flood from a landslide dam failure. The higher the water level and the greater the volume of water behind the dam, the higher the flood level will be downstream. Most dam-failure flood flows decrease rapidly downstream. However, if a flood incorporates a significant amount of easily eroded sediment into the flow, then it can turn into a debris deluge many times larger than the flow at the dam itself.

Why not construct a useful dam on a landslide dam? Actually, in the twentieth century, at least 167 dams were constructed this way. Most of these were built on top of rockfalls or rock slides because those are most stable. One outstanding example of a landslide-dammed reservoir that produces hydropower is Lake Waikaremoana, the largest landslide-dammed lake in New Zealand.

Only one of these modified landslide dams failed catastrophically. In 1928, the St. Francis high-arch concrete dam, north of Los Angeles, failed while the reservoir was being filled because it was built on the toe of a large Pleistocene landslide in schist bedrock. Four hundred and fifty people were killed. One seldom-addressed concern about constructing a dam atop a landslide is whether other slopes around the new reservoir are also prone to landsliding. If so, filling a reservoir would raise the water pressure in surrounding slopes, possibly leading to major slope failure that would drive water to surge over the dam.

Mitigation of Damages from Landslides

Landslides are widespread throughout the United States and Canada, and their damages can be extremely costly (**FIGURE 8-39**, p. 222). Few insurance policies cover them or any other type of ground movement. Landslides in the United States cost more than $2 billion and 25 to 50 deaths per year. Worldwide, landslides have caused an average of 7,500 deaths per year over the last century and $20 billion per year over the period from 1980 to 2000. As with other hazards, major landslide disasters increase with the growth of world population as people settle in less suitable areas; the vast majority of deaths occur in less-developed countries.

FIGURE 8-39 WHERE ARE THE LANDSLIDES?

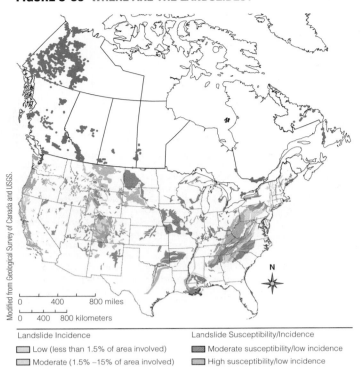

Modified from Geological Survey of Canada and USGS.

Landslide Incidence

- ▢ Low (less than 1.5% of area involved)
- ▢ Moderate (1.5% –15% of area involved)
- ▢ High (greater than 15% of area involved)

Landslide Susceptibility/Incidence

- ▢ Moderate susceptibility/low incidence
- ▢ High susceptibility/low incidence
- ▢ High susceptibility/moderate incidence

Landslides are widespread, not only in mountainous areas of the western United States and Canada and in the Appalachians, but also in the subdued terrain of the central United States and Canada. Canadian landslide areas shown in red.

People build on landslide-prone areas because they are attracted by scenic views or because the land is more affordable.

People often build at the base of cliffs to have their homes nestled in scenic environments (see FIGURE 8-15).

In some cases, they build among large boulders that provide convenient highlights for landscaping. They tend to not think much about where the boulders came from; if they did, they would quickly conclude that they came off the cliff above. These are truly dangerous places to live.

As with all hazards, however, understanding the processes that control downslope movement allows scientists to evaluate risks and implement strategies to reduce damages.

Record of Past Landslides

As with many hazards, a record of past landslides indicates that future landslides are likely. A large slide following a long period without them does not preclude soon having another. In fact, the existence of landslides in an area indicates favorable circumstances for slides.

A hummocky hillside, sometimes obscured by vegetation, may be the remnant of an old landslide. A stretch of road that cracks or exhibits broad waves in pavement, often requiring repaving, is also suggestive of sliding terrain (**FIGURE 8-40**).

Building a road across, constructing a building on, or removing material from the base of such hummocky topography would be unwise. Old landslides can be reactivated by any of the processes that initiate new landslides—that is, adding water, steepening the slope, undercutting the toe (or removing toe material), loading the upper part of a slope, removing vegetation, or earthquakes. Many old landslides are reactivated by removing material from the toe because that material encroached on a road, railroad, or construction site.

What about reduction in slope following landsliding? Does that make the slope less prone to future sliding? Not necessarily; preexisting slip surfaces and surface fractures that permit further water penetration both contribute to

FIGURE 8-40 EVIDENCE OF LANDSLIDES

Donald Hyndman.

A. This hummocky landslide terrain is near Gardena, north of Boise, Idaho. **B.** Wavy road across landslide terrain near Zion National Park, Utah.

further sliding. Building of roads or structures on an existing landslide merely aids additional slope movement. Clearly, if the conditions are appropriate for landsliding, preexisting slip surfaces can reactivate.

Landslide Hazard Maps

The best strategy is to avoid building in places prone to landslides. A Geographic Information System (GIS) can be used to build debris-flow and landslide-hazard maps. Such GIS maps can be used to prescribe restrictions in land use such as on road building, timber harvesting, or even housing subdivisions. High-risk landslide areas where development should be restricted include:

- Steep slopes and clearly mountainous areas.

- Local slopes that exceed the local angle of repose (30 to 45 degrees on a hillside).

- Areas with abundant loose debris on a slope.

- Slopes with low permeability, as in fine-grained soils.

- Areas where large amounts of rainfall or snowmelt enter the ground.

- Locations where shallow slides commonly develop at the interface between bedrock and the loose colluvium that covers it.

- Locations of previous shallow landslides of any size.

Shallow slides are more likely to develop on slopes with sparse vegetation and a lack of significant tree roots to hold shallow material in place. This factor is relevant only for debris flows and translational slides that are shallower than the depth of tree root penetration.

In the GIS approach, the area of concern is mapped—for example, on a scale of 1:20,000—and the area is divided into a set of polygons. Each polygon is chosen as having consistent internal attributes such as slope, concave-upward curvature, soil texture and depth, ease of slope drainage, slope-facing direction, type of vegetation, bedrock type, length of roads within the polygon, and presence of slope failures. Polygons with slopes of less than 15 degrees, for example, may be excluded from study because they typically lack evidence of landsliding and are less likely to slide.

Engineering Solutions

Because the relationship between forces that keep slopes from sliding is established, engineers can sometimes restore the balance among forces to keep a slope stable.

For a slope overloaded at its top, we can add load to the lower part of the slide to resist movement. To stop a slope from moving, highway engineers sometimes pile heavy boulders on the toe area to increase the resisting mass (**FIGURE 8-41**). The slope angle can also be changed by

FIGURE 8-41 WEIGHT THE BOTTOM

Heavy boulders are often piled on the lower part of a slide to resist movement. This road, cut through a landslide on U.S. Highway 101 near Garberville, in northern California, has been stabilized by loading its toe area.

removing the slope's top, adding weight to its base, or remaking the entire slope with a lower angle.

Rock cliffs or slopes can be sprayed with a cement mixture called *shotcrete*, or gunite, to restrict water access. Some are draped with heavy wire mesh to prevent falling rocks from reaching buildings or highways (**FIGURE 8-42**). Some are drilled and anchored in place by *rockbolts* (**FIGURE 8-43**, p. 224).

FIGURE 8-42 SCREENED ROCKS

Chicken wire and heavy cable mesh

Heavy cable mesh draped over crumbly road cut to protect U.S. 2, east of Wenatchee, Washington.

FIGURE 8-43 DRILL AND BOLT

A

B

A. Workers drill and bolt a rock cliff above Highway 1 at the mouth of Topanga Canyon, California, west of Los Angeles. **B.** Long rockbolts and concrete stabilize a fractured slope near Frigiliana, Spain.

Removing water from soil can increase its strength, making it less likely to slide. One of the most effective mechanisms for water removal involves trees and shrubs taking up water from the soil through their roots, a process called *evapotranspiration* that dries the soil. Water from roots reaches leaves, where it is transpired into the air. In addition, a portion of the rain falling on leaves or soil evaporates at the surface rather than percolating into the ground.

Some kinds of trees and shrubs take up water from the soil more eagerly than others. In general, those that grow prolifically adjacent to streams or lakes use great quantities of water. The most common of these are willows, cottonwoods, and aspens. Tamarisk trees may completely drain irrigation ditches in the arid southwest. Where those same plants grow well above a floodplain, the ground probably contains excessive water and may be in danger of sliding. Planting trees or shrubs that use large amounts of water can help stabilize a slope. People sometimes cover potential slide areas with plastic to prevent water penetration (**FIGURE 8-44**), although this also has the unintended consequence of shutting down evapotranspiration.

It is possible to artificially drain and thus stabilize many slopes. If pore spaces in the soil are very small or disconnected from each other, then the soil has low permeability; water does not flow easily through the small pores. One widely used method to stabilize slopes involves drilling holes inclined slightly upward into the slope and inserting *perforated pipes*. Water drains into the pipes and trickles out to the surface (**FIGURE 8-45**). A more expensive approach is to dig deep trenches in the slope with a backhoe,

FIGURE 8-44 LEAKY SWIMMING POOL?

After the 1999 Dana Point landslide north of San Diego, California, black plastic was spread on the ground in the lower left to prevent water infiltration. Note that the pale gray concrete pool (right center) was amputated by the headscarp. Did a leaking swimming pool add water to the slope?

line the trenches with *geotextile* fabric (cloth that permits water but not sediment to flow through), then backfill them with coarse gravel. Water will trickle out through the gravel for years. If the situation is truly desperate, it may help to drill wells into the slope and pump the water out. All these methods reduce the water pressure in the soil, which makes it less likely to slide.

FIGURE 8-45 DRAIN THE WATER

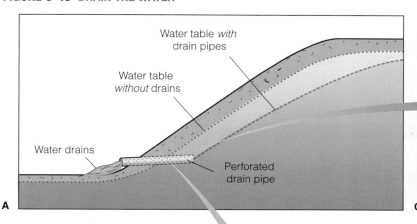

Water table *with* drain pipes

Water table *without* drains

Water drains

Perforated drain pipe

A

Granite

Water draining

C

Donald Hyndman photos.

Water draining

B

A. Installation of a perforated drainpipe can lower the water table and reduce the chance of sliding. **B.** Water drains from perforated groundwater-drainage pipes in shale road cut on U.S. Highway 101 north of Garberville, California. The low permeability of this shale necessitates many drainpipes. **C.** Even water in granite fractures can foster landsliding. Horseshoe Bay – Squamish Highway, British Columbia.

Cases in Point

Slippery Smectite Deposits Create Conditions for Landslide
Forest City Bridge, South Dakota ▶

Even the subdued topography of the Great Plains can be susceptible to landslides. Problems began when the U.S. Army Corps of Engineers built Oahe Dam across the Missouri River at Pierre in the early 1950s, and the new reservoir behind the dam rose about 30 meters. The steep river banks were marked by ancient landslides because

smectite that formed from ancient deposits of volcanic ash in the notorious Pierre Shale swelled and became extremely slippery. The slopes stabilized when the river deposited gravel in the channel at the end of the last ice age, but water rising in the reservoir and an increase in water pressure in the slopes again made the slopes more prone to

Martin McDermott and Roy Hunt, McKinney Drilling Co.

View to east

Landslide escarpment (elev. 1900 ft.)

Hwy. 212

Bridge elev. 1670 ft.

Lake Oahe (elev. 1600 ft.)

▶ *A huge landslide at the U.S. 212 Forest City Bridge across the Missouri River was reactivated by raising the reservoir level behind Oahe Dam.*

sliding. A 1.4-kilometer-long U.S. 212 bridge built across the new reservoir in 1958 crossed those old slides; its approaches immediately began to move slowly down into the reservoir. Drilling

showed the slide to be about 21 meters thick. At one point, repeated repaving of the highway approaches accumulated a total thickness of 2.1 meters of asphalt! Engineers helped stabilize the slopes by

excavating a broad area of material upslope—which lessened the slope angle and unloaded the head of the slide—and piling heavy rock riprap to load the toe of the slide.

Water leaking from a canal triggers a deadly landslide.
Logan Slide, Utah, 2009 ▶

In June 2009, Jacqueline Leavey and her children, ages 12 and 13, moved into a rented home at the base of a 30-meter-high terrace along the Logan River valley. The south-facing, steep, brush- and tree-covered slope behind the house consisted of unconsolidated dirt, sand, and gravel. Neighbors reported water flowing along the north side of Canyon Road for more than a week before the slide; the water was cloudy with dirt, suggesting it was carrying mud from the slope that later collapsed.

On the morning of July 11, Leavey called her sister and her landlord to say that water was flooding into the back bedroom; he came to clear out his small water-diversion ditch behind the house. She soon called back and said the water was coming in too fast; he called 911 to report water pouring down the hill. Hearing rumbling and branches breaking, she went outside to check on the water and mud pouring against the house. While her landlord went up the hill to inspect the canal, she checked the bedrooms. The next-door neighbor came home about 11:45 a.m.; soon afterward he heard cracking sounds and saw part of the hill collapse. About the same time, a mass of mud and debris slammed into the back of Leavey's house, lifted, collapsed, buried most of it, and carried it 6 meters off its foundation. Leavey and her children were killed—buried in the slide debris. Other neighbors ran to the back of the house and called out but heard no one inside. A power line was down, so they retreated when they smelled gas. Mud in the slide continued to spread across Canyon Road, into nearby yards, and along the road for a block.

Afterward it became clear that a 20-meter section of the water-filled irrigation canal had collapsed along with the hillside above and below it. The hundred-year-old concrete-lined canal about half-way up the slope was filled each spring to supply corn fields and other farms in the area. The canal is cut into the steep slope and filled on its downslope side by a 3-meter-wide dirt embankment that is also used for canal maintenance. The upslope side is held back in the steepest sections by hill-slope cribbing. Downslope creep of the hill above presses against the concrete wall of the canal, cracking and displacing it locally, and kinking the once-strong steel pipes (now rusting) that straddle the canal walls to keep them apart. It appears that the canal walls are bulging under the weight of the hillside above them. Although significant water must have been leaking into the porous ground from the canal, at least one sizeable spring just above the canal, and only a few meters from the failure point, was flowing six days after the slide.

According to neighbors, the ditch was not well maintained; it had cracks in the concrete liner and at least one hole larger than 15 centimeters across in the bottom. The irrigation canal has had recurring problems for many years; the Salt Lake Tribune reported that in more than 100 years, there have been dozens of cases of landslides along Canyon Road that have affected homes and endangered people. In most cases the canal was blamed. A mid-1980s resident of the destroyed home said the canal broke several times in the decade she lived there, once sending so much mud and water into the back bedrooms of her home that it "took weeks to clean up." In 1996, the canal failed above another street, and the county diverted the

water down city streets. In 2003, a study sponsored by the county concluded that the canal was deficient in several locations. Two years later, a slide near the July 2009 disaster site showed that slides were indeed likely. The report was presented to Logan City. In May 2009, the Logan City mayor, himself a professional contractor, commented in a city council meeting that "water is leaking from the canals due to cracks in the walls." He said he was not overly concerned by such cracks because all canals have fissures that are "going to cause some drips." Although the city repeatedly received warnings about the danger of the canal to nearby residents, little was done; they left oversight of the canal and decision making to the irrigation company that consists of 800 private owners who hold usage rights to the water. Monitoring of the canals is left to "water masters" who walk them to look for problems. They have few or no qualifications, no training program, and are paid low wages. The Director of Public Works in Logan once noted that "we have to assume [the canal operators] do their jobs." A reading of the City of Logan Land Development Code gives the impression that it focuses on the protection of the irrigation canals from new development rather than on the safety of homes from the irrigation canals.

Steel sheet piling of U.S. 89 along the top of the slope holds back at least some fill that supports the edge of the highway and loads the top of the slope; its presence suggests that it was placed there to hold back a locally filled / oversteepened part of the slope. Because the slide was so wet and was

(continued)

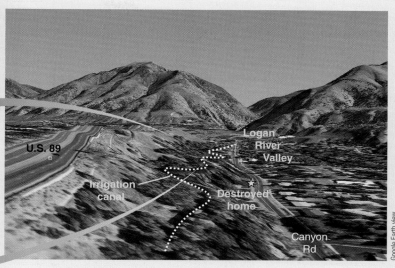

Google Earth view.

Hyndman.

▶ **A.** *An irrigation canal along a steep hillside in Logan, Utah, just east of the slide, with a retaining wall on the slope above and heavy steel pipes holding the canal walls apart. Note the kinks in many pipes, indicating the upper slope is pressing inward.* **B.** *The irrigation canal, cut into a steep hillside in Logan, Utah, is marked by the white dotted line; view is to the east. U.S. 89 is at the top of the terrace to the left; Canyon Road is at the bottom of the hillside to the right. The home buried in the landslide below the canal is marked by a yellow star at the base of the slope just left of Canyon Road.*

expelling a lot of water just before the main failure—and saturated soil easily loses cohesion—it seems likely that water in the slope had much to do with its failure. Given the sequence of events—abundant water emerging from the slope, followed by a small slide just above the house—the slide must have abruptly steepened the water-weakened slope. The small slide then grew upward and caused the collapse of a big section of the water-filled irrigation canal and failure of the main slide that swept down onto the house.

So what caused the landslide? In cases involving human modification of the natural environment, lawsuits are common following damages and deaths from natural hazards. More than one party is commonly targeted by these lawsuits. In this case there are several possible causes:

- Did the building of U.S. Highway 89 load the top of the slope with some fill?
- Was the water in the slope due to canal leakage prior to the slope failure?
- Did the water in the slope derive from springs in the slope above the canal and unrelated to it? What is the source of the water in the springs? Utah State University occupies much of the bench upslope from the irrigation canal. Did the spring water come from the university property or farther north?
- Did cut and fill from building the canal locally steepen and load the slope on its downward side?
- The state provides little or no control or oversight of the irrigation canals.
- The city permitted building on a dangerous site long after the construction of the canal.
- The developer built the house at the base of a dangerous slope and sold it to the owner.
- Did building the house involve local undercutting of the slope?
- The mortgage company provided a loan for purchase of the house in a hazardous location, long after the building of the canal.
- The city did not warn the resident of the hazard, despite a previous history of leakage and collapses of the same canal at this and other locations.
- People should be responsible for their own safety and their property—they should be aware of all potential hazards, including landslides.

If you were on a jury, who do you think should be held responsible for the damages and deaths? In such circumstances, damages are commonly distributed among the guilty parties in proportion to degrees of guilt. Given what you now know, what percentages would you assign among the possible culprits listed above?

A Coherent Translational Slide Triggered by Filling a Reservoir
The Vaiont Landslide, Italy ▶

In a classic case in the southern Alps of northeastern Italy north of Venice, filling a reservoir behind the newly completed Vaiont Dam caused a catastrophic mountainside collapse. Engineers completed the modern, 264-meter-high, thin-arch concrete dam in 1960 across a narrow,

rock-bound gorge. It was designed to provide both flood control and hydroelectric power. When the dam was ready for use, the engineers filled the reservoir behind it to 23 meters below the spillway.

Heavy summer rains in 1963 filled the reservoir to only 12 meters below the

spillway of the dam. The mountainside on the south side of the reservoir consisted of limestone and shale layers parallel to the 35- to 40-degree slope that was known

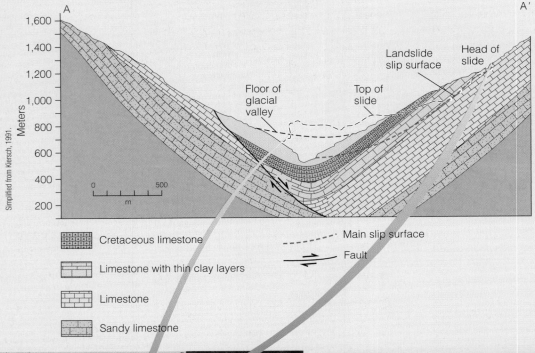

▶ The 1963 Vaiont slide moved catastrophically on weak layers of shale within limestone beds parallel to the mountain face. **A.** Cross section of the valley. **B.** The slip surface is in the upper right; the slide mass fills the center of the view, and the dam is just out of sight in the lower right. The highway looping around the toe of the slide mass in mid-view provides scale. **C.** The dam, viewed from downstream, survived even though a high-velocity wave of water 125 meters high swept over it (up to the yellow line) and killed more than 2,500 people downstream.

(continued)

to be unstable. An ancient slide plane that was partly exposed was not recognized (or perhaps not acknowledged).*

Engineers had monitored the slope just upstream from the dam for three years, and small landslides were expected. The slope had been creeping at 1 to 30 centimeters per week. By September 1963, the rate had increased to 25 centimeters per day; by October 8, it was up to 100 centimeters per day. At that point, engineers finally realized the size of the mass that was moving. They quickly began lowering the reservoir, but it was too late. Continued rain slowed reservoir draining and saturated the mountainsides. As often happens, grazing animals sensed danger a week before final failure and moved off that part of the slope.

At 10:41 p.m. on October 9, 238 million cubic meters of rock and debris collapsed catastrophically from the mountain face just upstream from the dam. It moved down at 90 kilometers per hour, filled 2 kilometers of

the downstream length of the reservoir, and moved 260 meters up the far mountainside. The displaced water, in a wave 125 meters high, swept over the dam and downstream, destroying Longarone and four smaller villages just 2 kilometers below the dam. The destruction occurred two minutes after the slide began. People had no chance. Another huge wave swept into a town on the upstream end of the reservoir. The two waves killed 2,533 people.

The resulting slide mass filled the reservoir. Amazingly, the well-built dam remained almost undamaged. The slide generated strong earthquakes and a violent blast of air that shattered windows and blew the roof off a house well above the final level of the slide mass.

Engineering failures contributed significantly to the disaster. Exploratory drilling before dam construction intersected zones in which little or no drill core was recovered, a condition suggesting broken

rock with much pore space. A tunnel excavated during dam building crossed a strongly sheared zone, but work continued without thorough examination of the implications. A study of the surrounding area would have shown that heavy surface runoff from higher slopes disappeared into innumerable solution fractures upslope of the slide plane. That water would dramatically increase the internal pore-water pressure in the rocks above the reservoir.

An important lesson from the disaster is that slopes that seem to be moving slowly may at some point fail catastrophically, and catastrophic landslides sometimes show precursory movement.

In addition to the lives lost, the cost of the landslide was considerable. Loss of the dam and reservoir cost $490 million; other property damage downstream came to a similar amount, and civil lawsuits for personal injury and loss of life cost even more.

*Sometimes, if we want something badly enough, we ignore significant negative aspects. The deep narrow rock gorge seemed an ideal place to build a dam.

A Rockfall Triggered by Blasting
Frank Slide, Alberta ▶

The small town of Frank mined coal just north of Waterton–Glacier National Park in the Front Ranges of the Canadian Rockies. At 4:10 a.m. on April 29, 1903, 30 million cubic meters of rock fell 762 meters

from the steep east face of Turtle Mountain, swept across the town, buried most of it, and killed 70 of the 600 people living there. Moving on bedding planes and approximately parallel fractures in crystalline

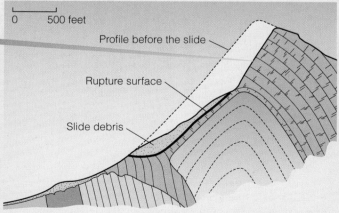

Hyndman.

Limestone boulder slide deposit

Modified from Cruden & Kahn, 1973.

0   500 feet

Profile before the slide
Rupture surface
Slide debris

▶ **A.** *The Frank slide peeled off the whole east face of Turtle Mountain and spread a spectacular bouldery deposit across the valley and up the slope to the west. Huge limestone boulders are part of the Frank slide deposit.* **B.** *This cross section shows the slope before and after the Frank slide. Fracture sets in some rocks and approximately parallel bedding surfaces in other rocks provided zones of weakness.*

(continued)

limestone, the event took less than 100 seconds. It began as a translational slide, which developed into a rock or debris avalanche as it gained speed. The giant pile of rock rubble continued north across the current route of Alberta Highway 3 and moved up the slope to the east to a height of 120 meters above the valley floor. Consensus is that the rubble moved as a "fluid" with compressed air, pulver-ized rock, and possibly water from the river that crossed its path. The immediate cause of the tragedy was mine cuts (low on the mountainside, now under the debris) that undermined tilted layers of sedimentary rocks. Although seventeen miners were actively working within the mountain, they were below the slide scarp. Rescuers were able to dig down to the shafts and save them; for once, a mine was a safer place to be than in the town below the mountain. Local tradition tells of several attempts to sink a shaft through the slide mass to recover a payroll in the vault of the buried bank.

A High-Velocity Rock Avalanche Buoyed by Air
Elm, Switzerland ▶

On September 11, 1881, a 360-meter-high mountain face at Elm, Switzerland, collapsed into a rockfall that quickly transformed into a debris avalanche. The problem began when amateurs with no mining experience were digging slate for use as chalkboards in classrooms. The quarry opened a 65-meter-deep notch at the base of a high cliff. When the excavation reached a depth of more than 50 meters, a large crack developed, and the cliff above began to creep slowly downward. Quarrying stopped only because small rocks were falling and injuring the workers.

Surviving eyewitnesses recalled that everyone expected the cliff to fall, but no one expected it to shatter into a flood of broken debris that would rush down to the main Sernf Valley, turn 60 degrees, then continue along that nearly horizontal valley floor for another 1.5 kilometers. Before it turned, part of the mass surged up the far slope of the main valley to a height of 100 meters, overwhelming those who were running uphill to escape it. Ironically, dogs and even cattle instinctively ran to the side and survived. The whole event took 40 seconds, no time to run far. Sixty-five people died.

An eyewitness watched the mass breaking up as it began to fall. It hit the floor of the slate quarry, completely disintegrated, and shot out horizontally. It did not flow along the ground but launched over the slope below and cleared a creek in the valley bottom. Witnesses saw houses, trees, fleeing people, and cattle under the flying debris. The underside of the rockfall was sharply bounded, but the upper side was a cloud of rocks and dust. One witness described its movement along the main valley as a torrential flood with a bulging head that tapered to the rear. It roiled as if boiling. The deposit is shaped like that of a debris flow with well-defined ridges at the sides and toe. Internal waves on the surface of the flow were convex downstream. No rubble sprayed beyond the toe.

The flood of rubble certainly did not ride on a cushion of air everywhere along its path. If it had, it could not have carved parallel grooves in the flat valley floor; in one place it unearthed a water pipe buried to a depth of one meter and carried it one kilometer downstream. Some houses near the leading edge of the slide were moved off their foundations. One was filled with rocks and broken boards. An old man standing inside was buried up to his neck but was otherwise uninjured. Apparently, the material flowed not as a dense mass but swirled, almost like a viscous fluid.

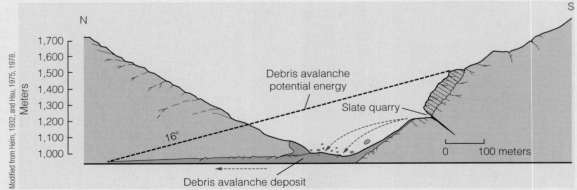

▶ *This cross section shows the Elm, Switzerland, debris avalanche and the location of the slate quarry that set it off.*

Critical View

A Road along stream, (right below road on right), southwestern Montana.

Home

Small slump

Slumping road

Donald Hyndman.

1. What would have caused slumping of the road at the right edge of the photo? Explain.

2. What would have caused slumping above the road? Explain.

C Campground near Yellowstone Park

Donald Hyndman.

1. Where did the giant boulders come from (in the fore-ground and in the campground)?

2. What is the process called that moved them to here?

3. What is the evenly sloping rocky surface behind the trees called?

4. Why are the largest boulders concentrated at the base of the slope?

B Washington Cascades

1. What would have caused the treeless path from the cliff at the top and down through the valley?

2. What processes would have promoted that movement?

Rocky cliff

Low brush

Donald Hyndman.

D Near north entrance of Yellowstone Park

Donald Hyndman.

1. What process produced the distinctive topography of the brownish grassy slope across the middle of the photo?

2. Would this be a good place to build a new highway to reach the area at the upper left of the photo? Why or why not?

E Recent subdivision

Donald Hyndman.

This expensive new home was built on a flat area excavated in an evenly-sloping soil hillside (below the trees in the background) in a dry climate. The brownish area in the foreground is another flat area excavated for another house. The green area below the house has been recently planted with grass.

1. What caused the patch of gully erosion in the lower left? Where, specifically, did the water come from to erode the gully?
2. What could be done to stop the process that produced the gullying?

G Barren cliffs with thick layers of sandstone in southern Utah.

Donald Hyndman.

The new houses at the base of the mountain are surrounded by some huge boulders that make for interesting landscaping.

1. Where did those boulders come from?
2. Is this a good place for houses—why or why not?

F Winter 1980 landslides in the Ventura area of southern California.

Housing subdivision

Douglas Morton, USGS.

The green, brush-covered slopes covering most of the photo have recent landslides covering most of their lower slopes.

1. Why are those slopes covered with brush and no trees? Given your answer, why specifically should that process lead to landslides?
2. What else likely contributed to the sliding?

H Highway along a very steep coastal cliff south of San Francisco, California.

Donald Hyndman.

1. This part of the roadway was evenly graded but now is wavy—up and down. What caused that waviness?
2. At the base of the roadcut and protruding from the concrete are many huge bolts. Why are they there—what do they do and how do they work?
3. Coarse screen covers the upper roadcut. What is its purpose?

Chapter Review

Key Points

Forces on a Slope

- A driving force pulls materials downslope, while a resisting force holds them in place. When the driving force is greater than the resisting force, a landslide occurs, bringing the two forces back into equilibrium.

- The relationship between slope angle and load is a key factor in slope failure. Steeper slopes and slopes with heavier loads are more likely to overcome frictional resistance and fail.

- Water is a key factor in slope stability. A little water coating holds grains together through cohesion. More water fills pore spaces and adds load to the slope—it pushes grains apart and makes a slope more prone to slide. **FIGURES 8-4** and **8-5**.

Slope Material

- Internal surfaces sloping in the same direction as the land surface provide weak zones that facilitate slip, especially if the surfaces daylight—that is, come back to the surface at their lower end—or if they are old landslide slip surfaces. **FIGURE 8-6**.

- Quick clays, consisting of clay flakes deposited in saltwater, can stand on edge like a house of cards. If the clays are raised, drained, and rinsed with freshwater, they are susceptible to collapse. **FIGURE 8-7**.

Causes of Landslides

- Landslides can be caused by adding soil moisture, slope material instability, or jarring by earthquakes.

- Changes in slope imposed by external factors—such as undercutting by a stream or building a road, loading of the upper part of a slope caused by construction, addition of water by various means, or removal of vegetation—can also destabilize equilibrium and promote sliding. **FIGURE 8-9**.

Types of Downslope Movement

- Rockfalls occur in areas with steep cliffs when rocks are broken loose by freezing or ground shaking. The distance a rockfall travels, its runout, is related to the mass of the rock and the distance from which it falls. Higher, more massive rocks fall farther. **By the Numbers 8-2**.

- Debris avalanches are similar to rockfalls except that their fragments disintegrate into tiny pieces that entrain air or water and flow at high velocity, much like a dense liquid. As with rockfalls, their travel distance depends primarily on their height of fall.

- Rotational slides and slumps are common in weak, homogeneous material. Sliding masses rotate on a curving surface. **FIGURES 8-22** and **8-23**.

- Translational slides are facilitated by slip surfaces inclined nearly parallel to the ground surface. **FIGURES 8-24B** and **8-25**.

- Soil creep is the slow downslope movement of soil and rock, which can be accelerated by alternating periods of freezing and thawing.

- Conditions for snow avalanches include slope steepness, weather, and temperature. Unstable layers in a snowpack can be caused by new snow fall or periods of melting and refreezing. Snow avalanches can occur spontaneously or they can be triggered by human activities, like skiing or snowmobiling.

Hazards Related to Landslides

- Earthquakes trigger many landslides, especially rockfalls.

- Loosely packed sandy soil saturated with water can "liquefy," settle into a smaller volume, and spread when shaken by an earthquake. **FIGURES 8-35** to **8-37**.

- Landslides can block waterways, creating a landslide dam, which can later fail and cause flooding. Failure of landslide dams depends on the size of the dam, its material, and the rate at which the lake behind the dam rises.

Mitigation of Damages from Landslides

- Hummocky landscapes and wavy roads are a sign of landslide activity.

- Landslide hazard maps can be used to restrict development in landslide-prone areas.

- Engineering solutions to landslides—including water removal—are geared toward restoring the balance between the forces that act on a slope, as well as reducing the penetration of water.

Key Terms

Questions for Review

1. Describe the forces exerted on a slope. Use a diagram to illustrate your description.

2. Why does a pile of dry sand have gently sloping sides whereas a pile of wet sand can have nearly vertical sides?

3. Why is the friction force on a gently sloping slip surface greater than that on a steeply sloping slip surface?

4. Explain why raising the water table may lead to slope failure.

5. Describe the sequence of events that leads to quick clay formation.

6. List the main factors that affect whether a slope will fail in a landslide.

7. What factors determine the distance a rockfall will travel laterally from its point of origin?

8. What is the difference between a rockfall and a debris avalanche?

9. Why does the top of a rotational slide tilt back into the slope?

10. What is the difference between a rotational slide and a translational slide?

11. What factors accelerate soil creep?

12. What conditions commonly lead to snow avalanches?

13. What steps should you take if you are caught in an avalanche?

14. What causes liquefaction of sediments? Briefly explain the process.

15. Under what circumstances is flooding a hazard related to landslides?

16. What parts of the United States are most susceptible to landslides?

17. Name several landscape features that indicate a record of past landslides.

18. Why might people want to build in landslide-prone areas?

19. List several ways in which old landslides are commonly reactivated.

20. What can be done to slow or stop the movement of a rotational slide?

21. List several distinctly different ways in which water can be removed from a wet slope that has begun to slide.

Discussion Questions

1. A developer in the steep coastal area northwest of Los Angeles requests a permit from the county to subdivide a large plot of land for expensive homes. Old landslides are documented in the vicinity. Should the county provide the permit? Considerations include personal freedom for use of one's property, road building, water and sewer lines, property taxes to the county, aspects of liability, and insurance. In case of damages because of ground movement, who is liable and who should pay?

2. A moderate-size earthquake (e.g. magnitude 6) triggers a large landslide that severely damages many homes in a large 15-year-old subdivision near San Francisco. Homeowners' insurance companies refuse to pay for damages. Who should pay for the damages? Is there shared liability and if so by whom and why?

3. Homes in a 5-year-old subdivision built on layers of swelling clays on flat ground near Denver, Colorado, begin to twist and break up. Who should be liable and why? If liability should be shared, what are the considerations?

Sinkholes, Land Subsidence, and Swelling Soils

9

Shrinkage and settling of ground near Apache Junction, east of Phoenix, Arizona. Excessive groundwater withdrawals for agricultural and urban uses caused differential subsidence. Fissures do not tend to open more than 2.5 centimeters (1 inch) per event, but they then widen as blocks slump down into the fissure.

Donald Hyndman.

Shrinking Ground

For more than 100 years, groundwater has been critical for agriculture, mining, and municipal uses in the Phoenix and Tucson areas of southern Arizona. With potential evapotranspiration of over 100 centimeters per year and desert-climate precipitation in some valleys as low as 7 centimeters per year, the natural replacement of groundwater supplies just does not add up. The rapidly growing metropolitan population demands ever more, and in some agricultural areas the groundwater level has fallen more than 180 meters (600 feet). Differential subsidence of the ground between adjacent areas from Phoenix to Tucson, an area of more than 7,700 square kilometers, has led to earth fissures as much as 6 meters deep and 9 meters across. Widening of the fissures has damaged highways, sewer lines, buildings, and other structures.

The region pumped almost all of its water from underground until demand led to building the massive, federally funded Central Arizona Project. That open, concrete-lined channel began supplying a large volume of water from the dwindling Colorado River in 1985. In an ironic twist of fate, the Central Arizona Project, which imports water through excessive pumping, has itself been affected by widening fissures and has required strengthening. Part of the imported water is used to recharge the groundwater, so that subsidence has slowed and water levels in some areas are slowly rising.

Types of Ground Movement

Ground movements may not be as dramatic as earthquakes and volcanoes, but they cause far more monetary damage in North America because they deform and effectively destroy roads, utility lines, homes, and other structures. **Sinkholes** develop when the overlying ground collapses into underground soil cavities over limestone. **Land subsidence** often occurs when large amounts of groundwater or petroleum are pumped out or when an earthquake causes closer repacking of sediment grains. **Swelling soils** typically form by the alteration of volcanic ash into clays that swell when wet.

Sinkholes

In some places, especially in limestone terrains of the eastern United States, the ground may suddenly collapse, leaving sinkholes that are tens to hundreds of meters across (**FIGURE 9-1**). Not only can sinkholes damage houses and roads, but they can drain streams, lakes, and wetlands.

In low areas, they can channel contaminants directly into underground aquifers, the main source of water used for drinking and other purposes in many parts of the country.

Processes Related to Sinkholes

Some common sedimentary rocks are **soluble** in water, allowing them to dissolve, which poses hazards to those living above them. Salt and gypsum (often called "evaporites") are highly soluble; limestone and other carbonate rocks are slowly soluble in acidic rainwater. Areas with underlying limestone can exhibit caves, springs, sinkholes, and streams that sink into the ground. **Caverns** form as carbonate rocks near the water table dissolve in groundwater (**FIGURE 9-2**). The roof of a cavern that was formerly supported by water pressure is then susceptible to collapse and the potential formation of a sinkhole. Where the soluble rocks are close to the surface (see FIGURE 9-2b), the solution cavities and caverns can grow large enough for the roof rocks to collapse.

Limestone dissolves when water droplets in the atmosphere take in carbon dioxide to form weak carbonic acid. Slightly acidic rain falls and percolates through soil and sediment down to the bedrock. Where the bedrock is limestone

FIGURE 9-1 SINKHOLE COLLAPSE

In May 1981, a large sinkhole developed in Winter Park, near Orlando, Florida. It swallowed a house, a couple of Porsches, a camper, and half of a municipal pool.

FIGURE 9-2 CAVERNS

A

B

A. A limestone cavern exposed along Interstate Highway 44 near Springfield, Missouri, shows sagging and fracturing of the rocks above it.
B. A deep hole in limestone near Austin, Texas, indicates the presence of a hidden cavern at depth. Plans for a large shopping center here were abandoned because of the danger of collapse.

(calcium carbonate), the acidic water slowly reacts with the limestone along fractures, dissolving it to widen cracks and leave cavities (**By the Numbers 9-1:** Formation of Calcium Bicarbonate). The rate of solution is slow with mildly acidic rain, on the order of millimeters per 1,000 years. Because the reaction between acidic water and limestone occurs more rapidly under warm, moist conditions, caverns and sinkholes are most common in tropical or subtropical climates. Dissolution is amplified by rain that is more acidic because of air pollution—acid rain. In both cases, water flows through fractures and cavities in the limestone and carries away the soluble calcium bicarbonate.

Limestone can dissolve above, at, and below the water table. Where limestone is above the water table, acidic water running down through fractures slowly widens them. Just below the water table, cavities can rapidly widen if horizontal bedding surfaces are open enough to conduct large amounts of water. In this case, the large flow past limestone surfaces brings fresh acidic water to these reactive surfaces and carries away the calcium bicarbonate products. This is the environment in which large caverns generally form. When the water table drops below the top of a cavern, water percolating through fractures above can precipitate calcium carbonate as the water locally evaporates and loses its carbon dioxide. These formations are called *stalactites* when they hang from the roof and *stalagmites* when they grow from the floor (**FIGURE 9-3**, p. 238). Caverns found high on hillsides above a current water table suggest that the water table has dropped, often the result of a nearby stream eroding its valley deeper.

Limestone bedrock near the water table dissolves along fractures to create an uneven and potholed upper surface. Later erosion of the soil cover may expose that surface as a limestone landscape called **karst** (**FIGURE 9-4**, p. 238). In extreme cases, deep solution of the limestone along dominantly vertical fractures can form an extremely ragged, toothy-looking, and often picturesque landscape. Perhaps the best known of these are in the Guilin and Kunming areas of southwestern China.

Types of Sinkholes

Three types of sinkhole formation are common, with gradations between them:

1. **Dissolution.** Where the soil cover is thin and highly permeable, acidic groundwater seeps through it and dissolves the underlying limestone along fractures. The upper parts of fractures widen to form a lumpy or jagged karst surface (see FIGURE 9-4). The overlying soil can slowly percolate or ravel down into the fractures

▶ By the Numbers 9-1

Formation of Calcium Bicarbonate

$$H_2O \quad + \quad CO_2 \quad = \quad H_2CO_3$$
(rain water) (carbon dioxide) (carbonic acid in water)

Carbonic acid reacts with limestone to form calcium bicarbonate:

$$H_2CO_3 + CaCO_3 = [Ca^{++} + 2HCO_3{}^-]$$

FIGURE 9-3 STALACTITES AND STALAGMITES

A

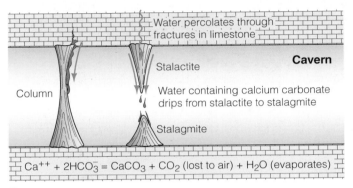

Water percolates through fractures in limestone

Cavern

Column

Stalactite

Water containing calcium carbonate drips from stalactite to stalagmite

Stalagmite

$$Ca^{++} + 2HCO_3^- = CaCO_3 + CO_2 \text{ (lost to air)} + H_2O \text{ (evaporates)}$$

B

A. This cavern near Tulum, Mexico, shows stalactites hanging from the ceiling and stalagmites growing from the floor. They may ultimately build into continuous columns. **B.** Water containing dissolved calcium carbonate seeps down through limestone into a cavern. There the evaporating water loses carbon dioxide and again precipitates calcium carbonate as stalactites and stalagmites.

to create a surface depression. Where the groundwater level is high or the fracture plumbing becomes clogged with sediment, the depression may fill to form a pond. These depressions are shallow and not generally dangerous.

2. **Cover Subsidence.** Where tens of meters of sandy and permeable sediment exist on top of limestone bedrock, numerous sinkholes can form as the soil slowly fills expanding fractures and cavities in the limestone. Depressions generally form gradually.

3. **Cover Collapse.** Where overlying sediments (called *overburden*) contain significant amounts of clay, this cover will be more cohesive and less permeable. As a result, it does not easily ravel into underlying cavities in limestone. This can allow a soil cavity to grow large and unstable, leading to the sudden collapse of its thinning roof (**FIGURE 9-5**). The lack of warning makes these steep-sided sinkholes destructive and dangerous.

Cover-collapse sinkholes open with little or no warning, taking with them roads, parking lots, cars, and occasionally even houses and other buildings, as in the February 25, 2002, collapse at a road intersection in Warren County, Kentucky (FIGURE 9-5b). In this case, storm water funneled underground at three corners of the intersection and caused further solution and instability in a limestone cavern underground. Before the road was built, karst experts who knew that the underground cave passage had an unstable roof, recommended a different, safe route for the road, but their advice was not followed. Individual collapses such as this cost more than $1 million each to repair.

FIGURE 9-4 KARST

A **B**

A. This exposed karst on the west coast of Turkey is typical of dissolved limestone bedrock on which covering soil has been eroded away. The lumpy appearance of the limestone emphasizes below-ground solution as the mechanism of erosion. **B.** Extreme karst weathering can be seen in the Stone Forest, south of Kunming, China. Individual spires are more than 10 meters high.

FIGURE 9-5 FORMATION OF A COVER-COLLAPSE SINKHOLE

Sediments spall into a cavity.

As spalling continues, the cohesive covering sediments form a structural arch.

The cavity migrates upward by progressive roof collapse.

The cavity eventually breaches the ground surface, creating sudden and dramatic sinkholes.

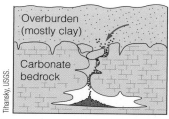

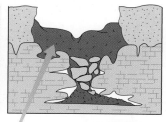

A

A. This sequence of events can lead to a cover-collapse sinkhole. **B.** The sinkhole on the right formed on February 25, 2002, by cover collapse over a limestone cavern in Warren County, Kentucky, where storm-drain water was funneled underground at a road intersection.

B

Areas That Experience Sinkholes

Slowly dissolving carbonate rocks underlie more than 40 percent of the humid areas of the United States east of Oklahoma (**FIGURE 9-6**, p. 240). Approximately 55 percent of Kentucky is on limestone weathering to karst, including one of the largest and most famous caverns in the United States, Mammoth Cave National Park. Slow solution of this limestone has left the landscape pockmarked with more than 60,000 sinkholes, some of which line up along lineaments.

As recently as 20,000 years ago, during the last ice age, when much ocean water was tied up in continental ice sheets, the whole carbonate platform of Florida was above sea level. The peninsula then was roughly twice as wide as at present. The groundwater level was much lower so many caverns sat above the water table. When the ice sheets melted, sea level rose and the water table rose to fill most of the caverns. Sand and clay up to 60 meters thick cover most of the cavernous limestone that is weathered to a karst surface. Sinkholes are prominent in central Florida, and many of the lakes and ponds in west-central Florida are water-filled sinkholes (see FIGURE 9-6b).

Fluctuations of the groundwater level can contribute to sinkhole formation. Periods of heavy rainfall tend to loosen soil, enlarge cavities, and promote sinkhole formation. More sinkholes tend to form during dry seasons or when excessive pumping drops groundwater levels. One circumstance that leads to unusual groundwater use and lowering of the aquifer is pumping during a prolonged spell of freezing weather—for example, when strawberry and citrus farmers spray warm groundwater on their plants to form an insulating coating of ice.

Areas with the greatest potential for sinkholes are those where surface water tends to percolate into the ground, recharging the aquifers below. This potential is greatest where the water table lies below the tops of limestone caverns, allowing for unsupported open space in cavities above the water level. Widespread limestone near Pittsburgh, Pennsylvania, is broken by numerous near-vertical fractures. Acidic rainwater percolating down through those fractures slowly dissolves and widens them. Soil above the limestone cracks may intermittently break apart to form a progressively enlarging cavity. Gradual expansion of that cavity eventually leads to formation of a cover-collapse sinkhole at the surface. Most Pennsylvania sinkholes are typically 1 to 6 meters in diameter, and fractures seldom enlarge to form underground caverns.

Areas with the least potential for sinkholes are those where water is being discharged to the surface. There the water is likely to fill the cavities and helps support their roofs; the flowing groundwater is unlikely to be corrosive to carbonates.

FIGURE 9-6 SINKHOLE DISTRIBUTION IN THE UNITED STATES

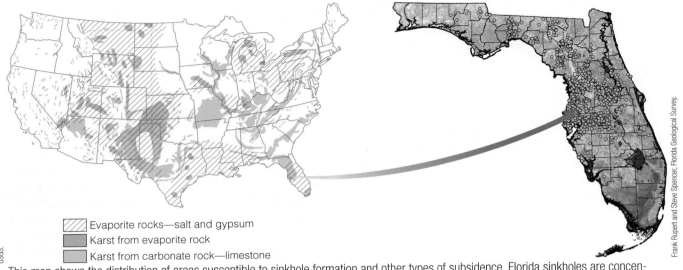

- ◿ Evaporite rocks—salt and gypsum
- ▓ Karst from evaporite rock
- ▒ Karst from carbonate rock—limestone

USGS.

Frank Rupert and Steve Spencer, Florida Geological Survey.

This map shows the distribution of areas susceptible to sinkhole formation and other types of subsidence. Florida sinkholes are concentrated in the central part of the state (blue dots).

In urban areas, large volumes of water carried by storm drains and leaking water mains can enhance the development of sinkholes. A large water flow can quickly flush soil from above fractured limestone to form large sinkholes. Roads and buildings above such sinkholes are vulnerable to severe damage (**FIGURE 9-7**). In Allentown, Pennsylvania, a nearly new eight-story office building broke up in 1994 after a pair of large sinkholes developed below major support columns. Damage required razing of the $10 million building.

In late May 2010, a steep-sided funnel-shaped sinkhole 60 meters deep and 16 meters wide swallowed a street intersection in downtown Guatemala City, Guatemala, taking with it a three-story building. Almost one meter of rain from Tropical Storm Agatha swamped the region, leading to failure of a large sewage pipe and undermining the street. A similar but larger sinkhole opened in 2007.

On August 13, 2006, an 18-meter wide by 23-meter deep sinkhole collapsed under one end of a home in Nixa, just south of Springfield, Missouri. The two-car garage, initially cantilevered over the hole, collapsed shortly afterward, and the house continued to break up (**FIGURE 9-8**). Sinkholes are abundant in the Karst Plain in southwestern Missouri, and most appear to be soil-collapse features. The Nixa sinkhole developed at the intersection of crossing fractures, which are typical for the area. Examination of the site suggests that sinkhole collapse may have been promoted by water seepage from an abandoned septic tank and its tile drain field; decayed roots of an old tree; roof downspouts draining into a gravel field on the gentle upslope side of the house; and well-drained soils that conveyed water to the fractured bedrock. Following destruction of the house, the sinkhole was

FIGURE 9-7 BUILDING DAMAGE DUE TO SINKHOLES

William Kochanov, Pennsylvania Geological Survey.

A sinkhole formed under this house in Palmyra Borough, Lebanon County, Pennsylvania.

filled with one- to two-meter boulders and then capped with soil. Collapse apparently occurred because of drainage water percolating through soil filling a cavity above a limestone cavern. Because the soil cavity was under the concrete floor of a two-car garage, it was not noticed until it grew beyond the edge of the garage and collapse occurred. More commonly, such sinkholes appear as small holes that progressively enlarge.

Construction activities can also lead to sinkhole development through increased loads on ground surfaces, dehydrating of foundation soils, and well drilling. Heavy

FIGURE 9-8 HOUSE COLLAPSES INTO SINKHOLE

Doug Gouzie.

2006 sinkhole collapse under a house in Nixa, Missouri.

FIGURE 9-9 HEAVY EQUIPMENT CAUSES COLLAPSE

Florida Geological Survey.

A truck-and-drill rig that was drilling a new water well near Tampa, Florida, overloaded the roof over a limestone cavern, causing it to collapse. The truck and equipment eventually sank out of sight in a crater 100 meters deep and 100 meters wide.

construction equipment or the weight of a new structure can impose loads (**FIGURE 9-9**).

Sediments such as salt and gypsum also produce karst landscapes and underlie large areas of the United States. Their relative ease of solution in water can lead to cavity formation in just days. They are widespread in the southeastern states, the Appalachians, and western Texas. In some cases, underground mining of salt and gypsum creates artificial cavities that can be enlarged by solution in water. Early on the morning of March 12, 1994, a section of shale roof rock 150 by 150 meters, collapsed into an underground mining excavation in the Retsof Salt Mine in the Genesee Valley of New York State. The hole in the shale barrier allowed groundwater to pour into the mine, gradually filling its cavities and destroying the largest salt mine in the United States.

Land Subsidence

Land subsidence occurs when the ground settles as a result of changes in fluid levels underground. Because subsidence occurs across large regions, its effects are less obvious and dramatic than those of sinkholes. Nonetheless, subsidence also causes considerable damage. Small faults cause some areas to drop more than others, damaging houses and utilities. In some areas, *earth fissures* form due to differential subsidence between adjacent areas (see chapter opener photo, p. 235). In coastal areas, subsidence can sink communities closer to sea level and leave them more vulnerable to flooding.

Subsidence of the ground is a serious problem throughout North America. This lowering of the ground surface is caused by a variety of human activities, including extraction of groundwater, drainage of organic or clay-rich soils, and thawing of permafrost.

Mining Groundwater and Petroleum

One of the most common causes of land subsidence is the pumping of water or petroleum out of the ground. As rainwater soaks into the ground and flows toward rivers and streams, it travels through porous rocks—*aquifers*. Many communities use these aquifers as a source of freshwater by pumping the groundwater. In areas where the amount pumped exceeds the recharge from precipitation, this is considered **groundwater mining**. Ground subsidence resulting from groundwater and petroleum extraction has been significant in many parts of the world, including the United States (**FIGURE 9-10**). **Table 9-1** lists a few of the more significant cases.

FIGURE 9-10 GROUNDWATER SUBSIDENCE MAP

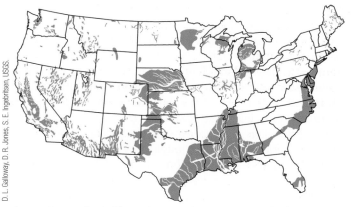

D. L. Galloway, D. R. Jones, S. E. Ingebritsen, USGS.

Areas of ground subsidence by groundwater pumpage in the United States.

Table 9-1		Examples of Ground Subsidence Resulting from Groundwater and Petroleum Extraction	
LOCATION	**AMOUNT OF SUBSIDENCE (m)**	**FLUID EXTRACTED**	**SOME CONSEQUENCES**
Sacramento–San Joaquin Valley, California	9 in some places	Groundwater for agricultural use	Fields in the delta-area sink below sea level
Phoenix to Tucson, Arizona	Up to 5	Groundwater for agricultural and municipal uses	Subsidence and earth fissures
Las Vegas, Nevada	Up to 2	Groundwater for municipal use	Subsidence, ground failure
Long Beach, California	9	Oil and gas	Pipelines and harbor facilities damaged; dikes needed to prevent seawater flooding
San Jose and Santa Clara Valley (Silicon Valley), California	>2–2.5	Groundwater for orchards, industrial, and municipal use	Tidal flooding of San Jose by San Francisco Bay
Houston–Galveston, Texas (a circular area 130 km across)	>2	Oil and gas, groundwater for municipal use	Frequent flooding, sometimes severe; submerged wetlands
Everglades, Florida	2–3.5	Drained for agricultural land and urban development	Wetlands now reduced by 50%; subsidence; saltwater intrusion into groundwater now slowed
Mexico City, Mexico	8–10	Groundwater for industrial and municipal use	Tilting buildings, broken water, sewer, and utility lines
Pisa, Italy	Several	Groundwater for municipal use	Danger to Leaning Tower from uneven subsidence
Venice, Italy		Groundwater for industrial and municipal use	Settling of buildings below sea level, severe flooding at especially high tides
Delta of the Rhine and Meuse rivers, The Netherlands	Large areas to 6 m, locally up to 6.7 m	Drainage of agricultural land	Dehydration and oxidation of peat-rich soils, and further ground subsidence

Many aquifers consist of thick accumulations of sand and gravel with water filling the spaces between the grains. Withdrawal of water from the pore spaces of loosely packed sand grains permits them to pack more tightly together and take up less space (recall FIGURE 8-3, p. 205). The original groundwater reservoir capacity cannot be regained by allowing water to again fill the pore spaces, because sand grains cannot be pushed apart into a more open structure once their pore space has collapsed.

Large water withdrawals can cause permanent reduction of aquifer capacity and subsidence of the ground. Prominent examples of subsidence due to long-term groundwater withdrawal can be found in Silicon Valley around the south-end of San Francisco Bay; the southern half of the Central Valley of California; the Long Beach area south of Los Angeles; southern Arizona; the Texas Gulf Coast region near Houston; and Venice, Italy (**Case in Point:** Subsidence Due to Groundwater Extraction—Venice, Italy, p. 252).

In the San Jose area at the southern end of San Francisco Bay, depletion of groundwater accompanying the rapid growth in population caused about 4 meters (13.1 feet) of ground subsidence from about 1920 to 1995 (**FIGURE 9-11**). Following rapid subsidence of 80 cm in 4 years around 1960, groundwater pumping was slowed and water imported, beginning in 1963. This has nearly stopped the subsidence.

In the Houston area, oil and gas were produced early along with groundwater. More than one meter of subsidence has been observed in parts of this area. As a consequence, coastal flooding that was formerly a problem only with the occasional major storm is now a significant problem with many much smaller storms (**FIGURE 9-12**, p. 244).

In the coastal areas of Wilmington and Long Beach just south of Los Angeles, subsidence began in the 1940s with pumping of groundwater at the naval shipyard and later with oil extraction. By 1958, subsidence occurred over 52 square kilometers by as much as almost 9 meters, warping rail lines and pipelines and damaging buildings and shearing off oil wells. (**FIGURE 9-13**, p. 244).

Mitigation of ground subsidence due to groundwater or petroleum extraction initially involves stopping the activity that causes the problem. In some cases, governments institute regulations that prohibit or limit groundwater or petroleum extraction or initiate programs to reinject fluid into the ground. This can slow or eventually stop the subsidence. In 1958, when the California Subsidence Act was

FIGURE 9-11 GROUNDWATER PUMPING CAUSES SUBSIDENCE

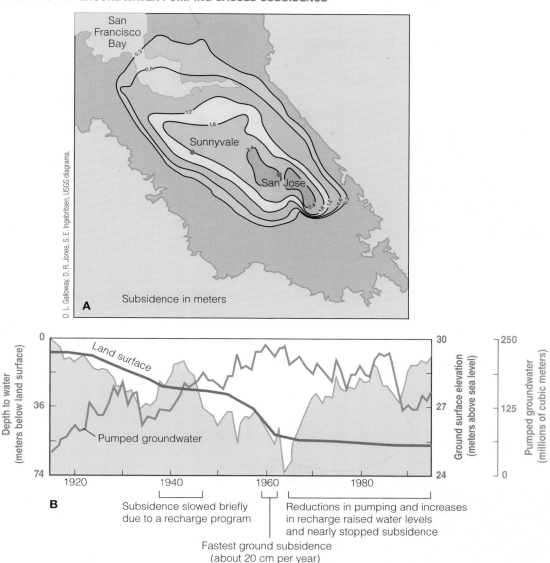

D. L. Galloway, D. R. Jones, S. E. Ingebritsen, USGS diagrams.

A. Map of subsidence in the Santa Clara Valley due to groundwater pumping. The area of greatest subsidence (dark orange) reached a total of 4 meters in about 70 years. **B.** Groundwater pumping in Santa Clara, California (shown in green), increased significantly from 1915 through 1960. As expected, the increases in pumping increased the depth to groundwater (shown in blue). It also had the unanticipated effect of dropping the land surface by subsidence (shown in brown), until water was imported beginning in 1963 to recharge the groundwater. This recharge and reduced groundwater pumping nearly stopped the subsidence.

passed, groups of property owners in the Los Angeles area injected water underground to repressurize the pore spaces and stabilize the rate of ground subsidence. However, most of the subsidence or compaction of sediments cannot be reversed once it has occurred. Reduction in aquifer porosity and permeability is also generally permanent.

Groundwater can be a renewable resource in many basins, provided that the amount removed does not exceed what is replenished, at least over many years. Sustainable-use

policies can help maintain this balance. Such policies must take into consideration a number of factors, including changes in climate and water quality. Available wet-year storage for water in pore spaces between sediment grains for use in dry years must be carefully monitored so that the subsurface reservoir capacity is not permanently reduced. The sustainable use, or "safe yield," for a basin may approach natural and artificial replenishment levels, but the climatic variability that affects both the natural and artificial

FIGURE 9-12 OIL AND GAS PUMPING CAUSES SUBSIDENCE

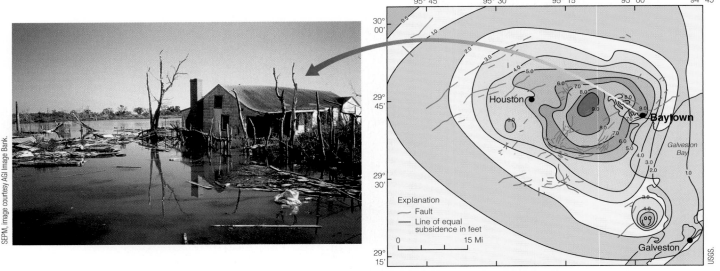

Subsidence of the Brownwood Subdivision of Baytown, Texas, has surrounded houses with brackish lagoon water. Subsidence in the Houston–Galveston region; contours in feet.

FIGURE 9-13 OIL PUMPING CAUSES SUBSIDENCE

Subsidence in the Long Beach area southeast of Los Angeles was caused by oil and water extraction from the basin sediments. Note that the ground settled around the well stems (white "posts"), leaving their well heads standing high. The levee (left edge of photo) around the subsided ground under the oil well area was installed to prevent seawater flooding from the adjacent marina.

restocking of the groundwater system must be taken into account. Stream flows also need to be maintained during the low-flow season to protect the health of stream ecosystems.

Drainage of Organic Soils

Subsidence can also be caused by draining organic-rich soils such as peat. In their natural state, these organic soils are saturated with anaerobic (without free oxygen) water,

which allows them to maintain stability. When groundwater levels drop because of activities such as municipal and agricultural pumping, the organic soils are exposed to aerobic (oxygen-rich) water that percolates from the surface down to the water table. This allows bacteria in the sediment to oxidize the organic matter, which causes decomposition largely into carbon dioxide, a gas linked with additional natural hazards through global climate change (as discussed in Chapter 10). This decomposition occurs at a rate as much as 100 times greater than that at which such organic material accumulates.

The semi-arid Sacramento–San Joaquin valley of central California is noted for its prolific agricultural products, from citrus fruits and wines to grains, cotton, and rice. The area produces a quarter of the food in the United States. Heavy pumping of groundwater for crop irrigation has lowered the water table, causing decomposition of organic matter and more than 9 meters of subsidence in some areas of the valley (**FIGURE 9-14**). That subsidence threatens agricultural production, large water-supply lines to major metropolitan areas, and important ecosystems. Compounding matters, the main river channels now flow well above the level of the agricultural ground, held there behind dikes, and leading to flood risk (**FIGURE 9-15**).

A vast area of the Florida Everglades was drained decades ago for farming, causing almost 2 meters of subsidence. An Everglades restoration plan is being implemented to reduce these and other impacts and take the Everglades back toward a natural state.

Sinking of the Mississippi River delta area, including New Orleans, has increased because of levees that now prevent the river from adding silt to its floodplain during floods. That sinking also promotes further saltwater

FIGURE 9-14 SAN JOAQUIN VALLEY SETTLES

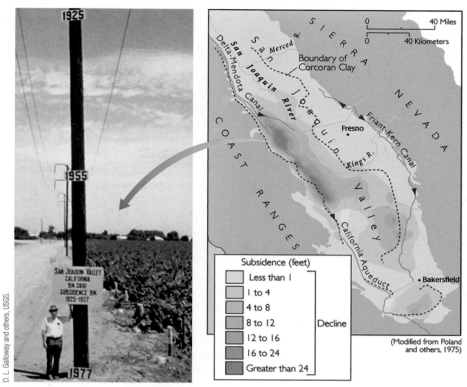

Parts of the San Joaquin Valley, California, settled almost 8 meters (more than 25 feet) in the 50 years between 1925 and 1977.

FIGURE 9-15 SACRAMENTO DELTA SUBSIDES

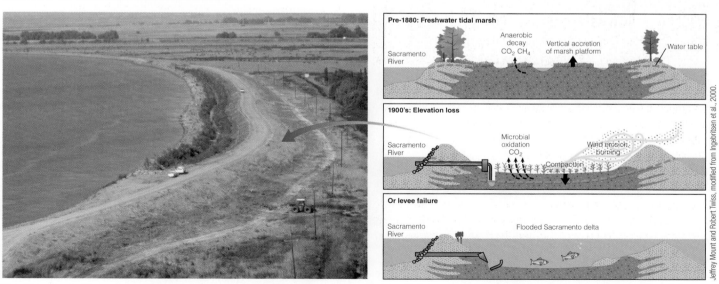

The Sacramento River stands above the now-subsided delta farmlands, held back by high levees. Pumping water from the Sacramento River delta lands (brown area) has caused compaction, leaving them below the level of the Sacramento River (blue, on left).

incursion that kills coastal marsh vegetation, including cypress forests. An additional large factor in delta subsidence is oxidation of organic-rich sediments.

In Europe, The Netherlands is most vulnerable to subsidence because this small country lies on the delta of the Rhine and Meuse rivers and one-quarter of the country lies below sea level. Artificial drainage of the peat-soil meadows on the delta began about 1,800 B.C. to permit farming. Those soils are now one to two meters below sea level; areas around the giant container-shipping center of Rotterdam lie as much as 6.76 meters below sea level. Large areas are sinking at one to eight cm per year. In many areas, groundwater lies at the surface or as much as 70 cm below the ground surface. During the dry season, the water table drops, so soils dry out and the peat oxidizes and settles.

Huge North Sea storms arriving at high tide have always eroded and flooded the coast, but disastrous floods in 1953 killed 1,800 people and destroyed thousands of homes and farms. The Delta Works commissioned at that time built a series of levees and dams to hold back the sea (**FIGURE 9-16**).

Drying of Clays

Clay soils can be particularly prone to subsidence when they dry out. This is especially true in randomly oriented marine clays that have a house-of-cards arrangement with extremely high porosity (see FIGURE 8-7). Collapse of that open arrangement can result in rapid land subsidence. The abundant inter-grain spaces of quick clays can collapse when jarred by an earthquake or heavy equipment or if they are flushed with fresh water.

The Leaning Tower of Pisa is a famous example of subsidence. The ground under the tower consists of 3 meters of clay-rich sand and more than 6 meters of sand over thick, compressible, and spreading marine clay—not the type of material that today's engineers would choose as a foundation. It was built as a bell tower for the adjacent cathedral of Pisa, near the west coast of Italy. Construction that began in 1173 was stopped in 1185 with only three levels completed, because the tower began settling and leaning. Construction resumed in 1274 after 100 years of inactivity. Ten years later, all except the top bell enclosure was complete, but again the rate of leaning accelerated, so construction was again stopped. The tower, including its bell chamber, was finally completed in 1372, but by this time it was 1.5 meters off vertical.

By 1975, the 55-meter-high Leaning Tower had sunk 2 meters into the ground and was nearly 4 meters off vertical. If the photo in **FIGURE 9-17** appears to show a slight curvature of the sides of the tower, that is not an illusion. The builders tried to straighten the tower as construction proceeded. Various attempts have been made to stabilize it. A 600-ton weight was added to the base of the tower on the side that had not subsided (left side of FIGURE 9-16). More recently, the tower was straightened by nearly a half meter, and a new foundation was built to keep it stable and safe for tourists.

One area of water-rich marine clays subject to subsidence is the Ottawa–St. Lawrence River lowland, where seawater invaded lakes that formed at the edge of the

FIGURE 9-16 STORM SURGE BARRIERS

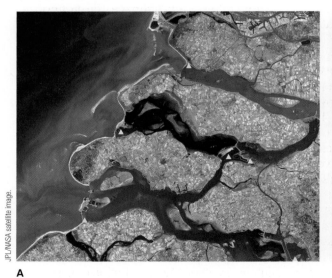

A **B**

A. Satellite view of southern Holland with storm-surge barriers marked with yellow triangles **B.** Maeslant barrier gates that close to block storm surge from reaching Rotterdam.

FIGURE 9-17 LEANING TOWER LEANS

Compression of marine clays by the Leaning Tower of Pisa's load has caused the tower to lean. The cables that help keep it from falling are just visible between the tower and the church on the left. The tower curves slightly because builders tried to straighten the upper floors during construction.

retreating continental ice sheet 10,000 to 13,000 years ago. Much of Ottawa, Ontario, is built on the notorious Leda clay, a marine layer up to 70 meters thick. It is especially weak because the natural salts have been leached out. The solid grains of the clay amount to less than one-third of the total volume so the clay dries and shrinks under old homes and buildings, causing differential settling and distortion (**FIGURE 9-18**). Landslides throughout the St. Lawrence River valley and its tributaries have involved the Leda clay. In an ironic twist, the headquarters building for the Geological Survey of Canada, completed in 1911, was built on a site underlain by more than 40 meters of Leda clay without consulting the geologists who worked there. The heavy outer stone walls immediately began to sink into the clay, causing cracking of the structure and separation of its tower from the main building. The tower was finally torn down in 1915 and the building and its foundations strengthened at considerable cost.

FIGURE 9-18 HOUSE DISTORTED BY CLAY SHRINKAGE

This is a telltale sign of a brick house (in this case, in Ottawa, Ontario) settling and being distorted by shrinkage of its clay substrate. Note the diagonal zigzag crack in the brick wall (see arrow).

An infamous case involving Leda clay began in April 1971 at St. Jean Vianney, Quebec, where a new housing development had been built on a terrace of Leda clay 30 meters thick. Heavy rains loaded the clay, and eleven days later it collapsed and flowed downslope toward the Saguenay River at 25 kilometers per hour for some 3 kilometers. The slide took 31 lives and 38 homes (compare FIGURE 8-8).

Thaw and Ground Settling

Geothermal gradient refers to the fact that the temperature at the surface of the earth increases with depth. Rocks deep in the earth are hotter. Near Earth's surface, the ground temperature increases by about 0.6°C per 100 meters of depth (6°/km). In temperate climates where average year-round temperatures are above freezing, the ground at depth is well above freezing. Arctic climates of Canada, Alaska, Northern Europe, and Asia have in-ground temperatures that remain below freezing year round. Water in the ground remains frozen, leading to a ground condition called **permafrost**. The ice occurs as coatings on grains or it completely fills the pores between grains and supporting strands of peat moss. Ice veins and even wedges, layers, and masses of more-or-less solid ice form near the surface.

In other northern climates or high mountain areas such as the Alps and Tibet, where average temperatures remain below freezing, the ground is frozen all winter but may thaw within a meter or two of the surface for two months in summer. Still farther north, in the Arctic, only a thin surface layer may thaw in the warmest summer weather. The distribution of permafrost around the North Pole is shown in **FIGURE 9-19**, p. 248.

Local environmental factors also affect ground temperature. Large areas of water in lakes and rivers help keep the

FIGURE 9-19 FROZEN GROUND IN THE ARCTIC

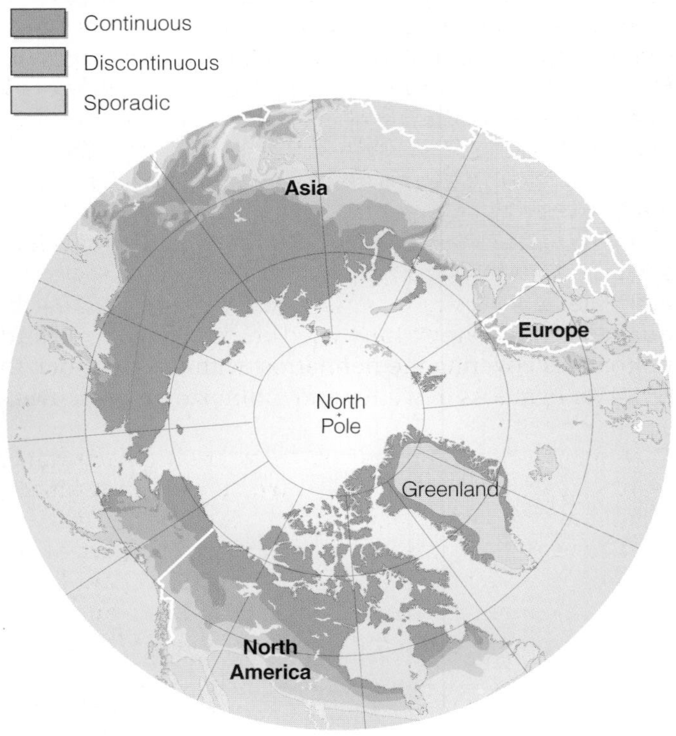

Permafrost zones

- ▢ Continuous
- ▢ Discontinuous
- ▢ Sporadic

Distribution of permafrost and thawing permafrost in northern North America and Asia.

FIGURE 9-20 THAWING PERMAFROST

Permafrost near Tuktoyaktuk, Northwest Territories of Canada.

(**FIGURE 9-20**). Landslides become more frequent both on slopes in soil and in bedrock as the once-solid ice turns to water between grains and in rock fractures. Most of the slides are shallow, a few meters deep, as the thawed layer slides off the remaining permafrost.

Structures such as buildings, roads, railroads, pipelines, and power poles, in permafrost areas have foundations built on or in the permanently frozen ground. The structural integrity of buildings and other structures can be compromised if the permafrost begins to thaw (**FIGURES 9-21** and **9-22**). Buildings that were solidly anchored on the ice settle and twist out of shape.

Those building in permafrost areas should take into acount the possibility of thaw and its consequences. The Trans-Alaska Pipeline is suspended above the ground surface because oil flow causes frictional heat in the pipe; that would thaw the permafrost, which could break the pipeline

ground warmer, as does vegetation that maintains deeper snow cover. However, forest fires or removal of vegetation for construction can also lead to thawing of permafrost.

When permafrost thaws, water runs out of the frozen ground, causing the ground to settle and flow downslope

FIGURE 9-21 BUILDING DAMAGE DUE TO PERMAFROST THAW

A **B**

A. This house near Fairbanks, Alaska, settled and deformed as permafrost thawed. **B.** Modern buildings are often raised off the ground so that (cold) air can circulate under a heated building, minimizing the possibility of permafrost thawing. Nome Alaska.

FIGURE 9-22 THAW DAMAGES RAILWAY

The Copper River Railway in Alaska was abandoned by 1960 because thawing permafrost severely deformed its line.

(**FIGURE 9-23**). The ground is also kept colder by an ingenious device called a *thermosiphon* (FIGURE 9-23B). The vertical posts that support the pipeline contain a gas that condenses to a liquid at cold air temperatures. The liquid settles to the bottom of the pipe within the frozen ground. Heat from the permafrost is sufficient to vaporize the liquid; that vapor rises to the top of the pipe where it again cools and condenses and sinks. This process dissipates the heat at ground level and helps keep the ground frozen so structures on it remain stable.

The pipeline is also built in a manner that permits it to slide back and forth on supports; that allows for ground movements, thermal expansion of the pipe, and earthquakes. One strategy long used in building roads and railroads is covering the permafrost with crushed rock to promote heat loss in winter and minimize heat access in summer.

Damages due to permafrost thaw are increasing as a result of global climate change. With a warming climate, ground temperatures increase, the permafrost thins, and the active zone of freeze-thaw thickens. The permafrost in some areas is now beginning to thaw to deeper levels during the summer, causing damage to more and more structures and contributing to climate change (discussed in Chapter 10).

Thawing permafrost not only wreaks havoc with the ground surface, but meltwater flowing into the Arctic Ocean promotes the melting of Arctic sea ice. Destruction and decay of surface vegetation and thawed peat can release immense amounts of methane and carbon dioxide, fostering even more global warming. These topics are discussed in more detail in Chapter 10.

FIGURE 9-23 PIPELINE EXPANSION

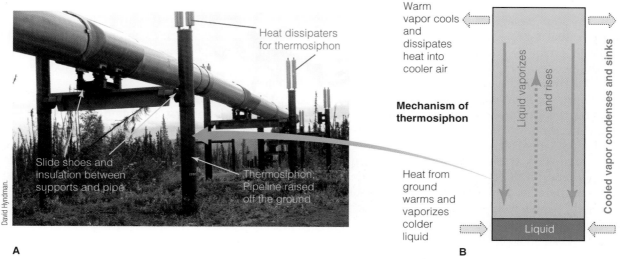

A. The Trans-Alaska pipeline carries crude oil from the North Slope of Alaska to the ice-free port of Valdez. Above-ground segments of the pipeline are built in a zigzag geometry with slide shoes designed to allow the pipe to freely accommodate ground movements and the thermal expansion and contraction of the pipe. **B.** Thermosiphons dissipate heat in the air so that it will not melt the ground.

FIGURE 9-24 DISTRIBUTION OF SWELLING SOILS

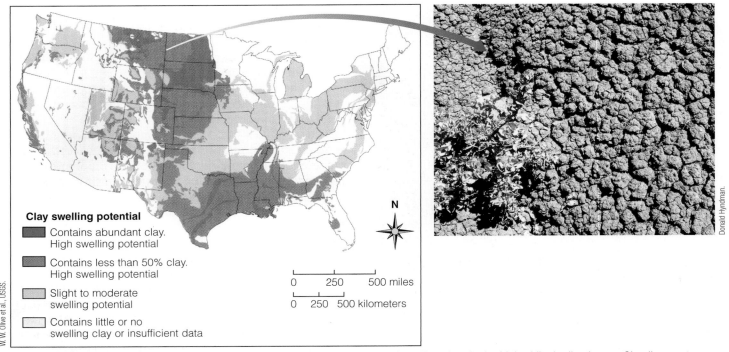

Clay swelling potential

- Contains abundant clay. High swelling potential
- Contains less than 50% clay. High swelling potential
- Slight to moderate swelling potential
- Contains little or no swelling clay or insufficient data

Areas of swelling soils in the United States. Popcorn weathering is a telltale sign of swelling clays in the Makoshika badlands near Glendive, eastern most Montana. View is 60 centimeters across.

Swelling Soils

Swelling soils can be just as damaging to structures as sinkholes, subsidence, and permafrost thaw. Much of the United States is affected by swelling soils, including most large valleys in the Mountain West region; all of the Great Plains; much of the broad coastal plain area of the Gulf of Mexico and the southeastern states; and much of the Ohio River valley (**FIGURE 9-24**). Even some areas not normally associated with swelling clays can have local problems.

Many homes in a new subdivision at the base of the Sierra Nevada Range, just east of Sacramento, California, began cracking soon after construction. The clay-bearing Ione formation swells during rainy winters, causing cracks in walls and twisting of window and door frames. The annual cost in the United States from swelling soils exceeds $4.4 billion; highways and streets account for approximately 50 percent of this total, homes and commercial buildings for 14 percent. When buying in an area subject to swelling soils, people would be well advised to have an expert examine a home before purchase.

The same process that causes subsidence in some clays when they dry out is responsible for swelling when water soaks into the interlayer spaces of the clay mineral structure. During long dry periods, the soil shrinks; in long wet spells, it expands.

Swelling soils and shales are those that contain **smectite**, a group of clay minerals that expands when they get wet. You can often spot the presence of swelling clays where the expansion and cracking of surface soils forms *popcorn clay* (FIGURE 9-24b). These minerals form by weathering of both aluminum silicate minerals and glass in volcanic ash (their behaviors are discussed in Chapter 8). Driving on a dirt road built on smectite-rich soils can also be hazardous when roads are wet and the clay becomes extremely slippery. Mud builds up on tires and shoes—the "gumbo" that is so familiar to those who live in such areas (**FIGURE 9-25**).

The expansion and contraction of soils causes the cracking of foundations, walls, chimneys, and driveways. Damage also results when adjacent areas swell at different rates. Swelling is problematic for roads and structures when layers of clay are interspersed with layers of sedimentary rocks. Swelling of the clay layers creates differential expansion, deforming houses and roadways. Parts of the southwestern suburbs of Denver, south of U.S. Highway 285 and east of state Highway 470, were built from the 1970s through the 1990s over the upturned edges of these formations. Where horizontal, smectite-bearing layers under the High Plains tilt up to reach the surface, those layers create differential expansion under houses and streets (**FIGURE 9-26** and **Case in Point:** Differential Expansion over Layers of Smectite

FIGURE 9-25 GUMBO AND SWELLING CLAY

Gumbo that builds up on shoes and tires on wet dirt roads is a good indication of smectite-rich soil.

FIGURE 9-26 SWELLING NEAR DENVER BULGES ROAD

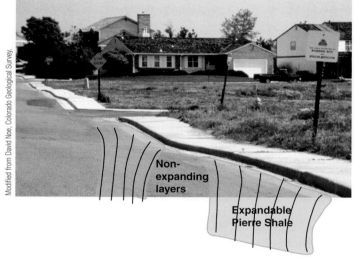

Non-expanding layers

Expandable Pierre Shale

Houses in parts of Denver are built over inclined beds of smectite-bearing Pierre shale. Wetting of the shale beds (schematically drawn in under the road) causes swelling that bulges the ground and deforms both the road and houses. Sangre de Cristo Road in southwestern Denver has since been leveled and repaved, eliminating the evidence.

Clay—Denver, Colorado, p. 255). The problem appears when rainfall, snowmelt, yard watering, and, sometimes, a leaking underground pipe wets the clay. Long ridges parallel to the mountain front, and separated by swales, grow 10 to 20 meters apart. A ridge rising under a house stretches

FIGURE 9-27 DIFFERENTIAL WETTING CAN DEFORM HOMES

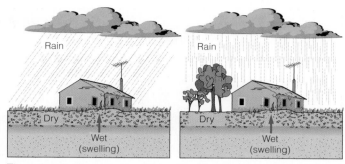

The soil under the wet side of a house, either because of a prevailing rain direction or lack of vegetation, can swell or deform the house.

and pulls it apart. Walls buckle and crack, chimneys fracture, windows pull apart, and doors no longer close.

Homes built on smectite clays can experience problems because of different moisture conditions on separate parts of a property. For example, the area under a house remains drier, whereas the ground surrounding the house is open to moisture from rain, snow, lawn watering, house-gutter drainage, or a septic tank or drain field. Or prevailing wind-driven rain against one side of a house can cause the ground around that side to swell (**FIGURE 9-27**). Vegetation concentrated along one side of a house can dry out the soil on that side. Even the removal of vegetation can increase moisture in the soil by decreasing evapotranspiration. In each case, the ground receiving more moisture expands, while the ground receiving less moisture contracts, deforming and cracking buildings.

Swelling mostly occurs within several meters of the ground surface because the greater pressure at depth typically prevents water from entering a clay's mineral structure and lifting everything above it. Thus the problem of swelling and deformation of buildings does not extend to larger, heavier structures such as large commercial buildings and large bridges.

The obvious solution to damage from swelling soils is to build elsewhere. Where this is not feasible, near-surface effects can be minimized by preventing water access to the soil, removing the expanding near-surface soils, or sinking the foundation to greater depth. The addition of hydrated lime—calcium hydroxide, $Ca(OH)_2$—to an expandable clay can reduce its expandability by exchanging calcium in molecular interlayer spaces so water cannot easily enter. However, this should not be done with quick clay soils, because that would markedly reduce soil cohesiveness, and foster soil collapse and landslides.

Cases in Point

Subsidence Due to Groundwater Extraction
Venice, Italy ▶

Venice, along the northeastern coast of Italy, is built at sea level on the now-submerged delta of the Brenta River. The delta and the city have, for 1500 years, been subsiding for a variety of reasons, the most important being extraction of groundwater from the delta sediments. The load of buildings squeezes water out of the soft delta sediments, and pumping groundwater for domestic and industrial use collapses the pore spaces in the sediments. That decreases the volume of the sediment, which sinks the original surface of the delta below sea level.

The high tides that accompany winter storms compound the problem. They submerge most of the few normally dry walkways under as much as a meter of water. Maps on the water taxis advise people of the few dry routes available for walking. Pumping groundwater from under the city is now prohibited, and sinking has all but stopped. But the high winter tides that formerly flooded the city twice a year now come several times annually. Ancient drainage pipes installed to carry off rainwater now carry seawater back into the city during times of high winter tides, along with whatever effluent they normally carry.

As the delta subsides, so does Venice. Some parts have now sunk as much as several meters. The ground floor of Marco Polo's thirteenth century house is now two meters below present ground level. Below that are older floors from the eleventh, eighth, and sixth centuries.

St. Mark's Square

Drain backflow in St. Mark's Square

BAR MERICANO

Raised walkway

St. Mark's Square during storm tide

NASA

Donald Hyndman

Donald Hyndman

▶ *Increasingly frequent high tides reverse the ancient drainage system of Venice such that lagoon water backflows, bubbling up through drains, and floods St. Mark's Square. In the satellite view, the twisting green channel through the center of Venice is the "Grand Canal," part of the original Brenta River delta.*

(continued)

▶ **A.** *At high tide, some streets and sidewalks are flooded, creating problems for deliveries.* **B.** *Buildings sinking into Venice's mud are flooded and deteriorating. First-floor doorways and flooded ground floors are permanently abandoned.*

To accommodate the sinking, Venetians continued to raise the street levels. Streets that were above sea level when they were built are now canals, and doorways and the lower floors of homes are now partly submerged, a condition that prompted Ulysses S. Grant to remark that the place appeared to have a plumbing problem.

Tides in most of the Mediterranean Sea range less than 30 centimeters, but where the tidal bulge sweeps north up the Adriatic Sea near the east end of the Mediterranean, they average one meter. About ten times per year, tides reach 40 cm above normal levels and flood most sidewalks. Once a year, on average, tides reach 65 to 70 cm above normal levels and flood about two thirds of the city's buildings; every 3 to 4 years, high tide reaches 8 centimeters higher and floods 90 percent of the city. When the low atmospheric pressure of a storm that raises sea level coincides with high tide, sea level rises even higher; coinciding with a sirocco, a warm winter wind blowing north out of Africa that pushes the tidal bulge ahead of it, the tidal rise can be especially high. Such a series of coincidences on November 4, 1966, when three days of rain had already raised both river and lagoon levels to cause catastrophic flooding, tides reached about 130 cm above normal high tide.

Merely raising the sea walls around the city will not cure this problem. Climate change and the concurrent rise in sea levels undoubtedly contribute. A few historical highlights illustrate the political background that makes any solution to the problem difficult:

- **1534** A major dike is built to divert the mouth of the Sile River to the east, away from Venice.

- **1613** Venice, on the delta of the Brenta River, diverts the Brenta around the city. New sediment is no longer added to that part of the delta.

- **1744–1782** Seawalls are constructed along the offshore barrier islands to protect the Venice lagoon from the Adriatic Sea.

- **1925** Large-scale pumping of groundwater from under Venice begins for industrial use.

- **1966** A major storm on November 3, compounded by an especially high tide during a full moon, pushes Adriatic water into the Venice lagoon, which is normally about 60 centimeters deep. The water reaches nearly two meters above normal sea level. The city floods for 15 hours, causing severe damage to buildings and works of art. Heating-oil tanks float, breaking attached pipes and leaking oil into canals and homes. Sinking permanently floods the first floors of most buildings; most families now live on their second floors.

- **1970** After 45 years of pumping and gradual sinking, the government begins to completely prohibit industrial pumping of groundwater from under Venice.

- **1981–1982** A technical team of the Public Works Ministry and Venice City Council approves plans for a major flood-control project to protect the shallow lagoon from flooding. The plan is to build four major mobile barriers across the three barrier island inlets to the lagoon that can be raised at times of potential high water.

- **1984** The Italian Parliament appropriates the equivalent of billions of dollars to build the barriers and raise parts of the urban center.

Since 1908, the combination of sea-level rise and natural and man-induced subsidence has reached 24 centimeters. Although the Italian government agreed decades ago to spend hundreds of millions of dollars on a solution, nothing happened until recently.

The controversial Mose project, a system of giant, moveable gates across the barrier island inlets, began in 2003 and is intended to prevent floods during especially high tides. The gates, like a layer of hollow boxes, lie in a concrete trough on the seafloor. With the arrival of an especially high tide, pumped air floats one edge of a gate above sea level, forming a

(continued)

wall to block the incoming tide. As of the end of 2009, the project was 63 percent done, with an expected completion date in 2012. However, given Italian politics, governments that change hands every year or two, and internal bickering, the budget for this now 6-billion-dollar project continues to be debated. There are those who argue that it is more important to spend available funds on building industrial capacity and promoting new jobs rather than on preserving Venice as one of the world's historic and artistic centers. If this sentiment wins out, development would involve a third industrial zone at the landward edge of the lagoon in Marguera.

During floods, the impermeable marble bases of most of the old buildings are submerged, and seawater seeps into the porous bricks of their walls, which then begin to crumble. More frequent submergence of drainpipes that carry rainwater from the city into the lagoon creates another major problem.

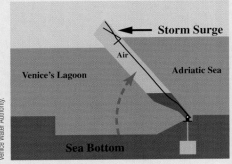

▶ *The new Mose flood barrier, pneumatically raised during storm high tides, floats a wall across the incoming tide.*

These pipes act as the city's sewers; they carry untreated sewage, including laundry suds and toilet effluent, into the canals, then into the lagoon, where the tides periodically flush it out to sea. During especially high tides, of course, the drains carry the effluent back into the

canals and walkways of the city. Some of the controversy concerning the proposed movable barriers at the mouths of the lagoon involves restriction of the tidal action that flushes the lagoon.

One of the highest Venice floods in decades reached a peak on December 1, 2008 (but it still ranked below both the 1986 level and the record level of 193 cm reached in 1966). Flood level is designated as 101.6 centimeters, and although the city erects wooden platforms that permit pedestrians to cross flooded areas around St. Mark's Square, the water rose so fast in 2008 that the platforms could not be emplaced in time. Ground-floor homes and shops flooded. Alarms sounded to warn both residents and visitors to stay indoors or use high wading boots. Such flooding has become more frequent and generally higher in recent years. Since groundwater pumping has been halted, the higher tides are blamed on climate change.

Subsidence Due to Groundwater Extraction
Mexico City, Mexico ▶

Mexico City, in south-central Mexico, was settled in the early 1500s when Spanish invaders conquered the Aztecs who earlier settled there at the edge of broad shallow Lake Texcoco. As their capitol city grew, they used water from the lake faster than it was replenished from the surrounding mountains, in turn extracting water from the now-dry lake bed and building on its surface. Mexico City, with a population of over 20 million, grew on the old, dry lakebed about 2,200 meters above sea level. Because it has limited surface water supplies, and relies primarily on groundwater that is replenished by summer rains and snowmelt from the surrounding mountains, its aquifer has been depleted since the early 1900s.

Sand and gravel interlayered with and below smectite-rich swelling clays provide a good source of water but the ground settles as water is withdrawn and

the slippery clays spread and deform. By 1948, it became apparent that subsidence was caused by withdrawal of groundwater for municipal use; by 1973, widespread subsidence had reached 5 meters over large areas and in some places 9 meters. It continues to settle 0.3 m/year in some places. Since then, the population has grown from 9 million to 20 million. Some of the subsidence has been in the gravels but most has been in the clays because of water extraction and to a lesser extent loading by buildings. Many buildings have cracked, settled, and tilted. Collapse of the porous lakebed sediments has caused subsidence of as much as 8.5 meters in the central part of the city. Because the subsidence is somewhat variable, buildings tilt, and sewer and water lines crack or break. Damage to sewer lines and cracking of impervious clay layers permits sewage and other contaminants

to infiltrate to the aquifer below. In some places sinking is so rapid that children mark their heights on a pole to see if they grow faster than the ground sinks. Since the city has sunk below the overall lake floor, 24-hour pumping is needed to prevent summer rains from flooding the city.

Water and sewer lines have been deformed and broken; well casings, anchored at depth, now protrude from the ground. Leakage losses of 30 to 40 percent of potable water from broken lines compound the problem.

Following major floods in 1951, new reservoirs and water lines were built to bring water from outside the Mexico City basin; a massive aqueduct diverts water from a drainage area 120 kilometers away,

(continued)

Ground settled

Dave Rudolph.

▶ **A.** *Mexico City house severely distorted by settling of its left side.* **B.** *200-year-old Basilica de Guadalupe tilts to left; new basilica to right was built to replace it.* **C.** *Water-well casing protrudes almost 8 meters from ground as the ground settles around it.*

lifting it more than 1,200 meters. Ground-water decline and settling have lessened but settling and pipe breakage continue; some poor areas receive drinking water for only 2 hours per week. Although that water is safe when it leaves a treatment plant, it is often contaminated by old plumbing and storage tanks. Industrial plants are now required to treat sewage and recir-culate water for further use. Unfortunately almost all of the rainfall flushes down the sewers where it is lost to further use because sewage is pumped over a nearby divide and out of the basin. Authorities are experimenting with collection and use of rainwater for flushing toilets and washing. In 2009, water supplies were shut down for one weekend per month to conserve supplies diminished by an especially dry 2008.

In a related event in Guatemala City in February 2007, formation of a 100-meter deep sinkhole consumed a dozen homes when blockage and rupture of a major underground sewage line and heavy rains caused subsidence.

Differential Expansion over Layers of Smectite Clay
Denver, Colorado ▶

Denver is built on the central High Plains, which are underlain by flat-lying 70-to-90 million-year-old sedimentary rocks that curve up to nearly vertical at the Earth's surface just east of the Front Range of the Rockies. Among these rocks is the Pierre shale in the Denver–Boulder area, a subtle villain 1.6 to 2.4 kilometers thick. The shale contains smectite-rich clay layers formed by the weathering of volcanic ash, blown in from volcanoes far to the west. When wet, the clays swell upward and outward, lifting parts of houses and twisting them out of shape

Within a few years of construction, more than 0.6 of a meter of differential movement has created broad waves in roads and cracked and offset under-ground utility lines. Parts of houses rise, causing twisting, bending, and cracking of foundations, walls, and driveways. Windows fracture and chimneys sepa-rate from houses.

Until 1990, engineers attributed the problem to swelling soils. Their mitiga-tion plans called for drilled piers under foundations or rigid floating-slab floors. Most proved unsuccessful. More recent research by the Colorado Geological Survey and the U.S. Geological Survey demonstrated that the problem is the Pierre shale bedrock. That "bedrock" is fairly young and about as soft as the overlying soil. Recent work in the Denver area shows that excavation and homogenization of both clay- and non–clay-bearing layers to depths of less than 10 meters in preparation for new subdivisions can increase uniformity of expansion and minimize problems. Regulations enacted since 1995 require new subdivisions to excavate to a depth of at least 3 meters and replace this with homogenized fill before pouring founda-tions. Significant reduction in damage has resulted.

(continued)

Donald Hyndman.

A

KLP Consulting Engineers.

B

▶ **A.** *The driveway and garage door deformed in response to the swelling clays on which this home was built in a southwestern Denver suburb. The brick wall pulled away from the garage door, leaving a gap at the bottom (see arrow).* **B.** *Swelling clays in southwestern Denver deformed this house's basement foundation in spite of heavy steel pipes utilized to stabilize it.*

Critical View

A South of Denver, Colorado

Donald Hyndman.

1. Note the skewed garage door. What happened to the house to do that to the garage door?
2. What could be done to minimize further damage here?
3. What could have been done before the house was built to avoid most of this type of damage?

C Arizona

Michael C. Carpenter, USGS.

1. What probably caused this huge crack in the ground?
2. What likely caused that process?
3. What could be done to stop further damage of this type?

B Pennsylvania

William Kochanov, Pennsylvania Geological Survey.

1. What is this strange landscape called?
2. What kind of rock, visible here, develops into this type of landscape?
3. How does that landscape form?

D Southern Texas

Donald Hyndman

The roof of this cavern collapsed to leave a large opening. Large blocks in the lower right collapsed from the roof in the upper right.

1. How would such a large cavern form and in what type of rocks would it form?
2. The small lake in the lower left is the level of the groundwater table. Explain how various levels of the groundwater might affect collapse of the roof and collapse of the slabs of rock in the lower right.
3. What are the downward-pointing "teeth," at the edge of the opening in the upper left, called? Explain how such features form?

Air view, west of Houston, Texas

Donald Hyndman.

1. This landscape is characterized by thousands of small ponds tens to hundreds of meters across. How do the depressions form?

2. What type of bedrock lends itself to formation of these depressions?

3. Sinkholes may form in such areas. Are the collapse sinkholes likely to form in very wet weather or very dry weather? Why?

G Mammoth, Kentucky

USGS.

F Carlsbad, New Mexico

USGS.

1. What are the long, thin pencil-shaped formations descending down from the roof called?

2. What are these long formations made of?

3. Sometimes you can see a drop of water dripping from the end of one of these features. Where does that water come from?

4. What is in the water that adds to the lengths of these formations?

5. What process permits that dissolved material to precipitate on the ends of these long features?

1. Water is pouring out from this underground cave. Where does the water come from?

2. What kind of bedrock forms this type of cavern?

3. How does such as cavern form?

4. What is in the water that helps it to dissolve the rock and how does that substance get there?

Chapter Review

Key Points

Types of Ground Movement

- Sinkholes, subsidence, and swelling soils cause far more monetary damage in North America than other natural hazards.

Sinkholes

- Rainwater, naturally acidic because of its dissolved carbon dioxide, can dissolve limestone, forming large caverns underground. **FIGURES 9-2** and **9-3** and **By The Numbers 9-1.**

- Soil above limestone caverns can percolate, leaving a cavity that can collapse to form a sinkhole. **FIGURE 9-5.**

- Sinkhole collapse can be triggered by lowering the level of groundwater that previously provided support for a cavity roof; loading a roof; or an increasing water flow that causes the rapid flushing of soil into an underlying cavern.

Land Subsidence

- Removal of groundwater or petroleum from loose porous sediments can cause compaction, resulting in subsidence of the land surface.

- Subsidence of an area below sea level can result in flooding, submergence, and saltwater invasion of freshwater aquifers.

- Land subsidence can also be caused by drainage of peat or clay-rich soils; oxygen-bearing surface water invades the peat, causing it to oxidize and decompose.

- Clays contract as they dry out, leading to subsidence.

- In cold climates, water in the ground is frozen as permafrost for all or part of the year.

- When permafrost thaws, water runs out and the ground settles, causing fractures and landslides.

- Global climate change has increased the depth of permafrost thaw in some areas and may lead to greater structural damages.

Swelling Soils

- Smectite soils, formed by the alteration of soils rich in volcanic ash, swell when water invades the clay mineral structure. Such soils are abundant in the midcontinent region, the Gulf Coast states, and western mountain valleys **FIGURE 9-24**.

- Swelling soils can crack and deform foundations, driveways, and walls. Damages can occur when soils swell at different rates—for example, when one area of a property experiences more or less moisture than another due to vegetation or protection from rain.

Key Terms

caverns, p. 236
cover collapse, p. 238
cover subsidence, p. 238

dissolution, p. 237
groundwater mining, p. 241
karst, p. 237

land subsidence, p. 236
permafrost, p. 247
sinkholes, p. 236

smectite, p. 250
soluble, p. 236
swelling soils, p. 236

Questions for Review

1. What causes sinkhole collapse—that is, where is the cavity that it collapses into, and how does that cavity form?

2. Explain how caverns form in limestone.

3. Where in the United States are sinkholes most prevalent? Why?

4. What is karst and how does it form?

5. What are the main characteristics of soil above a limestone cavern that lead to the formation of sinkholes?

6. What weather conditions are most likely to foster the formation of sinkholes? Why?

7. Other than limestone, what other types of rocks are soluble and can form cavities that collapse?

8. What types of human behavior typically lead to widespread ground subsidence?

9. Name three areas in the United States that have seen major subsidence and in each case explain what caused it.

10. How can subsidence contribute to flooding?

11. Explain how permafrost thaw causes ground settling.

12. What is the relationship between permafrost thaw and global climate change?

13. What kind of soil swells, and in what regions of the United States is this soil common?

14. What causes swelling soils to swell?

15. How do swelling soils lead to structural damages?

16. Give an example of how a property might experience differential swelling. Make a sketch to illustrate your explanation.

Discussion Questions

1. Given that lowering groundwater levels in an area of sinkholes removes support from the tops of underground caverns, what restrictions, if any, should be placed on pumping groundwater out for crop irrigation?

2. In case of collapse of a sinkhole that damages a home or commercial building, who should pay for the damages? Why?

3. If a housing subdivision is built on ground that damages homes when the soil swells, who should be held liable? Why?

4. What restrictions should be placed on building in areas subjected to swelling soils damage? Why?

Climate Change and Weather Related to Hazards

10

1941

2004

Muir Glacier in Glacier Bay National Monument, Alaska, shows a dramatic record of melting in the last 70 years. Late summer images.

Bruce Molnia , USGS.

Rapid Melting in the Arctic

Glaciers worldwide have been melting for more than a century. Much of Glacier Bay in southeastern Alaska was a long valley glacier in the 1940s. Since then the glacier has melted back so far that the ocean now occupies that fjord. Cruise ships now follow a deep valley for tens of kilometers up valley to the toe of the ice. Glaciers in Glacier National Park, Montana, were once numerous; now within 20 years or so, the park will be "glacier" in name only. From the many glaciers in the Alps of Switzerland to the Southern Alps of New Zealand, Patagonia in southern South America, and even Mount Kilimanjaro in central Africa, glaciers are rapidly melting. Melting dates back to the end of the last ice age 10,000 years ago, with minor reversals since then in various parts of the world at different times. The so-called "Little Ice Age" saw advancing glaciers in different regions between about 1650 and 1850 but they were regional in extent. Since then, the picture appears to have changed. Continental ice is retreating worldwide.

Basic Elements of Climate and Weather

Weather refers to the conditions and activity in the atmosphere at a particular place and time—temperature, air pressure, humidity, precipitation, and air motion. **Climate** is the weather of an area averaged over a long period of time. Both have significant effects on the hazards discussed in the following chapters involving coastal erosion, hurricanes, thunderstorms, tornadoes, and wildfires.

Hydrologic Cycle

Water in the oceans covers more than 70 percent of Earth's surface. Unfortunately, the oceans are salty; humans need fresh water for life, and less than three percent of the water on the Earth is fresh. This fresh water is derived from evaporation of ocean water, some of which moves over continents and falls as rain or snow. Most of the rain that is not taken up by plants soaks in to become groundwater. That groundwater seeps back to the surface into lakes and rivers and ultimately flows back to the sea. This is called the **hydrologic cycle** (**FIGURE 10-1**).

Air can dissolve water vapor. The sun warms the ocean surface and water evaporates into the air above it. Under most conditions, we do not see the water vapor because it is colorless and invisible, like most gases. The amount of water vapor that air can hold depends on its temperature; cold air can hold little water vapor, whereas warm air can hold a lot. When air contains the maximum that it can hold, it is saturated with water vapor; its relative humidity is 100 percent. If it contains half as much as it can hold, its **relative humidity** is 50 percent.

If air cools to the temperature at which its relative humidity is 100 percent, the water-saturated air is still invisible. This is the dew point. If it cools further, it will have more water vapor than it can hold, causing some of the water to condense into tiny water droplets and form clouds or fog. If enough water droplets form, they coalesce into larger droplets and eventually fall as rain or snow.

As breezes blow across the oceans, the air picks up water vapor. Air cannot hold significant amounts of solids, so water vapor evaporates into the air, but the salt stays in the oceans. Much of the water vapor in the air drifts over the oceans with prevailing winds until it reaches a continent where some will drop out as rain or snow.

Rain or snow falling on the continents can take different paths. Some falls on leaves of trees or other vegetation; some of that evaporates directly back into air. Some of the moisture reaching the ground may run off the surface into streams or lakes, but a large part soaks into the ground to become **groundwater**—the water in spaces between soil grains or in cracks through rocks. The roots of trees and other vegetation draw on some of that moisture, taking it up to carry nutrients to the leaves. The leaves, in turn, transpire water back into the atmosphere because the air generally has less than 100 percent humidity. Direct evaporation and transpiration by vegetation are both included in **evapotranspiration**.

Groundwater slowly migrates in the direction the water table slopes, to reach streams, lakes, and wetlands; it is the main source of water for streams except in dry climates. Streams in turn carry the water downslope into larger rivers and ultimately back into the ocean to complete the cycle. Variations in the hydrologic cycle can lead to droughts and floods.

Adiabatic Cooling and Condensation

Adiabatic cooling occurs when rising air expands without a change in heat content. Whenever an air mass expands, the available heat is distributed over a larger volume, so the air becomes cooler (**FIGURE 10-2**). The rate of cooling as an air mass rises is called the **adiabatic lapse rate**.

During condensation, the water releases the same amount of heat originally required to evaporate liquid water into water vapor. In fact, this heat of condensation is large enough to reduce the adiabatic lapse rate from 10°C per 1,000 meters of rise for dry air to 5°C per 1,000 meters for water-saturated air (**FIGURE 10-3**). In other words, very humid air cools at half the rate of dry air as it rises. When an air mass falls, its temperature rises at the dry adiabatic lapse rate because it is below saturation and can thus hold more moisture.

When this moist air reaches land and is forced to rise over a mountain range, it expands adiabatically and cools, in what is called an **orographic effect**. Because cool air can dissolve less water, the moisture separates as water droplets that form clouds and in turn rain, hail, or snow (**FIGURE 10-4**). When an air mass moves down the other side of a mountain after much of the moisture has fallen as rain or snow, the air is generally warm and dry. This rain shadow effect is the reason deserts commonly exist on the downwind side of mountain ranges.

FIGURE 10-1 HYDROLOGIC CYCLE

Ahrens, 2009.

The hydrologic cycle shows how a small but important part of water from the vast oceans evaporates to cycle through clouds and precipitation to feed lakes and streams, vegetation, and groundwater, before much of it again reaches the oceans.

FIGURE 10-2 ADIABATIC COOLING

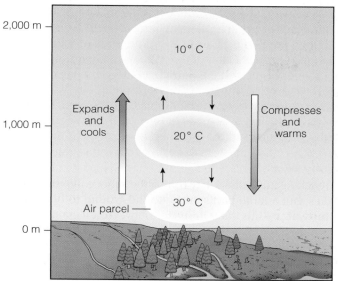

There is less atmospheric pressure at higher altitudes, so rising air expands and cools, while falling air compresses and warms. Under dry conditions, an air parcel will change temperature by 10°C per 1,000 meters of altitude change.

FIGURE 10-3 SATURATION REDUCES ADIABATIC RATE

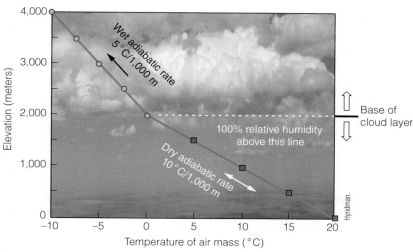

The wet adiabatic lapse rate is half the dry adiabatic lapse rate for rising air. Once a rising air mass reaches 100 percent relative humidity, condensation occurs; this releases heat back to the air mass. Note that the wet adiabatic rate applies only to rising air, as indicated by the upward arrow; the dry adiabatic rate can apply for rising or falling air.

FIGURE 10-4 OROGRAPHIC EFFECT

A. A schematic diagram of the orographic effect shows how warm moist air rises, expands, cools, and dumps precipitation. Descending on the other side of the mountains, it contracts, warms, and dries out. **B.** Moist air streaming over Mt. St. Helens from the west, descends to the east, warms, and dries out so the clouds evaporate. **C.** Humid air rising from the Pacific Ocean south of Monterey, California, cools and condenses to form clouds over the Santa Lucia Range.

CLIMATE CHANGE AND WEATHER RELATED TO HAZARDS 263

Atmospheric Pressure and Weather

Air pressure is related to the weight of a column of air from the ground to the top of the atmosphere. When an air mass near the ground is heated, the air molecules vibrate faster and have more collisions, causing it to expand, decreasing its density. The lower-density air rises to create a low-pressure region (a "low") (**FIGURE 10-5**). It has lower pressure because rising air exerts less downward pressure than still or falling air. Air near the ground surface is pulled in toward the low-pressure center to replace the rising air. The opposite occurs in high-pressure systems, where cool air in the upper atmosphere has a higher density than its surrounding warmer air and thus falls. As it moves toward the ground, it has to spread out or diverge, so air moves away from high-pressure systems near Earth's surface.

Air flows from high- to low-pressure areas, causing winds; high pressure pushes outward and low pressure pulls inward. Where these horizontal pressure differences are large, winds are strong.

Coriolis Effect

Because Earth rotates from west to east, large masses of air and water on its surface tend to lag behind a bit. It completes a rotation around its axis through the North and South Poles once every 24 hours; thus a point at the equator has to travel roughly 40,000 kilometers in a day (the approximate circumference of the planet), but Earth's rotation causes no movement for points on either the North or South Poles. This large rotational velocity of Earth's surface makes both oceans and air masses near the equator move from east to west because they are fluid and thus not pulled as fast as the rotating solid Earth below them. Because water and air move above the planet, which rotates faster under them, they lag behind Earth's rotation. Thus in the northern hemisphere, for water or air currents moving south, the Earth rotates faster to the east below them—the lag causes them to veer off to the west. The opposite occurs as water or air currents move toward the poles. In this case, the air or water is moving over areas of Earth with slower surface rotation, and the resulting lag causes a curve off to the east. This forced curvature due to Earth's rotation is called the **Coriolis effect**.

Thus, in the northern hemisphere, large currents of air and surface water tend to curve to the right; they curve to the left in the southern hemisphere. Ocean currents move westward as they flow toward the equator and eastward as they head toward the poles. Thus ocean currents circulate clockwise in the northern hemisphere and counterclockwise in the southern hemisphere. Similarly, hurricanes, tracking west across the Atlantic Ocean, turn in a clockwise direction as they approach North America.

Rising and falling air currents also rotate (see FIGURE 10-5) because of converging or diverging winds. When air rises in a **low-pressure system**, air near Earth's surface has to converge toward this low to replace the rising air. The Coriolis effect causes these winds to shift to the right of straight in the northern hemisphere, which causes these low-pressure systems to rotate counterclockwise. Falling air diverges away from a **high-pressure system** as it pushes down on the planet. Again the Coriolis effect shifts these northern hemisphere winds to the right, which causes a clockwise rotation of these high-pressure systems. The opposite is true in the southern hemisphere: low-pressure systems rotate clockwise, and high-pressure systems rotate counterclockwise. A simple way to remember this rotation is to use the **right-hand rule** in the northern hemisphere. That rule says that if you point your right thumb in the direction of rising or falling air, your bent fingers point in the direction of air rotation (in the southern hemisphere, use your left hand instead). For example, with low-pressure air rising, winds rotate counterclockwise (as viewed looking down on a map).

Air that rises faster will also rotate faster. Because storms are localized in zones of much lower pressure, the fast-rising air there rotates rapidly counterclockwise (in the northern hemisphere). On a weather map of air pressure, the closer together the pressure contours, the higher the winds. Circulating storms such as thunderstorms are small-diameter cyclones; tornadoes have still smaller diameters with higher-velocity winds. Severe thunderstorms accompanying fronts can spawn rapid circulation and tornadoes. These smaller low-pressure systems also rotate counterclockwise (in the northern hemisphere) unless spun off by some uncommon storm effect.

FIGURE 10-5 HIGH AND LOW PRESSURE SYSTEMS

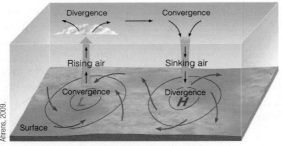

Differences in atmospheric pressure drive air movements and winds. Cold dry air descends to create high atmospheric pressure areas and spreads out as it sinks to the land surface; skies are clear. Relatively warm, humid air rises to create low atmospheric pressure areas. Near the land surface, the air is drawn in from the sides, rises, and eventually condenses to form clouds and rain. In the northern hemisphere, descending air in high-pressure cells rotates clockwise (viewed downward); rising air in low-pressure cells rotates counterclockwise. The opposite rotation occurs in the southern hemisphere.

Global Air Circulation

Atmospheric heating is especially prominent near the equator, while cooling occurs at the poles. Thus less-dense warm air rises near the equator, and nearby cooler surface air moves toward those low-pressure regions (**FIGURE 10-6**).

Higher in the atmosphere above the tropics, the rising air spreads both to the north and the south. At approximately 30 degrees north and south latitude, the cooled air sinks to form subtropical high-pressure zones before returning to the equator to complete the circulation. The air warms adiabatically where it sinks; because warm air can hold more moisture, the humidity in these regions is generally low; the climate is dry or even desert-like. At these latitudes lie the deserts of southern Arizona and northern Mexico, the Sahara and Arabia, central Australia, and northern Chile.

The combination of these global air movements with the Coriolis effect gives rise to the southwest-moving **trade winds** between the equator and 30 degrees north latitude and the northeast-moving **westerly winds** between latitudes 30 and 60 degrees north (see FIGURE 10-6). In North America, the westerly winds bring moist air from the Pacific Ocean to make the wet coastal climate of coastal Oregon, Washington, and southern British Columbia. To the south, the trade winds bring moist air from the Atlantic Ocean to make the wet climate of the Caribbean islands, eastern Mexico, and northeastern South America.

Regional winds can be locally important. During winter, strong westerly winds from the west cool and lose their moisture as they rise across the Rocky Mountains in Alberta, Montana, Wyoming, Colorado, and New Mexico. Continuing east, they form strong downslope winds that become warmer adiabatically as the air compresses into a smaller space at lower elevations. At Boulder, Colorado, such **Chinook winds** have reached speeds of 225 km/hr (140 mph). These warm, dry winds can rapidly melt any snowpack and cause flooding (see Chapters 11 and 12). In southern California, strong trade winds from the east develop in the fall when dry continental air flows southwest around a high-pressure system in Nevada. The air mass flows west, through mountain passes, and down toward the coast. In doing so, its temperature rapidly rises due to adiabatic compression, and its humidity drops. These hot, dry **Santa Ana winds** create extremely dangerous conditions for wildfires in southern California (discussed in Chapter 16).

Weather Fronts

Severe weather is often associated with **weather fronts**, or the boundaries between cold and warm air masses. Thunderstorms develop at such fronts. They are initiated by moist, unstable air rising, cooling, condensing, and raining. In a **cold front**, a cold air mass moves more rapidly than an adjacent warm air mass, causing rapid lifting and displacement of the warm air (**FIGURE 10-7A**, p. 266). In a **warm front**, a warm air mass moves more rapidly than the adjacent cold air mass, and thus it rises over the adjacent cold air mass (**FIGURE 10-7B**, p. 266).

Advancing cold fronts rapidly lift warm air to high altitudes, causing instability, condensation, and precipitation. Air in advancing warm fronts rises over a flatter wedge of cold air, causing widespread clouds and storms. Counterclockwise, midlatitude, low-pressure winds form where the jet stream dips to the south to form a *trough*. Cold and warm fronts sometimes intersect, such as at a low-pressure center. The rising air mass in the low-pressure center cools, causing condensation and rain. When cold air moves toward the equator, it collides with warm air, which is forced to rise.

A low-pressure center can develop as warm moist air to the south moves more rapidly to the east than cold dry air to the north. That counterclockwise shear between air masses can initiate a low-pressure cell where part of the warm air begins to move to the north, such as along a flex in the jet stream (**FIGURE 10-8**, p. 266). As this occurs, air is forced upward along both the cold and warm fronts, with the strongest upward motion occurring where these fronts come together. Such a system can continue to develop so that the angle between the cold and warm air masses will become tighter and tighter. The low-pressure

FIGURE 10-6 GLOBAL AIR CIRCULATION

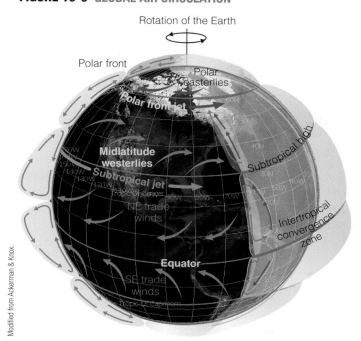

Global air circulation is dominated by the prevailing westerly winds that come from the southwest or west between 30 and 60 degrees north latitude and the trade winds that come from the northeast between the equator and 30 degrees north latitude.

FIGURE 10-7 COLD AND WARM FRONTS

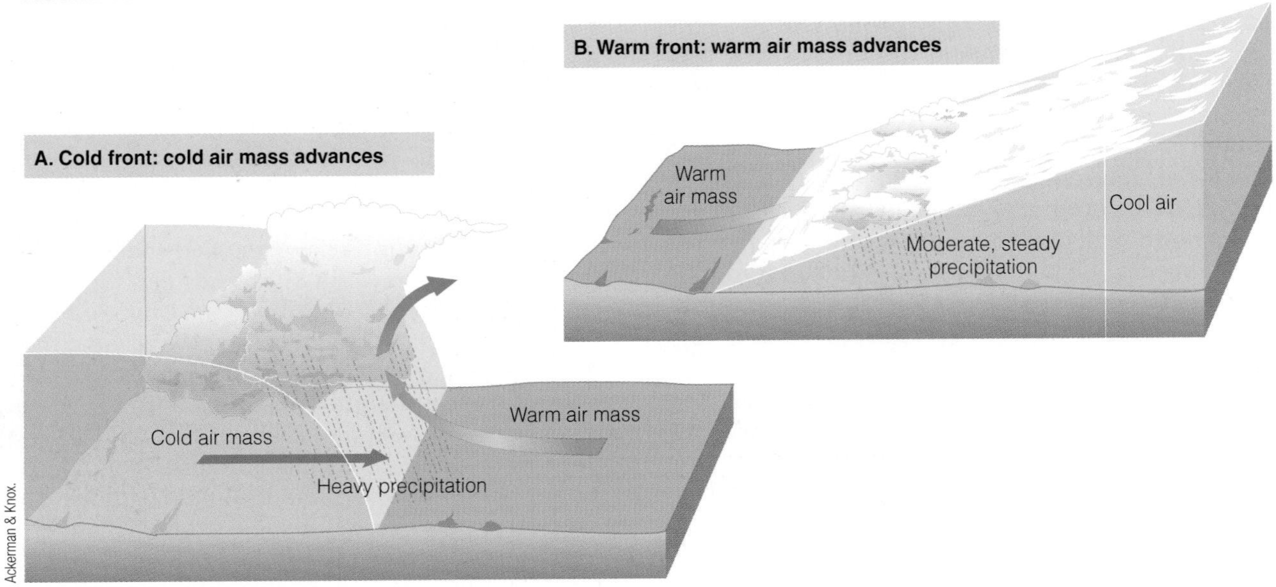

B. Warm front: warm air mass advances

A. Cold front: cold air mass advances

Warm air mass

Cool air

Moderate, steady precipitation

Cold air mass

Warm air mass

Heavy precipitation

A. A schematic diagram of a slice through a cold front shows how the steep front of a cold air mass advances and pushes the warm air mass upward. **B.** A slice through a warm front shows a warm air mass advancing over a cold air mass.

FIGURE 10-8 LOW PRESSURE DEVELOPMENT

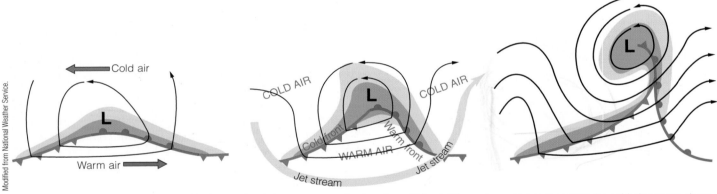

Cold air

L

Warm air

COLD AIR

L

COLD AIR

Cold front

Warm front

WARM AIR

Jet stream

Jet stream

L

A. A low-pressure cell can be initiated by interaction between a warm front and a cold front. Warm air moving east against the cold air causes shear rotation counterclockwise. **B.** The warm air pushes northeast over the front to the east, causing development of a warm front. Cold air moving southeast under the front to the west similarly forms a cold front. **C.** More rapid counterclockwise rotation strengthens the low, which can in some cases spin off from the intersection of the cold and warm fronts.

cell may finally spin off as a cutoff low that can remain in one place, dropping rain for a long period (**FIGURE 10-9**).

In eastern North America, coastal mountain ranges such as the Appalachians increase atmospheric convection (wind). Cold near-shore air can push warmer onshore air upward, causing expansion, cooling, and rain. Storms moving onshore tend to move northward toward the pole. If they do not move far offshore, they can persist for many days, dumping heavy rain along the coast and causing significant

storm surges. Because such higher-latitude storms can have similar wind strengths and rainfall as tropical cyclones, they can be just as destructive.

Jet Stream

Meteorologists often use weather maps that show the subtropical **jet stream** meandering across North America (**FIGURE 10-10**). This narrow, 3- or 4-kilometer-thick ribbon of

FIGURE 10-9 WEATHER RELATED TO FRONTAL SYSTEMS

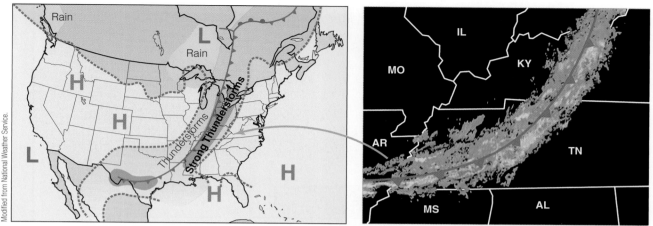

A weather map for June 26, 2003, shows a prominent cold front (blue) intersecting a warm front (red in upper right of map), with a cutoff low in Canada north of Lake Superior.

FIGURE 10-10 JET STREAM

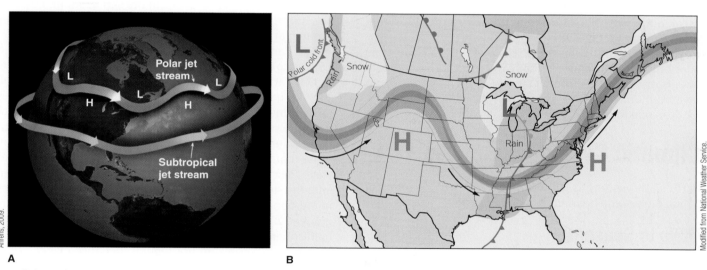

A. Polar and subtropical jet streams meander from west to east across North America. Shown here as continuous bands, they are actually discontinuous and variable in daily position. **B.** An example weather forecast for December 16, 2003, shows the meandering jet stream with a large polar low off western Canada that funnels rain into coastal California. Another Arctic low over the Great Lakes funnels rain into the Ohio River valley.

high-velocity winds blowing from west to east across North America sits at altitudes near 12 kilometers (that of many jet aircraft). The jet stream moves east along the interface between circulating high- and low-pressure cells (**H** and **L** on FIGURE 10-10B); the high-altitude rotation of these cells helps propel the stream.

When a meander in the jet stream shifts farther south, a polar low-pressure area in the North Pacific often lies off western Canada. The counterclockwise circulation of the low carries warm moist air from the Pacific Ocean directly into California. Subtropical storms initiated in the mid-Pacific tend to follow the south edge of the polar cold front into California, an event often called the *pineapple express*. These storms carry heat and moisture to the Pacific coast, where they can dump intense rainfall for a short time. In some cases, back-to-back storms every day or two provide heavy rainfall over broad areas and cause widespread flooding. Such frequent storms saturate the soils; runoff is magnified and landslides are common. **Table 10-1** lists some extreme rainfall events.

Table 10-1	Some Extreme Rainfalls*		
LOCATION	**DURATION**	**AMOUNT**	**CAUSE**
Cherrapunji, India, 1861	1 month	9.3 m	Monsoon
	1 week	3.39 m	Monsoon
	1 day	4.13 m	Monsoon
Cilaos, Reunion, 1952 (Indian Ocean)	1 day	1.87 m	
Alvin, Texas, 1979 (south of Houston)	2 days	1.09 m	Tropical storm
Guangdong, China, June 2010	6 hours	60 cm	
Rapid City, South Dakota, 1972	6 hours	37 cm	
Holt, Missouri, 1947	42 minutes	99 cm	
Curtea de Arges, Romania, 1889	12 minutes	21 cm	
Guadeloupe, West Indies, 1970	1 minute	3.8 cm	

*Compiled from various sources.

With more warm-air production to the south in the summer, the jet stream shifts northward over Canada; in winter, it moves down over the United States. Because the jet stream marks the boundary between warm, moist subtropical air and cold, drier air to the north, storms develop along weather fronts in this area.

The jet stream moves at speeds of 100–400 kilometers per hour. Long-distance aircraft flying east across the continents or oceans use the jet stream to their advantage to save both time and fuel. Those aircraft flying west avoid it for the same reason.

Climatic Cycles

Earth's atmosphere is subject to cyclic changes, most of which have some bearing on weather-related hazards.

Days to Seasons

Heat building during daytime hours can develop afternoon thunderstorms and, in some cases, tornadoes. Earth rotates once per day around its north–south axis (or poles), causing every place on Earth to face most directly toward the sun once a day at noon. At that time, the sun's energy is more intense than at any other, in the same way that the beam of a flashlight shines more brightly in a smaller area when it is pointed directly at a flat surface rather than at an angle.

Longer cycles are known as **seasons**, and many climate-controlled hazards are seasonal. For example, hurricanes occur from summer midway through fall, while tornadoes occur from spring midway through summer. Annual changes from winter to summer and back to winter occur because Earth's axis is tilted with respect to its annual orbit around the sun (**FIGURE 10-11**). When Earth's axis is perpendicular to the sun on both the vernal equinox (spring, usually March 20) and the autumn equinox (fall, usually September 22),

the sun is directly overhead at the equator. At these times, the sun's radiant heat strikes Earth's equatorial region directly, warming it more than at any other time. We are not, as people may believe, closer to the sun in summer. If that was truly the reason for seasons, the entire Earth would have summer at the same time.

The Earth's axis is tilted 23.5 degrees to the plane of the Earth's orbit around the sun; thus, when the sun points directly at 23.5 degrees north latitude on the summer solstice (usually June 21), the northern hemisphere receives its maximum solar radiation and thus the maximum heat from the sun. Although this is midsummer in the northern hemisphere, the warmest temperatures lag the solstice by more than a month. It takes time to heat up the land and water, just as it takes time to heat up an egg in a frying pan after the stove is turned on. When the axis tilts 23.5 degrees in the opposite direction, the sun is directly above 23.5 degrees south latitude, and it is midsummer in the southern hemisphere.

Some variations are not regular, because cycles of different periods often overlap to cause irregular periods (see FIGURE 1-4). Many theories have been proposed to explain significant observed cyclic changes in Earth's northern hemisphere climate, and these overlapping processes contribute to long-term climatic shifts. These include changes in ocean currents, cyclic changes in solar radiation because of changes in Earth's orbit, plate tectonic movements, and changes in atmospheric composition. Climate cycles occur across a broad range of time periods, from days to 100,000 years and more.

El Niño

Oceanic circulation in the equatorial Pacific Ocean normally flows westward, pushed by the trade winds (see FIGURE 10-7). A high-pressure cell exists over Tahiti in the central Pacific, and a major low-pressure cell resides over

FIGURE 10-11 SEASONS

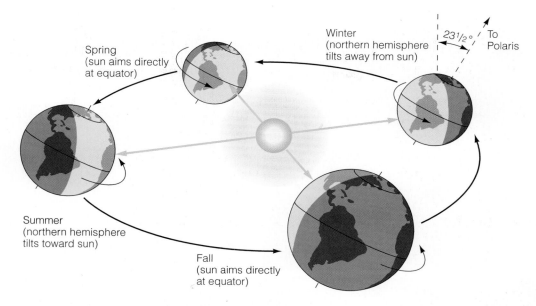

Earth rotates around its north-south axis, inclined at 23.5 degrees to Earth's orbit around the sun. Because Earth rotates around the sun once per year, the northern hemisphere is tilted toward the sun in summer, with the longest northern hemisphere day on the summer solstice (usually June 21). It then tilts away from the sun in winter, with the shortest northern hemisphere day on the winter solstice (usually December 21). The arrows from the sun show the solar energy striking the Earth from directly overhead—for example, on the northern hemisphere in summer.

the warm ocean water of Indonesia. Descending dry air over South America (**FIGURE 10-12A**, p. 270) and rising moist air over Indonesia circulate strong trade winds that pull warm ocean currents westward to Indonesia. Farther east, the trade winds blowing to the west across the Andes Mountains of South America descend to the coast, warming and severely drying the coastal desert environment.

Every six years (on average), especially during winter from November to March, the Pacific Ocean circulation reverses in a pattern called **El Niño**. The name is Spanish for "boy child," given by Peruvian fisherman because the weather phenomenon typically comes every few years around Christmas. The opposite extreme in weather patterns is called **La Niña**, the "girl child."

During El Niño, a major high-pressure area develops in equatorial latitudes over Indonesia and northern Australia in the western Pacific, and a major low-pressure cell develops near Tahiti in the central Pacific. Rising moist air over Tahiti, along with descending dry air over Indonesia and northern Australia, weakens the trade winds, allowing warm ocean currents to flow eastward toward Ecuador and Peru (**FIGURE 10-12B**, p. 270). The accompanying warming of the Pacific Ocean occurs along the equator between South America and the International Date Line in the mid-Pacific.

El Niño results in extreme weather and weather-related hazards, but it affects weather patterns differently around the world. In contrast to rains and floods in western South America, Mexico, and California, droughts and widespread fires affect Australia, Indonesia, and Southeast Asia (**FIGURE 10-13**, p. 271).

El Niño years bring increased rainfall to a normally dry area of Peru. A big low-pressure cell in the Pacific pulls the high-altitude, meandering winds of the subtropical jet stream south before it loops north as it approaches western North America. The warm water piles up against the west coast of South America, rises, evaporates, and causes heavy rainfall in the deserts of Peru. Late 1997 to early 1998 was a particularly intense El Niño for Peru (FIGURE 10-12c). Incessant rains—as much as 12 centimeters in a day—caused big problems for peasant farmers. Fields were soggy; rivers rose and flooded. In contrast, during La **Niña**, the usually lush Amazon rain forest is in the rain shadow of the Andes and is unusually dry and subjected to widespread wildfires.

If some areas like Peru are abnormally wet during an El Niño year, others are abnormally dry, because El Niño simply involves a large-scale redistribution of moisture rather than a change in the worldwide average. In the United States, part of the warm water sweeps north along the coast to bring warm subtropical rains to the

FIGURE 10-12 WARM WATER OF EL NIÑO

Normal conditions / La Niña

El Niño conditions

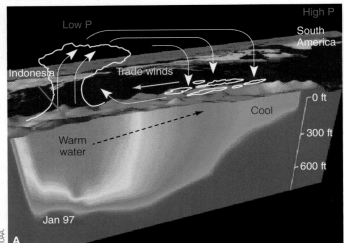

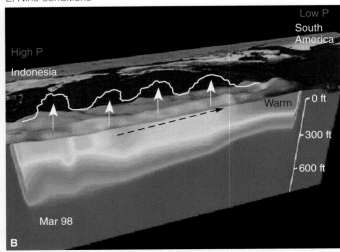

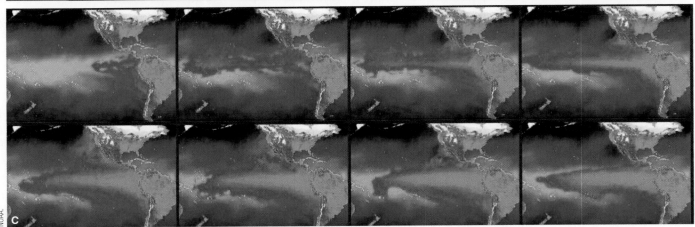

A slice deep into the Pacific Ocean from Indonesia east to South America. **A.** Under "normal" conditions, the near-surface trade winds blow toward the west, keeping the pool of warm water in the western Pacific Ocean, around Indonesia. There the warm, moist air rises and cools, and rains are abundant. **B.** During El Niño, the trade winds weaken and the pool of warm water sloshes back across the Pacific Ocean to the east. **C.** A time series of sea-surface temperatures during the very strong El Niño of 1997–98 shows the warm water (red) sloshing slowly eastward to pile up against the west coast of South America.

southwestern region (**FIGURE 10-14**). This keeps cold Arctic air farther north. Eastern Pacific weather systems off Mexico repeatedly hammer southwestern North America, especially coastal parts of California and northern Mexico. El Niño years in the generally dry Southwest bring more local flooding at lower elevations. In contrast, some northern parts of the United States tend to be drier than normal.

El Niño can also change patterns of other weather-related hazards. In eastern North America, hurricanes are less frequent and weaker, as in 2009, when the jet stream shears off the tops of westward-bound Atlantic storms. Although El Niño minimizes hurricanes, the strong jet stream flow over

Florida tends to stir up strong tornadoes. In February 1998, they killed 41 people near Orlando, Florida. Coastal erosion was also active in 2009–2010, another El Niño period.

Although the recurrence interval for strong El Niño events since 1930 averages six years, such events have been more frequent since 1976 (**FIGURE 10-15**). Some researchers blame climate change, while others disagree. The 1982–83 El Niño killed 2,000 people worldwide, with damage totals of $14.4 billion in 2010 dollars.

El Niño events can be predicted in advance, based on monitored sea-surface temperature (SST) variations in the equatorial Pacific Ocean. The issue is obviously

FIGURE 10-13 WEATHER DURING EL NIÑO

El Niño conditions

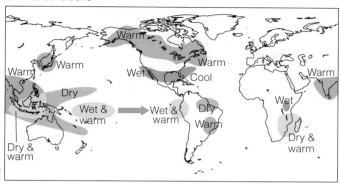

El Niño conditions see warm water of the equatorial Pacific Ocean moving east to Peru and sweeping north to California. That leaves the western Pacific areas around Indonesia drier than normal.

important in planning for both the likelihood of flooding in southern California during an El Niño event and the severity of the hurricane season in the southeastern United States during normal years. It is also important in planning for droughts and crop losses and for regional changes in heating-oil usage and costs. The difference can amount to billions of dollars; even the stock market responds to such predictions.

Tropical atmosphere and ocean buoys anchored in the equatorial Pacific Ocean monitor sea-surface and deeper water conditions, atmospheric temperature, humidity, and winds. Satellites monitor the ocean surface elevations, circulation patterns, and the huge back-and-forth rhythmic sloshing of the Pacific Ocean between El Niño events and normal conditions. When sea-surface elevations and temperatures rise in the central Pacific near Tahiti and atmospheric pressures fall, past events show that more rain is to be expected in Peru and southern California—a new

FIGURE 10-14 WEATHER CHANGES DURING EL NIÑO

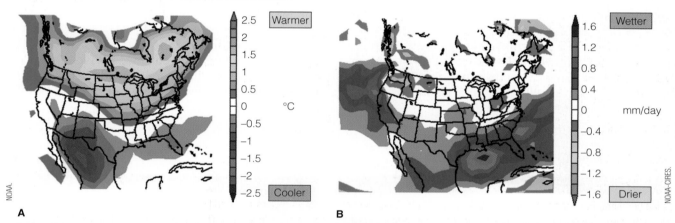

A. temperature change and **B.** precipitation change (average millimeters per day).

During El Niño, the southwestern United States is wet while the southeastern states are cooler than normal. Most of Canada is distinctly warmer. Average, compared with normal, of eight recent El Niño (November to March) effects, 1958 to 1998, for North America:

FIGURE 10-15 ENSO INDEX

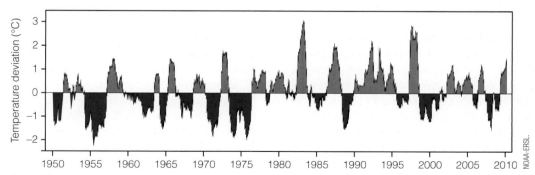

Sea-surface temperature anomalies in the mid–Pacific Ocean are shown from 1950 to 2009. El Niño shows abnormally high sea-surface temperatures. The horizontal 0 line is average.

El Niño. The models of how the process operates were successful in predicting the massive 1997–98 El Niño event a year before it occurred.

North Atlantic Oscillation

As with El Niño in the Pacific Ocean, there is a recurring atmospheric pressure pattern in the northern Atlantic Ocean, although there is no clear correlation between patterns in the two oceans. In contrast to El Niño, which is recognized by variations in Pacific Ocean water conditions, the **North Atlantic Oscillation (NAO)** is defined by variations in *winter atmospheric pressure* over the northern Atlantic Ocean, especially from December to March. Summer effects are much weaker. The variation in strength and position of the high- and low-pressure centers defines the NAO. There is, of course, interaction between the North Atlantic atmosphere and the ocean, but changes in the ocean lag behind those of the atmosphere by two years or more.

In a positive NAO, a strong high-pressure cell is generally centered west of southern Portugal, near the Azores in the eastern Atlantic Ocean, while a strong low-pressure cell is centered over the North Atlantic southeast of Greenland (**FIGURE 10-16**). Driven by the large pressure differences, the prevailing westerly winds are strong. In central Europe summers are cool and winters are wetter and warmer. In northeastern North America, southwesterly flow around the Icelandic low prevents cold Arctic air from moving south; winters are wetter and milder.

In a negative NAO, a weaker high-pressure cell forms farther south in the central Atlantic Ocean, closer to the equator, and a weaker low-pressure cell intensifies farther south, east of Newfoundland. The westerly and trade winds are both weaker. In central Europe, winter storms are fewer and weaker. Cold air moves into northern Europe and moist Atlantic air moves into the Mediterranean. Winter along the east coast of the United States is colder with more snow. Periodicity is not regular, but positive NAO comes along every three to ten years.

Atlantic Multidecadal Oscillation

Longer-term changes in the sea-surface temperature (SST) of the northern Atlantic Ocean, such as the **Atlantic Multidecadal Oscillation (AMO)** have recently become apparent. The SST appears to oscillate from slightly cooler to slightly warmer over several decades, typically about 70 years. The total temperature swing is only about 0.8°C (**FIGURE 10-17**), but because higher temperatures drive more frequent and stronger Atlantic storms, warmer periods may lead to wetter summers in the southeastern United States and more or stronger hurricanes. Warm phases occurred during the periods of about 1860–1890 and about 1930–1960. A new warm phase began around 1995 and, if past cycles are a guide, should continue for a few more decades (in other words, it will be multidecadal). This is not good news for those who live in the hurricane-prone southeastern United States.

FIGURE 10-16 NORTH ATLANTIC OSCILLATION

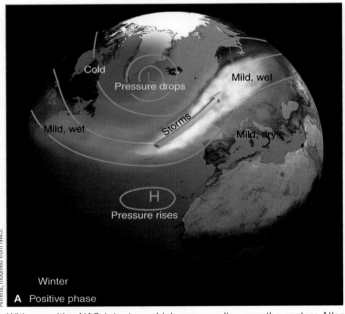
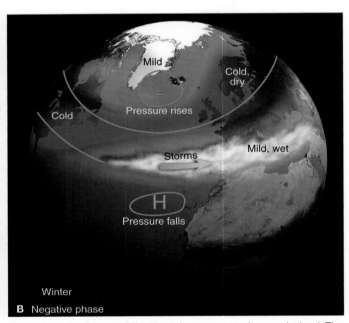

With a positive NAO (+), strong high pressure lies over the eastern Atlantic Ocean near the Azores while strong low pressure sits near Iceland. The westerly and trade winds are strong. Warm weather develops in the western Atlantic and northwestern Europe. With a negative NAO (-), both systems are weaker and lie farther to the southwest; the westerly and trade winds are weak. Darkest blue colors are unusually cold water.

FIGURE 10-17 MULTIDECADAL OCEAN TEMPERATURE

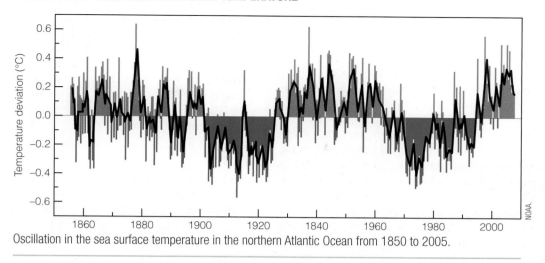

Oscillation in the sea surface temperature in the northern Atlantic Ocean from 1850 to 2005.

The multidecadal changes appear to correlate with the strength of the Atlantic Ocean's large-scale thermohaline circulation (see FIGURE 10-55), drought in the American Midwest (warm Atlantic Ocean), abnormally wet weather in southern Florida, rainfall in the Sahel (central Africa south of the Sahara Desert), and numbers and strengths of Atlantic hurricanes. If the warm phase of the AMO continues, we may see continuing droughts in the Midwest, creating major problems for water supply and agriculture, and strong hurricanes in the southeastern United States.

Long-Term Climatic Cycles

Longer-term climatic cycles are less obvious, but they play a large role in Earth's history. A 26,000-year cycle of the planet's axis acts somewhat similar to a spinning top with a wobble (precession) in the orientation of its axis. Over this cycle, Earth's axis of rotation points at different positions in space, which results in different amounts of solar radiation reaching *parts* of Earth. However, this does not affect the *total* radiation reaching the Earth. A 41,000-year cycle of Earth's axis tilt to the sun ranges from roughly 22 degrees to 24.5 degrees (**FIGURE 10-18**, p. 274), while the present tilt is 23.5 degrees.

Still longer cycles include the **ice ages** of the Pleistocene epoch, roughly the last two million years. These cyclic ice periods average 100,000 years apart but vary somewhat in length. The typical beginning of an ice age shows somewhat irregular advances to a maximum ice cover, followed by a rapid retreat to minimum ice cover over a few decades to a few thousand years. The atmosphere was 5°–10°C cooler during the last ice age relative to the present.

Various explanations for the cyclic nature of the Pleistocene epoch include changes in Earth's orbit and rotation. In the 1920s, Yugoslavian astronomer Milutin Milankovitch calculated that Earth's orbit around the sun slowly changes from nearly circular to elliptical over a period of 100,000 years. This corresponds well with the twenty ice ages recorded in the two million years of Pleistocene time. Milankovitch also calculated the angle at which the sun strikes Earth at every latitude and for each of Earth's orbital variations. He inferred that climate varied in response to these orbital changes—for example, that the ice ages developed when the Earth was tilted so that the latitude of northern Europe received less sunlight in summer. Thus, less snow melted each spring; snow gradually built up year after year, and the climate cooled finally into an ice age. He predicted that ice ages would occur every 100,000 and 41,000 years. More recently, studies of ocean-bottom sediment cores, fossil coral reefs that died off in cold periods, and oxygen-isotope temperatures in Greenland and Antarctic glaciers show a record of such intervals.

Global temperatures of the past are often estimated from studies of oxygen-isotope ratios ($^{18}O{:}^{16}O$) in sedimentary materials (e.g., the O in the mineral quartz, SiO_2) or glacial ice (the O in H_2O) formed at the time. During cooler periods, more water evaporated from the oceans is stored on the continents as ice and snow. Because evaporation preferentially takes the lighter ^{16}O into the air, the oceans become slightly enriched in ^{18}O during such periods. Because marine organisms incorporate oxygen from the seawater into their shells as they grow, the shells preserve the $^{18}O{:}^{16}O$ ratio of the seawater at the time. This permits estimates of the water temperature over long periods. Higher ^{18}O in a clamshell indicates lower northern hemisphere temperatures at the time the shell grew, because more of the water had evaporated from the oceans to form the continental glaciers. The snow that falls also preserves the record of this ratio for the ocean water at any given time; thus ice cores can give us a long-term record of climatic variations.

There are still-longer-term variations in climate. Throughout geologic time, for at least the last two billion years,

FIGURE 10-18 THREE FACTORS CONTRIBUTE TO LONG-TERM CLIMATE CHANGE.

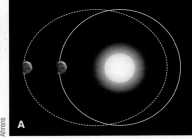
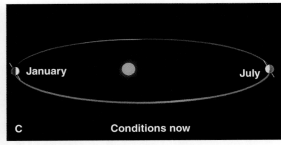
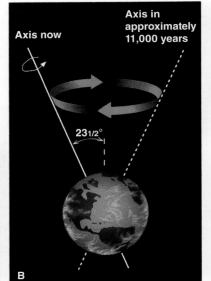
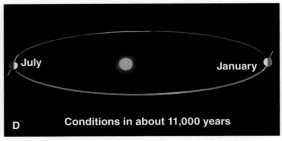

The difference is much less dramatic than the seasons. **A.** Earth's orbit changes from nearly circular to elliptical. **B.** Earth's rotation axis slowly changes its orientation to the sun. **C.** Now Earth is slightly closer to the sun in January. 11,000 years from now, Earth will be slightly closer to the sun in July.

there have been cycles of warm versus cold and wet versus dry. Warm most commonly coincides with wet, though the correlation is not perfect. Major cycles can last for tens of millions of years or even longer. Regardless of the origins of warm and cool or wet and dry cycles, some cycles are unrelated to others. Cycles of different lengths can overlap to give extreme highs or lows that do not come at well-defined regular intervals. These very long-term climate cycles, although important in the vast span of Earth's history, are not important to natural hazards that affect present-day humans nor climate change that affects us; we will not discuss them further.

Hazards Related to Weather and Climate

Many of the hazards related to weather are addressed in the following chapters of this book, including floods (Chapters 11 and 12), waves and beaches (Chapter 13), hurricanes and nor'easters (Chapter 14), thunderstorms and tornadoes (Chapter 15), and even wildfires (Chapter 16). Other weather-related hazards, like drought, heat waves, and snow or ice, may be less dramatic, but their effects can also be devastating. Natural events such as volcanoes can also trigger changes in weather patterns that pose hazards to humans.

Drought

Drought involves a distinctly lower-than-usual amount of precipitation in a region. Drought is a prolonged dry climatic event in a particular region that dramatically lowers the available water below levels normally used by humans, animals, and vegetation. It can involve a drop in the usable water available in reservoirs, groundwater storage, or stream flow (**FIGURE 10-19**).

Unlike many other natural hazards and disasters, drought is not an abrupt or dramatic event; it proceeds slowly and progressively. Drought may not be sudden or spectacular, but its effects can be disastrous. It can affect broad areas and last for months, years, or even decades, inflicting horrendous damage to soils, crops, and people's lives. In fact, annual U.S. losses from drought average more than $6 billion, more than twice that from floods or hurricanes. Widespread drought covers more than 40 percent of the United States every few years (14 times from 1900 to 2002), or about once every seven years on average (**FIGURE 10-20**). More than one-third of the United States suffered moderate to extreme drought in July 1988, compared with about 17 percent in July 2009. Losses from the 1988–89 U.S. drought totaled almost $40 billion.

Meteorologists commonly refer to drought as an extended period with much-less-than-average precipitation for a region. Farmers, ranchers, and foresters recognize it

FIGURE 10-19 DROUGHT

Donald Hyndman.

California Department of Water Resources.

A

B

A. A prolonged decrease in available water can dry ponds and reservoirs, leaving only mudcracks, such as in this former lake bottom.
B. Oroville reservoir on the lower Feather River, just north of Sacramento, California, had virtually dried up in February, 2009. The bridge over the reservoir spanned the river during the drought.

FIGURE 10-20 U.S. DROUGHT THROUGH TIME

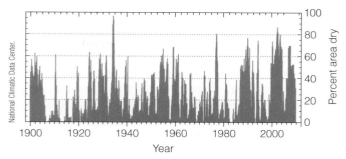

Percentage of the United States that was in moderate to extreme drought, January 1900–January 2010, based on the Palmer Drought Index.

as an extended period of insufficient moisture for plant growth in an area. The amount of precipitation required to sustain local crops depends on the moisture needs of those crops and the agricultural techniques used. Rice and cotton, for example, require tremendous amounts of water; open range grasses used by freely roaming cattle need very little. In some regions, farmers must use irrigation water to allow intensive farming of crops that could not be sustained by rainfall alone. Drought is often aggravated by overuse of the long-term naturally available supply of water. In the Great Plains and Midwest, the widespread drought from 1987 to 1989 was made worse by depletion of groundwater in many areas through heavy pumping and farming of marginal lands.

Farmers also compete for water with residential users, and drought conditions can lead to increased value for water. In a dry period during the fall and winter of 2008–09, some Northern California farmers found it more profitable to sell their water rights to cities and towns that needed it rather than use it to irrigate their crops.

Timing of precipitation can also be a factor leading to drought conditions. Winter snowpack tends to melt slowly, soaking more water into the ground and keeping streams flowing year round. Moisture falling in warmer seasons can rapidly run off or evaporate in the sun and wind, leaving less for groundwater and streams. Evaporation induced by high temperatures and winds can also contribute to drought conditions.

In North America, drought often begins with changes in the jet stream flow that normally moves westward from the warm tropical Atlantic Ocean, over the Gulf of Mexico, and then curves north into the Great Plains of North America. That northward trend, around a major high-pressure cell over the northern Atlantic Ocean, generally carries abundant moisture from the Gulf of Mexico into the Great Plains. During periods of extreme drought, such as in the Dust Bowl period of the 1930s, much of the jet stream continues west over Mexico rather than curving north to bring rain to the Great Plains. Such drought appears to correlate with the warm phase of the AMO.

Humans have been plagued by droughts for thousands of years (**Table 10-2**). The consequences of

Table 10-2			Some Severe Droughts	
WHEN	**WHO**	**LOCATION**	**EFFECT**	
~2,200 B.C.	Egypt, Old Kingdom; Akkad empire	Egypt, north Africa, Arabia; Iraq	civilization collapse	
1,300–700 B.C.	Late Bronze Age	Greece, Turkey, Middle East	civilization collapse	
~900 A.D.	Mayan civilization	S. Mexico, Guatemala, El Salvador, Honduras	civilization collapse	
~1,100 A.D.	Tiwanacu civilization		civilization collapse	
end of 1200s	Anasazi civilization	Arizona-Utah-Colorado-New Mexico	civilization collapse	
1362–1392, 1415–1440	Angkor Wat	Cambodia	civilization collapse	
mid-1970s to early 1980s	Darfur; overpopulation	North-central Africa	desertification conflict between farmers & nomadic herders, 300,000 lives lost	
2000s		Cusco, Peru	decreased runoff from receding glaciers	
2006, 2008–2009		Central and south Texas, eastern Mexico, Mexico City	water shortages, very low reservoir levels; $billions of crop and livestock losses	
2008		Cyprus (Mediterranean)	drought	
2008–2009		southeastern Australia	drought, severe wildfires	
2009		California	drought, water shortages, followed by heavy rains from El Niño in early 2010	

drought include famine, migration of large populations, social and political upheaval, war, and even the total collapse of civilizations. Developing countries are especially vulnerable to famine and political unrest during times of drought. In the more-affluent areas, the consequences of drought can be severe when crops fail, causing financial distress to farm families and the businesses that depend on them. In cases of prolonged drought, farmers are forced to abandon their land and move elsewhere to look for work.

Drought is not always a result of lower-than-normal rainfall; it can also be caused by human interventions, such as river damming or increased use of water by upstream areas. Dams built to retain water for irrigation often severely affect downstream flow, leading to crop failure, thirst, and disease from decreased water quality and downstream water contamination. When a country upstream builds a dam, the drop in downstream flow can initiate violent international conflict. Downstream countries such as Iraq and Bangladesh that depend heavily on irrigation have been strongly affected by other countries' upstream dams. Overuse of water in the upper parts of a stream drainage basin, either by excessive irrigation or water diversion, may leave downstream areas without water. Former wetlands dry up, and formerly irrigated lands are left dry. The previously fertile marsh and lake areas of the Tigris and Euphrates rivers in Iraq are now mostly dry

and saline. Similar water issues have also caused many intense negotiations and lawsuits between states in the western United States over the rights to water from rivers that flow between them.

Persistent drought takes a toll on soil moisture, groundwater levels, and reservoir levels. The consequences of drought can take months to years of average or above-average precipitation to restore pre-drought moisture levels. Drought years from 1998 to 2002 in much of eastern Colorado, Wyoming, and Montana left soils dry down to 60 centimeters depth in some areas. Low moisture levels in the soil can contribute to related hazards such as dust bowls (discussed in the following section) and wildfires (discussed in Chapter 16).

Drought also makes plants more susceptible to disease and infestation. Crops such as grains and corn, weakened by drought, can be overrun and destroyed by swarms of invading locusts. Weakened forests can be invaded and killed by pine-bark beetles (**FIGURE 10-21**). The risk of major wildfires is dramatically increased by all of the dead and dry pine needles. Areas of widespread infestation tend to be prime candidates for wildfires, including those in southern California in 2003, 2006, 2007, and 2009.

Could we experience the kind of severe drought that devastated the American Midwest during the Dust Bowl of the 1930s? Many experts believe we could. By early summer of 2007, drought was widespread throughout the

FIGURE 10-21 PINE BEETLES DEVASTATE FORESTS

Thousands of originally green pine trees weakened by drought and killed by pine-bark beetles in central British Columbia.

FIGURE 10-22 BOAT HOUSES LEFT STRANDED

Boat houses rest high and dry along the Pedernales River, August 2009, during the severe drought in central Texas.

United States, from southern California, Arizona, Nevada, and western Colorado to most of the southeastern states. Many experts expect that drought in the southwest will become the norm, an especially serious problem with their rapidly growing populations. A severe drought in Texas and Oklahoma suddenly ended with massive floods in April, 2007. By 2008 and 2009, however, the drought had returned and worsened (**FIGURE 10-22**). The weather pattern in the southeast was partly caused by the placement of the Bermuda High (with clockwise air circulation) in the Atlantic east of the Carolinas, which pulled moisture from the Gulf of Mexico and dumped heavy rain on Texas and the plains states. This huge high-pressure system moved far enough west to keep the southeastern states dry.

Dust Storms

A significant reduction of soil moisture permits **dust** to be carried in the air. Dust becomes a hazard when winds can pick up loose, dry dust or soil, creating a **dust storm** (**FIGURE 10-23**, p. 278). Such storms not only result in soil loss from the area where they originate, they can also obscure visibility, reduce sunlight and cause health problems in humans.

Dust storms most commonly develop in deserts, but they can occur in any area where soil moisture has been reduced. A reduction of soil moisture leading to dust can occur because of drought or poor land management, such as cultivation or overgrazing of marginally arid or semiarid land. In many poor countries, this type of cultivation is almost unavoidable, as people struggle to avoid starvation, but the practice occurs in developed nations as well. Elsewhere, tropical hardwood forests on marginal soils are stripped for the value of their timber, cut for firewood and cooking, or slashed and burned for cultivation. During much of the 1950s to 1990s, China cut most of its forests to fuel its factories and clear land for crops. When much of the new farmland on marginal soils failed, frequent choking dust storms enveloped Beijing (FIGURE 10-23b) and other parts of eastern China, and forced the country to make changes. The government now pays many farmers to leave the ground fallow and embarked on a massive campaign to plant extensive swaths of trees to help retain ground moisture.

Dust storms decrease visibility on the ground and in the air, affecting transportation by aircraft and automobile. They can also affect human health when people breathe in dust particles. Recent studies show that significant quantities of bacteria and fungi travel with dust over large distances. For years, many people in the San Joaquin valley of central California have come down with flu-like symptoms known as Valley Fever. The condition is not normally serious except for those with weakened immune systems. Fungal spores

FIGURE 10-23 DUST STORMS

A. A dust storm invades Lubbock, Texas on June 18, 2009. **B.** A cyclist rides through choking dust covering Beijing in 2001.

from soil generally lie dormant in the dry season. In the wet season, they form a mold that can be circulated when soil is disturbed by farming or construction, which caused some residents to contract the fever. The same disease is endemic in most of the southwestern states and northern Mexico. Much of the pollution comes from small garden plots in impoverished areas where people fertilize the soil by burning garbage that now includes not only waste from plants and animals but plastics and tires.

Dust storms are a global problem because winds carry dust around the world. Dust from eastern Asia travels across the Pacific Ocean, creating haze in North America, and even out into the Atlantic Ocean. Dust plumes can cross the Pacific Ocean to reach North America in only one or two weeks. Dust from North African storms often covers Europe and even the Caribbean, affecting the health of people, coral reefs, and other organisms (**FIGURE 10-24**). In 2006, a Gobi Desert sandstorm, just north of China, dumped 270 million kilograms of sand on Beijing.

Dust storms can carry pollutants as well as soil particles. Giant dust plumes in spring and summer from deserts in western China and Mongolia also carry smog, black carbon particles, industrial smoke and fumes, sulfates, and nitrates. The plumes can be 500 kilometers wide and almost 10 kilometers thick. The sulfates actually reflect more than 10 percent of the sunlight they are exposed to, but the black carbon or soot absorbs sunlight, increasing warming.

One of the most severe natural disasters in American history involved drought and dust. Extreme drought hit North America in the mid-1930s, turning the U.S. plains states into a "Dust Bowl" (**FIGURE 10-25**). Coinciding with the economic collapse of the Great Depression, huge areas of the Great Plains were abandoned as people migrated west seeking

FIGURE 10-24 DUST STORM

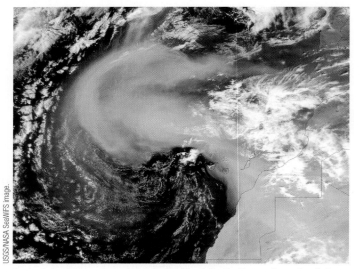

A giant dust storm sweeps out over the Atlantic Ocean from the Sahara Desert in northwestern Africa, on February 26, 2000.

work. The Dust Bowl crisis of the 1930s was brought on by a combination of extended drought and disastrous farming practices on the Great Plains. Advances in mechanical plowing permitted removal of natural grass sod to facilitate planting of wheat. Unfortunately, when the drought hit, the fertile soils dried out and were picked up by the wind and blown away to the east and northeast as immense dark clouds of dust. At its peak in July 1934, the drought affected more than 60 percent of the area of the United States (**FIGURE 10-26**).

FIGURE 10-25 DUST BOWL

A. A huge wall of dust bears down on Stratford, Texas, on April 8, 1935. **B.** Wind-blown dust buried farms and their machinery in Dallas, South Dakota, May 13, 1936.

FIGURE 10-26 DUST BOWL AREA

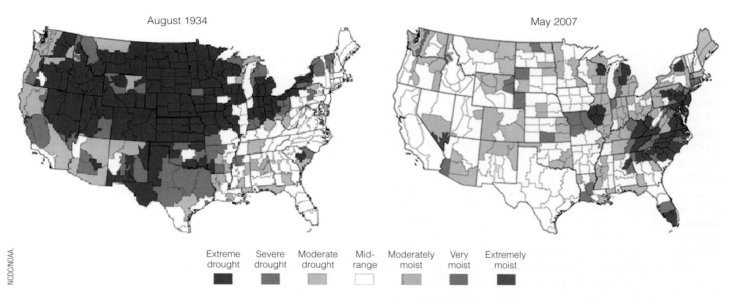

Extreme drought during the Dust Bowl in August 1934 spread over much of the central United States, compared with drought in the eastern United States in September 2007.

Financial hardships imposed by the Great Depression of the early 1930s forced farmers to grow more crops, which drove down prices, which forced them to further increase production to pay for their equipment and land. Unable to farm and unable to pay their debts, hundreds of thousands of people left the southern plains states of Texas, Oklahoma, Arkansas, Colorado, and Kansas. Most had no skills beyond farming; many were uneducated and worked menial migratory jobs for little pay. Living in roadside tents or tin shacks, with little food, they barely survived. Lessons learned during the Dust Bowl years led to better farming practices and less impact from later droughts. Much of the land was reseeded with native prairie grass rather than tilled and planted with wheat.

Much of the Great Plains of North America is similarly more suited to moderate grazing than cultivation. Semiarid areas affected by drought can recover when rains return if the ground has not been misused. Even prior to European arrival in North America, there were times of extreme drought and large areas of desert. The Sand Hills of Nebraska are giant sand dunes, active less than 1,000 years ago but now grown over with grass and brush (**FIGURE 10-27**, 284). Apparently the rains returned and vegetation sprouted, beginning in the low areas. Some of these areas are again cultivated, but the hills, with their thin, fragile soils, are not able to sustain cultivation without again reverting to open areas of sand. Grazing cattle often break the fragile cover, leaving persistent trails.

FIGURE 10-27 SAND HILLS

A

B

A. Small patches of flat ground below the giant dunes were once small lakes; they are now filled and grown over with grass. **B.** Air view of the Nebraska Sand Hills clearly shows the form of their original giant sand dunes, each more than a kilometer crest to crest.

Dust storms may also have some positive effects. Some areas such as the eastern Washington Palouse, may benefit from the addition of nutrient-rich soil particles deposited in a dust storm. Dust may even play a role in reducing hurricanes. Recent studies by NASA and NOAA scientists suggest that the super-dry air of major dust storms blowing out into the Atlantic from the Sahara Desert may prevent intensification of tropical waves in the atmosphere from forming storms that can develop into hurricanes. Atmospheric dust reflects sunlight, keeping the ocean water cooler.

Desertification

The vegetation cover of parts of semiarid scattered-tree grasslands, such as the Sahel region along the southern edge of the Sahara Desert in Africa, has lost its cover of grass, some shrubs, and patches of trees, mostly since the late 1960s, because of increased population, intensive cultivation in marginal areas, and overgrazing. Rainfall ranges from about 20 to 60 cm (8 to 24 inches) per year, equivalent to most of the Rocky Mountain West, east to the High Plains of central North Dakota and down to West Texas. Decades of drought in the semiarid Sahel has prevented regrowth of grass and fostered the evolution of new desert environments, or **desertification**. This is a major problem in parts of Chile, Brazil, Mexico, Morocco, Ethiopia, Pakistan, Afghanistan, Kazakhstan and other central Asian countries, Mongolia, China, and elsewhere. In China 150,000 km^2 per year of cropland is damaged by drought, and desertification; in 2009 it was costing $42 billion per year. Studies suggest that warmer SSTs of the Indian Ocean weaken the monsoon movement over Africa and promote drought in a belt across the continent. The United Nations Convention to Combat Desertification defines its concern as "land degradation in arid, semi-arid, and sub-humid areas resulting from . . . climatic variations and human activities." Desertification affects about 70 percent of Earth's dry lands, which equates to 30 percent of Earth's land surface (**FIGURE 10-28**).

Most people in the Sahel were once semi-nomads that raised livestock and farmed in the north in wetter seasons. They learned to live with their environment; in the dry seasons they migrated to grasslands farther south for grazing. More-permanent settlements grew and conflicted with traditional herder nomads. Periodic droughts caused widespread famines, including severe droughts in 1914 and 1968–74 that disrupted both grazing and farming.

Overgrazing, sometimes fostered by fences that prevent animal migration, breaks up surface soil, causes compaction of deeper soil, and reduces water infiltration. Artificial watering holes or watering tanks can have the same effect because animals don't migrate. Cultivation dries out the soil. If topsoil is exposed during high winds, dust storms remove its most productive parts or raindrop impacts loosen and wash it away. In Africa, the effect is severe due to overgrazing and over-cultivation, especially on marginally productive lands, and worsened by climate change. Two belts across Africa, the Sahel, south of the Sahara Desert,

FIGURE 10-28 DESERTS

Grazing of sheep and goats in Morocco, at the west end of the Sahara Desert, contributes to desertification. The High Atlas Mountains are in background.

and the other south of the equator, show growing drought. It has been estimated that one-half to two-thirds of productive non-irrigated land has been at least moderately subjected to desertification, as in the Dust Bowl of the 1930s.

Heat Waves

In contrast to dramatic or violent weather hazards—such as floods, hurricanes, and tornadoes—the effects of excessive temperatures build more slowly but can be even more deadly, killing greater numbers of people than floods, hurricanes, and tornadoes combined. Part of a heat wave's danger is that people tend to view hot weather more as a discomfort or inconvenience than a health emergency. Deaths during heat waves can occur when a person's core body temperature, normally 37°C (98.6°F), reaches 40°–41.1°C (104°–106°F), resulting in heat stroke. High summer temperatures in urban areas also accelerate the chemical reactions that create ground-level ozone and smog, further increasing stress on people's health. In some cases, heat adds stress to frail, elderly people, who die of heart attacks. Adequate water intake is especially important.

Urban areas bear an increased risk of high temperatures because of the **heat-island effect**. Extreme summer temperatures in a major city can be as much as 5°C (9°F) hotter than in the adjacent countryside. Cities and airports with many buildings and vast areas of dark pavement collect and absorb heat from the sun. Exhaust from cars, trucks, air conditioners, and factories trap more heat because the dark particles absorb heat from the sun in the same way. Urban areas also cool slower than rural areas at night because buildings and pavement retain heat longer, often for three to five hours after sunset. Although in the winter warmer temperatures reduce heating needs and melt snow

and ice on roads, the adverse summer heat-island effects much outweigh the winter benefits.

July of 1995 brought Chicago a period of high humidity and extraordinarily high temperatures that stayed above 37.8°C (100.2°F) for days and reached 41°C (106°F) on July 13. Some 600 to 700 people died of heat-related causes during this heat wave. Most were elderly, poor, and living in the inner city, where they were afraid to open their windows at night because of crime. Contributing to the problem was the heat-island effect, power failures, lack of warning, and inadequate ambulances and hospital facilities. In 1966, hundreds of people in St. Louis died from heat waves, and dozens more succumbed in 1980 and 1995. In July of 1993 and 1995, dozens died from heat waves in Philadelphia.

High humidity makes a considerable difference by limiting people's ability to cool by perspiration. The effect is dramatic at temperatures near or above the core temperature of the human body. For example, at 35°C (95°F) and 80 percent relative humidity, it feels like a very dangerous 56.1°C (133°F). Direct sunlight can add 15°F to these temperatures. In very dry climates, however, such as parts of the Mountain West, where the relative humidity can be as low as 20–30 percent, the apparent temperature can be lower than the temperature on the thermometer.

In August 2003, Western Europe sweltered in unusually high temperatures, in many places above 38°C, some 8–10°C above normal. More than 35,000 deaths in Europe were attributed to the heat; about 14,800 of those occurred in France, with thousands more in Italy and Spain. Although some parts of the world have long periods of high temperatures, people in such areas are familiar with these conditions and know to keep hydrated, stay in shady and well-ventilated areas, and use fans and air conditioners. Most of the victims in Europe were elderly people, in poor health and living alone, and small children, left behind when parents or other relatives went on August vacations. Most homes in France and other moderate climates do not have air conditioning.

Another heat wave in Europe in the summer of 2006 left southern England even drier than in 2003. Record temperatures throughout Western Europe climbed above 35°C in July. In the same month, temperatures in parts of India reached 49.6°C (121.3°F), and at least 884 people died. The western United States experienced an extended heat wave in July of 2006 that caused at least 164 deaths in southern California, mostly elderly people without air conditioners. Temperatures in Fresno exceeded 43.3°C (110°F) for six consecutive days, and Sacramento went over 37.8°C (100°F) for eleven days. Compounding the problem were electrical system failures that left more than a million people without power and air conditioning.

Snow, Ice, and Blizzards

Winter at northern latitudes, including Canada and most of the United States, brings cold temperatures and sometimes snow and ice. We bundle up to stay warm, especially when

FIGURE 10-29 LAKE-EFFECT SNOW

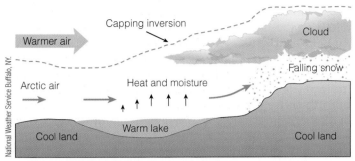

Cold arctic air moving across a warm lake picks up moisture evaporating from the lake, rises, cools, and condenses to form tiny ice crystals that fall as snow.

the wind blows. The wind makes a difference because evaporation of skin moisture causes heat loss, known as **wind chill**, which makes us feel colder. Wind chills can reach -51°C (-60°F) in Midwest or Canadian Prairie blizzards. Such bitter cold can lead to **frostbite**, the freezing of body tissue on noses, ears, or fingers. Frostbite is indicated when the affected areas turn pale or white and lose feeling. Victims need immediate medical attention; the frostbitten areas should be warmed slowly. When a person's core body temperature drops below 35°C (95°F), 2°C below normal, the person is experiencing **hypothermia**. Symptoms include uncontrollable shivering, drowsiness, disorientation, slurred speech, and exhaustion. Victims need immediate medical attention, and they should be warmed **slowly**. Wearing a hat, hood, and coat in cold weather is important in order to minimize heat loss through exposed skin. To keep fingers warm, mitts are better than gloves.

Compared with land nearby, areas downwind of large lakes often have unusually large snowfalls, called **lake-effect snow**, in winter. In the westerly wind belts of North America, the prevailing winds come from the west. Thus northeastern, eastern, and southeastern sides of most of the Great Lakes can receive extraordinary amounts of snow in a single storm. This can amount to one-third to one-half of the total annual snowfall. Factors that affect the amount of snow (**FIGURE 10-29**) include: expanse of ice-free water under the storm track (for evaporation); duration of strong wind blowing over the water (for greater evaporation); amount of moisture in the atmosphere (source of the snow); and topographic rise (hills) downwind of the water (to lift moisture, cause condensation, and snowfall).

Cold, arctic air moving over the broad expanse of the relatively warm water of one of the Great Lakes picks up large amounts of moisture evaporating from the lake. The cold air freezes the moisture. Reaching land, it rises and cools even more, to precipitate the moisture as tiny ice crystals—snow (**FIGURE 10-30**). As long as the cold wind keeps blowing across the water, snow continues to form and fall downwind. Snowfalls as heavy as 15 centimeters per hour have been recorded. Often the low-pressure cell, around which the wind is circulating counterclockwise, is off to the northeast. If that low-pressure cell remains stationary, the wind may continue to blow for days, bringing heavy snows to the eastern and southeastern sides of the lakes. Some of the cities most heavily affected are shown in FIGURE 10-29b. Will global warming lead to more ice-free winters on the Great Lakes and more lake-effect snow? The concern is real because recent winters have seen less ice and more open water, leading to more evaporation. In early February 2007, lake-effect

FIGURE 10-30 LAKE-EFFECT SNOW

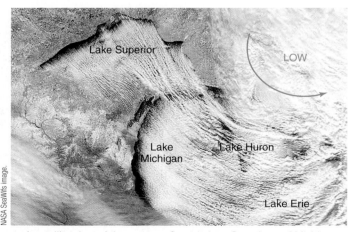

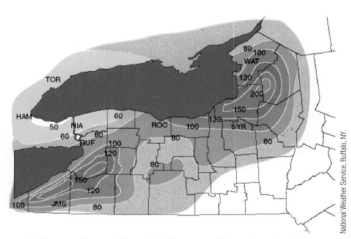

A. A satellite view of the western Great Lakes, Superior and Michigan, shows winds sweeping southeast, picking up moisture from the lakes, and dropping heavy snows on the southeast to east sides of the lakes. **B.** The average annual snowfall around Lake Erie and Lake Ontario shows three to four times as much downwind of the lakes. Contours in inches of snow.

FIGURE 10-31 ICE STORM DEVELOPMENT

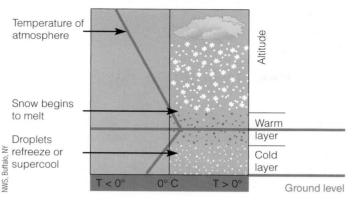

An ice storm can form when snow in the cold upper atmosphere falls through a warmer layer, melts, then refreezes as it continues down through a cold layer near the ground.

snows from the east end of Lake Ontario buried part of upstate New York in 3 to 4 meters of snow. People shoveled off roofs to prevent collapse, leaving snow piled above their eaves.

If the weather has been unusually cold for a long period, a lake may largely freeze over, so that little water evaporates from its surface. Under that circumstance, the cold, dry wind remains dry, and it doesn't snow. In the winter of 2005–06, the Great Lakes remained warmer and mostly free of ice. Cold winds formed ice near shore, but waves offshore broke up thin ice and stirred up warmer water from below, preventing ice from spreading. The winds picked up moisture from the water, which led to heavy snowfall downwind.

A similar phenomenon of snow falling downwind of a large body of water sometimes occurs during a Nor'easter, an offshore low-pressure cell off the coast. In this case the result would be **ocean-effect snow**. Cape Cod, Massachusetts, sometimes sees such storms. A classic Nor'easter blew north through the eastern states on February 12, 2006. It arrived in New York with claps of thunder and lightning. Dropping snow at rates of 7 to more than 12 centimeters per hour, it left about 30 to 76 centimeters on the ground from parts of North Carolina to near Washington, D.C., to New York City, to parts of Connecticut. It snarled traffic in cities and on highways and shut down buses, railroads, and airlines. In some areas, it downed electric power lines, cutting power to hundreds of thousands of homes. Another major Nor'easter struck the northeastern states in mid-April of 2007.

Weather events that cause **ice storms** depend on humid air and warmer temperatures above cold ground-level air (**FIGURE 10-31**). Such storms cause widespread power outages and can cripple transportation (**FIGURE 10-32**). Ice storms are especially common in New England and eastern Canada, where the humidity is relatively high. However, winter ice storms occasionally occur in southern areas such as Texas. A major ice storm in early December 2006 moved across Missouri and Illinois and into the northeast. Heavy snow and ice collapsed roofs and downed power lines, knocking out electricity for hundreds of thousands of people. At least 16 people died and 43 were hospitalized by carbon monoxide poisoning because they were heating with unventilated combustion heaters. Another ice storm in mid-January 2007 left ice coatings on everything from Texas to Maine. At least 39 perished, including people in Texas, Oklahoma, Missouri, Iowa, New York, and Maine. In January 2009, an ice storm wrecked havoc in Arkansas and Kentucky (**Case in Point:** A Massive Ice Storm in the Southern United States, p. 313)

FIGURE 10-32 ICE COATS EVERYTHING

A. A classic ice storm in the northeast coats power lines.

B. An ice storm in Geneva, Switzerland, on January 28, 2005, coats everything with a thick layer of ice.

Winter winds can be fierce, especially on the Great Plains of the upper Midwestern United States and the Prairie provinces of Canada, east of the Rockies. Blowing snow with winds greater than 56 km/hr in the United States (35 mph) or 40 km/hr in Canada can be classified as a **blizzard**. Such high winds often develop on the northwest side of a winter storm front, where the air pressure gradient is between a high to the northwest and a low to the southeast. In some cases, visibility can be very limited or even vanish, making travel extremely dangerous (**FIGURE 10-33**). Snow-packed roads become even more slippery and visibility may be very limited. High winds can blow already fallen snow in a ground blizzard. Severe conditions may develop into a whiteout with blowing dry powdery snow, fog, or ice droplets—a condition where objects and the horizon disappear and people lose depth perception and their sense of direction. People can even lose their sense of up and down, so even though they are standing on level ground, they lose balance and fall. Extreme cold with blizzards and whiteout conditions can be very dangerous, often resulting in frostbite, hypothermia, and death to those outside. In rural areas, where communication may be limited, people caught in rapid temperature drops and poor visibility have lost their way and frozen to death on ranches or even trying to run from a rural school to a nearby house less than 100 meters away.

Heavy snowfalls often create problems. In one example, snow falling in the Midwest on January 20, 2005, intensified, moved east to the Great Lakes region the next day, and continued into southern New England on January 22. The low-moisture content of the snow built to depths of 30–38 cm in Philadelphia and New York City and more than 90 cm in Boston and southeastern Massachusetts. Travel was severely curtailed and most schools in the northeast closed for two days.

Short-Term Atmospheric Cooling

The atmosphere is cooled when particles suspended in the air reflect sunlight back into space, reducing the amount of light that reaches Earth's surface, and thereby decreasing temperatures. The atmosphere can be temporarily cooled by anything that puts particulates such as ash, soot, and dust in the air. Forest fires, both natural and human-caused, and dust blowing off deserts and drought-affected croplands are major sources of particulates. Atmospheric cooling can also be caused by natural disasters such as volcanic eruptions or the potential impact of an asteroid striking Earth (see Chapter 17). Cooling can be a result of human pollution, such as from industrial smoke stacks. Atmospheric cooling can result in crop failure and famine, or even lead to the extinction of sensitive plants and organisms.

VOLCANIC ERUPTIONS The release of gas and ash during a volcanic eruption can trigger atmospheric cooling by blocking sunlight from reaching the surface of the Earth. Because the wind carries ash and particles around the globe, an eruption can have a significant impact on global temperatures.

Volcanic eruptions produce large amounts of ash and sulfur dioxide (SO_2), along with other aerosols that cool the Earth by reflecting sunlight back into outer space (**FIGURE 10-34**). The SO_2 dissolves in water droplets in clouds to form reflective sulfuric acid droplets that can remain in the atmosphere for a few years until rain washes them out. The eruption of Mount Pinatubo in the Philippines in 1991

FIGURE 10-33 WHITEOUT

A blowing ground blizzard or whiteout can eliminate ground-level visibility for even nearby objects such as these farm buildings. Goodland, Kansas, February, 1999.

FIGURE 10-34 VOLCANIC COOLING

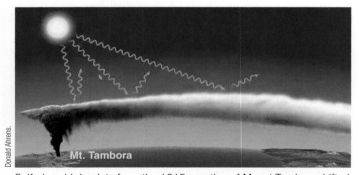

Sulfuric acid droplets from the 1815 eruption of Mount Tambora drifted around the northern hemisphere, reflecting some of the incoming radiation of the sun and lowering the atmospheric temperature to cause local crop failures.

lowered the northern hemisphere's temperatures by less than 1°F for a few years.

Because weather is confined mainly to elevations below 15 kilometers, there is no rain to wash out ash and it remains in the upper atmosphere. This causes spectacular sunsets, which are especially notable for their vivid streaks of green, a color not ordinarily observed in sunsets. Although beautiful, these sunsets indicate the continued presence of particles in the air that are reflecting sunlight away from Earth. Atmospheric cooling of a few degrees for a year or more can also lead to crop failures and, in severe cases, famine.

A significant example of atmospheric cooling as a result of volcanic activity occurred in April 1815, when a gigantic eruption reduced the peak of Tambora volcano on the densely populated Indonesian island of Sumbawa from an elevation of 4,300 meters to 2,900 meters. The eruption (with a Volcanic Explosivity Index of 7) produced 40 cubic kilometers of ash and pumice, an even larger volume than was produced in the eruption that reduced Mazama to Crater Lake almost 8,000 years earlier (see Chapter 7). The eruption killed some 10,000 people directly by pyroclastic flows and another 80,000 in the famine and epidemics that followed. Brightly colored orange and red sunsets were seen worldwide in the summer and fall of 1815. A persistent "dry fog" with reddish sunsets continued into the summer of 1816 because of sulfate aerosols in the stratosphere.

The environmental aftermath developed into a global catastrophe of famine and misery. The rhyolite ash that Tambora injected into the upper atmosphere blocked enough sunshine throughout the northern hemisphere to reduce the average temperature about 0.5°C (0.9°F) within a few weeks. That seems a small drop, but it was enough to inflict agricultural havoc. Freezing temperatures in New England throughout the middle of 1816, "the year without a summer," caused widespread famine. On June 6, 1816, snow fell in parts of New York and Maine. Summer frosts ruined crops as far south as Virginia, including, by some accounts, Thomas Jefferson's corn. Meanwhile, abnormally cold and rainy weather caused widespread crop failure and famine in Europe.

As the example of Mt. Tambora shows, any huge eruption of rhyolite ash is likely to develop into a global climatic catastrophe. No warning or evacuation scheme can mitigate this type of danger. People everywhere should dread a major eruption of any large resurgent caldera such as Yellowstone Volcano or Long Valley Caldera.

POLLUTION Since the industrial revolution, pollution has been a major source of atmospheric particulates, which block sunlight and affect weather patterns. Industrial smokestacks contribute particulates, especially in underdeveloped countries. Soot levels in some of the world's megacities, such as Beijing, Cairo, Mumbai, New Delhi, Seoul, Shanghai, and Tehran, are very high and rising, in some cases reducing sunlight by 25 percent. Pollution drifting east from coal-fired plants in China has even shifted normal rainfall patterns, causing more rain in the south and worse droughts in the north. Burning wood, coal, and peat provide significant amounts of particulates in poor countries. Sooty brown clouds form over much of southeastern Asia.

Just as with other types of particulates, as air circulates around the globe, pollution becomes a global climate concern. A 3-kilometer-thick brown cloud of smoke, soot, and dust roughly the size of the United States was recently discovered over part of the Indian Ocean. Some of that pollution is due to the huge population in India who use dried cow dung as a cheap fuel source for cooking.

Pollutants can have complex affects on atmospheric conditions. Dark soot in the cloud of pollutants over the Indian Ocean absorbs heat and warms the upper atmosphere. However, it also reduces heat to the lower atmosphere and cools the surface of the ocean, which reduces evaporation. This in turn increases the intensity of regional droughts because less moisture is available to fall as rain.

The Greenhouse Effect and Global Warming

The Greenhouse Effect

In a phenomenon called the **greenhouse effect**, Earth's atmosphere traps heat in about the same way a greenhouse permits the sun to shine through its glass but prevents much heat from escaping (**FIGURE 10-35**, p. 286). Sunlight is short-wave radiation that passes easily through Earth's atmosphere (or the glass of a greenhouse), heating the planet's surface. The heat that radiates from Earth outward is long-wave radiation. Some of this radiation escapes into space, but some is absorbed by gases in the atmosphere. As the gases absorb this radiation, air temperature increases. Without the greenhouse effect, Earth would be uninhabitable, as the average temperature would fall below freezing.

The greenhouse effect explains why temperature tends to drop more quickly on a clear night than on a cloudy one: heat radiates to outer space unhindered by clouds. The water vapor of clouds absorbs huge amounts of heat, preventing its loss from the atmosphere. Actually, during the day, different types of clouds have quite different effects. Thin, high-altitude clouds, above the flight paths of commercial aircraft, permit the sun to shine through unhindered to heat the ground. But they prevent heat loss back to outer space. Thick, low clouds, however, have white tops that reflect the sun's rays away from Earth; they prevent sunlight from reaching the ground, so those cloudy days stay cooler (**FIGURE 10-36**, p. 286).

FIGURE 10-35 GREENHOUSE EFFECT

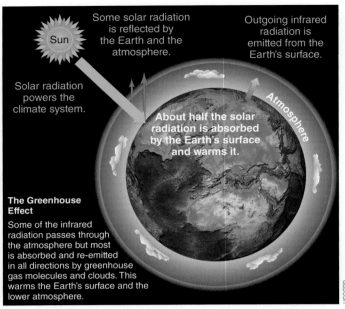

Outgoing long-wave radiation blocked by glass

Incoming short-wave radiation passes through glass

A

Some solar radiation is reflected by the Earth and the atmosphere.

Outgoing infrared radiation is emitted from the Earth's surface.

Solar radiation powers the climate system.

About half the solar radiation is absorbed by the Earth's surface and warms it.

Atmosphere

The Greenhouse Effect

Some of the infrared radiation passes through the atmosphere but most is absorbed and re-emitted in all directions by greenhouse gas molecules and clouds. This warms the Earth's surface and the lower atmosphere.

B

A. A greenhouse heats up inside because the short-wavelength radiation—incoming sunlight—passes through glass but most of the longer-wave return radiation—heat—cannot pass through glass and is trapped inside. **B.** Carbon dioxide and other greenhouse gases in Earth's atmosphere act in a similar manner as the glass covering a greenhouse.

FIGURE 10-36 CLOUDS SHADOW SURFACE

Reflective cloud tops

Cloud shadow on ground

White cloud-tops reflect much of the incoming sunlight, leaving a shadow on, and cooling, the ground over the Great Plains.

GREENHOUSE GASES AND THE CARBON CYCLE Gases in Earth's atmosphere that absorb radiation and contribute to the greenhouse effect are called **greenhouse gases**. Water vapor is the most important greenhouse gas. Others include carbon dioxide (CO_2), methane (CH_4), ozone (O_3), and nitrous oxide (NO_2). The total greenhouse effect of individual gases depends on their effectiveness in blocking the outgoing long-wave radiation (heat) and their abundance in the atmosphere. CO_2 has the greatest total effect, followed by CH_4, but the others are also significant. The most dramatic effect on temperatures occurs in the polar regions because polar air is drier than tropical air and is thus more sensitive to the concentrations of such greenhouse gases.

Many greenhouse gases are produced by natural processes. Most atmospheric water vapor has evaporated from the oceans, which cover about two-thirds of Earth's surface. Carbon dioxide and methane are emitted by erupting volcanoes, animals, decaying vegetation, and forest fires. For example, the 2007 Southern California wildfires generated about 8.7 million tons of CO_2 (actually a small amount compared with other sources). NO_2 is generated by oxidation of nitrogen in the atmosphere during lightning storms. Greenhouse gases are also byproducts of human activities, such as the burning of fossil fuels, which we discuss in more detail later.

The release of carbon dioxide into the atmosphere is part of a cyclic carbon exchange called **carbon cycle** (**FIGURE 10-37**). Carbon occurs naturally in rocks, the ocean, and the atmosphere. CO_2 dissolves both in the atmosphere and in ocean water in proportions that maintain a long-term equilibrium between them. If more CO_2 is loaded into the atmosphere, some of that dissolves into ocean water; if more were loaded into the ocean water, some of that would dissolve back into the atmosphere.

FIGURE 10-37 THE CARBON CYCLE

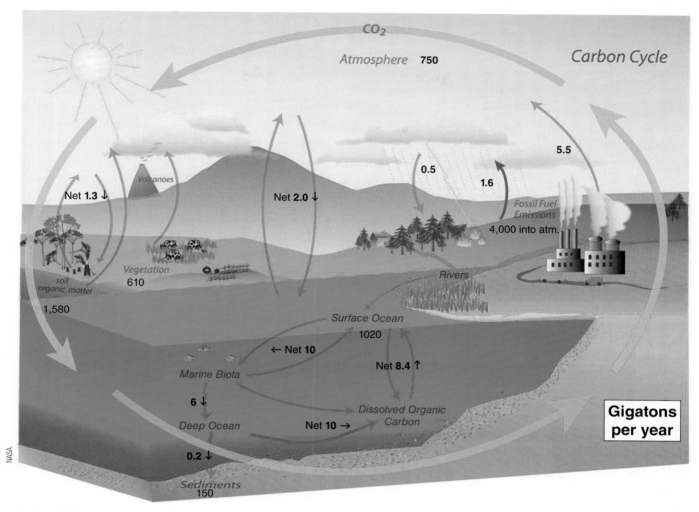

Carbon dioxide, vented into the atmosphere from burning hydrocarbons, mixes with water, and falls in rain. Dissolved in water, it may be precipitated as limestone or dissolved in the ocean. Taken in by plants, it is stored in their tissues until they are either buried and ultimately turned to peat or coal, or burned in wildfires to again return to the atmosphere.

In the ocean, dissolved CO_2 is used by microscopic plants and animals. Microscopic plants in the ocean, such as algae in coral reefs, take up CO_2 and release oxygen just as plants do on land. Oceanic animals, including microscopic forms that drift with ocean currents, and clams, corals, and other coastal forms with carbonate shells, take in oxygen and give off CO_2, to form carbonate-shell skeletons or protective coatings. When microscopic near-surface animals die, their soft-body organic-carbon parts and hard-carbonate parts both sink to the sea floor, thereby removing carbonate from the ocean-atmosphere part of the system.

On land, carbon is held as biomass, in live plant material and in dead material on the ground and in the soil. In photosynthesis, plants take up CO_2 from the atmosphere and release oxygen while growing. This annual cycle can be seen in the CO_2 concentration measurements in Hawaii, where the atmospheric levels decline in summer as the Northern hemisphere biomass (where most plants on Earth exist) takes significant levels of this gas out of the atmosphere through photosynthesis. Most of the carbon, by far, is locked in rock, primarily in limestones $(Ca,Mg)CO_3$, with smaller amounts stored in plants, coal, and oil. Erosion and solution of limestone carries some to the ocean; from there the carbon can be exchanged with carbon dioxide in the atmosphere.

Since the Industrial Revolution, human activities have been another important contributor to CO_2 in the atmosphere. Human-produced CO_2 emissions result from transportation, industrial and residential energy consumption, and land use changes such as deforestation (**FIGURE 10-38**, p. 288).

FIGURE 10-38 SOURCES OF EMISSIONS

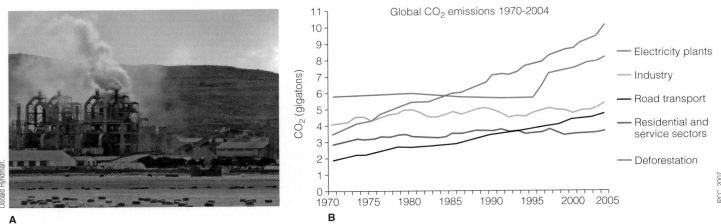

A. This cement plant in Peru emits a huge amount of CO_2 and other pollutants. Cement plants are major contributors of CO_2 because they expel it by turning limestone into the CaO in cement ($CaCO_3 \rightarrow CaO + CO_2$). **B.** Total worldwide human-produced CO_2 emissions by use. Note that electricity production involves burning coal, oil, and natural gas.

FIGURE 10-39 THIN ATMOSPHERE

In this image to the west over the Pacific Ocean, Earth's atmosphere appears as a very thin translucent line covering Earth's surface. All of the carbon dioxide and other pollutants that we expel into the atmosphere are trapped in that thin zone.

By far, the largest proportion of world energy use involves burning fossil fuels, which generates large amounts of CO_2. Fossil fuels, including coal, natural gas (methane), and petroleum oil, are all hydrocarbons, meaning that they are made of various combinations of carbon (C), hydrogen (H), and oxygen (O). Of those fuels, CH_4 generates less CO_2 when burned because it has a smaller proportion of carbon to hydrogen (1 C to 4 parts H) compared with about 1.8 for liquid petroleum, about 5.6 for soft coal, and 10.7 for hard coal. Soft coal, although far more abundant than other fossil fuels, contains much more moisture, so it generates less heat per unit of weight. For the same energy, burning CH_4 releases 30 to 40 percent less CO_2 than burning coal or oil.

Coal-fired power plants are the largest contributors of CO_2 into our thin atmosphere (**FIGURE 10-39**), more than gasoline-powered cars and trucks combined. Coal burning also typically generates more of other pollutants such as sulfur and mercury. Unfortunately, coal, the most-polluting source, produces most of the energy in China, India, and many other developing countries. Coal produced half of the electricity used in the United States in 2005, followed by

19 percent for nuclear power and natural gas, 6.5 percent for hydropower, and 3 percent for oil.

Millions of motor vehicles that travel roads across the world generate large amounts of CO_2 and other greenhouse gases. Thousands of jet aircraft burn fuel that generates large amounts of CO_2 and water vapor in the upper atmosphere. Jets contribute only about 3 percent of the total human-produced CO_2 emissions, but they contribute much more than this when the water vapor they produce is factored in. Methane and nitrous oxide are produced from farm animals and decay of agricultural materials. Decay in landfills also generates methane, although some of that gas is now being captured, cleaned, and used to generate electricity.

Some of the human-produced greenhouse gases are absorbed by natural buffers as part of the carbon cycle. However, these buffers cannot absorb CO_2 at the rate that it is currently produced. In fact, new studies show that plants take up significant CO_2 only when provided with sufficient soil nitrogen to aid in that growth. Large-scale deforestation has also reduced the plant matter available to absorb CO_2.

FIGURE 10-40 TEMPERATURE AND GREENHOUSE GASES

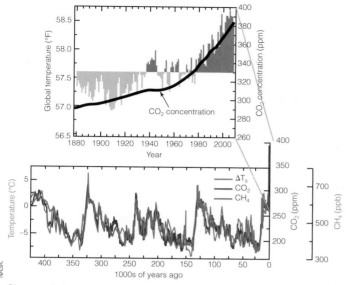

Changes in temperature correlate with changes in CO_2 and CH_4, for at least the past 400,000 years.

Some CO_2 also dissolves into ocean water, which means that the oceans also serve as a natural buffer to soak up some of the greenhouse gases that we emit. However, warm water dissolves less CO_2 than cold water, so as water temperatures rise, less CO_2 is absorbed by the ocean. A 2007 study published by the National Academy of Sciences showed a much faster rate of increase in atmospheric CO_2 concentrations than expected—a 35-percent increase since 1990, not only because of coal burning but because of decreased absorption of CO_2 by the oceans.

The increase in greenhouse gases has a cumulative effect, because almost everything expelled into the atmosphere stays there (FIGURE 10-38). Some of the larger particles are washed out in rain. Very few escape into outer space.

Scientists have established a record of historical levels of greenhouse gases over the last 400,000 years by analyzing tiny amounts of air trapped in Greenland and Antarctic ice (**FIGURE 10-40**). This record reveals that some variation in greenhouse gases occurs naturally. CO_2 levels in the distant past appear to be cyclic, showing variation from less than 200 to almost 300 parts per million (ppm). Initially, some people argued that the current increase in CO_2 was merely natural variation, not a result of human activity. However, the levels of CO_2 are now far higher than at any time during the past 650,000 years. The pre-industrial value of about 280 ppm skyrocketed to about 388 ppm by 2010, with current increases of almost 2 ppm per year. Concentrations of CH_4 and NO_2 have also risen far above their pre-industrial values. The CO_2 concentration has increased some 30 percent since before the Industrial Revolution, less than 300 years

ago. The vast majority of scientists now agree that this increase is a result of human activities.

Global Warming

As greenhouse gases become more abundant in Earth's atmosphere, they absorb more radiation and increase the surface temperature on Earth. Earth's average surface atmospheric temperature has been rising since the Industrial Revolution began in the late 1700s. The temperature increased by about 1°C (~1.8°F) over the past century, with the most dramatic increase since 1970 (**FIGURE 10-41**, p. 290). Earth's surface temperatures have been rising at an alarming rate, especially since about 1920. The changes are even more dramatic at high latitudes in the Arctic. In the past 90 years, they have gone up about 8°C (more than 14°F). As shown in FIGURE 10-41, this period of increasing temperatures corresponds to the rising level of CO_2 in the atmosphere, as well as increased carbon emissions from human activities.

A few people have argued that recent atmospheric increases in temperature are part of the cyclic changes in the sun's energy. Those energy cycles climax about once every 10 to 11 years. However, a comparison of levels of solar energy to surface temperatures shows that the surface temperature continues to rise even during a drop in the sun's energy (**FIGURE 10-42**, p. 290). In fact, we appear to be at about the minimum of a decade-long cycle of solar radiation, which means the temperature will soon rise as we enter a period of increased solar radiation.

Scientists can use the current understanding of the relationship between carbon dioxide emissions and global temperatures to project future **global warming**. The Intergovernmental Panel on Climate Change (IPCC) in 2007 estimated that Earth's average surface temperature will likely rise by 1.6°–3.4°C over the next century and possibly more than 5°C (9°F) in some places. This global warming, a significant part of climate change, compares with a rise of 4°C since the peak of the last ice age 20,000 years ago. The rate at which temperatures continue to rise depends on how much CO_2 humans emit in future years. This could be affected by economic conditions, population growth, and greater use of alternative energy, among other things (**FIGURE 10-43**, p. 291). Unfortunately, even if no additional greenhouse gases were added to the atmosphere after the year 2000 (the scenario indicated with the orange line in FIGURE 10-43), temperatures would continue to increase because of heat slowly released from the warmed oceans. The most likely scenario, indicated with the green line, envisions rapid world economic growth, decreased economic differences among world regions, and a balanced use of fossil and non-fossil energy sources.

Based on the most likely scenario for future fossil fuel consumption, by 2080 in North America, temperatures in the Arctic are expected to rise as much as 7°C (12°F) in winter and 3°C in summer (**FIGURE 10-44**, p. 291). Temperatures in the Great Lakes region and eastern Canada

FIGURE 10-41 CARBON EMISSIONS AND TEMPERATURES

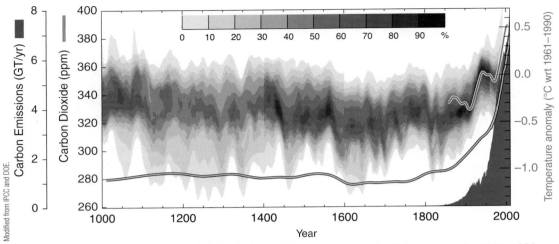

Global surface temperature anomalies (°C) for the last 1,000 years compared with the average for 1961–1990 [IPCC, 2007] and compared with changes in carbon dioxide [ice core data from Etheridge et al., 1988] and annual carbon emissions [Oak Ridge National Lab, Boden et al., 2009]. The red line represents measured temperatures while the orange shades are temperature reconstructions that have higher uncertainty.

FIGURE 10-42 SOLAR RADIATION AND SURFACE TEMPERATURES

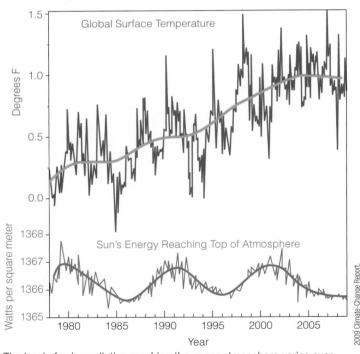

The level of solar radiation reaching the upper atmosphere varies over cycles of ten to eleven years. Because the increase in global surface temperatures has continued to rise even during periods of reduced solar radiation, the long-term rise of global temperatures cannot be attributed to changes in solar radiation.

2009 Climate-Change Report.

are expected to rise by 4°C in winter and 6°C in summer. Temperatures in most of the central United States, including the already hot desert southwest, are expected to rise by more than 3°C in winter and about 4°C in summer. Cities such as Phoenix, Arizona, are well known for extreme summer temperatures, but major cities such as Sacramento, California, could before long experience heat waves for 100 days per year.

Effect of CO₂ Levels on Oceans

Rising levels of CO_2 in the atmosphere also have consequences for the ocean. Recall that in the carbon cycle, a significant proportion of CO_2 expelled into the atmosphere is soaked up by dissolving into shallow ocean waters. Waves and currents stir some of the CO_2-rich water to depth but it remains concentrated near the surface. Dissolved CO_2 is controlled by near-surface water temperature and used by microscopic plants and animals. Cold water can dissolve more CO_2 so cold ocean temperature promotes solution of CO_2 from the atmosphere; warm ocean water causes release of the CO_2 back into the atmosphere.

As the ocean takes up more CO_2, the water becomes more saturated and can tolerate less additional CO_2. Because warm water can hold less CO_2, the heating of water caused by global warming dramatically slows the mixing rate of CO_2 to deeper levels, further lessening the ability of surface waters to take up additional CO_2. The rate of increase in CO_2 taken up by oceans between 1992 and 1999, for example, dropped by half between 1999 and 2007.

FIGURE 10-43 PROJECTED GLOBAL WARMING

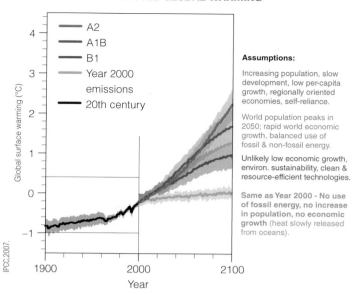

Assumptions:

Increasing population, slow development, low per-capita growth, regionally oriented economies, self-reliance.

World population peaks in 2050; rapid world economic growth, balanced use of fossil & non-fossil energy.

Unlikely low economic growth, environ. sustainability, clean & resource-efficient technologies.

Same as Year 2000 - No use of fossil energy, no increase in population, no economic growth (heat slowly released from oceans).

Projected increases in temperature for the twenty-first century depend on various scenarios for amounts of fossil-fuel burning and various feedback mechanisms. The orange line refers to the unlikely case of no further emissions after the year 2000.

Another consequence of further addition of CO_2 into the oceans is that CO_2 reacts with water to form carbonic acid (H_2CO_3) which makes the ocean more acidic (**By the Numbers 10-1:** Ocean Acidity). The significance for oceanic organisms such as corals and clams that use $CaCO_3$ for their shells or skeletons is critical. As pH falls, $CaCO_3$ can no longer precipitate and may even dissolve. Coral reefs also have a symbiotic relationship with reef algae, whereby the survival of one depends on the other. Warming of seawater tends to stress the coral and promote coral bleaching as the symbiotic algae die off. Major bleaching events have been recorded in the last few years. These changes reflect the extent to which climate change upsets the balance of life and food-chain links in the oceans.

Consequences of Climate Change

The warming of Earth's surface brings changes to weather patterns and other aspects of climate. The term "global warming" does not mean that all parts of the Earth warm; it refers only to the warming aspects of "climate change." However, the consequences of climate change are complex and may even bring colder weather to some areas. Significant changes include greater variability of weather, with more extremes of temperature, winds, and precipitation.

The 2007 IPCC report paints a bleak picture of what may lie ahead in some areas. Abrupt changes in regional climate, in either cooling or drying, would have disastrous consequences for today's world populations. Extreme weather events such as torrential rainfalls are expected to become

▶ By the Numbers 10-1

Ocean Acidity

Acidity of the oceans depends in large part on the amount of carbon dioxide dissolved:

$$CO_2 + H_2O = H_2CO_3$$

(H_2CO_3 breaks down into charged atoms or ions that are acidic)

Acidity is expressed as pH in numbers that range from 1 (extremely acid) to 14 (extremely alkaline), with 7 as neutral—neither acidic nor alkaline. Normal sea water prior to the industrial revolution (1700s) had a pH of 8.18; by the 1990s it had dropped to 8.10. By 2050 it is expected to drop to 7.95. These numbers may not seem dramatic, but the increase in acidity appears to be occurring faster than predicted.

FIGURE 10-44 PROJECTED WARMING

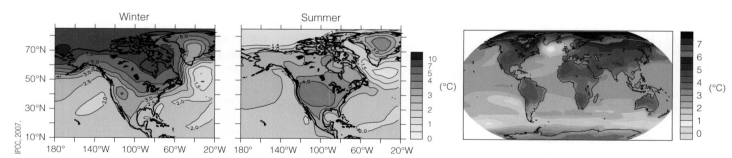

These maps indicate how much temperatures are projected to increase in different parts of the world over the next century. The temperature scale indicates how much the average temperature for the period from 2080 to 2099 will increase over the average temperature from the 1980–1999 period. Note that the largest increase in winter temperature is in northern latitudes.

more frequent because atmospheric heat, the engine that drives storms, gets stronger. Intense precipitation events would couple with longer and more severe drought periods. In large areas, the drop in temperature and rainfall would limit agricultural production, and water supplies would be severely curtailed. Food availability would be disrupted, resulting in spreading malnutrition; public health would suffer from the spread of insect-borne diseases such as malaria, cholera, and dengue fever to higher latitudes and elevations. Economic disruption, population migration, political upheaval, and conflicts over resources seem likely.

The sources and amounts of greenhouse-gas emissions and the broad-ranging consequences for life on Earth are pervasive and dramatic. In December 2009, climate scientists developed a "Climate Change Index" to illustrate these changes (**FIGURE 10-45**). The index integrates the key indicators of climate change: sea-level, average temperature, atmospheric carbon dioxide, and arctic sea-ice cover. An index value of 0 refers to no change, and 100 refers to the maximum recorded annual change since 1980. The change for each year is added to that for the previous year to show the cumulative effect. The index decreased only in 1982, 1992, and 1996. Those were the years of major volcanic eruptions (El Chichon, 1982; Pinatubo, 1991–92; Montserrat and Ruapehu, 1996) that blasted huge amounts of ash into the upper atmosphere. Because ash reflects sunlight, thus temporarily decreasing atmospheric temperature, those eruptions likely contributed to the dips in temperature.

FIGURE 10-45 CLIMATE CHANGE INDEX

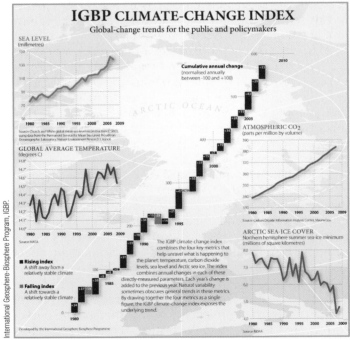

The Climate Change Index, and its four contributing factors, shows a dramatic and increasing rate of increase since the first measurements in 1980.

Warming Oceans

The increase in atmospheric temperature has been moderated by the fact that most of our planet is covered with oceans that soak up more than 80 percent of the potential effect of global warming. The problem, however, rests in the capacity of water to soak up heat. It takes a large amount of heat to warm a pot of water on the stove. By contrast, it takes a much smaller amount of heat to warm the same volume of air by the same amount. The oceans are slowly soaking up heat from the warming atmosphere. The average change in ocean temperature over the past 40 years was 0.3°C in the top 300 meters and 0.06°C in the top 3.5 kilometers.

This small warming of the seas is more ominous than it appears because ocean heat contributes significantly to the energy that drives storms. Hurricanes, for example, form and strengthen where SSTs are above 25°C (77°F). Warmer seas may cause these storms to be stronger and more frequent in the future. In addition, water expands as it warms. As a result, sea level will rise with global warming even if no glaciers and ice sheets melt (discussed further in Chapter 13).

Warming of the atmosphere also causes more evaporation from the oceans, increasing its water vapor content and thereby causing still more global warming—an unfortunate positive feedback effect. The feedback associated with water vapor is thought to roughly double the warming effect from carbon dioxide increases alone.

Warming of the oceans is a trend that cannot easily be reversed. Warm oceans can cool only if the atmosphere above them is significantly cooler. Unfortunately, the oceans are such a huge heat sink, covering about two-thirds of Earth's surface, that we can't easily cool the atmosphere enough to lower ocean temperatures. Studies by NOAA, reported in January 2009, show that even if human-caused CO_2 emissions were completely halted now, the amount of CO_2 in the atmosphere would decrease about 20 percent in the first century but then level off. Temperature of the lower atmosphere would decrease only very slightly over hundreds of years because although human-caused CO_2 addition would cease, the huge added amount already in the oceans would slowly release back into the atmosphere. The slower loss of atmospheric heat to the oceans would prevent significant atmospheric temperature drop for at least 1,000 years—thus much of the human-caused warming effect is essentially irreversible.

Precipitation Changes

An increase in Earth's surface temperature also has an impact on precipitation, including greater extremes of wet and dry, as well as more storms. Higher atmospheric temperatures cause more evaporation from the oceans and the ground and more water to dissolve into the atmosphere. In general, areas closer to the poles and near the equator

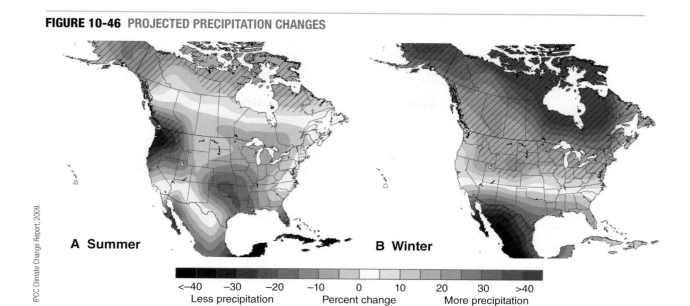

FIGURE 10-46 PROJECTED PRECIPITATION CHANGES

IPCC Climate Change Report, 2009.

A Summer

B Winter

<-40 -30 -20 -10 0 10 20 30 >40
Less precipitation   Percent change   More precipitation

The relative expected change in precipitation to the end of the century, between 2010 and 2100 (higher emissions scenario), **A.** for summer (June to August) and **B.** for winter (December to February). The absolute amount of change depends on the actual emissions, but the trends are clear. Note that most of North America will become drier, especially in the summer growing season. The consequences for crop growth, water supply, wildfires, and other natural hazards are a looming problem.

will become wetter, and the warmer, mid-latitude regions will become drier (**FIGURE 10-46**). Thus many areas that are now dry will get drier, and many wet areas will get wetter. By 2080, much of Canada is expected to be wetter in winter, whereas Mexico is expected to be 15–50 percent drier. In summer, the precipitation across most of the United States may not change much, but the Pacific Northwest is expected to be 20–30 percent drier.

Many regions in trade-wind latitudes will become drier during the dry season because the descending dry air at about 30 degrees north and south latitudes will expand toward the north and south poles. If atmospheric CO_2 concentrations rise from the current level of about 385 ppm to a peak of 450–600 ppm before the year 2100, long-term dry-season rainfalls in some areas will drop to conditions similar to those of the American "dust bowl" of the 1930s (**FIGURE 10-47**, p. 294). Southern Europe and perhaps the southwestern United States would become even more arid; a long-term decrease in rainfall has already been documented in these regions. In contrast, higher-latitude areas, including the northern plains of the United States, Canada east of the Rockies, and eastern Asia north of China, are expected to become wetter.

Most of Earth's near-surface heat is concentrated in the oceans, and warmer oceans lead to more evaporation and thus greater rainfall. More energy in the atmosphere often leads to more storms. Atmospheric pressure would show higher highs and lower lows. That would cause winds to

flow faster between them, with associated stronger and more frequent storms. Natural disasters such as hurricanes and tornadoes would likely be more frequent and stronger. Warming sea temperatures may be responsible for the recent rise in frequency and strength of especially stormy weather in the southwestern United States.

Although populations in some northern climates might welcome modest increases in temperature, detrimental effects would include more droughts and floods. It might seem strange that there could be more drought with more precipitation, but drought depends on the balance between precipitation and evaporation; higher temperatures cause more evaporation, which may have a larger effect than an increase in precipitation.

Rising temperatures, as well as other effects of pollution, mean that future floods are likely to come earlier in the season with more frequency and severity. In the western United States and Canada, the spring snowpack has been melting as much as a month earlier than four decades ago. This is due in part to rising temperatures, but also because black soot from industrial smoke stacks and vehicles settles on mountain snow packs and turns them darker. Instead of their white surface reflecting most of the sun's energy, more energy is absorbed by the darker surface and the snow melts faster. As the surface snow evaporates or melts, the soot remains and accumulates at the surface, collecting more heat and causing still faster melting. Thus snow packs melt faster and earlier in spring,

FIGURE 10-47 GLOBAL CHANGE IN PRECIPITATION

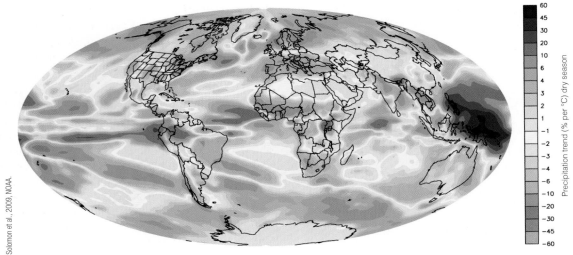

Solomon et al., 2009, NOAA.

Projected average changes in precipitation percent per °C of warming in the dry season (compared with the period between 1900–1950).

leading to earlier and higher floods. Higher flood waters threaten levees not built for those levels. Consequences also include more severe mudflows and coastal erosion.

Despite the increased flooding, these areas are also at risk for more dryness in the summer months. The West depends on the melting snow pack in late spring to retain moisture in groundwater for the dry summers. Higher temperatures mean that more precipitation falls as rain rather than snow, washing away quickly rather than slowly seeping into the ground as snow melts. Without the stored moisture from the later melting of the snowpack, summers are drier, leading to more thunderstorms and more fires.

Earlier snow pack melting also impacts the generation of hydropower and irrigation because water used for such purposes is often needed—but less available—in late spring and summer. Reservoir levels would decline earlier in places like the Sierra Nevada Range of California, and some reservoirs might not have sufficient storage to release needed water for the remainder of the year.

More-rapid melting also accelerates the shrinking of our remaining glaciers. Warmer winters in the Midwest would reduce the seasonal frozen surface area of the Great Lakes, which would lead to more winter evaporation from the open water surface and thus lower the lake levels. By 2007, such predictions had been realized. Lakes Superior, Michigan, and Huron were at record low levels. The snowpack is expected to be smaller with runoff occurring earlier, thus summer stream flow is likely to be lower on average.

Decreased rainfall in already dry areas could severely reduce the water supply for drinking and other municipal and industrial uses, farming, and ecosystems. Areas that already

have water shortages would likely be in desperate shape during similar periods in the future. Water access in places such as the Middle East would spark even more conflict, if not war. Upstream countries, such as Turkey and Iran, have already diverted irrigation water with dams, depriving arid Iraq and Syria, which are plagued by still lower river levels.

Increased shortages of potable water could be alleviated by desalination of sea water in areas close to the coast but not without cost, increased energy expenditures, and thus additional production of greenhouse gases. Desalination can produce fresh drinking water in areas close to the sea but not for those far inland. For example, in the desert area of San Jose del Cabo, at the south end of Baja, Mexico, a large hotel complex with its own desalination plant produces 400 m³ (about 106,000 gallons) of drinking water per day for the equivalent of $0.64 per m³ (2.4¢ per gallon) (**FIGURE 10-48**). The clean, tasty water is used for drinking, cooking, bathing, and laundry. Treated waste water is used for irrigation. Seawater is pumped from wells in near-shore beach sand, run through settling tanks and filters to remove sand and fine particles, then pushed under high pressure through semi-permeable membranes to remove salt and any other contaminants. Forty percent of the water passes through the membranes and the remaining sixty percent, now even more salty, is returned offshore by pipes under the ocean floor.

Arctic Thaw and Glacial Melting

The surface of the Arctic Ocean has been frozen for as long as anyone can remember, preventing ships from using the Arctic as a summer pathway between the Pacific and Atlantic oceans. Since the late 1970s, however, Arctic sea

FIGURE 10-48 DESALINIZATION

Semipermeable membrane-lined tubes

High-pressure seawater fed into tubes

Fresh water out

Pump

Donald Hyndman.

A small desalination plant that produces 400 m³ per day of fresh water at the south end of the Baja Peninsula, northwestern Mexico.

portion of energy reflected away from Earth's surface is called its **albedo**. Once some ice has melted, however, the dark surface of the ocean only reflects 10 percent of the sun's energy, leaving 90 percent of the original energy to heat the water and in turn melt more ice.

In August of 2005, a giant mass of ice, 66 square kilometers in area and 37 meters thick, cracked off the ice shelf at the coast of Ellesmere Island, 800 kilometers south of the North Pole, and drifted west for 50 kilometers until it froze into floating sea ice. The mass formed more than 3,000 years ago and was the largest loss in 30 years. Ice shelves of northern Canada are now more than 90 percent smaller than in 1906 when explorer Robert Peary first surveyed the region. Strong winds and warm water temperatures in 2005 probably made the ice shelf more vulnerable. If the ice drifts southward into the Beaufort Sea, it could endanger active shipping routes and significant oil and gas drilling platforms. Other ice shelves also show significant new cracks.

From 2005 to 2008, the volume of multiyear sea ice declined from about 11,000 to less than 5,000 km³. Similarly, from that period, older sea ice had thinned from an average of 4.1 to 2.7 meters. As more ice melts and more open water is exposed, the dark surface of the water absorbs approximately nine times the solar energy relative to snow and ice, which causes more melting. This positive feedback mechanism doesn't bode well for future impacts of global warming.

ice has dwindled to approximately 50 percent of its previous area (**FIGURE 10-49**). Arctic ice, about three meters thick in the 1960s, has thinned to about one and a half meters today. Unfortunately, there is also a feedback mechanism that accelerates melting of Arctic Ocean ice. White ice reflects 90 percent of the sun's energy back into space. The

FIGURE 10-49 ARCTIC THAW

Alaska

Canada

1979 minimum

Greenland

NASA Satellite images.

Northwest Passage **Closed**

Alaska

Canada

2003 minimum

Greenland

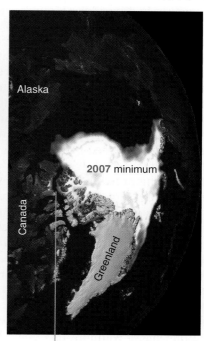

Alaska

Canada

2007 minimum

Greenland

Northwest Passage **Open**

The Arctic Ocean in 1979, 2003, and 2007 shows dramatic decrease in sea-ice cover with time. Note especially the increase in dark open water near Alaska and adjacent Siberia and Russia (top of image).

FIGURE 10-50 PERMAFROST THAW

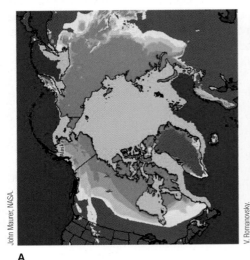

A   B   C

A. Permafrost distribution, northern hemisphere. Red and orange colors are, respectively, continuous and discontinuous permanently frozen ground, Still paler colors are sporadic occurrences of permafrost. **B.** Permafrost ice under a parking lot at the University of Alaska, Fairbanks, is melting, leaving cavities in the ice and potholes in the ground. **C.** This house in Alaska is settling and breaking up as a result of permafrost thaw.

PERMAFROST THAW Frozen ground (permafrost) that covers much of northern Canada, central Alaska, and northern Asia is thawing (**FIGURE 10-50**). Buildings that were once solidly anchored in the ice have to be abandoned because they are settling and deforming. Permafrost that once remained frozen through the summer at a depth of 0.5 meter finally began thawing in the early 1990s. Roads and railroads deform and become unusable (see FIGURE 9-22), spruce trees tilt at odd angles in forests, and sinkholes form in ice thawed beneath the ground surface (see FIGURE 10-50b, photo of permafrost thawing). Permafrost around Fairbanks, in central Alaska, averages about 50 meters thick, increasing to about 200 meters in northern Alaska, and more than 600 meters near the Arctic coast. Studies suggest that more than one-half of the top three or more meters may thaw by 2050—and the rest by 2100.

Compounding the problem, the thawing of permafrost releases large amounts of additional greenhouses gases into the atmosphere. Wind-blown dust and organic materials found in permafrost in Siberia and Alaska contain about three-quarters as much carbon as is tied up in living vegetation on Earth. When permafrost thaws, its organic material is warm enough to decay, and some of its carbon returns to the atmosphere as CH_4 and CO_2. Much of this carbon, sealed in permafrost for more than 10,000 years, may be released into the atmosphere over the next century. This would cause a huge increase in global warming, especially since CH_4 has about 20 times as much of a warming effect as CO_2 for an equivalent amount of gas. About one million square kilometers of western Siberian permafrost has begun to thaw, along with huge areas of northern Canada and central Alaska.

Studies in 2009 showed that permafrost contains about twice as much carbon as is presently in our atmosphere.

Permafrost melting also feeds more water into north-flowing rivers and into the Arctic Ocean, accelerating warming and melting there. Northern rivers are thawing several weeks earlier than in past decades.

SEA FLOOR THAW Sea ice melting in the shallow waters of the Arctic Ocean continental shelves appears to be initiating release of methane because the shallow sea floor is also thawing. **Methane hydrate** is frozen methane-ice compound trapped in layers at depths of 0.5 to 3 kilometers under the seafloor of many of the world's continental slopes (**FIGURE 10-51**). Just as methane and carbon dioxide are released on land by permafrost melting, which causes decay of frozen peat, the same thing is happening to previously frozen decayed vegetation under the sea floor. That organic matter formed long ago when the sea level was lower and that surface was the vegetation-covered land.

Most of the hydrate, though not all, sits in the sediment wedge of the subduction zones, where soft sediments are scraped up against continent edges. Locally it lies in seafloor venting "chimneys" where methane rises from depth, pools under the impervious methane hydrate layers, and finally reaches the stable zone of methane hydrate layers. Some methane escapes into the overlying ocean. Present estimates indicate that about three percent of global methane emissions now naturally come from the seafloor.

Stability of methane hydrate depends on pressure (depth of burial) and temperature (thawing of the hydrate at temperatures above 0° to 15° or 20°C). As the Arctic Ocean is

FIGURE 10-51 METHANE HYDRATE

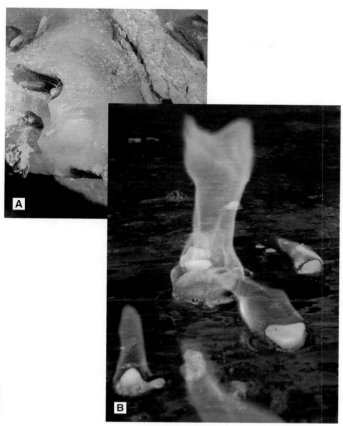

Ian MacDonald.

A. Methane hydrate (pale orange) layer on the ocean floor. **B.** Pieces of white methane hydrate decomposing and burning on the deck of a research ship.

now melting and its temperature rising slowly, methane will be released from the thawing sea floor. Recent studies show that the continental shelf north of Russia and Siberia, an area of about 1.5 million square kilometers, now shows gas bubbles rising from the thawing sea floor.

Based on projected temperature increases, huge amounts of methane may be released in coming decades. The amounts are comparable to burning much of our available resources of coal and petroleum. The impact of this gas on the greenhouse effect is compounded by the fact that methane is twenty times more potent than CO_2 in blocking outgoing thermal radiation.

A possible trigger for release of large amounts of methane involves the thinning of hydrate layers as sea levels rise with global warming—but perhaps even more important would be an increase in underlying methane "pools" as the hydrate stability zone thins; those pools may permit increased venting of methane through the sea floor. Major earthquakes, hurricanes, seafloor slumping, and drilling activities that initiate landslides on the continental slope would also destabilize methane hydrate in the affected zones. Small landslides triggered by earthquakes can destabilize larger areas of gas

hydrate that in turn weaken the slope and cause much larger areas of landsliding. A few giant submarine landslides, probably triggered by earthquakes that destabilized the methane hydrate layers in the past have caused major tsunami that inundated coastal regions (see Chapter 5).

GLACIAL MELTING Glaciers in the world's highest mountains have been melting rapidly, especially since about 1930. Unfortunately, as glaciers melt, their smaller area is more-closely surrounded by dark rocks that soak up more heat. That, in turn, causes more rapid melting. Muir Glacier in Alaska (chapter opener), for example, is rapidly retreating, as are the remaining small glaciers in northern British Columbia, the North Cascades of Washington, and Glacier National Park, Montana. These mountain glaciers, and those in the Himalayas, are especially important in maintaining fresh-water flow for downstream populations during dry seasons, including those of Pakistan, northern India, and Bangladesh. That flow provides municipal water, irrigation of crops, and hydropower. Even in the southern hemisphere, in the southern Andes of Argentina and in Antarctica, huge glaciers are dramatically shrinking (**FIGURE 10-52**). In Africa, Mount Kilimanjaro had large glaciers in 1912; by 2000, they were mostly open canyons.

Sea-Level Rise

Scientists now believe that global warming will cause the sea level to rise from between 0.8 and 2 meters by 2100. Global warming leads to sea-level rise from two primary factors—about half from water added due to ice melting on

FIGURE 10-52 GLACIAL RETREAT

1928

2004

Daniel Beltra, Greenpeace International.

Upsala Glacier in southernmost Argentina shows dramatic melting from 1928 (top) to 2004 (bottom).

land and half from the heating and expansion of sea water. Melting of Arctic Ocean ice does not raise the sea level because floating ice merely displaces the same volume of water as the submerged part of the ice. This is why melting ice cubes in a completely full glass of water will not make the water overflow.

Melting ice of continental glaciers, discussed previously, will continue. Antarctica also appears to be warming, especially in West Antarctica, at about half of the worldwide rate. Antarctic glaciers are now flowing more rapidly, apparently at least in part from breaking up of the Antarctic ice shelf. In April, 2009, rifts appeared in the 13,000 km^2 ice shelf and it appeared ready to break away. That floating ice shelf was helping to hold back the toes of flowing glaciers. Accompanying that melting, sea level has been rising at an average of 3.3 mm per year since 1990, or 33 cm per 100 yrs (about 13 inches). The rate of warming in Antarctica is likely to increase. Today the ozone hole over part of Antarctica permits reflected sunlight to escape, slowing warming. The hole is expected to heal by about 2050 now that the chlorofluorocarbons that caused it have been banned.

Melting of all Antarctic ice would raise sea level about 73 meters. If the entire Greenland ice cap were to melt, the added water would raise sea level by about 6.5 meters; the Greenland ice cap is now rapidly melting and spalling, three times faster in 2006 than in 1996. Fortunately, wholesale melting of the Greenland and Antarctic ice sheets, which would cause global sea level to rise a phenomenal 80 meters, is not likely in the foreseeable future.

As water warms, it expands, so the increase in ocean temperature raises sea-levels even further. Expansion occurs because heating causes the water molecules to vibrate faster and therefore take up more space. If CO_2 concentrations exceed 600 ppm before the end of the twenty-first century, thermal expansion of the ocean will cause sea-level to rise at least 0.4 to 1.0 meter.

The actual rise in sea level from both thermal expansion and glacier melting was about 19 centimeters between 1870 and 2000 (0.15 centimeter per year), but the rate doubled to about 0.3 centimeter per year between 1993 and 2005. The current rate of sea-level rise is roughly 30 centimeters per 100 years, and sea level is expected to rise by between 30 and 50 centimeters over the next century (**FIGURE 10-53**). New measurements by the USGS suggest, however, that sea level is likely to rise even more—about 1.2 meters (approximately 4 feet) by 2100.

EFFECTS ON HUMAN POPULATIONS People live at sea level for diverse reasons, sometimes because cities grew there for historic reasons. Some in wealthy, industrialized countries choose to live close to beaches for views and recreation. Others, in poor countries, live in places such as sea level deltas because fertile soils there facilitate growing crops to feed large numbers of people. The consequences of a major sea-level rise would be serious for low-lying

FIGURE 10-53 SEA LEVEL RISE

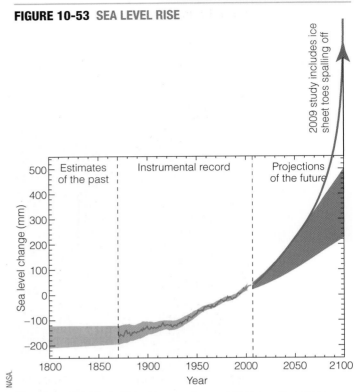

Rise of sea level from thermal expansion and glacial melting since 1800. Projection in blue arrow includes ice breaking off from toes of ice sheets.

areas on which five billion of the world's 6.8 billion people now live; this includes many of the world's largest cities (**FIGURE 10-54**). A rise of 3.5 to 12 meters, for example, would endanger New York City, Tokyo, and Mumbai, India (formerly Bombay). Even a moderate sea-level rise of 18 to 35 centimeters would accelerate erosion of coastal areas near Boston and Atlantic City.

A small sea-level rise along a gently-sloping coast translates to a very large landward migration of the shoreline. With 60 centimeters of rise, beaches in the southeastern United States would move about 60 meters landward, and almost 30,000 square kilometers would submerge. Coastal barrier islands would move shoreward much faster than today, making innumerable coastal communities even more vulnerable than they are now. Large areas of coastal Louisiana, Florida, and the North Carolina Outer Banks would be submerged (**FIGURE 10-55**).

Raising coastal river levees to protect low-lying areas would be expensive. Raising the levees protecting fields in the Sacramento River delta only 15 centimeters would cost more than one billion dollars. Proposals in the Netherlands, long noted for its levees and coastal storm barriers, suggest long-term spending of one billion Euros per year because of sea-level rise accompanying global warming.

In poor countries, without the resources to build and maintain protective barriers, coastal populations will be

FIGURE 10-54 RISING SEA LEVEL ON A DELTA

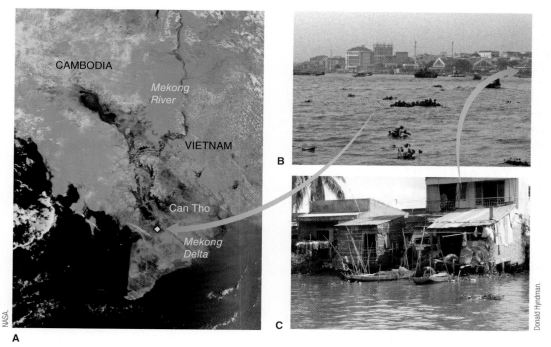

A. The Mekong River delta in Cambodia and Vietnam. **B.** Mekong River levels rise and fall with the tide at Can Tho, 120 km upstream from the South China Sea. **C.** Thousands living along its banks are endangered by sea-level rise.

FIGURE 10-55 POTENTIAL COASTAL FLOODING

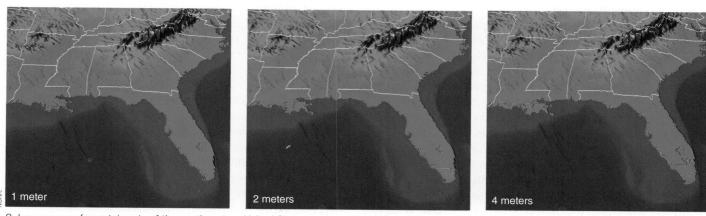

Submergence of coastal parts of the southeastern United States, with sea-level rise of 1 meter, 2 meters, and 4 meters, respectively. Hardest hit will be Louisiana, Florida, and the Outer Banks of North Carolina.

displaced. Researchers now predict that over the next four decades, tens of millions of people will be uprooted due to climate change and rising sea level. Bangladesh, which occupies the immense near-sea-level delta of the Ganges and nearby rivers, is already subjected to severe flooding during storms. Bangladesh would lose more than 17 percent of its land if sea level were to rise 1 meter. That would have a disastrous effect on its agriculture and the millions of people who depend on it. Deaths could run into the tens to hundreds of thousands during strong cyclones (**Case in Point:** Rising Sea Level Heightens Risk to Populations Living on a Sea-Level Delta—Bangladesh and Calcutta, India, p. 314).

Even if we are successful in the worldwide goal of stabilizing greenhouse gas concentrations in the atmosphere, it will take hundreds to thousands of years for temperatures to stabilize, as discussed earlier, while sea-level continues to rise.

FIGURE 10-56 THE GLOBAL CIRCULATION SYSTEM

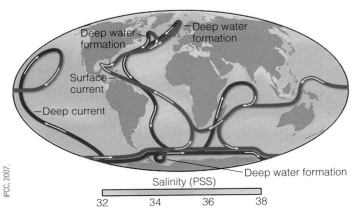

IPCC, 2007.

Salinity (PSS)

32   34   36   38

Red arrows represent surface warm currents and blue arrows deep cold currents.

Unfortunately, sea-level rise would take several centuries to stabilize, due to the long duration of heat uptake. It is clear that the world needs to act quickly and decisively to prevent further warming accompanied by additional rising sea levels.

Global Ocean Circulation

Changes in ocean currents resulting from changes in salinity could rapidly change climate in some areas. The large-scale circulation in the Atlantic Ocean (so-called thermohaline circulation) involves a current of warm, shallow water moving northward away from the equator. Closer to the Arctic, where it cools and sinks, the current pulls more warm surface water behind it (**FIGURE 10-56**). The sinking water moves south at depth. The warm, salty Gulf Stream, which moves northeast along the east coast of North America and across to Europe, is part of this circulation. Westerly winds carry heat from the ocean to keep Europe warm in winter, in spite of its northern latitude. Changes in local ocean salinity from factors such as melting glaciers and ice sheets could alter the path and strength of the Gulf Stream. Lower-density fresh water cannot sink in seawater, so circulation of the ocean current may slow or stop. Any reduction in the flow of the warm Gulf Stream to the north could thus cool Europe and some other northern hemisphere climates.

Global warming causes Arctic sea ice, Greenland glaciers, and frozen ground in Arctic lands to melt faster, pouring fresh water into the North Atlantic. In 2005, 221 cubic kilometers of Greenland's ice melted, compared with 91.7 cubic kilometers in 1995. As glacial ice melts, water pours down crevasses and holes in the ice: some refreezes; some flows below the glaciers, causing immense subglacial floods and faster flow of glaciers that calve more icebergs into the ocean (**FIGURE 10-57**). Kangerdlugssuaq Glacier in Greenland, which contains four times the water in the

FIGURE 10-57 GLACIER MELTS

Joel Harper.

A moulin, or melt hole, in a glacier carries large amounts of water to the glacier base, increasing buoyancy and lubrication of glacier flow.

Great Lakes, flowed downslope at 3.5 kilometers per year in 1996; it had accelerated to more than 16 kilometers per year by 2005.

Past climate studies show, however, that the warm oceanic conveyor belt has been affected before. It shut down 12,700 years ago in the cold period known as the Younger Dryas and then restarted 1,800 years later. Each of these changes occurred within only about a decade. Another, less severe abrupt cooling 8,200 years ago lasted only about 100 years. Lesser variation in temperature continued. A medieval warm period 1,000 to 700 years ago was followed by the "Little Ice Age" between the years 1300 and 1850. Winters were severe, glaciers advanced, crops failed, famine and disease were widespread, and large numbers of people migrated from northern Europe. Simultaneous drought in the southwestern United States has been linked to the collapse of the Mayan civilization in about 900 A.D. and the Anasazi just before 1300. Because it affects ocean circulation worldwide, cooling has also been linked to disruption of the monsoons of Southeast Asia that are critical to supporting the huge populations of the region. Monsoons are warm, with moist winds drawn off an ocean by warm air rising against a coastal mountain range. The moisture-laden air rises, cools, and dumps its moisture as heavy rains. Chilling of the continent prevents the mountain air from rising, and moist air is not pulled off the ocean.

Mitigation of Climate Change

Atmospheric CO_2 is expected to double in the next century, and there is no clear way to stop this trend. Given that carbon dioxide in the atmosphere has already reached levels such that the oceans are becoming increasingly resistant to further uptake of CO_2, and both atmosphere and ocean CO_2 would continue to rise even if we could immediately stop further production, how can we slow or stop global warming?

Policy changes, including those that reduce energy generated by burning fossil fuels, are necessary to reduce vulnerability to some of the effects of global warming. Mitigation could be accomplished by increased conservation, improved efficiency of power production and use. regulation of emissions carbon taxes, carbon trading, subsidies and tax credits, and capture and long-term storage of greenhouse gases, along with use of alternative fuels such as wind, solar, and possibly nuclear power. Conservation and efficiency are ideal because they cost almost nothing, are simple, and generate no pollution.

Reduction of Energy Consumption

If world population increases by 10 percent, greenhouse gas concentrations will also increase by about 10 percent, assuming no change in average per capita greenhouse gas production. The most direct way to slow the increase in greenhouse gases is to minimize population growth. Every person on the planet contributes to global warming, and world population now approaches 7 billion.

To complicate matters, the rate of population growth is higher in developing nations than in the developed world, and it is in the developing world where energy demands are increasing most. The total annual CO_2 emissions from fossil fuels in developing nations have now surpassed those of developed nations, and their rate of increase is much faster. For example, the increase in U.S. energy demand is one to two percent per year, but in China, where a growing consumer class is newly able to own cars, it is about ten percent per year. Of the total forty percent projected increase in demand for fossil fuels over the next 25 years, seventy-five percent will likely come from developing nations.

Minimizing population growth can be attempted through government enforcement, as implemented in China with the One Child Policy. However, this type of government intervention may infringe on people's religious beliefs as well as their freedom of choice; this can lead to decisions that are politically difficult, such as forced abortion or sterilization. Populations in the developing world often want large families because they want more people to contribute to household income or tasks, such as gathering food or firewood or even standing in line at community faucets that provide potable water. They also need children to care for their elders as they age.

Given the challenges of minimizing population growth, another method of cutting greenhouse gas emissions is to reduce the per-capita consumption of energy through conservation and greater efficiency. Conservation is an inexpensive, simple, and easy way to save large amounts of electric power and therefore reduce carbon output used in power production. Buildings use about half of all fuels in heat, light, and ventilation. Conservation mechanisms for buildings include insulation to minimize loss of heat or maintain cooler environments, more-efficient generation of light, and circulating and filtering air.

The design of more energy-efficient household electronics could significantly reduce emissions with little sacrifice of convenience on the part of consumers. An estimated four million tons of carbon is used per year in the United States to power electronic devices in standby mode—devices such as TVs, phones, computers, alarm clocks, and microwave ovens. The Department of Energy estimates that the "standby" portion of each person's electric bill will reach 20 percent by 2010!

More-efficient transportation would involve more use of mass transit such as railroads and buses. Automobiles, especially use of large vehicles with one or two people, make extremely inefficient use of fuel. The use of hybrid gas-electric vehicles is increasing, and more major auto companies are planning to produce plug-in hybrids in 2010 and several 100-percent-electric car models are slated for the market by 2011.

Alternative Energies

The current effort to mitigate climate change is largely focused on the exploration of alternative energy sources that have lower emissions. Power from hydro, wind, nuclear, solar, and geothermal sources generate no CO_2.

WIND POWER In 2008, wind accounted for only about one percent of total electricity generation, slowed in part by the need to build new transmission lines from wind turbine farms. However, about 35 percent of new electricity generated was from wind.

In 2008, all wind turbines were located onshore, especially in Texas and California (**FIGURE 10-58**, p. 302). By August, 2009, a rapid expansion of wind farms along the lower Columbia River valley east of Portland, Oregon boosted wind energy from Oregon and Washington to 2,000 megawatts, enough to light all of Portland and Seattle. The greatest onshore potential is in the windy Great Plains east of the Rockies: eastern Montana and Wyoming, the Dakotas, Nebraska, Iowa, Kansas, western Oklahoma, and north Texas.

FIGURE 10-58 WIND ENERGY

A. Wind turbines near Leipzig, Germany. The crop in the yellow field is rapeseed, which is used as a biofuel. **B.** Part of a giant, new wind farm in south-central Washington state.

Although wind power is in its infancy, the cost of wind-generated electricity has dropped dramatically since the 1990s with efficiencies in manufacturing larger numbers of towers and in newer technologies. By 2010, the cost of new wind-generated power was almost on par with that from coal and natural gas, especially when the costs of carbon dioxide sequestration are included. Once the infrastructure is built, the cost of providing electricity from wind drops virtually to zero. Because wind power produces no carbon dioxide, mercury, or other pollutants, it is being viewed more favorably by electric power companies. Electric power from existing coal-fired plants is still less expensive but the pollution they generate is increasingly an issue.

In the future, wind power may also come from offshore sources. Wind power may see more rapid growth in the United States because of new federal leases in parts of the outer continental shelf, areas that, according to the Interior Department, have sufficient potential to supply enough electrical energy to meet current demand. The greatest interest and potential offshore is on the continental shelf off the northeastern states, where electricity prices are high, water depths are shallow, and demand is huge from big coastal cities. Wind turbines off the west coast are disadvantaged by the deep water close to shore.

Arguments against wind as an energy source are that the giant turbines are unsightly and can kill birds and bats. Although noise was once an issue, modern wind turbines are very quiet. Another limitation is that it may cost as much as $100 billion to build enough transmission lines from suitably windy locations, especially in the Great Plains and Midwest, to the main transmission grids to carry the power to the big east-coast cities.

SOLAR ENERGY Some solar collectors generate electricity from silicon panels, and excess electricity can charge batteries for various domestic uses. These can be quite effective but are expensive to manufacture. Solar panels that plug directly into home circuit panels are now available at large home-improvement stores and are expected to become less expensive.

New *thin-film* solar materials, which can be used to coat windows and roofs, use little or no silicon, are much cheaper to manufacture, and promise to reduce costs significantly. Huge thin-film solar projects planned in Inner Mongolia are being subsidized by the Chinese government.

Solar thermal is an energy-collection method being built in deserts, such as those in southern California and Nevada, that direct sunlight from mirrors to liquid-filled tubes or to a central tower that stores the energy. The energy heats water to steam to drive an electric generator, and transmission lines carry the electricity to cities. Plants are operating in Spain, Egypt, and Israel. Other, much-less-expensive solar heating panels are merely water-filled tubes heated by the sun to provide hot water for washing or domestic heating (**FIGURE 10-59**).

NATURAL GAS Methane, a natural gas, is a cheap and efficient fuel used for heating homes and driving power plants. Although methane is a greenhouse gas, burning it generates much less CO_2 than other fossil fuels. Supplies of methane were dwindling, however new techniques permit drilling down to methane-bearing formations, turning the drill

FIGURE 10-59 SIMPLE SOLAR COLLECTORS

Simple but effective solar collectors that heat water for domestic use are almost ubiquitous in some countries such as Turkey.

horizontally within the shale, and fracturing the shale with high-pressure water. This allows gas to seep into the fractures and be recovered. Newly discovered huge reserves of natural gas include the Barnett Shale in Texas, the Haynesville Shale in Louisiana, the Marcellus Shale in Pennsylvania, and the Horn River basin in northern British Columbia. The big increase in supply of natural gas has driven down the price, making it cheaper to burn than coal. In mid-2008, its price rose to thirteen to fourteen dollars per million BTUs (British Thermal Unit), but by mid-2009, it had dropped to three-and-a-half to four dollars per million BTUs. Such low gas prices dampen interest in clean energy projects.

The vast majority of new power plants being built are designed to run on natural gas. Because it is cheaper and cleaner than gasoline or diesel, cities such as New York, Atlanta, and Los Angeles have converted some of their trucks and buses to run on natural gas. Many global warming experts suggest that cars should also use it.

Energy companies are interested in the possibility of tapping the vast reserves of methane frozen in the seafloor. Current estimates are that methane hydrate holds about twice as much energy as all other fossil fuels combined—gas, oil, oil shale, and coal. If only one percent of the methane hydrate were recovered, that would double the current methane reserves of the United States. Methane under U.S. waters is about 1,100 times the current U.S. reserves and 53 times the world's current land reserves. Unfortunately, heating to melt the hydrate at depth to release the methane uses much of the heat value that could be obtained.

Technology to tap those offshore reserves economically is not presently available.

Although methane is cleaner and cheaper than coal, unburned it is 20 times more potent than CO_2 as a greenhouse gas. As reported by the U.S. Environmental Protection Agency, methane is responsible for 75 percent as much warming as CO_2. Although CO_2 remains in the atmosphere for hundreds of years, CH_4 does so for only about a decade. Thus cutting the amount of methane released into the atmosphere has an almost immediate and very significant effect.

Methane is released from landfills, sewage, oil and gas drilling, coal mines, and feedlots. Channeling these sources into collection systems can remove it from the atmosphere. It can in turn be burned in community power plants as a source of relatively clean energy. In December, 2009, the United Nations Clean Development Mechanism pressed for setting up a Global Methane Fund to build methane-capture projects; these projects could remove 1.3 gigatons of annual CO_2-equivalent emissions for minimal cost.

CLEAN COAL Recent discussions about "clean coal" are misleading. Coal is cheap and abundant but generates about twice as much CO_2 as natural gas in generating the same amount of electricity. Proponents of clean coal assume that almost all of the CO_2 produced can be captured and permanently sequestered. The newest coal gasification plants, now in the experimental stage, are more efficient, would reduce pollution, use less water, and more easily collect CO_2 for underground disposal. However, coal burning not only pollutes with carbon dioxide, sulfur dioxide, and mercury, but the strip-mining that removes waste rock from above the coal typically lays waste to huge areas of landscape.

NUCLEAR ENERGY Japan and countries in Western Europe that have little fossil fuel generate significant amounts of nuclear energy—about 16 percent of worldwide energy supply. France generates about 78 percent of its power from nuclear fission—the breakup of uranium atoms—and it recycles much of the nuclear waste to make additional fuel. Nuclear power presently produces about 20 percent of electrical power in the United States, from 103 plants, all built before 1980.

Problems with nuclear power include limited deposits of uranium fuel, projected to last only about 50 more years, and disposal of nuclear waste, most of which has low levels of radioactivity. Nuclear power earned a bad name from the power-plant disasters at Three Mile Island, Pennsylvania, in 1979 and Chernobyl in the Soviet Union in 1986. Three Mile Island was caused by mechanical failures, poorly trained operators, and falsification of reactor leak rates. Chernobyl was a poorly designed, constructed, and monitored plant.

The problems of safety and disposal of nuclear waste now appear to be less of a hazard to mankind than further increases in greenhouse gases, and the power industry is under pressure to reduce reliance on coal-fired plants.

As a result, new nuclear fission power plants are again being planned in the United States, but the first will not come online before 2015. The U.S. Congress provided incentives in the form of risk insurance and production tax credits for the first few new reactors. As a result, several companies applied to construct new nuclear plants. However, the cost of building nuclear power plants has risen much faster than for other electrical-generation plants, placing the future of new plants in jeopardy. Only with federal loan guarantees will they consider new construction. Nuclear fusion—the combining of parts of atoms at temperatures of a hundred million degrees Celsius, as occurs in our sun—is feasible but presently very inefficient and likely at least 50 years away from commercial production.

TIDAL POWER SYSTEMS Tidal power systems are pollution free but expensive to build. Various schemes involve either tidal currents running underwater turbines in narrow fjords with large tidal ranges or capture of the potential energy difference between the heights of high versus low tides. Most existing installations are small, but one operating on the Rance River in France since 1967 generates about 68 megawatts of average power. Tidal power areas are under consideration in a few promising areas, from the Bay of Fundy, between Maine and New Brunswick, to Cook Inlet, near Anchorage, Alaska.

HYDROGEN FUEL Hydrogen fuel, touted as clean and pollution free, generates electric power by burning hydrogen with oxygen. Unfortunately, it must be generated by separating hydrogen and oxygen, which takes large amounts of traditional energy sources. The advantage is that hydrogen can be generated in centralized plants where CO_2 produced by burning can be captured and hopefully sequestered underground.

BIOFUELS Biofuels are used as gasoline supplements that burn cleaner than fossil fuels. Government support and subsidies have primarily focused on ethanol produced from corn as the most promising biofuel. However, recent studies question the efficiency of ethanol as a fuel, pointing to increased emissions during the production process. They show that ethanol producers that use natural gas in production produce 5 percent more emissions than gasoline; those that use coal produce 34 percent more emissions. Although ethanol burns cleaner than gasoline, the number of miles per gallon of fuel is lower. If you include the fertilizer and machine-fuel energy used to produce the crops, it takes at least 25 percent more energy to get ethanol or biodiesel from corn, soybeans, and other commonly used plants than is available in those fuels. Ethanol must be transported by truck or rail, rather than pipeline, thereby using more fuel. In addition, it would take an estimated 43 percent of all crop land in the United States to replace only 10 percent of gasoline and diesel fuel supplies.

Ethanol has other downsides. Its primary source is corn, which is important for animal feed and as a food export to poor countries. In a single year, the world price of corn quadrupled and that of wheat doubled, leaving poor countries without enough food for their people. In addition, the huge amounts of water used in producing ethanol are depleting worldwide aquifers and polluting water supplies. Also reducing the interest in production of ethanol is the increase in price, both because of higher corn prices and the economic recession of 2008–10, during which consumers cut back on driving. A feedback mechanism involving less driving—and therefore less fuel used—increased stockpiles of ethanol and drove down its price.

In future, instead of producing ethanol from corn, biofuels may be produced from algal farms that can generate 14 times as much fuel from the same size area of land. Pilot projects have shown the feasibility, but abundant sources of nutrients and water must be found. Although not yet commercially viable, a cellulose source such as grass or logging residues produces about four times as much ethanol with the same energy input. However, cellulose growth requires a lot of land. Hopefully, new technologies will develop to make this a net gain rather than a net loss.

GEOTHERMAL POWER Geothermal power generation is concentrated in California where "The Geysers" in the northern Coast Range produce more than 2,000 megawatts of geothermal power, about 80 percent of the country's total. One megawatt of electricity can power about 1,000 homes. Most of the remainder comes from crustal spreading zones around the Salton Sea of southern California and the thin, hot crustal areas of the Basin and Range of Nevada, Oregon, and Utah, areas with expanding potential.

COST OF ALTERNATIVE ENERGY A major factor in increasing alternative energy use is price to the consumer. With so much demand for energy, cost is what drives most usage. Economic considerations, time, and convenience drive people to use the least expensive modes of heating, lighting, and transportation.

The least expensive fuel sources today are coal and natural gas, costing about 5¢ per kilowatt-hour. However, if the hidden costs of health effects from air pollution were included, coal would cost about 6.3¢. Wind energies currently cost about 6¢ per kilowatt-hour, while nuclear costs about 6.5¢. Solar energies from solar panels currently cost about 22¢ per kilowatt-hour, although prices are expected to drop as solar technologies improve. As alternative energy technologies improve, costs should come down, making them more appealing to industry and consumers. If the real costs of fossil-fuel burning—including the costs of health problems associated with other emissions and sequestration of CO_2—are included, wind and other power sources are about as low as—or lower—than fossil fuels.

Carbon Trading

The implementation of government policies to raise the cost of fossil fuels, for instance, by taxing carbon emissions, could also make alternative fuels more affordable in comparison. Higher prices or policies are needed to push people to conserve fuel use, but taxes and other artificial increases in prices make for uncomfortable political issues. A few countries have implemented taxes based on carbon use. Studies show that in developed countries, a price of about $50 per ton of CO_2 would significantly reduce carbon use for power generation and heating—those make up 80 percent of energy use.

One proposal is to implement a **cap-and-trade** law to limit carbon emissions. A similar system helped the U.S. government reduce pollution by gases that cause acid rain in 1990. Limits placed on smokestack emissions required industry to either cut them or trade (and pay for) the right to emit additional pollution with companies possessing more efficient plants that produced much less pollution. The same system could be used to limit CO_2 emissions.

Most member countries of the European Union try to limit emissions by encouraging carbon trading. In this scenario, companies are limited to a certain level of carbon emissions based on past emission levels. A company that reduces its greenhouse gas emissions may sell its permitted emissions credits to one that generates more than it did in the past. The idea is to encourage companies to reduce greenhouse gas emissions without hindering growth. Unfortunately, the baseline for individual companies does not distinguish whether a company uses old, heavily polluting, "dirty" power generation or new, state-of-the-art, low-pollution power.

Some industrialized countries, such as the United States, have been reluctant to cut back on carbon emissions for fear of hindering economic growth. Arguments that point to the cost of a cap-and-trade system seldom take into account the potential costs of climate change. Although very difficult to estimate, current studies indicate that costs associated with climate change could range from 5 to 20 percent of gross domestic product (the market value of all goods and services produced in a country). For the United States, 10 percent of 2010 GDP was about $1.4 trillion.

Careful analysis of the emissions problem by NOAA and other climate scientists indicates that cap-and-trade would slow the increase in CO_2, but others argue that an outright tax on emissions would be more effective. Some countries, such as Sweden, use a carbon tax (a tax on burning fossil fuels); others use voluntary schemes.

Global Cooperation

Meaningful reductions in greenhouse gas emissions can only be accomplished if all major industrial producers of CO_2 commit to the effort. By 1997, world governments recognized the danger posed by the increase in atmospheric CO_2 and its correlation to global warming. One hundred and eleven countries, including most industrialized nations—but neither the United States nor Russia—signed the **Kyoto Protocol** in 1997 to reduce emissions of six greenhouse gases beginning in 2008. Russia signed the protocol in 2004, bringing the percentage of industrialized-nation signatories to 55 and thus meeting the treaty's threshold for activation. The United States declined to agree to such targets despite being the largest contributor to the emissions at that time (**FIGURE 10-60**), arguing that China, India, and other large contributors should also have to agree and that the costs would have a detrimental effect on U.S. economic growth. Instead, the United States said it would reduce the *rate of increase* of greenhouse gas emissions to less than the rate

FIGURE 10-60 EMISSIONS BY COUNTRY

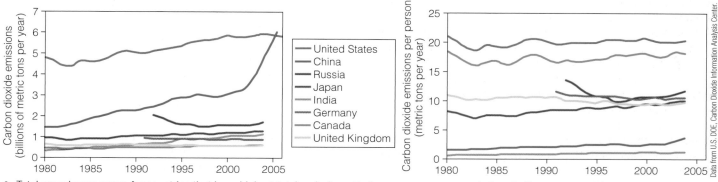

A. Total greenhouse gases for countries that have highest total emissions. **B.** Average per person emissions in the same countries.

of increase in U.S. economic growth. Canada followed a similar policy.

Several states, however, decided to do their part in spite of the federal government's stand. In 2006, California mandated a cap on global warming greenhouse gas emissions and intends to cut them to 1990 levels by 2020. In April 2007, the U.S. Supreme Court ruled that carbon dioxide from cars and trucks is considered a pollutant. In 2009, the U.S. Environmental Protection Agency (EPA) provided new fuel efficiency regulations for new motor vehicles. The EPA recently required fuel importers and refiners to reduce the sulfur content of diesel fuel by 97 percent, thereby permitting building of clean diesel cars and light trucks that can be 30 percent more fuel efficient. They also designated carbon dioxide and other greenhouse gases as pollutants. Large emitters putting out more than 25,000 tons of carbon dioxide per year, including 13,600 coal-fired power plants, oil refineries, and smelters, would be regulated. Those power plants generate about 40 percent of U.S. industrial greenhouse gases. New construction or modification would be required for permits that call for the "best available technology" to minimize emissions.

Nations are polarized over who is to blame and what to do about climate change. The countries that contributed most CO_2 historically, such as the United States, profited from unrestrained industrial development to become the wealthiest nations in the world. The countries that are today going through periods of rapid industrial development, such as China and India, feel that they too should be allowed to develop without environmental restraints. They argue that the industrialized nations must first reduce their emissions because they can more afford the associated costs. Huge populations in Southeast Asia are so poor that they cannot afford to purchase more efficient fuels. How do you tell a family that has too little money for food to stop cooking with the only fuel that it can afford?

Without the cooperation of large emitters like China and India, the prospect for significant reduction of emissions is bleak. China still generates far less greenhouse gases than the United States on a per capita basis, but their production is rapidly increasing because of their immense population and rapid development. China overtook the United States as the largest total emitter of greenhouse gases in 2006 due to minimal pollution controls, a dramatic increase in number of heavily polluting coal-fired power plants, cement production, and steel plants, in addition to very rapid economic growth. In early signs of a policy shift, some Chinese economists argued in 2009 that the country might soon be rich enough to afford some of the necessary changes. The average American produces 16 times as much CO_2 as the average person in India, but this too is changing as India's economy improves.

China's and India's only abundant energy source is coal, the least efficient fuel, which generates the most greenhouse gases; the prospects are alarming. Because hundreds of millions of individual families in India are so poor, they cannot afford $20 cooking stoves that use less than half the wood compared with the open fires that most use. Several non-profit groups are now trying to provide the stoves. Such low-tech solutions are often much simpler, less expensive, and more effective in reducing energy use than other mechanisms. Because federal and state tax incentives are often much higher for expensive retrofits, legislative proposals before congress now propose more generous tax breaks for projects that save more energy. Even with one of the most optimistic scenarios for reversing the increase in CO_2 with strong limits on emissions, the total CO_2 in the atmosphere is not expected to level off until after 2060 (at a level of about 475 ppm compared with 388 ppm in early 2010). Without immediate and drastic countermeasures, China's rate of emissions increase threatens to greatly exacerbate the greenhouse gas problem in the near future. China and India are determined to expand access to electricity, grow their economy, and reduce poverty, even if that involves increased emissions. China did, however, pass laws in late 2009 requiring its power companies to purchase electricity from renewable sources even if it is more expensive than coal-fired electricity.

On September 8, 2007, the leaders of 21 Asia-Pacific countries agreed to a 25-percent reduction by 2030 in their "energy intensity," the energy needed to produce a dollar of economic product. Thus countries are permitted to increase their emissions as long as they become more efficient and their emissions per dollar decrease. Unfortunately, although the rate of increase in emissions would slow under this agreement, the total greenhouse gas emissions would continue to increase. The agreement does little to solve the problem. In July, 2009, members of the G8 major-economic countries did commit to reduce their carbon emissions by 80 percent by the year 2050. Indonesia, the third largest emitter, contributes CO_2 by cutting down and burning forests for both lumber and the clearing of land for palm oil and acacia plantations. In spite of many countries' refusal to sign binding agreements at the 2009 Copenhagen climate-change conference, most countries pledged to reduce CO_2 emissions by 17 to 30 percent by 2020. However, for some, such as India and China, the reduction is in the emissions intensity—that is, they pledge to reduce the *rate of increase* in their emissions rather than their total emissions. Unfortunately, they set no short-term goals, and the agreement is not binding.

Sequestration of Greenhouse Gases

There are two obvious ways to minimize or reduce the addition of CO_2 to our atmosphere: (1) by dramatically reducing the burning of fossil fuels (including for heating, electricity, and transportation) and (2) by the physical removal and long-term storage, called **sequestration**, of large amounts of manmade CO_2 as it is produced.

The first step in sequestering CO_2 is carbon-capture. Carbon dioxide can be captured at its source in fossil-fuel-powered electrical generation plants that use fuel cells to ultimately produce CO_2 and H_2O. This requires retrofitting or replacing existing power plants with new-generation plants that first treat coal with steam. Carbon dioxide capture directly from the atmosphere is feasible, and two companies are now storing CO_2 in abandoned oil reservoirs, one in Canada and one in Algeria. Others are in the works (see **Case in Point:** CO_2 Sequestration Underground—The Weyburn Sequestration Project, p. 315).

Once the CO_2 has been captured it must be placed into long-term storage. Pumping CO_2 underground into the pore spaces of sedimentary rocks in oil fields has long been used to help push residual oil to the surface. In 2008, about 30 million tons of CO_2 were injected annually into old oil fields in the U.S. If all of the world's oil fields used such CO_2 reinjection, roughly half of the new CO_2 we produce could be kept underground.

Other underground sites may have even greater potential for long-term storage of CO_2; capacity in other sedimentary rocks is more than ten times that in oil fields. Anticlines (arches in sedimentary rocks) that were once sealed well enough to hold reservoirs of natural gas may be one of the more permanent places to lock away sequestered CO_2. Disposal of the displaced saline water in wells on land has also been demonstrated on a large scale, but it is not yet clear that it can be trapped for long periods. Disposal of the CO_2 would be below depths of 800 meters to increase its density almost to that of water.

Concerns about underground sequestration of CO_2 include fears of the gas escaping as well as the contamination of aquifers that have useable groundwater resources. Another concern is whether the pumping of large amounts of fluid under pressure could trigger earthquakes in some areas, as was the case when the injection of fluids for geothermal power production triggered several earthquakes at Basel, Switzerland, in 2006 (see FIGURE 4-3). If a significant amount of the CO_2 should escape to the surface along faults or fractures, it could collect in low areas to suffocate animals or humans. Such ponding occurred in Lake Nyos, when CO_2 escaped from a volcanic-region lake, and near Mammoth, California, when volcanic CO_2 collected in a winter cabin and almost suffocated a forest ranger (see FIGURES 7-17 and 7-18). A serious potential problem is the acidification of fresh groundwater supplies because CO_2 mixed with water produces carbonic acid (a weak acid). Potential side effects on groundwater supplies remain to be seen. Injection of sulfur dioxide (SO_2) at the same time as CO_2 would promote the formation of sulfuric acid.

The legalities of underground storage of CO_2 became clearer in 2008–09, when Wyoming, the largest coal-producing state, passed laws regarding sequestration. The laws specify that surface landowners own the right to store CO_2 under their land. In 2009, Wyoming ruled that anyone who injects the gas underground is legally responsible for it forever.

Sequestration under the ocean could be done in a deep sea environment, such as at offshore petroleum wells. Another option may be to pump liquefied CO_2 into deep ocean-floor sediments, where it would be trapped because of its higher density. Since 1996, a Norwegian company has used the process under the North Sea, disposing of about one million tons of CO_2 per year. To put this in perspective, one large coal-fired power plant produces about 8 million tons of CO_2 per year.

If CO_2 were piped directly to the ocean bottom from offshore drilling platforms, the liquid CO_2 could pool as a "lake" on the ocean bottom. Over a period of hundreds of years, much of that would gradually dissolve back into the ocean. Thus such disposal methods would not be permanent but would slow the rise of CO_2 in the atmosphere. Environmental concerns have yet to be evaluated.

Pumping CO_2 underground into the water-bearing pore spaces of immense basalt lava flows (discussed in Chapter 6) such as the Deccan Traps of India, Siberian basalts of central Asia, and the Columbia River Basalts of Washington state, provides the opportunity to store the gas and permit it to react with minerals to form insoluble carbonate minerals. The reactions result in an increase in volume, so the products could clog pores in the rock and prevent further reaction. Carbon dioxide and water react with olivine and pyroxene to form $MgCO_3$ (magnesite) and with plagioclase feldspar to form $CaCO_3$ (calcite). Whether this is feasible on a large scale may depend on initial reactions coating the mineral surfaces to slow further reaction and take up more CO_2. The test of the process in a natural environment is presently ongoing in Iceland.

A variant of the basalt-reaction proposal involves reacting magnesium-rich minerals such as olivine and serpentine in peridotite (see Appendix 2 online) with CO_2 to form $MgCO_3$ and storing it in the pits created by mining the peridotite. The energy required for both mining and crushing has to be part of the equation and it is not yet clear how such a process could be carried out commercially on a large scale.

Storage in above-ground tanks has been demonstrated as viable for decades but has not yet been implemented on a commercial scale. The volume needed to sequester CO_2 emitted from a single large coal-fired power plant would be about 10 million cubic meters per year, so the total volumes needed for all such power plants are extremely large.

Sequestration cannot be accomplished without the expenditure of additional energy and, generally, not without production of additional CO_2, generally about 25 percent of the amount of CO_2 removed. Carbon dioxide sequestration has only been shown to be feasible on a small scale and for relatively short times. It remains to be seen whether the

FIGURE 10-61 HOW DO WE STABILIZE CO$_2$?

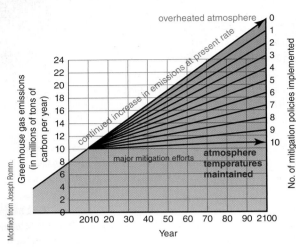

Minimum major mitigation policies to be implemented (sub-equal gains for each):
[Assumes policies implemented in 2010. Later implementation moves curves up]
- **Capture and sequester CO$_2$** from all large coal-fired plants
- 700 gigawatts of electricity from **new major nuclear-powered plants** (no **CO$_2$**)
- 4,000 gigawatts of **electricity from solar sources**
- 5 million gigawatt hours **power saved in more-efficient buildings**
- 5 million gigawatt hours **power saved in industry by heat recovery and cogeneration**
- 1 million large **wind turbines**
- **All cars average 60 miles per gallon** – including more hybrids and diesels
- **Cellulose-based biofuels from one-sixth of world's cropland**
- **End all tropical deforestation**
- Additional gains from **greater use of rail replacing trucks, more-efficient trucks, heating from natural gas (methane) captured** from unavoidable sources such as landfills and methane seeps

["giga" = 1 billion]

We have to do **all** of these things to stop increasing emissions entirely!

Necessary measures to stabilize the amount of CO$_2$ in the atmosphere. The uppermost sloping line represents the approximate increase in CO$_2$ with "global business as usual" without aggressive reduction in fossil fuel combustion. Beginning in 2010, each of the progressively gentler-sloping lines shows a ballpark reduction in the increase in CO$_2$ by using one of the alternatives to burning fossil fuels. Note that the figure assumes that we could have achieved these additional efficiencies by the year 2010; that did not happen.

process can be expanded to sufficient scale and whether storage can be viable over hundreds or thousands of years without significant negative consequences.

GEOENGINEERING Present efforts to curtail global warming focus on reducing production of CO$_2$ or removing some of the CO$_2$ that is being produced. An entirely different approach considers blocking some of the sun's incoming energy using geoengineering. Suggestions include injecting several megatons of sulfur dioxide into the atmosphere annually, either from jet aircraft or balloons. The approach would attempt to mimic the short-term cooling effects of volcanic eruptions like Mt. Pinatubo in 1991. Because the SO$_2$ would wash out of the atmosphere eventually, the injection would have to be done almost every year. Side effects might include damage to some plants and animals, including skin cancer from damage to the ozone layer, and possibly disruption of the monsoon rains in some areas. This could cause significant droughts with the only benefit being a minimal reduction in global temperatures.

Political concerns include questions of liability. Would any unusual weather be blamed on whichever country or organization engineers the change? Geoengineering has had unintended consequences elsewhere. In November 2009, the Beijing Weather Modification Office seeded clouds to provide moisture during a lingering drought. As a result, a huge blizzard dumped massive amounts of snow, wreaking havoc on highways, collapsing buildings, and causing $650 million in damages. Imagine the potential liability of engineering a substantial annual change in Earth's atmosphere!

COMBINING ALL STRATEGIES In order to avoid any further increase in CO$_2$ in the atmosphere, we must combine numerous mitigation strategies, including decreasing energy consumption, increasing use of alternative energy sources, and removing greenhouse gases from the atmosphere. Identification of the primary sources of human-caused greenhouse gases and reduction in those sources is difficult but achievable (**FIGURE 10-61**). All of these greenhouse-gas-reduction strategies are feasible. The main problems are political. The public demand economic growth, jobs, and improvement of their standard of living; government leaders are reluctant to hinder economic growth. Few seem to understand the future costs and consequences of inaction. With each additional year that passes without enacting these efficiencies, atmospheric CO$_2$ levels increase further and the task becomes more difficult. Even attainment of all of these reductions in greenhouse gas emissions will not reduce them below the current levels.

Without major decreases in human-caused emissions and some type of engineering intervention to remove CO$_2$ from the atmosphere, we may already have reached the point of irreversible climate change. So is that reason to say that the situation is hopeless, we have already gone too far, so we can stop worrying about CO$_2$? Certainly not, because with even higher amounts of CO$_2$ in the atmosphere, temperatures and their consequences will come sooner and be even more extreme.

Cases in Point

A massive ice storm in the Southern United States
Arkansas and Kentucky, January 2009 ▶

On January 26, 2009, rain ahead of a weather front in New Mexico and central Texas encountered cold air and light snow surrounding a large Arctic high centered on Iowa. Ahead of the front, in northern Texas and Oklahoma, the moist air over-riding the cold to the north turned to freezing rain and snow as it fell. In northwestern Arkansas and Kentucky, sleet, freezing rain and snow around midnight turned to heavy freezing rain by 6 am on the 27th. Freezing rain, thunder, and the cracking of tree limbs continued all day and into the evening. Many power lines can withstand only about 1.3 cm of ice buildup. Heavy ice coating up to 5 cm thick loaded, bent, and snapped tree branches, weighed down power lines, and coated roads.

By evening, large tree limbs and whole trees snapped, falling on homes and power lines. Power failed locally in mid-afternoon, then totally by 8 pm.

Five days later, thousands of Kentucky National Guard troops were still clearing branches, trees, and power lines from roads leading to remote communities and still-stranded residents. They went door-to-door to hand out food and bottled water to people still without power. More than 700,000 homes and businesses lost power in Kentucky, 1.3 million throughout the storm region that extended northeast through Ohio and beyond. Emergency shelters were set up in most towns, though with power out at some radio

stations as well, it was difficult to spread the word that shelters were available. Loss of power also shut down pumping stations and therefore water supplies. FEMA provided more than 300 high-capacity electric-power generators for emergency operations and other critical facilities. Convoys of utility trucks from around the region arrived to help re-store service. 260,000 remained without power a week after the storm.

42 deaths were attributed to the storm; most were from hypothermia, traffic accidents, or carbon monoxide poisoning from inside use of charcoal grills or improperly used space heaters and generators. One police officer died when an ice-laden branch fell on him.

Larry McGriff and Mel Coleman, NWS.

Bill Atchley.

▶ *January 2009 ice storm a. In Arkansas. b. in Kentucky*

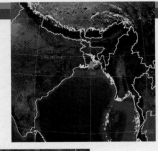

Rising Sea Level Heightens Risk to Populations Living on a Sea-Level Delta
Bangladesh and Calcutta, India ▶

The vast delta regions of Bangladesh and the northern coastal region of India around Calcutta are low lying and especially fertile—and one of the most densely populated regions on Earth, with some 130 million people in an area not much larger than New York State. People are exceptionally poor; their livelihoods are based directly on agriculture and therefore are intimately tied to weather and its associated hazards. The delta lands on which people live and farm are virtually at sea level. Most people live on farms rather than in the cities, and the few rail lines connect only the larger cities. Roads are narrow and crowded with heavy trucks, buses, and rickshaws. Bridges span only a few of the smaller river channels. People get around on bicycles, small boats, and ferries. Traffic jams in the cities are frequent on normal days. Imagine the chaos of trying to evacuate hundreds of thousands or millions of people with a couple of days' warning given such limited transportation.

Even when there is adequate satellite warning to evacuate populations in the region, most people do not leave because they believe that others will steal their belongings or because they have lived through one major cyclone. The average time between cyclones suggests to them that the next major one "will not come for many years," or that if they die, it is "God's will." The problem is enormous:

- In October 1737, 300,000 died in a surge that swept up the Hooghly River in Calcutta.

- In 1876, 100,000 died in another surge near the mouth of the Meghna River; thousands more perished in 1960 and 1965.

- In November 1970, cyclone winds of 200-kilometers-per-hour accompanied a 12-meter surge that swept across the low-lying delta of the Ganges and Brahmaputra Rivers in Bangladesh. 400,000 people died, many of them in the span of only 20 minutes. Whole

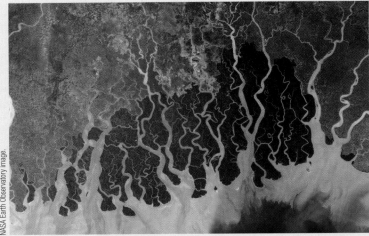

▶ *The nearly sea-level delta area of the Ganges and Brahmaputra rivers is laced with distributary channels and their tributaries. The dark central area is a preserve of mangrove swamps. The remainder is farmland.*

NASA Earth Observatory image.

villages disappeared, along with all of their people and animals.

- Again, on April 30, 1990, a cyclone with 233-kilometer-per-hour winds and a 6-meter surge swept into Bangladesh, drowning 143,000 people. The 1990 Bangladesh population of 111 million has been projected to double in 30 years, placing millions more on the delta.

- On June 6, 2001, nearly 100,000 people were stranded up to their chests in water following heavy monsoon rains. Villages were flooded when a flood protection embankment failed, and rail service was disrupted when the floods swept away a small bridge.

- In late November 2007, Tropical Cyclone Sidr, the worst to hit Bangladesh in many years, came directly north from the Bay of Bengal to strike the low-lying coast. More than three million people evacuated. A combination of 250-kilometer-per-hour winds, a 5-meter storm surge, huge waves, and flooding due to torrential rain swept at least 3,500

people to their deaths. Coastal fishing villages were especially hard hit.

Catastrophic floods in Bangladesh are caused by several factors. The heavy monsoon rains from April through October are carried by warm, moist Indian Ocean winds from the southwest and magnified by the orographic effect of the air mass rising against the Himalayas. The rains are torrential and widespread in the drainage areas of the Ganges and Brahmaputra rivers, which come together in the broad delta region of Bangladesh. Rivers swell annually to twenty times their normal width. Those same drainage areas have been subjected to widespread deforestation and plowing of the land surface, which causes massive erosion and heavy siltation of river channels. The decrease in channel capacity causes the rivers to overflow more frequently.

In the delta region of Bangladesh, gradual compaction and subsidence of delta sediments coupled with gradual rises in sea level compound the problem. Relative sea-level rise compared with the delta surface raises the base level of the rivers, reduces their gradient, and raises the water level everywhere on

(continued)

the delta. Cyclone-driven storm surges can put virtually the whole delta—75 percent of the country—underwater in a few hours. Other deltas in southeast Asia are subject to similar hazards (compare FIGURE 10-54).

Given that people live on the sea-level delta of two major rivers and in the path of frequent monsoon floods and tropical cyclones, their options are few. The government plans to dredge and channelize more than 1,000 kilometers of riverbeds to improve navigation and build homes for flood victims on the new levees. Unfortunately, the impetus appears to be to improve food production rather than ensure safety and will do little to prevent the problem. Ironically, it is those same floods that bring new fertile soils to the land surface. Dredging will increase the chances of flooding downstream; and the new levees, built from those fine-grained materials dredged from the delta's river channels, will easily erode during floods. Levee breaches will lead to avulsion of channels (formation of new river pathways), widespread flooding, and destruction of fields. One useful suggestion is to construct high-ground refuges and provide flood warning systems.

Donald Hyndman.

AFP/Getty Images.

▶ **A.** *Houses along near-sea-level ground in the Mekong delta of Vietnam are just above water level even without a storm surge.*
B. *An Indian boy reaches his house near Guwahati, India, in the Brahmaputra flood of July 8, 2003.*

CO_2 Sequestration Underground
The Weyburn Sequestration Project ▶

The world's largest-scale experiment to test the viability of permanent underground storage is ongoing in southern Saskatchewan. The Weyburn sequestration project of the International Energy Agency pumps CO_2 under pressure into depleted oil reservoirs 1,450 meters below, both to permit recovery of any remaining oil and to permanently remove CO_2 from the atmosphere. Initial injection occurred in September 2000. Six thousand tons were injected per day, and by September 2007, fourteen million tons of CO_2 had been injected from a North Dakota coal gasification plant, pumped to Weyburn by a 320-kilometer pipeline. Apparent faults cut the essentially horizontal sedimentary layers, but their potential as paths for leaking fluids is unknown. Local deformation caused by injection of the high-pressure fluids complicates the situation by widening fractures or opening new ones. Review of the results indicates that the Weyburn site may be suitable for long-term storage of CO_2.

Another experimental sequestration project is in the Colorado Plateau of southeastern Utah and adjacent Colorado, New Mexico, and Arizona, an area that surrounds many large coal-fired power plants that generate enormous amounts of CO_2. In fact, such power plants are the largest contributors of atmospheric CO_2. The area consists of horizontal sedimentary rocks with structural domes that naturally trap CO_2. Some of the natural CO_2 has reacted with rock minerals to form solid carbonate minerals, thereby trapping the CO_2 in the solid form. Preliminary data and modeling are encouraging, suggesting that after 1,000 years, 70 percent of the CO_2 should remain trapped in the rock.

A third experiment pumped CO_2 into sandstones sealed against a salt dome near the Gulf Coast of Texas. The CO_2 remained well sealed but reacted with mineral grains to mobilize iron, manganese, and toxic elements, displacing them in the saltwater that originally filled the spaces between the grains. Such toxic materials could contaminate groundwater in such regions.

Given the amounts of CO_2 being produced, it is clear that new cleaner, higher-efficiency, coal-fired plants now being built will need to capture the CO_2 directly at the point of generation, then transport it to sites for permanent burial underground. The amounts being sequestered at these and other sites are very modest, far from the amounts that need capturing in the future. The major concerns are how long CO_2 will remain trapped and the contamination of groundwater. Another potential problem is the triggering of earthquakes by pressure from the fluids (see FIGURE 4-3).

Chapter Review

Key Points

Basic Elements of Climate and Weather

- Water continuously evaporates from oceans and other water bodies, falls as rain or snow, is transpired by plants, and flows through streams and groundwater back to the oceans. **FIGURE 10-1**.

- Rising air expands and cools adiabatically, that is, without loss of total heat. The rate of temperature decrease with elevation—that is, the adiabatic lapse rate—in dry air is twice the rate in humid air. **FIGURES 10-2** and **10-3**.

- Cool air can hold less moisture; thus, as moist air rises over a mountain range and cools, it often condenses to form clouds. This is the orographic effect of mountain ranges. **FIGURE 10-4**.

- Warm air rises, cools, and condenses to form an atmospheric low-pressure zone that circulates counterclockwise in the northern hemisphere. Cool air sinks, warms, and dries out to form a high-pressure zone that circulates clockwise in the northern hemisphere. Air moves from high to low pressure, producing winds. **FIGURE 10-5**.

- West-to-east rotation of Earth causes air and water masses on its surface to lag behind a bit. Because the rotational velocity at the equator is greater than at the poles, the lag is greater near the equator. This causes air and water masses to rotate clockwise in the northern hemisphere and counterclockwise in the southern hemisphere. **FIGURE 10-6**.

- The prevailing winds are the westerlies north of about 30 degrees north (wind blows to the northeast) and the trade winds farther to the south (blowing to the southwest). Cells of warm air rise near the equator and descend at 30 degrees north and south. **FIGURE 10-6**.

- Warm fronts (where a warm air mass moves up over a cold air mass) and cold fronts (where a cold air mass pushes under a warm air mass) both cause thunderstorms. Such fronts often intersect at a low-pressure cell. **FIGURES 10-8** to **10-9**.

- The subtropical jet stream meanders eastward across North America, across the interface between warm equatorial air and colder air to the north, and between high- and low-pressure cells. **FIGURE 10-10**.

Climatic Cycles

- Earth's climate has cycles from days and seasons to those that come thousands of years apart. **FIGURE 10-11**.

- Equatorial oceanic circulation normally moves from east to west, but every few years the warm bulge in the Pacific Ocean drifts back to the east in a pattern called El Niño, bringing winter rain to the west coast of equatorial South and North America, including southern California. **FIGURES 10-12** to **10-15**.

- The North Atlantic Oscillation is a comparable shift in winter atmospheric pressure cells that affects weather in the North Atlantic region. It also shifts every few years, but the times do not correspond to those of El Niño. **FIGURES 10-16** and **10-17**.

Hazards Related to Weather and Climate

- Because flooding depends on water in the atmosphere and tropical air contains ten times as much water as cold polar air, tropical air masses bring the wettest storms. Warm, moist, tropical air moves to much higher latitudes during the summer.

- Vegetation is abundant in wet climates, so rain falls on leaves and soaks slowly into the ground to feed groundwater and year-round streams. Lack of vegetation in dry climates permits rain to fall directly on the ground, where most of it runs off the surface. Especially heavy rainfall can cause floods, as can prolonged rainfall that saturates surface soil to prevent further rapid infiltration.

- The atmosphere can be cooled by particulates from sources such as industrial smokestacks, forest fires, and volcanic ash eruptions.

The Greenhouse Effect and Global Warming

■ Greenhouse gases such as carbon dioxide and methane trap heat in the Earth's atmosphere much as the glass in a greenhouse permits the sun to shine in but prevents most heat from escaping. **FIGURE 10-35**.

■ Atmospheric carbon dioxide and temperatures are increasing, especially since 1970. **FIGURE 10-38,** and **10-41**.

Consequences of Climate Change

■ Consequences of climate change include more frequent and stronger storms, smaller snowpacks and earlier runoff, drier vegetation and more fires, and warming and expansion of the oceans that leads to rise of sea level. **FIGURES 10-45** and **10-52** to **10-54**.

Mitigation of Climate Change

■ Reductions in greenhouse gas emissions into the atmosphere could be accomplished by increased conservation, improved efficiency of power production and use, regulation of emissions, carbon taxes and trading, subsidies and tax credits, capture and sequestration of greenhouse gases, and use of alternative fuels such as wind, solar collectors, and possibly nuclear power.

Key Terms

adiabatic cooling, p. 262
adiabatic lapse rate, p. 262
albedo, p. 295
Atlantic multidecadal oscillation (AMO), p. 272
blizzard, p. 284
cap-and-trade, p. 305
carbon cycle, p. 286
Chinook winds, p. 265
climate, p. 265
cold front, p. 265
Coriolis effect, p. 264
desertification, p. 280

drought, p. 274
dust, p. 277
dust storm, p. 277
El Niño, p. 269
evapotranspiration, p. 262
frostbite, p. 282
greenhouse gases, p. 286
groundwater, p. 262
global warming, p. 289
greenhouse effect, p. 285
heat-island effect, p. 281
high-pressure system, p. 264
hydrologic cycle, p. 262

hypothermia, p. 278
ice ages, p. 273
ice storms, p. 283
jet stream, p. 266
Kyoto Protocol, p. 305
La Niña, p. 269
lake-effect snow, p. 282
low-pressure system, p. 264
methane hydrate, p. 296
North Atlantic oscillation (NAO), p. 272
orographic effect, p. 262
relative humidity, p. 262

right-hand rule, p. 264
Santa Ana winds, p. 265
seasons, p. 268
sequestration, p. 306
trade winds, p. 265
warm front, p. 265
weather, p. 262
weather fronts, p. 265
westerly winds, p. 265
wind chill, p. 282

Questions for Review

1. If a humid air mass has 100-percent relative humidity and is 20°C at sea level, what would the temperature of this same air package be if it were pushed over a 2,000-meter-high mountain range before returning again to sea level? Explain your answer and show your calculations.

2. Explain the orographic effect on weather.

3. An area of low atmospheric pressure is characterized by what kind of weather?

4. Why do the oceans circulate clockwise in the northern hemisphere?

5. Explain the right-hand rule as it applies to rotation of winds around a high- or low-pressure center.

6. What is the main distinction between a cold front and a warm front?

7. What causes Earth's seasons, such as winter and summer?

8. What main changes occur in an El Niño weather pattern?

9. Why do streams flow year-round in a wet climate? Explain clearly.

10. How much has Earth's atmosphere increased in temperature in the last 1,000 years? When did most of the increase begin?

11. About how much has carbon dioxide increased in the atmosphere? When did the increase begin? Why?

12. What are the main sources of important greenhouse gases?

13. Other than an increase in temperature, what would be the most prominent changes in weather with global warming?

14. Approximately how much is sea level expected to rise in the next 100 years? What country is expected to see the largest loss of life as a result of rise in sea level? Why?

15. What is the main contributor to the rise in worldwide sea level? Be specific as to what makes the level rise—not just "global warming."

Discussion Questions

1. Because a large proportion of carbon dioxide, blamed for much of climate change, comes from heavy industry such as burning coal for electric power and producing cement from limestone, to what extent should factories be required to incur major costs to reduce emissions, thereby increasing costs to the consumer?

2. Some people advocate conservation of energy by better insulating houses, using less heat and air conditioning, and turning off lights and electric devices when not in use. How important, how easy is this, and how likely are people to go along with such conservation?

3. The United States and Canada, the largest carbon emitters per capita, have been reluctant to reduce their carbon emissions at a rate greater than their rate of economic growth. Is this reasonable and why or why not?

4. The graph "How do we Stabilize CO_2" (Figure 10-61), outlines the steps necessary to stabilize atmospheric CO_2 at present levels. How plausible is each of these stabilization measures and are they worth the economic cost?

5. Because even if we manage to apply all of the stabilization measures listed in #4, we cannot reduce the total CO_2 in the atmosphere, then is it worth all of the cost and suffering, should we even try?

6. Instead of all of these mitigation efforts, should we merely adjust to climate changes as they come?

7. Because a large part of the climate-change problem is overpopulation, some people have suggested that people everywhere should be limited to having no more than one child, as in China's one-child policy—in order to slowly reduce world population and minimize climate change?

Streams and Flood Processes

Karl Christians, Montana DNRC.

No, this house was not built overhanging the cut bank of the river. When it was built, it was ten meters from the bank of the Clark Fork River near Plains, Montana. Authorities finally burned the house before it fell into the river.

Too Close to a River

A resident of Plains, Montana, had just finished building a new house in a picturesque location on the outside of a big meander bend, next to the Clark Fork River. It had a great view of the river and the surrounding mountains. Although the site was on the floodplain, he thought it would be amply safe because it was about ten meters from the river bank.

In late May of 1997, rising water in the spring runoff rapidly eroded the steep bank on the bend of the river next to the house. The fast-moving water caused progressive caving of the bank until it undercut the edge of the house. The owner hired a company to move his house, but he was too late; before they could move it, erosion had progressed so far that the moving company was not willing to risk loss of its equipment on the unstable riverbanks. Finally, local authorities decided to burn the house rather than find it floating down the river in pieces. The situation was especially embarrassing because the owner was also the local disaster relief coordinator.

Stream Flow and Sediment Transport

What the homeowner in the preceding account didn't consider is that a river is not a fixed structure like a highway but is subject to natural processes, including course change and flooding. The first step to understanding flooding is to understand the natural processes by which rivers and streams transport water and sediment.

Rivers are complex networks of interconnected channels with many small tributaries flowing to a few large streams, which in turn flow to one major river. They follow valleys that they have eroded over thousands to millions of years. Rivers respond to changes in regional climate and local weather through the amount and variability of flow and to the size and amount of sediment particles supplied to their channels.

Stream Flow

Streams and rivers collect water and carry it across the land surface to the ocean. Streams in humid regions collect most of their water from *groundwater seepage*—water percolating down through porous soils to reach the stream. Flow generally increases in the downstream direction as additional water from tributary streams and groundwater enters the channel. Streams accumulate surface water from their **watershed** (or drainage basin), the entire upstream area from which surface water will flow toward a channel (**FIGURE 11-1**).

The **discharge** of a stream, or total volume of water flowing per unit of time, is the average water velocity multiplied by the cross-sectional area of the stream (**By the Numbers 11-1**: Total Flow of Stream). Because there is no easy way to measure the average velocity of water in a stream, *point velocities* are measured at equal intervals and depths across a stream channel, and each point velocity is multiplied by a cross-sectional area surrounding that point (**FIGURE 11-2A**). New instruments called *acoustic doppler current profilers* have been developed to more accurately approximate stream flow by measuring water velocity at hundreds of locations based on the shift in sound frequencies due to moving particles (see FIGURE 11-2b).

Sediment Transport and Stream Equilibrium

Rivers and streams carry sediment downstream along with water, eroding material in one place and depositing it in another. Streams change to maintain a **dynamic equilibrium** in which the inflow and outflow of sediment is in balance. A stream that is able to maintain this equilibrium is called a **graded stream**.

The cross section of a stream adjusts to accommodate its flow, as well as the sediment volume and the grain sizes

FIGURE 11-1 WATERSHED OF A STREAM

The watershed of this stream near Boise, Idaho, is outlined in blue. This area includes all of the slopes that drain water to feed the stream.

▶ By the Numbers 11-1

Total Flow of Stream

The total flow of water in a stream depends on the average velocity of the water times the cross-sectional area through which it flows:

$$Q = VA$$

where:

Q = discharge or total flow (m³/sec)

V = average velocity (m/sec)

A = cross-sectional area (m²) = width (m) × depth (m)

supplied to the channel. The geometry of a channel cross section is controlled by flow velocities and the associated ability of a stream to carry sediment. Most streams are wide and shallow, with nearly flat bottoms, but the cross section of a stream adjusts based on the erodibility of the bottom and banks and the nature of the transported sediment. Streams flowing through easily eroded sand and gravel at low flow generally have steep banks and broad, nearly flat

FIGURE 11-2 MEASURING STREAM FLOW

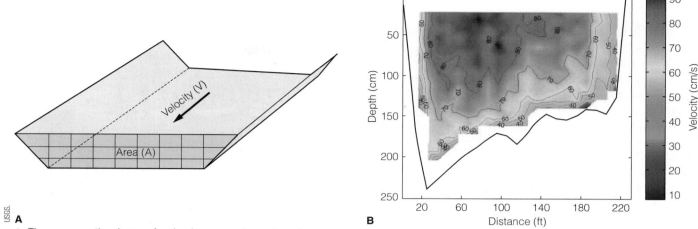

A. The cross-sectional area of a simple stream channel can be approximated by dividing it into a rectangular grid. With individual velocity measurements for each such box, the total flow would be the sum of $V_1 \times A_1 + V_2 \times A_2 + \ldots$, where A_1 and V_2 are the areas of individual boxes.
B. Stream flow through the San Joaquin River in California was measured using an acoustic doppler current profiler. The stream flow is then calculated by summing the velocity of each cell by the cross-sectional area of that cell.

bottoms. Streams flowing through bedrock or fine silt and clay tend to be narrow and deep because these materials are less easily eroded.

A stream also adjusts its gradient in response to water velocity, sediment grain size, and total sediment load in order to be able to transport its supplied sediment over time. The **gradient** of a stream, or its channel slope, is the steepness with which it descends from its highest elevation to its lowest, typically expressed in meters per kilometer. Most streams begin high in their drainage basins, surrounded by steeper slopes and often by harder, less easily eroded rocks. Coarser grain sizes, such as those supplied to a stream from the steeper slopes of mountainous regions, require a steeper gradient or faster water to move the grains (**FIGURE 11-3**). There the stream moves the rocks and sediment on down the valley. The greater discharges and smaller grain sizes that exist downstream lead to gentler slopes in those channels. Thus, the gradient generally decreases downstream as sediment is worn down to smaller sizes and the larger flow there is capable of transporting the particles on a gentler slope. Ultimately, the stream will reach a lake or the ocean, a **base level** below which the stream cannot erode.

Where a tributary stream descends from a steeper gradient in mountains onto a broad valley bottom, it leaves the narrow valley that it eroded to reach a local base level of a larger valley. The rapid decrease in its gradient causes it to drop much of the sediment it was carrying. Thus where the slope decreases, the stream changes from an erosional mode, where it is picking up sediment, to a depositional

FIGURE 11-3 GRADIENT CHANGE

Coarse gravel brought in by the small tributary on the left creates steepening and rapids in the Wenatchee River, Washington.

mode, where it is depositing sediment. The excess sediment may spread out in a broad fan-shaped deposit called an **alluvial fan**. The term *alluvial* implies the transport of loose sediment fragments, and *fan* refers to the shape of the deposit in map view. Similarly, where a river reaches the base level of a lake or ocean, the abrupt drop in stream velocity at nearly still water causes it to deposit most of its sediment in the form of a **delta**. The delta is like an alluvial fan except that the delta sediments are deposited underwater.

Sediment Load and Grain Size

Eroding riverbanks and landslides can supply a stream with particles of any size from mud to giant boulders. The velocity and volume of the flow limit both the size and the amount of sediment that can be carried by the stream. Empirical curves can be used to estimate the maximum particle size that can be picked up or transported for a given water velocity (**FIGURE 11-4**). Note that the velocity required to mobilize particles is appreciably greater than that needed to transport the same particles.

The relationship of larger grain sizes requiring higher velocities for movement does not hold where fine silt and clay particles make up the streambed. In that case, the fine particles lie entirely within the zone of smoothly flowing water at the bottom and do not protrude into the current far enough to be moved. Clay-size particles also have electrostatic surface charges that help hold them together.

In general, coarser particles in a stream channel provide greater roughness or friction against the flowing water (**By the Numbers 11-2:** Velocity in Channel). Thus, coarser particles also slow the water velocity along the base of a stream. For this reason, mountain streams with coarse pebbles or

▶ By the Numbers 11-2

Velocity in Channel

The velocity multiplied by the channel roughness is proportional to the average water depth of the channel multiplied by the square root of its slope:

$$V n = 1.49R^{2/3} s^{1/2}$$

where:

R = hydraulic radius is proportional to average water depth

n = Manning roughness coefficient:

  ≈ 0.03 for straight, small streams or grassy floodplains with no pebbles

  ≈ 0.05 for sinuous small streams with bouldery bottoms or floodplains with scattered brush

  ≈ 0.10 to 0.15 for brushy flood zones or floodplains with trees

s = slope of the channel

boulders in the streambed often appear to be flowing fast but actually flow more slowly than most large, smooth-flowing rivers such as the Missouri and Mississippi. Note that the water velocity depends on increases with water depth and slope (**FIGURE 11-5**).

Although a stream is capable of carrying particles of a certain size, there can still be a limit to the volume of sediment—or its **load**—a stream can carry. Large volumes of sand dumped into a stream from an easily eroded source or sediment provided by a melting glacier will overwhelm its carrying capacity. The excess sediment will be deposited in the channel.

The suspended load, or carrying capacity, of a stream depends upon the discharge (**By the Numbers 11-3:** Carrying Capacity of a Stream). As a result, the load of a stream increases during a flood.

Flooding is a natural part of the process by which rivers move sediment and maintain equilibrium. When peak flood velocity and depth are high enough to develop significant turbulence, the erosive power of a stream becomes very large, temporarily increasing both the size and volume of sediment it can carry.

At low water, virtually all of the material brought into the stream stays put, backing up the water behind it. When water rises in the stream, such as during flood, it has a higher velocity and thus can carry more sediment particles, flushing the accumulated sediment downstream.

Flooding also increases the size of sediment particles a stream can move. During floods, coarser particles are gradually broken down to smaller-sized particles in the channel and are then flushed downstream.

FIGURE 11-4 STREAM VELOCITY AND EROSION

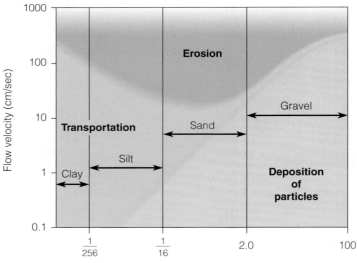

This diagram shows the approximate velocity required to pick up (erode) and transport sediment particles of various sizes. Note that both axes are log scales, so the differences are much greater than it seems on the graph.

FIGURE 11-5 STREAM SLOPE AND GRAIN SIZE

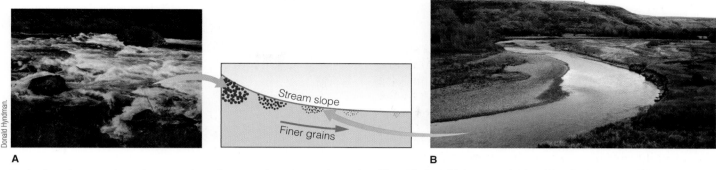

A

B

Grain sizes decrease downstream as slope decreases in a stream channel. **A.** The turbulent, high-energy Lochsa River in the mountains of northeastern Idaho has a steeper gradient and carries larger grains. **B.** The Smith River south of Great Falls, Montana, has a lower gradient and carries finer particles.

By the Numbers 11-3

Carrying Capacity of a Stream

Leopold and Maddock demonstrated that carrying capacity is proportional to discharge:

$$L \propto Q^n$$

where:

L = suspended load transport rate (cm³/sec)

Q = discharge (cm³/sec)

n = an exponent, generally between 2.2 and 2.5

Similarly, if coarser material is added to the channel, such as from a steeper tributary or a landslide, it accumulates in the stream channel until a large flood with sufficient velocity occurs or the gradient of the channel increases enough to move that size of material (**FIGURE 11-6**).

Channel Patterns

The way in which streams pick up and deposit sediment also determines the pattern of the channel and the way the channel moves over time, which in turn determines the type of flooding characteristic of each type of stream.

Meandering streams, which sweep from side to side in wide turns called meanders, are most common. Multichannel **braided streams** are much less common, and naturally straight streams are rare. Meandering streams are more typical of wet climates with their finer-grained sediments, and braided streams are common in dry climates with abundant coarse sediment. The patterns of most rivers fall somewhere in between meandering and braided. Even within one river, some reaches may meander and others

FIGURE 11-6 ACCUMULATION OF BOULDERS IN A STREAM

Giant granite boulders dumped by a landslide into the Feather River in the Sierra Nevada range of California can be moved only in an extreme flood.

may braid, depending on the erodibility of the banks and the amount of supplied sediment. **Bedrock streams** are fast moving, high energy streams that occur in steep, mountainous areas where the streambed is solid rock.

Meandering Streams

Meanders in a stream are created when an irregularity such as a boulder or tree root diverts water toward one bank. As the water swings back into the main channel, it sweeps toward the opposite bank, much like a skier making slalom turns. This deep and highest-velocity part of the stream is known as the *thalweg*. The flow erodes the banks along the outside of these turns to carve out the meanders. The flow preferentially erodes the outside of meander bends

FIGURE 11-7 PATTERNS OF EROSION AND DEPOSITION IN A MEANDERING STREAM

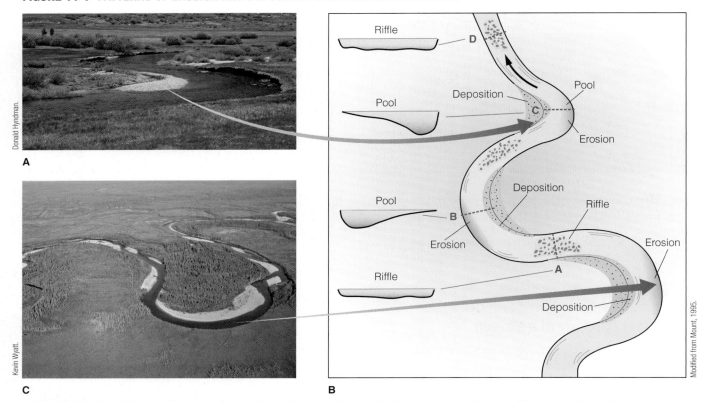

A. The Carson River in Nevada illustrates the eroding cut bank on the outside of meanders and the depositional gravelly point bar on a meander's inside. Flow is toward the right. **B.** Cross sections of a typical meandering stream channel from a riffle at A, downstream through pools at B and C, then finally through a riffle at D. Note that the river erodes on the outside of the meander bends where it has a deep channel, and it deposits on the inside of bends where it has a shallow channel. **C.** Deepwater channels on outside of meander bends and prominent point bar deposits on the inside of meander bends along Beaver Creek, north of Fairbanks, Alaska.

because the water has forward momentum that drives it into the bank, where the higher velocity water can mobilize more sediment. As the water rounds the corner of a bend, it slows, and sediment is then deposited as a **point bar** downstream along the inside corner of the bend. The flow commonly alternates between deep pools on the outside bends, where sediment has been eroded, and shallow riffles between there and the next outside bend, where sediment has been deposited (**FIGURE 11-7**).

Over a long time, continued erosion of the outside of meander bends causes the meanders to deepen and migrate. Meanders sometimes come closer together until floodwater breaks through the narrow neck separating them. The abandoned meander is called an **oxbow lake** (**FIGURE 11-8**). When a stream cuts across a meander bend, either naturally or by human interference, the new stretch of channel follows a shorter path for the same drop in elevation. The steepening of the channel slope increases water velocity, eroding finer particles and often leading to rapids. Engineers commonly add rock rubble, or **riprap**, along the banks of a straightened portion of a stream to diffuse the stream's energy and minimize erosion.

FIGURE 11-8 OXBOW LAKE

Meanders in this river near Houston, Texas, eroded the outsides of bends and migrated until one meander bend spilled over to one farther downstream, leaving an abandoned oxbow lake in the center of the photo.

Relative Proportions of Meandering Streams

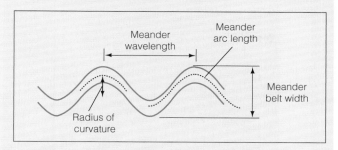

- **Meander wavelength** ≈ 12 × channel width or 1.6 × meander belt width or 4.5 × radius of curvature
- **Meander belt width** ≈ 2.9 × radius of curvature

The size and shape of river meanders follow some general relationships (**By the Numbers 11-4:** Relative Proportions of Meandering Streams). These relative proportions are maintained regardless of stream size, whether it is a small stream, two meters across, or the lower Mississippi River, 1,000 meters across. Thus, artificial attempts to change a channel by narrowing it or straightening it will be met by the river's attempts to return to a more natural equilibrium channel cross section and meander path. If we straighten a river channel, to accommodate a nearby road, for example, or if we riprap a channel bank to hinder bank erosion, the stream will adjust its path to maintain its natural proportions—often causing new problems for those who would make those changes.

Meandering streams flood in a typical way as a result of their patterns of erosion and deposition. Over the course of time, meanders erode outward and slowly migrate downstream. That process of erosion and deposition over a period of centuries gradually moves the river back and forth to erode a broad valley bottom. At high water, the flooding river spills out of its channel and over that broad area—its **floodplain** (**FIGURE 11-9**).

Think back to the vignette that opened this chapter, about the homeowner who built his house ten meters from a river but was surprised to find the river right at his doorstep one day. He lacked two key pieces of information about the behavior of meandering rivers when he built his home at the outside of a meander bend: he hadn't considered that meanders erode outward and move over time or that the rising water at spring runoff would speed up the erosion process.

Braided Streams

Braided streams do not meander but form broad, multi-channel paths. These streams are overloaded with sediment, which they deposit in the stream channel, locally

FIGURE 11-9 FLOODPLAIN

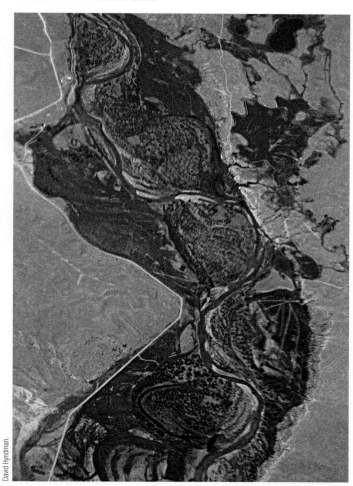

David Hyndman.

The darker floodplain of this northern Great Plains river was carved out as shifting meanders of the stream gradually widened the floodplain. A major flood would fill the floodplain wall-to-wall.

clogging it so that water must shift to one or both sides of the deposit. With further deposition, water shifts to form new channels. That behavior is promoted by dry climates in which little vegetation grows to protect slopes from erosion. Such braided streams are characterized by eroding banks, a steep gradient, and abundant stream *bedload*, which is the sediment carried along the stream bottom. With waning flow, bedload is deposited in the channel and flow diverges around it, splitting into separate channels. Depending on flow, some channels dry up, while others temporarily become dominant (**FIGURE 11-10**, p. 326).

Braided channels are characteristic of meltwater streams flowing from sediment-laden glaciers and of arid Basin and Range valleys in which heavy, intermittent rains from mountains carry abundant sediment across valley alluvial fans. When a stream channel moves from an area of high slope to one of lower slope, sediment will generally deposit

FIGURE 11-10 BRAIDED RIVERS

A. This braided river channel of the Wairou River is southwest of Blenheim, New Zealand. **B.** The strikingly braided Tanana River near Fairbanks, Alaska.

FIGURE 11-11 ALLUVIAL FAN

Eroding pile

Depositional fans

Water runs off this steep pile of sand in heavy rains despite its high permeability. Overland flow erodes gullies and carries sediment down onto depositional fans. Note that water drains into the gullies on the eroding area but spreads out over wider areas on the depositional zone.

FIGURE 11-12 HAZARD AREAS OF ALLUVIAL FANS

Moderate Hazard

Lower Hazard

The highest hazard area is at the apex of the fan, where the flow concentrates. The hazard becomes less as the flow spreads out, decreasing in depth and intensity downslope. However, the path of the flow may change to concentrate on different parts of the fan as it finds the easiest way down the slope. Shown here is Copper Canyon, Death Valley.

in an alluvial fan because the stream can no longer carry its full load (**FIGURE 11-11**).

Alluvial fans are somewhat arched, with higher elevations along their midlines and lower elevations toward their sides, because they are built by deposition of sediment flushed out of a mountain canyon at the apex of the fan. During a flood, all of the flow from a canyon concentrates at the apex of the fan and then spreads out, decreasing in depth and intensity downslope (**FIGURE 11-12**). At different times the flow might concentrate on different parts of the fan as it finds the easiest path down the slope.

Active alluvial fans are always marked by braided streams. Like other braided environments, most alluvial fans tend to be in dry climates and lack significant vegetation. People tend to build on alluvial fans because they are inviting areas, with gentle slopes at the base of steep, often rocky, mountainsides. They typically do not realize that the fan is an active area of stream deposition because the streams rarely flow.

In fact, alluvial fans are particularly dangerous flooding areas. Torrential floods can wash out of the canyon above with little or no warning. They destroy, bury, or flush houses and cars downslope. Deaths are common, especially during rainstorms at night, when people do not hear the telltale rumble of an approaching flood.

Bedrock Streams

A stream is called a bedrock stream when it has eroded away all of the easily transportable sediment to reach resistant rock. Sections of bedrock streams tend to abruptly steepen. These abrupt changes in gradient from gentle to steep are called *knickpoints*. Upstream, the gentle gradient has low energy for the river and may deposit sediment. Downstream, the steeper channel provides high energy, and sediment erodes.

High-gradient bedrock channels generally have deep, narrow cross sections that carry turbulent and highly erosive flows during floods (**FIGURE 11-13**). Their high energy

FIGURE 11-13 BEDROCK STREAMS

A. The turbulent, high-energy Colorado River in Grand Canyon, Arizona.
B. A steep, bedrock channel of the Shotover River in New Zealand cleans out any loose material during every flood.

FIGURE 11-14 POTHOLES

Potholes in the streambed of McDonald Creek in Glacier National Park were probably formed by rocks and boulders transported during a flood.

and turbulence allow them to transport all of the loose material in the channel. This can include large boulders, which impact and abrade the channel sides. Bedrock is resistant to erosion, so only major floods can scour or pluck fragments from the rock.

Floods in bedrock channels have excess energy that cannot be dissipated by increasing channel roughness or greater loads of sediment transport. Extreme turbulence is typical, and large-scale vortexes or whirlpools can appear. These can be effective in swirling rocks on the bottom to drill potholes in the bedrock (**FIGURE 11-14**). Recreational rafting or boating at high water in such channels, although exciting, is especially hazardous because of the extreme turbulence and vortexes. Flotation devices can be ineffective at raising a person caught in a whirlpool.

Groundwater, Precipitation, and Stream Flow

Atmospheric moisture in the form of rain is the main source of water for streams. Water stored in snow and ice is a secondary reservoir. After a rainfall, some rainwater evaporates from the ground and vegetation surfaces, some is taken up by vegetation, and some soaks into the ground. Rivers and streams collect the water that percolates down through the soil to groundwater.

The amount of precipitation on the land surface varies according to region and season, and there can also be differences year to year. Surface water interaction with groundwater is determined by these climatic factors.

FIGURE 11-15 GAINING AND LOSING STREAMS

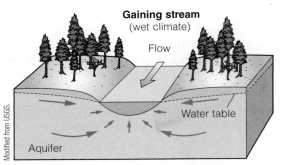
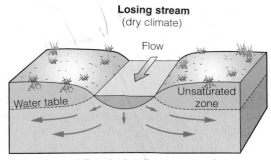

In wet climates, groundwater flows into gaining streams, ensuring year-round flow. In dry climates, water from streams feeds the groundwater. These losing streams may dry up between rainstorms.

Gaining streams are fed by groundwater, whereas **losing streams** lose water into the ground (**FIGURE 11-15**).

In areas of moderate to high annual rainfall, groundwater levels stand higher than most streams and thus continuously feed them, ensuring a year-round flow. Stream flow in more humid regions is high during wet weather periods and low during dry weather periods. During low-water periods, the stream surface generally sits at the groundwater surface, which is the exposed water table. The rate at which water flows from groundwater into a stream depends upon both the slope of the water table and the ease of flow through the water-saturated sediments or rocks. Storm discharge into a stream includes this groundwater flow plus any overland flow. New groundwater does not need to travel all the way from its infiltration source during a storm to a river before the stream flow increases—it merely has to raise the water table at the infiltration source. That increases pressure on the groundwater and displaces "old water" farther downslope into the stream. Groundwater in such areas responds slowly to changes in rainfall, and inflow to streams from groundwater is gradual and relatively constant. Thus, wet climates tend to provide streams that flow year-round and flood in prolonged heavy rains.

In semi-arid to arid regions, however, losing streams shed water into the ground and may even dry up between storms. When they do flow, they drain water into the ground and therefore raise the water table. In areas with low annual rainfall, little or no vegetation can grow to soften the impact of raindrops and slow their infiltration. Rain falls directly on the ground, packing it tightly, which permits less infiltration and causes more surface runoff. The rain also kicks up sediment from the surface, permitting it to be carried into streams. During dry seasons, less water gets into the ground, even during the less frequent rainstorms, so less of it feeds the groundwater (**FIGURE 11-15B**). Dry climates with no year-round streams can see flash floods after any major or prolonged rainfall.

Streams in deserts generally flow only during and shortly after a rainstorm but then dry up until the next storm.

Because more sediment is supplied to the dwindling amount of water, sediment deposits in the gullies. This progressively chokes the flow, causing some of the water to spill over and follow another path. This gully in turn fills with sediment, and so on. The result is a braided alluvial fan that continuously deposits sediment that builds with time.

Precipitation and Surface Runoff

Typically, rainwater slowly soaks into the ground and feeds streams through groundwater. During torrential rainfall, however, some water may flow across the ground as **surface runoff** directly into streams. The intensity of precipitation plays a significant role in the rate of runoff to streams and, in turn, floods. Light precipitation can generally be absorbed into the soil without surface runoff. Heavy precipitation can overwhelm the near-surface permeability of soils, leading to rapid runoff over the surface, a process called *overland flow*. The water that soaks into the soil raises the local groundwater level; that adds pressure to the groundwater and forces more water back downslope through the ground and into the streams.

The ability of the ground to absorb rainwater also depends on the permeability of the soil and the extent to which it is already saturated with water. Highly permeable soils may absorb heavy rainfall better than impermeable soils absorb much lighter rainfall. Rapid flood peaks during large storms are most common in areas with fine-grained soils or desert soils, especially tight clay hardpan or soils with a shallow, nearly impervious calcium carbonate-rich layer. The same is true of areas with near-surface bedrock or shallow groundwater that has little capacity to absorb rainfall (**Case in Point:** Heavy Rainfall on Near-Surface Bedrock Triggers Flooding— Guadalupe River Upstream of New Braunfels, Texas, 2002, p. 345). At the other extreme, decomposed granite soils dominated by coarse sand have high permeability and high infiltration capacity, which can cause stream levels to rise rapidly. Flat areas that contain many depressions can temporarily store enough water to delay runoff.

Even where the permeability of the soil is relatively high, heavy rainfall, especially over a long period or in multiple storms, can saturate near-surface sediments, thus forcing the water to flow over the surface to streams. The ground is also less able to absorb rainfall when it is frozen. In both of these cases, most of the water that reaches the ground surface runs off directly to a stream. For example, in June, 1972, when Hurricane Agnes dumped 5 to 7 centimeters of rain in Pennsylvania on ground already saturated with water, large floods occurred. In 1993, prolonged rains over-saturated soils in a broad area of the upper Mississippi River valley, leading to disastrous floods.

Some regions are more susceptible to heavy rainfall than others. Because rainfall depends on moisture in the atmosphere, heavy rainfall tends to develop where moist tropical or subtropical air over an ocean moves onto land or rises against a coastal mountain range. Or the moist air mass may collide with a cold front, where it rises and condenses. In North America, the Gulf Coast and southern Atlantic coast have those characteristics. On the west coast, from northern California through southern British Columbia, the westerly winds bring moist Pacific air eastward to collide with coastal mountain ranges, where these air masses rise and shed their rain. Coastal parts of southern California and adjacent Mexico, in the belt of trade winds, which blow from dry land areas out over the ocean, are generally very dry. During El Niño, however, the trade winds weaken, and moist Pacific air comes ashore. California then sees repeated heavy rainstorms that cause floods and landslides.

Intense precipitation often accompanies major storm systems. Tropical cyclones, including hurricanes, move westward in the belt of trade winds. In the Atlantic Ocean, they drift westward and then northward or eastward along the eastern fringe of the United States, where they interact with mid-latitude frontal systems. Reaching the Atlantic coastal plain, Hurricanes Camille in 1969 and Agnes in 1972, for example, caused significant flooding, with 25 to 50 centimeters of rain in many areas. Heavy rainfall accompanies thunderstorms, which last for a few minutes or an hour or two. A line of thunderstorms may prolong the deluge for several hours, whereas tropical cyclones may stretch out the rains for several days. Areas such as Southeast Asia, which have seasonal monsoons, may experience extreme rains for several weeks or even months.

At high elevations, much of the winter precipitation falls as snow. This stores the water at the surface until snowmelt in the spring. Regions north of the influence of warm, moist, tropical airflow are affected by snowpack on the ground for more than a month in most years. If the snowpack melts gradually, much of the water can soak into the ground. If it melts rapidly, either with prolonged high temperatures or especially with heavy warm rain, large volumes of water may flow off the surface directly into streams. When the snowpack warms to 0°C, a large proportion of meltwater remains in the pore spaces and the snowpack is "ripe." Water draining through the pack concentrates in channels at the base of the snow. Water has a large heat capacity, so dispersal of warm rainwater effectively melts snow. If the ground is already saturated with snowmelt, water from a rainstorm may entirely flow off the surface. If the ground is frozen, as is sometimes the case with minimal snowpack, heavy rains quickly run off the surface because the water cannot soak into the ground.

Flooding Processes

Flooding occurs when the amount of water entering a stream, for example from rising groundwater or surface runoff, causes the level of the stream to surpass the capacity of its channel.

The *bankfull* level of a stream is the height at which the water reaches the highest level of its banks, which typically occurs every 1.5 to 3 years. Larger, more infrequent floods fill the channel and spill out over its floodplain. It is not simply the size of the flow that makes a flood, but how the flow compares to the normal capacity of that particular channel. Large rivers can have large flows without being above flood level. Small streams can flood with fairly small flows.

Changes in Channel Shape during Flooding

Patterns of erosion in a streambed can change dramatically during flooding. An increase in discharge during a flood involves an increase in water velocity, water depth, and sometimes width of a stream. This happens because as water depth and velocity increase during a flood, shear or drag at the bottom of the channel increases. That extra shear picks up more sediment, increasing erosion (**FIGURE 11-16**, p. 330).

Channel scour, the depth of sediment eroded during floods, affects the shape of the stream channel and distribution of sediment. The grain size a stream can carry is proportional to its velocity; thus, rising water first picks up the finest grains, then coarser and coarser particles. Sediment is carried in suspension as long as grains sink more slowly than the upward velocity of turbulent eddies. Fine sediment is first picked up in eddies. At higher flows, pebbles or boulders may tumble along the bottom and even be heard as they collide with one another. This causes more erosion, deepening the channel.

As water velocity increases, the water drags much more strongly against the bottom. Increase in frictional drag on the stream bottom provides more force on particles on the streambed and thus more erosion. That friction also slows down the water (**By the Numbers 11-5:** Drag on Stream Bottom).

As a flood flow wanes, the coarser sand and gravel in suspension progressively drop out, thereby raising the streambed (**FIGURE 11-17**, p. 330). Thus, as water level rises, the stream begins eroding and the water gets muddy; as the level falls, the sediment in transit begins to deposit. For this

FIGURE 11-16 STREAM CHANNEL AT VARIOUS FLOWS

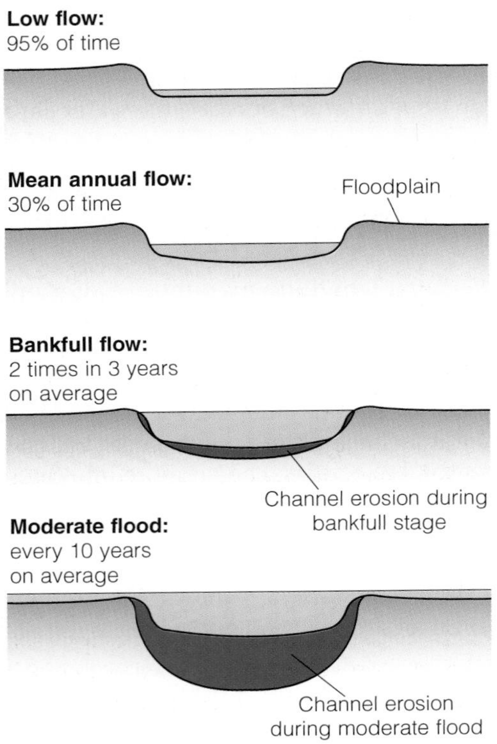

Low flow:
95% of time

Mean annual flow:
30% of time

Floodplain

Bankfull flow:
2 times in 3 years
on average

Channel erosion during
bankfull stage

Moderate flood:
every 10 years
on average

Channel erosion
during moderate flood

The shape of a channel changes with the level of flow. The greater the flow, the more erosion occurs to deepen the channel and increase its capacity.

► By the Numbers 11-5

Drag on Stream Bottom

Drag or total friction on the stream bottom is proportional to velocity squared:

$$\tau_0 \; \alpha \; v^2$$

where:

τ_0 = friction

v = velocity

---

reason, a cross section of a flood deposit shows the largest grains at the bottom, grading upward to finer sediments.

As a stream spills over its floodplain, it changes from a deep, high-velocity channel to a shallow, broad, low-velocity channel. As water velocity slows at the edge of the deeper channel, sediment deposits to form a **natural levee** (**FIGURE 11-18**). These features form a nearly continuous low ridge along the edge of the channel that may keep small floods within the channel. The floodplain, with its relatively

---

FIGURE 11-17 CHANNEL CHANGES DURING FLOODING

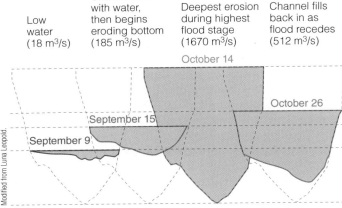

On September 9, 1941, the San Juan River near Bluff, Utah, was at its normal depth. With early water rise (September 15), sediment from upstream deposited to raise the channel bottom. As water rose further to the maximum level (October 14), the channel eroded to its deepest point. As the flood level waned (October 26), sediment again deposited to raise the channel bottom.

---

slow moving water, is part of the overall river path; it carries a significant flow during floods.

Stream channel changes are insignificant with normal flows or even small floods, regardless of their frequency. Significant changes occur only when flow reaches a threshold level that mobilizes large volumes of material from the streambed and channel sides. Streams in humid regions adjust their channels to carry the typical annual flows that fill them. Channels in semiarid regions adjust their channels to less frequent large floods because smaller flows do not significantly affect channel shapes.

Flood Intensity

The destructive effect of a flood depends primarily on its intensity. The intensity of a flood can be measured by the discharge of floodwater and the water's rate of rise. Flood intensity varies over time according to the rate of runoff, the shape of the channel, distance downstream, and the number of tributaries it has. For example, floods in small, narrowly confined drainage basins are typically much more violent than those along major rivers such as the Mississippi.

Rate of Runoff

The intensity of flooding is related to how much water flows in a stream over a period of time. Anything that increases the rate of runoff will cause more intense flooding. For example, there will be more runoff in areas where the ground is not very permeable and more where it is frozen

FIGURE 11-18 NATUREAL LEVEES

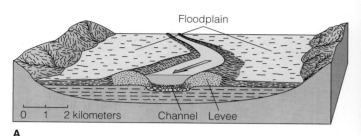

Floodplain

Channel Levee

0 1 2 kilometers

A

B

Marli Miller.

A. The main channel of a river has the coarsest gravels at the bottom, grading to finer grains above; the natural levees are still finer grains settled out in shallower water; the floodplain consists of very fine-grained muds that settled out from almost still water during floods. **B.** Houses built on natural levees along a channel and floodplain in the Mississippi River delta.

or already saturated with water. Rapid large runoff can develop in urban settings with large areas of pavement, houses, and storm sewer systems, or in natural areas that have been deforested. Deforestation can increase the volume of storm runoff by roughly ten percent.

We can depict flood intensity graphically in a **hydrograph**. A typical flood hydrograph rises steeply to the flood's crest, where the flood reaches its peak discharge, and then falls more gently (**FIGURE 11-19**). In areas where

FIGURE 11-19 URBANIZATION AND FLOODING

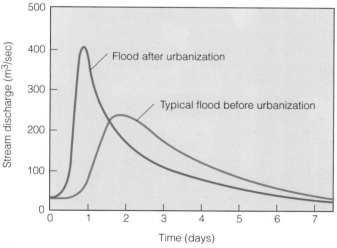

Flood after urbanization

Typical flood before urbanization

This hydrograph is a plot of stream discharge versus time for a similar eighteen-hour rainfall event for the same area before and after urbanization. Note that the area under the two curves is similar, that is, approximately the same total volume of water for both floods. Actually, because less water infiltrates, the flood volume after urbanization will be a little larger.

the rainfall saturates the soil and is forced to run over the surface to rapidly feed streams, the hydrograph peaks much more quickly and rises to a greater maximum discharge.

A **flash flood**, which comes on suddenly with little warning, is a flood with a very steep hydrograph. Any type of flood can be dangerous, but flash floods are especially so because they often appear unexpectedly, and water levels rise rapidly (**Case in Point:** A Flash Flood from an Afternoon Thunderstorm—Big Thompson Canyon, Northwest of Denver, p. 346). A map of flash-flood hazard tendency for the United States shows that high flash-flood danger areas are primarily located in the semiarid Southwest—Southern California to western Arizona and West Texas (**FIGURE 11-20**, p. 332). Moderate flash-flood dangers exist in areas such as the eastern Rocky Mountains, the Dakotas, western Nebraska, eastern Colorado, New Mexico, and central Texas. This is not to say that other areas are not prone to flooding; rather, the floods there tend to be less extreme compared with normal stream flows.

Deaths occur often in flash floods because of the little warning they provide and their violence. Even under a clear blue sky, floodwaters may rush down a channel from a distant storm. On many occasions, people have been caught in a narrow, dry gorge because they were not aware of a storm far upstream. At night, people in their homes have been swept away.

Stream Order

The number of tributaries of a stream (its **stream order**) has a significant effect on the rate of rise of floodwaters during and following a storm. Small streams that lack tributaries are designated *first-order streams* (**FIGURE 11-21**, p. 332). First-order streams join to form *second-order streams*; second-order streams join to form *third-order streams*; and so on.

Low-order streams tend to respond rapidly to storms with steep hydrographs because water has to travel only a short

FIGURE 11-20 FLASH-FLOOD HAZARD

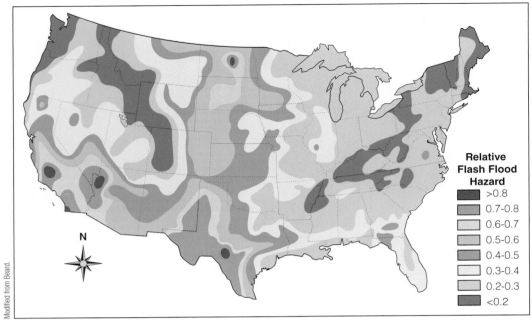

Modified from Beard.

Regions of the United States with higher numbers (orange areas) are most susceptible to flash floods. Notice that flash floods are most severe in the driest regions of the country.

FIGURE 11-21 STREAM ORDER

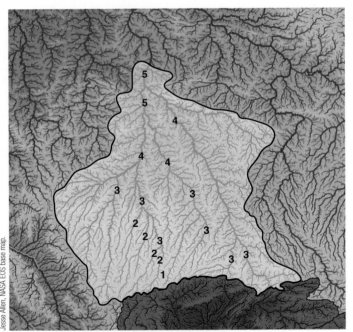

Jesse Allen, NASA EOS base map.

First-order streams have no tributaries but join to form a second-order stream, and so on. The watershed is outlined in black.

distance to the stream. Such streams provide less flood warning time for downstream residents. They have smaller drainage basins and carry coarser and larger amounts of sediment for a given area.

A storm in a headwaters area may cause flooding in several first-order streams. As the **flood crest** of each moves downstream, the length of time it takes to reach the second-order stream varies, so each first-order flood crest arrives at a somewhat different time. The flood peak for the second-order stream will therefore begin later and be spread over a longer period. The same goes for several second-order streams coming together in a third-order stream; its flood peak will again begin later and be spread over a still longer period. Thus, high-order streams with numerous tributaries have longer lag times between storms and downstream floods; their hydrographs are less peaked and cover longer time periods. Flood warning time for downstream residents is longer.

Downstream Flood Crest

Even with local intense rainstorms on a small drainage basin, there will be a lag between a storm and the resulting flood peak. A torrential downpour may last for only ten minutes, but it takes time for the water to saturate the surface layers of soil and to percolate down to the water table. More time is required for overland flow to collect in small gullies

FIGURE 11-22 DOWNSTREAM FLOODING

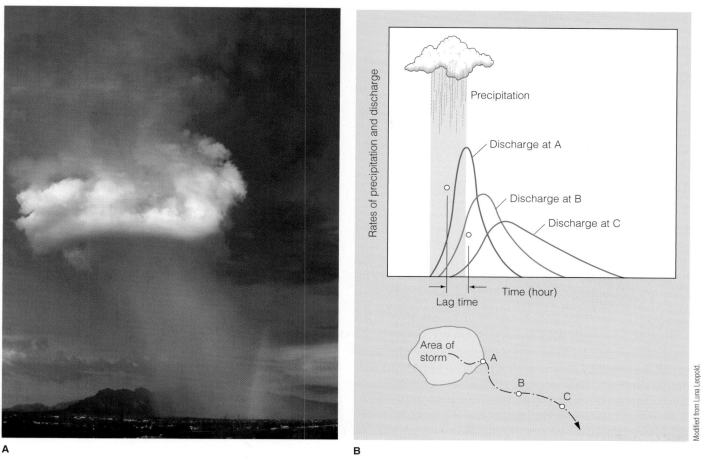

A. Localized afternoon rainfall over Tucson, Arizona. **B.** Storm rainfall entering a stream precedes the flood crest that it causes. The flood hydrograph nearest the rainfall area is highest and narrowest. Farther downstream at location B, the flood hydrograph crests at a lower level but lasts longer. Still farther downstream at C, the flood crests at an even lower level and lasts longer.

and for water in those gullies to flow down to a stream. In turn, it takes time for the water in small streams to combine and cause flooding in a larger stream. The length of the lag time depends on many factors, including slope steepness, basin area and shape, spacing of the drainage channels, vegetation cover, soil permeability, and land use.

Flood intensity depends on similar characteristics. If a storm occurs only in an upstream portion of a watershed, the peak height of the hydrograph will be lower and the flood duration will be longer farther downstream from the storm area (**FIGURE 11-22**).

At the downstream edge of the main rainfall area, water levels rise as water flows in from slopes, tributaries, and upstream. The level often continues to rise until about the time the rainfall event stops (location A in FIGURE 11-22b). At that time, water levels on the upstream slopes begin to fall, and flood level at that point in the stream peaks and starts

receding. Farther downstream, some of the earlier rainfall has already begun to raise the water level. Water continues to arrive from upstream, but the size of the stream channel is larger downstream because it has adjusted to carry the flow from all upstream tributaries. Thus, the flood flow fills a wider cross section to shallower depth (location B). Even farther downstream, the channel is still larger and the flood crest is still lower (location C). In cases where the precipitation occurs over the whole drainage basin, flows will increase downstream.

For flood hazards, we are normally more concerned with the maximum height of the flood crest than when the first water arrives from a storm. The flood crest moves downstream more slowly than the leading water in the flood wave. Note in FIGURE 11-22b that the first rise in floodwater from the storm appears downstream later at locations A, B, and C. Note that the flood crest, or peak discharge, also takes longer to move downstream. In general, the flood crest or flood wave moves downstream at roughly half the speed of the average water velocity.

Flood Frequency and Recurrence Intervals

Flood frequency is commonly recorded as a **recurrence interval**, the average time between floods of a given size. Larger flood discharges on a given stream have longer recurrence intervals between floods.

100-Year Floods and Floodplains

A **100-year flood** is used by the U.S. Federal Emergency Management Agency (FEMA) to establish regulations for building near streams. A 100-year flood has a one-percent chance of happening in any single year, although it also has a one-percent chance of happening in the year following a similar-magnitude event. A 100-year floodplain is the area likely to be flooded by the largest event in 100 years—on average.

Bridges and other structures should be designed to accommodate the runoff in at least the expected life of the structure, typically between 50 and 100 years. Those who design structures such as dikes, bridges, and buildings for a given life span consider the probabilities of large floods.

As with all calculations of recurrence intervals, however, the accuracy of the 100-year floodplain is not always reliable. The 100-year floodplain is based on extrapolation from a few large events over a short and incomplete flood record. The 100-year flood level does not account for huge probable

changes, including later upstream alterations to the drainage basin. FEMA maps show the limits of a 100-year floodplain, with no indication of uncertainty in this level. This is in spite of the fact that the level is a simple estimate based on limited data, always changing, and may not have been updated in decades. We know that almost all upstream human activities will decrease the average number of years for floods to reach the former 100-year level and raise the height of the average 100-year event. We also know that we cannot prevent flooding during major storm events, in spite of some people's belief in safety above the 100-year floodplain.

Complicating matters is the fact that floods due to extreme precipitation events are likely becoming more frequent, from climate change, upstream landscape changes, or some cyclic aspect of climate. All-time records were set, for example, in the Susquehanna River of Pennsylvania in 1972, the Santa Cruz River in Tucson in 1983, the upper Mississippi River in 1993, the Red River in North Dakota in 1997, and the American and San Joaquin Rivers of California in 1997. When an unusually large event occurs or several events occur more frequently than expected, researchers must go back and recalculate the 100-year flood flow.

Recurrence Intervals and Discharge

For a given stream, the statistical average number of years between flows of a certain discharge is the recurrence interval (T). The inverse of this (1/T) is the probability that a certain discharge or flow will be exceeded in any single year. To determine recurrence intervals for floods on a stream, the peak annual discharge is recorded for each year on record (**FIGURE 11-23**). The largest of these discharges is then ranked number 1, the next largest 2, and so on. Each discharge can

FIGURE 11-23 FLOOD FREQUENCY

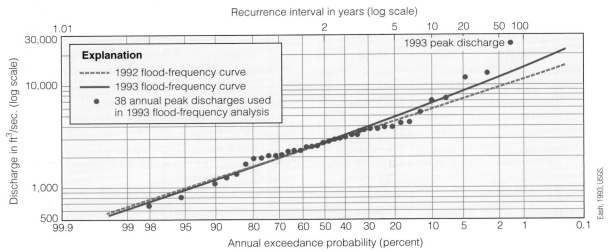

The flood frequency plot for Squaw Creek, a tributary of the Mississippi River at Ames, Iowa, is plotted for before and after the largest flood of historical record in 1993. The recurrence interval (plotted at top) and exceedence probability (plotted at bottom) are shown.

then be plotted against its calculated recurrence interval using the "Weibull formula." The calculated recurrence interval of a given flood depends on the total number of years in the flood record and the rank of the flood in question. That means that any new larger flood reduces the rank of the flood in question and can dramatically reduce its recurrence interval (**By the Numbers 11-6**: Recurrence Intervals).

For example, the 1993 Mississippi flood changed the recurrence interval for flooding in St. Louis. The adjusted recurrence interval for the 1993 flood at St. Louis is 100 years or less, rather than previous estimates of up to 500 years. Any new record flood changes previous recurrence intervals because the previous largest flood is now the second largest in the recurrence interval formula. This can have major consequences for 100-year-event floodplain maps. Properties that were outside the 100-year flood hazard area are now within it; flood protection structures that appeared to provide an adequate margin of safety no longer do so.

In general, floods with large recurrence intervals are more frequent in smaller catchment basins (**Table 11-1**). Such large floods are much more frequent in semiarid climates (such as the desert southwestern United States) and in monsoon climates (such as Sri Lanka) compared with tropical environments (like Congo and Guyana) because the semiarid regions have much larger variations in precipitation.

Paleoflood Analysis

A major problem in estimating the sizes and recurrence intervals of potential future catastrophic floods in North America is that we have only a short record of stream flow data; the measurement of flood magnitudes is not much more than 100 years, even less in parts of the West. The record is longer in Europe and much longer in civilizations such as China and Japan that have written records extending back a few thousand years. We can project graphical data on magnitudes and their recurrence intervals to less frequent larger events (see FIGURE 11-23). However, as outlined above, any record-size event can dramatically change the estimate of recurrence intervals.

In order to extend the record further into the past and recognize larger floods, **paleoflood analysis** uses the physical evidence of past floods—preserved in the geologic record—to reconstruct the approximate magnitude and frequency of major floods. Even where paleoflood magnitudes cannot be determined reliably from the evidence, the flood height can in some cases be fairly well determined. By itself, this can provide critical information on the minimum hazard of a past flood.

Table 11-1	Chance of a Flood of a given size in a Certain Period of Time (Occurrence %)				
FLOOD (RECURRENCE INTERVAL)	ANY SINGLE YEAR	10 YEARS	25 YEARS	100 YEARS
10 years	10			
50 years	2		40	86
100 years	1		22	63
1,000 years		1		9.5
10,000 years				1

FIGURE 11-24 HIGH-WATER MARKS AFTER FLOODS

A. A catastrophic flood on this small stream, Shoal Creek, in Austin, Texas, on November 15, 2001, followed an intense rainstorm that dropped 7.5 centimeters of rain in one and a half hours. It overtopped the bridge deck and left branches tangled in its railings, evidence of the flood height. The high water also deeply eroded the channel and undercut the supports, causing the bridge deck to sag. **B.** This organic drift line marks the high-water line of the same flood.

EARLY POST-FLOOD EVIDENCE: PALEOFLOOD MARKERS The nature and magnitude of a flood is most obvious immediately after it occurs. Debris, including leaves, twigs, logs, and silt, carried in floodwaters, tends to collect at the edges of the flow, including in back eddies. These provide perhaps the best evidence for maximum flood height, though they may not be preserved long after the flood, and driftwood is likely to end up well below the maximum water level. Debris may pile up on bridges, providing a flood's minimum height. More commonly, floods leave behind drift lines marking the high-water level for a short period of time until they wash away with the next rainstorm (**FIGURE 11-24**).

The best reaches of a stream for paleoflood analysis are those with narrow canyons in bedrock, slack-water sites, and areas with high concentrations of suspended sediment. Useful features that indicate the height, velocity, and size of a flood include the following (**FIGURE 11-25**):

- **High-water marks**, which can provide the elevation and width of the high-water surface
- **Cross-sectional area**, if a cross section is exposed (assuming no postpeak scour or channel fill)
- **Mean flood depth** (equal to cross-sectional area divided by water-surface width)

FIGURE 11-25 PALEOFLOOD MARKERS

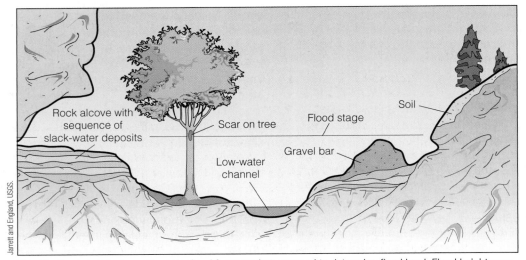

This idealized sketch shows the paleoflood features that are used to determine flood level. Flood heights can be estimated from the elevations of slack-water deposits, scars on trees, gravel-bar deposits, and eroded soil.

FIGURE 11-26 WATER VELOCITY AROUND A BEND

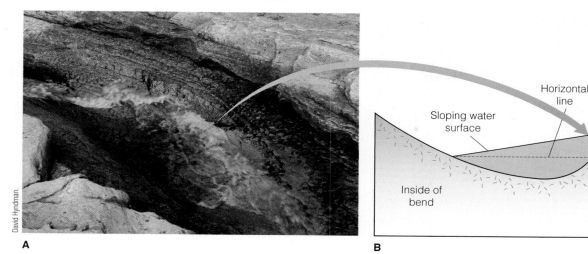

A. Fast-moving water banking up around a stream bend in the Grand Canyon, Arizona. **B.** This diagram shows water-surface banking up on one side of a channel as it races around a bend.

- **Estimated water velocity**, which can be estimated (±50%) from the inclination of the water surface as it flows rapidly around a bend (**FIGURE 11-26**)

- **Mean flow velocity**, which can also be estimated from the size of the largest boulders that were moved

- **Discharge** for high water within-bank flows, which can be estimated from cross-sectional flow area and water slope along the channel

Streams in different environments and different climates, however, are highly variable. Most usable evidence comes from meandering streams, not braided or straight streams. Studies have been conducted mainly in single regions, such as the west-central United States, where the evidence of floods is best preserved. Unfortunately, these results cannot easily be extended to other regions.

TREE RING DAMAGE Individual trees may preserve damage effects from a flood and indicate the number of years since flooding (**FIGURE 11-27**). Scars on a tree trunk or branch remain at their original heights during tree growth. The height of the damage generally indicates the minimum height of a flood, though it could be somewhat above the flood height if vegetation piles up.

The age of trees growing on a new flood-deposited sand or gravel bar indicates the minimum age since the flood that produced the bar. When all of the oldest trees on the deposit are of roughly equal age, then that age is probably close to the age of the deposit.

SLACK-WATER DEPOSITS During a flood, silt and fine sand can be deposited on sheltered parts of floodplains, the mouths of minor tributaries, shallow caves in bedrock

FIGURE 11-27 TREE-RING EVIDENCE OF FLOODING

The number of tree-growth rings after the point of damage (arrows) indicates the number of years since damage occurred.

canyon walls, or downstream from major bedrock obstructions (**FIGURE 11-28**, p. 338).

Organic material in silt and mud layers on floodplains, and occasionally in well-preserved siltlines, can be dated with radiocarbon methods to indicate the times of former floods. The elevations of these deposits provide bounds on

FIGURE 11-28 SLACK-WATER DEPOSITS

These Lake Missoula slack-water sands were deposited over coarse, darker, cross-bedded flood deposits near Starbuck in southeastern Washington State.

FIGURE 11-29 EFFECT OF URBANIZATION ON FLOOD PATTERNS

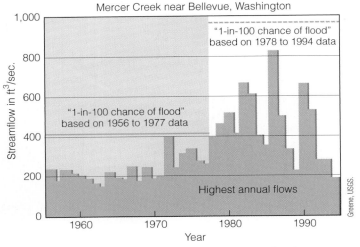

The 100-year flood for Mercer Creek, Washington, near Seattle, increased dramatically following rapid urbanization from 1977 to 1994.

the heights of those floods, which can be combined with data on channel geometry and slope to estimate flood flow using numerical models.

Boulders are often deposited where flood velocity decreases, such as where a channel abruptly widens or gradient decreases. These provide a minimum height for a flood.

Problems with Recurrence Intervals

Although recurrence intervals are perhaps the best estimate available of the likelihood of a flood, there are significant limitations to their accuracy.

First, for a recurrence interval to be accurate, the data on which it is based must cover a long enough time interval to be representative. (A quantitative example of how dramatic the change in recurrence interval can be with a single slightly larger record flood is shown in **By the Numbers 11-6**.) Second, the recurrence interval assumes that the upstream conditions that affect stream flow in the watershed must be similar through time. Although the use of paleoflood data can dramatically extend the total time interval on which a record is based, the conditions under which paleofloods occurred may have been significantly different than more recent conditions. Floods during the ice ages of the Pleistocene epoch, for example, originated in a colder, wetter climate; the conditions were different than at present, probably both for amounts and frequency of precipitation.

Human impacts, such as urbanization, channelization, the building of dikes and dams, deforestation by any process, and overgrazing, also change the conditions for flooding. Population growth clearly causes changes, especially where residential communities continue to expand the area of impermeable ground with pavement and concrete. Rapidly growing cities such as Bellevue, just east across Lake Washington from Seattle, removed vegetation and added huge areas

of pavement and rooftops. The estimated 100-year recurrence interval flow for Mercer Creek, which drains the area, was estimated in 1977 as 420 cubic feet per second. In 1994, following 17 years of rapid growth, the 100-year flood was estimated as 950 cubic feet per second (**FIGURE 11-29**)! The entire character of floods here changed due to urbanization of the region.

Finally, all of the floods plotted to determine recurrence interval or exceedance probability must belong to a single group that originates from similar flood causes so that the distribution of flood sizes should be random. They all must result from an El Niño event, for example, or from storms originating at a warm front. Combinations of causes, such as a hurricane colliding with a cold front, would create aberrations that would not fall in the same group and should ideally not be plotted on the same frequency distribution. Unfortunately, the data sets are so small—the period of record so short—that all floods on a river are typically lumped together, regardless of size or origin.

Although recurrence intervals are useful tools for studying floods, the complex interactions that lead to flooding mean that these intervals cannot be seen as predictions. Even worse is a simplistic perception of a 100-year flood as one that comes along every 100 years. In fact, the number is merely an average—such a flood may recur anytime, including in the year following a major flood.

Mudflows, Debris Flows, and Other Flood-Related Hazards

Where large amounts of easily eroded sediment are carried downstream by floodwaters, the nature of the sediment transport also changes. The amount of sediment

Table 11-2	Characteristics of Floods with Increasing Proportions of Sediment to Water		
	WATER FLOODS	HYPERCONCENTRATED FLOWS	DEBRIS FLOWS
Sediment concentration (% by volume)	0.4–20	20–47	47–77
Bulk density (g/cm³) depends on amount of sediment in transport	1.01–1.33 Note: Water has a density of 1 g/cm³.	1.33–1.80	1.80–2.3 Note: Most rock has a density of ~2.7 g/cm³.
Deposits, landforms, and channel shape	Bars, fans, sheets, wide channels (large width-to-depth ratio)	Similar to floods	Coarse-grained marginal levees, terminal lobes, trapezoid to U-shaped channels
Sedimentary structures	Horizontal to inclined layers, imbrication*, cut-and-fill structures, ungraded to graded	Massive or subtle layers, subtle imbrication, normal or reverse grading**	No layers, no imbrication, reverse grading at base, normal grading at top
Sediment characteristics	Well-sorted, clast-supported, rounded particles	Poor sorting, clast-supported, open texture, mostly coarse sand	Poor sorting matrix supported, extreme range of particle sizes, ± some megaclasts

*Imbrication is the overlapping of pebbles with their flat sides sloping upstream. The current has pushed them over.

**Normal grading has coarser grains at the bottom of a deposit, finer grains at the top.

transported in a flood varies widely, from less than one percent to 77 percent by volume (**Table 11-2**). As the proportion of sediment increases to more than 20 percent, a flood becomes a hyperconcentrated flow, and with more than 47 percent sediment, a **debris flow**. A debris flow is concentrated enough that you could scoop it with a shovel. In some cases, a single flow begins as a debris flow, and then with additional water downstream forms a hyperconcentrated flow, and perhaps then a water flood. If mud or clay dominates the solids choking a flowing mass, it is a **mudflow**; if volcanic material dominates, it is called a **lahar**. If the solids are more abundant, coarser, and highly variable in size, it is a debris flow. (Debris-flow hazards are described in Chapter 12.)

We can determine the type of a former flood from the characteristics of the deposits:

- Water floods have little internal shear strength and are dominated by the fluid behavior. Sediment moves in suspension and by rolling and bouncing along the channel bottom.

- Hyperconcentrated flows contain higher sediment concentrations and thus have larger overall densities and viscosities. Water and grains behave separately; grains colliding with one another may not settle at all or settle at rates only one-third of those in clear water.

- Debris flows contain still higher sediment concentrations; the water and grains move as a single viscous or plastic mass, with water and sediment moving together at the same velocity. Internal shear strength is high, so shear is concentrated at the base and edges

of a flow. As flow velocity slows, few particles settle. As water escapes from the spaces between fragments at the flow edges, the flow slows and finally stops moving.

All of these flow types are dangerous. Debris flows are most destructive because of their higher densities and the large boulders they carry; they can move at higher velocities than other floods, and their greater mass can destroy larger and stronger structures.

Mudflows and Lahars

Like landslides, mudflows are mobilized by abundant water in a slope and are often triggered by heavy or prolonged rainfall. The tiny pore spaces and low permeability of mud slurries retain water and keep the mud mobile. Boulders are rafted along near the surface of flows that have high density and resemble wet concrete. Earthquakes jar others loose.

Active volcanoes are notorious for spawning lahars, especially during eruptions, because volcanic ash supplies abundant mud-size material. However, lahars often contain rocks and boulders as well. The collapse of heavily altered ash on the flank of a volcano during heavy rains, or rapid melting of snow or ice, forms a dense slurry that collects loose volcanic debris as it races downslope (**FIGURE 11-30**). During volcanic eruptions, hot ash may mix with rain, lake, or stream water downslope to pour down nearby valleys. The rapid mixing of hot rocks and ash with ice and snow causes much more rapid melting than situations in which these hot materials simply fall on the snow surface. Even if

FIGURE 11-30 LAHAR

This is the flow front of one of the many fast-moving lahars racing right to left down a valley from Mount Pinatubo in the Philippines. Note that the surface of the flow is covered with rocks and pebbles, especially in the nose of the cresting flow.

the falling ash is too cool to melt snow or ice, large volcanic eruptions often generate their own weather. The heat of the eruption pulls in and lifts outside air, causing it to expand and cool. That causes moisture in the air to condense; locally heavy rains fall on the newly deposited ash, flushing it downslope.

Even long after major ash-rich eruptions, the amount of ash on a volcano's flanks provides ample material for catastrophic lahars. Such lahars often have extremely long runouts, especially where they are confined to valleys. They can continue to move on slopes as low as one percent—a one meter drop in 100 meters.

Debris Flows

Debris flows are common and extremely dangerous. They are widespread, begin without warning, move quickly, and have tragic consequences for both structures and people. They are most common in steep mountainous areas such as the U.S. Southwest (**FIGURE 11-31**). They are especially common along major active faults, where tectonic movements actively build mountain belts such as the Basin and Range of Nevada, the Wasatch Front at Salt Lake City, and the Andes Mountains of South America. The steep slopes maintain a continuing supply of broken debris. Most debris flows empty onto alluvial fans, making these a hazardous place for residences (**Case in Point:** Desert Debris Flows and Housing on Alluvial Fans—Tucson, Arizona, Debris Flows, 2006, p. 347).

As debris flows spread across a fan, they block old channels with piles of gravel and boulders. Subsequent debris flows then overflow to erode new channels. The details of the channels change, but the general style of channels on the fan remains the same. Flows continue moving on a broad alluvial fan at slopes less than 10 to 15 degrees but rarely at slopes less than 5 degrees or so. Because debris flows thin as they spread out on a fan, their energy dissipates, and their largest boulders drop near the head of the fan; progressively smaller ones drop downslope. Having deposited most of their bouldery load, water floods can continue to lower slopes.

Debris flows differ from stream flows in the amount of solid grains suspended within them. Because rocks are approximately 2.7 times as dense as water, a slurry of rocks with sediments in the pore spaces can be twice the density of water in a flooding stream. That permits these flows to

FIGURE 11-31 DEBRIS FLOWS

A. A debris-flow basin at the north edge of San Bernardino, California, filled and overflowed to surround houses in a winter event in 1980. It is being re-excavated in this photo. **B.** A series of debris flows from the huge fault scarp of the Wasatch Front behind Salt Lake City, Utah, in 1983 inundated homes built on alluvial fans at the mouths of steep canyons, including Rudd Creek (pictured here).

FIGURE 11-32 DEPOSITS FROM A DEBRIS FLOW

A. Bouldery natural levees formed in 1996 by a major debris-flow on a tributary to the Columbia River, near Dodson, east of Portland, Oregon. Note the large boulders concentrated at the top of the deposit. **B.** Finer-grained particles fill the spaces between the boulders and below them. It is the watery, finer-grained matrix of the flow that buoys the boulders on top. Sabino Canyon 2006 debris flow, Tucson, Arizona.

pick up and carry huge boulders, some as large as a car or even a school bus. The high density of debris flows in steep terrain can propel them at a higher velocity than clear water.

Debris flows are characterized by internal shearing, with some parts moving faster than others. Much of their movement is by the whole mass sliding on a stream bottom, with a lesser amount of jostling between fragments. Except for all of the boulders and internal shear, movement is something like wet concrete coming down the chute of a cement truck. Slippage at the base of a flow permits it to scour material from its channel, and to entrain more material in the flow.

Debris flows tend to move in surges. When particle sizes vary, jostling moves the largest pieces to the edges and front of a flow. The same thing often happens to a human body caught in a snow avalanche. Boulders bob along a debris flow's surface (**FIGURE 11-32A**); they are pushed to the sides and front, typically forming prominent bouldery natural levees.

Although water drains easily from large spaces between boulders, movement depends on having lots of fine-grained particles and water in the pore spaces (FIGURE 11-32b).

Movement slows when water drains from between fragments, especially at the toe and flanks of a flow (**FIGURE 11-33**). Fine material has small interspaces that drain water much more slowly, so movement continues in the fine-grained rear parts of a flow.

Many years of occasional rains tend to wash mud and assorted debris down slopes and into canyon floors. After decades or centuries, canyons contain enough accumulated debris to provide the raw material for a large mud or debris flow. That will happen as soon as one of the occasional heavy desert rains flushes the canyon floor and spreads its burden of debris out over the alluvial fan downslope. Then another long period may pass before the canyon again accumulates enough debris to repeat the performance. It recharges with loose debris from the sides of the canyon. Many geologists familiar

FIGURE 11-33 DISTRIBUTION OF SEDIMENT

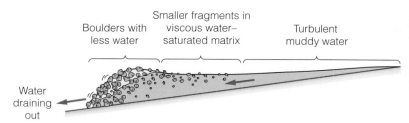

A. A schematic diagram shows the distribution of grain sizes and water in a debris flow. **B.** The steep, bouldery snout of a debris flow that flowed from upper right to left. East of Tucson, Arizona. Note geologists at right, for scale.

with desert canyons can tell just by looking whether a canyon is sufficiently charged to produce a new debris flow. In other cases, only the more recent loose deposits in the canyon flush out as a debris flow; older deposits remain available to move in another storm. These events are most likely in watersheds that have lost their cover of brush to a fire within the previous few years. Bare ground does not absorb rain as well as it does when covered with plants, so the hills shed water like a roof, and the heavy surface runoff flushes the canyons (discussed in Chapter 16, Wildfires).

Debris flows commonly begin with a heavy rainfall or rapid snowmelt that fills pore spaces above less permeable bedrock (**Case in Point:** Intense Storms on Thick Soils—Blue Ridge Mountains Debris Flows, p. 348). This increases the pore pressure that sets a mass in motion, especially if it is disturbed by an earthquake or even a strong gust of wind on a large area of slope. Sometimes the initial movement is a landslide that begins in a swale, where soils tend to become thicker and the groundwater

level is high. Sometimes they begin when water flowing downslope picks up sediment, then "bulks up" with more sediment farther downslope. Flows tend to keep moving as long as the positive pore-water pressure is maintained. A major six-day storm in November 2006, on the flanks of Mount Hood, Oregon, that produced 34 centimeters of rain, initiated debris flows, some induced by landslides. Others formed when the rain-saturated, loose pyroclastic sand material that had accumulated in upper reaches of the channels, bulked up and flowed downstream. Flows in the White River backed up behind, then crested over, the highway 35 bridge and washed out the highway in two places.

Flows tend to move in surges or waves, each one with a steep front of especially coarse boulders. The initial main surge is often followed by a series of smaller surges traveling faster than the overall flow. Individual surges may slow and stop, only to be remobilized by a subsequent surge. Although most debris flows are not witnessed, their velocities and peak discharges can be estimated from peak-flow mud lines left on channel sides and cross-channel banking angles at sharp bends (see FIGURE 11-26).

The lower ends of swales are especially risky places for homes. In coarse colluvium (loose, broken material over bedrock), landslides that develop into debris flows begin on slopes from 33 to 45 degrees, measured down from horizontal. Where a debris flow has occurred in the past, it will occur again in the future. Evidence of a previous occurrence indicates it is likely to happen again. Disturbances such as fire, logging, housing developments, road building, and volcanic activity can change a basin to make it more prone to debris flows. The maximum amount of loose material available in catchment basin channels provides an indication of the maximum size of a future debris flow. A large debris flow that removes most of the loose material in the basin will lessen the maximum size of future flows for years until the loose material again builds up. Evidence of former debris flows include:

FIGURE 11-34 EVIDENCE OF A DEBRIS FLOW

David Hyndman.

This debris flow in the Andes of northwestern Argentina was rich in rocks and boulders until it began losing water and slowing down. That permitted the water-rich finer-grained material (covering boulders in the middle left of the photo) to catch up and, in turn, flow over the surface of the head of the flow. Debris flows come down this channel every year, and in one case the flow eroded down to a buried natural gas pipeline, causing a massive explosion.

- A valley floor strewn with boulders that seem much too large for the modern stream to move. In exposed cross sections, huge boulders are perched on top of finer-grained deposits that are massive and unlayered and have large, angular boulders in a finer matrix (**FIGURE 11-34**)

- Levees of coarse, angular material next to a stream (FIGURE 11-34)

- Deep, narrow channels cut in those deposits (FIGURE 11-34)

- Fan-shaped deposits forming rounded lobes with coarser material at the outer edges

- Rocks lodged against trees, deposited in tree branches, or embedded in bark (**FIGURE 11-35**)

- Bark scars high on trees on their upstream sides (FIGURE 11-35)
- Lobes of even-age vegetation younger than the surrounding vegetation
- A drainage basin with large, actively eroding areas
- Active faulting, which helps supply broken rock to a slope

Glacial-Outburst Floods: Jökulhlaups

The toe of a glacier, where meltwater feeds a stream, can occasionally see sudden catastrophic floods. The water can originate either in a lake dammed by the glacier or in water pooled near the base of the glacier. In the former case, a glacier flowing down a mountain valley can cross a larger stream, damming it to form a lake. When the lake water gets high enough, the water can seep under the base of the glacier and rapidly enlarge a tunnel (**FIGURE 11-36**, p. 344). If the tunnel gets sufficiently large and the roof collapses, the resulting downstream flood may be catastrophic. In other cases, the lake water may float the glacial ice dam, leading to rapid failure and flooding downstream. Camping downstream from a glacier-dammed lake can be hazardous. Such floods are called **glacial-outburst floods**, or Jökulhlaups. On August 14, 2002, water dammed behind Hubbard Glacier burst through to drain the lake. In April 2010, Jökulhlaups during eruption of Eyjafjalla volcano under a glacier in Iceland forced evacuation of hundreds of people, severed the coastal road, and destroyed a huge area of farmland.

Meltwater pouring down through crevasses and holes in a glacier finds its way to the glacier's base where it can lubricate glacial movement (discussed further in Chapter 10). That water continues down-valley under the glacier, appearing at its toe as a meltwater stream. Sometimes large volumes of water pond under gently sloped glaciers, such as those in Antarctica. Elsewhere, glacial movement blocks a subglacial channel, leaving it vulnerable to a sudden outburst flood and potential catastrophe downstream. Active glaciers in many parts of the world pose glacial-outburst flood hazards to campers, hikers, sightseers, and some residents. Such floods are most common in very hot or rainy weather. If a meltwater stream rises rapidly or you hear a roaring sound up-valley, move quickly up the side of the stream bank.

Many glaciers on Mount Rainier, Washington, have experienced glacial-outburst floods, some of them triggering debris flows or lahars. Subglacial-outburst floods are also common in Iceland, where lava eruptions beneath a large ice cap may melt the ice. The meltwater flows downslope under the ice, causing bulging of the ice cap 15 kilometers downslope. Every few years the under-ice lake periodically bursts out as a flood.

Hubbard Glacier, the largest glacier reaching the sea in North America, crossed a fjord near Yakutat, at the north end of the Alaska panhandle to fill Lake Russell. Three months later, on August 14, 2002, the ice dam failed, releasing the second largest historic flood worldwide (FIGURE 11-36b). The peak flow was about 54,000 cubic meters of water per second in a river 100 meters

FIGURE 11-35 HEIGHT OF DEBRIS FLOWS

A. Huge boulders are lodged against the upstream sides of trees and the trees are debarked (arrows) along Tumalt Creek, east of Portland, Oregon. **B.** Rocks (above and at the right of the red pocketknife) are lodged in the branches of a tree, which was debarked to about that height by a debris flow along Sleeping Child Creek, south of Hamilton, Montana, in July 2001. The area burned in a massive wildfire the year before.

FIGURE 11-36 GLACIAL-OUTBURST

A. A lake formed by a tributary glacier blocking a valley drained when a tunnel eroded under it. Tulsequah area, northwestern British Columbia.
B. Advance of the Hubbard Glacier to raise Russell Lake in early May 2002 set the stage for the outburst flood on August 14.

across—30 times the peak flood flow of the lower Mississippi River. The lake took 36 hours to drain. Fortunately, Yakutat, at the mouth of the bay, was not damaged. The largest glacial-lake flood on record was at the same site in 1986.

During the last ice age, huge glacial meltwater floods appear to have spread southeastward from the continental ice sheet in southeastern Alberta. Some of these giant subglacial floods had volumes of tens of thousands of cubic kilometers. Similar gigantic floods, but from glacial lakes ponded behind ice-age glaciers, have been well documented. Gigantic floods from the Altai in central Asia and multiple drainages of glacial Lake Missoula that spread across southeastern Washington State thousands of years ago have also been well documented but pose no present danger.

Ice Dams

During winter and early spring, ice coating rivers can block stream channels; as the ice begins to melt, it breaks up and moves. A sudden warm spell that causes ice to break up can dam a channel at constriction sites, such as bridges (**FIGURE 11-37**). Large rivers flowing northward have the additional problem that ice upstream thaws before the ice downstream. As upstream meltwater flows north, it encounters ice jams that restrict downstream flow, causing floods (**Case in Point:** Spring Thaw from the South on a North-Flowing River—The Red River, North Dakota, 1997 and 2009, p. 349).

Other Hazards Related to Flooding

Different types of hazards are often interrelated, as discussed in Chapter 1, so it shouldn't surprise you to learn that flooding is related to a number of other hazards. Torrential rain can develop from hurricanes or thunderstorms (Chapters 14 and 15) or where humid, tropical air

FIGURE 11-37 ICE DAM

An ice jam built up at a river constriction threatens a bridge at Gorham, New Hampshire. The power shovel tries to get the river flowing again.

over an ocean is forced to cool and condense as it rises against a mountain range (Chapter 10). Coastal areas can flood as hurricane winds cause sea level to rise locally as it comes onshore. Heavy rains can be initiated by a volcanic eruption as hot volcanic gases pull in humid air and carry it to high elevations where it condenses (Chapter 6). Those same rains, falling on ash dumped on volcano flanks, can produce mudflows that race down nearby valleys and in turn gradually winnow into dirty floods. Wildfires often denude slopes, causing heavy rainfall to flow rapidly off their surfaces; they can cause debris flows or raise flood levels much faster than otherwise. Floods can, in turn, initiate other hazards. A flooding stream sweeping around the outside of a bend can undercut and steepen the stream bank or slope above, causing it to slide (Chapter 8).

Heavy Rainfall on Near-Surface Bedrock Triggers Flooding
Guadalupe River Upstream of New Braunfels, Texas, 2002 ▶

Like much of south-central Texas, the area around Austin and San Antonio is fairly dry except for heavy rainstorms delivered by warm, moist air funneled from the Gulf of Mexico. One such storm lingered over the Guadalupe River watershed upstream from New Braunfels for the first week of July 2002. It began as a typical Texas summer storm, a tropical low-pressure system fed by moisture from the Gulf. However, instead of moving east, it stalled over central Texas for almost a week because no other weather systems helped push it eastward. In that time, it dumped more than three-quarters-of-a-meter of rain on thin soils covering widespread limestone bedrock that prevents rapid infiltration. The amount of precipitation was unprecedented, surprising even the regional meteorologists.

The river 16 kilometers upstream from New Braunfels is dammed for a combination of flood protection, water supply, and recreation. Storm inflow from upstream

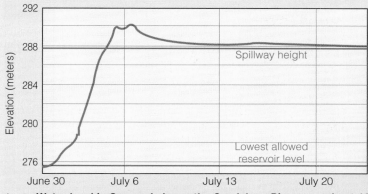

▶ **A.** *Water level in Canyon Lake on the Guadalupe River rose about 12 meters (40 feet) in four days during the flood in early July 2002 (blue line).* **B.** *The flood crested at 2.1 meters (7 feet) above the spillway.*

▶ **A.** *Water recedes in the 2002 Guadalupe River flood below Canyon Lake after washing out this bridge.* **B.** *A kitchen submerged during the 2002 flood retained heavy deposits of mud.*

raised the Canyon Lake reservoir level 12 meters in four days, then topped the spillway of the broad containment dam by more than 2 meters beginning on July 4.

The flood raced down the narrow, tree-lined valley of the Guadalupe River, gouging a new rock-bound canyon. It ravaged the homes of people who had built near the stream, thinking they were protected by the upstream dam.

The flood crest reached New Braunfels six hours after topping the dam and reached the Gulf of Mexico seven days later. So why wasn't the dam built higher in the first place, high enough to withstand a flood of this magnitude? Some large dams and bridges are built to withstand only a 100-year flood. To build them higher costs significantly more money, especially given that the dam

would have to be much wider if it were higher. Any dam may prove inadequate for future flood control because later upstream changes in land use, such as additional urban development, cause more rapid runoff and higher flood peaks. Climate changes such as global warming may also cause more moisture to be picked up from the Gulf of Mexico to cause more rainfall inland.

A Flash Flood from an Afternoon Thunderstorm
Big Thompson Canyon, Northwest of Denver ▶

July 31, 1976, was the beginning of a three-day celebration of Colorado's centennial. Motels and campgrounds in the otherwise sparsely populated canyon were filled with more than 3,500 people. Moist summer air masses from the east often rise into the Rocky Mountain Front from the High Plains; they cool and then dump their moisture in thunderstorms. That day, a cool Canadian air mass moved in to become stationary near the Big Thompson basin, while a warm front continued to funnel in moisture from the southeast.

Thunderstorms formed along the north-trending front just east of Estes Park, west of Fort Collins, and 80 kilometers northwest of Denver. It remained there, dumping approximately 30 centimeters of rain in four hours, 75 percent of the total for a typical

year. The heaviest rain fell directly over Big Thompson Canyon, just east of Rocky Mountain National Park. For its uppermost 34 kilometers through the park, the river descends more than 1,524 meters. At the town of Estes Park, it enters a narrow, rocky canyon for 40 kilometers and drops another 762 meters before spreading out on the Great Plains.

Heavy rain began to fall at 6:30 p.m., and the danger of a major flood soon became evident; police moved through the canyon telling people to leave. Unfortunately, because heavy summer thunderstorms in the area are common, many people did not believe they were in danger and remained. Rapid runoff from the mountain slopes developed into a flash flood, and by 7 p.m. a wall of water more than 6 meters deep was

racing down the canyon at roughly 6 meters per second (~22 kilometers per hour). Flows were between 860 and 1,512 cubic meters per second.

By 8 p.m., the flood was carrying dirt, rocks, buildings, and cars with people in them. Most of Highway 34 through the canyon was washed out. Of the people in the 400 cars on the canyon highway, those who abandoned their cars and ran upslope survived; the 139 who tried to outrun the flood in their cars died. More than 600 others were never accounted for. An ambulance crew that drove into the canyon to render aid reported that

▶ **A.** *This car was caught in the Big Thompson flood.* **B.** *A house in the bottom of Big Thompson Canyon was washed downstream, coming to rest on a concrete bridge, though somewhat the worse for the experience.*

R. J. Jarrett, USGS.

a huge, choking dust cloud led the wall of water, picked up the ambulance, and slammed it into a wedge on the canyon wall. The crew climbed out of the wrecked ambulance and up to a ledge 15 meters above the highway. The water surged to their level, but they survived and were rescued by helicopter the next morning, along with hundreds of others. In the debris at the mouth of the canyon was a motel register with 23 names. The motel disappeared in the flood; none of its guests were found.

The flood destroyed 400 cars, 418 houses, and 52 businesses and totaled $137 million (2010 dollars) in damage. The flow rate was four times larger than that of any previously measured flood in the 112 years of record. The recurrence interval for a flood of this magnitude is uncertain, but radiocarbon dating of old flood deposits suggests a recurrence interval of several thousand years, assuming no changes in climate. Some 30 years after the flood, the giant boulders carried downstream remained in the canyon, and the gorge was completely cleared of soil and vegetation. Could it happen again? Not much has changed. The winding highway still follows the canyon bottom, and the patches of narrow canyon bottom are again lined with houses.

Donald Hyndman.

▶ *Big Thompson Canyon, east of Estes Park, Colorado, still preserves the bouldery deposits from the July 1976 flood that killed hundreds of people.*

Desert Debris Flows and Housing on Alluvial Fans
Tucson, Arizona, Debris Flows, 2006 ▶

Following four days of rain, July 31 saw the largest of the early morning convective thunderstorms that dumped heavy rain on the Santa Catalina and adjacent mountain ranges around Tucson, Arizona. Upper-level winds from the northwest collided with lower-level winds from the west rising against the front of the range. The rainfall intensity was not unusual for Tucson's heavy summer thunderstorms, but saturation of the thin soils by the days of preceding rain appeared to be more important than the rainfall intensity.

More than 500 slope failures occurred on the flank of the Santa Catalina Mountains, and adjacent ranges experienced 100 additional slides. Twenty to 25 centimeters of rain fell on Sabino Canyon, with flood flows reaching 442 cubic meters per second, a new 100-year record.

The flows raced downslope, coalesced, and continued down valley in all of the canyons on the west side of the Santa Catalina Mountains. Soils are only about 1 meter thick over bedrock on the steep, rocky slopes. Debris flows began most commonly near the base of bedrock cliffs where excess surface water may have poured onto the already saturated soils to initiate failure. The flows quickly bulked up downslope as they collected more rocky soil material. Reaching gentler slopes at the main canyon floors, the flows rapidly thickened. In some places they severely eroded roads; in others they clogged large culverts and buried roads and small buildings.

▶ **A.** *Hydrograph for flood flows in Sabino Creek from July 27 to 31, 2006.* **B.** *One of numerous debris flows initiated on the steep slopes of Sabino Canyon that carried huge boulders down to the main valley.*

▶ *Rest room at the end of Sabino Canyon.* **A.** *Before debris flows.* **B.** *After. Arrows point to paired features.*

Intense Storms on Thick Soils
Blue Ridge Mountains Debris Flows ▶

Even relatively humid areas with higher rainfall, gentler slopes, and deeper soils can be subjected to debris flows. In August 1969, Hurricane Camille, one of only three Category 5 hurricanes to hit the United States in the twentieth century, came onshore in Mississippi, where it did severe damage. Before petering out, it stalled over the mountains of central Virginia, causing still more damage, primarily in debris flows. It dumped an amazing 71.1 centimeters of rain in eight hours, saturating the ground and causing shallow slides that quickly developed into debris flows.

At least 1,100 slopes slid, generally where they were steeper than a grade of 17 percent (10-degree slope). Most began where groundwater is closer to the surface, near the inflexion point between convex rounded hilltops and the concave segments above main river channels. Where slopes were covered with coarsely granular soil, they were generally stripped down to bedrock. Formations that were most vulnerable to sliding contained nonresistant, well-layered, and mica-rich rocks that dipped parallel to the downslope direction.

Years later, on June 27, 1995, an intense storm stalled over the Blue Ridge Mountains in Madison County, Virginia, where it dumped more than 76 centimeters (two-and-a-half feet) of rain in sixteen hours. It triggered more than 1,000 debris flows; one in Kinsey Run northwest of Graves Mill began as a landslide that developed into a 2.5-kilometer-long debris flow. It raced down the slope in surges at 8 to 20 meters per second, eroding and incorporating ground material to depths up to 0.6 meter. Velocities were calculated from the tilt of the flow surface where it poured around bends in the channel. Its deposits included coarse rocks with boulders up to 7 meters across in a matrix of clay and sand.

Although this flow did not impact people, others living elsewhere in the Blue Ridge Mountains are vulnerable. Debris flows are common in the region, and events of this magnitude have recurrence intervals of tens of years.

A decade later, Hurricane Ivan drenched the Blue Ridge Mountains of North Carolina on September 16, 2004, with nearly 30 centimeters (a foot) of rainfall. It initiated numerous landslides and debris flows, destroying many houses and killing 5 people.

Rick Wooten, North Carolina Geological Survey.

▶ *Debris flows from Hurricane Ivan destroyed this home in Starns cove, Buncombe County, NC.*

Spring Thaw from the South on a North-Flowing River
The Red River, North Dakota—1997 and 2009 ▶

The winter of 1996–97 brought numerous heavy snowstorms to the northern plains, a snowpack two or three times the depth of the previous record. In early April, a major blizzard dumped up to 1 meter of snow on parts of the area; this was the culmination of eight major blizzards that provided as much as 3 meters of snow in Fargo.

The Red River that flows through Fargo and Grand Forks, North Dakota, and Winnipeg, Manitoba, freezes in winter. Come spring, it thaws first in North Dakota but remains frozen farther north, a recipe for flooding beyond the control of local residents. As the thaw progresses, ice floes drift north to pile up in ice jams that cause widespread flooding. The very flat terrain around Winnipeg, Grand Forks, and Fargo is the old flat bottom of Glacial Lake Agassiz; it covered much of southeastern Manitoba, with a prominent tongue extending south up the Red River valley. That lake was dammed on the north by the continental (Laurentide) ice sheet until about 9,800 years ago. Grand Forks, North Dakota, is entirely within a 500-year floodplain; its highest elevation lies less than 10 meters above the Red River's flood level.

People in the area are used to floods, but not on the scale of 1997. On April 17, the Red River broke its 100-year record at Fargo by cresting at 6.9 meters above flood level. Discharge reached some 3,400 cubic meters per second (compared with this river's average of 75 cubic meters per second). On April 19, it crested at Grand Forks 7.9 meters above flood level and covered 90 percent of the city for several days. Flooding of the water and sanitation systems not only left the city without drinking water and sewage treatment but also contaminated the flood water, making the city uninhabitable. Fire broke out in downtown Grand Forks on April 19 and spread to eleven buildings, creating severe access problems for firefighters.

Sixty thousand people in the Grand Forks area were evacuated, and damage and cleanup costs exceeded $1 billion. Fortunately, no one died in Grand Forks. Overland flooding from snowmelt from farm fields on April 3 worsened the problem. To lessen damage to the larger population areas, authorities built a dike on the outskirts of town to block overland flooding from the fields. The Minnesota River also flooded, though it caused

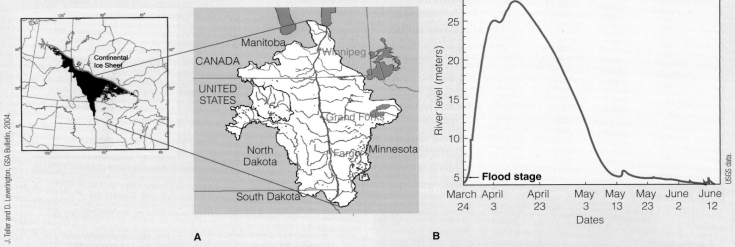

J. Teller and D. Leverington, GSA Bulletin, 2004.

USGS data.

A. *The Red River flows north along the border of North Dakota and Minnesota, through southern Manitoba, and into Lake Winnipeg. The extent of Glacial Lake Agassiz, about 9,800 years ago, controlled the area of the 1997 flood. Water draining northeast on a gentle slope was dammed by the south edge of the continental ice sheet.* **B.** *This hydrograph shows the flood level for the Red River at Fargo, North Dakota, March to June 1997.*

less damage than the Red River. Seven people died in North Dakota and four in Minnesota. In contrast to many rainfall floods, the hydrograph for this flood is rather broad, remaining above flood level from March 26 to May 20. The federal government purchased and moved many homes in the lower parts of Grand Forks and tore them down to minimize damage from future floods. By 2006, $410 million in federal, state, and local funds had facilitated rebuilding and raising 85 percent of levees and flood walls across the east side of the city. The new protection sits 1

meter above 1997 flood levels. The city is now considering either moving levees farther from the river or constructing a flood-diversion channel to bypass the city. The new floodway, a "Greenway," provides a broader pathway for the flooding river; it includes parks, gardens and golf courses, paths for bikes and walkers, and athletic fields. Breaks in the dikes or floodwalls provide pedestrian access to the Greenway. Fargo, a little farther upstream from Grand Forks, was not severely damaged by the 1997 floods.

So what does the future hold? Climate predictions suggest that the average precipitation and temperature in the northern Great Plains will both increase during the next century. If this is correct, floods on the Red River may become more frequent and larger in future.

After the 1997 floods, experts recommended building levees farther from the river in Fargo, but business owners along the river objected, and nothing was done. In late March, 2009, flooding began farther upstream; the situation in Fargo again looked ominous. Intense cold and an unusually snowy winter left snow blanketing the still-frozen ground, meaning snowmelt would run off instead of soaking in. Spring rains added to the runoff, causing rapid melting, and accelerating the flooding. The National Weather Service predicted that the Red River would crest between 11 and 12 meters, 6.7 meters above flood stage; it reached 11.9 meters in the 1997 flood. By March 27, the National Weather Service estimated a crest of 13.1 meters. To protect the city, 6,000 volunteers and 1,700 National Guard members filled sandbags in below-freezing temperatures to raise about 19 kilometers of levees. The freezing of moisture in the sand confounded these measures; one resident characterized the effort as stacking big frozen turkeys on the levee. This also

D. Schwert, North Dakota State University.

Dr. G. Brooks., Geological Survey of Canada.

A. *Ice jams that filled the channel of the Red River below Grand Forks, North Dakota, caused the April 1997 flood.* **B.** *The Red River in Manitoba also had extensive flooding during this event.*

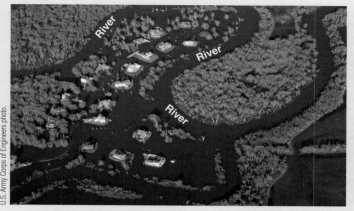

▶ **A.** *Homes in the area are built on mounds to keep them above floods, but the flood blocked access.* **B.** *East Grand Forks, in the 1997 flood. Homes in the rising water floated off their foundations.*

meant that the sandbags didn't sag to fill spaces between the underlying bags. The freezing temperatures did help, however, by preventing more snow from melting. In both Fargo and across the river in Moorhead, Minnesota, residents of some low-lying areas were asked to evacuate. Much to the relief of inhabitants, colder temperatures on March 28 slowed melting and the water dropped a little. But new heavy snows moved in and added

to meltwater later. The threat remained of the fast-flowing water breaching weak parts of the levees; a helicopter dropped 900 kilogram sandbags into the river to deflect the current away from vulnerable sections. Breaches swamped dozens of homes and the buildings of one private school campus that was also heavily flooded in the 1997 event. In spite of the severe damage in Grand Forks in 1997 and their nearness to the river, many

people had no flood insurance coverage for the 2009 event.

Decades earlier, farther downstream in Manitoba, a disastrous flood on May 8, 1950, caused damages of more than $790 million when eight dikes failed. In response the government built a new 47-kilometer-long flood channel for the Red River to divert it around Winnipeg; it was completed in 1968.

▶ **A.** *The 1997 Red River flood at Grand Forks, North Dakota.* **B.** *Downtown Grand Forks was submerged. The water level rose high enough to inundate some buildings to their rooftops as well as the bridge, which is visible near the right side of the photo.*

Chapter Review

Key Points

Stream Flow and Sediment Transport

- A stream's flow, or discharge, is proportional to the average water velocity and the cross-sectional area of the flowing part of the channel. **FIGURE 11-2; By the Numbers 11-1.**

- Stream equilibrium, or grade, is adjusted to accommodate grain size, amount of sediment in a channel, and the amount of water. Coarser grains and less water lead to steeper channel slopes. Similarly, larger particles can be moved only by higher water velocities. **FIGURE 11-4.**

- The sediment load that can be carried depends on the total flow, or discharge. Thus, floods can carry more sediment and erode more from a channel.

Channel Patterns

- Meandering streams are characterized by wide sweeping turns. This type of stream is more typical of wet climates. **FIGURE 11-7.**

- Meandering streams erode on outside bends and deposit sediment in point bars on inside bends in a pattern of deep pools and shallow riffles.

- Meanders gradually migrate over a whole valley floor. The area of this migration is the area that typically floods, called the floodplain. **FIGURE 11-9.**

- Braided streams are multi-channel streams that are overloaded with sediment because of either a dry climate with too little water or overabundant sediment supplied to the channel (below a melting glacier, for example). **FIGURE 11-10.**

- Braided streams deposit sediment in alluvial fans, which can be dangerous flooding areas. **FIGURE 11-12.**

- Bedrock streams are fast-moving, mountainous streams that have eroded the streambed down to bedrock. Flooding in bedrock streams is high-energy and highly erosive.

Groundwater, Precipitation, and Stream Flow

- In humid regions most streams are gaining, that is, they are fed by groundwater and flow year round. In drier climates, there are losing streams, which lose water into groundwater. These streams flow primarily after a rain and may dry up between rainstorms. **FIGURE 11-15.**

- When water cannot be absorbed into the ground, it runs directly into a stream as surface runoff. Surface runoff is generated when rainfall is particularly intense, such as during a storm, when the ground is impermeable, frozen, or water saturated.

Flooding Processes

- Streams commonly reach bankfull level every 1.5 to 3 years on average and spill over floodplains less often. **FIGURE 11-16.**

- A flood does not have to be a large flow; rather, it is an unusually large flow for the stream.

- During floods, not only does the water surface rise higher, but the channel also erodes more deeply. Channel scour strongly affects damage. **FIGURE 11-17.**

- Patterns of erosion and deposition during flooding create natural levees along a river's banks.

Flood Intensity

- Flood intensity is a measure of stream discharge over time. Flood intensity varies with the rate of runoff, the shape of the channel, the number of tributaries, and distance downstream.

- There is a higher rate of runoff where the ground is less permeable, which can be due to soil composition, dry climate, deforestation, or urbanization. These areas are prone to sudden, severe flash floods.

- In stream order, the smallest tributaries are designated first-order; those where first-order tributaries join are second-order; and so on. Low-order streams have short lag times between rainfall and flood and are more prone to flooding. Higher-order streams with many tributaries are less prone to flooding. **FIGURE 11-21**.

◾ Downstream flooding is characterized by the flood peak arriving more slowly but lasting for a longer time **FIGURE 11-22**.

Flood Frequency and Recurrence Intervals

◾ Flood frequency is commonly recorded as a recurrence interval, the average time between floods of a given size.

◾ Average flood frequency, or recurrence interval, is based on the past record for that stream. A 100-year flood has a one percent chance of occurring in any single year, regardless of the date of the last major flood. The recurrence interval for a stream is calculated as the total number of years of flood record for that stream +1, divided by the rank of the flood under consideration, where the largest flood on record has a rank of 1, the second largest has a rank of 2, and so on.

◾ The 100-year floodplain is the area flooded by the largest flood in 100 years on average.

◾ Paleoflood analysis, the study of the magnitude and timing of past floods, includes indications of high-water marks, cross-sectional area, and meander wavelength, among other factors. **FIGURE 11-25**.

◾ Recurrence intervals may be poor predictors of future flooding because of limited records of past flooding, changing climate conditions, and human impacts, among other things.

Mudflows, Debris Flows, and Other Flood-Related Hazards

◾ With more than 47 percent sediment, a flood becomes a debris flow. If mud or clay dominates the solids choking a flowing mass, it is a mudflow; if volcanic material dominates, it is a lahar.

◾ Glacial-outburst floods occur when a glacier moving downslope crosses a stream, damming it to form a lake. When the lake water gets high enough, it may float the dam and rush downstream in a flood. **FIGURE 11-36**.

◾ When rivers flow north, downstream portions of the river may remain frozen after upstream portions have melted. This can lead to an ice dam that blocks water from flowing downstream, backing up water and causing flooding. **FIGURE 11-37**.

Key Terms

100-year flood, p. 330	discharge, p. 316	gradient, p. 317	oxbow lake p. 320
alluvial fan, p. 317	dynamic equilibrium, p. 316	hydrograph, p. 327	paleoflood analysis, p. 331
bedrock streams, p. 319	flash flood, p. 327	lahar, p. 335	point bar, p. 320
braided streams, p. 319	flood crest, p. 328	load, p. 318	recurrence interval, p. 330
base level, p. 317	floodplain, p. 321	losing streams, p. 324	riprap p. 320
channel scour, p. 325	gaining streams, p. 324	meandering streams, p. 319	stream order, p. 327
debris flow, p. 335	glacial-outburst floods, p. 339	mudflow, p. 335	surface runoff, p. 324
delta, p. 318	graded stream, p. 316	natural levee, p. 326	watershed, p. 316

Questions for Review

1. A river slope or gradient adjusts to what three factors?
2. Does water move faster in a mountain stream or in a large smooth-flowing river like the Mississippi? Why?
3. If a large amount of sediment is dumped into a stream but nothing else changes, how does the stream respond? Why?
4. Characterize the differences between flooding patterns in meandering and braided streams.
5. Explain the patterns of erosion and deposition in a meandering stream. Use a sketch to illustrate your answer.
6. How does an alluvial fan develop? Why is flooding particularly hazardous in these areas?
7. Why does a stream bottom erode more deeply when its water level rises in a flood?
8. Stream power or the destructive capacity of a stream depends on what factors?

9. What aspects of weather cause a flood? (Be specific: not merely "more water.")

10. How often, on average, does a stream reach bankfull level just before spreading over its floodplain?

11. Why does removal of vegetation by any mechanism cause more surface runoff and thus more erosion?

12. In what climates are flash floods most common? (Provide an example location.)

13. Why do floods in first-order streams provide only a short warning time for people downstream but high-order streams may provide days of warning?

14. Does a flood crest move downstream at the same speed as the water flow? Explain why or why not.

15. What is the simple formula for calculating the recurrence interval for a certain size flood on a stream in 1999 if there are 69 years of record? Give a numerical example.

16. What specific evidence can be used to estimate the maximum water velocity in a prehistoric flood?

17. Why might a recurrence interval not be a good predictor of future flooding?

18. Characterize the differences between debris flows and water floods.

19. What events may lead to a lahar?

20. Describe the events that lead to a glacial-outburst flood. Use a sketch to illustrate your answer.

21. Why, in parts of Canada or the northern United States, do north-flowing rivers often flood in spring? Explain clearly.

Flood Discussion Questions are at the end of Chapter 12.

Floods and Human Interactions

12

Houses floated down river to pile up against a railroad bridge across the Cedar River, Iowa, on June 14, 2008.

Jeff Robertson, AP.

Lessons Learned?

A fter the "Great Flood" of the upper Mississippi River valley in 1993, many residents chose to stay, believing strengthened levees would protect them against future flooding. Local and regional governments continued to allow construction on floodplains in the pursuit of economic development and additional tax dollars. Near St. Louis, Missouri, about 30,000 homes were built on land that was flooded in 1993. Banks no longer required federal flood insurance, people dropped their coverage, and others were permitted to build new houses and businesses behind the levees.

But in June 2008, after months of above average precipitation, the region again experienced severe flooding (**Case in Point:** Upper Mississippi River Valley Floods, 2008, p. 380). Cedar Rapids, Iowa, a city of about 124,000, was particularly hard hit, suffering more than a billion dollars in damages.

After yet another destructive flood, have residents and government officials learned their lesson? For their part, government officials are taking steps to reduce future damages in some of the affected areas. By August, Cedar Rapids had made plans for the U.S. Army Corps of Engineers to build a new levee farther back from Cedar River, leaving more of the floodplain free to contain the river during floods. Federal money will also be used to buy out property owners on the river side of the levee. One hundred ninety-two homeowners were made offers, but more than twenty-four of them declined, even though they had suffered major flood damage two months before.

Development Effects on Floods

Natural stream processes cause disasters only when humans place themselves in harm's way. Before Europeans settled in North America, when rivers flooded, native residents merely packed up and moved to higher ground until the water subsided. Europeans, however, tended to establish towns along rivers because of the access that rivers provide to water and transportation. These settlers caused severe alterations of the landscape through urbanization, logging, and grazing, as well as forest fires. They changed the sediment load of rivers by adding loose gravel or excavating sand and gravel from river channels and floodplains, increasing the size and damaging nature of floods.

Modern civilizations continue to build near rivers because of the concentration of towns, jobs, and farm land along them. When modern residents build on a floodplain they often choose a site near the river, on the outside bend of a meander, where the house stands well above the river channel to provide a great view. Unfortunately, in doing so, they locate in the area of greatest bank erosion during a flood.

When settlers first established towns near rivers, they did not realize that rivers continually strive to minimize the energy that they expend; they alter their slope and path to accommodate the amount of flow and sediment supplied to them. If we make artificial changes in a channel, the river tries to adjust to minimize those changes. Today we understand how rivers change in response to human impacts and the increased flood risk these changes can bring.

Urbanization

Increasing urbanization in many parts of the world promotes increasing numbers of flash floods and higher flood levels. Urbanization involves transforming natural landscapes through logging, paving, and building. This affects flooding because water cannot soak into pavement or rooftops, so wherever roads or buildings cover the ground, the water is forced to run off rapidly into nearby streams, causing flooding. Dense urban areas often build artificial concrete channels to rapidly move this water downstream and minimize flooding (**FIGURE 12-1**).

Floods in urban environments cause special hazards when residents try to evacuate on flooded roads. Driving through a road that is under water can be dangerous or even fatal. Even though water may appear shallow, the force of its flow against the wheels or side of a car can wash it downstream. Fast-moving water only 30 centimeters deep is extremely dangerous (**FIGURE 12-2**). Water 60 centimeters deep and above the vehicle floorboards can push a vehicle off the road and potentially drown the occupants. Compounding this danger, shallow, fast-moving water can erode a deep channel underwater that may not be visible to a driver. Driving into apparently shallow muddy water that hides a washed-out roadway can drop a vehicle into deep water and cause drowning. Even where water is still or hardly moving, the settling of part of a roadway can cause unexpected deep water.

Most cars will initially float because their weight is generally much less than the volume of water required to fill them. Eventually, however, water seeps in and they sink. If you are in a sinking car, immediately climb out through a window or kick out the windshield. It is best to back out through a window so you are facing the car and can hang on to it. It is important to act quickly because once the car is partly submerged, it will be difficult or impossible to open the doors to escape.

Fires, Logging, and Overgrazing

Upstream human actions such as deforestation by fire, heavy logging, or overgrazing of a watershed also impact stream processes. These activities all cause excessive erosion, which dumps large amounts of sediment into channels, increasing their load. Recall from Chapter 11 the relationship between stream gradient and sediment transport. Streams choked with sediment develop a braided

FIGURE 12-1 CHANNELIZATION

A

B

A. The Los Angeles River runs through a straight section of concrete channel with angled energy dissipation structures. The channel reduces flood risk to residents by collecting and moving water rapidly out of the city during periods of intense rain. **B.** A channelized stream in New Hampshire.

FIGURE 12-2 DO NOT DRIVE THROUGH FLOODWATERS!

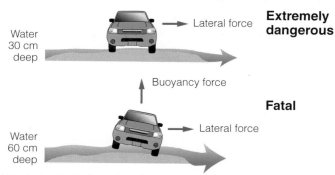

Vehicle begins to float when the water reaches its chassis, which allows the lateral forces to push it off the road.

Washed-out roadway can be hidden by muddy water, allowing a vehicle to drop into unexpected deep water.

A. Even shallow floodwaters can lift a vehicle and wash it downstream. **B.** Don't venture into this type of flooded road, even though it doesn't look very deep. Pisgah Road, Versailles, Kentucky, September 23, 2006.

channel pattern and become steeper in order to transport a greater sediment load.

In a recent example, torrential rains around Jakarta, Indonesia, in early February 2007 caused severe flooding, killing dozens and forcing the evacuation of more than 400,000 people. In about half of the city, hundreds of thousands were without clean water or electricity. Authorities blamed the severity of the flooding on deforestation

of hillsides near the city, inadequate urban planning, and trash-clogged storm drains and rivers.

Deforestation can significantly impact a stream environment, including rate of runoff and sediment load. In vegetated areas, rain droplets impact leaves rather than landing directly on the ground. Rich forest soils soak up water almost like a sponge, providing a subsurface sink for rainwater that maintains a long-term moisture source for the vegetation. Fire burns away vegetation that protects the soil, permitting rain droplets to strike the ground directly and run off the surface. Intense fire also tends to seal the ground surface by sticking the soil grains together with resins developed from burning organic materials in the soil. This decrease in soil permeability reduces water infiltration. As a result, direct overland flow carves deep gullies into steep hillsides and feeds large volumes of sediment to local streams (see Chapter 16 for further discussion of the impacts of fire).

Logging by clear-cut methods sometimes involves skidding logs along the ground, which removes brush and other vegetation and leaves the ground vulnerable to erosion. In a technique called *tractor yarding*, felled trees are skidded downslope to points at which they are loaded on trucks. The skid trails focus downslope to a single point like tributaries leading to a trunk stream. Additionally, logging roads cut through the forest tend to intercept and collect downslope drainage, leading to the formation of gullies, increased erosion, and the addition of sediment to streams.

Cattle and sheep grazing on open slopes or next to stream banks similarly remove surface vegetation that formerly protected the ground. Rainfall running off the poorly protected soil erodes gullies, thereby carrying more sediment to the streams. Once gullies begin, the deeper and faster water causes rapid gully expansion (**FIGURE 12-3**).

Whether as a result of overgrazing, deforestation, or fire, vegetation removal can also increase the contribution of sediment to streams by landslides. Some water is taken up by plants in the form of *transpiration*, the natural evaporation of rainfall from leaves and removal of water from soil via roots through the leaves. When vegetation is removed, more water soaks into the ground and runs off its surface. More water penetrating a slope tends to promote landslides, which in some areas contribute as much as 85 percent of the sediment supplied to a stream. That upsets the balance of the stream, causing increased sediment deposition and increased flooding downstream.

Mining

Mining can also change sediment load and disrupt the equilibrium of a river. During the California Gold Rush in the late nineteenth century, floods were triggered when miners added large amounts of loose gravel to stream channels as a by-product of gold mining (**Case in Point:** Addition of Sediment Triggers Flooding—Hydraulic Placer Mining, California Gold Rush, 1860s, p. 372).

Today large amounts of sand and gravel are used in construction materials for roads, bridges, and buildings. Much of that sand and gravel is mined from streambeds or floodplains in a process called **streambed mining**. Because the water flow in the stream is unchanged, the decrease in sediment load means the water has more energy and increased velocity, which leads to greater energy downstream. Downstream from the mining area the stream uses this excess energy to erode its channel deeper. Deepening a channel can severely damage roads and bridges. It also typically lowers the water table because more groundwater flows into the deeper stream channel; as a result, groundwater supplies away from the stream decline.

Where gravel is mined from pits on a floodplain, temporary gravel barriers are often used to channel the stream around the pits (**FIGURE 12-4**). Later, rising water may erode the barriers; water entering the pit from upstream slows and deposits sediment in the upper end of the pit, eventually filling it.

Although such mining may appear, therefore, to exploit a renewable resource, the gravel removed from flood flow

FIGURE 12-3 **EROSION CAUSED BY OVERGRAZING**

Heavy sheep grazing has encouraged deep erosion in a steep slope in New Zealand. The broader valley in the upper right is a more extreme stage of the same process.

FIGURE 12-4 **FLOODED GRAVEL PIT**

Now flooded, these gravel pits along the South Platte River near Denver, Colorado, are separated from the river by only thin gravel barriers. The pits are flooded because they are dug down to below the level of the river.

by deposition is not being carried farther downstream. The increased stream energy downstream amplifies the erosion. As water nears a deepened gravel pit in the channel, its velocity increases, in many cases eroding the upstream lip of the pit and washing that gravel down into the pit. That lip migrates upstream. The deepening channel undercuts roads, bridge piers, and other structures, often causing failure. Erosion concentrates near piers where there is increased turbulence. The cost to local governments to repair or replace such structures often exceeds the value of the mined gravel. In one prominent court case, it was shown that gravel mining from stream bars deepened the channel by as much as three meters for many kilometers downstream from Healdsburg, California (**Case in Point:** Streambed Mining Causes Erosion and Damage—Healdsburg, California, p. 373). The change threatened several bridges, destroyed fertile vineyard land, and lowered groundwater levels.

Removal of sediments from a stream often has consequences far from the site of removal, where a river deposits its sediment at the coast. Normally, much of the sediment deposited in the river delta is carried up or down the coast by longshore drift. When the sediment supply is reduced because of gravel mining or an upstream dam, the longshore movement of sediment along the coast continues but is not replenished. Beaches erode and may even disappear. Waves that normally break against the beach then break against the beach-face dunes or sea cliffs, causing severe erosion (discussed in Chapter 13).

Bridges

Bridges promote erosion in a channel when they increase water velocity by restricting the flow of water over the floodplain. The road or railroad approaches

FIGURE 12-5 BRIDGES AND EROSION

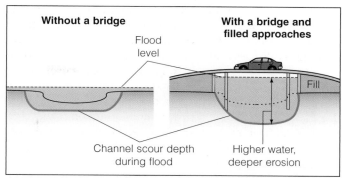

Without a bridge the river spreads over its floodplain during a flood (shown at left). With a bridge the river has a narrower and deeper channel, increasing channel scour (shown at right).

to bridges commonly cross floodplains by raising the roadway above the 100-year flood level. This typically involves bringing in fill that effectively creates a partial dam across the floodplain; the river flow is restricted to the open channel under the bridge. During a flood, when a river would usually spread over its banks and onto the floodplain, the bridge forces the water through a deeper, narrower channel (**FIGURE 12-5**). The deeper water flowing under the bridge flows faster, causing greater erosion of the channel under the bridge; that may undermine the pilings supporting the bridge, causing it to fail (**FIGURE 12-6**). Where more enlightened planners or engineers design the bridge—or where better funding is available—the approaches may be built on pilings that permit floodwater to flow underneath the roadway.

FIGURE 12-6 EROSION AT BRIDGE SUPPORTS

A. A flood scour hole eroded next to the downstream end of one of the Alaska Highway bridge support piers in the Johnson River.
B. High-water turbulence at the downstream end of a bridge pier indicates that the water is moving faster and eroding more. Clark Fork River, Missoula, Montana.

FIGURE 12-7 LEVEES ATOP OLD NATURAL LEVEES

A. Levees are almost always built on old natural levees, right next to river channels. Confidence in the protection afforded by the levee has encouraged industrial and residential development behind these levees in Moorefield, West Virginia. **B.** Fine materials making up this levee in the Sacramento River Delta of California is sealed with plastic and sandbags. The river side of the levee is protected from currents and wind-driven waves by coarse rocks.

Levees

With permanent settlement of floodplains, people feel the need to protect their structures from floods. The most common response to protecting an area from flooding is to artificially raise the riverbanks in the form of a **levee**. Individuals, municipalities, states, and the U.S. Army Corps of Engineers commonly try to protect floodplain areas from floods by building levees. However, as discussed here, the behavior of streams and rivers is governed by natural processes and fixed relationships between stream flow, channel slope, grain size, and meander shape. This means that human efforts to control a river are often frustrated by the river reverting to its natural course.

Recall from Chapter 11 that deposition of sediment during flooding creates natural levees along the banks of a river. Artificial levees are almost always built on top of original natural levees at the edge of the stream channel (**FIGURE 12-7A**). In the past, levees were built from locally handy materials in the floodplain, typically sand and mud from the floodplain itself or dredged from the river channel. Because these materials are often finer-grained than that carried by the river during floods, they are easily eroded, and do not make good barriers to fast-moving water. Higher-quality levees built by federal agencies are often mixed with coarser gravels or faced with coarse riprap to resist erosion. Different materials have different advantages (**FIGURE 12-7B**). Compacted clay resists erosion and is nearly impermeable to floodwater but can fail by slumping; crushed rock is more permeable but less prone to slumping.

Levee Failure

Another unintended consequence of levees is fostering a false sense of security and encouraging development in flood hazard areas. With levees in place, more people build homes and businesses behind them, where they feel safe from floods (FIGURE 12-7). However, even well-built levees can fail during major floods.

Although a levee may be initially of sufficient height to constrain a 100-year flood, the stream will eventually *overtop* the levee in a larger flood. Unfortunately levees are of varying quality, and, as noted in Chapter 11, major floods can come more frequently and at higher levels with time as regions become more developed. Communities can be faced with raising the level of a levee that is not high enough to contain a flooding river (**FIGURE 12-8A**).

When a flood breaks through the walls of a levee, it is called a **breach** (**FIGURE 12-8B**). When a large flood first breaches a levee, the water crossing the breach tends to be relatively clear and below its sediment-load capacity, so it erodes vigorously. As the breach erodes deeper, the floodwater carries more sediment and the water-surface slope decreases, ultimately bringing the breach flow into equilibrium and limiting further erosion. If floodplain flow stops moving locally, the breach flow will slow, sediments will deposit, and the breach will stop flowing, often in less than a day.

Sometimes a breach will lead to **avulsion**, where the main channel of a river is redirected through the breach. If the floodplain flow is unrestricted downstream and the flood level remains high, the breach flow may continue to erode, ultimately diverting the main channel through

FIGURE 12-8 LEVEES CAN FAIL

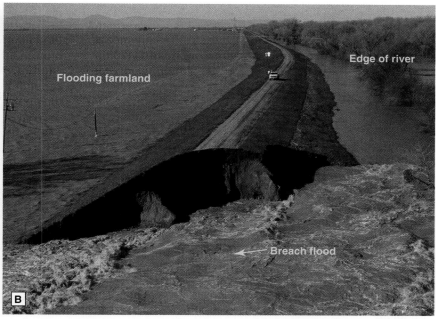

A. A barge and crane add coarse rock to shore up a levee protecting a town. **B.** Breach of a major levee on the Sacramento River, California, rapidly flooded surrounding farmlands and towns in January 1997.

the breach. The stream moves to a new lower-elevation path on the floodplain. Because the new path is lower, the stream does not return to its original path. The Mississippi River did this locally in the 1993 flood. The social and economic consequences of avulsion on such a major river can be severe (**Case in Point:** The Potential for Catastrophic Avulsion—New Orleans, p. 374; and **Case in Point:** A Long History of Avulsion—Yellow River of China, p. 375).

Common levee failures are caused by bank erosion from river currents or waves, slumps into the channel, or *seepage* (**FIGURE 12-9**). Water seeping rapidly enough

FIGURE 12-9 TYPES OF LEVEE FAILURE

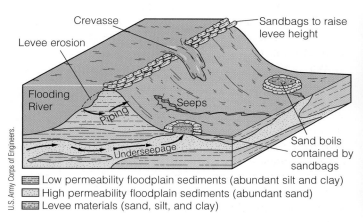

A. A levee may fail by overtopping, seeping through, or piping. Even where a levee is not overtopped, water may percolate through porous areas to cause seeps, landslides in its flanks, and ultimate collapse. Seepage from the river channel beneath the levee can rise behind the levee to form sand boils and, ultimately, flooding on the floodplain. **B.** A sandbagged levee at Granite Falls, Minnesota, in danger of being overtopped.

through parts of a levee erodes sediment at the side of the levee, which may then progress until the levee is in danger of failure. Cloudy water seeping out can indicate that soil is being washed out of the levee through a process called **piping**.

Levees can be more susceptible to these types of failures because of their composition from local material. The floodplain materials on which levees are built often are composed of old permeable sand and gravel channels surrounded by less permeable muds. The floodplain muds beyond the current channel are sediments that were left behind by the river where it spilled over a natural levee to flow on the floodplain. Recall from Chapter 11 that a river migrates across all parts of that floodplain over a period of hundreds or thousands of years. Under a mud-capped floodplain, the broad layers of sand and gravel deposited in former river channels interweave one another (**FIGURE 12-10**). These permeable layers provide avenues for transfer of high water in a river channel to lower areas behind levees on a floodplain. If a flood is prolonged, seepage beneath the levee can transmit enough water to flood surrounding areas behind the levee.

Even without overtopping or breaching the levee, water can also escape through the old gravels beneath it. Rising floodwater in a river increases water pressure in the groundwater below; this can push water to the surface on the floodplain, where it can potentially rise to nearly the water level in the river. With prolonged flooding,

FIGURE 12-10 MIGRATION OF MEANDERS

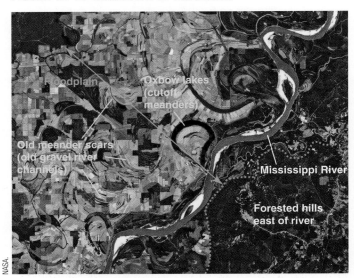

Because a river migrates laterally across its floodplain, sediment under the floodplain consists of old gravel river channels and old muddy floodplain deposits. The Mississippi River shows gradual migration of tight meanders before being cutoff to form an oxbow lake. The river flows from upper right to lower left (*modified from Farrell*).

FIGURE 12-11 SAND BOILS

A California Conservation Corps crew places sandbags around sand boils at the Sacramento River, at north Andrus Island, in the delta area southwest of Sacramento on January 4, 1997.

that water often reaches the surface as **sand boils** (**FIGURE 12-11**). The sandy water, under pressure, gurgles to the surface to build a broad pile of sand a meter or more across. Workers defending a levee generally pile sandbags around sand boils to prevent the loss of the piped sand. They leave an opening in the sandbags to let the water flow away and reduce water pressure under the levee.

Failure of artificial levees damages not only homes and businesses but also fertile cropland on floodplains. A flooding river without artificial levees spills slowly over its floodplain, first dropping the coarser particles next to the main channel to form low natural levees. Farther out on the floodplain, mud settles from the shallow, slowly moving water to coat the surface of the floodplain and replenish the topsoil. When artificial levees fail, the floodplain areas adjacent to a levee breach are commonly buried under sand and gravel from the flood channel and eroded levee. Farther away, the faster flow from the breach may gully other parts of the floodplain and carry away valuable topsoil.

Unintended Consequences of Levees

Levees do save some towns from flooding, at least for a while, but constraining a flooding river with a levee can have unintended negative consequences. Every time we build a new levee to protect something on a floodplain or to facilitate shallow-water navigation, we reduce the width of the flood-flow part of the river and raise the water level during flooding (see also **By the Numbers 11-1**). When a river floods under natural conditions, water flowing rapidly downstream spills over the riverbank, spreads

FIGURE 12-12 EFFECT OF LEVEES ON FLOW

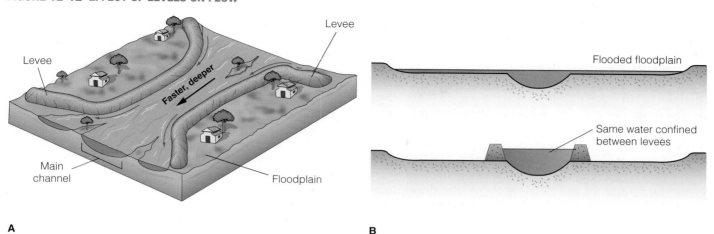

A

B

A. Levees that constrict the flow of a stream cause water to flow faster and deeper past levees. This causes flooding both upstream and downstream.
B. Levees confine at least the same amount of water between them as could spread out over the floodplain.

out over the floodplain, and slows down. Although that shallow water eventually continues downstream, its temporary storage on the floodplain lessens flood levels downstream. Constructed levees along stream banks eliminate that storage, causing higher flood levels downstream.

Additionally, all of the floodwater that should have spread over the floodplain is confined between the levees, causing the flood-flow to be much deeper and faster (**FIGURE 12-12**). Upstream of a levee section, water levels rise because the flow is constricted, causing flooding. Immediately downstream, the water level is higher because it is deeper within the constricted area. Flooding will thus commonly occur both upstream and downstream where it would not have occurred before the levees were built. For this reason, levees are sometimes intentionally breached during flooding to reduce flooding elsewhere (**Case in Point:** Repeated Flooding in Spite of Levees—Mississippi River Basin Flood, p. 377).

A study conducted for the U.S. Congress in 1995 showed that if levees upriver had been raised to confine the 1993 Mississippi River basin flood, the water level in the middle Mississippi would have been 2 meters higher than it was. In fact, the increase in flood heights, for constant river flow, has been 2 to 4 meters in the last century in the parts of the Mississippi River that have levees. This rise in flood level is mostly the result of levee building. The upper Missouri and Meramec rivers do not have levees and show no increase in flood levels.

Wing Dams

In addition to levees, navigational dikes or **wing dams** constrict river channels in areas such as St. Louis to increase river depth for barge traffic during low flow

(**FIGURE 12-13**). Even at high flow, when the river tops the wing dams, these structures increase resistance to flow near the river banks, which slows the velocity and raises the water level. As a result, for the same flood discharge, the water level or height increases. This artificial increase in river level clearly affects the inferred recurrence interval for any huge flood, such as that on the Mississippi in 1993. The 1993 peak level lies well above the recurrence interval curve adjusted for river flow without the wing dams.

Dams and Stream Equilibrium

Another major human intervention in rivers is the construction of dams across river channels. Dams are built for a variety of reasons, including flood control, hydroelectric power generation, water storage for irrigation, and recreation. These benefits must be weighed against significant cost to taxpayers as well as environmental impacts and increased flooding hazards. For example one benefit, such as water storage for irrigation, may lead to high reservoir levels behind a dam and too little remaining water storage in the reservoir to prevent downstream flooding at times of high rainfall or snowmelt.

Flood-control dams are built to stop downstream flooding. These dams are built high enough to contain a certain magnitude of expected flood, perhaps a 100-year flood. The ability of a dam to resist such a flood depends on a reservoir having a low-enough water level so it can fill up during a flood and avoid flooding downstream of the dam. This water can then be slowly released throughout drier parts of the year. Flood protection capacity will decrease as a reservoir fills with sediment; the rate of

FIGURE 12-13 WING DAMS

Wing dams on the Mississippi River provide a deepwater channel for shipping at lower water but narrow and slow the higher flows and thus increase flood levels.

FIGURE 12-14 EFFECT OF DAMS ON EROSION

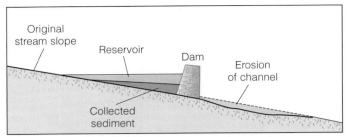

Trapping of sediment from the stream in the reservoir behind a dam causes erosion downstream.

sediment infill increases with upstream deforestation or urbanization.

As with levees, dams built for flood protection can actually amplify the danger of flooding. Dams across rivers remove sediment from streams because the velocity in reservoirs behind the dams drops virtually to zero. Downstream of a dam, a stream carries little or no sediment, so it erodes its channel more deeply during flooding (**FIGURE 12-14**).

The unintended impact of dams on river equilibrium is illustrated by the 200-meter-high Three Gorges Dam on the Yangtze River, completed in May 2006. The Chinese government relocated residents of at least ten major cities flooded by the dam, which was designed to provide hydroelectric power and prevent catastrophic floods. Since then the dam has led to a variety of major problems. Landslides developed along the reservoir upstream because rising groundwater levels destabilized adjacent slopes. Faster-moving, lower-silt-content water downstream of the dam is eroding river levees. In late 2007, the government persuaded four million people to resettle because they were endangered by river bank collapses. It appears that the giant port of Chongqing, at the upper end of the reservoir, will be silted up and closed by about 2016 because the river deposits its load where the water slows and no longer carries the sediment downstream.

Floods Caused by Failure of Human-Made Dams

The intent of dams is not to allow people to build on floodplains, but that is often the effect. People may feel protected by a dam, but that is a false sense of security.

At some point, the dam may not be adequate, and in extreme cases the dam may fail. Federal agencies or states own less than 8 percent of dams, local agencies and public utilities own about 19 percent, and private companies or individuals own 59 percent. Thus, care in siting, design, construction quality, and maintenance of dams is highly variable. When the U.S. Army Corps of Engineers studied more than 8,000 U.S. dams in 1981, they found that one-third of them were unsafe. More than 3,300 high and hazardous dams are located within 1.6 kilometers of a downstream population center. Few local governments consider the hazard of upstream dams when permitting development.

Dams can fail for a variety of reasons, including overtopping, erosion of the dam foundation, poor design, low-quality construction or maintenance, or natural disasters such as earthquakes. Like levees, dams can fail when water overtops or seeps through them. In the 1972 flood in Rapid City, South Dakota, the reservoir was overtopped and the dam failed after a prolonged rainfall. Although most dams could be built higher, the cost increases rapidly with height, in large part because a dam's length also increases rapidly with its height. Seepage of water under a dam or subsurface erosion along faults or other weak zones in the rock below it can lead to erosion of its foundation, resulting in catastrophic failure. The Teton Dam in eastern Idaho, which failed on June 5, 1976, took eleven lives and caused more than $3.2 billion in damage (**Case in Point:** Dams Can Fail—Failure of the Teton Dam, Idaho, p. 383).

Poor design and engineering standards can lead to dam failure, a fate suffered by a privately owned slag-heap dam at Buffalo Creek, West Virginia, which drowned 125 people in 1972. Improper maintenance, including failure to remove trees, repair internal seepage, or properly maintain gates and valves can also put a dam at risk of failure, as can negligent operation. In 1966 the failure to open gates early enough during high flows caused flooding on the Arno River in Florence, Italy (**Case in Point:** Catastrophic Floods of a Long-Established City—Arno River Flood, Florence, Italy, 1966, p. 381).

Other disasters can also trigger dam failure. Landslides into reservoirs can cause surges and dam overtopping, as happened when the Vaiont Dam in northeastern Italy was overtopped in 1963, killing 2,600 people. Filling the reservoir behind the dam increased pore-water pressure in sedimentary rocks sloping toward the reservoir. A catastrophic landslide into the reservoir displaced most of the water, which drowned more than 2,500 people downstream (see Chapter 8, **Case in Point:** The Vaiont Landslide, p. 232). Earthquakes can weaken earth-fill dams or cause cracks in their foundations. The Van Norman Dam, owned by the city of Los Angeles and situated less than 10 kilometers from the epicenter of the 1971 San Fernando Valley earthquake, is immediately upstream from thousands of homes. It is an earthen structure that was 30 years old when the earthquake struck. The quake caused a large landslide in its upstream face and so drastically thinned the dam that it seemed likely to fail. Authorities evacuated 80,000 people from the area downstream until operators were able to lower the water to a safe level and avert dam failure.

The hazard potential of a failing dam to people living downstream depends on the volume of water released, the height of the dam, the valley topography, and the distance downstream. In a broad open valley, the speed and energy of the water from the failed dam would spread out and quickly dissipate, whereas in a steep narrow valley the water could maintain high velocity for a great distance. Calculations show that a dam-failure flow rate in a broad open valley would likely drop to half of its original rate in 60 kilometers or so, but in a steep narrow valley half the original flow rate could be maintained for 130 kilometers downstream.

Flooding on Rapid Creek in the Black Hills of South Dakota in June 1972 provided dramatic evidence of why it is dangerous to live downstream from a dam (**FIGURE 12-15**). In just six hours, 37 centimeters of rain fell over the Rapid Creek drainage basin. Southeast winds carrying warm, moist air from the Gulf of Mexico banked up against the Black Hills, where it encountered a cold front from the northwest.

The creek's typical flow of a few cubic meters per second became a torrent of 1,400 cubic meters per second within a few hours. With rising water, authorities began ordering evacuation of the low-lying area close to the creek at 10:10 p.m., and the mayor urged evacuation of all low-lying areas at 10:30. The spillway of a dam just upstream from Rapid City became plugged with cars and house debris, raising the reservoir level by 3.6 meters. At 10:45 p.m. the dam failed, releasing a torrential wall of water into Rapid Creek, which flows through Rapid City. The dam had been built just 20 years earlier for irrigation and flood control after an earlier flood. Building of this and other dams made people feel secure, so they built homes along the creek downstream. The flood just after midnight killed 238 people, destroyed 1,335 homes and 5,000 vehicles, and caused $690 million in damages. More than 2,800 other homes suffered major damage.

In this case, the lesson was learned—at least for the time being. The city used $207 million in federal disaster aid to buy all of the floodplain property and turned it into a greenway, a park system, a golf course, soccer and baseball fields, jogging and bike paths, and picnic areas. Since then, building on the floodplain has been prohibited. However, decades later, pressure increased to build shopping centers and other structures in the greenbelt. The decision rested with a politically divided city council

FIGURE 12-15 DAMAGES CAUSED BY DAM FAILURE

A. Damages from the June 10, 1972, flood at Rapid City, South Dakota, were extensive. Cars were mangled, and the only thing left of a nearby house is the tangle of boards in the lower right. **B.** This house was carried from its foundation onto the road by the flood.

locked in the usual struggle between developers and the promise of additional jobs versus long-term costs, aesthetics, and safety.

Reducing Flood Damage

Increased development of floodplains has brought greater casualties and costs related to floods. Floods are among the deadliest weather-related hazards in the United States, with 84 deaths per year over a recent 10-year period. Floods are also one of the costliest hazards, causing one-quarter to one-third of the annual dollar losses from geologic hazards, an average of $4 billion a year. Between 1929 and 2003, the long-term average annual costs of flood damages has gone up by a factor of 10, from $400 million to about $4 billion (**FIGURE 12-16**). Individual annual losses vary widely. A significant part of this is because more and more people are living in dangerous areas.

Land Use on Floodplains

If people build homes on floodplains or developers succeed in obtaining changes in local ordinances to allow building, a change to years of heavier rainfall can lead to an unpleasant surprise. Rivers in some areas, such as north of the Gulf of Mexico, can remain well within their banks for many years and then flood frequently during wet years. Such a weather shift can occur rapidly (**FIGURE 12-17**).

People have various reasons for settling on floodplains, including cheap land, fertile soil, or scenic views. As a result, many floodplains in the United States are heavily settled. Many people do not realize that a floodplain is part of the natural pathway of a stream or river.

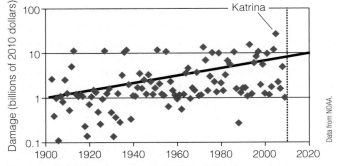

FIGURE 12-16 COST OF FLOODING

Data from NOAA.

Annual U.S. flood damages from 1929 to 2003, in constant dollars, show a strong increase, partly due to an increase in population but also from people living in more dangerous areas.

People often build their houses at the outside of a meander bend because the view of the river is best there or the trees are larger. Unfortunately, if you follow this strategy, you may find the stream running through your living room during the next flood. In mid-November 2006, major floods on the Skykomish River in northwestern Washington severely eroded river banks and destroyed homes along its banks. Many of these were carefully located to take advantage of the spectacular mountain and river scenery but were built too close to the river (**FIGURE 12-18**). At Mount Rainier National Park, about 45 centimeters of rain fell in 36 hours, washing out roads and campgrounds and destroying a major highway near Mount Hood, Oregon. A hunter who ignored road-closure signs was drowned when his pickup truck was swept into the flooding Cowlitz River.

After severe flooding in 1999 in the southern Mexican state of Tabasco killed more than 900 people, the government initiated a project to strengthen the levees that protect the more than half a million people of Villahermosa, the capital city. Unfortunately the work was never done. According to local reporting, flood-protection monies were used for other purposes and developers were allowed to build in high-risk areas. At the end of October and the beginning of November 2007, a cold front brought days of heavy rain. The low-lying state occupies the coastal ends of several Gulf Coast rivers that breached levees to affect 70 percent of the state and displaced about half of the state's two million people. For days, thousands of people remained stranded on rooftops or upper floors of houses surrounded by putrid, brown, debris-laden water. Food distribution, clean drinking water supplies, electric power, and public transportation came to a halt. Authorities were concerned about outbreaks of cholera and other waterborne diseases in the region's tropical climate. Flooding affected at least 50 of the state's medical centers. At least 30 people died.

People quickly forget most major floods that have affected an area and believe that similar or larger floods are unlikely to occur anytime soon. Similarly, if people have not seen their homes or businesses flooded in their lifetimes—or those of their parents—they also believe that such an event is not likely to happen to them. In both cases, people significantly underestimate their risk.

One way to reduce damages from flooding is to restrict development on floodplains. Natural rivers spill over their floodplains every couple of years. Because they often flood, these areas should be reserved for agriculture and related uses, not housing and industry. Parks, playing fields, and golf courses are reasonable uses in urban areas. Unfortunately, past government policy has failed to discourage development in flood-prone areas. The federal government pays millions to build and maintain levees to

FIGURE 12-17 CHANGING WATER LEVELS

A

B

Seasonal water levels in central Texas can change rapidly. **A.** Lake Travis, Austin, Texas, December 31, 2006, is so low that the lake is almost invisible in the distance. **B.** The same view six months later on June 28, 2007. Note the same "Public Notice" sign almost submerged at left. The same thing happened here in 2009–2010.

FIGURE 12-18 STREAM BANK EROSION

A 2006 flood on the Skykomish River, northeast of Seattle, Washington, undercut the stream bank underlying this homeowner's new house, which had spectacular views of the mountains and river.

protect such developments and to provide disaster relief following floods. The cycle continues when governments allow relief funds to be used to rebuild in the same unsuitable places.

Federal government policies began to change after the disastrous 1993 upper Mississippi River flood. The Federal Emergency Management Agency (FEMA) began buying up floodplain land to prevent people from rebuilding there and being flooded again. Disaster relief funds were provided only if people moved out of the floodplain. The government purchased many homes on floodplains with the requirement that building new structures there was prohibited. Unfortunately, some exceptions have been made. At St. Louis, for example, billion-dollar developments have been placed on land flooded in 1993. (**Case in Point:** Flood Severity Increases in a Flood-Prone Region—Upper Mississippi River Valley Floods, 2008 p. 380).

Some people feel that regulations on floodplain building infringe on their right to use their property as they choose. See, for example, the intensive public debate about development in flood-prone areas in some parts of California. However, an individual's choice to build on a floodplain often infringes on many other individuals. Huge sums of public tax dollars are spent every year to fight floods, build flood-control structures, and provide relief from flood damage for structures that should not have been built on the floodplain. Streams and rivers generally pass through many people's property. How one person or one town affects, restricts, or controls a stream commonly influences stream impacts to others both upstream and downstream. Developers and builders can make profits by building on floodplains, leaving homeowners and governments to pay the price of poor or insufficient planning. A coordinated approach is necessary to protect everyone. In many places, floodplains are not adequately zoned to minimize damages. Newly developed structures should

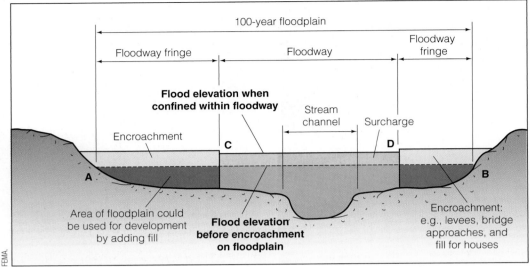

FIGURE 12-19 FLOOD INSURANCE DEFINITIONS

This schematic river cross section shows FEMA definitions for flood insurance purposes. Line AB is the flood elevation before any encroachment. Line CD is the flood elevation after encroachment. The surcharge is the rise in flood-water level near the channel as a result of artificial narrowing of the floodplain. The surcharge is not to exceed 0.3 meters (Federal Insurance Administration requirement) or a lesser amount specified by some states.

generally be prohibited, and in some cases entire towns should be moved. Expensive as this may be, it is less expensive in the long run.

Flood Insurance

Flood insurance is one way to mitigate costs of flood damage to individuals and also influence behavior to reduce future flood damages. The **National Flood Insurance Program (NFIP)** for the United States was established by the National Flood Insurance Act of 1968 and the Flood Disaster Protection Act of 1973. The NFIP made insurance available to those living on designated floodplains at modest cost. Insurance for floods is provided by the federal government but purchased through private insurance companies. Ratings and insurance premiums were intended to be actuarial, that is, they were based on flood risks and the existence of certain mitigation measures.

Guidelines for federal flood insurance stipulate several definitions. The 100-year floodplain is formally separated into a floodway and a flood fringe (**FIGURE 12-19**). The **floodway** includes the stream channel and its banks. During flooding, this zone carries deeper water at higher velocities. Most new construction is prohibited in this area, including homes and commercial buildings. Also prohibited are structures, fills, and excavations

that will significantly alter flood flows or increase 100-year flood levels. The **flood fringe** zone includes the stream channel and banks but is farther from the stream channel and still below the 100-year flood level. It is mostly floodplain, so during flooding it may be underwater; water there is generally shallower and flows more slowly.

Rates for this insurance depend on the likelihood and severity of flooding and are designated in mapped flood zones. Flood-hazard boundary maps and flood insurance rate maps (FIRMs) are available for communities under the regular FEMA program (**FIGURE 12-20**). To be eligible for flood insurance, a community must complete the required studies to designate floodplain zones and enforce its regulations. Larger amounts of insurance are available at actuarial—that is, "true risk"—rates.

Although flood insurance is clearly a good deal for those in flood-prone areas, it is not a good deal for U.S. taxpayers, who foot the cost of any losses. By 2006, the NFIP was $20 billion in debt, and insurance premiums are not high enough to pay it off. The long-term prognosis is worse; the NFIP insures $870 billion in homes and businesses in areas of high risk for flooding. Some policyholders have filed claims again and again. Losses will continue to mount, with most of the costs ultimately paid by taxpayers across the United States.

FIGURE 12-20 FLOOD HAZARD MAP

This example of a National Flood Insurance Program flood-hazard map is for part of East Lansing, Michigan.

FIGURE 12-21 REPEATED FLOOD DAMAGES

Davenport, Iowa, on the Mississippi River floodplain, was again underwater during the 2001 event.

For people's behavior with respect to a river to be appropriate, flood insurance premiums for a given location should be proportional to the risk of damage caused by flooding. Premiums are currently set too low to cover the actual cost of flood insurance. In 2003, participants in the NFIP were paying only 38 percent of actuarial risk rates. Clearly, premiums need to be raised to actuarial levels.

Another problem with flood insurance is convincing individuals to purchase it. Although insurance can be purchased up until 30 days before flood damage occurs, few people whose properties are damaged by flooding have purchased such insurance. Before the 1993 Mississippi River flood, only 5.2 percent of households in the flood hazard area had purchased flood insurance.

Some people may not purchase flood insurance because they assume that flood damage will be covered by their normal homeowner's insurance. In fact, homeowner's insurance only covers flooding events when the water source is inside the home, from a burst pipe, for example. Homeowners without separate flood insurance are liable for flood damages, such as water damage to walls, floors and furnishings; mud deposits; as well as the growth of mold that often occurs when warm water from outside stands in a home for more than a few days.

People often don't purchase insurance because they don't realize they are at risk. A survey of Missouri residents conducted seven months after the end of the catastrophic flood in 1993 found that approximately 70 percent of floodplain residents did not know they lived on a floodplain. Some communities flooded by the Mississippi River in 1993 were again flooded in 2001 (**FIGURE 12-21**). One method of making people aware of their risk is requiring flood insurance to sell a property or permits to modify or develop its land or buildings. This is often when the property loan must be secured by a flood insurance policy if it is within a floodplain. Most experts believe that real estate agents should be required to disclose flood risks when properties are offered for sale. Only since 2002 have virtually all banks and mortgage companies required that a residence be surveyed to see if it is on a floodplain. If it is, it needs to be covered by flood insurance before the mortgage company or bank will provide a mortgage on a property.

Flood insurance also provides an opportunity to encourage those making claims to relocate or rebuild homes with better flood protection. Recent, stricter controls require that where flood insurance funds are used for reclamation or rebuilding, the work has to conform to NFIP standards. For insured buildings, the rebuilt lowest floor must be on compacted fill and at least 2 feet above the 100-year flood level. After two floods in less than a decade, some 10,000 homes and businesses were approved for removal or non-rebuilding along more than 400 square kilometers of floodplain. As of 2003, flood insurance policyholders with a record of repeated losses receive funds only for the purpose of relocating outside the flood zone, elevating their homes above flood level, or doing flood proofing or demolition. Those who reject the mitigation offer are charged insurance rates based on standard actuarial costs for properties with severe repetitive losses.

A bill passed by the U.S. Senate in 2008 requires that people buy flood insurance if a levee breach will result in their homes being flooded. In spite of the hazard, many towns and their residents resist such regulations because they claim they curtail economic development and the cost of insurance premiums is prohibitive. 2008 flood insurance premiums for Illinois floodplain homes, for example, average about $400 per year.

Environmental Protection

Human alterations to the landscape can have significant effects on the magnitude of future floods. By protecting rivers and the watersheds that feed them, governments can also reduce the intensity of future floods. Current restrictions dictate that building or encroaching on a floodway must not raise the level of a 100-year flood by more than one foot (30 centimeters). Some states set more stringent restrictions of no more than a half-foot rise (15 centimeters). Bridge approaches and levees, however, commonly encroach on a channel. Waste disposal, storage of hazardous materials, and soil-absorption sewage systems, including septic tank drain fields, are prohibited in both floodways and flood fringes.

Changing government policy toward artificial river barriers also reflects a better understanding of natural river processes and may reduce future flood damages. A 1994 committee of federal experts recommended that levees along the lower Missouri River be moved back from the river by 600 meters to give the river room to meander and spill over its floodplain during high water. An ongoing federally funded wildlife-habitat project restored about 675 square kilometers of floodplain accessible to the river. It was a patchwork dependent on volunteer sales that continued into the early 2000s. However, they also authorized raising other levees.

Reducing Damage from Debris Flows

Debris flows are among the most dangerous of downslope movements because of their sudden onsets and high velocities (**FIGURE 12-22**). Debris flows can be highly destructive, even on slopes of less than 30 degrees that have thick brush or forest.

A broad alluvial fan spreading from the mouth of a desert canyon with picturesque boulders littering its surface is not a safe setting for residential development (**FIGURE 12-23**). The dangers include not only burial in heavy debris but also huge impact forces from fast-moving boulders. Hundreds of thousands of people live on gravelly alluvial fans in Los Angeles, Palm Springs, Phoenix, Tucson, Salt Lake City, Denver, and elsewhere. Most of these fans were built up from fast-moving slurries of sand, gravel, and boulders—debris flows. Even arid regions can have periods of intense or prolonged rainfall. People who live on such fans are at considerable risk from debris flows.

The best solution to minimize the impact of debris flows is to avoid building in vulnerable areas (**FIGURE 12-24**).

FIGURE 12-22 DANGEROUS FLOWS

A. In May 1998, a muddy debris flow from the steep hillside to the right of this house in Siano, Italy, east of Mount Vesuvius, had sufficient momentum to blow right through the walls of the house and out the other side. Debris on both balconies (arrows at left) provides an indication of the height of flow. **B.** It may not be obvious, but the car this geologist is standing on is at the roof level of a house buried by a bouldery debris flow (note the roof vent pipe above the car's roof).

FIGURE 12-23 DEBRIS FLOW HAZARD

People living in the modern housing subdivisions on the surface of alluvial fans in the Palm Springs area of California find that boulders make for great landscaping but seem unaware of how the boulders got there.

FIGURE 12-24 OBVIOUS HAZARDS

Debris-flow fan

Some debris-flow hazards are really obvious. New homes on a debris-flow fan in Georgetown, Colorado, west of Denver. Huge boulders among houses document previous debris flows.

Especially hazardous areas should be zoned as open space such as parks, golf courses, and agriculture. Building on the debris fan should be prohibited, and in some cases existing development should be bought out and removed to open pathways for future flows. Where development is necessary, buildings and streets should be oriented with their lengths parallel to the downslope direction of flow to limit building exposure to flows.

Although insurance cannot be purchased for landslides or other ground movement, debris flows and mudflows may provide some exceptions. The distinction for insurance purposes is generally that if the flow is too watery to be shoveled, then damages can be claimed under flood insurance. Thus, some fast-moving, watery debris flows may be covered.

Early Warning Systems

Where buildings predate recognition of a hazard, education and early warnings can minimize problems. Conditions can be monitored and alerts provided when values reach known thresholds for triggering an event. Warnings of increased hazard, during prolonged or heavy rainfall, for example, can help, but the steep terrain in which most debris flows occur leaves little time for evacuation once a flow begins moving. This is especially true at night in a heavy rainstorm. Storm sounds can drown out that of an approaching debris flow.

Detection devices can help warn people of an already moving debris flow or mudflow. The best are acoustic flow monitors that detect the distinctive rumble frequency of ground shaking caused by debris flows. The sensed motion is telemetered automatically to downstream sirens. Once alerted, people should immediately run to higher ground off to the sides of a debris flow or mudflow path. Such sensors are used, for example, at Mount Rainier and in Alaska, Ecuador, and the Philippines. Mudflows are most frequent around active volcanoes because of the abundance of volcanic ash. Prominent examples of recent volcanic mudflows include Mount Hood, Oregon, in November 2006; Mount St. Helens, Washington, in 1980; Mount Pinatubo, the Philippines, in 1991; and Nevado del Ruiz, Colombia, in 1985.

Where likely sources of debris flows are a short distance upstream and the channel gradient is high, the warning time may be too short for evacuation. A tripwire installed in the canyon three kilometers upstream from heavily populated areas on the Caraballeda fan in Venezuela after the December 1999 disaster would have provided only five minutes' warning for the large population there, almost certainly not enough for most people to move out of danger (**Case in Point:** Alluvial Fans Are Dangerous Places to Live—Venezuela Flash Flood and Debris Flow, 1999, p. 382). Inexpensive tripwire sensors are in use in many areas, but falling trees, animals, or vandalism can trigger false warnings.

FIGURE 12-25 DEBRIS FLOW PROTECTIONS

A. A modern debris-flow collection basin was built in Rubio Canyon, north of Pasadena, California. Downstream is to the upper left. **B.** A 1978 storm filled the basin behind a debris-flow dam and overtopped the dam at La Crescenta, near Pasadena. **C.** Owners of houses immediately below a debris-flow dam at the northern edge of Pasadena apparently trust the dam to block all debris flows and floods that may come down the canyon.

Trapping Debris Flows

Damage from debris flows can be limited through construction of barriers (**FIGURE 12-25**). Walls can be built to deflect large-volume flows to a part of a fan that has little development. Debris flows can also be channeled into a debris basin large enough to contain all loose material in the channel upstream. These must be cleaned out after each flow.

Structures used to trap debris flows in canyons upstream from alluvial fans include permeable dams that stop boulders but permit water to drain, that is, grid dams consisting of cross-linked steel pipes, horizontal beams, vertical steel pipes, or reinforced columns. Widely used in Canada, Europe, Japan, China, Indonesia, and the United States, they abruptly slow the progress of debris flows by draining the water. The grid is generally spaced to permit people, animals, and fish to easily travel through the structure.

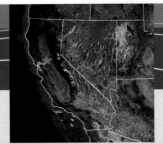

Cases in Point

Addition of Sediment Triggers Flooding
Hydraulic Placer Mining, California Gold Rush, 1860s ▶

The practice of hydraulic placer mining in California during the 1860s and 1870s provides a classic case history of how rivers respond to large volumes of added sediment load. Gold miners originally panned gold or separated it from sand and gravel in streambeds with small sluice boxes fed by water diverted from the stream. As those river gravels became depleted, the miners discovered gold in high-level terrace gravels above the streams. To separate that gold, they used high-pressure jets of water from higher elevations to hose down the gravels into large sluice boxes at stream level. The accumulated loose gravel was picked up by the streams during flood and washed downstream. The Bear River in the western Sierra Nevada, for example, built up its bed by as much as 5 meters in response to hydraulic gold mining upstream.

The first big flood from heavy rainfall in the Sierra Nevada in January 1862 flushed much of the placer gravel from tributaries into the main rivers and in turn out through the mouths of their canyons into the edge of California's Central Valley. The rivers became choked with sediment because they could not carry it all; channels filled with gravel, and the flood spread

▶ *Hydraulic gold mining in the Sierra Nevada in the 1860s.*

(continued)

far beyond where it should have—in some cases all the way down to San Francisco Bay. Previously productive farmland was covered with gravel, rendering it unusable. Cities were not much better off. The next catastrophic floods, in 1865, turned the Central Valley into an "inland sea" 20 miles wide by 250 miles long, submerging farms and towns. Similar floods occurred in the following 30 to 40 years. A total of 1.5 billion cubic yards of sediment spread out into the Central Valley.

Hydraulic mining was finally outlawed in 1884, but by then the damage was done. Landowners and governments tried to deal with the floods by the usual means of treating the symptoms: Build levees near the river to contain flooding; when those are topped during a subsequent flood, build them higher. Channelize and straighten the river to carry water through more quickly and prevent it from backing up to form a lake. Build dams to contain the floods. These actions, however, made matters worse downstream. Floodwaters raced right down the channel rather than spreading out across floodplains to slowly drain back into channels as the flood waned. Flood levels between the levees were much higher, so water flowed faster.

Downstream towns built levees to protect themselves, but the sediment-choked channels contained so much new sediment from upstream that their beds in some cases built up higher than the towns behind the levees. On January 19, 1875, a modest flood along the Yuba River breached a levee at Yuba City, north of Sacramento, sending a flood of gravel through the town, all but destroying it.

The hydraulic placer-mining fiasco may be past, but other landscape alterations such as overgrazing or deforestation by fire can lead to similar results, including, in some areas, braided streams.

Streambed Mining Causes Erosion and Damage
Healdsburg, California ▶

The Russian River downstream from Healdsburg, California, north of San Francisco, provides an excellent example of the consequences from river-channel gravel mining. Beginning in 1946, both the region's population and large construction projects increased the need for sand and gravel mined from the Russian River watershed. By 1978, private companies were mining 4 million metric tons per year of sand and gravel along the Russian River, mostly downstream from Dry Creek.

Of this, 80 percent came from stream terraces and 10 percent from draglines within the stream channel. They excavated

F. M. Mann Jr., courtesy of Byron Olson.

A

▶ **A.** *Dry Creek bridge supports near Healdsburg, California, have been severely undercut by stream-bed gravel mining downstream.* **B.** *Downcutting of part of the Russian River channel between 1940 (before gravel mining began) and 1972 as a result of gravel-mining operations downstream along the Russian River in northern California.*

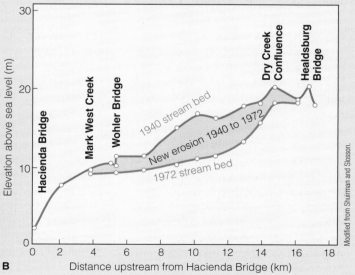

Modified from Shurman and Slosson.

B Distance upstream from Hacienda Bridge (km)

(continued)

gravel in pits as deep as 18 meters below water level in a stream that is typically less than 1 meter deep.

Recall that the slope of a stream strives to remain in equilibrium with the amount and grain size of supplied sediment and the erodibility and cross section of its channel (see "Dams and Stream Equilibrium," p. 363). If any of these factors change, the others will adjust to help bring the stream back toward equilibrium. In the early 1970s, Byron Olson and other fruit farmers along Dry Creek, upstream from Healdsburg, noticed that riverbanks

were eroding, steepening, and collapsing more rapidly in storms than they had in the past; the stream channel became wider at the expense of adjacent orchards and vineyards.

Local farmers blamed the gravel miners for the increased erosion and filed suit for damages. Detailed stream studies followed that eliminated large storms, fire or flood events, and land use changes as causes for the observed increase in discharge and erosion. By 1972 the total channel deepening of the Russian River upstream from Dry Creek ranged from

2 meters at the confluence to 5.7 meters some 5 kilometers upstream and 2 meters some 9 kilometers upstream. Dry Creek eroded its bed 3.3 meters deeper for 10 kilometers upstream from the confluence. It also undercut and threatened bridge supports.

The removal of so much gravel left Dry Creek with much less bedload to carry, locally increased the channel gradient into the deep mining pits, and increased turbulence. This resulted in more aggressive downcutting all the way from the mining areas into Dry Creek.

The Potential for Catastrophic Avulsion
New Orleans ▶

Originally settled in 1718 as a French colony on a natural levee of the Mississippi River 4.5 meters above sea level, New Orleans soon built low artificial levees to protect itself in times of flood. By 1812, the levees extended 114 kilometers upstream. New Orleans has battled the river ever since; with each large flood, the levees were built higher. Major levees built in 1879 to protect New Orleans broke in 1882 in 284 places. In 1927, with 2,900 kilometers of levees, the flood broke through in 225 places. In the 1973 record flood, the river submerged 50,000 square kilometers of "levee-protected" floodplain. Then in 2005, Hurricane Katrina surged over levees on the east side of New Orleans and submerged much of the city.

By 1900, the city began to spread north into the marshes of the floodplain by building canals and draining them with pumps. Peat on the floodplain began to compact because of dewatering and the weight load of buildings and roads, so the city began to settle. Parts of it are now almost 4 meters below sea level, even farther below the Mississippi River that flows along the levees right next to downtown. Ships on the Mississippi look down on the city. High-capacity pumps capable of pulling 1,100 cubic meters per second keep the groundwater at bay and the ground free of water, even after torrential rains.

Each additional levee further confines the river, which aggravates its tendency to flood downstream. Each time the river rises higher during flood, it flows faster. Instead of permitting the river to spread over its floodplain to shallow depths during flood, levees along the Mississippi and tributaries exacerbate the problem by raising the river level well above the natural floodplain.

As the Mississippi River carries sediment to the Gulf of Mexico, it builds its delta seaward, thereby decreasing the river's slope in the depositional delta area. At the same time, the river builds both its bed and natural levees higher. During a major flood, the river breaches the natural levees, and its water heads down a steeper, shorter, lateral path to the sea. Because water flows faster on the steeper slope, it begins to carry more of the river's flow, gradually taking over to become the new main channel. Sixty to seventy years ago, a new distributary channel, the Atchafalaya River, formed and began to sap some of the flow from the main channel. If the Atchafalaya were to take over the whole flow of the river, New Orleans would be left high and dry without its river; that would destroy its key role as a shipping center for the whole Mississippi basin.

Recognizing the problem, the U.S. Army Corps of Engineers in the 1960s

Natural levee

Mississippi River

Marli Miller.

▶ *New Orleans, the major shipping center on the lower Mississippi River, is protected by levees such that the river now stands four meters above the downtown area.*

built the Old River control structure to permit 30 percent of the flow to enter the Atchafalaya, keeping 70 percent in the main channel through New Orleans. The idea was to regulate flow so that floodwaters could be channeled into the Atchafalaya to save New Orleans and other towns from flooding. Some of the sediment carried by a flood could also be channeled off to minimize further siltation of the main channel through New Orleans. The 1973 Mississippi River flood almost destroyed the Old River control structure, thereby

(continued)

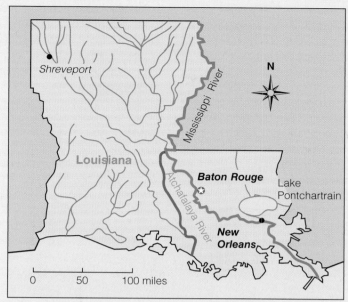

▶ *In Louisiana, the Mississippi River takes a long, gentle path to the Gulf of Mexico. The Atchafalaya River drains part of the flow of the Mississippi upstream from Baton Rouge, taking a shorter and therefore steeper path.*

channeling the main flow into the Atcha-falaya River.

New Orleans is now at the mercy of the river. If a catastrophic flood should breach Mississippi River levees in the area, much of the city would drown in as much as 7 meters of water, along with a thick layer of sand and mud brought in through the breach. Could it happen? It was not supposed to happen in the upper Mississippi in 1993. Nor was it supposed to happen in 2005 when Hurricane Katrina surged in from the east. If a catastrophic flood destroyed the Old River control structure, it could lead to dominance of the Atcha-falaya River over the current Mississippi course through New Orleans. Or the U.S. Army Corps of Engineers could decide to deliberately breach the barrier at the Old River control structure to save New Orleans. Either way, the city loses.

In addition to permanently flooding and eroding large floodplain areas, the disruptions to a city that depends on river shipping for its economic livelihood can be severe. Avulsion of the Mississippi River above New Orleans would be economically catastrophic for the city. The Yellow River avulsion in 1855 took only one day. Avulsion of the Saskatchewan River in the 1870s turned more than 500 square kilometers of its floodplain into a belt of braided channels and small lakes. How long can the U.S. Army Corps of Engineers keep the Mississippi confined and prevent it changing course to the straighter, steeper path to the ocean?

A Long History of Avulsion
Yellow River of China ▶

The 4,845-kilometer-long Yellow River (Huang Ho) drains most of northern China, an area approximately 945,000 square kilometers. That is a similar length but less than one-third the drainage area of the Mississippi River. The upper reach of the river, flowing generally east from the Tibetan highlands, carries relatively clear water through mountains and grasslands for more than half the river's length. The middle reach south from Baotou and the deserts and plains of Inner Mongolia drains a broad region of yellowish wind-deposited silt, or loess, originally blown from the Gobi Desert in Mongolia to the northwest.

Sediment supply to the river from the loess plateau is vigorous because of vast arid-to-semiarid hilly areas of easily eroded silt. Before heavy agricultural use of the loess plateau began in 200 B.C., it was mostly forested, and the sediment load fed to the river would have been one-tenth the current amount. After the tenth century A.D., agriculture had largely destroyed the natural vegetation. Silt supplied by sheet-wash and gully erosion is so abundant that floods carry hyperconcentrated loads of yellow-colored sediment that gives the river its name. Average sediment load during a flood is generally greater than 500 kilograms per cubic meter, which is 20 percent by volume or 50 percent by weight.

The lower reach of the Yellow River, downstream from Zhengzhou, flows

▶ *The largest remaining area of loess tableland at Dongzhiyuan in Gansu, China, is being rapidly eroded. The size of the view can be inferred from the road around the end of the deep canyon.*

(continued)

across a densely populated and cultivated alluvial plain, one affected repeatedly by flooding for more than 4,000 years. As the river gradient decreases and spreads out over a width of several kilometers, sediment deposits progressively raise the channel bottom; this in turn requires regular raising of the levees. Near Zhengzhou, sediment deposition accumulated on the river bottom by an average of 6 to 10 centimeters per year, aggravated flooding, to rapidly fill the reservoirs behind dams.

As early as 4,000 years ago, Emperor Yu dredged the channel and dug nine separate diversion channels to divert floodwaters. In 7 B.C., Rhon Gia advised evacuating people rather than fighting the river, but no one followed his wise suggestion. After a disastrous flood in A.D. 1344, people used a combination of river diversion, river dredging, and dam construction. After each flood, they plugged breached levees and raised existing ones. With the channel elevated by deposition, some breaches drained the old channel and followed an entirely new path to the sea—that is, by river avulsion. Unfortunately, as with most rivers, levees here are built from the same easily eroded silt

that fills the channel; the river erodes the levees just as it does the loess plateau. Downstream from a breach, the river bed between levees is left dry, a serious problem for those who are dependent on its water.

In 1887, the river topped 20-meter-high levees and followed lower elevations to the south to reach the East China Sea at the delta of the Yangtze River at Shanghai. The flood and resultant famine killed more than one million people. Official government policy since 1947 has been to contain floods by flood-control dams and artificial levees along the channel. However, these structures are designed to control a flood with a recurrence interval of only 60 years, clearly not a long-term solution. The bed of the river is now as much as 10 meters higher than the adjacent floodplain. Because the riverbed is well above the surrounding landscape, the lower 600 kilometers of the river receives no water from either surface runoff or groundwater.

In 1960, the Chinese completed the San-men Gorge Dam, 122 meters high and more than 900 meters wide. Its 3,100-square-kilometer reservoir,

designed for flood control and electric power generation, is now filled with sediment and is useless for both of its designed purposes. At the same time, a major effort was launched to plant trees and irrigate huge areas of the silt plateau to reduce the amount of silt reaching the Yellow River. Unfortunately, most of the trees died. As with droughts, construction of reservoirs and extraction of water for irrigation and other uses leads to low downstream flow, especially since 1985. Flood level in 1996 was at an all-time high, even though the discharge was much lower than the 1958 and 1982 floods.

The lower 800 kilometers of the Yellow River has repeatedly shifted its course laterally by hundreds of kilometers. China has lived with the Yellow River and its floods for thousands of years and has tried to control the floods with levees just as we do. It has not worked. Levees repeatedly failed and killed thousands of people. Avulsion dramatically changed the river course several times. With each flood, the people built levees higher and they still failed. There should be a message here; our levees fail just about as frequently.

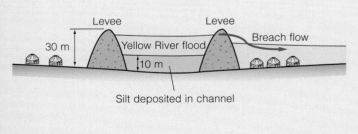

▶ **A.** *Levees of the Yellow River stand high above the surrounding floodplain.* **B.** *Aggradation of the channel of the Yellow River, between levees, has raised the channel some 10 meters. Even with 30-meter levees, breaches often lead to avulsion and abandonment of the main channel.*

Repeated Flooding in Spite of Levees
Mississippi River Basin Flood, 1993; a sequel in 2008 ▶

The Great Flood of 1993

It was a wet winter and spring in the northern Great Plains, so most of the ground was saturated with water from rain and melting snow, and the rivers were rising. Unfortunately, heavy rainstorms continued as flood waters arrived from upstream. The storms kept forming in the same area; it rained and rained, literally for months. Highways, roads, and railroads in the upper Mississippi River basin were submerged for weeks on end, along with homes, businesses, hospitals, water-treatment plants, and factories.

Almost 100,000 square kilometers, much of it productive farmland, lay underwater for months. Wells for towns and individual homes were flooded and contaminated, requiring the boiling of domestic tap water. Even large cities were affected. Des Moines, Iowa, went without drinking water and electric power for almost two weeks. A few towns on the floodplain, such as Grafton, Illinois, at the confluence of the Illinois River with the Mississippi, are not protected by levees. When the water rises, people merely move out, then clean up afterward. But after six big floods in twenty years, some people began to think

about moving the town. Small towns such as Cedar City and Rhineland, Missouri, accepted a governmental buyout of their flood-damaged homes and moved off the floodplain. The homes were bulldozed to create public park land.

In spite of extensive high levees, more than 70,000 homes in the Mississippi River basin flooded, predominantly those of poor people living on the floodplains, where the land is less expensive. Some houses lay for weeks in water as high as halfway up the second story. After the lengthy flood, putrid gray mud coated floors and walls, and plaster was moldy to the ceiling because it and the insulation wicked it up. Belongings that could not be moved in time had to be thrown out and carted away.

In all, fifty people died, and damages exceeded $26.5 billion (in 2010 dollars), the worst flood disaster in North American history—that is, until the Hurricane Katrina–driven flooding of New Orleans in 2005.

The first half of 1993 had twice the average precipitation. By late June, flood-control reservoirs in the upper Mississippi River basin were nearly full. The storms kept forming in the same area; it rained and rained, literally for months. Between April and August

most of the area was drenched with 60 centimeters of rain; regions in central Iowa, Kansas, and northern Missouri received more than one meter. On June 17 and 18, 7 to 18 centimeters of rain fell in southern Minnesota, Wisconsin, and northern Iowa. In late June, flooding in Minnesota was the worst recorded in 30 years. The floods killed a total of 24 people.

Weather systems generally move east, but 1993 was different, especially in late June and July. The jet stream moved farther south, bringing cool, dry air from Canada and circulating around a low-pressure system in southwestern Canada. That part of the jet stream swept northeastward from Colorado toward northern Wisconsin. At the same time, warm, moist air pulled into the central United States from the Gulf of Mexico collided with the cool, dry air to produce persistent low-pressure cells and northeast-trending lines of thunderstorms centered over Iowa and the surrounding states. Even this was not particularly unusual. What was unusual was its coincidence with a persistent high-pressure system that stalled over the southeastern coast. That high kept the storms from moving east as they would normally have done.

▶ *The Missouri River flooded, spilled over its floodplain, and submerged U.S. 54 near Jefferson City, Missouri, in late July 1993.*

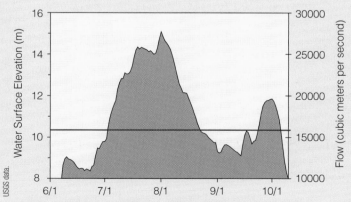

▶ *The 1993 flood hydrograph for the Mississippi River at St Louis. The river was above flood level for more than one month.*

(continued)

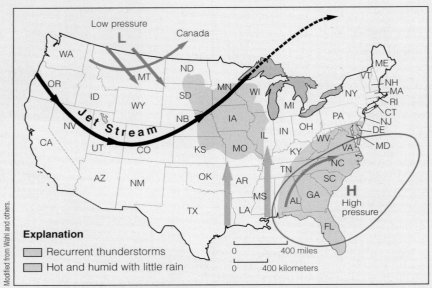

Modified from Wahl and others.

▶ *The dominant weather pattern for June and July 1993 that created the Mississippi River floods included a stationary low over western Canada and a persistent high (red oval) off the Southeast coast. The main flooding was in the blue outlined area.*

On the fifth of July, for example, a stationary weather front extended from northern Missouri to southeastern Wisconsin. A series of cold fronts rotated counterclockwise around a low and collided with the warm, moist air over Iowa. This collision lifted the warm air, causing condensation and thunderstorms. Strong upper-atmosphere winds of the jet stream generated a chimney effect that created additional updraft. Storms in Iowa between July 4 and 24 dumped a total of 82 centimeters of rain on various parts of the region.

Downstream Flooding

The flood crest moved downstream while it continued to rain. The flood wave moved downstream at approximately two kilometers per hour, so it was relatively simple to predict when the maximum flood height would occur at any location. In reality, however, other factors came into play. Tributaries added to the flow, and broad areas of floodplain that were not blocked by levees removed flow and then gradually released it to the river. Tributaries backed up and even flowed upstream. Most places experienced multiple flood crests. Between Minneapolis and Clinton, Iowa, high bluffs confine the river to a narrow floodplain. From Clinton down to St. Louis, the floodplain is wider, and levees of different heights had been built by various agen-

cies. From St. Louis down to Cairo, Illinois, they channelized the Mississippi River to maintain dikes for an average river depth of 7 meters. In St. Louis the river crested at nearly 6 meters above flood level; it was above flood level from late June through August, then again from September 13 to October 5. The U.S. Army Corps of Engineers halted all river-barge traffic on the Mississippi north of Cairo, Illinois, in late June because it could no longer operate the locks and dams along the river.

August 1993 had the largest flood on record for the upper Mississippi; in most places in this drainage, the recurrence interval of the flood was 30 to 80 years. In the lower Missouri River drainage basin in Nebraska, Iowa, and Missouri, the peak discharge was greater than a 100-year event. River flow reached 29,000 cubic meters per second at St. Louis, 160 percent of the average flow at New Orleans near the mouth of the river.

As floodwaters rose, locals, National Guard personnel, and others dumped loads of crushed rock and filled sandbags to raise the height of critical levees across the region. For weeks on end, it seemed the work would never stop. As the higher levees raised river levels, they became saturated, causing slumping. Crushed rock in the levees minimized that. People inspecting a levee would sound

an alarm if wave erosion became significant or if they found a leak. Cloudy water indicated that soil was being washed out of a levee through a process called piping.

The levee across the river in tiny West Quincy, Missouri, stands about 9 meters (the height of a three-story building) above the fertile farmland to the west. In 1993, the levees held until one night a young local man pulled a few sandbags from the top of one. Within a few minutes, a trickle grew into a torrent and then a massive breach that submerged West Quincy under 4.5 meters of water and flooded farmland for 10 kilometers to the west. He was sent to prison. He said he wanted the bridge to wash out so he could party and have an excuse to not go home. National Guardsmen then patrolled the levee at night; locals with guns supplemented the patrol to ensure no sabotage.

Rivers across the basin overtopped or breached numerous levees as the flood crest reached them. There were 1,576 levees on the upper Mississippi River. Over 75 percent of those built by local or state agencies were damaged, while less than 20 percent of the 214 that were federally constructed suffered damage. Levees failed from north of Quincy, Illinois, to south of St. Louis, Missouri, and on the Missouri River from Nebraska City, Nebraska, south through Kansas and Missouri to St. Louis. Once a river breached a levee, it flushed sand and gravel over previously fertile fields, flooded the area behind the levee, and flowed downstream outside the main channel. Each levee breach lowered the river level, sparing downstream levees, at least for a while. During the peak flooding, 25,000 square kilometers of floodplain was underwater.

Individuals, towns, and government agencies go to great lengths to maintain and raise levees to protect towns from floods. So why would levee district officials and the U.S. Army Corps of Engineers intentionally sever a Mississippi River levee? They did just that in southwestern Illinois on August 1, 1993, to save the historic town of Prairie du Rocher. Fifteen kilometers north of Valmeyer, Illinois, a levee failed, permitting Mississippi River floodwaters to flow onto the floodplain on the east side of the river. There it

(continued)

U.S. Army Corps of Engineers.

▶ **A.** *In the breach of this levee during the 1993 flood on the Mississippi, the river (on the left) spills through two breaches to cover the floodplain (right). A second levee at the right edge of the photo provides temporary relief.* **B.** *This 1993 Missouri River flood near Jefferson City, Missouri, left many homes deeply submerged in floodwaters for an extended period of time.*

began moving south, soon overtopping a pair of levees on a tributary stream and continuing south behind the main levee to flood Valmeyer.

Twenty-seven kilometers farther south, another tributary stream flanked by levees of its own would stop the floodplain water and protect the town of Prairie du Rocher unless they also were overtopped. However, water in part of a floodplain enclosed by levees will gradually rise to the level of the inflow breach upstream, because it behaves like a lake. The Mississippi River, of course, decreases in elevation downstream; levees protecting Prairie du Rocher were high enough to keep out the advancing flood but not as high as the Mississippi River at the breach 42 kilometers upstream. That means that the "lake" behind the main Mississippi River levee would rise well above the flood level on the Mississippi and flood Prairie du Rocher.

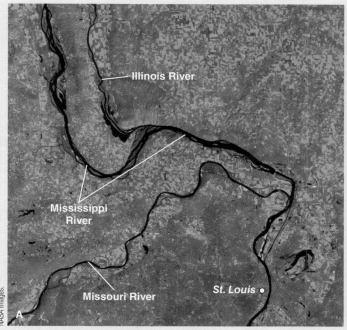
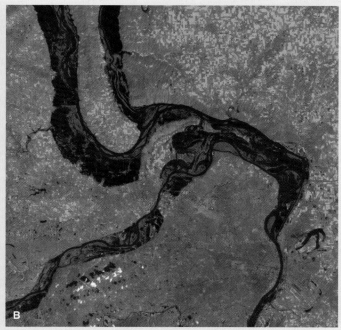

NASA images.

▶ *The Mississippi River flood of 1993 can be clearly seen by comparing satellite images taken* **A.** *before and* **B.** *during the flood. Note that the river fills its floodplain except in the channelized St. Louis reach.*

(continued)

The solution was to deliberately breach the main Mississippi River levee one kilometer upstream from Prairie du Rocher to permit the "lake" behind the levee to flow back into the Mississippi.

Officials and the Corps of Engineers breached the levee before the floodplain flood reached Prairie du Rocher so that the backflood of Mississippi River water would cushion the oncoming wall of

water on the floodplain. When the "lake" behind the floodplain rose higher than the Mississippi, it again flowed back into the river.

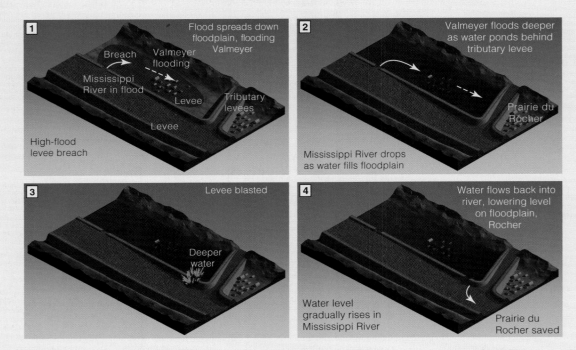

▶ *This three-dimensional drawing shows the Mississippi River, levees, and floodplain in the Valmeyer–Prairie du Rocher area of Illinois. The sequence of events proceeds with the breach of the main levee and the flow south along the floodplain, to the deliberate breach to release water back into the Mississippi River near Prairie du Rocher. The water on the floodplain (near Prairie du Rocher) is blocked behind the side-stream levee; it will gradually rise as a "lake" to the level of water at the upstream breach. That ponded water is higher than the water flowing in the adjacent Mississippi River (third image). Explosive creation of an artificial breach permits the ponded water to drain back into the Mississippi River, as shown in the fourth image. That drainage reduces pressure on the tributary levee, protecting Prairie du Rocher so it can be drained back into the river.*

Flood Severity Increases in a Flood-Prone Region
Upper Mississippi River Valley Floods, 2008 ▶

In June, 2008, persistent drenching rains fell in the upper and central Mississippi drainages of South Dakota, Iowa, Wisconsin, Illinois, southern Indiana, Missouri, Nebraska, Kansas, and eastern Oklahoma. Heavy winter snowfalls and heavy spring rains saturated the ground and prevented later rains from soaking in. The precipitation from January to April, 2008 was 33% above average. Severe flooding affected

cities throughout the region. Levees broke, roads, railroads, and towns flooded, bridges collapsed, sewers overflowed. Shipping locks on the Mississippi River were closed because of high water, halting commercial shipping on the river. Lives and businesses were disrupted and farm crops on floodplains were destroyed. An estimated 16,000 square kilometers of cropland flooded in the Midwest.

On June 17, the levee protecting Gulfport, Illinois, failed, submerging much of the city under three meters of water. FEMA contracts with outside engineers certified that a levee can hold back a 100-year flood. Unfortunately, the floodplain maps on which they depend were 20 years old and outdated—river conditions had changed, and levee quality varies. The continued building of homes, shopping centers, and

(continued)

parking lots on the floodplain caused more-rapid runoff into streams and higher flood levels. In addition, the flood at Gulfport was estimated to be roughly a "500-year event."

The Cedar River in Iowa rose to more than 3.6 meters higher than in 1993, breaking a 157-year-old record and causing more than $1 billion in estimated damages in Cedar Rapids, Iowa, a city of about 124,000. About 23.8 square kilometers, 1,300 city blocks, went under water. On June 12, 2008, the rising Cedar River in Cedar Rapids caused collapse of a railroad bridge that was weighted down with hopper cars full of rocks to help keep it from washing away. Three of the city's four main drinking-water wells were contaminated by floodwater fouled by petroleum and had to be shut down. The remaining well was protected by sandbags and pumps that removed water from around it. Flooding and power outages from storms forced temporary dumping of untreated sewage into different rivers. Some flooding of low-lying houses and basements was caused not by failed levees but by the rise of groundwater due to higher river levels between levees.

In some cases, sewer lines backed up and flooded basements with foul water. After two weeks under water, basements were often tainted by toxic black mold that ruined much of their contents. In late July, most of the 1,000 flooded businesses in Cedar Rapids were still closed. 5,000 homes were not much better off. In Iowa City, the Iowa River flooded a dozen buildings of the University of Iowa in spite of a week of sandbagging.

Hundreds of FEMA workers, state and local employees, national guardsmen, volunteers, and even prisoners helped with sandbagging. Some levees held; others did not. In Missouri, numerous levees failed, flooding hundreds of square kilometers of floodplain. In some places, topsoil was eroded from fields; in others, sand was deposited on them. Individual thunderstorms dumped up to 15 centimeters of rain on certain areas; one place in Iowa got 7.5 centimeters in an hour. Severe crop losses were reported for corn and winter wheat. By June 17, Iowa had lost almost 10 percent of its corn crop, but in July, weather conditions turned ideal, and

by early August, the remainder of the crop was in good condition.

Levees failed where flood levels were higher than in 1993 or saturated earth collapsed in small landslides. Hogs, flooded out from nearby farms, climbed onto roofs and levees to avoid drowning. Some were shot because of concern that their hoofs would puncture sandbags. Others were taken to safety in boats. Elsewhere a muskrat dug a hole through a private earth levee in eastern Missouri, causing mudslides that breached the levee. In some places, flood-levels dropped as nearby levees failed, then rose again. Lake Delton, a manmade, one-square-kilometer water body in Wisconsin, drained suddenly on June 10 after storms overfilled it and eroded its earthen dam. Four homes on its muddy banks collapsed into the lake.

When levees in the Midwest failed, leaving houses and businesses under meters of water, people again looked to FEMA for help. By June 25, the agency had approved more than $88 million in Wisconsin, Iowa, and Indiana, for immediate needs such as temporary housing and home repairs.

▶ *The Cedar River and downtown Cedar Rapids, Iowa, on June 13, 2008.*

Catastrophic Floods of a Long-Established City
Arno River Flood, Florence, Italy, 1966 ▶

As in many other old European towns, multistory buildings are constructed right next to the channel of the Arno River. The river itself is channelized through the city, with vertical walls raised only a little above street level. In

the 1600s, walls almost 9 meters high were built on the riverbanks, reducing the bankfull channel width from 300 meters to its current 150 meters. Even at low water, the river reaches both walls in many places. Although

(continued)

the surrounding region is hilly, the old part of town, including most of its famous museums and churches, is on a broad, flat floodplain one to two kilometers wide.

In Florence, the Arno is only 160 kilometers from its headwaters but has a record of catastrophic floods, including those in 1117 and 1333 that decimated the city and its bridges. Two years before a disastrous flood in 1547 that killed more than 100 people, Bernardo Segni pointed out that cutting so many trees for timber in the mountains upstream permitted water to erode the soil and to silt up the beds of the rivers. Thus, humans had contributed to the flood. In spite of major floods averaging one per 26 years, Florence remained unprepared.

The largest flood ever on the Arno River occurred on November 3, 1966. Following an exceptionally wet October that saturated soils, a heavy storm dumped 48 centimeters of rain on Florence and the Arno headwaters, a third of the average annual rainfall. Discharge reached 2,580 cubic meters per second and overfilled reservoirs. The flood tore through small towns upstream and continued downriver at almost 60 kilometers per hour into its narrow, concrete-lined river

channel within Florence. At 2:30 a.m. on November 4, floodwaters rapidly rose to a depth of 6.2 meters, 2 meters higher than the 1333 flood.

By 4 a.m., water invaded the main square and was soon 1.5 meters deep in the Piazza Duomo. The raging floodwaters rose to the roadway of the famed Ponte Vecchio, threatening the famous bridge that has spanned the Arno River since Roman times; it was previously destroyed by floods in 1117 and 1333 and rebuilt each time. A bus carried downriver by the raging torrent crashed into the bridge during the 1966 flood, opening a huge hole. Ironically, that permitted water to pass, probably saving it from complete destruction. By 7 a.m. on November 4, water one to two meters deep completely covered the central part of the city and was six meters deep in other parts of town—up to third-floor levels. Heating oil from thousands of basement tanks flushed to the surface and was carried along with the floodwaters, contaminating everything it touched. For a city whose claim to fame is being one of Europe's most valuable centers of culture and art, the effect was disastrous, with damage from mud and polluted water

and the destruction of priceless medieval and Renaissance paintings, sculptures, and books, many of which were stored in basements. Twenty-nine people died.

A significant part of the blame for these devastating floods was attributed to the residents. Since pre-Roman settlement in the region, they had stripped natural vegetation from the hills. That had produced an annual cycle of winter floods and summer droughts. The 1966 problem was compounded by the failure to gradually release water from two hydroelectric dams upstream from Florence during heavy October rains. Late on November 3, the dam operators realized they had a problem, so they released a huge mass of water from the upstream dam, which in turn required immediate opening of the downstream dam, unleashing a wall of water.

In the last few decades, the extraction of gravel from the Arno River channel for construction materials and the construction of reservoirs upstream have caused increased channel erosion. Little has been done to rectify the basic causes, but most vulnerable art works are now kept on the upper floors of buildings and out of range of future floods.

▶ **A.** *A street in Florence following the 1966 flood.* **B.** *Boulders and flood debris from upstream.* C. *Arno River and Ponte Vecchio now.*

Alluvial Fans Are Dangerous Places to Live
Venezuela Flash Flood and Debris Flow, 1999 ▶

The rainy season in coastal Venezuela is normally from May through October, so a storm in the first two weeks of December 1999 was unusual. A moist southwesterly

flow from the Pacific Ocean collided with a cold front to bring 29 centimeters of rain. Then on December 14 through 16, when soils were already soggy, torrential rains

(continued)

dumped 91 centimeters near sea level at Maiquetia International Airport in 52 hours! Imagine almost a meter of water on the landscape in a little more than two days, all headed downslope. Higher elevations received twice as much rainfall as areas along the coast. This was the area's greatest storm in more than half a century.

Flash floods and debris flows inundated the coastal towns of Maiquetia and La Guaira, 56 kilometers north of Caracas. Most homes and buildings in this area crowd large alluvial fans and narrow valley bottoms at the base of incredibly steep, unstable mountainsides. Floods began after 8 p.m. on December 15. Eyewitnesses who fled the river and watched the flood from nearby rooftops reported crashing rocks and debris flows. Massive mudslides and floods killed an estimated 30,000 people and left more than 400,000 homeless. Exact numbers are difficult to determine because muddy slides buried many people or carried them out to sea.

Mud and debris either swept away or buried shantytowns of tin-and-cinderblock shacks that covered extremely steep mountainsides. Although the potential costs for property damage were low in these barrios, the potential for loss of life was high.

Losses totaled $2.31 billion (in 2010 dollars), with heavy damage to homes, apartment buildings, roads, telephone and electric lines, and water and sewage systems. Because the only non-mountainside land along this coastal part of Venezuela is on alluvial fans, the large fan at Caraballeda was intensively developed with large multi-story houses and many high-rise apartment buildings. At the upper end of the fan, the flood flow was above channel capacity; on reaching the fan, the flood separated into several streams and spread debris throughout the city, up to six meters thick in some places. Outside the main channel, flows

destroyed many two-story houses. They impacted the second stories of several apartment buildings, leaving boulders more than one meter in diameter. The ends of several buildings collapsed. Residents described several high streamflows and debris flows that began the night of December 15 and continued until the next afternoon.

This was not the first disastrous debris flow in the region. Eight similar events are found in the historic record between 1798 and 1951. Examination of older debris-flow deposits shows that some flows were larger than the 1999 flow—thicker and with larger boulders.

Food, water, clothing, and antibiotics were in short supply. Evacuation and disaster recovery was especially difficult because many areas were hard to reach. The single highway along the coast was extensively blocked and cut by debris flows and flood channels.

A

B

▶ **A.** *The debris flow and flash flood in Venezuela destroyed most of the homes on the low-lying Caraballeda fan at the mouth of the canyon.* **B.** *The end of this apartment building collapsed when debris-flow boulders crushed key support columns. The largest boulder is more than two meters high.*

Matt Larsen, USGS.

Dams Can Fail
Failure of the Teton Dam, Idaho ▶

The Teton Dam, near Rexburg in eastern Idaho, was built by the U.S. Bureau of Reclamation to provide not only irrigation water and hydroelectric power to

east-central Idaho, but also recreation and flood control. After dismissal of several lawsuits by environmental groups, construction began in February 1972, and

(continued)

filling of the reservoir behind the completed dam commenced in October 1975. The dam was an earth-fill design, mostly using a fill of wind-blown silt, 93 meters high and 945 meters wide. It had a thin "grout curtain," or concrete core wall to prevent seepage of water through the dam. The intensely fractured rhyolite bedrock below contained large gas-vent openings; the largest openings were filled with concrete slurry along only a single line because the amount of concrete needed was more than expected. The fill material deep in the dam was poorly compacted. Filling of the reservoir behind the dam in mid-May was at almost one meter per day, three times the normally permitted fill rate.

On June 3, 1976, workers discovered two small springs just downstream from the dam. On June 5 at 7:30 a.m., a worker discovered muddy water flowing from the right abutment (viewed downstream). Although mud in the water indicated it was carrying sediment, project engineers did not believe there was a problem. By 9:30 a.m., a wet spot appeared on the downstream face of the dam and quickly began washing out the embankment material. The hole expanded so rapidly that two bulldozers trying to fill it could not keep up and were themselves lost in the hole. At 11:15 a.m., project officials told the county sheriff's office to evacuate the area downstream. At 11:55, the crest of the dam collapsed; two minutes later, the reservoir broke through and rushed downstream. The flood obliterated two small towns, spread to a width of 13 kilometers over Rexburg, with a population of 14,000, and continued downslope at 16 to 24 kilometers per hour.

The flood killed eleven people and 13,000 head of livestock and cost the federal government almost $1.2 billion (in 2010 dollars) in claims and other costs. The cause of failure was never settled, but numerous flaws came to light. After construction started, U.S. Geological Survey geologists expressed concern about pressures from a filled reservoir and loading that could cause movement around the dam, as well as internal shearing, endangering the dam. It was the highest dam that ever failed and marked the end of large dam building in the United States.

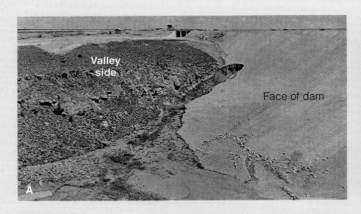

U.S. Army Corps of Engineers and U.S. Bureau of Reclamation.

▶ *Progressive failure of Teton Dam, eastern Idaho, June 5, 1976:* **A.** *At 11:20 a.m., muddy water pours through the right abutment of the dam.* **B.** *At 11:55 a.m., the right abutment begins to collapse and a large volume of muddy water pours through the dam.* **C.** *and* **D.** *Just after noon, the dam fails and the reservoir floods through it.*

Critical View

Test your observational skills in interpretation through the following scenes relevant to floods and human interactions.

A This is a dry-climate area in western Montana. A modern house and garage are built in the bottom of a canyon next to a tiny seasonal stream.

Donald Hyndman.

1. What natural hazard should have been considered before building?
2. Where would have been a better location for the house in the area visible in this photo?
3. What other natural hazard is apparent? Consider the subjects of preceding chapters.

B This vehicle camped overnight in a sheltered gravelly area in Death Valley, California.

D. Steensen, National Park Service.

1. What process destroyed the vehicle?
2. Describe the likely size and nature of the destructive process.
3. What clues are there in the photo that the dry channel could be prone to floods?

C This bouldery deposit severely damaged this house in western Colorado.

1. What process likely deposited the material?
2. A widespread event affected the forest about a year before deposition of the material. What kind of event or events would lead to that movement and deposition of the boulders?
3. Why are the larger boulders sitting at the top of the deposit, rather than sinking to the bottom?

S. Cannon, USGS.

D A foot-and-horse bridge across MacDonald Creek in Glacier National Park was destroyed in this flood in November 2006.

Shaun Bessinger, National Park Service.

1. What could the builders of the bridge have done to ensure it survived a flood like this?

2. What could lead to such an unusual flood well after the spring wet season and after a dry summer?

F This new, unfinished home with a great view was undercut and severely damaged along a river in northwestern Washington State.

M. Nauman, FEMA.

E Flooded home on the Salt River floodplain near Lebanon, Kentucky.

K. Crawford, USACE.

1. Would normal homeowner's insurance cover damages of this type? Why or why not?

2. Could the homeowner have purchased flood insurance to pay for damages on this home?

3. What type of damages would be expected in this environment?

4. Note the For Sale sign out front. Why would this home be a poor investment, even if offered at a bargain price?

1. What normal stream process led to the failure of the river bank under this home?

2. What conditions might have accelerated the collapse of this bank?

3. What would have been a better choice for a building site near the river?

Chapter Review

Key Points

Development Effects on Floods

- Urbanization aggravates the possibility of flash floods because it hastens surface runoff to streams.

- Cars driven into a flooded roadway with water above their floorboards are often pushed off the road, which can cause their occupants to drown. Most vehicles will float until they fill with water and sink. **FIGURE 12-2**.

- Deforestation and overgrazing increases erosion and adds sediment to streambeds, making streams steeper and more prone to flooding.

- Mining of stream gravel removes sediment so that the excess stream energy causes erosion downstream.

- Bridges with in-filled approaches prevent water from flowing over floodplains, causing a deeper, faster flow of water under the bridge. Increased erosion around bridge pilings puts bridges at risk of failure.

Levees

- Although people feel safe behind them, levees fail from overtopping or breaching, bank erosion, slumps, piping, or seepage through old river gravels below the levee. **FIGURE 12-9**.

- Avulsion occurs when a breach flow does not return to the river but follows a new path down a valley, flooding cities in its way.

- Because levees confine streams to main channels rather than permitting floodwaters to spread over floodplains, they dramatically raise water levels during a flood and increase flooding upstream and downstream from the levee. **FIGURE 12-12**.

Dams and Stream Equilibrium

- Dams are built on rivers to provide electric power, flood control, water for irrigation, and recreation. They can also cause a flood hazard to those living downstream.

- Dams slow water at their reservoirs, which accumulate sediment that would typically be transported downstream. Without that sediment, flow is more erosive downstream of a dam.

- Dams can fail during floods due to seepage and erosion under them, poor design and construction, and by major landslides into reservoirs upstream.

- Floods caused by the failure of human-made dams are worst in steep, narrow valleys.

Reducing Flood Damage

- Floods in the United States cause one-quarter to one-third of the monetary losses and nearly 90 percent of deaths by natural hazards.

- Flood damages could be reduced by restricting development on flood plains.

- Many people living on floodplains are eligible for national flood insurance but do not purchase it because they are not aware that they live on a floodplain or do not believe their risk is great.

Reducing Damage from Debris Flows

- Damage from debris flows can be reduced through land use planning, early warning systems, or structures to trap or divert flows.

Key Terms

avulsion, p. 356

breach, p. 356

flood fringe, p. 364

floodway, p. 364

levee, p. 356

National Flood Insurance
  Program (NFIP), p. 364

piping, p. 358

sand boils, p. 358

streambed
  mining, p. 354

wing dams, p. 359

Questions for Review

1. What non-natural changes imposed on a stream cause more flooding and more erosion?

2. How would a hydrograph for a drainage basin change if major urban growth were to occur upstream? Draw a sketch to illustrate your answer.

3. What are the negative effects of mining sand or gravel from a streambed?

4. Roughly what depth of flowing stream is dangerous to drive through?

5. What should you do if your car is floating in water?

6. What negative physical effect do most bridges have on the streams they cross? Draw a sketch to illustrate your answer.

7. Aside from protecting the adjacent stream bank, what effects do levees have on a stream?

8. What process can lead to the failure of a river levee?

9. What process can lead to flooding of the floodplain behind a levee (of a flooding river) even if the levee does not fail?

10. What is a common sign of seepage under a levee?

11. Under what circumstance (or for what purpose) might a levee be deliberately breached?

12. What negative physical effects do dams have?

13. Name three possible causes of dam failure.

14. What is the difference in the danger of a dam failure in a wide open valley versus a narrow closed valley?

Discussion Questions

1. People should not build homes on floodplains because of the danger of flooding. What are better uses for floodplains?

2. How should flood assistance be offered in situations where an area has had damaging floods multiple times in a few decades? Who should pay for such assistance (affected individuals, local, regional, national)?

3. Why are engineered structures such as levees and dams much more popular as a mitigation measure than moving people out of affected areas?

Waves, Beaches, and Coastal Erosion

13

Breaking waves at Rockaway Beach, near Pacifica, California.

Hyndman.

Coastal Cliff Collapse

Pacifica, California, a small coastal community just south of San Francisco, made the news in late 1997 as El Niño storms pounding the coast threatened a row of modest houses on a high terrace near the center of town. Vertical cliffs along that section of the coast consist of rocks that collapse intermittently to produce sand and gravel at the base of the cliff. Much of the narrow beach, however, is fed and maintained by sands that drift south along the shore. In recent decades, however, cliff erosion and collapse has accelerated because rivers in the Sierra Nevada and coastal ranges no longer bring as much sediment to the coast. Dams built for flood control, irrigation, and water supply on those rivers trap sediments that were once carried to the coast and distributed by coastal waves. Sediment that used to deposit in San Francisco Bay was removed by dredging, so tides did not wash it out to the Pacific.

Heavy rock boulders, or **riprap**, emplaced along the beach to protect the cliff may have done so temporarily, but by 1972 the block of houses facing the ocean had succumbed to cliff collapse; those facing landward remained. The section of cliff "protected" by riprap had actually eroded more rapidly

385

than the "unprotected" section immediately to the south (**FIGURE 13-1**). In the early 1980s new riprap was again placed at the base of the cliff, and after a 100-meter gap, more riprap was placed farther south. By 2002 riprap below the houses was again replaced.

Big winter storms during the strong El Niño of 1997–98 accelerated cliff collapse below the landward-facing houses, first eating away at back yards that were originally more than 30 meters from the coastal cliff. By March of 1998, the cliff had retreated landward of the coarse riprap below homes at the south end of the group; these houses had lost their yards and parts of their decks, the remains strewn at the base of the cliff (FIGURE 13-1). The corner of one home was left hanging over open space and had been condemned as unsafe. Cliff erosion continued so that by December 2003 only two of the original dozen landward-facing houses remained. One was sold to a new buyer in 2004! Those two still remained in March 2010, but their back yards had disappeared. How long before they're gone?

Living on Dangerous Coasts

People have always lived along the shores of inlets and bays and fished in nearby streams and lagoons, but their structures along the open coast were once temporary shelters that could be moved or low-value ramshackle summer cabins that could be replaced after storms. In the early years after European settlement of North America, coastal tourism was not important because of the difficult access through local brush and the incidence of malaria. By the 1700s, people began building more costly and permanent structures on the protected landward side of some barrier islands. The old-timers understood beach processes and built homes on stilts on the bay side of bars with temporary structures at the beach.

The advent of steam locomotion in the early 1800s, followed by railroads and a large increase in population in the continental interior, also led to deforestation, land cultivation, and overgrazing on a large scale. Invention of the internal combustion engine continued the trend. This removal of protective cover from the land led to heavy surface erosion and large volumes of sediment delivered to the coasts. Along the steep Pacific coast, longshore drift of these sediments caused widespread enlargement of beaches.

FIGURE 13-1 CLIFF COLLAPSE

A. Beach erosion and cliff collapse endanger homes in Pacifica in March 1998. Collapsed parts of houses litter the base of the cliff. **B.** Heavy riprap at the base of the sea cliff again attempts to protect houses above. Taken from the same viewpoint as in December 2003, this photo shows that the seven homes nearest the camera are gone. Only two of the original ten houses remain, and one of these sold in 2004 for $450,000! (See FIGURE 18-2)

By the 1850s, reduction in working hours, formation of an urban middle class with money, and expanded transportation via railroads and steamships changed the ground rules. Coastal tourism and resorts expanded, especially after the late 1940s. As populations grow in numbers and affluence, more people move to the coasts, not only to live but also for recreation. People installed utilities, paved roads, and built bridges to the islands, along with more expensive permanent homes, hotels, and resorts along the same beaches. More recently, second homes for summer use have become popular, some used for only a few weeks per year. Others have become year-round dwellings for urban retirees. By 2004, approximately 42 percent of the population of the continental United States lived in coastal counties. Coastal populations continue to grow. For example, the population of Florida grew by 75 percent from 1980 to 2003.

Beaches and sea cliffs constantly change with the seasons and progressively with time. When people build permanent structures at the beach, coastal processes do not stop; the processes interact with and are affected by those new structures. Instead of living with the sea and its changing coastline, they tried to hold it back and prevent natural changes to the beaches and sea cliffs.

Hurricanes and their dramatic aftermaths are often viewed as abnormal or "nature on a rampage." In fact, they are normal for a constantly evolving landscape. What is abnormal is how human actions and structures cause natural processes to impose unwanted damage. To be aware of coastal hazards, we need to understand wave processes and the formation of beaches and sea cliffs. We also need to understand how human activities affect wave action, beach response, and sea-cliff collapse.

Waves and Sediment Transport

Winds blowing across the sea push the water surface into waves because of friction between air and water. Wind-driven waves are described in terms of **wave height**, **wavelength**, and **period** (**FIGURE 13-2**). Gentle winds form small ripples. As the wind speed increases, ripples grow into waves that grow higher as they catch even more of the wind energy, somewhat like a sail. Two other factors that increase wave height are **fetch**, which is the length of water surface over which the wind blows, and the amount of time the wind blows across the water surface. Ocean waves are generally much larger than those on lakes, and prolonged storms often build huge, damaging waves. Big waves generated by storms far offshore can travel for hundreds or thousands of kilometers, so they can appear at a beach even when there is little or no wind. Because waves can move outward from a storm center at different times as the storm moves, waves generated can be different sizes and move at different speeds. When a faster wave overtakes a slower one, they can

FIGURE 13-2 WAVE FEATURES

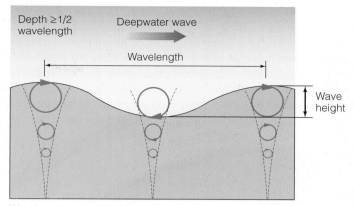

Wavelength is the distance between successive wave crests; wave height is the height from crest to trough; and wave period is the time it takes for two crests to pass a point.

interfere to increase or decrease the size of the combination (compare FIGURE 1-9). Such combination waves can be very large and even lead to occasional giant "rogue" waves.

The timing and size of waves approaching a shoreline vary by the location and size of offshore storms. Waves move out from major storm centers, becoming broad, rolling swells with large wavelengths (**FIGURE 13-3**). A constant, mild onshore wind will produce waves with smaller and shorter wavelengths.

Offshore, water in a wave itself does not move onshore with the wave; it merely has a circular motion within the wave, otherwise staying in place. You can see that motion by watching a stick or a seagull floating on the water surface. It moves up and down, back and forth, not

FIGURE 13-3 STORM WAVES

In addition to shore damage, huge storm waves can topple boats.

approaching the shore unless blown by the wind or caught in shallow water where the waves break. Waves in this circular motion are not damaging because the mass of water moves only slightly forward, then back. Watch waves moving past the pilings of a pier or against any kind of vertical wall in deep water. The waves have little forward momentum and do not splash against the vertical surface; they merely ride up and down against it. A steep "wall" of coral reef facing offshore from some tropical islands has a similar effect, thereby helping protect such islands from the impact of storm waves.

When waves approach shore, they begin to "feel bottom." These are conditions under which waves gain the potential for serious damage. Waves begin to feel bottom when the water depth is less than approximately half the wavelength (**FIGURE 13-4**). Because the size of the circular motion is controlled by wave size, the depth at which wave action fades out is controlled by wavelength. In shallower water, the crest of the wave moves forward as the base drags on the bottom. Waves slow in shallow water but rise in height, which causes them to break (see FIGURE 13-4 and **By the Numbers 13-1:** Wave Velocity). The momentum of the upper mass of water carries it forward to erode the coast.

Big waves are more energetic and cause more erosion, because **wave energy** is proportional to the mass of moving water. This can be approximated by multiplying the density of water by the volume of water in a wave, which is roughly the wave height squared times the wavelength (**By the Numbers 13-2:** Wave Energy). The fact that the height term is squared tells us that waves that are twice as high have four times the energy; those that are four times as high have sixteen times the energy.

Wave Refraction and Longshore Drift

Waves often approach shore at an angle. The part of each wave in shallower water near shore begins to drag on the bottom first and thus slows down; the part of the wave still in deeper water moves faster, so the crest of the wave curves around toward the shore. This is called **wave refraction** because waves bend, or refract, toward shore.

When wave crests approach a beach at an angle, the breaking wave pushes the sand grains up the beach slope at an angle to the shore. As the wave then drains back into the sea, the water moves directly down the beach slope perpendicular to the water's edge. Thus, grains of sand follow a looping path up the beach and back toward the sea. With each looping motion, each sand grain moves a little farther along the shore (**FIGURE 13-5**). The angled waves thus create a **longshore drift** that essentially pushes a river of sand along the shore near the beach. Over the period of a year or so, with high and low tides, large and small waves, and storms, most of the sand on the active beach moves farther along shore.

Along both the west and east coasts of the United States, longshore drift is dominantly toward the south. Longshore drift in parts of coastal California averages a phenomenal 750,000 cubic meters of sand past a given point per year, more than 20,000 cubic meters per day. If a large dump truck carried 10 cubic meters, this would be equivalent to 2,000 dump-truck loads per day. Along the East Coast, drift rates are much less but still average 75,000 cubic meters per year, more than 2,000 cubic meters per day.

FIGURE 13-4 CHARACTERISTICS OF WAVES CHANGE ON APPROACHING SHORE

At depths less than one-half wavelength, motion of particles in a wave are flattened. As waves approach shore, they drag on the bottom and their crests lean forward to break, as seen in this photo taken along the west coast. The water is brown from stirred sand.

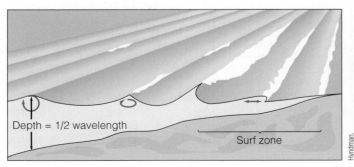

Wave Velocity

Wave velocity in deep water is proportional to the square root of wavelength:

$$v = \sqrt{\frac{gL}{2\pi}} = 1.25\sqrt{L}$$

where:

v = velocity (m/sec)

L = wavelength (m)

g = acceleration of gravity (9.8 m/sec^2)

π = 3.1416

Thus, waves with a wavelength of 5 meters move at 1.25 x 2.24 = 2.8 m/sec. or 10.1 km/hr. Waves with a wavelength of 100 meters move at 1.25 x 10 = 12.5 m/sec, or 45 km/hr. Note that if we measure the period (the time between wave crests), we can determine both the wavelength and wave velocity using the graph below.

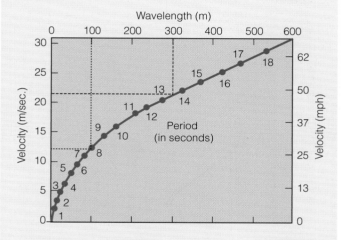

In shallow water, wave velocity is proportional to square root of water depth:

$$V = \sqrt{gD} = 3.1\sqrt{D}$$

where:

V = velocity (m/sec)

D = water depth (m)

Waves on Irregular Coastlines

Waves approaching a steep coast, such as those along much of the Pacific coast of North America or the coast of New England or eastern Canada, encounter rocky points called **headlands**, separated by shallower sandy bays, that reach into deeper water. Waves bend or refract toward the rocky points, causing the energy of the waves to break against the headlands (**FIGURE 13-6**). Thus, wave refraction dissipates much of the wave energy that would

Wave Energy

Doubling the wave height quadruples the energy:

$$E_w = 0.125\ r\ gH^2\ L$$

where:

E_w = energy of the wave

r = water density (g/cm^3: close to 1)

g = acceleration of gravity = 9.8 m/sec^2

H = wave height (m)

L = wavelength (m)

FIGURE 13-5 LONGSHORE DRIFT

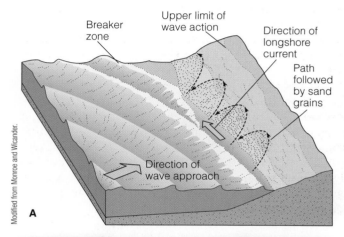

A. Sand grains are pushed up onto a beach in the direction of wave travel. Gravity pulls them back directly down the slope of the beach. The combination is a loop that moves each sand grain along the shore with each incoming wave. **B.** Waves in shallow water drag on the bottom and slow so their crests curve to approach nearly parallel to the shore.

otherwise have impacted sandy bays. Sand pounded off a rocky point migrates along the shore due to wave refraction and is dumped along beaches in bays. Currents on both sides carry sand into adjacent bays. Over time, the activity of waves thus tends to straighten coastlines by eroding the points that extend into the sea (**FIGURE 13-7**).

Rip Currents

As waves pile up water onshore, it streams back offshore to create a **rip current**. Some such high-energy beaches show a prominent scalloped shoreline with cusps five to ten meters apart, a suspicious sign of rip-current danger (**FIGURE 13-8**). Such rip currents can persist for long periods. Permanent rip currents can develop at groins, jetties, or rock outcrops, where water flowing against those structures piles up and is forced to flow offshore along the structure. A scalloped beach may form in an area of rip currents because each current erodes sand as it flows offshore, leaving the beach slightly indented at the point of offshore flow (**FIGURE 13-8**). Elsewhere, rip currents may appear unpredictably for a short period of less than 10 minutes after a large swell from a distant storm pushes water up onto a beach.

These streams of muddy-looking water can be dangerous to people not familiar with them because the currents are too fast to permit even strong swimmers from swimming directly back to shore. Many people drown at the coast when caught in rip currents, which are especially powerful and dangerous during strong onshore winds and big waves. Rip currents are sometimes called undertows, but this is a misnomer because these currents would not drag someone under the surface. The drowning danger comes from swimmers becoming overtired while fighting the current. You should escape a rip current by swimming parallel to shore and then back to the beach (see FIGURE 13-8c).

FIGURE 13-6 WAVES BREAK AGAINST HEADLANDS AND REEFS

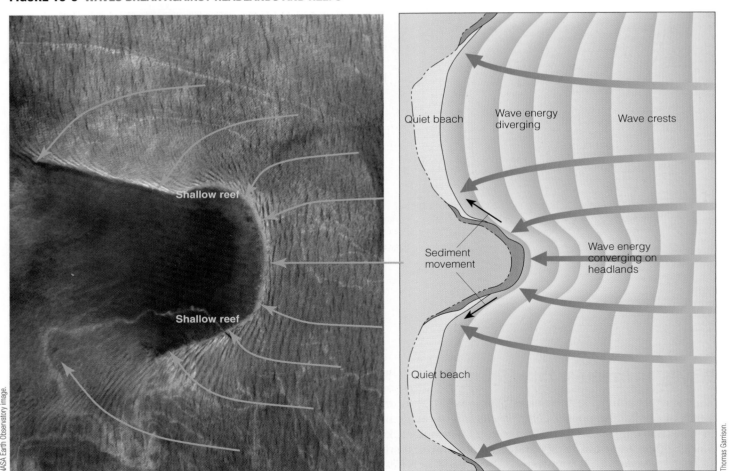

A. Waves approaching a shallow reef in the Caribbean Sea, from right, curve to almost parallel the reef. **B.** Wave crests bend to conform to the shape of the shoreline; wave directions bend to attack the shoreline more directly. Thus, rocky headlands are vigorously eroded, and bays collect the products of that erosion.

FIGURE 13-7 EROSION OF HEADLANDS

Wave refraction has eroded the former series of headlands along Drakes Bay, Point Reyes National Seashore, California, into a straighter coastline.

FIGURE 13-8 RIP CURRENTS

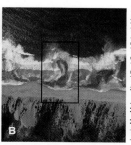

A. Prominent scallops in beach sand are a sign of dangerous rip-current action. Lima Peru. **B.** A prominent rip current carries muddy water offshore in Monterey Bay, California. **C.** To escape a rip current, swim parallel to shore. Don't try to swim directly towards the beach against the current.

FIGURE 13-9 WAVES UNDERCUT A CLIFF

This cliff south of Puerto Vallarta, Mexico, has been undercut by waves. Large chunks of rock break off and are pounded into sand by the waves.

Beaches and Sand Supply

Beaches are accumulations of sand or gravel supplied by sea-cliff erosion and river transport of sediments to the coast. The size and number of particles provided by sea-cliff erosion depend on the energy of wave attack, the resistance to erosion of the material making up the cliff, and the particle size into which the cliff material breaks. Waves often undercut a cliff that collapses into the surf (**FIGURE 13-9**) and then break it into smaller particles. The size and amount of material supplied by rivers depends similarly on the rate of river flow and the particle size supplied to its channel.

Most of the sand and gravel supplied to the coast is pushed up onto beaches in breaking waves; it then slides back into the surf in the backwash. Big waves during winter storms carry sand offshore into deeper water, leaving only the larger gravels and boulders that they cannot move. The gentle breezes and smaller waves of summer slowly move the sand back onto the beach. As a result,

FIGURE 13-10 ROCKY WINTER BEACH

A sandy summer beach covers the lower part of a bouldery upper beach left by winter waves north of Newport, Oregon.

FIGURE 13-11 GRAIN SIZE AND BEACH SLOPE

The beach slope steepens shoreward where the breaking waves reach their upper limit. The result is a ridge, or berm, as in this case at Positano, Italy.

some cliff-bound beaches show sand near the water's edge with gravel or boulders upslope. In such areas, the beaches are commonly sand in summer but more steeply sloping gravel or boulders in winter (**FIGURE 13-10**). The active beach extends from the high-water mark to some ten meters below sea level.

Beach Slope: An Equilibrium Profile

Grain size strongly controls the slope of a beach, called its **shore profile**. Just as in a stream, fine sand can be moved on a gentle slope—coarse sand or pebbles only on a much steeper slope. Whether the sediment moves shoreward or not depends on the balance (equilibrium) between shoreward bottom drag by the waves, size of bottom grains, and downslope pull by gravity. This balance is called the **equilibrium profile**. Thus, the slope of the bottom is controlled by the energy required to move the grains, which is related to water depth, wave height, and grain size (**FIGURE 13-11**). Shallower water, smaller waves, and coarser grains promote steeper slopes offshore, just as in rivers. In the breaker zone offshore, waves can easily move the sand, and the beach surface is gently sloping. As breakers sweep up onto the shore, the water is shallower, their available energy decreases, and the shore profile steepens. Sand there can move back and forth only on such a steeper slope.

As waves travel into shallow water and begin to touch bottom, they shift sediment on the bottom, stirring it into motion and moving it toward the shore. Most sediment moves at water depths of less than 10 meters. Long-wavelength storm waves, however, with periods of up to 20 seconds, reach deeper; they touch bottom and move sediments at depths as great as 300 meters on the continental shelf. Those large waves have the energy to spread the grains into a gentler slope; they erode the grains above water level and deposit them offshore below sea level.

On shallow, gently sloping coastlines, such as those in much of the southeastern United States, the beach both onshore and offshore becomes steeper landward. This is because the waves use energy stirring sand on the sea bottom, so that they slow as they ride up onto the beach. Most of the wave energy is used in waves breaking and moving water and sand upslope. The active beach slope is controlled by the grain size being moved and the amount of water carrying the grains. To reiterate, larger grains or less water requires a steeper slope to move the grains.

On beaches with coarser sand or gravel, much of the water soaks quickly into the ground rather than flowing back offshore over the surface; thus, less sand or gravel is eroded from the beach. Water flowing back off the beach carries finer sand with it. This effect is most significant with large storm waves that still have most of their water available to flow offshore and carry sand with them. Therefore small waves tend to leave more of their sand onshore. During low tide, winds pick up drying sand on the beach and blow it landward into dunes. The next strong storm may erode both the beach and the dune face and carry the sand back offshore.

Loss of Sand from the Beach

Sand in the surf zone moves with the waves; however, where it goes changes with the tides and with wave heights. Larger waves, especially those during a high tide or a major storm, erode sand from the shallow portion of a beach and transport much of it just offshore into less-stirred deeper water. Much of the eroded material comes from the surface of the beach, which flattens the beach's profile. More comes from the seaward face of dunes at the head of the beach, where waves either break directly against the dunes or undercut their face.

Storms bring not only higher waves but also a local rise in sea level—called a **storm surge**. High atmospheric pressure on the sea surface during clear weather holds the surface down, but low atmospheric pressure in the eye of a major storm permits it to rise by as much as a meter or more. The stronger winds of a storm also push the water ahead of the storm into a broad mound several kilometers across. The giant waves of a hurricane and the higher water level of storm surges take these effects to the extreme. They cause massive erosion and decrease the beach slope (see Chapter 14 for further discussion).

Along the Gulf Coast and southeastern coast of the United States, storms and heavy erosion are most likely to occur in the hurricane season of late summer to early fall. The most severe area of coastal erosion in the United States, with more than five meters of loss per year, is along the Mississippi River delta, where dams upstream trap sediment, and the compaction of delta sediments lowers their level. One to five meters are lost annually along much of the coasts of Massachusetts to Virginia, South Carolina, and scattered patches elsewhere. **Nor'easters**, the heavy winter storms that hit the northeastern coast of the United States with similar ferocity, do similar damage.

It might seem that continual erosion of sea cliffs and erosion by rivers would add more and more sand to beaches, making them progressively larger. This does happen in some areas. Beaches are actually gaining sediment near the mouth of the Columbia River between Washington and Oregon, between Los Angeles and San Diego, along much of the Georgia and North Carolina coasts, and along scattered patches in Florida and elsewhere on the Gulf of Mexico coast.

However, many coastlines are continuing to erode. What happens to the sand? Nature gets rid of some of it. Some is blown inland to form sand dunes, especially in areas without coastal cliffs. Other processes can permanently remove sand from the system. Some is beaten down to finer grains in the surf and then washed out to deeper water. Rip currents form when waves carry more water onshore than returns in the swash. That current flows back offshore in an intermittent stream that carries some of the beach sand back into deeper water. Huge storms such as hurricanes carry large amounts of sand far offshore.

Some sand drifts into inlets that cross barrier islands, where dredges remove it to keep the inlets open for boat traffic. Some migrates along coasts for a few hundred kilometers until it encounters the deeper water of a **submarine canyon** that extends offshore from an onshore valley (**FIGURE 13-12**). One prominent example of a submarine canyon extends offshore from the Monterey area of central California. These valleys extend across the continental shelf to where the sediment intermittently slides down the continental slope as turbidity flows onto the deep ocean floor. Thus, much of the longshore drifting sand of beaches is permanently lost to the beach environment.

Sand Supply

With European settlement of North America over the past 400 years, attempts to control nature have upset the natural sediment balance. The supply of river sediment to the coasts has been severely reduced from building dams that

FIGURE 13-12 LONGSHORE DRIFT AND SEDIMENT LOSS

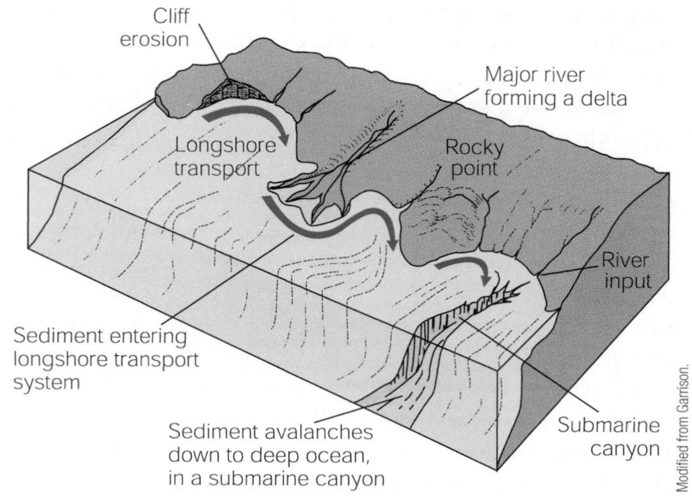

Longshore drift of beach sediment often leads to loss of the sediment in a submarine canyon.

trap sediment, and by sand and gravel mining from streams. With less sand and gravel, beaches shrink and the waves break closer to, and more frequently against, coastal cliffs. Waves undercut the cliffs, which collapse into the surf (**FIGURE 13-13**).

Anything that hinders sand supply to a beach from "upstream" on a coast or removes sand from the moving longshore current along a beach results in less sand to an area of coast and erosion of the beach. Although sand mining in many places is now permitted only offshore at depths greater than 18 to 25 meters, monitoring of such activity is sometimes lacking. And some communities tacitly condone mining by purchasing beach sand for use on roads. Dams on rivers trap sand, keeping it from reaching the coast, and mining sand from river channels or from beaches for construction has a similar effect. In many industrial countries, major dam-building periods on rivers began in the 1940s to generate electricity, store water for irrigation, and provide flood control. That and better land use practices caused dramatic reductions in the amount of sediment carried by rivers and supplied to beaches. Beach erosion again accelerated. The resulting erosion of

beaches and coastal cliffs is clear in California. Shoreline recession of 5 to 10 meters per year was common but in some places, for example, at the mouth of the Nile River in Egypt, was as much as 200 meters per year.

Erosion of Gently Sloping Coasts and Barrier Islands

The beach on a gently-sloping shoreline becomes steeper toward shore, thereby building an offshore **barrier island** or barrier sandbar. Farther landward, the area below sea level is a coastal lagoon (**FIGURE 13-14**). Offshore barrier islands and the lagoons behind them are products of dynamic coastline processes: erosion, deposition, longshore sand drift, and wind transport. Barrier island communities live within this constantly changing environment.

Barrier islands migrate with the gradual rise of sea level. Because the equilibrium profile of a beach and the position of a barrier island are linked to wave size and water depth, a beach and barrier island shift landward as the water level rises (**FIGURE 13-15**). Current rates of sea level rise are approximately 30 centimeters per century. On especially gently sloping coasts, like those of the southeastern United States, that 30-centimeter rise can move the beach and barrier island inland by 100 to 150 meters or more over the course of a century. Barrier island migration happens over decades, but all of the significant movement occurs in stages during hurricanes and other major storms.

Barrier islands form primarily along gently sloping coastlines such as those of the East and Gulf coasts of the United States, but they also form across the mouths of some shallow bays and estuaries along the West Coast. The sea level rise after the last ice age drowned the mouths of these river valleys.

During the ice ages of the Pleistocene epoch more than 12,000 years ago, sea level dropped as water was tied up in continental ice sheets; the shoreline receded far out onto the continental shelf. In fact, wave-base erosion and deposition that formed the continental shelves may be related to thousands of years at which sea levels stood some 100 meters lower than at present. Following the last

FIGURE 13-13 STORM EROSION

A. Storm waves undercut this East Coast parking lot, causing it to collapse into the ocean. **B.** During Hurricane Ike in 2008, almost two meters of sand were eroded from this beach southwest of Galveston. The concrete platform at author's head level was a concrete carport floor poured onto beach sand and now suspended after beach erosion.

FIGURE 13-14 BARRIER ISLAND

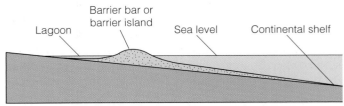

This cross section shows an offshore barrier bar with a sheltered lagoon behind it. The waves create a steeper profile for the sand than the overall slope of the coastline.

FIGURE 13-15 BARRIER ISLAND MIGRATION

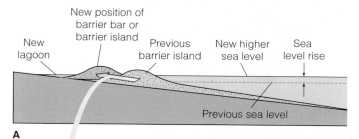

B

A. The barrier island migrates landward as sea level rises. B. The Morris Island Lighthouse, near Charleston, South Carolina, was on the beach in the 1940s. It is now 400 meters offshore because the sand of the barrier island on which it once stood gradually migrated landward. C. Storm waves reduced the level of sand by almost 4 meters during one storm at Westhampton, New York. The barrier island migrated landward, leaving these houses stranded offshore.

C

ice age, as continental ice sheets melted 12,000 to 15,000 years ago and sea level rose, waves gradually moved sands on the continental shelf landward, piling them up ahead of the advancing waves. The Atlantic and Gulf of Mexico coastal plains are gently sloping, creating the conditions necessary for the formation of barrier islands. The gently upward curving equilibrium profile of a sandy bottom produced by the waves is steeper than the coastal plain, so it tops off landward in a ridge, the barrier island.

Offshore barrier islands, also called barrier bars, are a part of the active beach, built up by the waves and constantly shifting by wave and storm action. Offshore barrier islands are typically 0.4 to 4 kilometers wide and stand less than 3 meters above sea level. Winds picking up dry beach sands may locally pile dunes as high as 15 meters above sea level. Where high tides or storms carry the sea through low areas in the barrier bars, the water returns from lagoons to the sea at low tide through the same inlets, eroding them and keeping the channels open (**FIGURE 13-16A**). Over time, inlets through barrier islands naturally shift position; some close and others open (FIGURE13-16b). Longshore drift of sand at times closes some gaps, and storms open others, so they intermittently change locations.

Oysters grow in the quiet waters of lagoons on the coastal side of barrier islands. If you find oyster shells on the seaward-side beach of a barrier island, one possibility is that the island gradually migrated landward, over lagoon mud.

Beach waves winnow out the fine mud of the lagoon, leaving the heavier oyster shells embedded in the beach sand. Migrating barrier islands also overwhelm trees growing along lagoons; their stumps reappear later along a beach's upper edge as the bar gradually moves landward after a storm.

Development on Barrier Islands

Although barrier islands help protect low-lying coastal areas from storm wave and flood damage, the islands themselves are hazardous places to live. Many barrier islands are now so crowded with buildings that they bear little resemblance to their natural states. Most distinctively, former broad beaches erode rapidly in front of buildings, especially during hurricanes (**FIGURE 13-17**).

There are a few barrier islands at the mouths of drowned river estuaries along the West Coast; people build on those just as they do on barrier islands on the East Coast. On gently sloping areas of some western beaches, as in parts of Southern California, people build right on the beach. Perhaps they purchase homes or build them in good weather, not realizing that big winter waves come right up to the house. Some homeowners pile heavy boulders on the beach in front of their homes, hoping to protect them. The protection is temporary because big waves reflect off the boulders, washing away the beach sand in front and steepening the remaining beach. Eventually, the boulders

FIGURE 13-16 INLETS BETWEEN BARRIER ISLANDS

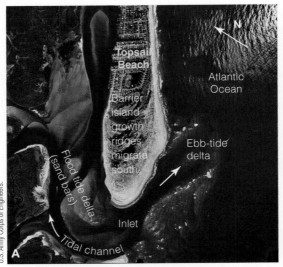

A. A vertical air photo of Topsail Inlet, near Wilmington, North Carolina, shows sand deltas deposited by the in-flowing flood tide and out-flowing ebb tide. **B.** A new breach from Hurricane Isabel, 2003, develops into a new inlet.

FIGURE 13-17 BEACH MIGRATION

Several hurricanes shifted the beach shoreward at North Topsail Beach, North Carolina, destroying beachside homes one by one as both beach and barrier island migrated landward. **A.** Use the colorful condominiums in the upper left here as a reference and note those in the center of the photo taken after Hurricane Bertha on July 16, 1996. **B.** This is the same location following Hurricane Fran on September 7, 1996. Finally, **C.** shows the same location after Hurricane Bonnie on August 28, 1998. Note that the series of three photos span only about two years.

will slide into the wave-eroded trough, leaving the homes even more vulnerable. But then the beach is gone.

Shifting sand closing an existing inlet commonly hinders access to marinas and boating in protected lagoons behind bars. It also hampers sea access to any coastal industrial sites on the mainland. Existing inlets are thus often kept open by dredging and building jetties along inlet edges. Where significant populations or large industrial sites are affected, the U.S. Army Corps of Engineers will often construct or maintain an inlet. Where significant settlement has occurred on or behind a barrier island, the maintenance of inlets severely hinders natural evolution of the island and beach.

When a storm overwashes and severs a beach-parallel road or cuts a new inlet across a barrier island, some homes and businesses are isolated on part of the bar (**FIGURE 13-18**). Generally, road engineers fill the new inlet and rebuild the road. Unless they do so immediately, the inlet typically widens over the following weeks or months as tidal currents shift sand into the lagoon and back out. Therefore, the scale of the repair project can quickly get out of hand. For example, a winter storm in 1992 at Westhampton, New York, opened a breach inlet 30.5 meters wide. Within eight months, the inlet widened to 1.5 kilometers.

Through past experience with hurricanes and other big storms, people living on barrier islands learn to build homes on posts, raising them above the higher water levels and bigger waves of some storms. In many areas, building codes require such construction. Codes also require preservation of beachfront dunes to help protect buildings from wave attack. A coastal construction control line (CCCL), established in the 1980s by the Department of Protection of Florida, imposed higher standards for land use and construction in high-hazard coastal zones. It restricts the purchase of flood insurance to those communities that adopt and enforce National Flood Insurance Program construction requirements governed by the standard building code. On the coastal side of the CCCL, requirements are more stringent for foundations, building elevations, and resistance to wind loads.

Dunes

At low tide, winds pick up drying sand and blow it landward to form **sand dunes**, which help maintain the barrier islands. Major storms or hurricanes wash sand both landward into lagoons and seaward from barrier islands. Sand moves back up onto the higher beach in milder weather, and wind blows some dry sand into the dunes again.

Unfortunately, most dunes along the coasts of the southeastern and Gulf states have disappeared as a result of development. People level the dunes to build roads, parking areas, and houses, or to improve views of the sea. They remove vegetation deliberately—to provide vistas—or inadvertently—by trampling it underfoot to reach beaches or by using off-road vehicles. They move sand onto lagoon-margin marsh areas for housing and marinas. They remove

FIGURE 13-18 BARRIER-ISLAND BREACHES

A. This breach of Hatteras Island on the North Carolina Outer Banks followed the arrival of Hurricane Isabel in September 2003. The ocean is to the right. **B.** This breach severed the only road to the west from Galveston to the mainland during Hurricane Ike. The arrow in **A.** and the yellow line in **B.** mark the centerline of roads.

sand from dunes for construction or for reclaiming beaches damaged by erosion elsewhere.

Adding structures to a beach can also affect the formation of dunes. Sand still blows off the beach, but it drifts around houses instead of forming dunes on back beaches. It piles against houses and covers sidewalks and streets (**FIGURE 13-19**). After storms, the sand is routinely removed from roads by either private citizens or municipal employees. Shoveling snow in some northern areas can be hard work, but imagine having to shovel sand, which weighs much more and will not melt. The sand is sometimes placed back on an upper beach, but often it is pushed onto vacant lots, where it is lost to beach surfaces. In areas of lower sand supply, the wind may funnel between houses to scour local areas.

When people modify dunes, they upset the natural equilibrium of sand movement. True, they improve views, but

FIGURE 13-19 DUNES AND DEVELOPMENT

These huge houses are right on the beach, along the former line of dunes, at Oxnard Shores, in southern California. Sand blown off the beach piles as drifts around houses and covers roads, sidewalks, and driveways. Shoveling snow in northern climates is heavy work, but this is ridiculous. "For Sale" signs are often numerous along this beach-parallel street.

FIGURE 13-20 SEVERE EROSION IN THE ABSENCE OF DUNES

With no line of dunes to protect it, the beach under this walkway at Holden Beach, North Carolina, was severely eroded by Hurricane Floyd, and the beachfront dune was completely washed away. The homes behind are next.

often build elevated walkways to cross dunes and reach a beach; visitors get to beaches using roads and raised paths over dunes (**FIGURE 13-21**). Without walkways, such paths to beaches can become deeply eroded. Unfortunately, where dunes are managed at local levels with strong input from landowners, the dunes are often low and narrow to permit easy access to the water and direct views of the ocean. Such low dunes provide minimal protection and storms easily remove them, exposing homes to direct wave attack (**FIGURE 13-22**).

Sand can be trapped by placing sand fences across the wind direction to slow blowing and promote sand deposition on a fence's downwind side. Fences can help keep sand on an upper beach or in dunes; they can also be used to prevent drifts from forming where they create problems for roads and driveways. Beach nourishment projects also often involve dune nourishment. The sand scraped from a beach or from overwash sediments may be used to partly rebuild artificial dunes or close gaps opened through existing dunes.

Where vegetation has been lost, it can be replanted to help stabilize the sand, though new vegetation may be difficult to establish if the sand is salty or mobile. Straw or branches from local coastal shrubs can be strewn on sand surfaces to slow movement and help establish the growth of dune vegetation. Dune vegetation that is diverse and native to an area is best for its likelihood of survival. Dune grass, or "Sea Oats," are often planted for that purpose, but it takes about three years without severe damage and root dehydration for the grass to establish roots in the sand and survive.

Natural re-vegetation of dunes may be accomplished using cuttings taken from nearby dunes, but this can be a slow process. European beach grass introduced along the Pacific coast of the United States has been even more successful in trapping sand than has native vegetation, but it creates dunes that are less natural and higher than the originals. It now dominates dunes in coastal Washington

with each storm a beach gets closer until the view gets too good (**FIGURE 13-20**). Houses are the next to go. Bigger storm waves will impact such structures because any protective dunes have long since disappeared.

As communities have become more aware of the importance of sand dunes in maintaining beaches, many have introduced measures to protect them. When buildings are elevated four meters or so on pilings, as is commonly the case across the southeastern and Gulf coasts, the effect on wind-blown sand is much reduced. In many places, walking on dunes or otherwise damaging dune vegetation is prohibited. Homeowners

FIGURE 13-21 PATHS THROUGH DUNES

A. This walkway at Sunset Beach, North Carolina, once reached over a protective beachfront dune, which has since been removed by hurricane waves. **B.** A vehicle and foot path cut through beachfront dunes on the coast south of Atlantic City, New Jersey. Such paths dramatically increase erosion during storms.

FIGURE 13-22 DUNE MANAGEMENT

Low beach dunes are partly stabilized by sand fences and beach grass south of Atlantic City, New Jersey.

state and parts of Oregon. Residents often plant exotic vegetation because of appearance, but such plants often require artificial watering. The resulting rise in the water table can lead to the formation of gullies and increase the chance that coastal cliffs will have landslides.

Sea-Cliff Erosion

Much of the West Coast would be eroding back more rapidly except for the ever-present beach cliffs. Where these cliffs consist of soft Tertiary-age sediments, they are commonly undermined, causing landslides and cliff collapse. Disintegration of the collapsed material supplies sand to beaches down the coast. That part is good for the beach, but if your home is perched at the top of that cliff, where there is a magnificent view of the ocean, it may not be so good. As the cliff erodes, your house gets closer to the edge, and the view gets better, but your house eventually collapses with the cliff. Severe erosion of sea cliffs at Martha's Vineyard in Massachusetts in 2007 prompted a group of affluent homeowners to try to protect their cliff-top property by moving some homes and importing large amounts of sand to replenish the beach. Unfortunately, the effort will succeed only temporarily.

Hard, erosionally resistant rocks—granites, basalts, metamorphic rocks, and well-cemented sedimentary rocks—mark some steep coastlines. Such coastlines often consist of rocky headlands separated by small pocket coves or beaches. Much of the coast of New England, parts of northern California, Oregon, Washington, and western Canada are of this type. The rocky headlands are subjected to intense battering by waves and drop into deep water. Sands or gravels pounded from the headlands are swept into the adjacent coves, where they form small beaches (see FIGURES 13-6 and 13-7).

Raised marine terraces held up by soft muddy or sandy sediments less than 15–20 million years old mark other coastlines, including much of the Oregon and central California coasts. These terraces, standing a few meters to tens of meters above the surf, are soft and easily eroded. They were themselves beach and near-shore sediments not so long ago. Some terraces rose during earthquakes as the ocean floor was stuffed into the oceanic trench at an offshore subduction zone.

Elsewhere they may rise by movements associated with California's San Andreas Fault.

Along these coasts, cliffs line the heads of beaches, except in low areas at the mouths of coastal valleys. The beaches consist of sand, partly derived from erosion of soft cliff materials and partly brought in by longshore drift. Waves strike a balance between erosion and deposition of beach sands. Larger waves flatten a beach by taking sand farther offshore, and smaller waves steepen it. Where streams bring in little sand or the cliffs are especially resistant, a beach may be narrow. Where rivers supply much sand or where the coastal cliffs are easily eroded, a beach may be wide. A broad, sandy beach hinders cliff erosion because most of the wave energy is expended in stirring up sand and moving it around.

In addition to rock strength, the main factors affecting cliff erosion are wave height, sea level, and precipitation. All three factors are heavily influenced by intermittent events, from El Niño to hurricanes. When storms come in from the ocean, their frequency and magnitude strongly affect the rate of erosion. Along the coast of Southern California, erosion is amplified during El Niño events, such as those of 1982–83 and 1997–98 (**FIGURE 13-23**). Farther north along the coasts of Washington and Oregon, storms are more common during the weakest El Niño years.

Unfortunately, especially in the West, people also choose to build houses on cliff-tops for sea views, typically not realizing how hazardous the sites are. Developers promote building in such locations, charging a premium for land that may be gone in only a few years. Vertical cliffs made of soft, porous sediments are a recipe for landsliding and cliff collapse. Homeowners themselves exacerbate the problem by clearing beaches of driftwood that help reduce wave force. They unwittingly become agents of erosion by making paths down steep slopes, cutting steps, and excavating for foundations next to the cliffs. They irrigate vegetation and drain water into the ground from rooftops, driveways, household drains, and sewage drain fields. Adding water to the ground further weakens it and promotes landslides.

When such cliff-top dwellers see their properties disappearing and recognize that part of the problem is waves undercutting the cliff, the typical response is to dump coarse rocks—riprap—at the base of the cliff (**FIGURE 13-24**) or build a wood, steel, or concrete wall there. Waves reaching such a resistant barrier tend to break against it, churn up adjacent sand, and sweep it offshore. The new deeper water next to the barrier undercuts the barrier, which then collapses into the deep water. The barrier has provided short-term protection to the cliff, but after a few years that "cure" has done more harm than good. A bigger slab of the cliff collapses into the deeper water in the next big storm. Some people spray **shotcrete**, a cement coating, on the cliff surface to minimize erosion (**FIGURE 13-25**). This might be effective for a few years, until the waves undercut the coating.

In a few places, people even build at the base of cliffs, on the beach itself (**FIGURE 13-26**). They must know something that we do not! CalTrans, the California highway department, had enough sense to build the coastal highway six or so meters above high tide; the houses, on the beach side of the highway, sit on pilings sunk into the beach but have their lower floors more than three meters below the highway.

FIGURE 13-23 EROSION INCREASES DURING EL NIÑO

This beach north of Point Reyes, California, was eroded by waves during the 1997 El Niño event, shown at left. By April 1998, shown at right, the beach had been naturally rebuilt.

FIGURE 13-24 RIPRAP TO SHORE UP A CLIFF

A. Heavy riprap boulders were piled at the base of the same rapidly eroding cliff shown in FIGURE 13-1 before the next high tide in January 1998. Note the house debris already at the base of the cliff. **B.** Collapsing sea cliffs at Ocean Beach near San Diego destroyed some homes and severely threatened those remaining in March 1998. Huge boulders were placed at the cliff base in an attempt to arrest the erosion.

FIGURE 13-25 SHOTCRETE ON SEA CLIFFS

These large homes sit atop a sea cliff without a beach in Pismo Beach, California. Note that a new house is being built on the right, even though established houses have lost most of their yards and spray-cemented shotcrete on their cliffs to slow cliff loss.

FIGURE 13-26 BEACH DEVELOPMENT

Donald Hyndman.

These homes were built on pilings on the beach west of Malibu, California.

Human Intervention and Mitigation of Coastal Change

People with beachfront or cliff-top homes are commonly affected by storms that cause beach erosion or threaten the destruction of their property. Until something bad happens to their beaches, beach cliffs, or homes, however, many do not really think about the constant motion of sand along the beach from waves and offshore sand movement in storms. Unless tragedy strikes near home, they do not realize that soft sediment beach cliffs gradually retreat landward as they erode at their base and collapse or slide into the ocean.

The typical response to this threat to their property is to build structures to protect communities, a process

called **beach hardening**. An extreme example of beach hardening is evident along the New Jersey coast (**Case in Point:** Extreme Beach Hardening—New Jersey Coast, p. 413). Unfortunately, the structures put in place to protect communities actually contribute to further erosion and, in the long run, make those communities more vulnerable. More recent developments in beach management policies include beach nourishment and zoning.

Engineered Beach Protection Structures

Before large-scale tourism, coastal residences and even small communities fell to the waves or moved inland as the beach gradually migrated landward. Once longer-term investments were made in shore properties, however, individuals and governments were motivated to protect those properties rather than move them. Shore-protection projects followed the catastrophic Galveston, Texas, hurricane of 1900. These included building massive **seawalls** (**FIGURE 13-27**). Individuals during this period tried to stop the erosion threat with riprap or walls built of either timber or concrete at the back of the beach in front of their property (**FIGURE 13-28**). Sometimes a promenade or boardwalk fronted the beach near the main access road and local business district. Seawalls and other supposed beach protection, often called beach hardening, tends to front boardwalks and spread out along beaches from there. The thought was that the waves would beat against the boulders or walls rather than erode the homeowners' property.

Construction of seawalls accelerated until the 1960s, when scientists and governments began to recognize that much of these activities had long-term disadvantages. Although a structure may slow the direct wave erosion of a beach cliff for awhile, the waves reflect

FIGURE 13-27 BEACH HARDENING IN GALVESTON

U.S. Army Corps of Engineers.

Hyndman.

A. The massive Galveston seawall was built after the catastrophic hurricane of 1900. Beach sand was brought in by regular truckloads from coastal dunes to replace sand lost to storms. **B.** Storm waves pound the seawall at Galveston, Texas. **C.** Sand previously covered the heavy riprap but was eroded by storm waves during Hurricane Ike, 2008.

FIGURE 13-28 BARRIERS TO WAVES

A. These beachfront homes are protected, albeit temporarily, by heavy riprap, northwest of Oxnard, California. **B.** These huge interlocking concrete pieces are designed to reduce wave energy and protect the road at Muscat, Oman.

back off the barrier, stir sand to deeper levels, and carry the adjacent beach sand farther offshore (**FIGURE 13-29**). A wave sweeps up onto a gently sloping beach, and the return swash moves back on the same gentle slope. A wave striking a seawall, however, is forced abruptly upward, so the swash comes down much more steeply and with greater force, eroding the sand in front of the seawall. The beach narrows and becomes steeper, the water in front of the barrier deepens, and the waves reach closer to shore before they break. Thus, instead of a protected beach, bigger waves approach closer to shore. The result often hastens erosion and removal of the beach. When the water in front of a barrier becomes sufficiently deep, the beach is totally removed, and the waves may undermine the barrier, which then topples into the surf (**FIGURE 13-30**). If the beach lies at the base of a sea cliff, the supposedly protected cliff succumbs more rapidly.

For those who understand that sand grains on a beach tend to migrate along the coast, another approach has been to try to keep the sand from migrating. **Groins**, the barriers built out into the surf to trap sand from migrating down a beach, do a good job of that. They collect sand on their upstream sides (**FIGURE 13-31**).

Riprap walls, or **jetties,** are sometimes used to maintain navigation channels for boat access into bays, lagoons, and marinas. Jetties that border such channels extend out through the beach but typically require intermittent dredging to keep a channel open (**FIGURE 13-32**). They also block sand migration along the beach. **Breakwaters**, built offshore and parallel to the shore, have a similar effect,

causing deposition in the protected area behind the barrier (**FIGURE 13-33**).

Although all of these measures can trap sand, they have the unintended effect of reducing the sand that continues along a beach, which causes beach erosion on the down-current side of a barrier. Effectively, they displace erosion sites to adjacent areas.

Without continued supply, these beaches are starved for sand and thus erode away. So once someone builds a groin or jetty, people down-current see more beach erosion and are inclined to build groins to protect their sections. New Jersey shows these effects to the extreme. Except where replenished, its once sandy beaches are now narrow or nonexistent and lined with groins and seawalls (FIGURES 13-30 and 13-31). The impacts of inlets are dramatic at Ocean City, Maryland, and St. Lucie, Florida.

Most knowledgeable people view groins as a bad choice, and in many cases they are. The state of Florida now requires removal of groins that adversely impact beaches (or are simply nonfunctioning). Groins should not stop sand migration permanently—this should be a remedy only until sand is fully deposited on the upstream side. The effects, of course, remain.

However, in some cases, groins can work where the drift of sand farther down the coast is undesirable. For example, to keep an inlet open, dredges must remove sand moving down the coast into an inlet. Sand drifting into a submarine valley is generally carried far offshore and lost permanently from the coast (see FIGURE 13-12). In such cases, well-engineered groins can be effective. They take a wide variety of forms and sizes that depend on the specific

FIGURE 13-29 SEAWALLS INCREASE EROSION

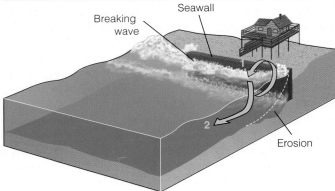

C

A. The Nor'easter of December 29, 1994, undermined this timber and steel wall at the right edge of photo and toppled it into the surf. Sandbridge Beach, City of Virginia Beach, Virginia. **B.** Waves crash against this seawall in Puerto Vallarta, Mexico. The downward force of the return wave undermines the sand at the base of the wall. **C.** This diagram shows how waves break against a seawall, causing erosion that may result in the collapse of the wall, as occurred in Virginia Beach.

FIGURE 13-30 ERODED BEACH

This is the sad ending for a mishandled beach north of Cape May, New Jersey. Building the coastal road led to protecting it with riprap, which hastened erosion of the beach. A concrete and rock seawall now lines the remains of the "beach."

FIGURE 13-31 GROINS

A. Groins at Cape May, New Jersey, trap sand. Longshore drift is from lower right to upper left. **B.** Houses crowd a narrow barrier bar at Manasquan, New Jersey. A groin in the foreground traps sand moving from left to right. The area downstream to the right, having lost its entire beach, is temporarily protected by a rock seawall.

FIGURE 13-32 JETTY

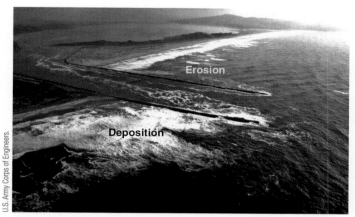

Longshore drift is interrupted at jetties just as it is at groins. The beach on the right side of this jetty near Tillamook, Oregon, has been eroded by longshore drift of sand from left to right.

FIGURE 13-33 BREAKWATERS

Breakwaters at Chesapeake Bay, Norfolk County, Virginia, in June 2002, shelter the beach behind them, causing deposition.

purpose they are intended for. Some even include artificial headlands.

Beach Replenishment

Are there better ways to protect a beach—or to repair the damage after its loss? Beginning in the 1950s, replacing sand on beaches became popular in the United States, in a process called **beach replenishment**, or

FIGURE 13-34 ADDITION OF SAND TO A BEACH

A truck dumps rusty, fine-grained sand excavated from the lagoon side of a barrier bar at Holden Beach, North Carolina, following a small storm in March 2001. Such sand is more easily eroded by smaller waves than the natural sand on the beach.

beach nourishment (**Case in Point:** Repeated Beach Nourishment—Long Island, New York, p. 413). To replace sand on a severely eroded beach, individuals sometimes contract to have sand brought in by dump truck (**FIGURE 13-34**). In fifty-six large federal beach projects in the United States between 1950 and 1993, the U.S. Army Corps of Engineers placed 144 million cubic meters of sand on 364 kilometers of coast. Enormous volumes have been placed in some relatively small areas, such as the 24 million cubic meters on the shoreline of Santa Monica Bay, California.

When the federal government agrees to foot a large part of the bill for a major beach replenishment project, the Army Corps of Engineers becomes the responsible agency. Engineers, geologists, and hydrologists with expertise in beach processes design a replenishment project and oversee the private contractors who actually do the work. Common sand sources include shore areas in which sand shows net accumulation or sources well offshore and below wave base. Sometimes sand is dredged off the bottom and transported to the beach area on large barges. Elsewhere sand is suction-pumped from the source and pumped through huge pipes as a slurry of sand and water. From there, the sand is spread across the beach using heavy earthmoving equipment (**FIGURES 13-35** and **13-36**).

In some cases, however, engineers choose to minimize costs by adding sand up the coast and permitting longshore drift to spread it into the area needing nourishment. Dumping sand at a cape, for example, can lead to sand migration into adjacent bays where it is needed. Similarly, sand drifting into inlets is often lost to the system if inlet dredges dump it far offshore. Some areas, such as Florida, now

FIGURE 13-35 BEACH REPLENISHMENT

A. To replenish a beach, the U.S. Army Corps of Engineers pumped a high-volume sand-and-water slurry from up the coast in 30-inch pipes. Heavy earthmoving equipment spread more than a meter of sand across the new beach at Ocean Isle Beach, North Carolina, in March 2001. **B.** As part of the massive beach replenishment at East Rockaway, New York, in March 1999, the groin at left minimizes the loss of the replenished sand by long-shore drift.

FIGURE 13-36 BEFORE AND AFTER REPLENISHMMENT

These views show Miami Beach, Florida, before and after beach nourishment by the U.S. Army Corps of Engineers.

require that dredged sand be dumped next to down-drift beaches (**FIGURE 13-37**). The operation thus permits the sand to bypass the inlet. Other bypass operations use fixed or movable jet pumps, most commonly in conjunction with conventional dredges.

Along cliff-bound coasts, sand, sediment, or easily disintegrated fill material is sometimes merely dumped over an eroding bluff to permit breakup by waves. Instead of rapidly eroding the cliff base, the waves gradually break up the dumped sediment to form new beach material.

One problem with beach replenishment is that natural coastal processes involve the migration of sand away from beaches. Unfortunately for replenishment projects, sand added to a beach in one place will quickly migrate elsewhere. A moderate storm the next month or next year could remove all of the added sand. In many cases the replenished beach is eroded away in the next big storm or several smaller storms—often after only a year or two. Because subsequent storms remove sand from the beach, part of the cost of a sand renourishment project involves

FIGURE 13-37 PIPING IN DREDGED SAND

Jetties bordering an estuary block southward drift of beach sand. That starves the beach to the south, leading to its erosion. Note that the beach south of the estuary is much recessed from the straight coastline. Lighthouse Point, Pompano Beach, Florida. Arrow shows location of discharge pipe. Dredging of sand from a river outlet between jetties at Lighthouse Point. The dredge is the yellow boat in the middle left. Sand from this dredging is piped to the down-drift side of the jetties to replenish the beach seen in the foreground.

continued small-scale nourishments at intervals of two to six years. Small trucking operations may need to be repeated every year, especially if the added sand is finer grained than that on the original beach. Sand finer than 0.5 millimeters is too fine; small waves will typically wash it offshore. At Carolina Beach, North Carolina, for example, the beach, lined with private homes, was replenished eight times in the 26 years between 1965 and 1991.

Another concern is that sand placement can change a shore profile (slope perpendicular to the beach) and actually increase erosion. Most people want to see the sand they pay for placed on the upper dry part of a beach rather than offshore. It is also easier to calculate the volume of sand added to the upper beach and therefore the appropriate cost. But the active beach actually extends well offshore into shallow water. If that part of the beach is not also raised, waves will move much of the onshore sand offshore to even out the overall slope of the beach. The next storm will carry much of the new sand offshore, where people view it as lost to the beach and think the replenishment was a waste of money. If half of the sand added is offshore in shallow water where it provides a more natural equilibrium profile, people have a hard time understanding that it is not being wasted. Either scenario leads to criticism of beach replenishment as a viable solution to beach erosion. As a result of such problems, some beach experts suggest using the expression "shore nourishment" rather than "beach nourishment" to emphasize that the shallow, underwater part of the beach is equally important.

Identifying the source for the sand is another hurdle for beach replenishment projects. For obvious reasons, mining sand from other beaches is generally not permitted. In some areas, significant sand is obtained from maintenance dredging of sand from navigation channels. Mining sand from privately owned sand dunes well back from a beach or dredging sand from a lagoon or other site behind a barrier island is sometimes possible, though expensive. A significant drawback to such sources is that dune and lagoon sand is generally finer grained than that eroded from the beach. Because storm waves were able to move the coarser-grained sand from the beach, slightly finer grained sand can be removed by even smaller waves.

Another solution used in many areas, especially along the southeastern coasts, is to dredge sand from well offshore and spread it on a beach. Because much larger equipment is required, regional or federal governments, often under the direction of the Corps of Engineers, normally undertake such projects. Taking sand from near shore deepens water and makes the beach steeper. However, waves shift sand into an equilibrium slope, a process that depends especially on the size of both sand grains and waves. An artificially steeper beach will erode down to the equilibrium slope. If the sand is taken from well offshore, the bigger waves of storms erode the beach down to a lower slope farther offshore.

Sources of usable sand in many areas are being rapidly depleted. Florida's sources of economically recoverable sand, for example, were almost exhausted by 1995. They are now looking at tens of billions of tons available from the Great Bahama Banks.

The scale and cost of beach nourishment projects is huge. A large dump truck carries roughly 10 cubic meters of sand. If a person's lot is 30 meters wide and the beach extends 65 meters from the house to the water's edge at mid-tide (roughly 2,000 square meters), one dump-truck load would cover that part of the beach to a depth of only half a centimeter or so; it would take 200 loads to add about a meter of sand to the beach in front of one house. Thus, if sand costs $30 per cubic meter, it might take 200 truckloads in front of every home to replace the sand removed in one moderate storm—for a cost of roughly $60,000 for each home. More than 28,000 kilometers of coast continue to erode.

As always, the role of government in the cost of such projects is a matter for debate. Sometimes, groups of residents, towns, or counties on barrier islands lobby the local, state, and federal governments to replenish the sand on a severely eroded beach. Because large sand replacement projects typically run into millions of dollars and home owners do not want their taxes to increase, local governments lobby state and federal governments to foot most or all of the bill. Politicians want to be reelected and bring as much money back to their communities as possible, so they lobby hard for state and federal funding. What this means, of course, is that the cost of replenishing sand to benefit a few dozen beachfront homeowners is spread statewide—or more commonly countrywide—among those who generally derive no benefit from the work. For example, the federal government paid approximately fifty percent of the cost of beach replenishment projects in Broward County, Florida, from 1970 to 1991, while local communities paid only four percent. To add insult to injury, beachfront communities often try to inhibit access by numerous mainlanders who flock to beaches on warm summer days. Although beach areas below high tide are legally public, access routes are often poorly marked or illegally posted for no access. Some communities also make beach approaches difficult by severely restricting parking along nearby roads.

Since the 1990s, federal and state funds for replenishment have dwindled. Environmentalists, inland taxpayers, and government officials often argue against spending millions of public dollars for major projects to benefit a few beach-front homes when the sand is likely to wash away in the next significant storm. Lacking state or federal support, local counties sometimes try to fund their own beach projects, but these often fail because even those living on the beach sometimes argue that they cannot afford the annual assessments. Since many of those who own beach-front homes live elsewhere, local voters often reject tax increases. Even where local groups successfully fund multimillion dollar projects, they often find that the new sand erodes in a few years.

Zoning for Appropriate Coastal Land Uses

A less expensive and more permanent alternative to beach hardening or beach replenishment is advocated by many coastal experts. They suggest moving buildings and roads on gently sloping coasts back landward to safer locations after major damaging storms. Sand dunes behind a beach can be stabilized with vegetation in order to provide further protection for areas behind them. The cost of moving buildings may be high, but it is less than the continuing long-term cost of maintaining beach-hardening structures that tend to destroy their beaches or of continually bringing in thousands of tons of sand that gets removed by following storms. And it does provide a way of living with the active beach environment rather than forever trying to fight it.

On cliff-bound coasts, buildings should not be placed close to cliff edges. Those that are too close should be moved well back, and foot traffic and other activities should be restricted to areas away from cliff tops and faces. Beaches should not be cleared of natural debris such as driftwood. Mining sand and gravel from beaches and streams should be prohibited. Additional dams should not be built on rivers that discharge in coastal regions with erosion problems; removal of old dams would eventually bring more sediment back to beaches and help protect cliffs.

Cases in Point

Extreme Beach Hardening
New Jersey Coast ▶

One of the most extreme cases of beach hardening and severe erosion is along the coast of New Jersey. Early development was promoted by extending a railroad line along more than half of the New Jersey shoreline by the mid-1880s. Buildings first clustered around railroad stations, dunes were flattened for construction, natural dense scrub vegetation was removed, and marsh areas were filled. The arrival of private automobiles accelerated development. Marsh areas along the back edge of the offshore bars were both filled and dredged in the early 1900s to provide boat channels. Inlets across barrier islands were artificially closed, and jetties built after 1911 stabilized others. People did not understand then that barrier islands were part of the constantly evolving beaches and that nature would resist human attempts to control it.

The large population nearby, and the demand for recreation, stimulated the building of high-rise hotels and condominiums in Atlantic City; it is now so packed with large buildings next to the beach that they strongly affect wind flow and sand transport. Large-scale replenishment of sand on beaches is widespread in some areas. In other areas, narrow and low artificial dunes are built in front of separate detached houses, as are barriers to provide backup protection from erosion. Groins are numerous. A prominent seawall and continuous groins bear no resemblance either to the original barrier island environment or to the beaches that attracted people in the first place.

Hyndman.

▶ **A.** *New Jersey has permitted condos and large houses to be built on the artificially replenished beach just south of Atlantic City. The dunes are gone; their only protection is riprap and steel walls.* **B.** *Giant sandbags buried in an artificial dune were exposed as the replenished beach was eroded to a flatter slope during a minor storm in October 2006. Wildwood area, south of Atlantic City.*

Repeated Beach Nourishment
Long Island, New York ▶

One of the most vulnerable coastlines in the northeastern United States is the south coast of Long Island, New York, which protrudes east into the paths of some strong Atlantic storms. As in any low-lying coast facing big waves, its southern portion is marked by a nearly continuous line of sand bars and barrier islands. The unincorporated community of Westhampton Dunes, about halfway east along the barrier island, has become a "poster child" for problems of beach renourishment and attempts to maintain a constantly shifting barrier island.

A chronology of events illustrates evolution of the problem:

1950s: Erosion of the beach threatens the town of Westhampton Beach to the

(continued)

▶ **A.** *The late-1992 storm breached the barrier bar at Westhampton Dunes and left houses stranded in the surf. The same houses are shown in Figure 13-15.* **B.** *The main breach was just down drift from the westernmost groin.* **C.** *A recent satellite view of part of Westhampton Dunes (right half of image) includes Moriches Inlet where tides flush the lagoon to the north. The west end of the former groin field is visible near the right edge of the image. Although the narrow barrier bar has escaped serious damage in recent years, the lack of major storms leaves the area vulnerable to catastrophic breaches and overwash in many areas.*

north. The Army Corps of Engineers proposes renourishment and construction of 23 groins "if necessary."

1962: A storm breaches the barrier island and, following political and legal pressure, 11 hastily constructed groins immediately restrict east-to-west longshore drift of sand. Beach erosion to the west, at the 4-kilometer-long community of West Hampton Dunes, abruptly increases from 0.5 to 4.5 meters per year.

1970: Four more groins are built to protect large homes and a condominium. In the late 1970s more houses are lost to erosion in severe Nor'easters.

1973 and 1984: Residents with eroding beaches file lawsuits against the government because the groins accelerated erosion, but they are not compensated. Various efforts to complete the groin field stall because of local, state, or federal concerns with cost-sharing to specific

state requirements for shore protection projects.

1988: New York's Sea Grant Program convenes a group of coastal experts who consider available options and predict formation of a breach at one site just west of the westernmost groin, but nothing is done because lawyers are reluctant to move forward in the midst of a lawsuit.

1991 and late November 1992, Nor'easter: Storms destroy many houses and breach the barrier island in several places; one breach widens to almost one kilometer, destroying dozens of houses. A total of 190 out of 246 homes have now been lost.

1993: The breach at the predicted site is filled in 1993 for $8.8 million and the groins are shortened to permit more sand to drift along the beach.

1994: Residents, now as a municipality, again file a lawsuit, this time successfully.

The Corps of Engineers, State, and County agree to maintain the beach for the next 30 years.

1997: The $53 million (adjusted to 2007 dollars) renourishment project is completed with 3.4 million cubic meters of sand. Since no major storms struck Long Island in the following five years, the effort is pronounced a success.

Most of the cumulative cost was paid by taxpayers of the State of New York. In return, they are provided with limited parking for 200 cars at two locations along 3 kilometers of road length and entry to the beach via seven public access points. The history of this short stretch of barrier bar emphasizes the hazards and pitfalls of trying to control the natural processes that shape and move a beach on a gently-sloping coastline.

Critical View

Test your observational skills in interpretation through the following scenes relevant to waves, beaches, and coastal erosion.

A This beach is on a large lake in the northern Rockies.

Donald Hyndman.

1. Why is the front of the beach so steep?
2. Why is it steeper at its top?
3. Why are pebbles larger at the top of the beach front rather than the bottom?

B This beach runs along the central California coast next to a highway.

Highway

USACE.

1. Why are huge boulders piled at the back of a nice sandy beach?
2. What are the consequences to the beach from piling these boulders here? Why?

C These houses sit right on the beach along the California coast.

Highway

D. Krohn, USGS.

1. What hazards can you identify in this photo?
2. What measures might these beachfront homeowners take to better protect their properties?
3. What is the direction of longshore drift? How can you tell?

D This fence runs between these houses and the beach on Long Beach Island, north of Atlantic City, NJ.

Hyndman.

1. Why is there a steep cliff in the beach sand in the foreground?
2. What effect does the slat fence have on the distribution of sand on the beach?
3. How does this fence reduce the hazard to these houses during a storm?

E These houses are near the New Jersey coast, south of Atlantic City. The ocean is 100 meters toward the lower left corner of the photo.

Hyndman.

1. Why are these houses built up on posts?
2. How do the posts help protect the home?

G This beach is along the central California coast.

David Hyndman.

1. What dangerous swimming conditions are present?
2. What evidence of those conditions is apparent in the pattern of wet sand on the upper part of the beach?
3. What evidence is there of those dangerous swimming conditions in the water just offshore?
4. If you were swimming in the water here, what action should you take to get to safety?

F This beach is along the Lake Michigan coast near Chicago.

Donald Hyndman.

1. What are the dark lines called that are protruding into the water from the beach?
2. Why is the beach shaped like giant saw teeth?
3. In what direction is the sand moved along the beach by waves?

H This steep coastline is along the central California coast. A house sits about four meters to the right, in the foreground.

Hyndman.

1. What hazard is apparent in the photo?
2. What has the owner of the home in the foreground done to minimize that hazard and why?
3. What has been done to protect the apartment building at the back of the photo (in distance) and why?

Chapter Review

Key Points

Living on Dangerous Coasts

- Beaches and sea cliffs constantly change with the seasons and progressively with time. When people build permanent structures at the beach, coastal processes do not stop; the processes interact with and are affected by those new structures.

Waves and Sediment Transport

- Most waves are caused by wind blowing across water. The height of the waves is dictated by the strength, time, and distance of the wind blowing over the water.

- Water in a wave moves in a circular motion rather than in the direction the wave travels. That circular motion decreases downward to disappear at an approximate depth of half the wavelength. Waves begin to feel bottom near shore in water less than that depth. **FIGURES 13-2** and **13-4**.

- Wave energy is proportional to the wavelength multiplied by wave height squared, so doubling the wave height quadruples the wave energy. **By the Numbers 13-2**.

- Waves approaching the beach at an angle push sand parallel to shore as longshore drift. **FIGURE 13-5**.

Beaches and Sand Supply

- The grain size and amount of sand on a beach depend on wave energy, erodibility of sea cliffs and the size of particles they produce, and the size and amount of material brought in by rivers. Larger winter waves commonly leave a coarser-grained, steeper beach.

- A beach at its equilibrium profile steepens toward shore. Larger waves erode the beach and spread sand on a gentler slope. **FIGURE 13-11**.

- Sand is lost from a beach through erosion, particularly during storms. Anything that removes sand supply from a beach, for example, damming rivers upstream, results in greater erosion.

Erosion of Gently Sloping Coasts and Barrier Islands

- Offshore barrier islands develop where the landward-steepening beach profile is steeper than the general slope of the coast. Wind blows sand shoreward into dunes.

- Barrier islands maintain their equilibrium profile by migrating shoreward with rising sea level. Because many barrier islands are now covered with buildings, islands cannot migrate landward but are progressively eroded. **FIGURE 13-15**.

- Tidal currents keep open estuaries and inlets that lead into lagoons behind bars. Storm surges and waves open, close, and shift the locations of such inlets.

- Dune sand spreads over a beach, helping to protect it from erosion during storms, but most dunes have disappeared by excavation, trampling underfoot, or wave action.

Sea-Cliff Erosion

- Sea cliff erosion is determined by the strength of the rock, wave height, sea level, and precipitation. As a result, erosion is amplified during storms.

- Homeowners can attempt to protect their sea-cliff property from erosion by placing riprap at the base of a cliff or covering it with concrete.

Human Intervention and Mitigation of Coastal Change

- Beach hardening to prevent erosion or stop the longshore drift of sand includes seawalls, groins, jetties, and breakwaters. All have negative consequences either after a period of time or elsewhere along the coast. In some cases, they result in the complete loss of a beach. **FIGURES 13-7 to 13-33**.

■ Beach replenishment or nourishment involves replacing sand on a beach, an expensive proposition that needs repetition at intervals. **FIGURES 13-34 to 13-37**.

■ A less-expensive and more permanent solution to the effects of beach erosion is to move threatened structures back from a beach and implement environmental protection strategies.

Key Terms

barrier island, p. 394	groins, p. 403	rip current, p. 390	storm surge, p. 393
beach hardening, p. 402	headlands, p. 389	riprap, p. 385	submarine canyon, p. 393
beach replenishment, p. 405	jetties, p. 403	sand dunes, p. 397	wave energy, p. 388
breakwaters, p. 403	longshore drift, p. 388	seawalls, p. 402	wave height, p. 387
equilibrium profile, p. 392	Nor'easters, p. 393	shore profile, p. 392	wavelength, p. 387
fetch, p. 387	period, p. 387	shotcrete, p. 400	wave refraction, p. 388

Questions for Review

1. Sketch the motion of a stick (or a water molecule) in a deepwater wave.

2. What force creates most waves?

3. How can there be big waves at the coast when there is little or no wind?

4. What factors cause growth of wind waves?

5. Why are ocean waves generally larger than those on lakes?

6. Where does the sand go that is eroded from a beach during a storm?

7. Where would a sand spit form on a barrier island relative to the direction of longshore drift? Draw a sketch to illustrate your answer.

8. Draw the shape of a barrier island before and after a significant rise in sea level.

9. How do sand dunes affect stability of a beach?

10. Which side of a beach groin collects sand? Why is this the case?

11. What happens to wave energy and erosion when riprap or seawalls are installed?

12. What can be done to prevent building in hazardous regions in coastal areas?

13. What can be done to minimize erosion in a coastal area, and what are some positive and negative aspects of those methods?

Discussion Questions

1. Who should pay for the costs of beach nourishment projects (local residents, state taxpayers, national taxpayers)? Should beach-user fees play a role in financing these projects?

2. If you owned a house near the edge of an eroding seacliff on the Oregon or California coast, what would you do? Keep in mind that you are morally and legally obligated to inform a buyer of any damages or imminent hazards that may affect the house.

3. If you were a FEMA administrator in charge of government aid to people living on a barrier island and whose homes are now cut off by hurricane breach of their barrier island, what would you do and why?

4. Where should beach-replenishment sand come from? What are the consequences of taking sand from that location?

Hurricanes and Nor'easters 14

Hyndman.

Sinking these posts more than a meter in the beach was not sufficient to prevent Hurricane Ike from toppling this house northeast of Crystal Beach on the Bolivar Peninsula, Texas.

Hurricane Ike made landfall at Galveston, Texas, just after midnight on September 13, 2008, as a category 2 storm with hurricane-force winds extending an incredible 190 kilometers out from its eye. Given dire warnings, most people evacuated, but many stayed to ride it out based on past false alarms and confidence in their own structures. The 5.2-meter-high concrete seawall built at Galveston after the disastrous 1900 hurricane spared the city from much of the wave effect, but the strongest onshore winds, directed at the Bolivar Peninsula northeast along the coast, obliterated most of the houses there. Although most structures along this part of the Texas Gulf Coast are raised on high, sturdy posts so that storm waves will pass under them, the big waves and surge also stir up and erode the sand that provides the foundation for the posts.

On Galveston Island southwest of Galveston, however, the high winds and surge initially came from the east before landfall, dramatically raising a mound of water in the lagoon behind the island. At landfall, winds turned south and southeast, pushing the lagoon surge back offshore, stripping meters of sand off the beach, and carrying it offshore. Here too the destruction was almost complete, especially in the first two rows of houses closest to the beach. Roofs and walls shredded in the winds; whole houses collapsed near the beach and were swept away in the waves as their posts snapped off or were undermined by loss of sand.

The aftermath was shocking. Northeast of Galveston where the onshore winds and surge were strongest, almost all of the houses, for ten rows back from the beach, were completely obliterated. Hundreds of meters from the beach, a tangled ridge of boards, plywood, roofing, furniture, cars, appliances, and power poles was all that remained. Hours of 3- to 4-meter-high surge with huge waves eroded more than two meters of sand from the beaches, carrying it inland, and demonstrating that the posts were either not sunk deeply enough in the sand or strong enough to withstand the winds and waves. Most of the houses were blown apart by the winds and flying debris, lifted off their support posts, or toppled and smashed in the surf when their posts either snapped or were undermined. Loss of sand also lifted and smashed roads, driveways, sidewalks, and palm trees. Septic tanks, formerly buried in the sand, were upended and emptied.

Power went out when power lines and poles were downed or toppled, water lines were severed and contaminated as sewage remained uncontained, and roads were either torn up or covered with sand and debris. The whole coast for almost 200 kilometers northeast and southwest of Galveston was uninhabitable for weeks. For the few who refused to evacuate and survived, there was no food, water, electricity, telephone, or evacuation after the storm. Refrigerators and freezers could not be kept cold and food rotted in the late-summer heat. In spite of being only a mid-category 2 hurricane at landfall, damages from Ike reached more than $27 billion along the coast, the third costliest tropical cyclone ever recorded.

Hurricanes

A **tropical cyclone** is a storm system consisting of a low-pressure center surrounded by strong rotating winds. These storms have wind speeds of more than 120 kilometers per hour and can exceed 260 kilometers per hour. They can also bring torrential rain, thunderstorms, tornadoes, and a high surge of water from the sea along the coast.

When a large tropical cyclone occurs in the North Atlantic and eastern Pacific, it is called a **hurricane**, a term derived from a Caribbean native Indian word meaning *big wind*. The same type of storm is called a **typhoon** in the western Pacific, Japan, and Southeast Asia. In the Indian Ocean these storms are simply called cyclones. In our discussion in this chapter, we will refer to all such storms as hurricanes.

Historically, tropical cyclones create some of the world's deadliest disasters. The worst disasters have occurred in heavily populated poor countries in Southeast Asia, where single storms can result in hundreds of thousands of deaths.

Formation of Hurricanes

Hurricanes begin to develop over warm seawater of at least 25°C (77°F), commonly between 5 and 20 degrees north latitude. In these tropical latitudes, temperatures are high and air-pressure gradients are weak. Air rises by localized heating, which causes condensation that can build into towering convective "chimneys" with frequent thunderstorms. Strong winds can shear off the tops of the convective chimneys and cause them to dissipate.

In the absence of winds, convection may strengthen when the air rises to high elevations. The warm, moist air over the ocean rises and spreads out at the top of the chimney. That warm air expands, cools, and releases latent heat; the air rises faster as the storm strengthens.

At the center of the storm is a low-pressure zone called the **eye**, clearly visible as a small, dark area in many satellite views (**FIGURE 14-1**). The eye of a hurricane is as much as 20°C warmer than the surrounding air. Rise of air near the eye pulls more moist air into the eye from low elevations at the periphery of the storm. Coriolis forces initiate rotation in the rising air, the highest winds and lowest air pressures focusing toward the core of the storm. The winds rotate counterclockwise in the northern hemisphere and clockwise in the southern hemisphere, with the highest wind speeds along the edge of the eye wall.

The whole storm may be from 160 to more than 800 kilometers in diameter. Once formed, storms move across the ocean with the prevailing trade winds, their forward motion averaging 25 kilometers per hour.

FIGURE 14-1 FORMATION OF A HURRICANE

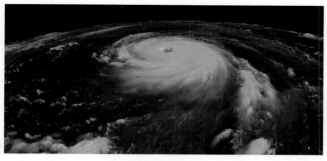

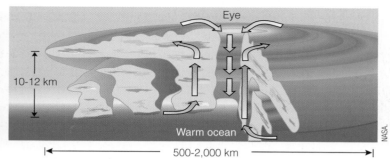

A. An oblique astronaut view of Hurricane Katrina looking north before landfall. The storm circulates counterclockwise, with its trailing winds spreading farther out (lower right). **B.** An idealized cutaway view of a hurricane as seen from the side. The top of the diagram corresponds to the astronaut view on the left. Air drawn in at sea level from outside the storm rises near the eye and spreads out; cold dry air sinks in the eye.

Classification of Hurricanes

Hurricanes are classified by their wind speed. Hurricane-force winds have sustained velocities greater than 119 kilometers per hour (74 mph). The highest winds in a hurricane exist along the edge of the eye wall. Inside the eye—generally 20 to 50 kilometers in diameter and bounded by a wind "wall"—the winds drop abruptly from, for example, 220 to 15 kilometers per hour (**FIGURE 14-2**). The air pressure drops from normal atmospheric pressure of approximately 1,000 millibars (1 bar) to 960–970 millibars in the eye. The air in the sharply bounded eye sinks, causing skies to clear as the dry air from above warms and can hold more moisture.

The **Saffir-Simpson Hurricane Scale** divides hurricanes into five categories based on barometric pressure and average wind speed, with Category 1 storms being the weakest and Category 5 the strongest. The lower the barometric pressure and the higher the wind speed, the stronger the hurricane (**Table 14-1**). A hurricane typically changes category as it develops, either intensifying or weakening. For example, Hurricane Katrina was a Category 5 storm at sea, but its winds slowed as it approached land, weakening to a Category 3 storm.

Although the Saffir-Simpson Scale is designed to indicate the intensity of a hurricane, it can be misleading as an indicator of the level of expected damage. Though people often focus their attention on the strongest hurricanes, Categories 4 and 5, lower-category storms can sometimes do almost as much damage and in some cases cost even more lives (**Table 14-2**). In addition to the wind speed, the amount of damage from a hurricane is heavily dependent on the height of water rise from the storm surge, how large an area is covered, the duration of high water and high winds, and how recently another storm has affected the area (discussed in greater detail in the next section).

FIGURE 14-2 WIND SPEED AND ATMOSPHERIC PRESSURE

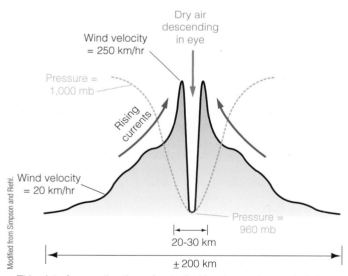

This plot of properties through a typical hurricane shows what atmospheric pressure (tan dashed line), wind velocity (heavy black line), and air motions would be for measurements taken from a plane flying through the center of a hurricane.

Areas at Risk

Tropical cyclones rotate counterclockwise but track clockwise in northern hemisphere ocean basins, the same direction as the ocean currents (**FIGURE 14-3**, p. 419). Cyclones rotate clockwise and track counterclockwise in the southern hemisphere. Northern hemisphere tropical storms begin in warm waters off the west coast of Africa,

Table 14-1

The Saffir-Simpson Hurricane Scale

CATEGORY**	EXAMPLE	BAROMETRIC PRESSURE*		STORM SURGE		WIND SPEED		DAMAGES***
		Mbar	In.	M	Ft	Kph	Mph	
Normal (no storm)		1,000	29.92	0	0	0	0	
Tropical storm				<1.2	<4	62–119	39–74	
1	Danny, 1997	>980	>28.94	1.2–1.7	4–5	120–153	75–95	Minor to trees and unanchored mobile homes.
2	Bertha, 1996; Isabel, 2003; Ike, 2008	965–979	28.5–28.93	1.8–2.6	6–8	154–177	96–110	Moderate to major damage to trees and mobile homes, windows, doors, some roofing. Low coastal roads flooded 2 to 4 hours before arrival of hurricane eye.
3	Alicia, 1983; Fran, 1996; Katrina, 2005	945–964	27.91–28.49	2.7–3.3	9–12	178–209	111–130	Major damage: Large trees down, small buildings damaged, mobile homes destroyed. Low-lying escape routes flooded 3 to 5 hours before arrival of hurricane eye. Land below 1.5 meters above mean sea level flooded to 13 kilometers inland.
4	Hugo, 1989	920–944	27.17–27.9	3.4–5.5	13–18	210–249	131–155	Extreme damage: Major damage to windows, doors, roofs, coastal buildings. Flooding many kilometers inland. Land below 3 meters above mean sea level flooded as far as 10 kilometers inland.
5	Camille, 1969; Gilbert, 1988; Andrew, 1992; Mitch, 1998	<920	<27.17	>5.5	>18	>249	>155	Catastrophic damage: Major damage to all buildings less than 4.5 meters above sea level and 500 meters from shore. All trees and signs blown down. Low-lying escape routes flooded 3 to 5 hours before arrival of hurricane.

*Standard atmospheric pressure at sea level = 29.92 inches of mercury = 1,000 millibars (mb) = 1 bar or 1 atmosphere pressure.
**Wind is the primary control used to categorize hurricanes—pressure and storm surge height were formerly used, but are now just reference.
***These are highly variable and depend on many factors as discussed in the text.

Table 14-2	Comparison of Hurricane Categories 5 and 2	
HURRICANE		
CHARACTERISTIC	CATEGORY 5	CATEGORY 2
Surge	High	Lower
Winds	High	Lesser
Diameter	Smaller (affects smaller area)	Larger (affects larger area)
Speed	Crosses coastline rapidly (shorter life span on coast; dumps less rain on smaller area and less flooding)	Crosses coastline slowly (longer life span on coast; often dumps more rain on larger area and more flooding)
Examples	Gilbert, 1988; Andrew, 1992; Camille, 1969	Agnes, 1972; Isabel, 2003; Frances, 2004

FIGURE 14-3 HURRICANE TRACKS

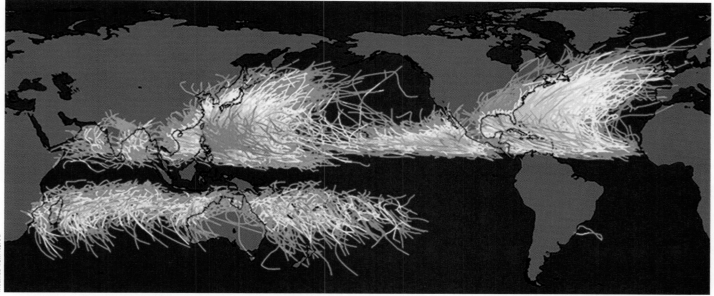

Modified from NOAA.

Colored lines indicate the paths of all hurricanes and tropical storms. Storm intensity is indicated by color, from weaker storms in blue to category 5 hurricanes in red.

then move westward across the Atlantic Ocean with the trade winds. They warm, pick up wind speed and energy, and often develop into hurricanes before they reach the Americas. They generally track west, northwest, and then north, either off the southeastern United States or sometimes into the continent. Typhoons in the western Pacific and cyclones in the north Indian Ocean have similar tracks.

Some 80 to 90 tropical storms and 45 hurricanes occur worldwide every year (130,000 such storms per thousand years). An average of six named hurricanes form annually in the Atlantic Ocean and Gulf of Mexico. Thus, in a short period of geologic time, cyclones are likely to have a

major effect on erosion, deposition, and overall landscape modification, especially in exposed areas such as barrier islands.

For the United States, subtropical cyclones or hurricanes are a major concern along the Gulf of Mexico and southern Atlantic coasts. More than 44 million people live in coastal counties susceptible to such storms, roughly 15 percent of the total population of the United States. Because of the rapid growth of these areas, a large proportion of the population has never directly experienced a hurricane and is poorly informed about the risks. These residents, and equally inexperienced developers and builders, choose to live and work in hazardous sites such

FIGURE 14-4 HURRICANE PATHS AND OCEAN TEMPERATURE

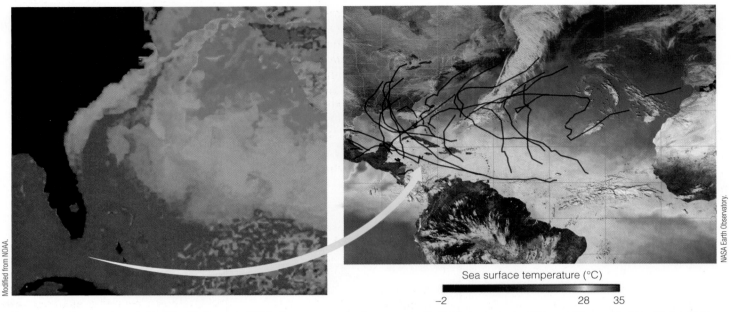

Modified from NOAA.

NASA Earth Observatory.

Sea surface temperature (°C)

-2   28   35

Sea-surface temperatures in the Atlantic Ocean in 2005 show warm water (orange colors) off the west coast of Africa, drifting westward with the trade winds to energize hurricanes impacting the southeastern United States (black paths indicate hurricanes in 2005). Streaky white clouds are also visible.

as beachfront dunes and offshore barrier bars. Rising property values amplify the monetary damage when a major storm does hit.

Hurricanes generally curve northward as they approach the southeastern coast of the United States because of the Coriolis effect. Some track fairly straight; others take erratic paths (**FIGURE 14-4**). They commonly strengthen as long as they remain over the warm water of the Gulf Stream that flows northeasterly along the east coast of Florida and continues up the coast past eastern Canada. Westward-moving storms therefore tend to track in arcs to the northwest and then north as they approach the coast. If they move over especially warm ocean eddies, they may strengthen dramatically, as in the case of Hurricane Katrina in the Gulf of Mexico in 2005 (**FIGURE 14-5**). As they move over cooler waters or continue over land, they gradually lose energy and dissipate.

Although most U.S. hurricanes affect lower-latitude areas, many of those that remain off the coast with northward trajectories reach as far north as New England. Only a few of these have intensities greater than Category 3, but those few can be highly destructive, especially where Long Island, Rhode Island, and eastern Massachusetts protrude into the Atlantic Ocean and across the hurricane paths. Based on historic accounts and prehistoric overwash deposits of sand covering layers of peat, at least 27 hurricanes of Categories 1 and 2 have hit the same area in the last 400 years, an average of one every 15 years.

FIGURE 14-5 HURRICANE STRENGTH AND OCEAN TEMPERATURE

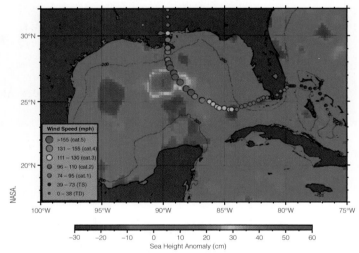

NASA.

Wind Speed (mph)
>155 (cat.5)
131 – 155 (cat.4)
111 – 130 (cat.3)
96 – 110 (cat.2)
74 – 95 (cat.1)
39 – 73 (TS)
0 – 38 (TD)

-30 -20 -10   0   10   20   30   40   50   60
Sea Height Anomaly (cm)

Sea-Surface temperature for the Gulf of Mexico in August, 2005, show an eddy of warm water (measured by sea-surface height, indicating thermal expansion of the water). The strength of Hurricane Katrina, as indicated by wind intensity, increased dramatically where it passed over the eddy of warm water (red).

Large hurricanes of Categories 3 to 5 strike most locations from Florida through the Texas Gulf Coast once every 15 to 35 years on average. Especially vulnerable are southern Florida, the Carolinas, and the Gulf Coast (**FIGURE 14-6**).

FIGURE 14-6 AREAS AT RISK FOR HURRICANES

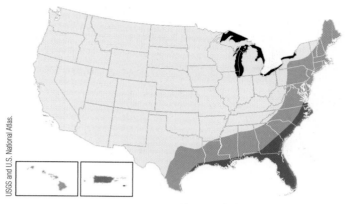

Potential number of hurricane strikes (and floods from hurricanes) per hundred years based on historical data. Red = more than 60; dark blue = 40–60; light blue = 20–40.

Although hurricane season spans from late June or July to November, most hurricanes develop in August or September, because the sun spends all summer warming ocean surface temperatures. On average, nine hurricanes develop in the eastern Pacific basin each year, four of them major.

Storm Damages

Over the past century, the costs related to hurricanes in the United States have dramatically increased, while the number of deaths has decreased (**FIGURE 14-7**). In fact, six of the ten most costly hurricanes occurred in 2004 and 2005 (**Table 14-3**). Increased costs reflect the rapidly growing populations along the coasts, more construction in unsuitable locations, and more expensive buildings. The number of deaths from hurricanes decreased considerably from its peak in the 1920s until a recent rise with Hurricane Katrina. The earlier decline is at least partly due to the improved ability of scientists to predict the locations of landfall and the coordinated ability to evacuate populations at risk.

Deaths and damages from hurricanes can be caused by storm surge, high winds with flying debris, waves, torrential rainfall, and flooding.

Storm Surges

Coastal areas experience **storm surges**, also called storm tides, as the sea rises along with an incoming hurricane, as a result of low atmospheric pressure and high winds. Under normal conditions, atmospheric pressure pushes down on the water surface, keeping the sea at its usual height. When a low-pressure system moves over an area of ocean, the height of the storm surge rises with lower atmospheric pressure (**By the Numbers 14-1**: Atmospheric Pressure Affects Surge Height, p. 422).

Prolonged high winds contribute to a storm surge by pushing seawater into huge mounds as high as 7.3 meters and 80 to 160 kilometers wide. The crest of a surge wave moves more or less at the speed of the storm because it is pushed ahead of the storm (**FIGURE 14-8**, p. 423). The pileup of water before the wind is higher, with greater wind speed and **fetch**, the distance the wind travels over open

FIGURE 14-7 COSTS AND DEATHS DUE TO HURRICANES

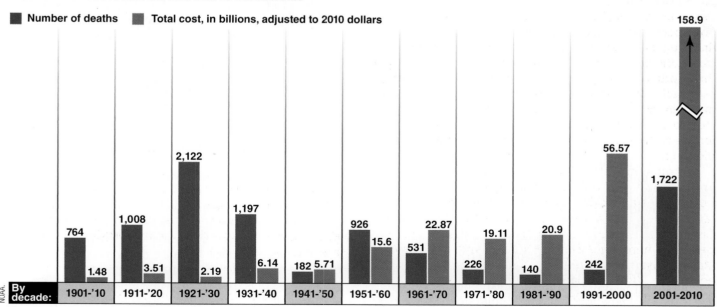

The death toll due to United States hurricanes has dropped steadily from its peak in the 1920s, except for a recent rise due to Hurricane Katrina. In contrast, the damage costs from hurricanes have risen rapidly over this same period.

Table 14-3			Costliest Hurricanes in the United States, 1900–2010			
RANK	HURRICANE	DATE	MAIN LOCATIONS	CATEGORY	TOTAL ESTIMATED LOSS*	
1	Katrina	2005, Aug.	New Orleans, Gulf Coast	3	91.4	
2	Andrew	1992, Aug.	Florida, Louisiana	5	51.9	
3	Ike	2008, Oct.	Galveston, Texas Gulf Coast	2	29.2	
4	Wilma	2005, Oct.	Southern Florida	3	23.2	
5	Charley	2004, Aug.	Florida, Carolinas	4	17.6	
6	Ivan	2004, Sept.	Alabama, Florida	3	16.7	
7	Hugo	1989, Sept.	Georgia, Carolinas, Puerto Rico, Virgin Islands	4	14.6	
8	Agnes	1972, June	Florida, northeast United States	1	13.4	
9	Betsy	1965, Sept.	Florida, Louisiana	3	12.8	
10	Rita	2005, Sept.	Louisiana, Texas, Florida	3	12.8	
11	Camille	1969, Aug.	Mississippi, Louisiana	5	10.6	
12	Frances	2004, Sept.	Florida, Georgia, Carolinas, New York, Virginia	2	10.5	
13	Diane	1955, Aug.	Northeast United States	1	8.3	
14	Jeanne	2004, Sept.	Florida, Pennsylvania, Georgia, S. Carolina, Puerto Rico, Virginia, Florida	3	8.1	
15	Frederic	1979, Sept.	Mid-Atlantic, Northeast United States	3	7.5	
17**	Allison	2001, June	North Texas	TS	6.9	
18**	Floyd	1999, Sept.	Carolinas, New Jersey	2	6.8	
24**	Isabel	2003, Sept.	North Carolina, Virginia	2	4.3	

* In billions of 2010 dollars. Figures from NOAA; Swiss Re Insurance Co., respectively.

** Well-known hurricanes Floyd and Isabel and Tropical Storm Allison are included here because they are featured prominently in this chapter TS = Tropical Storm.

▶ By the Numbers 14-1

Atmospheric Pressure Affects Surge Height

$$h = 43.3 - 0.0433\,P$$

where:

h = height of surge (in meters)

P = pressure in eye of hurricane (in millibars = thousands of the pressure of the atmosphere at the Earth's surface)

water, and with shallower water (**By the Numbers 14-2: Surge Height Depends on Wind Speed, Fetch, and Water Depth**, p. 423).

The mound of water ahead of a storm slows down and piles up as it enters shallow water at a coast. Shallower water of the continental shelf forces the offshore volume of water into a smaller space, causing it to rise. A bay, inlet, harbor, or river channel that funnels the flow of water into a narrower width also causes the surge mound to rise. Even the Great Lakes are large enough to build storm surges one to two meters high.

The level of surge hazard depends on a variety of factors. Because sea level rises rapidly during an incoming storm surge, low-lying coastal areas are flooded and people drown. In fact, about 90 percent of all deaths in tropical cyclones result from storm-surge flooding.

Contrary to many people's assumptions, the highest surge levels are not at the center of the hurricane but in its north to northeast quadrant (**FIGURE 14-9**). Because hurricanes rotate counterclockwise, winds in that "right-front" quadrant point most directly at the shore and cause the greatest surge and wave effect there. The forward movement of the storm enhances these winds because they blow in nearly the same direction that the storm is moving. Winds south of the eye of the hurricane are moving offshore and have no onshore surge or wave effects.

The path of a hurricane compared with shore orientation also has an effect, as does the shape of a shoreline. A hurricane arriving perpendicular to the coast can lead to a higher storm surge because the whole surge mound affects the shortest length of coast. One arriving at a low angle to the coast is spread out over a greater length of shoreline but will remain along the coast for longer. Although people may feel more protected by living along the side of

FIGURE 14-8 STORM SURGES

A. A five-meter storm surge during Hurricane Eloise attacks the west end of the Florida panhandle in September 1975. **B.** Katrina's storm surge at the Michoud-Entergy power plant at the I-510 bridge.

▶ By the Numbers 14-2

Surge Height Depends on Wind Speed, Fetch, and Water Depth

$$(height)^2 \propto \text{wind speed} \times (\text{fetch length} - \text{water depth})$$

an inlet rather than along the open coast, the surge height in such a location may actually be higher.

The forward speed of a storm center can have mixed effects. A faster storm movement pushes the storm surge into a higher mound that submerges the coast in deeper water, but slow-moving hurricane cells often inflict more overall damage because they remain longer over a region, dumping

---

FIGURE 14-9 WIND DIRECTION AND STORM SURGE

A. Winds in the northeastern quadrant of the storm are directed toward the shoreline and inflict the greatest damage; southwestern-quadrant winds are directed offshore and inflict least damage. Hurricane Isabel, 2003, is shown here at the North Carolina coast, as an example. State boundaries are superimposed on this natural color image taken at 11:50 a.m. EDT on September 18, 2003. **B.** The storm surge for Hurricane Ike, 2008, as its eye arrived at Galveston, Texas, was highest to the right of the storm track, shown as a dashed black line. For the orange-colored area in the upper left, there is no surge or no information.

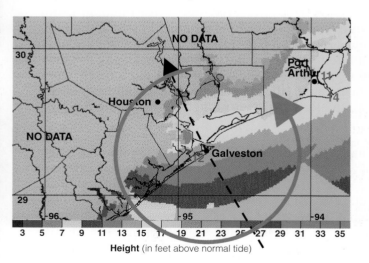

greater rainfall and causing more landward flooding. Similarly, a larger-diameter (often lower category) hurricane typically drops more rain because it impacts a larger area. It can also build higher waves because of longer fetch and form a wider surge mound because it pushes the water over a larger area.

The inland reach of storm-surge waters depends on a variety of factors. Most coastal areas of the southeastern United States and Gulf of Mexico stand close to sea level, so there is little to prevent surge flooding inland. However, inland progress of storm-surge waters is slowed dramatically by the presence of vegetation and dunes. Because surge waters flow inland like a broad river, they are slowed by the height of coastal dunes, especially those covered with brush, and by near-shore mangroves or forest.

Areas where dunes are low or absent or where vegetation has been removed permit surge waters to penetrate well inland. Thus, people who lower dunes or remove vegetation to improve their view or ease of access to the sea invite equally easy access from any significant storm surge. In many cases, pedestrian paths across dunes foster sites of erosion and overwash. A storm surge, finding a low area through dunes or an area of little or no vegetation, will flow faster through the gap, eroding it more deeply and causing more damage. This includes the destruction of buildings, roads, bridges, and piers, along with contamination of groundwater supplies with saltwater, agricultural and industrial chemicals, and sewage. Saltwater can invade aquifers and corrode buried copper electrical lines.

Buildings, bridges, and piers can be washed away by surge currents and waves. They can also float away if not well anchored to foundations or if the foundation is undermined by waves. Because larger waves feel the bottom at greater depth, they stir bottom sand and erode more deeply, undermining pilings and foundations. For this reason, most low-lying coastal homes are raised high on posts above the most frequent storm-surge heights, typically determined by local building codes (**FIGURES 14-10**, 14-17). Coastal houses are often built four meters off the ground on posts to raise them above storm surge and wave levels. Unfortunately, storm surges can rise higher than usual, and unless set deeply into the sand, wave erosion may undermine the posts and topple houses.

Because a storm surge locally raises sea level, moves inland as a swiftly flowing current, and raises the height of wave attack, it amplifies all of the erosional aspects of a storm, moving not only sand and sediment, but boats, cars, and houses (**FIGURE 14-11**). Anything already floating in the water will be readily moved in the direction of a surge and deposited in bays or lagoons. Houses are essentially big boxes full of air, so whole houses can be floated off their foundations and transported inland, often breaking up in the process. The nature and quality of construction are important in minimizing

FIGURE 14-10 UNSTABLE SUPPORT POSTS

Hyndman.

This house on the beach northeast of Galveston, Texas, lost its support posts during Hurricane Ike.

damage. A building's foundation should be well anchored to the ground, through deeply embedded strong piles. Floors and walls should be well anchored to the foundation and the roof securely attached to the walls with "hurricane clips."

When a surge comes at high tide, the resulting sea level rises still higher (**FIGURE 14-12**), with correspondingly greater reach inland and greater damage (**FIGURE 14-13**). Although the circumstance may be unusual, the additional load of a large surge mound on Earth's crust may be sufficient to trigger an earthquake in an area already under considerable strain—as in the Tokyo earthquake of 1923, in which 143,000 people died, mainly in the cyclone-fanned fire that followed the rupture of gas lines.

SURGES AND BARRIER ISLANDS Barrier islands are particularly vulnerable to storm surges during hurricanes. As discussed in Chapter 13, heavy erosion during storms is part of the natural process through which barrier islands are built and gradually migrate. When a storm surge moves inland over a low-lying coast, it raises water levels in lagoons and adjacent land areas, a huge volume of water that must return to the ocean after the storm passes. Much of that water reaches inland either as overwash of eroded dunes or through tidal inlets between segments of barrier islands. That overwash carries sand across an island, helping build island height, thereby making it less susceptible to erosion from later storms. Sand carried landward also builds the landward edge of the island as the storm removes sand from the seaward edge—so the island moves landward.

When sea level drops after the storm, that higher water level rushes back offshore across the already eroded

FIGURE 14-11 HEAVY OBJECTS MOVED BY STORM SURGE

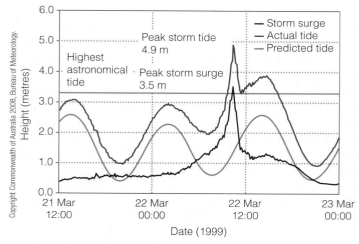

A. After Hurricane Hugo, pleasure boats moored in the lagoon behind the barrier island lay stacked like fish in a basket. **B.** Hurricane Camille's surge lifted this sailboat and carried it inland to collide with a house. Apparently the boat sustained less damage than the house. **C.** This house, at Carolina Beach, North Carolina, was not well anchored to its slab. The surge accompanying Hurricane Fran (Category 3) in 1996 carried it off its slab and onto a road. **D.** A roof, not well anchored to the walls, was lifted off a house in Crystal Beach, Texas, by the storm surge from Hurricane Ike in September 2008.

FIGURE 14-12 STORM SURGE AMPLIFIED BY TIDE

During Cyclone Vance, which struck Australia in March 1999, the storm surge arrived just before high tide. As a result, the surge was higher and lasted longer.

FIGURE 14-13 SURGE AT HIGH TIDE

Hurricane Katrina came ashore at high tide, raising the storm surge level by 0.6 meter, worsening the flooding with polluted water in New Orleans.

FIGURE 14-14 BARRIER ISLAND CUT OFF FROM MAINLAND

Cynthia Hunter, FEMA.

Hurricane Isabel severed Highway 12 along Hatteras Island, North Carolina. The continuation of the highway is visible directly across the breach to the right of the power poles.

beaches and through the widened inlets and breaches. New breaches can form in low or narrow areas of a barrier island depending on the storm surge height, wave size and direction, forward storm speed, and other factors. Other existing breaches can close.

When barrier islands are developed, buildings get in the way of the natural cycle of erosion and deposition that governs them. Breaches caused by storm surges can destroy bridges and access roads, leaving no road access to the mainland and isolating homes and towns (**FIGURE 14-14**).

Where a new breach severs the only road, government agencies are pressured to close the breach. However, closing such breaches decreases the number of tidal breaches along a barrier island, upsetting the natural equilibrium of the barrier island–lagoon system. Closing a breach can decrease the tidal circulation between lagoon and ocean, with a variety of negative consequences. Much of the erosion that we see after a hurricane occurs during that surge outwash, or "ebb tide." Sand carried back offshore can end up in deeper water below wave-base depth where it is lost to the beach system.

During Hurricane Katrina, barrier islands were severely eroded. For example, almost all of the homes on the beach side of the road running along Dauphin Island, Alabama, east of the hurricane's eye, disappeared, leaving only posts (**FIGURE 14-15**). In some instances, even the posts were snapped off. A large volume of sand eroded from the beach swept across the island during the storm surge.

Dauphin Island is an example of a barren offshore sand bar sprinkled with expensive homes repeatedly protected with federal taxpayer dollars; the fix is often temporary. A berm destroyed by Hurricane Georges in 1998 was rebuilt in 2000 for $1 million, only to be washed away by Tropical Storm Isidore two years later. Again rebuilt, Hurricane Katrina in 2005 washed away not only the berm but also many of the houses. A new sand berm 3 meters high and 5.7 kilometers long was completed in May 2007 at a cost of $4 million; it was severed two weeks later by waves during a high tide.

Waves and Wave Damage

The higher waves generated by a hurricane impact the coast with much more energy. They are able to move sand on a lower slope than smaller waves. They also stir up sand to greater water depths both offshore and onshore, eroding beaches and moving sand farther offshore to form more gently

FIGURE 14-15 SEVERE EROSION TO BARRIER ISLANDS

NOAA.

Marvin Nauman, FEMA.

During Hurricane Katrina, Dauphin Island, Alabama, lost most of the three rows of homes on its beach; all that remains are some of the posts on which they once stood. A portion of the sand eroded from the island was carried to the right into Mississippi Sound.

FIGURE 14-16 BEACH EROSION DURING A STORM

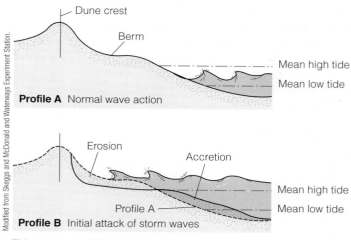

Profile A Normal wave action

Dune crest
Berm
Mean high tide
Mean low tide

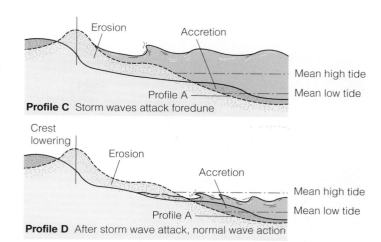

Profile C Storm waves attack foredune

Erosion
Accretion
Profile A
Mean high tide
Mean low tide

Profile B Initial attack of storm waves

Erosion
Accretion
Profile A
Mean high tide
Mean low tide

Profile D After storm wave attack, normal wave action

Crest lowering
Erosion
Accretion
Profile A
Mean high tide
Mean low tide

Modified from Skeggs and McDonald and Waterways Experiment Station.

This sequence of diagrams shows the effects of a storm wave attack on a beach and dune. Note the progressive erosion of original shore Profile A, in part because of waves on higher sea level of storm surge.

sloping beaches (**FIGURE 14-16**). Because dunes dissipate wave energy and protect homes placed behind them, erosion of the sand from beaches and dunes frequently undermines any structures there (**FIGURE 14-17**). Loose debris such as boards, branches, logs, and propane tanks carried by waves act as battering rams against buildings, amplifying damage.

The level of damage at a particular site can be dramatic, as shown in photographs taken before and after individual hurricanes (**FIGURE 14-18**, p. 428). Many beachfront homes are completely destroyed when the sand beneath them is washed offshore. Others nearby sustain major wind and wave damage (**FIGURE 14-19**, p. 428).

The width and slope of the continental shelf has a significant effect on wave damage. On a wide, gently sloping continental shelf offshore, the waves drag on the bottom and stir up sand; this uses up more wave energy before the waves reach land, decreasing the damage. Large amounts of moving sand on a shallow bottom offshore reduce the wave energy available for coastal erosion. On a narrow continental shelf, as in the Outer Banks of North Carolina, or a steeper slope offshore, larger waves maintain their energy as they approach closer to shore, thereby causing more damage. (**Case in Point:** Wind, Waves, Beach Erosion, Flooding—North Carolina Outer Banks, p. 452)

FIGURE 14-17 HOMES ON POSTS

Dave Gatley, FEMA.

Hyndman.

A. Hurricane Dennis on September 1, 1999, eroded the beach out from under these beachfront homes in Kitty Hawk, North Carolina. **B.** Homes raised on posts to be above storm surge heights can still be battered by hurricane winds and severe erosion of beach sand under them. This home was damaged during Hurricane Ike, 2008, southwest of Galveston, Texas.

FIGURE 14-18 BEFORE AND AFTER A HURRICANE

These photos show part of North Topsail Beach, North Carolina, **A.** after Hurricane Bertha on July 16, 1996, and **B.** after Hurricane Fran on September 7, 1996. The area is hardly recognizable. Compare individual houses in the second pair of photos. **1.** Is pushed off its foundation and houses to right are demolished. **2.** Is pushed off its foundation into new inlet to its left. **3.** Is now gone.

FIGURE 14-19 WAVE DAMAGE TO HOMES

A. Hurricane Alicia (Category 3) in October, 1983, removed the beach sand from under the concrete parking area below this house on the Texas Gulf Coast, leaving it suspended on the pilings that raised the house above surge level. It also peeled off much of the roof and the front half of the house. **B.** Hurricane Ike (high Category 2) in September, 2008, destroyed most houses near this beach along the Texas Gulf Coast. The house at top, just left of center, survived well because its roof slopes in four directions, few windows, and tall, sturdy posts.

It might seem that low-lying, subtropical islands such as Grand Cayman in the Caribbean and Guam in the western Pacific would be especially vulnerable to storm surge and wave damage. However, fringing coral reefs force storm waves to break well offshore, minimizing those effects. Water depths offshore of the reefs tend to be deep, and absence of a wide continental shelf prevents the storm surge from rising as high. However, those characteristics do not make these islands safe. Because many of them are low-lying, they are vulnerable to severe wind damage and storm surge.

Winds and Wind Damage

Although flooding compounded by water rise from storm surge can be catastrophic, wind damage is often ten times as great. Wind velocity has a major effect on wave height. Wind can also wreck weak buildings and blow down trees, power lines, and signs. It can fan fires and destroy crops; blow in windows, doors, and walls; and lift the roofs off houses. Wind damage is greatly magnified by flying debris (**FIGURE 14-20**). Significant harm also results from rain entering a wind-damaged structure.

FIGURE 14-20 WIND-BORNE DEBRIS

A. Yes, hurricane winds can be strong! The danger of wind-blown debris is vividly illustrated by a palm tree near Miami, which was impaled by a sheet of plywood during Hurricane Andrew in August 1992. This may have been caused by an embedded tornado. **B.** A 14-cm-diameter branch impaled the side of a house in Punta Gorda Isles during Hurricane Charley, August 2004.

Wind in some hurricanes, does far more damage to buildings than flooding. To understand the force of wind, the lowest-level hurricane winds at 119 kilometers per hour apply approximately 73 kilograms per square meter (15 lb/ft²) of pressure on the wall of a building. Thus, a force of 1,360 kilograms (3,000 lbs) would press against a wall 2.4 meters high and 7.6 meters long. If the wind speed doubled, the forces would be four times as strong; so in a Category 4 hurricane at 240 kilometers per hour, the force on the same wall would be 5,400 kilograms. Reduced pressure caused by the same winds on the downwind side of a building adds to the problem because the winds pull against the wall (**FIGURE 14-21**).

Roofs often fail before walls because of additional factors. The largest wind forces are caused by suction. Where a roof slopes toward the wind, the air is forced up and over the roof, lifting it in the same way that air flowing over an arched airplane wing lifts the plane (see FIGURE 14-21). A steeper-sloped roof actually performs better in high wind than one sloped more like an airplane wing. The lifting forces under overhanging eaves tend to cause failure there first. Roofs that slope to all sides without overhangs deflect the wind best. Larger roof spans are more vulnerable.

Roof material also makes a difference; shingles fare poorly because they can easily blow off during high winds. Metal, slate, or tile roofs are good; a single-membrane roof is better; and a flat concrete-tile roof is much better. Concrete or steel beams supporting roofs are better than wood. An extra nail or two in each shingle can save a roof.

Because of pressure differences inside and outside a building, sidewalls, windows, and roofs are commonly sucked out rather than blown in. When a house is breached during a storm by flying debris, for example, pressure

FIGURE 14-21 WIND FORCES ON A BUILDING

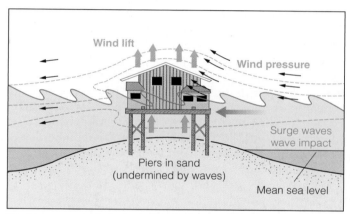

Wind pressure exerts force on the windward side of a building, while simultaneously pulling upwards on the roof. At the same time, reduced pressure on the sheltered side also pulls on the house. Wave and surge effects are also shown.

increases inside the building, exerting outward force on windows, walls, and the roof. If a garage door fails, the wind can inflate a house and blow its roof off and windows out.

Shutters or plywood well anchored over windows and exterior doors helps protect the integrity of a house more than anything else because those are the vulnerable points of entry. Impact resistance is also important for windows and doors. Tempered and laminated glass windows are significantly better than ordinary window glass. When windows or exterior doors make up more than fifty percent of a wall, it is especially vulnerable. Skylights are especially susceptible to penetration. Double-wide garage doors, particularly if the doors open overhead, are unusually susceptible because tracks holding doors may fail or the wind may force a door out of its track.

Much of the damage in homes from hurricanes, tornadoes, and earthquakes arises from the repeated flexing of wood structures nailed together. Nails pull out, or their heads pull through walls or roof sheathing, or they shear off. Recent changes in nail design have specifically addressed those failings: barbed rings around the nail shaft, larger nail heads, and higher-strength carbon-steel alloy shafts. Such nails can double resistance to high-wind damage and add 50 percent to resistance to hurricane damage.

Unreinforced brick chimneys fail in large numbers, even when situated against walls, because of poor attachment to the walls. Brick, stone, or reinforced-concrete block walls, however, are much stronger than wood-sheathed walls. Because much damage comes from flying debris, gravel or roof-mounted objects on nearby buildings are dangerous.

Rainfall and Flooding

Hurricanes cause significant flooding because of storm surge as well as heavy rainfall. Heavy rain and flooding during a hurricane can wash out structures, drown people, contaminate water supplies, and trigger landslides.

The characteristics of a particular storm, as well as the local topography, determine the severity of flooding. As mentioned earlier, Category 1 or 2 hurricanes or even smaller tropical storms can actually cause more flooding damage than larger storms. Lower-category, large-diameter or slow-moving hurricanes spend much more time over an area and typically drop large amounts of rain on large parts of a drainage basin, and for a longer period, so they cause more extensive and more prolonged flooding. Hurricane Agnes, only a tropical storm when it came back on land in Pennsylvania and New York in 1972, spread over a diameter of 1,600 kilometers and provided the largest rainfall on record in that area.

In another example, late on June 5, 2001, Tropical Storm Allison drifted slowly inland to 200 kilometers north of Houston and, over the next 36 hours, dropped heavy rain on southeastern Texas and adjacent Louisiana. Strengthening of a high over New Mexico caused Allison to loop east, then southwest to Houston a second time. In total, the storm dumped almost one meter of rain on the Port of Houston in less than one week, in an area that has notoriously poor drainage (**FIGURE 14-22**). Damages in the Houston vicinity alone reached $4.88 billion. The Texas Medical Center was especially hard hit. Its below-ground floors flooded. Backup generators were above ground, but unfortunately switches between the two systems were below ground and destroyed by the flooding.

More recently, Hurricane Ike in 2008, a strong category 2 storm, inflicted $29.2 billion in damages, making it the third costliest storm ever to affect the United States (**Case in Point:** Extreme Effect of a Medium-strength Hurricane on a Built-up Barrier island, Galveston, TX. Hurrican Ike, September 13, 2008, p. 449).

Local topography also influences flooding severity. Several centimeters of rain dumped over a few hours collects rapidly and runs off slowly because of the gentle slopes of coastal plains. Draining downslope, the water accumulates to even greater depths to cause catastrophic floods in low areas. A good example is Hurricane Floyd in 1999, following on the heels of Hurricane Dennis which had already saturated the ground (**Case in Point:** Back-to-Back Hurricanes Amplify Flooding—Hurricanes Dennis and Floyd, 1999, p. 455).

FIGURE 14-22 FLOODING DAMAGE

A. Tropical Storm Allison flooding in Houston. **B.** Water flowing into a below-street mall.

Deaths

Deaths in hurricanes depend not only on the strength of a hurricane, through the wind velocity and surge height, but also on patterns of development and the size and education of the local population. Obvious problems include buildings too close to the coast, buildings too close to sea level, and buildings, such as mobile homes, that are too weakly constructed. Less obvious is people's lack of awareness of the wide range of hazards involved in such events. Finally, a large population in the path of a storm cannot evacuate quickly or efficiently, especially if an accident blocks a heavily used highway.

Hurricanes have much higher rates of deaths in poor countries, such as in the Caribbean and Central America. Poverty, culture, and disastrous land use practices put such populations at a much higher risk of hurricane-related damages. Much of Central America is mountainous, with fertile valley bottoms that are mostly controlled by large corporate farms. Many people have big families that survive on little food and marginal shelter; unable to afford land in the valley bottoms, they decimate the forests for building materials and fuel for cooking, leaving the slopes vulnerable to frequent landslides. Others migrate in search of work to towns in valley bottoms where they crowd into marginal conditions on floodplains close to rivers subject to flash floods and torrential mudflows fed from denuded slopes. Hurricane Mitch, which was Category 5 offshore but weakened rapidly to a tropical storm onshore, provides a dramatic and tragic example of what can happen (**Case in Point:** Floods, Landslides, and a Huge Death Toll in Poor Countries—Hurricane Mitch, Nicaragua and Honduras, p. 458).

Poor countries in which low-lying coastal areas attract and provide food for large populations also have severe problems. The delta areas of big rivers in Southeast Asia support large numbers of people because the land is kept fertile by the frequently flooding rivers. As those populations grow, heavy land use in the drainage basins of the rivers leads to more frequent and higher floods from both rivers and tropical storm surges. (**Case in Point:** Floods, Rejection of Foreign Help, and a Tragic Death Toll in an Extremely Poor Country—Myanmar (Burma) Cyclone, May 2–3, 2008, p. 456).

The deadliest Atlantic hurricanes have affected the Caribbean islands with 64 percent of total storm losses. Before Mitch in 1998, Mexico and Central America accounted for 15 percent of total deaths due to storms, with the U.S. mainland at 20 or 21 percent. Major catastrophes in Central America include large death tolls:

- About 10,000 in Nicaragua and Honduras from Hurricane Mitch from October 22 to November 5, 1998

- Approximately 8,000 in Honduras from Hurricane Fifi, September 15–19, 1974

- 7,192 in Haiti and Cuba from Hurricane Flora, September 30 to October 8, 1963

- More than 6,000 at Pointe-a-Pitre Bay, Haiti, September 6, 1776

The largest numbers of deaths in U.S. hurricanes from 1900 to 2009 are shown in **Table 14-4**. Note that Katrina is the only recent hurricane to make the list, due to improved forecasting and warning systems, better transportation facilities, and the lower incidence of hurricane activity between 1970 and 1995.

Table 14-4	Hurricane Deaths in the United States, 1900–2009			
RANK	HURRICANE LOCATION	YEAR	CATEGORY	DEATHS
1	Galveston, Texas	1900	4	>8,000–12,000
2	Southeast Florida	1928	4	2,500–3,000
3	New Orleans, Gulf Coast (Katrina)	2005	3	1,500
4	Louisiana	1893	4	1,100–1,400
5	South Carolina, Georgia	1893	3	1,000–2,000
6	South Carolina, Georgia	1881	2	700
7	Southwest Louisiana, North Texas (Audrey)	1957	4	416
8	Florida Keys	1935	5	408
9	Louisiana	1856	4	400
10	Northeastern U.S.	1944	3	390
11	Florida	1926	4	372
12	Grand Isle, Louisiana	1909	3	350
13	Florida Keys, South Texas	1919	4	287
14	New Orleans, Louisiana	1915	4	275
15	Galveston, Texas	1915	4	275

Social and Economic Impacts

Because hurricane damage can spread over large regions, their social and economic impact can be significant. Disruptions from Katrina shut down 95 percent of petroleum output from the Gulf of Mexico, the biggest domestic source. Pipelines inland from the area shut down for lack of power. Several major petroleum refineries in Katrina's path were closed, almost all workers from offshore drilling platforms were evacuated, and 29 of the platforms were destroyed (**FIGURE 14-23**). When the giant anchors that held some platforms in place broke loose, they dragged, twisted, and sometimes severed seafloor pipelines that carried the crude oil to the mainland. The same thing happened with Hurricane Ivan. As a result, gasoline prices rose dramatically throughout the United States.

Dramatic disruptions in an unexpected evacuation sometimes separate immediate family members who have not made contingency plans to find each other in such a circumstance. Family members in other states can find themselves with no means to contact their kin in stricken areas because communication via telephone, email, and even snail mail may be impossible due to storm destruction. Specific plans, such as having a contact relative in another state, should be made before a major storm's arrival. Otherwise, local authorities or national services, such as those provided through the American Red Cross, may not be able to help people reach one another.

Many people in hurricane- and related flood-prone areas lack flood insurance. For a large proportion, the financial impact of a major event is long-term, if not permanent. Job losses are widespread. Companies cannot reopen because utilities are damaged for weeks or months. Without business, they have no income and cannot afford to pay their regular workers. Those workers need the income not only for day-to-day expenditures but for ongoing expenses such as health insurance and housing. Many move to other areas for work and may never return.

By September 1, 2006, one year after Katrina, $44 billion of federal money had been used to pay flood-insurance claims and accommodate immediate needs after the storm—paying federal workers and providing temporary housing for victims and operating funds to local governments. Only a small amount was used to repair water and sewer lines and electric power grids. This resulted in a major dilemma—the government won't spend on infrastructure until it is sure people are going to move back, but people can't move back until there is usable water, sewer, and power.

One year after the storm, thousands of New Orleans residents said that they would rebuild their damaged homes, even though their property remains in flood zones. Thousands of others are either uncertain or have decided to not return. By late 2009, the city's population had slowly climbed back to 72 percent of its original 462,000. A significant percentage of the present population resides in the 20 percent of the city that was not flooded. In one of the poorest and hardest-hit neighborhoods, the Lower 9th Ward, however, fewer than 2,000 out of more than 13,000 people have returned. On September 1, 2006, the New Orleans City Council ruled that any house being rebuilt had to be raised 3 feet (almost one meter) above ground level; unfortunately, many areas were flooded with 5 to 6 meters of water. A $6 billion upgrade of the more than 500 kilometers of levees around New Orleans will not be completed until 2010 at the earliest. Even then the levee system will not provide much more protection than in 2005. The effect of Hurricane Katrina on New Orleans provides many lessons in this regard. (**Case in Point:** City Drowns in Spite of Levees—Hurricane Katrina, 2005, p. 441).

Climate Change and Hurricane Damage

Hurricane damages can be amplified by weather and climate conditions. Higher ocean temperatures or warm-water eddies add energy. Storms can be stalled by other weather systems, thereby dropping more rain in a region. Two or more hurricanes hitting the same region one after another can also amplify damages.

The number and severity of hurricanes appears to be increasing. An average of five hurricanes develop in the Atlantic Ocean every year, two of them major (Category 3 or greater). The number naturally varies, but the correlation with the Atlantic Multidecadal Oscillation (discussed in Chapter 10) and with annual sea-surface temperatures (SSTs) suggests that the next few decades will see not only more hurricanes but more severe hurricanes. Although 2004 was a record year for hurricanes, 2005 provided an even larger number, including Katrina, the costliest hurricane of all time, and Wilma, the strongest ever measured (**FIGURE 14-24**).

FIGURE 14-23 OIL PLATFORMS IN THE PATHS OF HURRICANES

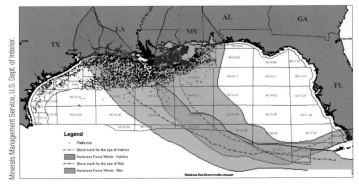

6,357 oil platforms are located in the Gulf of Mexico within 230 kilometers of the U.S. coast. 2005 hurricane paths are shown as pink and orange lines. The colored shading indicates areas that experienced hurricane-force winds from Katrina (brown) and Rita (pink). Note that the hurricanes veered and tracked clockwise.

FIGURE 14-24 HURRICANE WILMA

Hyndman.

Hurricane Wilma destroyed much of the waterfront property of Cancún, Mexico, in 2005. If a large hurricane were to impact coastal New Jersey, would this be the result?

In more than 150 years of records, only 1960, 1961, 2005, and 2007 had more than one Category 5 storm. Both Dean, which struck the Yucatan Peninsula of Mexico on August 21, 2007, and Felix, which hit eastern Honduras on September 4, 2007, were still Category 5 hurricanes at landfall.

Hurricane strength appears to be increasing as a result of the increase in ocean temperatures due to climate change (**FIGURE 14-25**). Warmer SSTs provide more energy to the

FIGURE 14-25 CORRELATION OF HURRICANE STRENGTH AND OCEAN TEMPERATURE

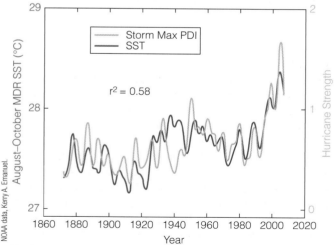

Hurricane strength in any given year appears to correlate quite well with Atlantic Ocean sea-surface temperature. Both are rising with time, suggesting that global warming will increase the strength of hurricanes as ocean water warms. PDI = Power Dissipation Index.

atmosphere, thus increasing the likelihood of more catastrophic hurricanes such as Katrina.

What impact might weather and climate changes have on the severity of future hurricane damages? If several of these weather and climate influences were to overlap in time (see FIGURE 1-4) and if such a storm were to hit a major city such as Savannah, Miami, or New Orleans (again), we might have truly incredible damages. When Hurricane Katrina decimated New Orleans in late August of 2005, it became North America's costliest natural disaster ever.

Hurricane Prediction and Planning

One reason that there have been fewer deaths from hurricanes over time is the great success in predicting and planning for hurricanes.

Hurricane Watches and Warnings

Storms are monitored by weather satellites and "hurricane hunter" aircraft that make daily flights into storms to collect data on winds and atmospheric pressures. Within two to four days of expected landfall, they drop dropsonde sensing instruments into a storm from 9-to-12-kilometer altitudes, their fall slowed by small parachutes. These transmit wind, temperature, pressure, and humidity information. The path of the storm is commonly controlled by nearby high- and low-pressure systems.

Hurricane predictions include time of arrival, location, and magnitude of the event. The National Hurricane Warning Center tries to give twelve hours warning of the hurricane path. Alerts come in two stages. A **hurricane watch** indicates that: "A hurricane is possible in the watch area within 36 hours." A **hurricane warning** is provided when: "A hurricane is expected in the warning area within 24 hours." On average, 640 kilometers (400 miles) of coastline is warned of hurricane landfall within 24 hours. Of that, 200 kilometers (125 miles) may actually be strongly affected by a storm. Thus, some $275 million in costs are borne by people ultimately not in the storm's path. Clearly, more accurate forecasting could save lives and significant evacuation costs.

One future prospect for predicting hurricane behavior is to monitor thermal anomalies in the ocean with satellite sensors. We know that a hurricane's energy is drawn from the heat in tropical ocean water. A newly recognized effect on hurricane strength is the presence of large eddies of warm water 100 or more kilometers across, which can apparently spin off larger oceanic currents and boost the energy of hurricanes passing over them, sometimes dramatically (compare FIGURE 14-5).

Uncertainty in Hurricane Prediction

Hurricane prediction, like any weather prediction, involves significant uncertainty. Hurricanes can quickly change paths or increase in intensity. Although forecasters can provide some indication of a hurricane's time of arrival, the exact landfall location and storm strength is less predictable (**Case in Point:** Unpredictable Behavior of Hurricanes—Florida Hurricanes of 2004, p. 459).

The National Hurricane Center's current 24-hour lateral forecast error for the path of a hurricane is 80 kilometers. Even a small shift in a hurricane's path could make a dramatic difference to its impact on coastal communities. A shift of as little as 80 kilometers in the Galveston hurricane in 1900 could have resulted in far fewer deaths. A shift of 1992's Hurricane Andrew just 32 kilometers to the north near Miami would have caused two or three times as much damage. Much less monetary loss would have resulted from a strike 64 kilometers to the south in the minimally populated Florida Keys.

Smaller storms can also quickly gain energy and become powerful hurricanes. Hurricane Andrew developed as a thunderstorm in warm waters off the coast of Africa, gradually strengthening as it moved westward in the trade winds, and becoming one of only a few Category 5 storms to make landfall.

Planning for Hurricanes

Planning for a hurricane should be among the first things people do when they move to a hurricane-prone coast. They should prepare their house for all of the potential hazards discussed in this chapter, and do so in anticipation of having little or no warning. Many of these preparations take days and should be an ongoing effort. If any cannot be done long before, specific plans should be in place to do them quickly. Don't assume that stores or workmen will be available after a hurricane watch or warning has been announced.

AS SOON AS A HURRICANE WATCH IS ANNOUNCED:

- Because winds cause the greatest damage, bring in or securely anchor anything that can become a damaging projectile in a high wind. Remove damaged or weak limbs, along with extra branches, to minimize wind force on trees. Remove outside antennas.

- Board up windows securely with plywood and reinforce garage doors. Close and secure storm shutters.

- Store as much clean drinking water as possible in plastic bottles, sinks, and bathtubs. Public water supplies and wells often become contaminated, and electric pumps do not work without electricity. A small generator with ample fuel can be helpful, but stores sell out quickly when such crises arise.

- Move any boats to a safe place, preferably above storm-surge height. Securely fasten a boat to its trailer with a rope or chain. Anchor the trailer to the ground, house, or large tree with secure tie-downs.

- Stock up on any prescription medications, food, and water—a one-week supply at home in addition to at least a three-day supply for evacuation.

- Fill your vehicle's gas tank. Gas station attendants may evacuate before you do or stations may run dry as everyone fills up on the way out of town.

IMMEDIATELY BEFORE LEAVING:

- Turn the refrigerator to the coldest setting. The power may be out for a long time.

- Turn off other appliances. The power surge while electricity is restored may damage them.

- Turn off a main natural gas line or propane tanks to your home, and anchor them securely.

- Evacuate when authorities say to do so.

Evacuation

Early warning in the United States, using weather satellites, allows people to either evacuate by road or take refuge in reinforced high-rise buildings, such as those in Miami Beach. Because forecasters refer to the statistical likelihood that a storm will strike a given length of coast and there is significant uncertainty in the direction of the storm track, many more people are warned to evacuate than those in the final path of the storm. The costs of evacuations are also large. The usual estimate is roughly $1 million per coastal mile evacuated—exclusive of any damages from the hurricane. This is an unfortunate but necessary cost for the safety of coastal populations.

One problem is convincing people to evacuate promptly. With recent heavy development in coastal areas, most of the 40 million people living in hurricane-prone localities have never been involved in an evacuation. Large seasonal tourist populations make matters even worse. Municipalities view hurricane publicity as bad for tourism and property investment, so they sometimes delay warnings or minimize the danger.

Some people who live through one hurricane do not leave, believing that if they have done it before, they can do it again. Many delay their evacuation because of inconvenience and cost or because they spend time purchasing and installing materials to help protect their homes from a storm. Others stay because they feel their homes are strong enough to survive. Still others delay evacuation until the last minute, thinking it will take them only an hour or two to drive inland, or to minimize lodging costs, or hoping the storm will miss their part of the coast.

Most people believe it will take less than one day to evacuate, but studies show that because of traffic jams and related problems, most evacuations take up to 30 hours. Only single two-lane bridges to the mainland serve most barrier islands. Those bridges and roads become snarled with traffic, and accidents can cause further dangerous delays.

The huge increase in population along the coasts in recent years has outgrown highway capacity. The roads are not able to cope with evacuating populations. Changing freeways to single direction traffic away from a coast merely moves bottlenecks inland. It also creates severe safety problems at off-ramps, where people traveling against the normal direction of traffic try to exit for lodging or fuel (**FIGURE 14-26A**). If people fail to evacuate before a storm arrives, high surge levels can flood roads and freeways under meters of water, making escape impossible (FIGURE 14-26b, c). Fallen trees, power lines, and other debris can block roads, or a surge may cover roadways, making escape impossible and rescue difficult. (**Case in Point:** Déjà vu? Hurricane Gustav, Gulf Coast, August 29–September 2, 2008, p. 448)

In part to combat these problems with evacuation, the latest approach is to move people 32 kilometers (20 miles) from the coast, beyond the limit of surge, to temporary protection, not hundreds of kilometers from the coast, as was done in the past. Even once residents successfully evacuate, however, lodging is completely inadequate to cope with an entire coastal population.

The most difficult area to evacuate in the United States is the Florida peninsula, which is only a few meters above sea level and less than 200 kilometers across. A hurricane can easily cross the whole state without losing much strength.

Population growth has been rapid, especially along the coasts, and it is difficult to predict the precise path of a hurricane.

When you hear a Hurricane Warning, the American Red Cross recommends the following:

- Plan to leave if you live on the coast, an offshore island, a floodplain, or in a mobile home. Tell a relative outside the storm area where you are going.

- If you evacuate, take important items: identification, important papers (e.g., passports, insurance papers), prescription medicines, blankets, flashlights, a battery-operated radio and extra batteries, a first aid kit, baby food and diapers, and any other items needed during a week or two in a shelter. Lock your house.

- Listen to the radio for updates. Use telephones only for emergency calls.

- Do not drive through floodwaters. Roads may be washed out, and two feet of water can carry away most cars. Stay away from downed or dangling power lines; report them to authorities.

FIGURE 14-26 EVACUATION

A. One-way traffic evacuation near Beaumont, Texas, before Gustav. **B.** Surges quickly cover escape routes. This warning sign is on the barrier-bar road to Sunset Beach on the offshore barrier island of South Carolina. **C.** Some traffic still tried to evacuate after Tropical Storm Francis in 1999 began to move in.

- If trapped at home before you can safely evacuate, be aware that flying debris can be deadly and hurricanes may contain tornadoes or spawn them. Stay in a small and strong interior room without windows; stay away from windows even if they are shuttered. Close and brace exterior and interior doors. Use flashlights for emergency light; candles and kerosene lamps cause many fires.

- If the eye of the hurricane passes over you, the storm is not over; high winds will soon begin blowing from the opposite direction, often destroying trees and buildings that were damaged in the earlier winds.

Managing Future Damages

Governments, communities, and insurance companies are finally reacting to damages and costs from hurricanes. They are beginning to push for "disaster-resistant communities" that are less vulnerable and incur reduced costs from coastal hazards. Policies used include land-use planning, building codes, incentives, taxation, and insurance. However, because tourism is the largest source of income for most coastal areas, most governing bodies are reluctant either to publicize their vulnerability or to place many restrictions on anything related to tourist income.

Natural Protections

Beachfront sand dunes absorb the energy of waves and advancing storm surges and can reduce damages to coastal communities. Cypress forests or thickets of mangroves limit the shoreward advance of waves and dramatically slow a surge's landward advance (**FIGURE 14-27**). Unfortunately, many of these areas have been modified or damaged over the years through development. Some communities have now recognized the importance of these areas and are working to protect or restore them.

Human alterations of dunes and mangrove stands that would increase potential flood damage are prohibited. Even walking on dunes is generally prohibited because it disturbs the sand and any vegetation covering it, thereby making the dune more susceptible to erosion by both wind and waves. To protect the dunes, people build elevated walkways over dunes to beaches (see FIGURE 13-21).

The damage inflicted on New Orleans by Hurricane Katrina was likely increased because decades of dredging channels for shipping and the emplacement of onshore oil-drilling sites introduced saltwater and killed off nearby cypress forests and marshes, which had previously provided natural protection from storm surges and waves. At the same time, the delta has been sinking because the levees of the Mississippi River and its tributaries have deprived it of sediment that was once replenished annually. A four-year, $500 million federal grant to restore coastal wetlands that was awarded in July 2005 should help somewhat in the future.

Building Codes

After major hurricanes obliterate a stretch of coast, you would think that people would have second thoughts about rebuilding in the same vulnerable locations. Unfortunately, the opposite seems to be true (**FIGURE 14-28**). Developers immediately descend on devastated areas and snap up coastal building sites, often paying as much or more than the value of the property prior to its destruction. Many want to build large hotels or condominium complexes. Some have been waiting to build their dream beachfront property; others are merely real estate speculators betting on a rapid comeback. Waterfront lots in the Florida Panhandle, for example, go for over a million dollars; those in formerly depressed areas several blocks back from the beach go for more than $100,000.

Most people agree that buildings in hurricane-prone areas should be built stronger, and in some areas legislation

FIGURE 14-27 COASTAL FORESTS PROVIDE STORM PROTECTION

A. Cypress forests near New Orleans and **B.** Mangroves along southeastern coasts present a formidable barrier to an advancing storm surge.

FIGURE 14-28 COASTAL DEVELOPMENT

A. A few months after Hurricane Ike completely annihilated most of the homes on the Bolivar Peninsula, northeast of Galveston, Texas, new houses were being built on the same beach with no protective dunes. **B.** Many of the hotels along Cancún's main beach sustained major damage after Hurricane Wilma struck in 2005, but the entire area has since been rebuilt.

has been passed to regulate such development. Studies by the research center of the National Association of Home Builders indicate that homes can be made much more resistant to hurricane damage at a cost of only a 1.8 to 3.7 percent increase in the total sales price of a property.

In the 1980s, the state of Florida designated a coastal construction control line (CCCL), seaward of which habitable structures were permitted only with adherence to certain standards of land use and building construction. The CCCL designates the zone that is subject to flooding, erosion, and related impacts during a 100-year storm. Hurricane Opal on October 4, 1995, provided a Category 3 test of the system. None of the 576 major habitable structures built to CCCL standards and seaward of the CCCL suffered substantial damage. Of the 1,366 preexisting structures in that zone, 768 (56 percent) were substantially damaged.

Unfortunately, developers, builders, local governments, and many members of the public often oppose such increased standards, especially for winds. Their argument is that these regulations unnecessarily increase housing costs and limit economic development. Often increased standards fail to pass at all levels of government until a major disaster and huge losses make the need obvious to everyone. Even then, the issue of whether the state or counties are required to foot the cost of enforcing new regulations often thwarts new laws. When adequate building standards are not enacted and enforced, the general public is eventually forced to pay for the unnecessary level of damage. People's federal and local taxes could be lower if it were not for such unnecessary costs.

With no national building code, state and local governments are responsible for enforcing their own codes. Some southeastern states do not have universal building codes.

When Hurricane Hugo hit South Carolina in 1989, half of the area of the state, including parts of the coast, had no building codes or enforcement at all. This lack of building regulations results in much more widespread and severe destruction of property.

Flood Insurance

Flood insurance is one way to reduce monetary damages from hurricanes and potentially influence people's behavior to keep them from settling in inappropriate areas. Significant destruction of homes and other structures in historic hurricanes has prompted some states such as North Carolina to dictate a setback line based on the probability of coastal flooding. Seaward of this line, insurance companies will not insure a building against wave damage. In spite of many challenges, the courts have so far upheld the building prohibition.

The National Flood Insurance Program (NFIP) requires federal mapping of areas subject to both river and coastal floods. The purpose is to implement a Flood Insurance Rate Map (FIRM) provided by the Federal Emergency Management Agency (FEMA). The designated special flood-hazard areas have a one-percent chance of being flooded in any given year, that is, they are comparable to the 100-year floodplain of streams. Coastal communities wishing to participate in the program must use maps of these flood-hazard areas when making development decisions. Coastal areas subject to significant wave action in addition to flooding are more vulnerable to damage. Significant waves are considered those higher than one meter. Therefore, construction standards are more stringent in such coastal zones. National flood insurance premiums are higher for those who live in more vulnerable areas.

Flood insurance costs in 2007 for a single-family dwelling built after 1981, with no basement, were generally $1,520 per year for a $250,000 replacement value. For a home in a high-risk coastal flood-risk zone, the rates rise to $3,275. Contents coverage adds an additional $2,000, and insurance is capped at $250,000 for a building. Land loss is not covered. As with stream flooding, there is a 30-day waiting period after purchase before the insurance takes effect. Given that hurricanes are not predictable for a specific location for more than a day or two in advance, it would seem prudent to maintain flood insurance if you live in an area of possible storm surge or coastal river flooding.

Some undeveloped areas were protected by the Coastal Barrier Resources Act of 1982, which prohibited federal incentives to development and prohibited the issuance of new flood insurance coverage. However, in areas not covered by that act, flood insurance remained available for elevated structures located as far seaward as the mean high-water line, regardless of local erosion rates.

Coastal building standards for the NFIP require the following:

- All new construction must be landward of mean high tide

- All new construction and major improvements must be elevated on piles so that the lowest floor is above the base flood elevation for a one-percent chance of flooding in any year

- Areas below the lowest floor must be open or have breakaway walls. Fill for structural support is prohibited

Raising the lowest floor above a 100-year flood level does nothing, of course, to prevent erosion. Piers may be undercut, and an eroding beach will eventually move landward from under a structure, causing its collapse.

With global warming and rising sea levels, coastal landforms including beaches and barrier islands will continue to migrate landward; it would seem wise to move coastal homes, roads, and railroads back farther from the coast. The alternative is to suffer increased damage and destruction along the coasts or require extremely expensive beach replenishment with disappearing supplies of replenishment sand.

Homeowners Insurance

After the four disastrous hurricanes in 2004, some small insurance companies left Florida entirely; several large companies stopped writing new policies or dropped certain policyholders. Almost all insurance companies significantly increased the cost of coverage after the storms. Recent major hurricanes including Ivan, Katrina, Rita, and Ike have led insurance companies to dramatically increase premiums for people living in susceptible areas, impose deductibles as high as $20,000 (before a company pays anything), and cap replacement and rebuilding costs. Many now exclude wind damages from policies and require additional premiums for wind coverage. In some areas of south Florida, insurance rates were three to four times higher in 2007 than in 2005; in South Carolina, they were seven times higher.

In the Mid-Atlantic States and New England, insurance companies have cancelled about one million homeowners' policies since 2004. Although most people have found other coverage, it comes with higher rates and larger deductibles. Some companies no longer provide insurance on Long Island and in New York City because of hurricane risk. Long Island protrudes directly into common hurricane paths. Exclusive summer homes on the southeast corner of Nantucket Island, Massachusetts, with an average 2007 price of $1.8 million, line a beach cliff that is rapidly eroding, endangering the homes. To slow the erosion, a group of wealthy homeowners put up $23 million of their own money to replenish the beach using sand dredged from two-and-a-half kilometers offshore.

A few years ago, governments in states such as Florida, Mississippi, and Louisiana created state-run insurance programs to cover homeowners who are unable to get insurance from private companies; private insurers in the state are billed part of the cost, which is then passed on to policyholders. Now, given the severe damage from the hurricanes of 2004 and 2005, the states feel the need to dramatically raise their premiums—to the forceful complaints of residents. Surprisingly, Florida's state-run insurance pool still provides coverage to people building expensive coastal homes on sites destroyed in recent hurricanes. A recent actuarial (actual cost) analysis indicated that state-run rates should be increased by an average of 80 percent—and in some areas more than double that. Politicians, however, resist raising premiums on state-run policies to avoid alienating policyholder voters. Thus, if the state has insufficient funds to cover losses, either the policyholder isn't paid for a loss or the state puts pressure on Congress to step in and provide the remainder from federal funds—that is, from taxes paid by everyone in the country. The overall problem is that too many people live in dangerous coastal areas.

Costs paid by homeowners insurance commonly include physical damage to homes caused directly by winds, flying debris, falling trees, and rain penetration after wind damage. In fringe areas of a storm surge, insurance companies often argue that water came from the surge, which is not covered, rather than from wind followed by rain.

Flood insurance is not included as part of homeowners insurance, but it is available through the NFIP. In low-lying areas, flood insurance is expensive and capped at $250,000 per home. In New Orleans and its vicinity, more than half of the eligible homes were not insured for floods, either because people could not afford the additional expense or didn't believe they would be flooded. In Mississippi, less than 20 percent of eligible homes had flood insurance. Since a large part of the damage in low-lying areas was from storm-surge flooding, homeowners insurance did not cover the damage. Flooded homes as far as 3 kilometers inland lacked flood insurance. Many were not in the floodplain zone designated after Hurricane Camille in 1969, which came ashore in Mississippi as a Category 5 storm, with a 7-meter surge.

Following Katrina, arson fires sprang up in a number of places, leading to the suspicion that some people may have set fire to their own flood-damaged homes, hoping that fire insurance would pay for rebuilding.

Extratropical Cyclones and Nor'easters

Extratropical cyclones are cold-weather storms that behave much like hurricanes and can cause as much damage. **Nor'easters** are extratropical cyclones that strike the northeastern parts of the United States. The most prominent Nor'easters struck in 1723, 1888, 1944, 1953, 1962, 1978, 1991, and 1993. They differ from hurricanes in several ways:

1. Named for the direction from which their winds come, nor'easters bring heavy rain and often heavy snowfall. Benjamin Franklin noted that precipitation begins in the south and spreads northward along the coast.

2. They are most common from October through April, especially February (rather than late summer for hurricanes).

3. They build at fronts where the horizontal temperature gradient is large and the air is unstable. They often form as low-pressure extratropical cyclones on the east slopes of the Rocky Mountains, such as in Colorado or Alberta, when the jet stream shifts south during the winter months. Similar storms can arise in the Gulf of Mexico; near Cape Hatteras, North Carolina; or near the Bahamas or east coast of Florida.

4. They lack distinct, calm eyes and are not circular in form but can spread over much of the northeastern United States and eastern Canada. As recognized in the late 1700s by Benjamin Franklin, smaller, counter-clockwise-rotating cyclonic weather systems are embedded in the broader overall flow.

5. They are cold-core systems that do not lose energy with height. If jet stream winds move an air mass away from the center of a storm, this drops surface pressure and increases storm strength.

6. Damage is concentrated along the coast, whereas much of the damage from hurricanes occurs farther inland. With strong winds from the northeast, they typically batter northeast-facing shorelines.

Nor'easters can build when prevailing westerly winds carry these storms over the Atlantic Ocean and if the jet stream is situated to allow the storms to intensify. The annual number of Nor'easters ranges from 20 to 40, of which only one or two are typically strong or extreme.

In addition to direct damage from high winds, a Nor'easter also generates high waves and pushes huge volumes of water across shallow continental shelves to build up against the coast as a storm surge. Low barometric pressure in these storms also permits the water surface to rise, creating higher storm surges. Especially high tides or surge movements into bays can amplify flood heights. Like those that accompany hurricanes, these surges flood low-lying coastal plains and overwash beaches, barrier islands, and dunes. As with hurricanes, the greatest damages occur when a major storm moves slowly at the coast or hits a coast already damaged by a previous storm.

A classification scale for Nor'easters by Davis and Dolan (1993) approximately parallels that of the five-category Saffir-Simpson Hurricane Scale, except that the emphasis is on beach and dune effects rather than wind speeds and surge heights. It infers a "storm power index" based on the maximum deepwater significant wave height (average of the highest one-third of the waves) squared multiplied by storm duration (**Table 14-5**).

Table 14-5	Dolan-Davis (1993) Nor'easter Scale			
CLASS	MAXIMUM DEEPWATER SIGNIFICANT WAVE HEIGHT (METERS)	AVERAGE DURATION (HOURS)	MOST COMMON SITE OF FORMATION	EXAMPLE STORMS
I (Weak)	2	10		
II	2.5	20		
III	3	35		
IV	5	60	Bahamas or Florida	December, 1992; March 1993 ("Storm of the Century")
V (Extreme)	6.5	95	Bahamas or Florida	March 7, 1962 (Ash Wednesday); Halloween 1991 ("The Perfect Storm")

Nor'easter wave heights are commonly 1.5 to 10 meters, with energy expended on a coast being proportional to the square of their height. Thus, a 4-meter wave expends four times the energy of a 2-meter wave. Wave height depends on fetch, or the distance the wind travels over open water. Waves with a long fetch and constant wind direction in a slow-moving storm can therefore be much more destructive than those in a stronger, fast-moving storm with variable wind directions. Where a storm center is well offshore, the highest waves may reach the coast after the clouds and rain have passed. When high waves are stacked on top of a storm surge, the effects are magnified. A severe Nor'easter can remain in place for several days and through several tide cycles. Examples of Class V storms include the Ash Wednesday storm of 1962 and the Halloween storm of 1991.

The Ash Wednesday Storm of March 7, 1962, was a high-latitude Nor'easter along the Atlantic coast of the United States; it stayed offshore approximately 100 kilometers, paralleling the coast for four days. The storm began east of South Carolina and migrated slowly north and parallel to the coast before moving farther offshore at New Jersey. Its slow northward progress was blocked by a strong high-pressure system (clockwise rotation) over southeastern Canada. It affected 1,000 kilometers of coast and caused more than $2.2 billion in damages (in 2010 dollars). Sustained winds over the open ocean blew 72 to 125 kilometers per hour and produced waves as high as 10 meters. Storm waves of 4 meters on top of a 1- to 2-meter storm surge washed over barrier sandbars that had been built up for years. A series of five high tides amplified the height of the surge. Beaches and dunes were extensively eroded and dozens of new tidal inlets formed. Almost all of heavily urbanized Fenwick Island, Delaware, was repeatedly washed over by the waves. The coastline moved inland by 10 to 100 meters.

The huge Halloween Nor'easter of October 1991, also called "The Perfect Storm," had 10.7-meter-high deepwater waves and lasted for almost five days. Its wave crests were especially far apart, with intervals ranging from 10 to 18 seconds between crests, so they moved much faster than most storm waves. Accompanied by a major storm surge, it caused heavy damage from southern Florida to Maine, especially in New England (**FIGURE 14-29**).

A huge Nor'easter ravaged the eastern states and Canadian Maritime Provinces in mid-April, 2007. It produced six tornadoes across northern Texas, then others farther east to South Carolina. It strengthened as it moved up the East Coast. Several people died in Texas, Kansas, and South Carolina. Wind gusts of 100 kilometers per hour were common from New Jersey to Maine and reached 250 kilometers per hour at Mount Washington, New Hampshire, an area notorious for high winds. The storm dumped up to 22 centimeters of rain and wet snow, which produced flash floods in the Carolinas, West Virginia, New Hampshire, and Nova Scotia. Winds pushed high waves that eroded beaches through New York, New Jersey, and New England.

FIGURE 14-29 NOR'EASTER

Peter Shugert, U.S. Army Corps of Engineers.

Storm waves batter this seawall at Sea Bright, New Jersey, during the 1991 Halloween Nor'easter.

Snow depth is one measure of the severity of a winter storm because of snow loading and accompanying building collapse. Snow depths, occasionally exceeding a meter, are typically greatest in Maine, northern Michigan, Wisconsin, and western New York. Heavy snow can even fall in southern states such as Alabama and Georgia. Hazards initiated by both tropical cyclones and extratropical storms include the following:

- Storm surges are well above normal tides and cause coastal flooding, salinization of land and groundwater, coastal erosion, damaged crops and structures, and drowning.

- Huge storm waves may overwash dunes and impact coastal structures.

- Heavy rain causes river flooding, flash floods, landslides, structural damage, and overflow of storm sewers and sewage systems. Nearly 60 percent of the people who die in hurricanes drown because of river floods near the coast. One-quarter of hurricane deaths are of people who drown in their cars or while trying to abandon them during floods. The heaviest rainfalls are from slower-moving storms and those with larger diameters. The National Hurricane Center predicts the total rainfall in inches by dividing 100 by the forward speed of the storm in miles per hour. For example, a storm approaching at 20 miles per hour would be expected to have 5 inches of rain (100/20 = 5 inches).

- High winds, including tornadoes, damage windows, roofs, and entire buildings. They disrupt transportation and utilities and create large amounts of debris. Tornadoes that accompany many hurricanes form as higher winds blow over slower surface winds hindered by friction against the ground. The combination forms a horizontal rolling motion that tilts up to become a twister rotating around a vertical axis (discussed in Chapter 15). Tornadoes can form near an eye wall or well away from the center of a hurricane; however, they are difficult to confirm because hurricane winds can reach speeds similar to those in tornadoes.

■ Immediately following a storm, communication lines are lost, power lines are down, roads are out, and a broad range of urgent needs overwhelm local governments. Salt, sewage, various chemicals, and bacteria contaminate surface water and groundwater. Many local officials are inexperienced in dealing with large-scale disasters and unfamiliar with the programs available for assistance. Coordination among all levels of government and teamwork is critical for recovery from such disasters.

Cases in Point

City Drowns in Spite of Levees
Hurricane Katrina, 2005 ▶

Hurricane Katrina set a new standard for damage levels and for questions raised about the potential effects of hurricane flood surge, coastal development patterns, and the need for far better disaster preparedness. The results underscore the expression "Whatever can go wrong will go wrong."

As Hurricane Katrina bore down on the Louisiana and Mississippi coasts, most people complied with evacuation orders, but many thousands of residents in the very poor, predominantly African American, eastern parts of New Orleans did not. Most of these people had no cars or other means of transportation and no money for travel even if transportation had been available. Many were tired of the time and expense of evacuating only to find that a storm did not strike their area. Many of them felt that New Orleans and the surrounding communities, especially areas not right on the coast, had survived hurricanes before and would do so again. Others felt that the brunt of the storm would miss them, just as Hurricane Ivan spared New Orleans when it roared into the Gulf Coast of Alabama and adjacent Florida a year earlier, wreaking havoc there.

However, many of these people were not there or perhaps did not recall some of the other storms that had struck the area. In 1965, Hurricane Betsy, a Category 3

▶ Hurricane Katrina with its well-developed eye moved north from the Gulf of Mexico into Louisiana and Mississippi. Southeastern Louisiana and the Mississippi River delta with the path of Hurricane Katrina and direction of the surge and winds.

(continued)

storm, left almost half of New Orleans under water, up to 7 meters deep in some places. Hurricane Andrew, in August 1992, passed just west of New Orleans. In 1969 Hurricane Camille, a Category 5 storm at landfall, caused major damage in Mississippi and Louisiana.

Although pre-Katrina planning predicted that at least 100,000 residents would not have transportation, few buses were sent to shuttle them out of the area. Hundreds of school buses were left unused because the city could not find drivers and because FEMA apparently asked that school buses not be used in evacuation efforts because they were not air-conditioned, and evacuees might suffer heatstroke. FEMA said that it was providing suitable buses but failed to tell the governor that they would come from out of state and not be immediately available.

Many residents who sought food, water, and shelter found their own way to the giant Louisiana Superdome stadium and to the New Orleans Convention Center; others were brought there by rescuers. By the time the storm arrived the next morning, 9,000 residents and 550 National Guard troops were housed in the Superdome. FEMA did arrange for eighteen medical disaster teams, as well as search-and-rescue teams, medical supplies, and equipment, but relief for most refugees did not come until four days after the storm.

Ten thousand National Guard troops were finally ordered to the area days after the storm to help with rescue, public safety, and cleanup efforts. They brought water, ice, tarps, and millions of ready-to-eat meals, but it was not enough. The hurricane-force winds blew off part of the roof lining of the Superdome, causing it to leak. The power went out and the air-conditioning failed; then water pressure dwindled, so toilets backed up. People received water and two meals a day. Some individuals (particularly the elderly and very young) were sick or went without their prescription medicine; some people suffered heatstroke, and several died. Ultimately more than 20,000 people ended up in the Superdome.

Approach and Landfall

The National Weather Service predicted that the hurricane would cause catastrophic damage and "human suffering incredible by modern standards," and it was absolutely correct. In the Gulf of Mexico, Katrina strengthened to a Category 5 hurricane, the maximum strength, with a central pressure of 902 millibars and sustained winds of 280 kilometers per hour. It weakened to a Category 3 in shallow waters before landfall in coastal Mississippi on Monday, August 29, at 7 a.m., where it obliterated nearly everything within hundreds of meters of the beach.

The Wind, Storm Surge, and Flood

However, with its counterclockwise rotation and landfall near the Louisiana–Mississippi line, the high near-shore winds on Katrina's north flank were directed westward toward New Orleans. Trees and power lines fell; the winds blew out windows, including those in hospitals, office buildings, and hotels; and as with other hurricanes, tornadoes did some of the most severe wind damage. Much of the city was under water, so rescuers had to use small boats and helicopters to reach people.

Most of the damage to New Orleans was caused by the storm surge that raised the water level in Lake Pontchartrain, other low tidal lagoons, and the large canals that typically drain water from the

▶ *Abandoned school buses at Port Sulphur, Louisiana, after the water receded.*

city into the lake; giant pumps pull water from the city into those canals. Pushed by fierce winds, the surge moved northwest through the deeply dredged Mississippi River Gulf Outlet. Even on a normal day, Lake Pontchartrain lies meters above the lower parts of New Orleans. In both cases, the eastern suburbs of New Orleans were rapidly inundated, with water rising as much as a meter every three minutes. Residents described a "river" with 2-meter waves rushing down streets.

Surges came over many levees on the eastern fringes of the city but did not breach them. However, canal walls failed

▶ *By August 31, the Superdome was surrounded by water, and sections of its roof covering had blown off. View is northeast; the Mississippi River is to the right.*

(continued)

● Pumping station   ⇨ Breach direction   △ Main water plant

▶ **A.** *This map of New Orleans for September 2, 2005, shows shallow flood depths on the natural levee areas at the north bank of the Mississippi River, increasing to 3 or 3.5 meters over much of the city. The flood area abruptly ends on the west at the 17th Street Canal; levees breached to the east.* **B.** *Almost the entire city lies below the average annual high-water level of the river and below the level of Lake Pontchartrain, so these giant pumps are used to drain the city.*

in several places, causing inundation of New Orleans that began about 18 hours after landfall. The canal walls generally consist of a ridge of dirt and rock topped by a concrete and steel floodwall 30 centimeters thick and 5 meters tall. Two long sections of floodwall failed, flooding homes that had been built just behind the levee and floodwall, where people felt protected from the water high above them. The water poured directly into the basin occupied by the city, drowning it in places with 4 meters of water. Only the homes on the high-standing natural

levee along the Mississippi River were largely spared. Engineers dammed the breached 17th Street Canal with steel sheet pilings 15 meters long to stop the flow of more water from Lake Pontchartrain through the breach into the city. It took two weeks to fill the breaches, using helicopters to drop giant sandbags. One plan involved deliberately breaching some lower levees to let water drain back out and to use smaller replacement pumps that were available. Levees along the canals were supposed to be over 5 meters high, enough to withstand a moderate

(Category 3) hurricane, but some had settled; in some cases, water surging over the walls likely undermined their bases, and soil used to build some of the levees included layers of easily eroded sand and shell fragments.

As the city filled with water, cars, houses, debris, and marsh grass floated with the surge, coming to rest in odd locations—sometimes houses ended up on top of cars or in streets. Even brick houses well anchored to 40-centimeter-thick concrete slabs floated and were swept down streets.

▶ **A.** *The 17th Street Canal crosses much of New Orleans south from Lake Pontchartrain (bottom edge of photo). The breach and flooded homes are visible on the far side of the canal to the right of the bridge. The Corps of Engineers are driving sheet pilings at the bridge to block water flowing down the canal from the lake while they also drop giant sandbags into the breach.* **B.** *This canal wall in the Gentilly area of New Orleans collapsed outward from the canal. The Corps of Engineers erected the corrugated steel wall as a temporary barrier against further flooding.*

(continued)

▶ **A.** *Modest houses in the Lower 9th Ward were floated off their cinder-block posts and deposited on cars, other houses, or streets.*
B. *This brick home on a 40- to 50-cm-thick concrete slab in the Chalmette area at the east edge of New Orleans floated in the surge and was carried five blocks and deposited in the middle of a residential street.*

The Pumps Fail

With most of the city under several meters of water, the pumps failed, either for lack of electricity or from overheating. Electricity soon failed throughout New Orleans and east along the coast through Mississippi and Alabama to the Florida panhandle. Because of power failures, destruction of base stations, and breaks in lines, telephones and cell phones wouldn't work. The lack of communication greatly hampered rescue efforts. Broadcasts from reporters in nearby cities and communication over the Internet became important. Ten major hospitals were forced onto backup power.

The water came in so fast that within minutes it was over people's knees, forcing them up to second floors and attics or onto roofs. A few of those who retreated into attics thought to grab an axe or a saw so they could break through the roof if the water continued to rise. They waited there, sometimes for days, but had no means to contact authorities or rescuers. Searchers rescued at least 1,500 people from rooftops and heard others beating on roofs from inside attics. Rescuers doing careful house-to-house searches found a few bodies, but fortunately not many.

Contamination, Disease, and Mold

The floodwaters were littered with pieces of houses, old tires, garbage cans, all manner of trash, sewage, coatings of oil and gasoline from ruptured tanks, and even bodies. A few people were fortunate to grab hold of larger pieces of trash or drifting boats to stay afloat. Many of those rescued had problems related to the polluted water, especially gastrointestinal illnesses, dehydration, and skin infections. Aircraft sprayed pesticides to kill mosquitoes because of the danger of malaria, West Nile virus, and St. Louis encephalitis. Even after September 16, some neighborhoods were still flooded.

Mold grew in most buildings in contact with the warm, contaminated water. Despite being structurally sound, many buildings were so affected that they had to be bulldozed. As the water receded, dark gray muck coated everything. In French Quarter buildings on the natural levee of the Mississippi, mold was growing a month later in rooms that were not flooded but still lacked air-conditioning. Wallboard, insulation, rugs, bedding, and almost anything else that had gotten wet had to be discarded. Even the bare studs of walls needed to be sanded, disinfected with bleach, and then dried with fans. Any wood frame structure standing in water for more than two or three weeks had to be demolished because mold was impossible to remove from deep in the wood. The effects of mold on people with allergies, asthma, or weak immune systems can be serious.

Relief Came Slowly, Many Victims Died

After Katrina, people waited for four days before anyone brought food, water, medical supplies, or vehicles for evacuation. Clearly emergency response to the catastrophe was a dismal failure. Day after day, federal Homeland Security and FEMA officials promised National Guard troops, supplies, and buses for evacuation—help that rarely materialized.

Distribution of aid for disasters is logistically complex, but giant building materials and grocery companies have become very efficient at distribution on a global scale. Since the Bam, Iran, earthquake disaster of 2003, aid organizations such as the International Red Cross have used those techniques to distribute aid quickly to victims of major disasters. FEMA presumably could have used those companies or such procedures.

Confirmed deaths totaled 1,810, including 1,464 in Louisiana and 238 in Mississippi. Thirty-nine percent of the people who died were more than 75 years old. As water rapidly rose, people were swept away by surge flow and drowned because they couldn't swim. Or they drowned in their houses after retreating to a higher floor or an attic and becoming trapped. Five people reportedly died from illness caused by bacteria related to cholera.

(continued)

▶ *Homes in the Chalmette area at the east edge of New Orleans were trashed by Katrina.*

Some patients in hospitals died when electricity necessary to power equipment such as respirators and dialysis machines failed and backup generators ran out of fuel; doctors and nurses squeezed hand-held ventilators for patients who couldn't breathe on their own. There was no running water or ventilation; seriously ill patients died in the 41°C (106°F) heat. Thirty-four nursing-home patients died in the flood; the owners said they never received the mandatory evacuation order and that re-locating would have killed some of the frail patients. Some hospitals were so damaged by flooding and mold that they will never reopen. Truly, anything that could go wrong did go wrong. These should be prominent lessons for any future potential disaster.

▶ *Some neighborhoods were still flooded weeks after the storm. Many already-damaged homes floated off their foundations and collided with other homes. Fetid water was everywhere.*

Impacts Farther South and East: The Hurricane Winds, Surge, and Waves

News media focused on the storm damage in New Orleans because its flooding and destruction was so dramatic and catastrophic. However, elsewhere down-river and along the coast to the east, the storm's effects were no less disastrous. Southeast of New Orleans, near the main dredged Mississippi shipping channel, the high winds, huge surge, and waves floated houses like matchboxes, dropping them on roads or other houses. Some buildings were moved with their con-crete foundation slabs still attached. The surge lifted shrimp boats of all sizes and dumped them onto nearby levees and roads. Nests of poisonous water moc-casins and other snakes swept in from the bayous added to the dangers. Rotting animal carcasses were scattered near the roads. Northeast of New Orleans, the high surge and waves from Katrina lifted seg-ments of a 10-kilometer-long Interstate 10 causeway and dropped them into Lake Pontchartrain. The U.S. 90 cause-way, across St. Louis Bay and the east end of the Back Bay of Biloxi, Mississippi, collapsed in a similar series of tilted road panels.

Katrina's eye tracked almost due north, making landfall on August 29 at 7 a.m. at the border of Louisiana and Mississippi, causing collapse of an apartment complex and killing dozens of people. Some people survived the fast-rising surge by climbing into treetops. Given its counterclockwise rotation, the strongest onshore winds, as high as 224 kilometers per hour at the deadly eastern edge of the eyewall, ham-mered the Mississippi coast near Bay St. Louis, where the surge reached its highest level of about 9 meters, the height of a three-story building!

On the mainland, the surge, extreme winds, and high waves crushed houses, toppled trees, and severed power lines; transformers exploded, and sailboats broke loose and were thrown across the coastal highway. Cars were scattered like toys. A few homes, specially built to withstand major storms, survived even where all of the neighboring houses were obliterated.

(continued)

▶ *A pair of 100-foot-long oil service vessels from the Mississippi River shipping channel ended up on Highway 23. The U.S. 90 bridge across St. Louis Bay, near the western end of the Mississippi coast, collapsed in the massive surge and giant waves as Katrina arrived. The surge and waves must have lifted the bridge deck segments and then dropped them either onto their supports or into the bay.*

U.S. Army Corps of Engineers.

John Fleck, FEMA.

A

B

Buildings all along the beachfront, from the border of Louisiana, through Mississippi, from Bay St. Louis, Gulfport, and Biloxi to Pascagoula, were leveled, leaving only tall posts and concrete pads to show where they once stood. Many beachfront houses were washed out to sea or undermined and toppled; others over large areas were reduced to kindling, even as far as 2 kilometers from the beach. Two years later, one-third of Bay St. Louis residents had not returned and many said they would not—it is too much to risk everything again.

In Gulfport, the surge rose 3 meters in a half hour; fierce winds tore the roofs off eight schools that were being used as shelters, and a hospital was heavily damaged, as were two huge casinos. In Biloxi, seven giant casinos, floating just offshore, were wrecked, including the state's largest; it was carried more than a kilometer inland.

The effect of the storm surge on houses in Gulfport is clear where the lumber that once made up the houses was swept up and stacked against the remaining heavily damaged houses. Buildings lifted from their foundations and slammed into nearby buildings are clear evidence of storm surge damage, as are huge piles of building debris banked up against one side of buildings with none on down-current sides.

In Alabama, a huge oil-drilling platform moored at a shipyard floated away and slammed into a suspension bridge across the Mobile River. Downtown Mobile saw severe flooding, not only from the surge but from heavy rainfall as the storm moved north. Following damage from each previous storm in coastal Mississippi and Alabama, developers took advantage of the destruction to build larger and more expensive structures right to the edge of the beach. Although many of the high-rise hotels survived the storm, some of the beaches that attracted them and on which they were built disappeared, as did most homes behind them.

Predictions, Preparation, and Response

In 2004 FEMA prepared a simulation of a major flood in New Orleans. The results, although unfinished, were unnervingly accurate: The simulation left much of the city under 3.5 meters of water and showed that transportation would be a major problem.

Batteries in emergency radios used by the mayor's staff, police, and firefighters would quickly drain and could not be recharged because the power was out. These were unlike radios used by teams fighting wildfires, which can be powered by ordinary disposable batteries.

A big contributor to the poor response to this disaster was lack of coordination among government groups with different responsibilities. Clear-cut lines of authority and communication were not in place. In some cases, the head of an agency said to proceed with a plan, but lower-level employees wanted signed papers to protect themselves from later criticism. Some governmental organizations were afraid of being sued if they stepped beyond their authority or made a mistake.

Different federal agencies, such as FEMA, charged with separate duties, communicated poorly with one another. Some did nothing because it was "not part of their jurisdiction." Interagency squabbles on the federal level and quarrels among

(continued)

▶ **A.** *A nearly new subdivision in a coastal area of Gulfport, Mississippi, was leveled by the surge, leaving its remains piled up against the battered houses farther inland.* **B.** *Two-story apartment buildings were lifted from their foundations and crashed into adjacent buildings.*

federal, state, and local governments appeared to involve protecting each organization's turf. Much of FEMA's problems originated when the federal government reduced its funding and relegated it to a role of responding to disasters instead of preparing for or preventing them. It couldn't even do that; it tragically neglected to stage adequate water, food, medical supplies, and transportation nearby in preparation for post-storm response. Individual medical and emergency organizations were tragically under-prepared.

FEMA, charged with handling response to disasters, proved tragically unprepared and inept. Five hours after landfall, FEMA's director decided to send 1,000 federal employees to deal with the storm's effects, but supplies were very slow to arrive. With thousands of people sheltered by the Red Cross in the Convention Center, FEMA said it had no "factual knowledge" of its use as a shelter until September 1.

Arranging temporary housing for an estimated 300,000 displaced people in the wake of Katrina was an immediate and enormous task. By September 4, 220,000 refugees were sheltered in Houston, San Antonio, Dallas, and other cities across the country. Outside New Orleans, FEMA provided army-style wood-frame tents and some travel trailers and mobile homes but insisted that before the tens of thousands of trailers could be moved, the sites that would receive them had to have water, sewer, and electricity hooked up, services

not available in many areas for months. In most of the city, even a month after the hurricane, power lines still dangled, tree branches and other debris still clogged the streets, and no stores or gas stations had reopened.

The Future of New Orleans?

An important question is whether New Orleans should be rebuilt in essentially its previous form. Two years later, little had been accomplished in most of the city. Should people be permitted to rebuild in a huge sinking depression several meters below sea level and below the Mississippi River, or should aid for reconstruction come with the requirement that any new homes be situated above sea level and outside the floodplain? The latter was one of FEMA's main requirements for people and companies seeking funds for rebuilding structures along rivers; this rule was put into effect after the disastrous 1993 upper Mississippi River flood.

Certainly the higher-elevation areas of New Orleans, those on the natural levees of the Mississippi River, should be restored. These areas provide the shipping and industrial facilities that serve not only the Mississippi River basin but much of the rest of the country. Some port facilities could be moved upriver about 150 kilometers to Baton Rouge, which is also a dredged deepwater port. The prospect of a permanent relocation of much of New Orleans' population to Baton Rouge

brings to mind the migration of people and businesses from Galveston, Texas, to the then small town of Houston after the disastrous 1900 Galveston hurricane. Most of those people never returned to Galveston.

New Orleans' natural-levee areas, including the lightly damaged, famous French Quarter, also make up the cultural and historical part of the city frequented by tourists, who provide a large portion of the city's income. Even in areas that should be rebuilt, where do you start? Months after the storm, large devastated areas had no gas stations; no open grocery, hardware, or building-supply stores; no schools; and no funds from property or sales taxes with which to pay the city or parish employees needed to repair roads and utilities. For people with jobs in rebuilding or at restarted refineries, where will they live, get groceries and gas, or find schools for their children? Without people in the area, there are no jobs; without jobs, people cannot return.

Katrina was the costliest natural disaster to strike North America to date. Insured costs reached $60 billion as of August 2006; federal government appropriations reached $71 billion within three months following the storm. After several months, FEMA had not allocated most of the federally appropriated funds and the federal government seemed to be backing off on initial promises. A total of 80 percent of $114 billion in allocated funds had been

(continued)

disbursed two years after the storm. Some estimates suggest total costs may top $200 billion, including payments to businesses and individuals. Four years after the storm, the population was estimated to be about 355,000, still down 100,000 from before the storm.

The Corps of Engineers plans to repair 60 kilometers of the 480-kilometer levee system to withstand a Category 3 storm. Improved sections will be 5.2 meters high rather than the previous 3.8 meters high. Rebuilding the system to withstand a Category 5 storm would cost more than $32 billion—that is $66,000 for each of the 485,000 original residents, $264,000 per family of four, much more than that for the many fewer who are expected to return!

Hurricanes have affected New Orleans before. Betsy, a Category 3 hurricane, submerged almost half of New Orleans in 1965; some places were under as much as 6 meters of water. The storm left 60,000 people homeless. Congress then authorized a gigantic construction project to raise the levees and link them to those of the Mississippi River to prevent such flooding ever again. As the city and its levees continue to sink, it will be just a matter of time before the next catastrophic flood.

Some of those levees along Lake Pontchartrain were built years ago by local governments or private groups and were not well engineered. Following a disastrous Mississippi River flood in 1927, the levees were built higher and strengthened. Additional levees were constructed in the 1940s and 1950s, and shipping canal walls were added in the 1960s. Unfortunately, the city on the floodplain continues to slowly sink, as groundwater is withdrawn for municipal and industrial uses, and buildings continue to compress the underlying peat. A total of 148 giant pumps remove water from the spongy sediments. Now much of the city lies 3 meters below Lake Pontchartrain's normal level and 4 meters below the Mississippi River—not a good place for hundreds of thousands of people to live.

Parts of New Orleans continue to sink; the Lakeview and Kenner parishes will continue sinking—a total of 25 cm over the next 10 years—and the Lower Ninth Ward will sink to about 3 meters below sea level in 50 to 100 years.

Déjà vu?
Hurricane Gustav, Gulf Coast, August 29–September 2, 2008 ▶

Almost three years to the day after Hurricane Katrina ravaged New Orleans, a similar hurricane, Gustav, reminded residents of the Gulf Coast that their coastal location is a dangerous one. Tropical Atlantic Ocean temperatures warming all summer finally spawned a series of strong westward-moving and strengthening storms in late August of 2008. Hurricane Gustav weakened to a tropical storm temporarily on August 27 as it passed over Haiti but it dumped 30 cm of rain on its deforested southwestern peninsula, causing landslides that killed 51 people.

Since Katrina, $2 billion has been spent on more than 350 km of rebuilt floodwalls, new and more flood-resistant pump stations, and other improvements, but the Corps of Engineers estimates that the whole flood-protection system is only one-third the level needed for a 100-year storm. However, only about one-quarter of the $12.8 billion in federal dollars authorized has been spent. The weakest link in the system is a planned barrier across the mouth of the "Mississippi River Gulf Outlet" canal on the east; that work is not planned until 2011. The area near that shipping canal sank about 6 cm between 2002 and 2005 and continues to sink; some of the levee failures were in the

Ronnie Simpson, FEMA.

▶ *Water begins to slosh over the Industrial Canal floodwalls in New Orleans.*

(continued)

areas of greatest subsidence. FEMA was much more prepared for a hurricane, and residents were more prepared to evacuate. In fact about 1.9 million people left southern Louisiana on August 30. The storm strengthened from a tropical storm to a category 4 hurricane in 24 hours as it entered the Gulf. A mile-long line of residents waited for buses in New Orleans to take them to shelters in northern Louisiana.

Petroleum companies raised and rebuilt their offshore production platforms to stronger standards and secured them to withstand stronger storms and damage there was limited. However, about 80 percent of homes lost power during Gustav because of downed transmission towers and lines. Trees fell and roofs were blown off some houses. When residents tried to return home after the storm on September 4, there was no electricity; without it sewage treatment stations failed and gas stations could not pump gas or power hospital generators. Eighteen people died along the U.S. Gulf Coast.

Extreme Effect of a Medium-strength Hurricane on a Built-up Barrier Island
Galveston, Texas: Hurricane Ike, September 13, 2008 ▶

Hurricane Ike provided a vivid reminder of the damage that can be wreaked by a moderate-category hurricane on a developed barrier island. A category 4 storm in the Atlantic Ocean, Ike ravaged Haiti and Cuba, weakened to category 1 in the Gulf of Mexico, then curving northwest and north, it strengthened to category 4, finally easing to category 2 before landfall at Galveston, Texas. Hurricane Rita in August 2005, a category 5 offshore, weakened to category 3 at landfall, near the Louisiana border east of Galveston, but the ordered full evacuation of Galveston and much of Houston did almost as much damage as the hurricane. Again before Ike, authorities ordered complete evacuation; about one million complied, but the lower wind speeds encouraged several thousand people to ride out the storm.

Following the tragic hurricane of 1900 that killed about 8,000 people, Galveston raised the level of the city and built a high sea wall to protect it from further hurricanes. However, there was a downside. Aggressive waves in front of the seawall stirred up and removed sand where there was once a nice beach, the main reason most people moved there in the first place. For many years, they dumped truckloads of sand—removed from dunes elsewhere along the coast—over the wall to provide a narrow beach for recreation and protect the base of the seawall from being undermined. Continued erosion and shortages of readily available sand, however, made it necessary to "pave" the remaining sand with huge blocks of granite to minimize further erosion and undermining of the wall. During Ike, the wall did in fact protect the city from the worst of the incoming storm surge and waves, but areas beyond the ends of the wall were not protected.

As in the majority of coastal areas, the most vulnerable homes sit on offshore barrier islands, especially those on beachfronts with no protective dunes. Natural dunes, originally present on the barrier islands, both northeast and southwest of Galveston, gradually blew landward during past hurricanes and other storms. Coastal houses built on the islands, however, stayed put, leaving them standing on the beach itself as the sand moved inland. Unfortunately, with so many houses built at the beach and with little zoning to control building activity, protective coastal dunes were also removed for construction or to improve views or beach access.

At landfall, Ike's low atmospheric pressure and winds of about 160 km per hour (about 100 miles per hour) drove a 3.6 meter-high surge onshore northeast of Galveston, over the Bolivar Peninsula, an offshore barrier island heavily populated with homes. Most were built 3 to 4 meters above the sand, on top of sturdy posts, to keep them above storm waves, as required by coastal flood insurance and coastal ordinances. Although low artificial dunes were piled on the beach in some places, by the time Ike arrived, the front rows of houses were on the beach itself, essentially unprotected. Ike had a huge diameter, with hurricane-force winds extending out from the eye for 193 km (120 miles) and tropical-storm winds extending out to 442 km. This made it the largest-diameter hurricane ever measured, a formidable storm with more kinetic energy (energy of motion) than any other Atlantic hurricane recorded by modern instruments. Winds and waves pounded the Texas coast for hours before the fierce winds of the eyewall arrived. The storm finally came ashore at Galveston with 176 km/hr winds, a high category 2, almost category 3.

Twenty-four hours before landfall, on the morning of September 12, the huge diameter of Hurricane Ike pushed a 3.6-meter-high surge (12 feet) northeast of the eye, on top of a 2-foot high tide. That pushed sea level and high waves over the low-lying barrier islands. Because of the counterclockwise rotation of the storm, winds and waves ahead of it blew to the west, so southwest of Galveston, winds angled toward the shore carried the high surge, or storm-tide, in over the lagoon landward of Galveston Island. Thus although destruction near the beach northeast of the eye was almost complete, damage

(continued)

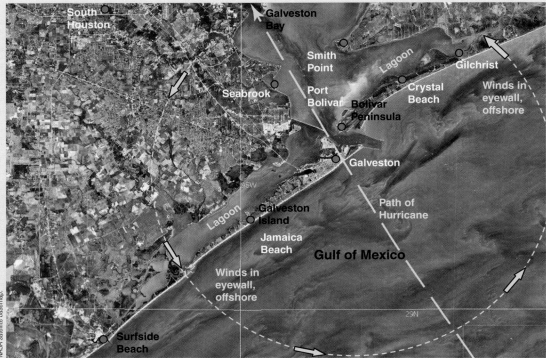

NASA satellite basemap.

▶ *At landfall, the high winds of Hurricane Ike's eyewall tracked northwest, directly over Galveston and just east of Houston. Northeast of Galveston, the winds and their accompanying waves and storm surge blew directly onshore, while southwest of the town they blew offshore. As the hurricane approached from the southeast (dashed yellow path), the storm surge was approaching a broad area of the Texas coast.*

southwest of the eye was also severe. At about 4 p.m., the storm surge began sloshing over the 5.2-meter seawall. Even though the level of the city was raised by more than a meter after the 1900 hurricane, the ground behind the beach slopes down toward the lagoon and water rose quickly.

Although Ike was only a category 2 storm at landfall, its huge diameter, covering most of the Gulf of Mexico, spread destruction over a broad area of the coast. Northeast of the eye, for about 190 km, the onshore-directed winds pushed a huge storm-surge mound of seawater far onto land. At Galveston the surge raised tide levels almost 3 meters 24 hours before landfall. Onshore, it swept inland about 50 km, reaching about 3.6 meters (almost 12 feet) along the west side of Galveston Bay. It also reached that height on the Louisiana border to the east; 32 km inland, it peaked at about 3 meters, primarily near the heads of bays that focused it. Elsewhere a wave of seawater more than 1 meter deep swept inland for more than 50 km, across

range and farmland and beyond Interstate 10. It lifted low barrier-island bridges off their supports, carried cars and boats inland, overturned, smashed, and sometimes buried them in sand.

Destruction at and near the barrier island beaches was nearly complete, with only a few especially strongly built houses remaining. The winds took a heavy toll. At 176 km/hr, they lifted off shingles and some complete roofs blew over mobile homes, shredded and flattened signs, and carried all kinds of debris that acted as missiles, impacting other structures. The high winds against the walls of some poorly-built houses simply blew them over. Houses on elevated posts that were not sunk deeply enough into the ground were blown over and shredded in the surf. Support posts that were weak or partly rotted, snapped. High winds and waves wrecked oil platforms offshore, damaged storage tanks, and ruptured fuel pipelines; oil and gas spills were widespread. The country's largest complex of petrochemical plants and

gasoline refineries, just inland from Galveston Island, came through largely unscathed because they were all shut down and secured before the storm arrived. However, chemical plants and most refineries could not be restarted for weeks after the storm because of lack of power.

Much of the damage occurred as the huge waves and storm surge swept onshore. Sand eroded from the beach, both in shallow water offshore and onshore, and was swept landward across the barrier island and into the lagoon behind. The huge mound of water not only carried boats, houses, parts of wrecked buildings, cars, grass, and debris onshore, but the seawater killed grass, crops, and brush that cannot tolerate the salt. Southwest of Galveston, as the wind turned south and subsided, the massive mound of water, standing several meters above normal sea level, swept back across the barrier island and offshore. In doing so, the fast-moving water further deeply eroded the beach, carrying as much as two meters or more of

(continued)

NOAA.

▶ *Most of the houses on the west end of Bolivar peninsula were completely obliterated, leaving only remnants of their ground-floor slabs.*

It remains puzzling why 13,000 people in the coastal area did not evacuate, in spite of dire warnings. Some were caught in the frustrating and chaotic evacuation from Hurricane Rita, category 3 at landfall, in 2005; evacuations can be expensive and most had never experienced a major surge like this. Some thought the storm would miss them or that they would be safe in their homes on tall stilts. Finally, some have an intense distrust of government and refuse to be told what to do. A new Texas law that took effect September 1, 2009, may help to convince them to leave. Police can now arrest people that don't leave under a mandatory order. In addition, they can now be forced to pay for any rescue during or after the storm.

In spite of the mandatory evacuation before Ike's landfall, 48 people died in Texas and 8 in Louisiana, and 3,500 people needed rescue. Coastal damages reached $27 billion and only hurricanes Andrew in 1992 ($51.9 billion) and Katrina in 2005 ($91.4 billion) cost more (in 2010 dollars). In the Houston area, 9 people died while cleaning up after the storm, from carbon monoxide poisoning from generators or from house fires. Months later, 202 remained missing, mostly along the coast.

FEMA handed out ice, millions of liters of water, and millions of meals at a dozen distribution centers in Houston; they also provided about 11,000 motel rooms for temporary evacuees. In contrast to the

sand offshore and undermining the posts of many remaining homes. In many cases flawed construction contributed to the destruction of houses. In some cases posts were not strong enough to support a house in the face of high winds, waves, or surge. Some untreated posts were weakened by rot. In many cases post tops were not well anchored to the main floors of houses, walls were not well anchored to floors, or roofs were not well anchored to walls. In many cases failure of one or two critical connections was enough to destabilize entire structures and cause them to collapse.

Since many barrier island homes are not connected to municipal sewer lines, the sewage typically goes into septic tanks. Because liquid outflow from the tanks flows into a drain field, in this case beach sand, the outflow must be above high-tide level for drainage to occur. Thus the septic tanks for homes on the beach are only shallowly buried. During Ike, numerous septic tanks were uncovered, their plumbing severed, lids torn off, and their contents spilled; in many cases the tanks were upended and floated to the surface.

Hyndman.

A

Hyndman.

B

▶ **A.** *Support posts failed under this house so it collapsed and broke up. Crystal Beach, northeast of Galveston.* **B.** *Support posts for this house were not strong enough to withstand the force of the surge flow. Some had rotted below sand level.*

(continued)

► **A.** Well away from the beach, the storm surge floated this house off its foundation and dumped it into nearby brush, along with debris from other houses. Northeast of Crystal Beach, Bolivar Peninsula. **B.** One-and-a-half to two meters of sand were eroded from this beach southwest of Galveston.

debacle after Hurricane Katrina three years before, they received praise for their efforts after Ike. In addition, the American Red Cross provided shelter and emergency food for tens of thousands of people. Days after the storm, many gas stations remained either out of gas or unable to pump it due to electrical outages; some people waited in line for 3 hours. People bailed water from fish tanks or swimming pools to flush toilets. With most stores, businesses, and restaurants closed, people couldn't work or purchase food and supplies. A week after the storm, most roads were still closed because of debris or washouts, only a single gas pump was operational in Galveston, and mosquitoes were rampant because of standing water.

► Severe beach erosion lowered the beach level uprooting and emptying many septic tanks. Near Gilchrist, on the Bolivar Peninsula northeast of Galveston.

Wind, Waves, Beach Erosion, Flooding
North Carolina Outer Banks ►

North Carolina's Outer Banks forms an endless stretch of white sandy beach that draws millions of visitors each summer. The beach is actually an offshore barrier sandbar resting on a shallow-water area of especially gently-sloping continental shelf that protrudes east into the Atlantic Ocean. That form places it farther into the path of many Atlantic hurricanes than anywhere else on the east coast between Florida and New York (compare hurricane strike potential, FIGURE 14-6). Although much of the Outer Banks is protected as national seashore, the northern two-thirds is accessible by road and a northern part, especially around Kitty Hawk and Nags Head, is heavily built up with expensive beach homes.

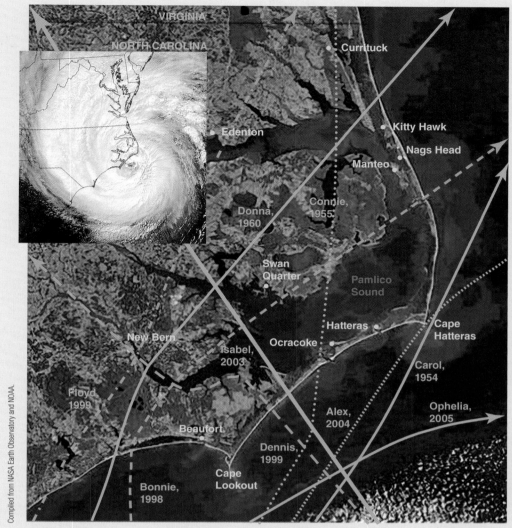

Compiled from NASA Earth Observatory and NOAA.

▶ *Some of the many hurricanes that have battered the North Carolina Outer Banks. Total Width of the view is about 210 km. Inset shows Hurricane Isabel for comparison; its size covered all of the area of the main map—compare shape of coastline to map.*

U.S. Army Corps of Engineers.

▶ *Beach front home in Kitty Hawk toppled into the surf after being undermined by Hurricane Isabel.*

Mark Wolfe, FEMA.

▶ *This beach-front home on stilts at Nags Head, North Carolina, was left stranded, then toppled on the beach after Hurricane Isabel in 2003.*

(continued)

A map of some of the hurricanes that have ravaged the Outer Banks makes clear that anyone who lives in the vicinity needs to be aware that this is a dangerous place to live. The whole map is less than the diameter of many hurricanes (compare hurricane-Isabel inset) so the winds, waves, and surge of many hurricanes would extend to more than half the width of the map from the tracks plotted here. Recall also that the storm surge impact is concentrated in the right-front quadrant of a storm (see FIGURE 14-9),

so for Hurricane Isabel, 2003, for example, most of the barrier-island surge impact from offshore would lie northeast of the track of the eye shown on the Figure below. However, for Hurricane Alex, 2004, the right-front quadrant surge impact would hammer the barrier island from Cape Lookout to Cape Hatteras and beyond. Then as they eye continued northeast, the opposite left-rear quadrant would hammer the barrier island with wind and waves from behind—from Pamlico Sound.

With every significant storm, beaches are eroded and sand swept offshore; in major storms where the storm surge and waves overtop barrier bars, sand is also swept across barrier bars into back-bar lagoons. In such cases, barrier islands literally migrate landward. If we examine the record of shoreward migration just north of Cape Hatteras on the North Carolina Outer banks, an area with few houses and little beach hardening, we see that in the past 150 years, the island has migrated landward about 800 meters to the west. That migration of more than 5-meters-per-year average does not bode well for houses built on a narrow barrier bar. Another area a little farther north at South Nags Head, which has beach-front houses, shows beach migration that has progressively destroyed rows of houses, mostly during major storms. The slow but progressive rise of sea level, now about 42 cm per century in this area, heightens the problem of barrier island migration. One hundred years ago it was less than half of that.

Drawn from information in S. R. Riggs and others, 2008.

PAMLICO SOUND

ATLANTIC OCEAN

2003 1999

1974

1955

rebuilt positions of Hwy 12

1852 shoreline

0   0.8 km

▶ *Landward migration of part of the North Carolina Outer Banks, just north of Cape Hatteras. In the last 50 years, highway 12 has had to be moved landward several times. Plotted on 1993 USGS aerial photograph when the beach was approximately along the 1955 position of Highway 12 (green line). The average erosion rate in the last 150 years is 5 meters per year!*

U.S. Army Corps of Engineers.

▶ *A Nags Head home on stilts is now out on the beach front after Isabel.*

Mark Wolfe, FEMA.

▶ *The Corps of Engineers fills a breach from Hurricane Isabel at Hatteras, North Carolina. The pipe brings sand in a water slurry to fill the breach.*

Back-to-Back Hurricanes Amplify Flooding:
Hurricanes Dennis and Floyd, 1999 ▶

Hurricane Dennis, a Category 3 storm, moved up the East Coast offshore beginning August 30, wandered back and forth erratically for a few days, then made landfall as a much weaker tropical storm on September 4. Unlike many hurricanes that move quickly across the shoreline, minimizing the time available for wave damage, Dennis remained 125 kilometers off the North Carolina coast for days, generating big waves that progressively eroded the beaches through a dozen high tides. Sand overwashed the North Carolina Outer Banks at numerous points,

and erosion was equivalent to that of a Category 4 hurricane. A large frontal dune built in the 1930s for erosion control along 150 kilometers of beachfront had been progressively eroded by storms in the previous decade, pushing much of its sand landward during overwash. Hundreds of buildings now rest on the seaward-sloping beachfront, where they are directly exposed to storm attack, but the houses would have to be removed to replace the old dune. Hurricane Dennis removed sand from beneath some buildings and stranded others below high-tide level.

Ten days later, from September 14 to 18, Hurricane Floyd hit the coast from South Carolina to New Jersey. However, it was only a strong Category 2 hurricane in September, when it reached the mainland near Cape Fear, south of the Outer Banks in North Carolina. Buildings on Oak Island, largely destroyed in a hurricane in 1954, had been rebuilt in their original locations well back from the beach. After 45 years of landward beach migration, some 20 to 45 meters in total, half of them were again destroyed by Floyd. Its big waves came on top of a 1.5- to 2.5-meter storm surge.

At least 77 people died in the hurricane, most of those in flooding. Military helicopters and emergency personnel in boats rescued thousands of people from flooded houses, rooftops, and even trees. Damages were in the billions of dollars. Some 2.6 million people evacuated from Florida to the Carolinas, the largest such evacuation in U.S. history. At the peak of the evacuation, almost every east-to-west highway was jammed with traffic, some almost at a standstill. Some people took two-and-a-half hours to cover fifteen miles. Most gas stations and restaurants were closed.

USGS.

▶ *The early September landfall of Hurricane Dennis brought huge waves on top of a high storm surge that unmercifully battered the North Carolina Outer Banks.*

Dave Gatley, FEMA.

▶ *After Hurricane Floyd, this family returns to its almost completely submerged home in a community near the Tar River just north of Greenville, North Carolina.*

Dave Gatley, FEMA.

▶ *Sand eroded from under the shallowly anchored posts under this house, causing it to topple during Hurricane Floyd on September 17, 1999, at Long Beach, Oak Island, North Carolina.*

(continued)

Floyd covered a larger area and lasted longer than many stronger hurricanes, so its heavy rains lasted much longer. The storm dumped more than 50 centimeters of rain on coastal North Carolina, an area where Dennis had saturated the ground only two weeks earlier. Floyd's torrential rains had nowhere to go, so the water ran off the surface. Flood levels from eastern North Carolina to New Jersey rose above the 100-year flood stage. Twenty-four-hour rainfall totals reached 34 centimeters in Wilmington, North Carolina. Total rainfall along part of the coast reached 53.3 centimeters. As if that were not enough, with rivers still high, heavy runoff carried sediment, organic waste, and pesticides from farms; hazardous chemicals from industrial sites; and raw sewage into coastal lagoons and bays and onto beaches.

According to FEMA, this was the worst flood disaster ever recorded, to that date, in the southeastern states. For many people, the trauma and mess of the storms were just the beginning. Thousands of dollars of appliances, computers, stereo equipment, clothing, and other belongings were ruined beyond repair. Far more than half of the people flooded out did not have flood insurance. Common reactions were, "No one told me I was in a flood area," and, "I didn't know my insurance didn't cover floods."

Floods, Rejection of Foreign Help, and a Tragic Death Toll in an Extremely Poor Country
Myanmar (Burma) Cyclone, May 2–3, 2008 ▶

Atlantic hurricanes mostly lie in the belt of trade winds that carry them from east to west. In the Pacific, they reach land in Southeast Asia—the Philippines, Vietnam, Taiwan, southern Japan, and southeastern China. On occasion, the southern fringe of the cyclone belt crosses Cambodia, Thailand, Myanmar (Burma until renamed by the military dictatorship in 1989), or storms even begin in the Bay of Bengal in the northeastern Indian Ocean. Such was the case on April 27 when a cyclone grew and began moving northwestward away from Myanmar. Encountering dry air it weakened; then confined between weather ridges to the north and south, it began to move northeastward and strengthened. By May 2, 2008, Cyclone Nargis was a category 4 storm with peak winds of 215 km/hr, making landfall in the south coast of Myanmar. On land, it turned northeast to just northeast of Yangon (Rangoon), the largest city in Myanmar, where its winds slowed to 130 km/hr. Although Indian meteorologists warned the Myanmar government more than two days before landfall, the time available was much less than the five to seven days necessary for most people to evacuate.

 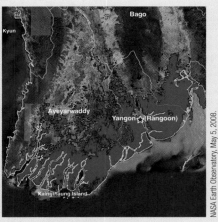

▶ **A.** *Cyclone Nargis path (warmer colors, higher intensity storm) and* **B.** *flooded area (blue).*

The southern, hardest-hit part of the country occupies the immense delta of the Irrawaddy River, an area about 200 km across. Because of the agricultural fertility of the delta, this, unfortunately, is also the area with the greatest population. In addition to the fierce winds, torrential rains and up to a 5-meter-high (13-foot) storm surge flooded most of the delta, with 2-meter storm waves on top of that. The surge moved as much as 50 km inland. Destruction was severe in this desperately poor country. Sewage mains broke, contaminating water supplies and destroying more than half of the rice crop. Most businesses and markets were closed in Yangon and food prices doubled or tripled. More than 138,000 died, making this the deadliest storm-related event since 1970; about 1.5 million were left homeless.

(continued)

Damages and deaths were made worse by loss of mangroves that were removed for firewood and to make room for agriculture. People had not experienced such a storm in their lifetimes, so there was no evacuation plan. Because all of the land is almost at sea level and transportation is minimal, people had no escape.

Almost immediately, help in the form of food, medicines, water purification systems, and shelter was offered from many countries, but Myanmar's military rulers, suspicious of outside influence, especially from the United States, initially refused aid. Finally on May 7–8, India was permitted to bring in more than 136 tons of relief supplies, including tents, blankets, and medicine, but Myanmar denied India's search and rescue teams and media people access to the hardest hit areas. A few days later, many other countries, especially the United States, United Kingdom, and Australia, sent large amounts of relief supplies. Of the people left homeless—in danger of disease or in need of food—a week after the storm, only one in ten had received any aid; after a month only 50 percent had received aid. Diarrhea was widespread. With all of the standing water, people worried about getting dengue fever. The only relief to the delta area was often provided by local Yangon volunteers and Buddhist monks. Many who lost their homes had no shelter from the continuing incessant rain; a few were fortunate to be offered a small ration of food and a bare floor to sleep on under the shelter of a monastery roof. The food that was available in delta towns was often provided by private businessmen and volunteers, without knowledge of the military authorities. The persistent rain was a mixed blessing; it provided the only source of clean drinking water.

Local activists were critical of the minimal government response to the tragedy. State-controlled media presented an upbeat message of widespread government relief efforts and near completion of rescue and aid. In contrast, images smuggled from the delta showed thousands of villagers that had received no aid, badly injured villagers, and numerous bloated bodies of people and animals still floating down rivers and in standing water. Relief workers accused the military of confiscating some international relief shipments and re-selling some to local shops, who in turn sold it to those in need.

▶ *A Myanmar woman in her destroyed house.*

The price of rice increased by 50 percent in Yangon, the largest city. The price of roofing tiles went up by a factor of 10. The military took credit for clearing roads and providing food and water, even though the work had actually been done by local people and monks, and supplies had been provided by foreign governments.

Finally, on May 23, three weeks after the cyclone, following protracted and widespread condemnation of resistance to aid by the military rulers, negotiations between the United Nations Secretary General and the Myanmar generals reached agreement to permit relief teams to enter the country, regardless of their nationality. But no military helicopters or personnel were permitted to distribute supplies. Much of the foreign aid was being confiscated and re-labeled as being provided by the Myanmar government. Some of it was sold on the black market and then appeared in local markets at extravagant prices. By the end of May, certain foreign-aid workers, mostly from Southeast Asian countries and from the U.N., were being allowed access to the hard-hit delta region. Access was also difficult because of problems with roads and too few trucks and boats. Much of the military government's concern about access to the delta stemmed from the fact that many inhabitants were ethnic minorities who had aggressively fought the army and that naval ship-based assistance could be used to support the insurgents. For years, the U.S. has imposed trade restrictions and sanctions against

Myanmar and been involved in efforts to help destabilize the military rule. Myanmar's army brutally put down pro-democracy demonstrations by Buddhist monks in September 2007. The military government fears that increased foreign presence would further resistance to their control.

With easing of government resistance to foreign distribution, the U.N. World Food Program and other aid organizations used lessons from past national disasters to organize distribution hubs, much like those of FedEx and UPS. They used local staff to set up about five improvised distribution centers in rented buildings in the affected area and utilized radio contact to Yangon and the internet to organize distribution of foreign aid from nearby Bangkok, Thailand. After the chaotic, competitive, and inefficient response to the 2004 Sumatra earthquake and tsunami, the U.N. organized a new system in which the World Food Program would be responsible for logistics and UNICEF (U.N. International Children's Emergency Fund) would be in charge of water and sanitation. Locals in Myanmar would provide distribution within the country.

By July 22, damage losses in Yangon and the Irrawaddy delta area were estimated at $1.7 billion for infrastructure and $2.3 billion in lost income. Damage estimates one year later were about $10 billion. Although the totals were not high by western standards, they were extreme for such a poor country.

Floods, Landslides, and a Huge Death Toll in Poor Countries
Hurricane Mitch, Nicaragua and Honduras ▶

Mitch formed as a tropical storm in the Caribbean Sea on October 21, 1998, then rapidly strengthened to Category 5 from October 26 through 28. Maximum sustained surface winds reached 290 kilometers per hour. Weakening, Mitch hovered near the north coast of Honduras before moving southwest and inland as a tropical storm on October 29. Waves north of Honduras probably reached heights of more than 13 meters. For the next two days, the storm continued westward over Honduras, Nicaragua, and then Guatemala, producing torrential rains and floods. Some mountainous areas received 30 to 60 centimeters of rain per day, with storm-total rainfalls as high as 1.9 meters. After several days, a weakened Mitch turned northeastward across Yucatán and over the Gulf of Mexico, where it strengthened to tropical storm status before pounding Florida's Key West on November 4–5.

Larger towns throughout Central America were concentrated in valleys near rivers, where businesses and homes were flooded, buried in mudflows, or washed away. Mitch was the deadliest hurricane in the western hemisphere since 1780; it killed more than 11,000 people in Central America, most in floods and mudflows. Whole villages disappeared in floodwaters and mudflows, and 18,000 people were never found. More than 3 million others lost their homes or were otherwise severely affected.

During and after the storm, there were critical shortages of food, medicine, and water. Dengue fever, malaria, cholera, and respiratory illnesses were widespread in the warm climate. Roads were impassable, and there were so few helicopters for distributing relief supplies that some areas did not receive help for more than a week. Survivors surrounded by mudflows had to wait days for the mud to dry enough for them to walk to rescuers.

NOAA.

▶ Hurricane Mitch, October 26, 1998, neared Honduras with winds at 289 kilometers per hour (180 miles per hour). Its cloudless eye is well defined. The storm track is shown as a green line that starts on the lower right and progresses to the upper right over this two-week period.

A large percentage of all crops were destroyed, including most of the banana and melon crops, as were shrimp farms in Honduras. Coffee and other crops in Nicaragua, Guatemala, and El Salvador suffered severe damage. Big agricultural companies that employed many of the people lost huge areas of farmland, so they had to lay off many employees. Damages came to more than $6.6 billion (in 2010 dollars) in Honduras and Nicaragua, two of the poorest countries in the world. It will take decades for these nations to get back to where they were before Hurricane Mitch.

USGS.

▶ Many homes in Honduras were buried by mudslides.

(continued)

Many people live and work on the floodplains of large rivers and are used to flooding during hurricane season almost every year. When a river rises, they pull their furniture onto the roof, leave their house, and set up temporary shelters along a nearby raised highway until the water recedes in a couple of weeks. This time, Mitch took away nearly everything and left behind several feet of mud.

There was one surprising outcome from Hurricane Mitch. Near the Pacific coast of Honduras, roads, bridges, adobe huts, sugar cane fields, and dairy farms that were damaged or destroyed by the hurricane's winds and floods in the formerly destitute area have been rebuilt and upgraded. Less than ten years later, the attention deriving from the destruction brought entrepreneurs that planted melon fields and ultramodern seafood farms and

built new concrete bungalows, motels, and textile factories. The profits, however, go to the foreign capitalists, lured by government and national labor union agreements that keep wages below the legal minimum for several years. The poor still struggle to survive.

Was the devastation of Hurricane Mitch a freak event that is not likely to repeat itself? Unfortunately, the combination of rapid population growth, widespread poverty, and lack of access to usable land makes Central America increasingly vulnerable to natural disasters. In 1995, 75 percent of Guatemala's and 50 percent of Nicaragua's populations were living in poverty, defined as living on less than $1 per day at 1985 prices. Those levels were worse after Hurricane Mitch. In 1974, 63 percent of Honduran farmers had access to only 6 percent of the farmable

land. Large corporate farms took over most of the fertile valley floors and gentle slopes for growing cotton, bananas, and irrigated crops and for raising livestock. Peasants were forced onto steep slopes where they cleared forests for agriculture, building materials, and firewood; this caused increased soil erosion and the addition of sediment to rivers. Those who cannot survive by growing crops on those slopes move to cities in search of jobs. Lacking access to safe building sites there, they build shelters on steep, landslide-prone slopes or flood-prone riverbanks.

Improved hurricane forecasts are of limited use for these poverty-stricken people. Even if warned of approaching storms, most are reluctant to abandon what little they own. In addition, they lack the resources to leave and have no way to survive if they do evacuate.

Unpredictable Behavior of Hurricanes
Florida Hurricanes of 2004 ▶

The 2004 hurricane season showed that even with our greatly improved prediction abilities, hurricanes still behave unpredictably. This was the first time that four hurricanes came onshore in a single state since 1886, when Texas was

the unfortunate victim. Charley arrived on August 13 as a Category 4 storm, Frances arrived on September 5 as a Category 2, Ivan made landfall on September 15 as a Category 4, and then finally Jeanne dealt the final blow as a

Category 3, following virtually the same track through Florida as Frances. In less than a month and a half, four storms wreaked havoc across much of Florida. In some places, later storms hammered areas destroyed from the earlier storms

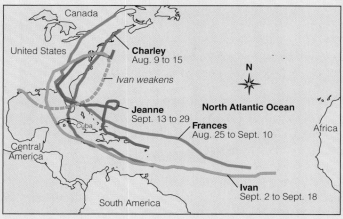

▶ **A.** *As Hurricane Ivan roared into the Gulf Coast on September 15, 2004, its eye crossed the west end of the Florida panhandle, while its strongest onshore winds, waves, and storm surge pounded areas just east of the eye (compare orange track of Ivan on map to right).* **B.** *Four major hurricanes decimated Florida in August and September 2004, striking many areas more than once.*

(continued)

FEMA.

▶ *Pre-engineered long-span metal building designed for use as a storm shelter proved ineffective during Hurricane Charley in Florida. Civic Center in Arcadia.*

FEMA.

▶ *This manufactured home provided no protection from a falling tree. Pine Island.*

as inhabitants were in the middle of repairs. This was the second costliest hurricane season on record; Charley inflicted $17.55 billion in damages, Frances $10.41 billion, Ivan $16.61 billion, Jeanne $8.07 billion (in 2010 dollars). Those hurricanes were, respectively, the 4th, 11th, 5th, and 13th costliest.

The first storm, Charley, began on August 9 as a tropical storm just north of South America, gradually strengthening as it curved from west to north. It approached the Gulf Coast of Florida as a Category 2 storm and was not expected to do severe damage. Much to the

surprise of forecasters, before landfall on August 13, just north of Fort Myers, Florida, it rapidly strengthened to a Category 4 hurricane with surface wind speeds of 232 kilometers per hour. In Florida, 22 people died, 12,000 buildings were destroyed, and 2 million customers lost electric power. It weakened to a Category 1 as it crossed Florida into the Atlantic, then made landfall again north of Charleston, South Carolina. It weakened over land to a tropical storm and then moved northeast up the coast.

Hurricane Frances reached Category 4 strength before hitting the Bahamas, then

paused for a day, long enough for 2.5 to 3 million people in Florida to evacuate. It weakened to a Category 2 before crossing Florida on September 5. It weakened further as it turned north through Georgia. In North Carolina and Virginia, it dumped about 37 centimeters of rain and caused heavy flooding. It spun off 137 tornadoes, and six people died in the United States.

Hurricane Ivan, locally dubbed Ivan the Terrible, grew in the warm waters of the Atlantic Ocean and strengthened to a Category 4 as it grazed the north coast of South America, then strengthened to a Category 5 with a core pressure of

J. Augustino, FEMA.

▶ *Many beachfront homes in Orange Beach, Alabama, vanished in Hurricane Ivan in September 2004, along with the beach that had attracted their owners. Even some behind the first row of buildings were destroyed; note the ground-level house to the right and behind the larger building on the right, in these before and after photos.*

(continued)

▶ *Ivan wrecked beachfront homes in Pensacola, Florida.*

910 millibars and winds of 260 kilometers per hour. It heavily damaged Grenada, Jamaica, and western Cuba before striking the Gulf Coast of Alabama at 2 a.m. on September 16, as a Category 3 storm.

In response to warnings that the hurricane could hit New Orleans, more than half of the city's residents evacuated. In Mobile, Alabama, about one-third of the residents evacuated, but many people were delayed up to 12 hours because of traffic congestion on highways. There

it caused major damage before turning northeast, dumping heavy rains, and weakening. Looping east into the Atlantic, it turned south and then west across the southern part of Florida, into the Gulf of Mexico, and again headed inland as a weak tropical storm near the Louisiana–Texas border. Tornadoes spun off from the leading edge of the hurricane, causing severe localized damage in northwestern-most Florida—more than 100 tornadoes in all of the eastern United States. Thirty-nine people died in

Grenada, 17 in Jamaica, 14 in Florida, and 8 in North Carolina. Damages in the Caribbean area reached $3 billion. Much of the eastbound lanes of the Interstate 10 bridge across a wide bay near Pensacola, Florida, were destroyed. Rains brought major flooding to Georgia, the western Carolinas, and Pennsylvania.

Jeanne hammered the same area of Florida only three weeks after Frances. At least 3,132 people died, mostly from Jeanne in Haiti.

Critical View

This building was damaged during Hurricane Charley in 2004.

FEMA.

1. What process or processes led to destruction of this building?
2. How can you tell that a storm surge was not the primary cause?
3. What can you see that made the building especially vulnerable to this type of destruction?

B Damage on the Texas Gulf Coast from Hurricane Ike.

Hyndman.

1. What process led to removal of the sand from under this concrete foundation?
2. What process or processes led to destruction of the house?
3. What would have made the house less susceptible to destruction?

C This building was damaged during Hurricane Charley in 2004, in Punta Gorda, Florida.

Andrea Booher, FEMA.

1. What happened to this large beach-front building?
2. What could have prevented the damage?

D This home in New Orleans was damaged during Hurricane Katrina in 2005.

Andrea Booher, FEMA.

1. Why is beach grass and other debris piled on the edge of this roof? What does it indicate?
2. What indicates that it was not carried there by the wind?

E Hurricane Isabel, 2003, Nags Head, North Carolina.

U.S. Army Corps of Engineers.

1. Why was this walkway and stairs elevated above the beach?
2. What happened to leave the walkway where it is?
3. What does the current state of this beach mean for future hazards to these homes?

F This house is located tens of kilometers south of New Orleans, just after Hurricane Katrina.

Donald Hyndman.

1. What are the main damages to this house and from what cause?
2. This house appears much less damaged than most of those in New Orleans, even though it is much closer to the Gulf of Mexico. Explain why that makes sense.

G This photo was taken on the Texas Gulf Coast after Hurricane Ike.

Hyndman.

1. Why are all of these boards left here in a long pile?
2. The large gray surface is the roof of a house. What is it doing there?
3. What does it indicate about the house?

H Hurricane Charley, 2004. Punta Gorda, FL.

Florida Division of Emergency Management.

1. What process or processes led to destruction of this building?
2. How can you tell that a storm surge was not the primary cause?
3. What can you see that made the building especially vulnerable to this type of destruction?

Chapter Review

Key Points

Hurricanes

- Hurricanes, typhoons, and cyclones are all major storms that circulate around a low-pressure system with winds from 120 to more than 260 kilometers per hour.

- Clear, calm air in the low-pressure eye is surrounded by the highest winds and stormy skies. **FIGURES 14-1** and **14-2**.

- Hurricanes that affect the southeastern United States form as atmospheric lows over warm subtropical water. They grow off the west coast of Africa, then move westward with the trade winds. Most hurricanes occur between August and October because it takes until late summer to warm the ocean sufficiently. They strengthen over warmer water and weaken over cool water or land.

- The strongest hurricanes, Category 5 on the Saffir-Simpson Hurricane Scale, have the lowest atmospheric pressure (less than 920 millibars) and the highest wind speeds (more than 249 kilometers per hour). **TABLE 14-1**.

- Hurricanes' impacts are greatest from North Carolina to Florida to Texas. Annual damages run from hundreds of millions to billions of dollars.

Storm Damages

- Hurricanes and other major storms cause severe coastal damage. Beaches and dunes are eroded, and buildings are severely damaged by wind, waves, and flooding. In addition to buildings and bridges, damages include the deaths of farm animals, agricultural damage, contaminated drinking water, and landslides.

- Damages from a hurricane depend on its path compared with shore orientation, presence of bays, forward speed of the hurricane, and the height of dunes and coastal vegetation.

- Storm surges, as much as 7.3 meters high and 160 kilometers wide, result from a combination of low atmospheric pressure that permits the rise of sea level and prolonged winds pushing the sea into a broad mound. **By the Numbers 14-1** and **14-2**.

- Surge hazards include significant rise of sea level and waves on top of the higher sea level. The surge and associated winds are concentrated in the northeast forward quadrant of a hurricane because of wind directions. **FIGURE 14-9**.

- High waves have much more energy and erode the beaches and dunes lower and to a flatter profile, especially if the waves are not slowed by shallow water offshore. **FIGURE 14-16**.

- Wind damages include blown-in windows, doors, and walls; lifted-off roofs; blown-down trees and power lines; and flying debris. **FIGURES 14-20** and **14-21**.

- Rain and flooding from hurricanes can be greater with large-diameter, slower-moving storms.

- Thousands of people in poor mountainous countries such as in Central America die in floods and landslides triggered by hurricanes because their homes are poorly constructed and situated on floodplains and unstable steep slopes.

- Thousands of people living on low-lying deltas of major rivers die in floods and storm surges.

Hurricane Prediction and Planning

- Hurricane predictions and warnings include the time of arrival, location, and event magnitude. The National Hurricane Warning Center tries to provide a 12-hour warning, but evacuations can take as many as 30 hours.

- Many people think they can evacuate quickly and leave too late. Storm surges and downed trees and power lines often close roads.

Managing Future Damages

- The National Flood Insurance Program requires that buildings in low-elevation coastal areas be landward of mean high tide and raised above heights that could be impacted by 100-year floods, including those imposed by storm surges.

- The nature and quality of building construction have a major effect on damages. Floors, walls, and roofs need to be well anchored to one another, and buildings should be well attached to deeply anchored stilts.

Extratropical Cyclones and Nor'easters

- Nor'easters and other extratropical cyclones are similar to hurricanes except that they form in winter, lack a distinct eye, are not circular, and spread out over a large area. Like hurricanes, they are characterized by high winds, waves, and storm surges.

Key Terms

extratropical cyclones, p. 439

eye, p. 416

fetch, p. 421

hurricane, p. 416

hurricane warning, p. 433

hurricane watch, p. 433

Nor'Easter, p. 439

Saffir-Simpson Hurricane Scale, p. 417

storm surges, p. 421

tropical cyclones, p. 416

typhoon, p. 416

Questions for Review

1. What causes a tropical cyclone or hurricane? Where does a hurricane get all of its energy?

2. Where do hurricanes that strike North America originate? Why there? Why do they track toward North America?

3. Why are coastal populations so vulnerable to excessive damage (other than the fact that they live on the coast)?

4. Where in a hurricane is the atmospheric pressure lowest, and approximately how low might that be?

5. When is hurricane season (which months)? Why then?

6. What two main factors cause increased height of a storm surge?

7. What effects does the wind have on buildings during a hurricane?

8. What effects do the higher waves of hurricanes have on a coast?

9. If the forward speed of a hurricane is greater, what negative effect does that have? What positive effect does it have?

10. If the eye of a west-moving hurricane were to go right over Charleston, South Carolina, where would the greatest damage be?

11. What is the difference in damage if a hurricane closely follows another hurricane—for example, by a week? Why?

12. Why is there more coastal damage during a hurricane if sand dunes are lower?

13. What shape of roof is most susceptible to being lifted off by a hurricane? Why?

14. Why is it so important to cover windows and doors with plywood or shutters?

15. What factors contribute to uncertainty in hurricane predictions?

16. What steps should you take to prepare for a hurricane?

17. Characterize the differences between a hurricane and a Nor'Easter.

Discussion Questions

1. How is climate change expected to affect hurricane damages? How can this play a role in the discussion of costs to reduce carbon emissions?

2. Examine FIGURE 14-4b and 14-6 and explain why the Atlantic coast of northern Florida and Georgia has fewer hurricane strikes than coastal areas farther north and south.

3. Why do developers, builders, local governments, and many members of the public oppose higher standards for stronger houses?

15 Thunderstorms and Tornadoes

The EF-5 tornado that ripped through Greensburg, Kansas, destroyed everything in 2007.

Michael Raphael, FEMA.

Twister Demolishes Kansas Town

The small town of Greensburg in southwestern Kansas was effectively wiped off the map by a powerful tornado just before 10 p.m. on May 4, 2007. It was an EF-5, the strongest tornado category, the first since the Oklahoma outbreak of May 3, 1999. The National Weather Service managed to provide 30 minutes of warning that a major tornado was headed for Greensburg; then with a more precise track, they activated tornado sirens and an emergency message ten to fifteen minutes before it struck, for people to take shelter immediately. In spite of this, nine were killed in Greensburg. The 2.7-kilometer-wide tornado decimated the town in fifteen to twenty minutes (**FIGURE 15-1**), following a northeastward path on the ground for 35 kilometers. Nearly an hour later the twister touched down again 46 kilometers to the northeast, causing fatal injuries to a sheriff in his patrol car. Winds were estimated at between 320 and 335 kilometers per hour. The storm developed when a major low-pressure system moving east from northern Nevada

Tornadoes

and Utah collided with a stream of moist air moving north from the Gulf of Mexico into Oklahoma and Kansas. The resulting cold front spun off days of severe storms.

The May 4 tornado demolished every building on the main street in Greensburg, a town of 1,600. It stripped the branches from trees and flattened most homes, churches, and other buildings. Even solid buildings were destroyed, including the high school, which was built in 1939 using two layers of bricks and mortar on the outside, with concrete blocks and mortar inside. Greensburg's emergency vehicles lay buried under their collapsed building (**FIGURE 15-2**). Searchers dug through the debris to find people trapped in their basements. One survivor was pulled from under rubble two days later. A tank of anhydrous ammonia, used in fertilizer, ruptured during cleanup, requiring further evacuation of part of the town. Many residents suggested that the town itself may not survive.

Rebuilding first required repairing utility poles, lines, and meters, along with water and sewer lines. First they had to be dug out from under the piles of debris. Damage estimates reached $153 million.

FIGURE 15-1 TORNADO DAMAGE

A deadly tornado swept through and destroyed nearly all of Greensburg, Kansas.

FIGURE 15-2 DESTRUCTION OF KEY BUILDINGS

Greensburg's emergency vehicles lie buried in rubble from the tornado of May 4, 2007.

Thunderstorms

Thunderstorms, as measured by the density of lightning strikes, are most common in latitudes near the equator, such as central Africa and the rain forests of Brazil (**FIGURE 15-3**). For its latitude, the United States has an unusually large number of lightning strikes and severe thunderstorms. These storms are most common from Florida and the southeastern United States through the Midwest because of the abundant moisture in the atmosphere that flows north from the warm waters of the Gulf of Mexico.

Thunderstorms form as unstable, warm and moist air rapidly rises into colder air and condenses. As water vapor condenses, it releases heat. Because warm air is less dense than cold air, this added heat causes the rising air to continue to rise in an updraft. This eventually causes a region of falling rain in an outflow area of the storm when water droplets get large enough through collisions. If updrafts push air high enough into the atmosphere, the water droplets freeze in the tops of **cumulonimbus clouds**; these are the tall clouds that rise to high altitudes and spread to form wide, flat, anvil-shaped tops (**FIGURE 15-4**). This is where lightning and thunder commonly develop.

FIGURE 15-3 THUNDERSTORM HAZARD AREAS

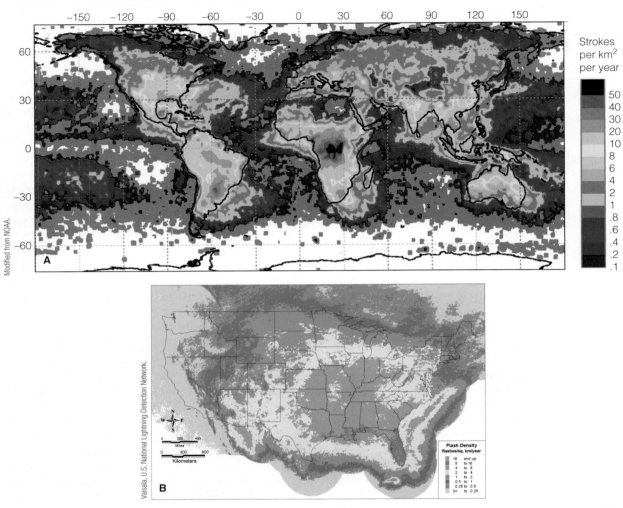

A. This worldwide map shows the average density of annual lightning flashes per square kilometer. **B.** North American lightning distribution from 1996–2005.

FIGURE 15-4 CUMULONIMBUS CLOUD

This type of classic cumulonimbus anvil buildup over Utah on August 26, 2004, is where thunderstorms and lightning typically develop.

Cold air pushing under warm, moist air along a cold front is a common triggering mechanism for these storm systems, as the warm humid air is forced to rapidly rise over the advancing cold air. Isolated areas of rising humid air from localized afternoon heating, or warm and moist air rising against a mountain front or pushing over cold air near the ground can have similar effects. Individual thunderstorms average 24 kilometers across, but coherent lines of thunderstorm systems can travel for more than 1,000 kilometers. Lines of thunderstorms commonly appear in a northeast-trending belt from Texas to the Ohio River valley. Cold fronts from the northern plains states interact with warm moist air from the Gulf of Mexico along that line, so the front and its line of storms move slowly east.

Thunderstorms produce several different hazards, including lightning, strong winds and hail. Strong winds can down trees, power lines, and buildings. Severe thunderstorms cause numerous wildfires, and sometimes large damaging hail and tornadoes. Of the more than 100,000 thunderstorms in the United States each year, the National Weather Service classifies about 10,000 as severe. Those severe storms spawn up to 1,000 tornadoes each year. The weather service classifies a storm as severe if its winds reach 93 kilometers per hour, it spawns a tornado, or it drops hail larger than 1.9 centimeters in diameter. Flash flooding from thunderstorms causes more than 140 fatalities per year (see Chapter 11).

Lightning

Lightning results from a strong separation of an electric charge that builds up between the top and bottom of cumulonimbus clouds. Atmospheric scientists commonly believe that this **charge separation** increases when updrafts carry water droplets and ice particles toward the top of cumulonimbus clouds where they collide with downward-moving ice particles or hail. The smaller, upward-moving particles tend to acquire a positive charge, while the larger, downward-moving particles acquire a negative charge. Thus, the top of the cloud tends to carry a strong positive charge, while the lower part carries a strong negative charge (**FIGURE 15-5**). This process is similar—though much larger—to the effects of static electricity a person can generate by dragging one's feet on carpet during dry weather—this action produces a charge that is discharged as a spark when one gets near a conductive object.

The strong negative charges near the bottom of these clouds attract positive charges toward the ground surface under the charged clouds, especially to tall objects such as buildings, trees, and radio towers. Thus, there is an enormous electrical separation, or potential, between different parts of the cloud and between the cloud and ground. This can amount to millions of volts; eventually, the electrical resistance in the air cannot keep these opposite charges apart, and the positive and negative regions join with an electrical lightning strike.

Because negative and positive charges attract one another, a negative electrical charge may jump to the positive-charged cloud top or to the positive-charged ground. Air is a poor conductor of electricity, but if the opposite charges are strong enough, they will eventually connect. Cloud-to-ground lightning is generated when charged ions in a thundercloud discharge to the best conducting location on the ground—locations such as tall trees, power poles, and people standing, for example in open areas, fields or golf courses. Negatively charged **step leaders** angle their way toward the ground as the charge separation becomes large enough to pull electrons from atoms. When this occurs, a conductive path is created that in turn causes a chain reaction of downward-moving electrons. These leaders fork as they find different paths toward the ground (**FIGURE 15-6**). Depending on the location and strength of negative and positive charges in a cloud and in the ground, lightning may strike at a considerable distance from the storm cloud.

If the hairs on your head are ever pulled upward by what feels like a static charge during a thunderstorm, you are at high risk of being struck by lightning. When one of the pairs of leaders connects, a massive negative charge follows the conductive path of the leader stroke from the cloud to the ground. This is followed by a bright *return stroke* moving back upward to the cloud along the one established connection between the cloud and ground. The enormous power of the lightning stroke instantly heats the air in the surrounding channel to extreme temperatures approximating 28,000°C (50,000°F). The accompanying expansion of

FIGURE 15-5 OPPOSITE CHARGES CAUSE LIGHTNING

In a thunderstorm, the lighter, positive-charged rain droplets and ice particles rise to the top of a cloud while the heavier, negative-charged particles sink to the cloud's base. The ground has a positive charge. In a lightning strike, the negative charge in the cloud base jumps to join the positive charge on the ground.

FIGURE 15-6 FORKED LEADERS

Lightning often follows a very irregular path between clouds and the ground, as shown by these lightning strokes in Tucson, Arizona.

the air at supersonic speed causes the boom that we hear as **thunder**.

The lightning channel itself appears to be only 2 or 3 centimeters in diameter, based on holes produced in fiberglass screens and long narrow tubes fused in loose sand. In fewer cases, lightning will strike from the ground to the base of a cloud; this can be recognized as an upwardly forking lightning stroke (**FIGURE 15-7**) rather than the more common downward forks observed in cloud-to-ground strokes. Lightning also strikes from cloud to cloud to equalize its charges, although there is little hazard associated with such strokes.

Major insurance companies reported in 2007 that lightning-associated claims in the previous six years rose 77 percent. Costs escalated because of the growing number of electronic devices in people's homes. The voltage surge in a home's wiring from a lightning strike can destroy personal computers, HD TV sets, DVD players, game consoles, other devices, and heating systems. U.S. lightning strikes cause about $144 million in direct property damage and 6,100 house fires every year. Lightning-protection systems for a whole home or a surge suppressor installed at the main circuit-breaker panel can reduce the excess voltage. Individual surge suppressors that you plug into an electrical outlet should have a Suppressed Voltage Rating of 330 or less.

Downbursts

As well as lightning, thunderstorms can also bring destructive winds. Small areas of rapidly descending air, called **downbursts** or *microbursts*, can develop in strong thunderstorms (**FIGURE 15-8**). Downbursts associated with thunderstorms can be wet or dry. A falling column of air driven by heavy rainfall reaches the ground and spreads out in winds of up to 240 kilometers per hour that move in straight lines radially outward from the center of the downburst. Downburst winds as fast as 200 kilometers per hour and microburst winds (a downburst with less than a 4-kilometer radius) are caused by a descending mass of cold air, sometimes accompanied by rain.

These severe downdraft winds pose major threats to aircraft during takeoffs and landings because they cause **wind shear**, which results in a plane losing the lift from its wings and plummeting toward the ground. A plane flying into the vertical wind shear at the edge of a downburst experiences an increase in headwind and a rapid increase in lift on its wings. If the pilots are not aware of wind shear as the cause, a common reaction is to decrease engine power and airspeed. Then beyond the plane of shear, the plane experiences downdraft and reduced lift. During takeoff or landing, when the plane is close to the ground and its

FIGURE 15-7 GROUND-TO-CLOUD LIGHTNING

David Hyndman.

This ground-to-cloud lightning stroke was observed near East Lansing, Michigan, in spring 2004.

FIGURE 15-8 DOWNBURSTS

A. An invisible, dry microburst kicks up dust as the mass of descending air strikes the ground and radiates outward (downward air trajectories marked in red). **B.** An aircraft flying into a microburst as it approaches an airport may reduce speed in response to sudden lift, possibly leading to a crash.

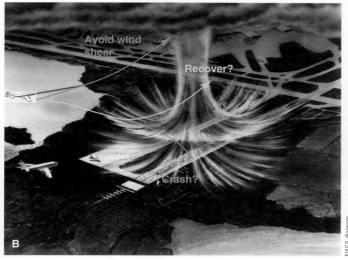

Avoid wind shear

Recover?

Crash?

NASA diagram.

airspeed is low, there may not be time to increase speed enough to escape the downdraft, and the plane may crash.

Fatal microburst aircraft crashes include a July 9, 1982, Pan American flight in New Orleans; an August 2, 1985, Delta flight in Dallas; a July 2, 1994, U.S. Air flight in Charlotte, NC; and a September 16, 2007, One-Two-Go flight at Phuket, Thailand. Once Dr. Tetsuya (Ted) Fujita proved this phenomenon and circulated the information to pilots and weather professionals, the likelihood of downbursts causing airplane crashes has been greatly reduced—though not eliminated.

When these descending air masses hit the ground, they wreak damage people sometimes mistake as having been caused by a tornado. On close examination, downburst damage will show evidence of straight-line winds: trees and other objects will lie in straight lines pointing away from the area where a downburst hit the ground (**FIGURE 15-9**). This differs from the rotational damage observed after tornadoes, where debris lies at many angles due to the inward flowing, rotating winds.

Other "straight-line winds" can accompany the gust front of a thunderstorm or where a mountain range forces winds to rise against stable layers of air where they must squeeze into a thinner zone and speed up. Mount Washington, in the Presidential Range of northern New Hampshire, is such a place. From November to April, two-thirds of the days are likely to have hurricane-force winds. The record there was 372 km/hour, higher than a category 5 hurricane and equal to the wind gusts in a strong EF 5 tornado. Such winds are partly in response to converging storm tracks from the Gulf of Mexico, Pacific Northwest, and the Atlantic Ocean.

Hail

Hail is another weather phenomenon associated with thunderstorms. Hail causes $2.9 billion in annual damages to cars, roofs, crops, and livestock (**FIGURE 15-10**). **Hailstones** grow when warm and humid air in a thunderstorm rises

FIGURE 15-10 HAIL

A. A violent storm over Socorro, New Mexico, on October 5, 2004, unleashed hailstones, many larger than golf balls and some 7 centimeters in diameter. **B.** Most cars caught out in the open suffered severe denting and broken windows. In some cases, hailstones went right through car roofs and fenders.

FIGURE 15-9 DOWNBURST DAMAGE

Downburst winds in Bloomer, Wisconsin, blew these trees down on July 30, 1977.

rapidly into the upper atmosphere and freezes. Tiny ice crystals waft up and down in the strong updrafts, collecting more and more ice until they are heavy enough to overcome the updrafts and fall to the ground. The largest hailstones can be larger than a baseball and are produced in the most violent storms. Hailstorms are most frequent in late spring and early summer, especially April to July, when the jet stream migrates northward across the Great Plains. The extreme temperature drop from the ground surface up into the jet stream promotes the strong updraft winds. Hailstorms are most common in the plains of northern Colorado and southeastern Wyoming but rare in coastal areas. Hail suppression using supercooled water containing silver iodide nuclei has been used successfully to reduce crop damage; however, this practice was discontinued in the United States in the early 1970s because of environmental concerns.

Safety during Thunderstorms

Thunderstorms can be deadly. Among weather-related events, only heat and floods cause more deaths. In 1940, lightning strikes in the United States killed about 400 people per year. The number dropped continuously to an average of 44 people per year from 1997 to 2006, likely due in part to increased awareness of the hazard and forecasts. Twice as many deaths occurred in Florida than in any other state. The maximum number of deaths from lightning strikes occurs at around 4 p.m., with a significant increase on Sunday—presumably because more people, such as golfers, are outside, engaging in leisure activities.

When someone is fatally struck by lightning, the immediate cause of death is heart attack, with deep burns at lightning entry and exit points. Seventy percent of lightning strike survivors have residual effects, including damage to nerves, the brain, vision, and hearing.

To reduce your risk of being struck, gauge your distance from the lightning. Lightning is visible before you hear the clap of thunder. The speed of light is almost instantaneous, whereas sound takes roughly three seconds to travel a kilometer. Thus, the time between seeing the lightning and hearing the thunder is the time it takes for the sound to get to you. Every three seconds means a kilometer between you and the lightning; if the time difference is 12 seconds, then the lightning is about 4 kilometers away. Even if there is blue sky above and the threatening cloud is several kilometers distant, lightning can arc to the side, well away from its dark origin cloud. The National Weather Service recommends that you take cover if you hear thunder within 30 seconds of a lightning strike and stay in a safe place until you do not see lightning flash for at least 30 minutes—the "30-30 rule."

Danger from lightning strikes can be minimized by observing the following precautions during a thunderstorm:

- Take cover in an enclosed building; its metal plumbing and wiring will conduct the electrical charge around you to the ground. A picnic shelter is not safe.

- Do not touch anything that is plugged in, including video games. Do not use a phone with a cord; cordless phones and cell phones are okay. One of us was struck by a lightning charge through a corded phone—not something you want to experience!

- Water conducts electricity. Do not take a shower or bath or wash dishes. Stay away from open water.

- Stay away from open fields. Lightning can travel along the ground for about 20 meters and kill or burn you severely.

- Do not take refuge in a cave. People on high peaks during thunderstorms sometimes take refuge in a cave or under a deep rock overhang—bad move. Although it may provide some protection from rain or hail, an electrical charge from lightning can surround the inside of a cave, placing an occupant in greater danger than they would have been outside on an open slope.

- Stay away from trees, power poles, or other tall objects. If there are tall trees nearby, staying under low bushes well away from the trees is a better plan.

- Stay away from metal objects, such as fences, golf clubs, golf carts, umbrellas, farm machinery, and the outsides of cars and trucks (**FIGURE 15-11**). Be aware of overhead power lines; people are sometimes electrocuted when they move lengths of sprinkler pipe or transport farm equipment.

- If you are driving, pull over and stop in a safe spot. Stay inside a car with the windows rolled up and do not touch any metal, including the steering wheel, gearshift, or radio. The safety of a car is in the metal shield around you, not in any insulation from the tires. The rubber tires of a motorcycle or bicycle do not insulate you from the ground; water on them conducts electricity.

FIGURE 15-11 DEADLY LIGHTNING

NOAA.

Reality can be gruesome. These cows were probably spooked by thunder and ran against the barbed wire fence, where they were electrocuted by a later lightning strike. Note that they were at the base of a hill but out in the open.

There are signs that you may be in imminent danger of being struck by lightning. The sky may turn a strange shade of olive gray. Storm clouds several kilometers away may indicate a lightning strike about to happen near you. People report that shortly before a strike, the surrounding air begins to hiss, buzz, and sometimes crackle, like you sometimes hear when standing under high-tension lines; strange blue sparks may leap from rocks and people's heads; hair may stand on end. Even if it strikes a few meters away, the charge through the ground may, if it does not kill you, do serious damage. People report a deafening explosion as the sky turns blinding white and a blast of energy exploding through their chest as they pass out.

If you are trapped in the open, and your skin tingles, or your hair stands on end, you are in immediate danger of being struck. Immediately crouch low, on the balls of your feet with your heels tightly together to minimize contact with electrified ground; much of the charge from the ground that enters your feet will pass from one heel to the other and back into the ground without passing through your body. Do not lie down because that increases your contact with the ground. Remember that the ground conducts electricity.

Most lightning-strike victims can survive with medical help such as the application of CPR. A lightning-strike victim does not carry an electrical charge or endanger others. However, rescuers should not put themselves in danger of another strike.

Tornadoes

Thunderstorms and other storm systems sometimes generate **tornadoes**—narrow funnels of rapidly rotating intense wind. Tornadoes are nature's most violent storms and the most significant natural hazard in much of the Midwestern United States. They often form in the right-forward quadrant of hurricanes, in areas where the wind shear is most significant. Even weak hurricanes spawn tornadoes, sometimes dozens of them.

The United States has an unusually high number of large and damaging tornadoes relative to the rest of the world, over 1,000 per year on average. Canada is second, with only about 100 per year, though some are strong; an F4 tornado touched down west of Winnipeg, Manitoba, in 2007. Europe experiences a moderate number.

The storms that lead to tornadoes are created through the collision of warm, humid air moving north from the Gulf of Mexico with cold air moving south from Canada. Because there is no major east-west mountain range to keep these air masses apart, they collide across the Southeastern and Midwestern United States. These collisions of contrasting air masses cause intense thunderstorms that sometimes turn into deadly tornadoes.

The average number of tornadoes is highest in Texas, followed by Oklahoma, Kansas, Florida, and Nebraska. **Tornado Alley**, covering parts of Texas, Oklahoma, Arkansas, Missouri, and Kansas, marks the belt where cold air from the north collides frequently in the spring with warm, humid air from the Gulf of Mexico, forming intense thunderstorms and tornadoes (**FIGURE 15-12**). Tornadoes are rare in the western and northeastern states. An individual **tornado outbreak**—that is, a series of tornadoes spawned by a group of storms—has killed as many as several hundred people and covered as many as thirteen states (**Table 15-1**).

The largest known tornado outbreak to date started just after noon on April 3, 1974. This **superoutbreak** lasted more than seventeen hours, killed 315 people, and injured 5,484 others. The map of the storm tracks (**FIGURE 15-13**, p. 474) shows that several of these tornadoes ended in downbursts.

One of the most severe tornado outbreaks in recent years occurred on May 3, 1999, in central Oklahoma

FIGURE 15-12 TORNADO RISK IN NORTH AMERICA

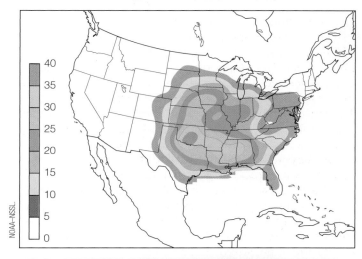

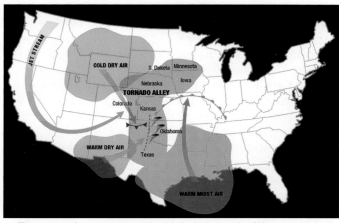

A. The areas of greatest tornado risk include much of the eastern half of the United States (for F2 and greater; data from 1921–1995). The scale indicates the number of significant tornado days per year.
B. Tornado Alley is influenced by warm, moist air from the Gulf of Mexico that collides with warm, dry air to the west and cold, dry air to the northwest.

Table 15-1		Deadliest U.S. Tornado Outbreaks On Record*		
NAME OR LOCATION	DATE	NUMBER OF TORNADOES (NUMBER OF STATES AFFECTED)	DEATHS	ESTIMATED DAMAGE IN MILLIONS (2007 $)
Tristate: MO, IL, IN	March 18, 1925	7 (6)	695	196
Tupelo-Gainesville (MS, GA)	April 5–6, 1936	17 (5)	419	242
Northern Alabama	March 21–22,1932	33 (7)	334	49
Superoutbreak, E. US and Ontario	April 3–4, 1974	148 (13)	330	1,860
LA, MS, AL, GA	April 24–25,1908	18 (5)	310	53
St. Louis, MO	May 27, 1896	18 (3)	306	380
Palm Sunday	April 11–12,1965	51 (6)	256	695
Flint: MI; MA, OH, NE	June 8-9, 1953	10 (4)	247	554
AR, TN	March 21-22, 1952	28 (4)	204	57
Easter Sunday	March 23, 1913	8 (3)	181	105
PA, OH, NY, Ontario	May 31, 1985	41 (3)	75	1,142
Carolinas	March 28, 1984	22 (2)	57	507
AL, AR, KY, MS, TN	Feb. 5–6, 2008	62 (5)	59	Unknown
OK-KS (F5)	May 3–4, 1999	76 (2)	49	1,600
Southeastern United States	March 27, 1994	2 (2)	42	271
Jarrell, TX (F5)	May 27, 1997	1 (1)	30	203
KS, MO, AL, GA	Feb. 28-March 1 2007	57 (4)	20	520

*From FEMA, NOAA, and other sources.

FIGURE 15-13 SUPEROUTBREAK OF 1974

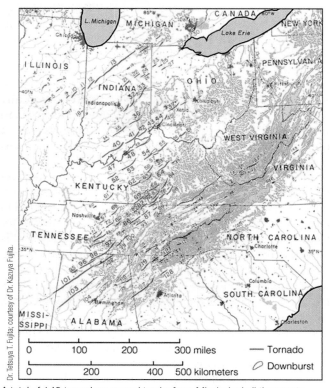

Dr. Tetsuya T. Fujita; courtesy of Dr. Kazuya Fujita.

A total of 148 tornadoes scored tracks from Mississippi all the way to Ontario and New York, with an overall storm path length of 4,180 kilometers.

(**FIGURE 15-14**). Eight storms producing 58 tornadoes moved northeastward along a 110-kilometer-wide swath through Oklahoma City. Eighteen more tornadoes continued up through Kansas. Individual tornadoes changed in strength as they churned northeast. Fifty-nine people were killed, and damage reached $800 million.

Tornado season varies, depending on location. The number of tornadoes in Mississippi reaches a maximum in April, with a secondary peak in November (**FIGURE 15-15**). An unusual tornado outbreak on February 5–6, 2008, killed at least 55 people in southeastern states (see TABLE 15-1). Farther north, the maximum occurs in May, and in Minnesota it is in June. These are the periods that people should be particularly vigilant for tornadoes. At these northern latitudes, tornadoes are virtually absent from November to February.

Even late-season tornadoes can be deadly, especially for those in mobile homes. Just after 6:30 a.m. on November 16, 2006, and without warning, a thunderstorm spawned a tornado that killed eight people in mobile homes west of Wilmington, North Carolina. At about 10:30 p.m. on October 19, 2007, a tornado just east of Lansing, Michigan, blew a mobile home into a nearby pond, drowning its two occupants. On December 29, 2006, a line of severe storms east of Waco, Texas, spawned tornadoes that killed one man, and on February 1, 2007, tornados killed 19 northwest of Orlando, Florida.

Most, though not all, tornadoes track toward the northeast. Storm chasers, individuals who are trained to gather storm data at close hand, know to approach a tornado

FIGURE 15-14 OKLAHOMA CITY TORNADO OUTBREAK

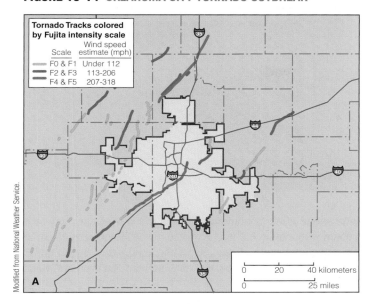

Modified from National Weather Service.

Tornado Tracks colored by Fujita intensity scale

Scale	Wind speed estimate (mph)
F0 & F1	Under 112
F2 & F3	113-206
F4 & F5	207-318

A

A. Booher, FEMA.

B

A. This map of the May 3, 1999, tornadoes shows their paths and intensities around Oklahoma City. **B.** One of these tornadoes threw these cars into a crumpled heap.

FIGURE 15-15 TORNADO SEASON

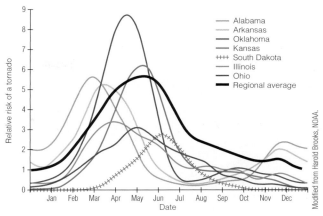

Modified from Harold Brooks, NOAA.

Tornado occurrence rises quickly in springtime to a peak in May and then gradually falls off through early winter (data from 1980–1999). The heavy black line is the average for the United States.

---

from the south to southwest directions so they will not be in its path. They also know it is safer to chase them on the flat plains rather than along the Gulf Coast, where the lower cloud base can hide the funnel from view.

Tornado Development

Tornadoes generally form when there is a shear in wind directions, such as surface winds approaching from the southeast with winds from the west higher in the atmosphere. Such a shear can create a roll of horizontal currents in a thunderstorm as warm and humid air rises over advancing cold air (**FIGURE 15-16**, p. 476). These currents, rolling on a horizontal axis, are dragged into a vertical rotation axis by an updraft in the thunderstorm to form a *rotation cell* up to 10 kilometers wide. This cell sags below the cloud base to form a distinctive slowly rotating **wall cloud**, an ominous

sight that is the most obvious danger sign for the imminent formation of a tornado (**FIGURE 15-17A**, p. 476). **Mammatus clouds** can be another potential danger sign, where groups of rounded pouches sag down from the cloud base (**FIGURE 15-17B**, p. 476).

A *supercell*, a giant thundercloud characterized by a rapidly rising and rotating mass of warm air that spreads out at its top as a broad mushroom, can produce severe thunderstorms, torrential rain, and tornadoes (**FIGURE 15-18**, p. 476). A slowly rotating, steep-sided wall cloud may descend from the nearly horizontal base of a main cloud. A smaller and more rapidly rotating funnel cloud, or *mesocyclone*, may form within the wall cloud or, less commonly, adjacent to it. After the heaviest rain and hail has passed, rotation within the mesocyclone may accelerate and tighten, typically rotating counterclockwise. If a funnel cloud descends to touch the ground, it becomes a tornado. Supercells are not common but often remain in place for extended times and produce the most violent storms.

Tornadoes derive their energy from the **latent heat** released when water vapor in the atmosphere condenses to form raindrops. Since latent heat is the amount of heat required to change a material from solid to liquid or from liquid to gas (for example boiling water to steam), the same amount of heat is released in the opposite change from gas to liquid. In a storm, when high humidity or a large amount of water vapor (gas) in the atmosphere condenses to liquid, the amount of heat released is the same as that required to boil and vaporize the same amount of liquid. Since the heat released is localized in the area of condensation, it creates instability in the atmosphere, in some cases fueling tornadoes. Note that in FIGURE 15-17a, the storm is moving to the right; the rainstorm passes over an area on the ground before the arrival of the tornado.

Tornadoes generally form toward the trailing end of a severe thunderstorm; this can catch people off-guard. Someone in the path of a tornado may first experience

FIGURE 15-16 WIND SHEAR

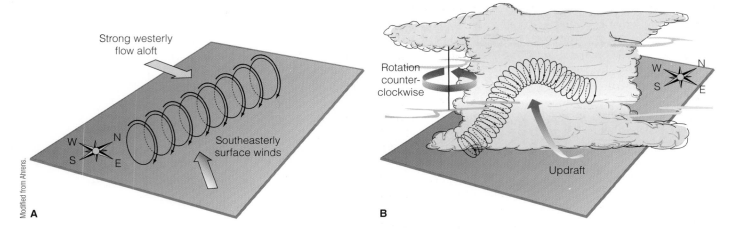

A. Wind shear, with surface winds from the southeast and winds from the west aloft. **B.** This slowly rotating vortex can be pulled up into a thunderstorm, which can result in a tornado.

FIGURE 15-17 CLOUD FORMATIONS INDICATING TORNADO CONDITIONS

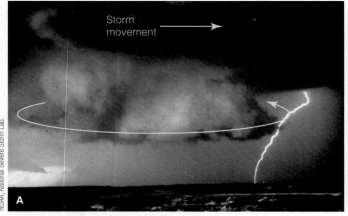

A. A slowly rotating wall cloud descends from the base of the main cloud bank, an ominous sign for production of a tornado near Norman, Oklahoma, on June 19, 1980. **B.** Mammatus clouds are a sign of the unstable weather that can lead to severe thunderstorms and potentially tornadoes. These formed over Tulsa, Oklahoma, on June 2, 1973.

FIGURE 15-18 SUPERCELLS AND TORNADOES

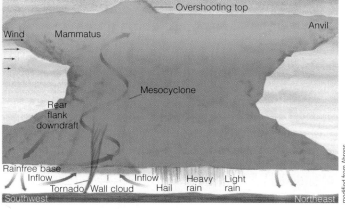

During a supercell storm over Utah, a wall cloud descends from the main anvil cloud. The rotation of the mesocyclone at the center of the cloud tightens to form a tornado.

wind blowing out in front of the storm cell along with rain, then possibly hail, before the stormy weather appears to subside. But then the tornado strikes. In some cases, people feel that the worst of a storm is over once the strong rain and hail have passed and the sky begins to brighten, unless they have been warned of the tornado by radio, television, or tornado sirens that are installed in many urban areas with significant tornado risk. Some tornadoes are invisible until they strike the ground and pick up debris. If you do not happen to have a tornado siren in your area, you may be able to hear an approaching tornado as a hissing that turns into a strong roar that many have likened to that of a loud oncoming freight train.

Conditions are favorable for tornado development when two fronts collide in a strong low-pressure center (**FIGURE 15-19**). This can often be recognized as a **hook echo**, or hook-shaped band of heavy rain, on weather radar. This is a sign that often causes weather experts to put storm spotters on alert to watch for tornadoes.

A strong tornado can form rather quickly, within a minute or so, and last for ten minutes to more than an hour. A tornado rarely stays on the ground for more than 30 minutes, generally leaving a path less than one kilometer wide and up to 30 kilometers long. Typical tornado speeds across the ground range from 50 to 80 kilometers per hour, but internal winds can be as high as 515 kilometers per hour—the most intense winds on Earth. In a major research project, VORTEX2, on the Great Plains in May, 2009, scientists were fortunate to be able to see up into the core of a tornado as it rolled over to nearly horizontal. The highest-velocity winds of the funnel spun around an open core reminiscent of the quiet eye of a hurricane.

Comparison of the winds of tornadoes with those of hurricanes (compare TABLE 15-2 with TABLE 14-1, p. 424) shows that the maximum wind velocities in tornadoes are twice those of hurricanes. Wind forces are proportional to the wind speed squared, so the forces exerted by the strongest tornado wind forces are four times those of the strongest hurricane winds. In many cases, much of the localized wind damage in hurricanes is caused by embedded tornadoes.

As a tornado matures, it becomes wider and more intense. White or clear when descending, tornadoes become dark when water vapor inside condenses in updrafts, which pull in ground debris (**FIGURE 15-20**, p. 478). In its waning stages, a tornado then narrows, sometimes becoming ropelike, before finally breaking up and dissipating (**FIGURE 15-21**, p. 478). At that waning stage, tightening of the funnel causes it to spin faster, so the tornado can still be extremely destructive.

Classification of Tornadoes

The severity of tornadoes is based on their internal wind speeds and the damage produced. In his research, Dr. Ted Fujita of the University of Chicago separated estimated

FIGURE 15-19 COLLIDING COLD AND WARM FRONTS

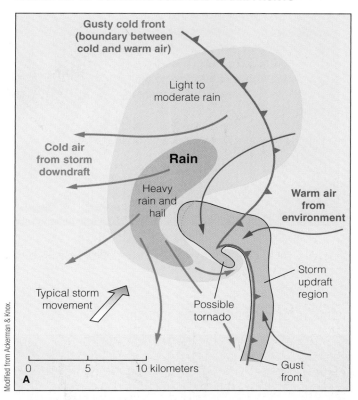

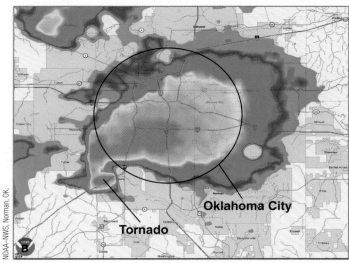

A. A common situation for tornado development is the collision zone between two fronts, commonly seen in a "hook echo" of a rainstorm. A pair of curved arrows indicates horizontal rotation of wind in the lower atmosphere. **B.** Radar hook-echo image of a classic supercell with an embedded tornado during the 1999 Oklahoma City tornado outbreak.

tornado wind speeds into a six-point nonlinear scale from F0 to F5. F0 causes minimal damage and F5 blows away even strong-frame homes. Early in 2007 the scale was updated as the Enhanced Fujita (EF) scale, based on detailed

FIGURE 15-20 TORNADO PICKS UP DEBRIS

A big tornado south of Dimmitt, Texas, on June 2, 1995, sprays debris out from its contact with the ground. This tornado tore up 300 feet of the highway where it crossed. Debris is drawn into the vortex of the tornado, lifted, and sprayed out.

FIGURE 15-21 DISSIPATION OF A TORNADO

This thin, ropelike tornado was photographed at Cordell, Oklahoma, on May 22, 1981, just before it broke up and dissipated. A strong tornado has descended from the cloud base, picked up and violently spewed out everything loose or breakable. As the cloud and head of the tornado moved to the left, its foot on the ground lagged behind. Finally the weakening tornado thinned to a narrow rope and lifted. Note that the foot of the tornado on the ground can sweep well away from its position at the cloud base.

wind measurements and long-term records of damage. The new scale uses three-second wind-gust estimates at the site of damage and is considered more reliable than the old scale. The main difference between the scales is that the new EF scale shows that particular damages occur at much lower wind speeds.

In addition, Dr. Fujita compiled an F-scale damage chart and photographs (known as the **Fujita Scale**)

corresponding to these wind speeds (**Table 15-2**). Reference photographs of damage are distributed to National Weather Service offices to aid in evaluating storm intensities (**FIGURE 15-22**).

You can see from TABLE 15-2 that, as with other hazards, moderate tornadoes are more frequent than severe ones. Although EF5 tornadoes are the most severe and most destructive, they kill fewer people because they happen rarely. EF5 tornadoes are infrequent, but on May 4, 2007, one ripped through Greensburg in south-central Kansas and flattened 95 percent of the town. Although there was a 20-minute warning, winds reached 328 kilometers per hour and 10 people were killed (see chapter-opener photo).

Tornado Damages

In his research on tornadoes, Dr. Fujita examined damage patterns. He noticed that there were commonly swaths of severe damage adjacent to areas with only minor damage (**FIGURE 15-23**, p. 480). He also examined damage patterns in urban areas and cornfields, where swaths of debris would be left in curved paths. This led him to hypothesize that smaller vortices rotate around a tornado, causing intense damage in their paths but allowing some structures to remain virtually unharmed by missing them. Such vortices were later photographed on many occasions, supporting this hypothesis. The most intense winds are within these embedded vortices, so the pattern of damage can vary greatly over short distances.

If a building is unlucky enough to be in the path of one of these destructive vortices, the resulting damage is dependent on the severity of the tornado as well as the building construction. Wind speeds and damage to be expected in buildings of differing strengths are shown in

Table 15-2		Fujita Scale of Deaths and Damages				
FUJITA SCALE VALUE	WIND SPEED ENHANCED kph [ENHANCED mi. / hr.]	NUMBER OF TORNADOES (1985–93)	% PER YEAR	% OF DEATHS	DAMAGE	
F0	104–137 kph [65-85 mph]	478	51	0.7	Light: Some damage to tree branches, chimneys, signs.	
F1	138–177 kph [86–110 mph]	318	34	7.5	Moderate: Roof surfaces peeled, mobile homes overturned, moving autos pushed off roads.	
F2	178–217 kph [111–135 mph]	101	10.8	18.4	Considerable: Roofs torn off, mobile homes demolished, large trees snapped or uprooted. Light objects become missiles.	
F3	218–265 kph [136–165 mph]	28	3	20.5	Severe: Roofs and some walls torn off well-constructed houses, trains overturned, most forest trees uprooted, heavy cars lifted and thrown.	
F4	267–322 kph [166–200 mph]	7	0.8	36.7	Devastating: Well-constructed houses leveled, cars thrown, large missiles generated.	
F5	323–374 kph [over 200 mph]	1	0.1	16.2	Incredible: Strong frame houses lifted and carried considerable distances to disintegrate. Auto-size missiles fly more than 100 yards; trees debarked.	

FIGURE 15-22 FUJITA SCALE OF TORNADO DAMAGES

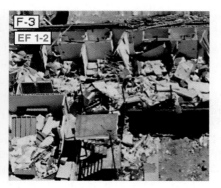

Dr. Ted Fujita developed the "F scale" for tornadoes by examining damage and evaluating the wind speeds that caused it. He used this set of photos as his standard for comparison. Conversion of his damage to the EF scale is approximated in the red numbers. EF5 would level virtually everything except for the strongest reinforced buildings.

FIGURE 15-23 DAMAGE PATTERNS FROM TORNADOES

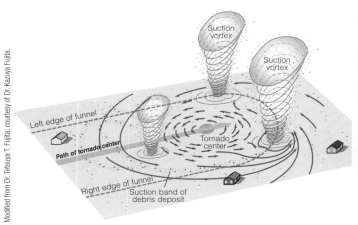

A. Tornadoes are composed of multiple vortices that rotate around their center. **B.** The Lake County, Florida, tornado of February 3, 2007, shows how selective tornado damage can be. The hook of debris distribution, in this case clockwise, is clear in the photo.

Table 15-3. Note that walls are likely to collapse in an F3 tornado in even strongly built frame houses; and in an F4, the house is likely to be blown down. Brick buildings perform better. In an F5 tornado, even concrete walls are likely to collapse.

Damage to a home generally begins with the progressive loss of roofing material, followed by glass breakage from flying debris. Then a roof may lift off and garage doors may fail. Exterior walls collapse, then interior walls, beginning with upper floors. Small interior rooms, halls, and closets fail last. The ease of destruction depends on how well roofs are attached to walls and walls to floors. Metal "hurricane straps" connecting horizontal and vertical members make a big difference. As with many other circumstances, damage and destruction are controlled by the weakest link.

Most susceptible to damage and destruction are farm outbuildings, followed by mobile homes, apartment and condo buildings, and then family homes. Winds of 225 kilometers per hour (km/hr) will cause severe damage to a typical home or apartment building and endanger people; 260 km/hr will effectively destroy such structures. Winds of 145–160 km/hr will severely damage a typical mobile home and endanger the occupants; 169 km/hr may destroy it, likely killing any occupants. Other than a reinforced tornado shelter, a basement with interior walls seems to be the safest place, but the variable directions of swirling winds leave no preferred location. Although the southwest corner of a basement was once thought to be safest; studies show that a house may be blown off its foundation and the house may collapse into that corner of the basement.

High-rise office buildings and hotels framed with structural steel are less susceptible to structural damage, but their exterior walls may be blown out. Downtown areas of cities with tall buildings are not immune to tornado damage. For example, instead of dissipating from turbulence among the buildings, an EF2 tornado moved into Atlanta, Georgia, at

Table 15-3	Expected Damages for Different Types of Buildings Dependent on Tornado Strenghth*					
EXPECTED DAMAGE BY EF-SCALE TORNADO						
TYPE OF BUILDING	**EF0**	**EF1**	**EF2**	**EF3**	**EF4**	**EF5**
Weak outbuilding	Walls collapse	Blown down	Blown away			
Strong outbuilding	Roof gone	Walls collapse	Blown down	Blown away		
Weak frame house	Minor damage	Roof gone	Walls collapse	Blown down	Blown away	
Strong frame house	Little damage	Minor damage	Roof gone	Walls collapse	Blown down	Blown away
Brick structure	OK	Little damage	Minor damage	Roof gone	Walls collapse	Blown down
Concrete structure	OK	OK	Little damage	Minor damage	Roof gone	Walls collapse

*Simplified and modified from Fujita, 1992.

FIGURE 15-24 URBAN TORNADO DAMAGE

Jeff Olson.

Tornado damage in downtown Atlanta, March 14, 2008. Note the high-rise building on left.

9:38 p.m. on March 14, 2008, causing about $250 million in damages (**FIGURE 15-24**). Its strength may have been intensified by the heat-island effect of the huge city.

Although many people believe that the low pressure in a tornado vacuums up cows, cars, and people, and causes buildings to explode into the low-pressure funnel, this appears to be an exaggeration. Most experts believe that the extreme winds and flying debris cause almost all of the destruction. Photographs of debris spraying outward from the ground near the base of tornadoes suggest this (**FIGURE 15-25**). However, even large and heavy objects can be carried quite a distance. The Bossier City, Louisiana, tornado ripped six 700-pound I-beams from an elementary school and carried them from 60 to 370 meters away. Another I-beam was carried to the south, where it lodged in the ground in someone's back yard

FIGURE 15-25 DESTRUCTIVE DEBRIS

© Eric Nguyen/CORBIS.

Debris sprays outward from a tight tornado rope in Mulvane, Kansas, June 12, 2004.

(**FIGURE 15-26**, p. 482). In another documented case, several empty school buses were carried up over a fence by a tornado before being slammed back to the ground.

Safety During Tornadoes

As with other hazards, tornadoes can be predicted with some success. Prediction and identification of tornadoes by the National Weather Service's Severe Storms Forecast Center in Kansas City, Missouri, uses Doppler radar, wind profilers, and automated surface observing systems. A **tornado watch** is issued when thunderstorms appear capable of producing tornadoes and telltale signs appear on radar. At this point, storm spotters often watch for severe storms. A **tornado warning** is issued when Doppler radar shows strong indications of vorticity, or rotation, or if a tornado is sighted. Warnings are broadcast on radio and television, and tornado sirens are activated, if they exist, in potential tornado paths. In contrast to the 12- to 24-hour warning often available before a hurricane's landfall, the warning time for a tornado at a particular location is 15 minutes or less in some eastern states, somewhat longer in the central states.

When a tornado warning is issued, people are advised to seek shelter underground or in specially constructed shelters whenever possible (**Case in Point:** Tornado Safety—Jarrell Tornado, Texas, 1997, p. 489). If no such area is available, people should at least go to some interior space with strong walls and ceiling and stay away from windows. The main danger comes from flying debris. People have been saved by taking refuge in an interior closet, or even lying in a bathtub while holding a mattress or sofa cushions over them. Unfortunately, in some cases a strong tornado will completely demolish houses and everything in them (**FIGURE 15-27**, p. 482).

Mobile homes are lightly built and easily ripped apart—certainly not a place to be in a tornado. Car or house windows—and even car doors—provide little protection from high-velocity flying debris, such as two-by-fours from disintegrating houses. Those in unsafe places are advised to evacuate to a strong building or storm shelter if they can get there before a storm arrives. If you cannot get to a safe building, FEMA recommends that you lie in a ditch or culvert and cover your head to provide some protection from flying debris.

It is as yet unclear whether vehicles provide more protection than mobile homes or lying in a ditch. Although cars are designed to protect their occupants in case of a crash, they can be rolled or thrown or penetrated by flying debris. If you are in open country, the tornado is far enough away, and you can tell what direction it is moving in, you may be able to drive to safety at right angles from the tornado's path. Recall that the path of a tornado is most commonly from southwest to northeast, so being north to east of a storm is commonly the greatest danger zone. However, some twisters, such as the F5 Jarrell tornado, moved southwest. Remember also that

FIGURE 15-26 TORNADOES MOVE HEAVY OBJECTS

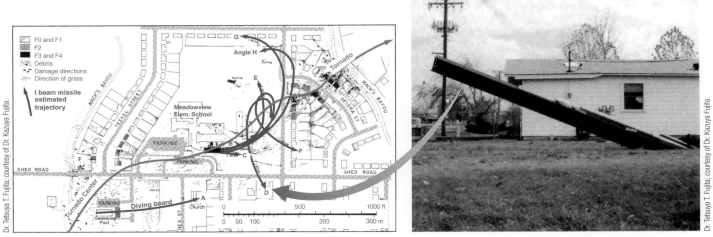

Six 700-pound I-beams were pulled from an elementary school in Bossier City, Louisiana, and carried by a tornado along these paths. Other objects, such as a diving board and a car, were also carried significant distances. The beam labeled "D" ended up stuck in the ground.

FIGURE 15-27 TORNADO DAMAGES

A steel-beam warehouse is not a safe place in a tornado; nor is a school bus. **A.** Roanoke, Illinois, 2004. **B.** Jackson, Mississippi, January 10, 2008.

the primary hazard associated with tornadoes is flying debris, and much to people's surprise, overpasses do not seem to reduce the winds associated with a tornado. Do not get out of your car under an overpass and think you are safe. In fact, an overpass can act like a wind tunnel that focuses the winds and you are higher above ground where winds are faster. Most overpasses lack hanging girders to hold onto in extreme winds. Once a few people park under an overpass, this can cause the additional problem of a traffic jam, thereby endangering others.

NOAA's weather radio and television network provides severe weather warnings. Typically, these warnings can furnish up to ten minutes lead time before a tornado's arrival. General guidelines include the following:

- Move to a tornado shelter, basement, or interior room without windows. In some airports, such as Denver International, the tornado shelters are the restrooms.

- Flying debris is extremely dangerous, so if your location is at all vulnerable, protect your head with a bicycle or motorcycle helmet.

- Although cars can overturn, and flying debris can penetrate their windows and doors, they still provide some protection—
especially below the window line.

Case in Point

Tornado Safety
Jarrell Tornado, Texas, 1997 ▶

On May 27, 1997, around 1 p.m., a tornado watch was issued for the area of Cedar Park and Jarrell, 65 kilometers north of Austin, Texas. Many people heard the announcement on the radio or on television, but most went on with their daily work. Storms are common in the hill country. This case seemed familiar: A cold front from the north had collided with warm, water-saturated air from the Gulf Coast to generate a line of thunderstorms. A tornado warning was issued at 3:25 p.m.

Just before 4 p.m., a tight funnel cloud swirled down from the dark clouds 8 kilometers west of Jarrell, a community of roughly 450 people. This tornado moved south-southwest along Interstate 35 at 32 kilometers per hour rather than taking a more typical easterly track.

When trained spotters saw a tornado on the ground ten to twelve minutes before the funnel struck, they sounded the alarm, and everyone who could took shelter. Some sought protection in interior rooms or closets; few homes had basements due to limestone bedrock close to the ground surface. People in this area were advised to take shelter in closets and bathtubs with mattresses for cover, but in this case it did not matter. Within minutes, the F5 tornado wiped 50 homes in Jarrell completely off their foundation slabs. Hail the size of golf balls and torrential rain pounded the area. Wind speeds were 400 to 435 kilometers per hour for the 20–25 minutes the twister was on the ground. At least 30 people died.

One woman had hidden under a blanket in her bathtub. Her house blew apart around her, and both she and the tub were thrown more than 100 meters. She survived with only a gash in her leg. Some people watched the tornado approach and decided to outrun it by car. They survived; but in other tornadoes, people have died trying to do this when they would have survived at home. Eyewitnesses reported that the Jarrell tornado lifted one car at least 100 meters before dropping it as a crumpled, unrecognizable mass of metal.

This was the second tornado to strike Jarrell; the first was only eight years earlier, on May 17, 1989. One of several tornadoes during the same event moved south through the town of Cedar Park, demolishing a large Albertson's supermarket, where 20 employees and shoppers huddled in the store's cooler. One of the authors happened to be a few kilometers south of Cedar Park playing golf that hot and humid Texas morning. Thunderstorms began to build on the horizon, and the sky took on a greenish gray cast. Early in the afternoon, golf course attendants quickly drove around the course warning players that two tornadoes had been spotted in the area. Because thunderstorms and tornadoes are fairly common there, many people become complacent; several people thought about finishing their rounds. Reaching the car in a drenching downpour, we realized that there was no safe place to go. Our cell phones were useless because all circuits were busy. Fortunately, the tornadoes were north of us, so we drove south into Austin to wait out the storm.

AP Images/Ron Heflin.

▶ *The F5 Jarrell tornado stripped many homes in the Double Creek Estates subdivision down to their concrete slabs.*

Critical View

A towering cloud formation

D. Bradford, NOAA.

1. What kind of a cloud is this and what does it indicate?
2. What other weather phenomena are often associated with it?

B This storm occurred over Norman, Oklahoma.

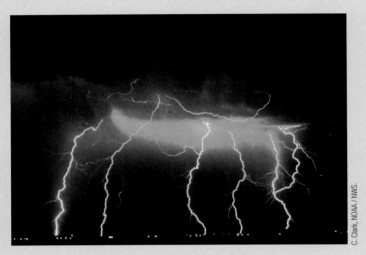

C. Clark, NOAA. / NWS.

1. Indicate relative positions of positive and negative static charges that lead to lightning.
2. Why is Oklahoma a likely location for thunderstorms?
3. If you were outside when this type of storm occurred, what should you do to keep safe?

C This destruction occurred in Dyersburrg, TN.

L. Skoogfors, FEMA.

1. What would have caused this kind of damage?
2. What category of storm does this damage suggest?
3. What should you do to survive this type of event?

D These clouds formed over St. Joseph, Missouri.

NOAA.

1. What is this type of cloud formation called?
2. What does this type of cloud formation suggest?

E This damage was the result of a tornado in Paris, Missouri.

NWS, Memphis.

1. What was the likely severity of the tornado that caused this damage?

2. What could the builders of this house have done to increase the chances that it would survive this type of event?

3. If you were living in such a place and a warning for this type of event were announced, what should you do to survive?

F This car was in Birmingham, Alabama.

A. Lannare, USACE.

1. What would have caused this kind of damage?

2. If you are in your car and see an approaching storm of this type, what should you do?

3. If you are in open country with lots of good, straight roads, what direction should you drive to avoid being in the path of such a storm?

G Sheet of plastic embedded in column, near Yazoo City, western Mississippi. April 24, 2010.

NOAA.

1. Imagine what this piece of plastic would have done to a person. How would you protect yourself from such flying debris if you were in a house lacking a tornado shelter?

2. If you were caught outside, out of range of a tornado shelter, and see a tornado coming, what would you do?

H Concrete slab once under a house, near Yazoo City, western Mississippi. April 24, 2010.

NOAA.

1. What strength of tornado (what EF level?) likely did this amount of damage?

2. How might you prepare for such a tornado, months or years in advance?

3. How might you retrofit a house on this slab so 2 or 3 people might survive inside it?

Key Points

Thunderstorms

- Thunderstorms are most common at equatorial latitudes, but the United States has more than its share for its latitude.

- Storms form most commonly at a cold front when unstable warm, moist air rises rapidly into cold air and condenses to form rain and hail. Cold fronts from the northern plains states often interact with warm, moist air from the Gulf of Mexico to form a northeast-trending line of storms. **FIGURE 15-3**.

- Collisions between droplets of water carried in updrafts with downward-moving ice particles generate positive charges that rise in the clouds and negative charges that sink.

- Because negative and positive charges attract, a large charge separation can cause an electrical discharge—lightning—between parts of the cloud or between the cloud and the ground. If you feel your hairs being pulled up by static charges in a thunderstorm, you are at high risk of being struck by lightning. **FIGURE 15-5**.

- Thunder is the sound of air expanded at supersonic speeds by the high temperatures accompanying a lightning bolt. Because light travels to you almost instantly and the sound of thunder travels one kilometer in roughly three seconds, if the time between seeing the lightning and hearing the thunder is three seconds, then the lightning is only a kilometer away.

- You can minimize danger by taking refuge in a closed building or car, not touching water or anything metal, and staying away from high places, tall trees, and open areas. If trapped in the open, minimize contact with the ground by crouching on the balls of your feet.

- Larger hailstones form in the strongest thunderstorm updrafts and cause an average of $2.9 billion in damage each year.

Tornadoes

- Tornadoes are small funnels of intense wind that may descend near the trailing end of a thunderstorm; their winds move as fast as 515 kilometers per hour. They form most commonly during the collision of warm, humid air from the Gulf of Mexico with cold air to the north. They are the greatest natural hazard in much of the Midwestern United States. The greatest concentration of tornadoes is in Oklahoma, with fewer to the east and north. **FIGURE 15-12**.

- The Fujita tornado scale ranges from F0 to F5, where F2 tornadoes take roofs off some well-constructed houses and F4 tornadoes level them. **TABLES 15-2, 15-3**.

- Tornadoes form when warm, humid air shears over cold air in a strong thunderstorm. The horizontal rolling wind flexes upward to form a rotating cell up to 10 kilometers wide. A wall cloud sagging below the main cloud base is an obvious danger sign for the formation of a tornado. **FIGURES 15-16** to **15-18**.

- On radar, a hook echo enclosing the intersection of two fronts is a distinctive sign of tornado development. **FIGURE 15-19**.

Safety During Tornadoes

- The safest place to be during a tornado is in an underground shelter or the interior room of a basement. Even being in a strongly built closet or lying in a bathtub can help. If caught in the open while driving a car, you may be able to drive perpendicular to a storm's path. If you cannot get away from a tornado, your car may provide some protection, or lying in a ditch and covering your head will help protect you from debris flying overhead.

Key Terms

Questions for Review

1. When is the main tornado season?

2. How are electrical charges distributed in storm clouds and why? What are the charges on the ground below?

3. What process permits hailstones to grow to a large size?

4. Why do you see lightning before you hear thunder?

5. List the most dangerous places to be in a lightning storm.

6. What should you do to avoid being struck by lightning if caught out in the open with no place to take cover?

7. In what direction do most midcontinent tornadoes travel along the ground?

8. How fast do tornadoes move along the ground?

9. What is a wall cloud, and what is its significance?

10. Why does lying in a ditch provide some safety from a tornado?

11. How do weather forecasters watching weather radar identify an area that is likely to form tornadoes?

12. What is the greatest danger (what causes the most deaths) from a tornado?

Discussion Questions

1. Explain why there are more tornadoes in the United States than in any other country. Be specific about the weather conditions that lead to tornado development and why.

2. Tornado experts say that if you encounter a tornado while in your car, you should get out and lie down in a ditch. Discuss the advantages and disadvantages of both locations in terms of your risk. Consider what about a tornado kills people.

3. How would you design a house to be safe from tornadoes, while keeping the cost to a minimum and having a design that would not look out of place in a typical residential neighborhood?

16 Wildfires

The Storm King Mountain fire races upslope toward fleeing firefighters. Eleven died here when the fire overwhelmed them.

Sonny Archuleta, USFS.

A Deadly Wildfire

As the Storm King Fire spread out below and west of them, firefighters recognized they were in danger. One group of eight smokejumpers began hiking up toward a burned-out patch identified as a "safety zone." They abandoned chain saws used to fight the fire and hiked up a ridge ahead of the fire as fast as they were able. they deployed their fire shelters; air tankers dropped fire retardant around them and rvived.

r group of firefighters backtracked up the fire line toward the main ridge, but not all of reach safety. The fire moved up the canyon at about one meter per second, up open meters per second. A few managed to reach the ridge and followed a steep drainage st and eventually down to the highway. Some surviving firefighters received first- and burns on their backs, necks, and elbows.

ghters, racing upslope ahead of fast-moving flames, were overtaken and died 60 to w the main ridge crest that would have slowed the fire's advance. Another firefighter

Wildfires

died 35 meters below it. Two others reached the main ridge and then were forced northwest toward a rocky face. Unfortunately, that led to a deep dead-end gully where they died when hot air and smoke engulfed them (**Case in Point:** Debris Flows Follow a Tragic Fire—Storm King Fire, Colorado, 1994, p. 501).

Fire Process and Behavior

Wildland fires are a natural part of forest evolution. They benefit ecosystems by thinning forests, reducing understory fuel, and permitting the growth of different species and age groups of trees. It is only when fires encroach on human environments that they become a hazard.

Factors that affect wildfire behavior include fuel, weather, and topography. In some cases, the fuel and conditions foster a firestorm that is virtually impossible to suppress without a change in the conditions.

Fuel

A fire requires three components—fuel, oxygen, and heat—and cannot progress without all three (**FIGURE 16-1**). The type of fuel available, its distribution, and its moisture content determine how quickly a fire ignites and spreads, as well as how much energy it releases.

Fuel loading refers to the amount of burnable material. Trees and dry vegetation are the primary sources of fuel for a wildfire. They burn at high temperature by reaction with oxygen in the air. The main combustible part of wood is cellulose, a compound of carbon, hydrogen, and oxygen.

When it burns, cellulose breaks down to carbon dioxide, water, and heat. Shrubs and trees also contain natural oils or saps that add to the combustibles.

Surface fires spread along the ground fueled by grasses and low shrubs. Vegetation with large relative surface areas accelerates ignition and burning because heat that promotes ignition begins at the outer surface of the fuel, then works its way inward. The large relative surface area of dry grasses, tree needles, and to a lesser extent shrubs, allows them to burn more easily than trees. Fires in dry grass spread rapidly. With a small total mass of burnable matter on the ground—that is, grass, dry leaves, shrubs, and litter—a surface fire may move through an area fast enough to consume ground cover and understory brush but not ignite trees. Heavy fuels such as tree trunks have a small surface area compared with their total mass; they ignite with difficulty and burn slowly. Some tree species such as ponderosa pine have evolved with time to survive fires that burn ground vegetation.

Low brush and branches are known as **ladder fuels** because they ignite first and then allow the fire to climb into higher treetops (**FIGURE 16-2**, p. 490). The ability of fire on the ground to easily reach tree crowns has a major effect on the rate of spread and intensity of a wildfire. Once flames reach into trees and then treetops, the smaller branches and fine needles easily ignite to form a **crown fire** ten or more meters high. A large crown fire will burn out at a single location, such as a cluster of trees, in less than a minute.

The intensity of a wildfire depends heavily on different types of vegetation. Lightweight flammable materials, such as dry grass and leaves, ignite easily but burn up quickly without generating very much heat. At the other extreme, heavy materials, such as tree trunks, are more difficult to ignite but burn much longer and generate much more heat. These include heavy coniferous (softwood) forests. By contrast, deciduous (hardwood) forests burn at somewhat lower intensities.

Ignition and Spreading

Fires can be naturally started by lightning strikes, intentionally set for beneficial purposes, or set accidentally or maliciously. Between 2001 and 2009, only 1.7 percent of wildfires in the United States were caused by lightning but 9.5 percent were human-caused, both by arson and

FIGURE 16-1 THE FIRE TRIANGLE

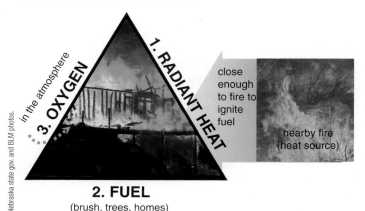

1. RADIANT HEAT

3. OXYGEN
in the atmosphere

2. FUEL
(brush, trees, homes)

close enough to fire to ignite fuel

nearby fire (heat source)

Nebraska state gov. and BLM photos.

A fire requires fuel, oxygen, and heat. Without any one of these, a fire cannot burn.

FIGURE 16-2 LADDER FUELS

A. Dense ladder fuels in Glacier National Park, Montana. **B.** Ladder fuels permit wildfire to climb into the trees; Yellowstone Park fire, 1988.

unintentionally. During the same period in California, causes included lightning 5 percent, arson 7 percent, trash burning 10 percent, vehicles 14 percent, campfires and power lines 3 percent each, smoking 2 percent, and unsafe use of power equipment about 30 percent. Various unintentional causes make up the remainder. A major heat wave in southern Europe in the summer of 2007 fostered catastrophic wild-fires. Some fires in Greece and Italy were ignited by arson-ists in protected forests to create new areas for construction. In 2000, Italy passed a law that bans construction for ten years after a wildfire, but the rules are not always enforced.

Investigators searching for clues to a fire's origin look for evidence left behind, including remnants of a campfire, matches, cigarettes, and accelerants such as gasoline. They also search for any glassy residue or elongated or straw-like groups of fused mineral grains (fulgurite) from a lightning strike. They can reconstruct the path of a fire by taking into account wind direction when the fire began and indica-tions of the direction in which the fire moved. Fire-source indicators include post-fire aerial photo and satellite burn

distribution; weather records and wind patterns during a fire; topography, such as slopes and canyons; more-severe scorching on the sides of trees facing a fire source; lower height and degree of defoliation toward a source; and greater degree of scorching and wood charring on tree trunks facing a source (**FIGURE 16-3**). Effects at the heel or back of a fire nearer the source tend to be less intense. Unburned grass stalks sometimes fall backwards toward a fire's source; burned stalks commonly fall forward in the direction of fire movement. Large non-combustibles, such as metal signs and rocks, can leave unburned patches in the direction of fire movement.

Fire ignites and progresses primarily by three mecha-nisms: radiation, convection, and burning embers. Radi-ant energy decreases with the square of the distance from a heat source, that is, at twice the distance the energy on a surface drops by four times (**FIGURE 16-4**). With only ra-diant heating—the energy transferred by thermal radia-tion, without air movement or contact with flames, 50 seconds of intense crown-fire burn is sufficient to ignite

FIGURE 16-3 DETERMINING THE PATH OF FIRE

Lawrence Berkeley Lab.

The more severely burned white side of this tree indicates the direction from which the fire came; in this case, the fire moved upslope to the right.

FIGURE 16-4 RADIANT HEAT AND DISTANCE

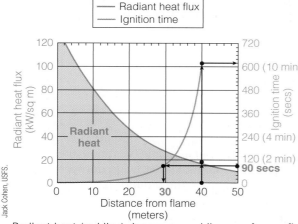

Jack Cohen, USFS.

Radiant heat (red line) decreases rapidly away from a flame that is 20 meters high and 50 meters wide. The time it takes for flammable material, such as wood, to ignite (blue line) increases farther from the flame.

dry fuels at a distance of 20 to 25 meters. Convective transfer of heat is more direct and efficient; it involves direct movement of heated air or flames to ignite fuel. Fire also spreads when burning embers, called **firebrands**, are carried with the wind and then deposited, igniting

FIGURE 16-5 FIREBRANDS

Wally Skalij, L.A. Times.

Burning embers carried by the wind start spot fires. Here firebrands blow across Angeles Crest Highway in August, 2009.

spot fires, which burn ahead of the main fire but can quickly spread (**FIGURE 16-5**).

Topography

Local topography, especially canyons, can funnel air, accelerate a fire, and cause more rapid spreading. In a canyon, as sometimes initiated by an untended campfire, fire can race up the valley because of this chimney-like funneling effect. Flames rise because their heat expands air, making it less dense. Thus, fires generally move rapidly upslope but slowly downslope (**FIGURE 16-6**). On a hillside, especially a steep one, flames and sparks rise with their heat, promoting the upslope movement of fires into new fuel. Although

FIGURE 16-6 FIRE MOVES UPSLOPE

National Park Service.

Heat rising from a fire moves upslope, and flames attack new fuels above them. Yellowstone Park fire, 1988.

flaming branches and trees can fall downslope, advance in that direction tends to be slow. To be upslope from a fire is to be in serious danger—as were the firefighters whose story introduced this chapter.

At a ridge top, fire often slows dramatically because the rising flames run out of fuel. A fire reaching a ridge crest can create an updraft of air from the far side that fans the flames, but the same updraft can reduce the chance that the fire will move down the other side.

Weather Conditions

Weather conditions can both ignite a fire and determine the rapidity of its spread. Lightning wildfires are ignited by thunderstorms triggered at weather fronts, especially after a period of dry weather (**FIGURE 16-7**). Lightning strikes in an area of heavy timber, at lower elevations or during high winds, can ignite a fire that progresses rapidly. Lightning strikes on rocky mountain peaks often have limited effect because fire progresses slowly downslope, and fuels there are often limited.

Fires start more easily and spread rapidly during dry weather because of low **fuel moisture**. High humidity makes it more difficult for fires to start and continue burning, whereas high temperatures and winds progressively reduce moisture over time and increase fire hazards. A few years of less-than-normal moisture dehydrates soil and lowers the water table, providing less moisture for the growth of trees and other vegetation. It may take several wet seasons to replenish the water shortfall below the surface. Thus, surface vegetation during a drought may green up in the rainy season but dehydrate quickly as dry weather returns.

Wind conditions dramatically affect the spread and direction of fires. An extreme example is the Santa Ana winds of Southern California, the strong trade winds that blow southwest from Nevada in the fall. The wind descending through California's mountain passes compresses and

FIGURE 16-7 LIGHTNING IGNITES FIRES

This fire was started by lightning in the Selway-Bitterroot Wilderness area of the northern Rockies on August 9, 2005.

heats up, causing the humidity to drop, and providing prime conditions for fire. Winds accelerate fires both by directing flames at new fuels and bringing in new oxygen in the air that facilitates burning. Once fires reach high into trees and crown, firebrands—burning embers carried downwind from flaming treetops—can ignite fuel as far as a kilometer away. High winds or extreme amounts of dry fuel can initiate **firestorms** intense enough to generate their own winds from a **convective updraft** of heat, which draws in new air from all sides and fans flames. Some types of vegetation, such as eucalyptus trees, contain oils that produce tall, intense flames that in turn can spawn firestorms.

With continuing effects of climate change, more frequent and larger wildfires emit greater amounts of carbon into the atmosphere—primarily carbon monoxide (CO) and carbon dioxide (CO_2). Longer, drier summers contribute much larger amounts of burnable vegetation and more fires. Earlier snowmelt in the spring leads to a reduction in soil moisture, earlier and more-complete dehydration of vegetation, and hotter fires. Weakened forests also become more susceptible to insect attack (pine-bark beetle infestations, for example) and tree deaths, so that forests are even more susceptible to fire.

Deliberate deforestation by fire also contributes. The largest areas of such deforestation in recent years have been in places like Indonesia, where forests are burned to make space for growing commercial crops, such as palm oil. The fires burn not only forest trees and shrubs but peat that has stored carbon in the ground for hundreds of years. Studies estimate that 13 to 40 percent of fossil fuel emissions come from such fires.

Secondary Effects of Wildfires

A major fire can lead to other hazards, such as floods and landslides, which are sometimes more disastrous than the fire itself. Major fires also produce air pollution.

Erosion Following Fire

Major fires often result in severe slope erosion in the few years following the fire. In normal, unburned areas, vegetation has a pronounced effect in reducing the runoff after rainfall and snowmelt. Tree needles, leaves, twigs, decayed organic material, and other litter on the ground soak up water and permit its slow infiltration rather than turning it into surface runoff. Evapotranspiration by trees and other vegetation also decreases the amount of water soaking deep into the soil.

Intense fires burn all of the vegetation, including litter on the forest floor (**FIGURE 16-8A**). The organic material burns into a **hydrocarbon residue** that soaks into the ground. This material fills most of the tiny spaces between fine soil grains in the top few centimeters of soil and sticks them together (**FIGURE 16-8B**). In some cases, water can no longer infiltrate the surface soils, thus most of it runs off. Such impervious soils are called **hydrophobic soils**.

FIGURE 16-8 SOIL CONDITIONS AFTER A FIRE

Surface of soil sealed by resins formed during fire

A. This steep slope on the O'Brien Creek drainage at the western edge of Missoula, Montana, had been ravaged by fire one month earlier, leaving it vulnerable to erosion. **B.** In hydrophobic soil such as this produced by the Banning fire in Southern California in 1993, the dry layer lies beneath the dark gray hydrophobic layer, which is composed of soil and ash. The view is about 20 centimeters across at the base.

The hydrophobic nature of soils is often attributed to intense burning, but the relationship to fires is not entirely clear. What is clear is that water more readily runs off the surface of charred soil.

Big rain droplets beat directly on unprotected ground. When rainfall is heavy and vegetation and ground litter have been removed by fire, the water cannot infiltrate fast enough to keep up with the rainfall. Raindrops splash fine grains off the surface, and the runoff carries them downslope as sheetwash, or overland flow, which easily carves out gullies and channels in the unprotected soils (**FIGURE 16-9**). All of this surface runoff quickly collects downslope in larger

FIGURE 16-9 FLOODING AFTER A FIRE

A. These gullies were eroded by a short-lived rainstorm one year after the 2000 fires in the southern Bitterroot Valley, Montana. **B.** This mudslide followed the Hayman, Colorado, fire on July 5, 2002.

gullies and small valleys. Water collecting in a valley with a large-enough watershed quickly turns into a flash flood or debris flow. These destructive and dangerous floods rip out years of accumulated vegetation, soil, and loose rock, carrying everything down valley in a water-rich torrent (see **Case in Point:** Debris Flows Follow Fire—Storm King Fire, Colorado, 1994, p. 501).

Rapid accumulation of water from overland flow following a fire creates a severely shortened and heightened stream hydrograph. This can lead to flash floods that wash out bridges, roads, and buildings. After a region has burned, it is unsafe to drive or walk in small valley bottoms during intense rainstorms because swift runoff funnels most of the precipitation into adjacent gullies and then quickly into sequentially higher-order streams. Even when storms are several kilometers away in the headwaters of a burned drainage, a flash flood downstream can appear under a clear blue sky. Homes that may have survived a fire in usually dry valley bottoms are vulnerable to debris flows from all of the new sediment washed downslope after a fire.

Mitigation of Erosion

Erosion following a fire cannot be prevented, but it can be minimized. Federal and state agencies and individuals can plant vegetation, grass, shrubs, and trees that prevent the direct impact of raindrops on bare soil. Over large areas, slopes are often seeded soon after a fire. Straw can be packed into tubes or laid out in bales to provide a barrier to overland flow, thus reducing slope erosion. Felling dead trees across a slope can have a similar effect. Drains can direct water flow laterally to valleys to minimize surface gullying. On small steep patches of particularly vulnerable soil, sheets of plastic can be spread to prevent water from reaching the soil.

Air pollution with wildfires

Major wildfires can spread smoke, soot, and ash over areas more than 1,000 kilometers away and, in some cases, around the planet. Pollutants include carbon monoxide, hydrocarbons, and nitrogen oxides. Sunlight can react with these pollutants to produce ozone at low elevations. In the stratosphere, ozone helps protect us from ultraviolet radiation, but close to the ground it can destroy living tissue and cause respiratory problems in some people. Southern California wildfires have repeatedly increased ground-level ozone to unhealthy levels.

Wildfire Management and Mitigation

Wildfires are necessary to the evolution of forest ecosystems, but out-of-control fires pose hazards to humans. Wildfires occur in most regions of the United States, where they affect wildlands and often encroach on suburban or

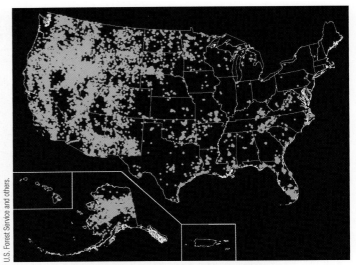

FIGURE 16-10 U.S. WILDFIRES

U.S. Forest Service and others.

A map of large wildfire locations from 1980 to 2003.

built-up areas **(FIGURE 16-10)**. From 1985 to 1994, an average of 73,000 fires per year burned on federal lands, consuming more than 121,000 km^2 and costing $411 million for fire suppression. From 2000 to 2007, annual firefighting costs exceeded $1.3 billion in four of the seven years. Additional millions were lost in timber values, and indirect costs resulting from landslides and floods are unknown.

Hazards to the public increase as more people move to forest-, range-, and other wildlands. Over ten days in October 2003, catastrophic fires destroyed 3,452 homes in Southern California (**Case in Point:** Firestorms Threaten Major Cities—Southern California Firestorms, 2003 to 2009, p. 502). From October 25 to November 3, 1993, Santa Ana winds fanned 21 major fires in Southern California, burning 76,500 acres (309.6 km^2) and 1,171 buildings. Three people died.

As with all hazards, a clear understanding of the natural processes underlying fire behavior allows governments and individuals to take steps to protect lives and property.

Forest Management Policy

The U.S. Forest Service was formed in response to a catastrophic wildfire that burned thousands of square kilometers of northern Idaho and western Montana in the summer of 1910. For most of the twentieth century, their policy was to aggressively fight wildfires. Until a few years ago, the Forest Service and its Smokey Bear mascot maintained that fire was bad and all fires should be extinguished as quickly as possible.

Because the United States spends upward of $1 billion per year on fighting fires, you might expect some significant long-term benefit. However, many experts argue that we get no long-term benefit. In fact, preventing fires leads to buildup of wildland fuels, which ultimately leads to

FIGURE 16-11 BITTERROOT VALLEY FIRES

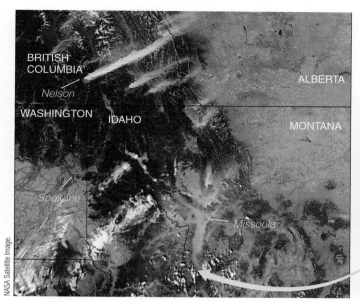

A. Forest fires spread throughout the northern Rockies on July 15, 2003, as shown by the red areas on the image. Fires in Glacier National Park burned just south of the Alberta-Montana border. Fires in the mountains farther south surrounded Missoula, Montana. Smoke plumes, clearly visible as bluish-gray areas, are quite different from the white clouds across the region. **B.** Fires in western Montana on August 6, 2000, chased these two elk into the relative safety of the Bitterroot River's upper drainage.

worse fires. Forest tree densities and fuel loads have risen to critical levels, creating ideal conditions for fires. Many open, park-like forests became choked with closely spaced smaller trees and dense underbrush, leading to disastrous wildfires, as in the case of the Bitterroot Valley fires of westernmost Montana in 2000 (**FIGURE 16-11**). Decades of fire suppression have led to a buildup of wildland fuels. Very large wildfires in recent years include the 1988 Yellowstone fire that covered 3,200 square kilometers, the 2002 Biscuit fire in Oregon, the 2002 Rodeo-Chediski fire in Arizona, the 2006 Darby fire in western Montana, and the 2007 Milford Flat fire in Utah, each covering more than 400 square kilometers.

Federal agencies now recognize that fire is a natural part of wildland evolution, a process necessary for the health of rangelands and forests. They recently shifted to mitigation and management of fires, including setting controlled fires intentionally in **prescribed burns**. The Forest Service now permits many wildfires to burn and consume excess fuel in wilderness areas as long as they don't endanger buildings or important resources.

Fighting Wildfires

When wildfires approach human developments, several approaches are used for fire suppression. Highly trained firefighters, called **smokejumpers**, parachute into remote areas of lightning-caused spot fires to exterminate them before they spread out of control. Other firefighters hike in from the nearest roads to hand-dig meter-wide paths, **firelines**, down to bare mineral soil to slow or stop a fire's spread. Since many low-intensity wildfires mostly burn along the ground, even a narrow firebreak can provide a defensible line. Bulldozers are used to cut firebreaks, both to slow the spread of active fires and as preemptive measures in fire-prone areas long before fires break out. Often they are located along ridge crests where wildfire activity naturally weakens because of the terrain. Aircraft drop giant loads of water or fire retardant along the lateral and advancing edges of fires to help direct and contain them (**FIGURE 16-12**, p. 496). Retardants include ammonium sulfate, ammonium phosphate, borate, or swelling clay. Helicopters scoop huge buckets of water from nearby streams or lakes and dump them on the leading edges of fires.

For the largest fires that threaten towns or critical facilities, firefighters sometimes deliberately set burnouts (**FIGURE 16-13**, p. 496). In a **burnout**, a large area in the path of a big fire is burned under controlled conditions and suitable weather, generally beginning at a road or river; the burnout fire burns back toward the first fire, eliminating fuel and stopping the progress of the main fire. Before lighting a burnout, the forest or range between the burnout and the area needing protecting is thoroughly drenched using planes and helicopters. The technicalities of a burnout are complex and dangerous, but they can work remarkably well. In July 2003, firefighters set a large burnout that saved the town of West Glacier at Glacier National Park from a huge fire bearing down on the area.

FIGURE 16-12 FIREFIGHTING

A. A helicopter fills up at a local pond in Sonoma Valley, California, before dropping its load on a fire nearby. **B.** A "slurry bomber" drops retardant on a fire. Bright orange dye is added to show pilots coverage from previous drops.

FIGURE 16-13 BURNOUT

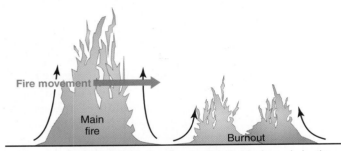

In a burnout, firefighters intentionally set a fire that is drawn in by the updraft of the main fire. This causes the intentional fire to burn back toward the main fire, which consumes fuels and hinders the wildfire's progress.

Risk Assessments and Warnings

Several government agencies study patterns in wildfire occurrence and spreading to assess the risk for regions of the United States at different times. Closely monitoring risk levels allows officials to reduce hazards and prepare emergency plans.

The vegetation, and in some regions topography, can increase the risk of fire (**FIGURE 16-14**). These include grasslands east of the Rocky Mountains and agricultural lands in the Midwest. Heavy coniferous (softwood) forests in the Northwest, the Rocky Mountain states, northern New England, and eastern Canada ignite more slowly but burn at high intensities. By contrast, deciduous (hardwood) forests of the southeastern states burn at somewhat lower intensities.

Typical weather patterns affecting wind direction or moisture for a region also influence fire risk. Fires tend to advance in the direction of prevailing winds, so fires in the northern United States and southern Canada tend to burn toward the northeast with westerly winds. In the trade-winds belt of the southern United States and northern Mexico, fires tend to burn toward the southwest. Thus, homes in the prevailing downwind direction from fire-prone or high-fuel areas are more at risk than those upwind from such areas. The desert regions in the Southwest, from northwestern Texas to southern California and northward to southern Wyoming are at higher risk because of their low moisture levels. Areas with more moisture are less at risk. Along the west coast, prevailing westerly winds carry moisture off the Pacific Ocean, and in the southeastern states, storms bring abundant moisture north off the Gulf of Mexico.

Agencies also monitor shorter-term weather conditions and alert the public of rising risk levels. The favorability of weather conditions for wildfire is called its fire weather potential (see FIGURE 16-14b). **Fire weather potential** is posted on fire danger-level signs along roads by the U.S. Forest Service and Bureau of Land Management. For a local area, it depends primarily on the number of high-temperature days, the air's relative humidity, moisture in available fuels, and wind speed. For large regions, it depends on similar factors over periods from weeks to months.

Drought maps provide a record of longer-term moisture deficits. A high-resolution infrared sensor is used to measure greenness of vegetation and inter-fuel moisture content. This greenness is compared with the typical value for each map area. Such estimates are used to infer fire danger and inform the public through news media and roadside signs.

Fire-detection methods have evolved from the fire lookout towers located on mountaintops decades ago, but some manned towers are still used, especially in remote areas. Ongoing detection tactics include the use of weather radar to detect suspect areas after lightning storms; light aircraft are then sent in to spot smoke. Aircraft are also used to detect new fire starts off of large fires caused by wind-carried

FIGURE 16-14 FIRE HAZARD AREAS

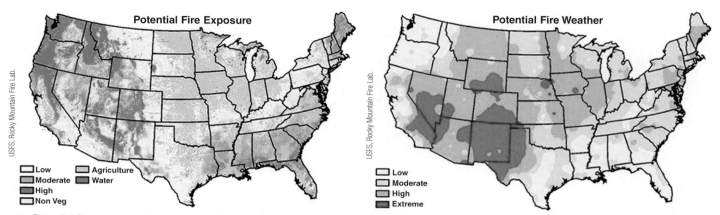

A. Potential fire exposure depends upon different vegetation types, how quickly fire spreads, and with what intensity. **B.** Extreme fire-weather potential in the continental United States averaged over 16 years.

firebrands. Another approach involves automated smoke-detection systems, mounted on tall towers, with multiple cameras powered by solar panels that communicate with central offices via microwave or satellite. False alarms from automated systems remain a problem.

Mounted in light aircraft, high-resolution infrared heat-detection cameras combined with visible-light cameras with GPS location equipment can now detect fires as small as 30 centimeters (~1 foot) across, even in bright sunlight. NOAA uses automated imaging by Earth-observing satellites, but results are dependent on hot fire temperatures, large fire extent, and subjective interpretation by fire analysts.

Protecting Homes from Fire

Many people who live in the woods do so to escape urban congestion and experience the beauty of nature (**FIGURE 16-15**). They like to be surrounded by trees and shrubs, but for a wildfire, trees and shrubs are fuel. Not only do forest dwellers put themselves and their property in danger, but they also endanger firefighters and compel fire-management officials to expend limited resources protecting individual properties rather than managing an overall fire.

The risk to flammable structures, especially people's houses, depends on the fire weather potential, the regional type of vegetation—how fast fire will spread and how hot it will burn—and the housing density. At one extreme, fire in a wilderness area or unpopulated desert endangers no houses. At the other extreme, a firestorm in a metropolitan area can become a catastrophe because it affects numerous homes and people.

Although individual homes scattered throughout a forest might constitute a low-risk area overall, the solitary homes could be in severe danger, especially if they are built from flammable materials and surrounded by ground fuels. Even those who live in an urban interface near a forest are in danger, particularly in dry, hot weather, where high winds can drive fires.

FIGURE 16-15 DENSE TREES INCREASE RISK TO HOMES

Fire consumes homes surrounded by heavy forest northwest of Payson, Arizona, in the Dude Fire, June 25, 1990.

Firebrands carried from treetop crown fires can easily ignite flammable rooftops and decks (**FIGURE 16-16**, p. 498). Wood-shingle roofs, which are favored by many people for their natural appearance, are often ignited by falling firebrands, even if a wildfire is not particularly close and there is no nearby vegetation. Once started, fire spreads rapidly across the roof and into the house. Wooden walls will not ignite from this process because the burning embers do not remain in contact long enough to ignite them. Composition shingles are better than wood shakes because, although they may burn if they get hot enough, the fire will not easily spread to other areas.

Those who live in the forest can minimize danger by building with flame-resistant materials. With a nonflammable roof and deck and no fuel within 30 meters, a building has a good chance of surviving, even with falling firebrands. Burning embers landing on a metal roof will not ignite it, although if the roof gets sufficiently hot the wood

FIGURE 16-16 ROOF FIRES

These fires were ignited by firebrands landing on roofs. The homes might otherwise have been saved from low-intensity ground fires nearby.

framework supporting it will ignite. Metal roofs, though not as natural, are flame resistant.

Trees or ground fuels around a home can also be a danger, either from the spread of fire along the ground or from falling firebrands. Low-intensity fires often burn forest homes, even though surrounding trees and fences may survive, because low-intensity fires burn ground fuel, such as needles, leaves, and twigs, without reaching higher into trees. In the aftermath of many wildfires, investigators frequently find homes burned to the ground but the surrounding trees still mostly green (**FIGURE 16-17**).

The best protection for homes in forests is fuel reduction before fires start. Homeowners should remove underbrush, low-hanging ladder fuels, and especially dry ground fuels (**FIGURE 16-18**).

Even once ground fuels are cleared, radiant heat from nearby burning trees can ignite a home. Since radiant heat decreases rapidly away from a fire, clearing fuels out to about 30 meters from a home will generally protect it from ignition by that mechanism. In fact, dry wood structures may not ignite even when fire is close enough to burn

FIGURE 16-18 REMOVAL OF LADDER FUELS

Ladder fuels have been removed from this slope to prevent fire from easily reaching the tree branches, but dry grass could still provide a path to the house.

FIGURE 16-17 SURFACE FIRES

A. Missionary Ridge fire, Durango, Colorado, June 2002. Surface fire spread through ground needles and ignited the house but did not burn part of a wooden fence, plastic garbage cans, ponderosa pines, and some low vegetation. **B.** In the Cerro Grande fire at Los Alamos, New Mexico, in May 2000, houses on one side of a street in a forested community burned to the ground, whereas those on the other side remained unscathed. The ground fire was stopped by the street, indicating that it spread through ground fuels.

FIGURE 16-19 WOOD STRUCTURE WITHSTANDS A FIRE

A wall of flame at the edge of a forest clearing provides enough radiant heat to severely burn skin but does not burn either wood posts or the siding on this home during the 2003 Wedge Canyon fire along the North Fork, Flathead River, Montana. During the fire (left) and after the fire (right).

people severely (**FIGURE 16-19**). Exposure to heat that will produce a second-degree burn on skin in 5 seconds will take more than 27 minutes to ignite wood!

In many cases, government regulations on building materials can mitigate damages from future fires. On October 20, 1991, a firestorm in the hills upslope from Oakland, California, destroyed or damaged 3,354 single-family homes and 456 apartments (**Case in Point:** Firestorm in the urban fringe of a Major City, Oakland-Berkeley Hills, California Fire, October 19, 1991 p. 504). It killed 25 people, and damage amounted to $2.2 billion. Eighty percent of the homes were rebuilt in the same risky hillside locations with their views of San Francisco Bay and the Golden Gate Bridge. However, new restrictions dictate that roofs be fire resistant, decks and sheds be built with heavier materials that don't burn as quickly, and landscaping use fire-resistant plants.

Evacuation before a wildfire

Preparation for evacuation in a fire-prone area is important because too little time is available when a fire approaches. Forest Service guidelines say to evacuate all family members and animals who are not essential to preparing the property well in advance, designating a contact person and safe meeting place beforehand. Wear cotton or wool clothes that do not burn easily, long pants, gloves, and a kerchief to cover your face. Tune into a portable radio to hear instructions. Place your vehicle in a garage with the garage door closed but an electric garage door opener disconnected so the door can be opened manually if power fails. As with other imminent disasters, put essential items in the car, including important documents. Crucial items might include bank

and insurance records, birth certificates, credit cards, medications, drivers licenses and passports, computer backup files, house photographs and an inventory of its contents, cell phone and charger, and family heirlooms. Close all interior doors, remove non-fire-resistant coverings from windows, turn off pilot lights, leave a light on in each room, and move stuffed furniture to the center of the room. Shut off gas lines, close exterior windows and vents, prop a ladder against the roof for firefighters, unlock exterior doors, connect garden hoses to faucets, and leave trash cans and buckets full of water for firefighters. If there is time, cover windows and vents with plywood at least one-half inch thick and wet down wood roofs.

Forced Evacuation

Under circumstances in which local authorities believe resident's lives are threatened, they may dictate mandatory evacuation of an area. In general, federal policy is to defer to the laws of individual states; local officials work with states to enforce those laws. FEMA and the American Red Cross may facilitate the evacuation and shelter of evacuees. California law authorizes officers to restrict access to hazardous areas, but since no court has upheld a law requiring a person to evacuate their own residence, "mandatory evacuation" apparently means that authorities believe strongly that people should evacuate for their own safety.

What to Do If You Are Trapped by a Fire

Of course, no one ever wants to be trapped by a wildfire, but if it ever should happen, keep in mind how fire behaves, and act accordingly. Light fuels such as grass and dry brush burn fastest, especially uphill and in the direction of the wind. Fuels burn uphill faster, so try to get upwind and downslope of a fire. You are not likely to outrun a wildfire up a slope—it may advance at more than a meter per second. If you can see the extent of a fire below, you may be able to move laterally, then downslope to avoid the advancing fire front. Even on flat areas it may move at 100 to 150 meters per minute. Since **convective updrafts** concentrate in swales or gullies on slopes, creating a chimney effect, avoid those areas—stay on outward-rounded (convex) parts of a hillside.

Try to reach flat, moist, grazed areas, ponds, or streams where vegetation is short, wetter, and not subject to updrafts. If trapped by a closely advancing fire, lie in a ditch, cover yourself with non-combustible material—such as wet clothes or wet dirt—and let the fire burn past you. If you can reach a car, park it over bare ground. Roll up the windows—covering them with opaque material to minimize access of radiant heat, if possible—and lie low. Even if the car catches fire, try to let the intense wall of fire pass before getting out. Except in movies, gasoline tanks rarely explode. Don't try to out-drive a closely approaching fire and don't drive through dense smoke—your chances of making it are slim. The safest place in an uncontrolled fire is in an area already burned. A burning line of short fuels such as grass may have a gap—a trail or road—that may provide access to already burned areas. Studies have shown that even running through a flame front 3 meters high (about 10 feet) and 20 meters (65 feet) deep may be survivable if you can maintain a speed of four meters per second (about 15 miles per hour). You would also need to hold your breath, because inhalation of hot gases causes severe and potentially fatal respiratory-system damage. See "A deadly Wildfire" at the beginning of the chapter and **Case in Point:** Debris Flows Follow a Tragic Fire: Storm King Fire, Colorado, 1994, p. 501.

Public Policy and Fires

As with all hazards, the best way to mitigate damages from fire is to keep people from building in high-risk areas. Unfortunately, with the fiercely independent attitudes of many rural residents, zoning restrictions are considered unacceptable infringements on their freedoms. They insist on their right to do with their property as they wish. Some even build in so-called **indefensible locations**, such as narrow canyons that are too dangerous for firefighters.

Who pays for the costs of fighting fires and trying to save the homes of those who choose to live in fire-danger zones? Insurance companies generally set premiums at levels based on risk and replacement costs. But fire-insurance rates, though rising, pay only a small part of the cost because federal and state governments pay most of the firefighting and cleanup expenses and generally provide a portion of rebuilding costs. When governmental agencies spend millions of dollars fighting fires, and most of those efforts are expended in protecting homes, much of the real cost is borne by the general public rather than the people who choose to live in wooded areas.

Although fire insurance for homes in wildland-urban interfaces is not as high as might be expected given the greater hazard, some insurance companies are now requiring wildland homeowners to clear ground duff and brush, and cut down trees to create a "defensible space" or install fireproof roofs. They are also dramatically increasing insurance premiums. The country's second-largest insurer announced in May 2007 that it would no longer provide new homeowner policies in California because of risks from wildfires and earthquakes.

One needs to ask whether public funds should be used to help people rebuild in places where wildfires—and associated floods and erosion—are ever-present. Our conscience tells us that we should help those in need, but such help merely encourages them and others to live in vulnerable areas. Perhaps, as FEMA has finally learned, we should provide more help to people who are willing to relocate to less-vulnerable places.

Should people accept more responsibility for their own safety and property? Until 2009, Australia, a country having equal problems with major disastrous wildfires, responsibility was largely left to the individual. The Australian government did not focus on protecting private wildland property, but it educated people on how and when to evacuate on their own, before fires reached them, and provided programs that teach ways to make property less vulnerable to fires. Government policy mandated either evacuating early or staying and defending your house from fire. Most deaths occurred during last-minute evacuations.

Australian summers are dry and hot, especially when heat waves push temperatures into the mid 40s°C (110°F) and humidity drops to about five percent. On February 7, 2009, a blistering summer day, those temperatures accompanied 100-kilometer-per-hour winds that fanned bush fires in areas near Melbourne, Victoria. One of several huge, fast-moving firestorms incinerated whole towns, killing 120 of the 173 people that died on that single day. Most of the fires were ignited by fallen power lines, others by lightning, cigarettes, power tool sparks, and arson. The fires followed a decade-long drought and gradual warming that has been associated with human-caused climate change. Following that catastrophe, the Australian government changed its stance and planned to create a nationwide fire alert system involving telephone and text messaging.

An alternative to zoning is to stipulate that if people do build in a vulnerable location, they must not expect public help in times of crisis. Such people should not count on public assistance for fire fighting, stabilizing streams or hillsides before or after floods, or rebuilding after catastrophes. If people still insist on living in fire-prone areas, they could create special fire-prevention districts that would tax their members—for example, a few thousand dollars per year—to create a pool of funds to pay the costs of future fire protection for their homes.

Cases in Point

Debris Flows Follow a Tragic Fire
Storm King Fire, Colorado, 1994 ▶

On the afternoon of July 2, 1994, lightning ignited a fire on a ridge flanking Storm King Mountain seven miles west of Glenwood Springs, Colorado. Dry lightning storms had ignited 40 fires in the general area two days earlier; firefighting teams were assigned to the fires threatening lives, residences, other buildings, and power lines. Firefighters were assigned to the Storm King fire two days later, when it appeared to be spreading. It moved slowly downslope in leaves, twigs, and grass under open pinyon-juniper and oak forest. Bureau of Land Management and U.S. Forest Service firefighters, including smokejumpers, supported by helicopter water drops, had been cutting fire lines flanking the fire and around spot fires.

On July 6, as a cold front moved in from the west, the humidity dropped, and winds from the Colorado River Gorge below were partly redirected up a side canyon to the north. The combined winds of 65 to 80 kilometers per hour created a shear vortex that lifted and spread burning embers well beyond the fire lines. Individual trees began to torch as the ground fire spread into trees. Strong, erratic winds and heavy smoke hampered the effectiveness of helicopter water drops. Firefighters working on fire lines along one flank of the fire and at the ridge above were endangered as the winds increased in speed and varied direction. Fourteen died when they were overrun as they ran upslope ahead of the fast-moving fire

(see the photo at the beginning of this chapter).

In the weeks that followed the fire, rains washed large amounts of ash and mineral soil downslope into drainage channels. Up to a meter of loose, silty sand and ash accumulated along the sides of most drainages, adding to coarser fragments of colluvium collected by slow downslope movements over the years. This set the stage for erosional events that followed.

On the night of September 1, 1994, torrential rains flushed most of this loose material downslope into larger canyons and mobilized debris flows. With no vegetation to slow progress over the ground surface, heavy rain led to rapid runoff. Rills and gullies formed on the bare hillsides, indicating rapid overland flow. These efficiently carried surface runoff to larger channels. The flow—composed of mud, boulders, and other debris—came down across Interstate Highway 70 and engulfed 30 vehicles, pushing some into the adjacent Colorado River. Fortunately, no one died in these flows. Some debris fans blocked almost half

▶ *Wind patterns controlling the Storm King Mountain fire. Regional westerly winds (light blue) and strong winds in the Colorado River Gorge (dark blue) were partly redirected up a side canyon by rising fire currents in this canyon. Turbulent fire eddies carried burning embers that spread spot fires behind the firefighters.*

of the river. Every major drainage, in which the headwaters burned, supplied debris flows that reached canyon mouths. Calculated velocities for many flows reached 4.6 to 8.5 meters per second, and discharges were estimated to be from 73 to 113 cubic meters per second.

Firestorms Threaten Major Cities
Southern California Firestorms, 2003 to 2009 ▶

The first of eleven major fires of the season began in Southern California on October 21, 2003. Hot, dry Santa Ana winds that blew westward out of the high deserts to the east gusted to 100 kilometers per hour. The air compressed and heated up as it descended toward the Pacific, rapidly dehydrating soils and vegetation. Sparks from downed power lines, careless campfire handling, barbeques, cigarettes, and even arson ignited the fires. Ten days later, the rapidly spreading fires had burned more than 300,000 hectares, or 3,000 square kilometers, about the area of Rhode Island. They leveled 3,600 homes and killed 24 people. Nine hundred of the homes were in the San Bernardino Mountains east of Los Angeles, caught in a fire likely started by arson on October 25. The largest, the Cedar fire in San Diego County, burned 2,207 homes and caused 15 deaths. A lost hunter trying to signal rescuers started that fire.

In May, 2009, someone clearing brush with a power tool sparked a wildfire that burned 33 square kilometers and destroyed 77 homes at the base of the Santa Ynez Mountains behind Santa Barbara. In that case, the fires were fanned by local winds, called "sundowners," that sweep downslope as evening air cools high in the mountains, becomes denser, and flows downward. Thirty thousand people required evacuation.

Many of those who died in 2003 had ignored evacuation orders, waiting until the last minute to leave. By then, the fast-moving flames overtook their lone evacuation road, cutting off all escape. The worst fires were in San Diego County, where the high cost of available land encouraged developers and individuals to build in areas vulnerable to brush fires. Some neighborhoods consisted of closely spaced wooden houses surrounded by pine trees, with some trees wedged against flammable roofs and wooden decks. California state law now requires strict standards for building with fireproof materials and clearing brush around homes. However, hundreds of thousands of people ignore the rules, and enforcement has been limited.

Four years of drought in the San Bernardino Mountains northeast of Los Angeles left the trees vulnerable to bark beetle infestation that killed large numbers of them. Although there are tight controls on fire-resistant building materials and brush removal, the cost of tree removal can be as

▶ **A.** *The Southern California fires of October 2003 and 2007 can be seen in these satellite photos. The largest area of fire at the right edge of both images is on the eastern outskirts of San Diego. The largest area of fires in the north-central part of both images is on the outskirts of Los Angeles. The Santa Ana winds carried the fires and smoke toward the southwest.* **B.** *Fires bear down on San Diego in 2003.*

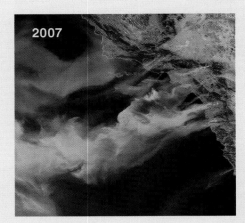

2007

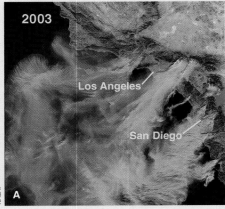

2003

Los Angeles

San Diego

NASA.

A

© John Gibbons / San Diego Union-Tribune / ZUMA Press.

B

▶ *Many homes were closely surrounded by trees and brush, which made it impossible for firefighters to save them from the Cedar fire east of San Diego.*

high as $850 per tree, so few people complied. As one resident put it, "Those dead pine trees are just gasoline on a stick." Such beetle-killed trees make for running crown fires that are almost impossible to stop. Fifteen thousand firefighters fought the fires, along with helicopters equipped with giant water buckets and air-tankers that dropped massive volumes of fire retardant. Insured losses from these fires amounted to more than $1.25 billion. A change in weather on November 1 calmed the Santa Ana winds and brought moisture from the Pacific. Two or three centimeters of white, powdery snow fell in the Big Bear Lake resort area near San Bernardino, slowing the fires and permitting firefighters to construct fire lines. The only drawback was that the rain and snow caused mudslides that closed highways.

Two months later, on December 25, torrential rains fell, with 4.39 inches on the San Bernardino Mountains and 8.57 inches on the San Gabriel Mountains. On the south side of Cajon Pass between them, the route of Interstate Highway 15, the unprotected soils of the burned area unleashed heavy mudslides that raced down canyons and through a trailer camp and recreation center. Fifteen people died in mud as much as 4 to 5 meters deep.

Costs of firefighting and cleanup are estimated to reach another $2 billion. Fire insurance is still available; however,

insurance companies build the heightened risk into their premiums. In some areas, insurers are becoming less willing to write coverage for fire. In others, homeowners are told to replace roofs and clear brush if they want to be covered.

In Hook Canyon on the southeast side of Lake Arrowhead, 350 homes burned. Some homeowners' associations in scenic areas banned property owners from cutting trees, as in the same area in the San Bernardino Mountains. They lifted the ban less than a year before the fire, but by then it was too late to do much even for those inclined to remove trees.

Population increases in outlying areas compound the problem. In recent years, as the population in the Sierra Nevada nearly doubled, so did the number of fires and acres burned, including large fires in 2007. Property damage in the same period went up 5,000 percent.

In this region's Mediterranean climate, more than 90 percent of the rainfall comes during the winter and early spring. Chaparral or scrub brush that covers many Southern California hillsides has lots of resins that help fires burn and spread. This vegetation burns in wildfires every 35 to 65 years, primarily during late summer or early fall, when hot, dry Santa Ana winds blow seaward off the deserts to the east. The hydrophobic soil layer produced by fires (see FIGURE 16-8b), along with the lack of vegetation, lead to greater overland flow and high stream flows. Studies suggest that erosion increases by a factor of 10 following chaparral fires. The frequency of fires and the damage produced are likely to increase dramatically as more people move to the woods.

Exactly four years later, on October 21, 2007, the hot, dry, Santa Ana winds picked up and again ignited catastrophic fires in nearly the same areas as in 2003. A wet winter in 2004–05 that permitted abundant growth of trees and brush was followed by a dry 2006 and an extremely dry 2007 with only about one-fifth the normal rainfall, which left the brush-covered hills tinder-dry. Some fires started when high winds downed power lines that sparked them; arson appears to be responsible for a few others. One fire near

▶ *Ponderosa pine trees killed by pine-bark beetle, northwest of Merritt, B.C.*

Los Angeles was started by a child playing with matches. Once ignited, high, shifting winds reaching gusts of more than 110 kilometers per hour, and essentially no humidity, fanned the intense fast-moving flames. Abundant dry brush, dense groves of eucalyptus trees, and water-starved and insect-killed trees provided abundant fuel. Under intense conditions even air-drops of water and fire retardant are ineffective because they evaporate before reaching the ground. When flames are longer than about three meters, a fire is considered unstoppable, and firefighters must retreat until conditions ease.

Four days later when the winds died down, 2,000 square kilometers had burned, more than 800,000 people had been evacuated, and more than 2,300 homes and businesses were destroyed. Damages totaled more than $1 billion in San Diego County. Only seven people died in the fires—a fact partly credited to aggressive evacuation and to a "reverse 9-1-1" calling system that warned people of a dangerous fire's approach. The worst fire in the northeastern outskirts of San Diego was dubbed the Witch Fire. The Witch Fire and as many as seven other major southern California conflagrations may have been sparked by high winds that caused arcing between high-voltage power lines. Investigations by the California Public Utilities Commission led to millions of dollars of lawsuits against the power companies. In one case a wood power pole snapped

(continued)

off, causing a fire. Those same winds carried burning embers downwind to ignite about two-thirds of the 74 homes that were destroyed in that fire. Many of those fires started on wooden decks.

Farther north in the Santa Monica Mountains west of Los Angeles, the Canyon Fire burned into the center of Malibu. Erratic winds carried firebrands downwind for more than half a kilometer, igniting new fires. This, along with heavy smoke, made conditions extremely difficult for both firefighters on the ground and for firefighting aircraft. By early November, and even with weakening winds, firefighters were still unable to contain two stubborn fires.

On August 26, 2009, an arsonist ignited what became the Station Fire in the San Gabriel Mountains above Burbank and Pasadena, California. Before it could be completely contained on October 16, the fire in the Angeles National Forest had blackened 650 square kilometers (250 square miles) of the southern, forested part of that steep, rugged mountain range. Two firefighters died. By September 23, suppression costs reached $87 million. For weeks afterward, helicopters made water drops on hot spots. In Mid-January, 2010, mandatory evacuations were ordered for the area, this time because heavy rains increased the danger of mud- and debris flows from the burned areas.

Compounding the problem in regions like southern California are two factors: Thousands of people live back in the wooded hills and rugged canyons on narrow, winding roads and federal policies call for firefighters to protect homes, even when they are built in high-risk, poorly accessible areas. Firefighters are under tremendous pressure to protect wildland homes in areas that, if safety were the operative priority, would be left to burn. Yet in spite of major wildfires in 1970, 1993, 2003, 2007, and 2009, even more people rebuilt in the burned areas and with still larger homes. Something needs to change!

Firestorm in the Urban Fringe of a Major City
Oakland-Berkeley Hills, California Fire, October 19, 1991 ▶

The steep hills behind Oakland and Berkeley, California, provide spectacular views west over San Francisco Bay to the City and its Golden Gate Bridge. Land so close to the cities with such views and with short commutes commands premium prices and promotes building of closely spaced, expensive homes. Steepness of the terrain dictated that streets are narrow and winding. In this Mediterranean-type trade-winds climate, summer and early fall weather is hot and dry with breezes blowing offshore. Homes are shaded by big Monterey pine and dense groves of eucalyptus trees; privacy is enhanced by lots of undergrowth. Most people built with wood, including untreated wood shake roofs to enhance the feeling of living in harmony with nature. Unfortunately, all of these characteristics make for a prime wildfire environment.

Overlapping this hazardous environment was a five-year drought that left the plants thoroughly dry and a freeze in December that killed and further dehydrated a lot of the vegetation.

Just before the fire: On Saturday and Sunday, October 19 and 20, temperatures were well above 33°C (90°F), as they

tend to be in October, with low humidity. Firefighters were doing final mop-up from a small grass fire (about 100 × 200 m) that had burned on Saturday but was beaten down and completely soaked by that night. It was also monitored during the night. The next morning, equipment and hoses were still in place for about 2 hours as firefighters searched for any hot spots and began picking up equipment.

The wind: Then just before 11 a.m., winds picked up an ember and blew it into a tree just outside the burn area; it exploded into flame and embers blew all over. Winds in October often come from the east. Locally called Diablo winds because they come from the direction of Mt. Diablo to the east, they are equivalent to the Santa Ana winds of southern California. These winds gusted to 80 to 100 kilometers per hour. Within a few minutes, the wildfire spread to surrounding vegetation and homes. Within the first hour it burned 790 homes, and firefighters were completely overwhelmed. The wind blew the fire downslope to the west and across ridges but the heat from the fire created its own winds, the defini-

tion of a **firestorm**. Heat from the fire also created cyclonic swirls that spread burning embers and carried the fire upslope.

Fighting the fire: Fire officials immediately called for reinforcements and air drops from as far away as the Oregon and Nevada borders and Bakersfield in southern California. More than 370 fire engines and 1,500 firefighters fought the fire. Within half an hour, the fire had reached the Parkwoods Apartments next to the Caldecott Tunnel on Highway 24, an eight-lane freeway under the Oakland-Berkeley hills. The fire moved so fast that it bypassed firefighters, forcing them to retreat to defensible spaces. Hot winds preheated and dehydrated everything ahead of the fire, so when reached by the flames buildings just exploded into flame. Ground crews were virtually helpless to slow the fire's rapid advance. They could not move hoses fast enough, Oakland's hydrant fittings were not the same size used by regional firefighting units, and hoses burned up in the fire.

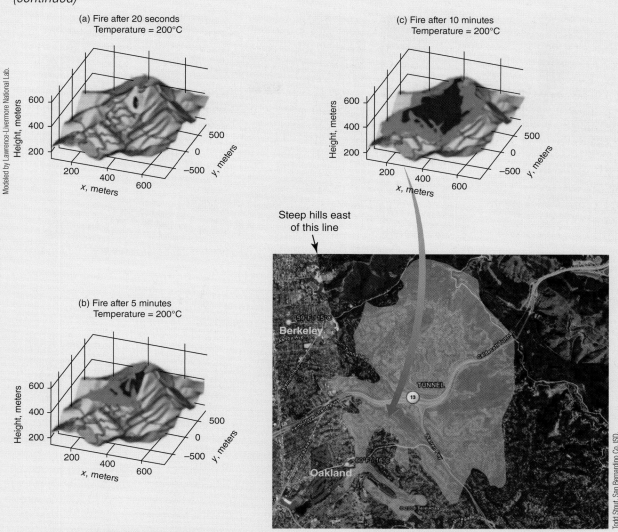

(a) Fire after 20 seconds
Temperature = 200°C

Modeled by Lawrence-Livermore National Lab.

(b) Fire after 5 minutes
Temperature = 200°C

(c) Fire after 10 minutes
Temperature = 200°C

Steep hills east
of this line

Todd Stout, San Bernardino Co. ISD.

▶ **A.** *In its initial stages, the fire burned rapidly, pushed down-valley and over ridges by winds from the east. It moved downslope more than 1500 meters in 10 minutes.* **B.** *Area of the 1991 Oakland-Berkeley hills fire.*

At the fire's height, supply tanks and reservoirs ran out of water due to fire suppression efforts, residents hosing down roofs and shrubs, and leaving sprinklers running when evacuating. In addition, the fire destroyed power lines to 17 water pumping stations used to lift water from Oakland. Department of Forestry pilots complained that as soon as they released their huge buckets of water, the winds just blew it away into a mist. Large air tankers dropped hundreds of loads of fire retardant. By mid-afternoon, the wind slowed and came from the west. Finally that evening, the wind died down to 8 kilometers per hour, permitting firefighters to stop and ultimately contain the fire's spread.

Evacuation: Thirty-degree slopes and narrow, winding streets created serious problems with panicked residents trying to evacuate and firefighters rushing in. Both were hampered by sightseers. Trapped people abandoned cars and fled on foot. More than 10,000 people were evacuated, many by car through smoke and flowing debris on the steep, narrow streets. Problems arose, not only with water supply but with communication between fire-control supervisors and firefighting units and

between units that were brought in from other areas. Radio frequencies were overwhelmed with the heavy use and there were "dead spots" caused by the rugged terrain.

The aftermath: To that date, the fire was the worst in California history: 25 people died, 150 were injured, and 3,810 homes and apartment units were destroyed. Over half the deaths were people trapped in cars. About 610 hectares (6.1 km²) burned within an overall area of about 13.6 km². Total physical damages reached about $2.3 billion (2010 dollars).

This was not the first big fire there, nor will it be the last. A major fire in 1923

(continued)

▶ *Little remained of expensive houses on the steep slopes of the 1991 Oakland-Berkeley hills fire but their chimneys and foundations, along with shells of burned-out cars.*

burned large areas and more than 550 houses in the same general region. A somewhat smaller one struck in 1970.

Fire safety codes were strengthened and fire hydrant couplings are now standardized. Water cisterns were added and radio communications improved and standardized. The Oakland Fire Department now requires brush clearing 9 to 30 meters from buildings and chimneys must have a spark arrestor and no vegetation within 3 meters. Highly flammable building materials, including untreated wood walls, wood roofs, small-dimension lumber for decks, and firewood stacked next to houses, are not permitted. Ground duff, such as dry pine needles and small-dimension lumber, are more like kindling; they ignite quickly. Since flames generally move upslope, fire-resistant materials should cover deck undersides and overhanging eves. Fire entering a house will burn it from the inside, regardless of external fire-proofing; double-pane windows slow its entry.

On June 20 and 21, 2008, a violent, dry lightning storm ravaged northern California with about 8,000 lightning strikes and started more than 2,100 wildfires. The storm, arriving from offshore to the west, after the two driest months on record, was not forecast. In less than a month, about 4,450 square kilometers burned. Two days later, lightning sparked more than 500 additional fires, including one burning in dry grass and brush southwest of Napa that grew to 16 square kilometers in the Coast Range to the northwest. An orange sky choked with smoke covered most of the state north of San Francisco. Up to 21,000 firefighters from the state and U.S. Forest Service, local government fire departments, California National Guardsmen, and teams from Canada, Mexico, Australia, and New Zealand battled the fires. They concentrated on fighting fires threatening lives and buildings. Almost 20,000 buildings were in danger of burning. One resident and two firefighters died.

Further storms ignited more wildfires and stoked others. Fires raged all over northern California, especially west and northwest of Redding and in the Sierra Nevada mountains southeast of Redding. On July 8, 3,800 homes and 10,000 residents of small towns north of Oroville, California, were told to evacuate when temperatures suddenly rose, humidity dropped, and winds drove fire over fire lines. One person who did not leave was found dead in a burned-out house.

Critical View

A This 2006 wildfire burned in the Kaibab National Forest, Arizona.

Jackie Denk, USFS.

1. Why is this fire only on the ground?
2. What could lead to a more intense fire in which the trees burn?

B This wildfire is burning on a hillside.

National Park Service.

1. Why is there so much flame in this fire?
2. In what direction is the fire moving and how can you tell?
3. How does the topography of this location affect the direction the fire is moving?

C This community in the Los Alamos National Forest in New Mexico was destroyed by fire.

Andrea Booher, FEMA.

1. Describe what has burned and what was most susceptible to burning.
2. What would have helped the fire move from one burn area to the next?
3. How would the fire have crossed the paved roads?

D This fire occurred in the Las Padres National Forest in July, 2006.

USFS.

1. This fire appears to be burning green vegetation. How can that happen?
2. Why does the fire appear to be intense?

Chapter Review

Key Points

Fire Process and Behavior

- Fire requires fuel, oxygen, and heat. Lacking any one of these, fire cannot burn. **FIGURE 16-1**.

- Ground fires are fueled by low-growing vegetation such as grass and low shrubs. The higher surface area of dry grass and needles makes them burn faster and more easily. Ladder fuels such as the low branches of trees help fire climb into treetops where it burns as a crown fire. **FIGURE 16-2**.

- Fires can be naturally started by lightning strikes, intentionally set for beneficial purposes, or set accidentally or maliciously.

- Fires spread through radiation, convection, or wind-borne embers called firebrands.

- The spread of fire is affected by local topography. Fire can be accelerated by air funneled through a canyon. A fire tends to accelerate upslope as heat rises.

- Fires start more easily and spread more quickly during dry seasons or droughts, when fuel moisture is low. Winds accelerate fires by directing the flames to fuels and by supplying more oxygen.

Secondary Effects of Wildfires

- Hydrocarbons formed in a fire seal soils, making them hydrophobic, and force water to run off the ground surface rather than soak in; this can lead to flash floods and mudflows. **FIGURES 16-8, 16-9**.

- White ash produced during some fires forms a calcium-sodium hydrate that swells up when wetted. This can seal pore spaces in the soil, leading to runoff and flash floods.

- Black-ash particles lie parallel to the ground and promote water runoff.

- Reduction in evapotranspiration from vegetation after a fire can increase the amount of water in the ground and thereby promote landsliding.

Wildfire Management and Mitigation

- U.S. Forest Management policy has changed over the years from stopping all fires to letting fires burn—as long as they are under control—in order to consume dry fuels.

- Techniques for fire suppression include cutting firebreaks down to bare ground, having helicopters dump large buckets of water, and having air tankers dump huge loads of fire retardant. In out-of-control fires that threaten critical facilities, firefighters sometimes set burnouts to burn back toward an advancing fire and thus deprive it of fuel. **FIGURE 16-12, 13**.

- People who live in the woods are surrounded by fuel for fire. Many build their homes out of wood, which provides more fuel for any wildfire.

- Prolonged dry weather reduces fuel moisture and increases fire danger. Thus, satellite imaging for greenness compared with normal conditions provides an indication of regional fire danger. **FIGURE 16-14**.

- Much or most of the cost of protecting a few who choose to live in dangerous places is borne by the vast majority of the public who do not. Zoning restriction or other policies may be needed to discourage building in fire-prone areas. **FIGURES 16-15, 16-17**.

Key Terms

burnout, p. 495
convective updraft, p. 492
crown fire, p. 489
fire weather potential, p. 496

firebrands, p. 491
firelines, p. 495
firestorm, p. 492
fuel loading, p. 489

fuel moisture, p. 492
hydrocarbon residue, p. 492
hydrophobic soils, p. 492
indefensible locations, p. 500

ladder fuels, p. 489
prescribed burns, p. 495
spot fires, p. 491
surface fires, p. 489

Questions for Review

1. Wildfires are beneficial to forests in what two ways?

2. What are the two main causes of forest fires?

3. What two conditions lead to more fire-prone forests?

4. Why do winds accelerate fires? Give two specific reasons.

5. Why do fires typically advance faster upslope than downslope?

6. What natural conditions and processes of a fire lead to new fires well beyond the burning area of a large fire?

7. Why is hill-slope erosion more prevalent after severe wildfires? Be specific.

8. What main techniques are used to minimize post-fire erosion? Name two specific and quite different techniques.

9. What are the main differences in a stream hydrograph after a large fire?

10. In addition to pouring water or retardant on fires, what techniques are used to fight fires? Name two specific techniques.

11. On which side of a forest is a home at greater risk to fire, and why?

12. What can people living in the forest do to minimize fire danger to their houses?

Discussion Questions

1. Most wildfires cost far more in damages and suppression than any individual could afford. Homeowner's insurance premiums are generally much lower than actuarial costs because state and federal firefighters are paid by their respective governments. Who should pay the overall costs of fire suppression and damages? Why?

2. Many people live in wildlands such as brushy areas and forests that are highly susceptible to wildfire. Should these areas be zoned as off limits to homes? Would such restrictive zoning tread on individual rights? Would that be necessary or appropriate?

3. Firefighters are often required to first defend private residences from wildfire, even when the occupants have evacuated, while neglecting the overall progress of the fire. Is this appropriate? Should homeowners who choose to live in such fire-prone environments be responsible for defense of those homes or is it the responsibility of public agencies to defend those homes?

4. Public agencies such as the U.S. Forest Service spend funds to thin forest areas in the wildland-urban fringe to minimize fire danger. Should public or private entities do such work and who should pay for it?

17 Impact of Asteroids and Comets

Simulated impact of a giant asteroid on the Earth.

NASA.

The Ultimate Catastrophe?

An asteroid 10 to 15 kilometers in diameter struck the Yucatán peninsula of eastern Mexico 65 million years ago, opening a crater about 80 to 110 kilometers in diameter. The energy released from such an impact would have been equivalent to that of 100 trillion tons of TNT or a million 1980 eruptions of Mount St. Helens. The impact would have had dramatic consequences—giant tsunami that swept over the continental margin, widespread acid rain, and wildfires. Prolonged darkness from all the atmospheric dust would have triggered abrupt cooling of Earth for many years, resulting in extinctions of plant and animal life. It appears that the impact of this asteroid killed the dinosaurs and the majority of other species on Earth at that time. What would an impact of this size do to Earth today, and what are the chances of such an impact occurring in our lifetimes?

Asteroids

Projectiles from Space

Space objects that cross Earth's path include asteroids and comets, and the smaller pieces of rock that cause meteors.

Asteroids

Asteroids are chunks of space rock orbiting the sun just like Earth. Astronomers believe that asteroids are remnants of material that did not coalesce into planets when the other planets formed around our sun. This theory is founded on the fact that most asteroids are found in the **asteroid belt** between Mars and Jupiter, where you would expect another planet to be orbiting. Most asteroids stay in the asteroid belt, where they pose no danger to Earth.

The gravitational influence of planets perturbs the orbits of some asteroids, causing them to migrate toward the inner solar system, potentially crossing Earth's path. The dangerous few that may strike Earth within a few days of discovery are difficult to spot because their trajectories toward Earth leave them nearly stationary in the night sky. They are recognized on sequential telescope images as changing position and their approximate paths are then calculated. The majority of these asteroids are less than two kilometers in diameter, and most are less than 100 meters to one kilometer in diameter (**FIGURE 17-1**). Smaller bodies, called *meteoroids*, are more abundant.

Comets

Comets are similar to asteroids but consist of ice and some rock; they are essentially giant, dirty snowballs (**FIGURE 17-2**). They do not come from the asteroid belt but range far beyond the planetary part of our solar

FIGURE 17-1 ASTERIODS

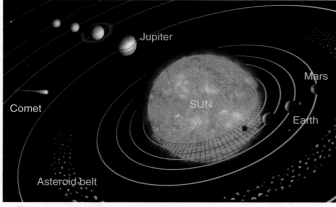

A. This photo compilation shows Asteroid Ida from four different sides.
B. The asteroids lie in a belt between Earth and Jupiter, in the same orbital plane around the sun as all of the planets.

FIGURE 17-2 STRUCTURE OF A COMET

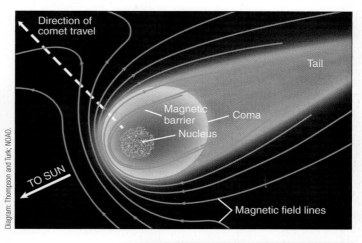

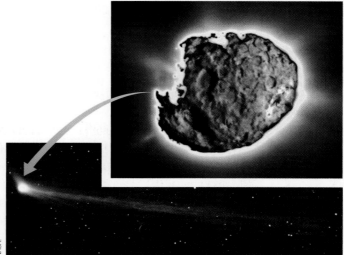

A comet consists of a solid nucleus of a rock and ice mixture surrounded by a coma of dust and gas. The tail is a mixture of water, other volatiles, and dust that the solar wind sprays away from the direction of the sun. This close-view 2004 image of Comet Wild 2 shows its heated surface spraying incandescent dust and gas into space.

system, where they make up the Oort cloud. The Oort cloud forms a vast spherical region around the sun that extends to more than 100,000 times the distance between Earth and the sun. It contains billions of comets. An inner doughnut-shaped zone of millions of comets, the Kuiper belt, lies in the plane of the solar system and extends to 55 times the distance between Earth and the sun.

When they come within the influence of the sun's solar wind, comets spray off water, other volatiles, and dust that form their glowing tails (FIGURE 17-2). The visible tail of a comet does not indicate its direction of travel; rather, the tail points away from the sun. As comets spray off virtually all of their water, they gradually become dehydrated, leaving only rocky material. At that point, they are not easily distinguished from asteroids. In fact, there may be a continuous gradation between rocky comets and icy asteroids.

Most comets have heads less than 15 kilometers in diameter, but they travel at speeds up to 60 or 70 kilometers per second. At those velocities, impact with Earth would be a catastrophe. Some comets traverse our solar system as frequently as once every 10 years. Of all space projectiles, comets have the highest chance of coming close to Earth, or even colliding with it.

Meteors and Meteorites

When small pieces of rock from space enter the atmosphere at high speeds, friction with the air molecules heats the surrounding air to white-hot incandescence, forming **meteors**, or what we sometimes call *shooting stars*. Most space rocks that enter Earth's atmosphere originate in the asteroid belt between Mars and Jupiter (FIGURE 17-1b).

Earth's atmosphere shields us from the impact of most of these space projectiles, because small ones burn up in the upper atmosphere. When a large rock enters Earth's atmosphere, it heats up due to friction with the air and forms a fireball that glows for a period of time before it either disintegrates or survives to strike the Earth. When an object is large enough to survive its passage through the atmosphere without burning up and colliding with Earth, it is called a **meteorite**.

Upon entering Earth's atmosphere, the air around a large fragment of rock heats to become incandescent, but its core typically does not get especially hot. Meteorites are cool enough when they reach the ground that they fall on buildings or dry grass without starting fires. A meteorite that fell in Colby, Wisconsin, on July 14, 1917, was cold enough to condense moisture from the air and become coated with frost.

Because meteorites travel much faster than the speed of sound, we hear only those that are relatively close. If we hear no sound, the meteorite is probably more than 100 kilometers away.

Unless a meteorite is witnessed falling from the sky, it can be difficult to distinguish one from any other rock on Earth. Meteorite fragments are hard to collect, since relatively few meteorite falls are witnessed; fewer than 1,000 have ever been seen in the United States. Only 20 or 30 witnessed falls lead to meteorite finds worldwide each year. Sometimes, large rocks break up in Earth's atmosphere and fall as a **strewn field**, spread out around the main impact site. The Allende meteorite that fell in Chihuahua, Mexico, in 1968 scattered fragments over an oval-shaped area 50 kilometers long and up to 10 kilometers wide.

Identification of Meteorites

Meteorites come in several types, all of which are somewhat similar to rocks thought to make up the deeper interior of the Earth. **Iron meteorites** make up six percent of all meteorites (**FIGURE 17-3**). Because metallic meteorites consist mostly of a nickel-iron alloy, they are attracted to magnets. Even stony meteorites often contain some iron, so they are often magnetic. Most meteorites have a *fusion coating*, a very thin layer of dark glass, formed when friction against Earth's atmosphere heats it above its melting temperature. A meteorite's coating is sharply bounded and quite different than its interior; in contrast, the weathering rind on an Earth rock is generally not sharply bounded. Weathering of the meteorite may later result in rusting that turns the coating reddish brown. Metallic meteorites probably crystallized slowly in the deep interior of a large, solid body in our solar system. Collisions between such bodies, and their breakups, lead to some collisions with Earth. Iron

FIGURE 17-3 IRON METEORITE

Dave Hyndman, at Smithsonian.

The Henbury Iron Meteorite, found in northern Australia, has characteristic small depressions on its surface due to partial melting upon passage through Earth's atmosphere approximately 10,000 years ago.

meteorites are extraordinarily heavy, with densities of 7.7 to 8 grams per cubic centimeter. By comparison, Earth's surface rocks have densities of 2.6 to 3 grams per cubic centimeter and water has a density of 1 gram per cubic centimeter.

Stony-iron meteorites make up less than one percent of all meteorites (**FIGURE 17-4**). They consist of nearly equal amounts of magnesium and iron-rich silicate minerals, such as olivine and pyroxene, in a nickel-iron matrix. They probably come from a zone between the deeper, iron-rich parts of a large asteroid and the outer stony parts.

Chondrites are stony meteorites that make up 93 percent of all meteorites. They consist primarily of olivine and pyroxene, magnesium-iron–rich minerals, along with a little feldspar and glass, similar to the overall composition of Earth's mantle. They have densities of approximately 3.3 grams per cubic centimeter. Millimeter-scale silicate spheres called chondrules enclose nickel-iron inclusions within or surrounding the chondrule. **Achondrites** are stony meteorites that are similar to basalt, a common rock on Earth. They consist of variable amounts of olivine, pyroxene, and plagioclase feldspar.

Iron meteorites are distinctive. They differ from other continental rocks and are generally black unless oxidation over many years has turned their surfaces brown. Iron meteorites are extremely hard, virtually impossible to break with a hammer. Over time, when exposed to the weather, they rust to iron oxides. Pieces of manufactured iron are more abundant and may look similar, but meteorites can be distinguished by polishing a sawn surface. Iron meteorites show intersecting sets of parallel lines marking the internal structure of the nickel-iron minerals. When these lines are accentuated by acid etching, they show the distinctive patterns characteristic of iron meteorites.

Even stony meteorites are distinctively heavy. They are made of peridotite that is 15 to 20 percent more dense than most common rocks. Most contain enough metallic iron to be still heavier. Stony meteorites may be broken, exposing the meteorite's fresh interior. If the broken surface is unaltered, you may see small inclusions of metallic, silver-gray-colored nickel-iron that strongly suggest a meteorite.

Some meteorites are rounded by their passage through the atmosphere. Iron meteorites are commonly more angular and sometimes twisted-looking. Some show rounded thumb-sized depressions. Others have an orientation related to their direction of travel, with a smooth leading end and a pitted rear end (see FIGURE 17-4, bottom).

Evidence of Past Impacts

Asteroid impacts have been recognized worldwide (**FIGURE 17-5**, p. 514). The largest proportion presumably fell into the oceans that cover two-thirds of Earth's surface. Most of those have remained undetected because details of the ocean floor are not well known. Because older parts of the ocean floors have subducted into oceanic trenches, evidence of early impacts is no longer available. On continents, impact sites are broadly distributed but have been found concentrated in areas of greater population or well-exposed, vegetation-poor regions. As with all natural hazards, recognizing and analyzing evidence of past impacts helps scientists determine how often meteorites strike and what consequences they have.

Impact Energy

The record we have of past collisions with space objects is a result of the tremendous energy of these impacts. Because most asteroids travel at velocities of 15 to 25 kilometers per second, they have incredible energy. Recall that the energy of a moving object is equal to its mass times the square of its velocity (**By the Numbers 17-1**: Energy, Mass, and Velocity). Thus, the energy doubles for every doubling of the mass of the asteroid but quadruples for every doubling of

FIGURE 17-4 STONY-IRON METEORITE

This meteorite that crashed through the roof of a home in Auckland, New Zealand, in 2004 is approximately 13 centimeters long.

Meteorite Magazine / Grant Christie.

▶ **By the Numbers 17-1**

Energy, Mass, and Velocity

$$ke = 0.5mv^2$$

where:

ke = kinetic energy

m = mass

V = velocity

FIGURE 17-5 RECOGNIZED IMPACTS WORLDWIDE

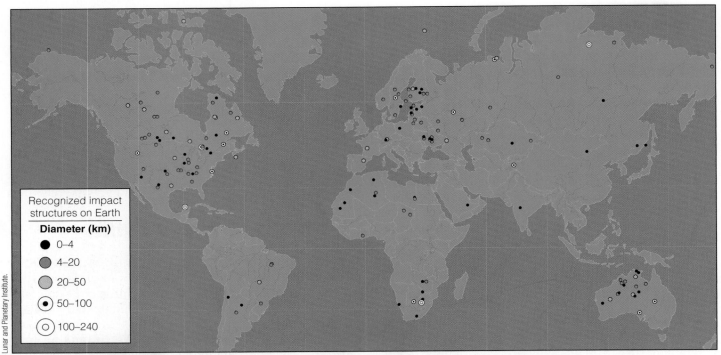

Lunar and Planetary Institute.

Recognized impact structures on Earth

Diameter (km)
- 0–4
- 4–20
- 20–50
- 50–100
- 100–240

Meteorites of various sizes have peppered Earth for billions of years. The uneven distribution of recognized impacts is related mainly to how easily they are identified because of population density and land cover.

the velocity. Comets typically have less mass than asteroids but greater velocity. Because comets are mostly ice, with a density of 0.9 grams per cubic centimeter, their overall densities, including their rock component, tend to be similar to that of water, 1.0 gram per cubic centimeter. However, comets tend to travel at much higher velocities than asteroids—for example, 60 to 70 kilometers per second. Because the energy quadruples for every doubling of the velocity, comets can have extremely high energy in spite of their lower densities.

On impact, the kinetic energy of the incoming object is converted to heat and vaporization of the asteroid and the target materials. This melts more rock, excavates a crater, and blasts out rock and droplets of molten glass. The result is a huge fireball that heats and melts rock and burns everything combustible.

Impact Craters

One might imagine that an asteroid striking Earth would create a big hole in the ground that has a shape dependent on the incoming angle of the object. However, the incredible velocity, and therefore the energy, of the impactor requires that it explode violently on impact. The effect is more like a missile being fired into a sand surface. It blasts a nearly round hole regardless of the impact

angle (**FIGURE 17-6** and **Case in Point:** A Round Hole in the Desert—Meteor Crater, Arizona, p. 523). The largest identified open crater is the 100-kilometer-diameter Popigai Crater of Siberia.

If the impactor is large enough, the explosion violently compresses material in the bottom of the crater, accelerating it to speeds of a few kilometers per second and ejecting material outward at hypervelocity. The center of the crater rebounds rapidly to form a central cone; that cone and the outer rim almost immediately collapse inward to form a wider but shallower final crater (**FIGURE 17-7**). The older Manicouagan Crater of eastern Canada is preserved as a striking ring of lakes in the basement rocks of the Canadian Shield (**FIGURE 17-8**, p. 516).

Impact craters provide evidence about the size and date of past impacts and also help us speculate about the damage those impacts would have caused. For example, the age and size of the Chicxulub crater basin in Mexico, mentioned at the beginning of this chapter, makes that impact the most likely cause of the mass extinction that killed the dinosaurs. That crater is 195 kilometers in diameter, with a pronounced central uplift (**FIGURE 17-9** , p. 516). An impact large enough to have such a devastating effect should theoretically create an initial crater 200 kilometers in diameter—with a much

FIGURE 17-6 OPEN IMPACT CRATER

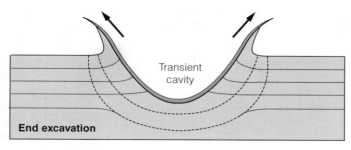

End excavation

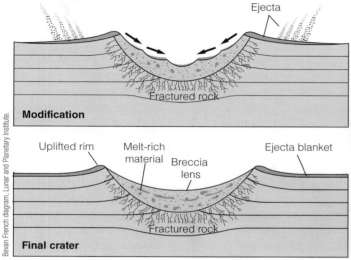

Ejecta

Fractured rock

Modification

Uplifted rim — Melt-rich material — Breccia lens — Ejecta blanket

Fractured rock

Final crater

Bevan French diagram, Lunar and Planetary Institute.

This panoramic view of Meteor Crater in Arizona shows its relatively small size for an impact crater, 1.2 kilometers across. First, a transient crater is excavated, compressed, and fractured; the base of the cavity melts; and the rim is raised. Then the ejecta blanket spreads around the cavity, and the part of the rim slumps back into the cavity. Finally, fallback material partly fills the cavity, along with some melt-rich material.

David Roddy, USGS.

FIGURE 17-7 LARGE IMPACT CRATER

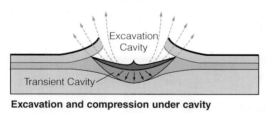

Excavation Cavity

Transient Cavity

Excavation and compression under cavity

Uplift, excavation and collapse

Central basin — Crater rim

Megablock zone — Central peak — Melt/allochthonous breccia

Final crater — Stratigraphic uplift

Modified from Morgan.

First, a transient crater is excavated and compressed, and the base of the cavity melts. Second, the base of the transient cavity rebounds as excavation continues. Then, the raised rim of the transient crater and central uplift both collapse to form a larger and shallower crater basin partly filled with inward-facing scarps, large blocks, smaller fragments, and melt rocks. The final crater is much broader and shallower than the initial transient crater. "Autochthonous" refers to materials more or less in their original position. Some large craters on the Moon have this central uplift (see below).

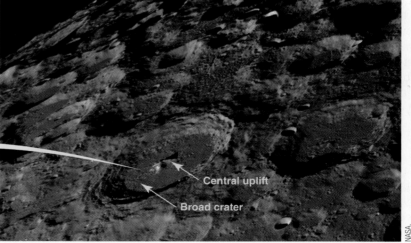

Central uplift

Broad crater

NASA.

FIGURE 17-8 LARGE IMPACT CRATER

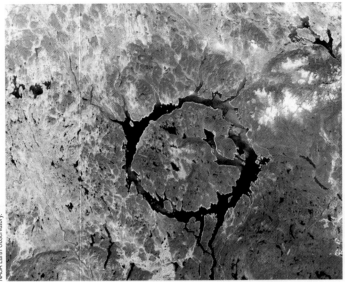

The giant Manicouagan Crater of the eastern Canadian Shield is one of the largest impact craters observable on the surface of the Earth. The crater shows as a dark ring of lakes 65 kilometers in diameter in this Space Shuttle image.

larger final diameter—but Chicxulub is the largest crater of the right age found so far. The crater was slowly buried later by the quiet accumulation of limestones on the continental shelf, so it is not exposed at the surface. However, this burial also preserved some features that tend to

erode with time. It has been studied through drilling and geophysical methods.

Aside from the final crater shape, the record of the impact is preserved in minute particles and chemical traces in sedimentary rocks deposited around that time. There is also evidence of huge tsunami waves formed in the Gulf of Mexico in this period. At the Brazos River, Texas, these waves left debris 50 to 100 meters above sea level. Massive submarine slope failures were common around the Gulf of Mexico and along the east coast of North America at that time.

The Manson impact structure in central Iowa was once thought to be 65 million years old but more recently determined to be 74 million. At 35 kilometers in diameter it is much too small to be the main impact site for the K-T extinction event. Because concurrent impacts are known for other events, breakup of the asteroid may have caused multiple impacts. This impact structure is buried under ice-age glacial deposits but has been studied through drilling and geophysical methods. Basement granite and gneiss under the center of the crater have been raised at least four kilometers above their original position. As with most other well-documented asteroid impact sites, shocked mineral grains are present in the target rocks.

Another major impact structure is evident on the Virginia continental margin near the mouth of Chesapeake Bay. From its crater shape as well as other evidence, we know that a major asteroid or comet about 3 km in diameter and travelling about 110,000 kilometers per hour impacted this site about 35 million years ago (**FIGURE 17-10**). It initially blew open a crater 35 kilometers in diameter, followed by

FIGURE 17-9 THE CHICXULUB IMPACT SITE

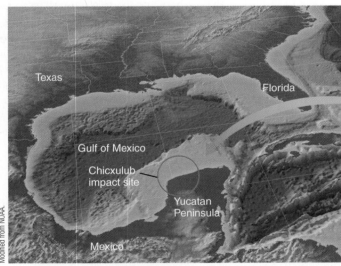
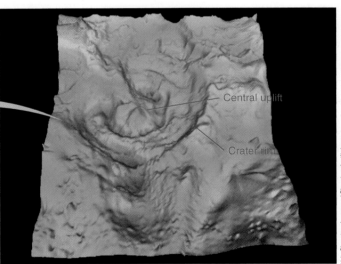

The Chicxulub Crater at the northern edge of the Yucatán Peninsula of eastern Mexico is thought to be the site of the impact that caused the extinction of the dinosaurs 65 million years ago. The crater was imaged using geophysical methods because it is buried below the sea floor and partly filled with sediment. The low-lying areas around the current Gulf of Mexico were a shallow continental shelf 65 million years ago.

FIGURE 17-10 CHESAPEAKE BAY IMPACT

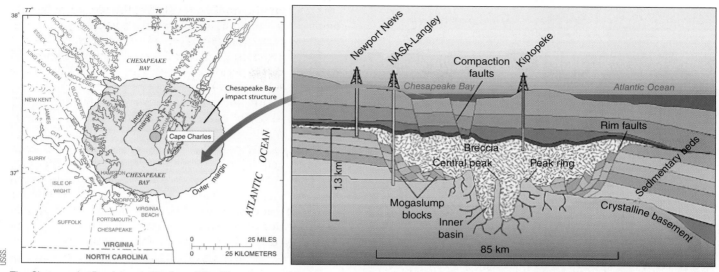

The Chesapeake Bay impact site, from 35 million years ago, is now covered by a few thousand feet of sediments and water of the bay.

collapse of its near-vertical walls to form a wider, shallower crater about 85 kilometers in diameter, with its floor two kilometers below sea level. Millions of tons of fragments and dust would have been ejected tens of kilometers into the upper atmosphere. Pieces of those fragments, with shreds of impact melt that fell back into the crater, were found by drilling at the site between 2004 and 2006. The impact would have had dramatic consequences—giant tsunami that swept over the continental margin, widespread acid rain, and wildfires. Prolonged darkness from all the atmospheric dust would have triggered abrupt cooling of Earth for many years, resulting in the extinction of many plants and animals.

Shatter Cones and Impact Melt

The same energy that blasts out a crater also has an effect on rocks in the area. Rocks on the receiving end of an impact show distinctive characteristics, especially **shatter cones**. These cone-shaped features, with rough striations radiating downward and outward from the shock effect, range from 10 centimeters to more than a meter long and are considered proof of asteroid impact (**FIGURE 17-11**, p. 518). They form most readily in fine-grained massive rocks. Apexes of shatter cones point upward toward the shock source. Cones directly under the impactor should be vertical; those off to the sides flair down and outward from the source in "horsetail" fashion. Thus, the distribution of shatter cone orientations provides evidence for the location of the center of an impact site. Individual cones are often initiated at a point of imperfection, sometimes a tiny pebble, as in FIGURE 17-11b. The Sudbury structure

north of Toronto, Ontario, has some of the best-known shatter cones (**Case in Point**: A Nickel Mine at an Impact Site—The Sudbury Complex, Ontario, p. 524). They are distributed over an area 50 kilometers by 70 kilometers around the intrusion.

Sometimes melting of the asteroid produces the glass; sometimes the glass forms by melting the target material. Molten glass droplets sprayed out during impact form tiny, hollow *spherules* (**FIGURE 17-12**, p. 518). These droplets of glass, generally a millimeter or two in diameter, and often altered to green clay, are typically the most obvious signatures of an impact feature found in sediments. In silicate target rocks, the impact melt may also form sheets, dikes, and ejecta fragments, and be disseminated in breccias. The melt develops under extremely high impact pressures. Impact-melt compositions are a mixture of the compositions of target rocks shocked above their melting temperatures. Because the impacting meteorite is often melted as well, impact melt may contain small but extraordinary amounts of nickel, iridium, platinum, and other metals that are abundant in iron meteorites.

Fallout of Meteoric Dust

Although the impact site itself is the best source for clues about past impacts, other evidence can be found far from such sites. Fragments and dust sprayed out from a large site can drift around Earth. The enormous impact at the Chicxulub site blew out enough material to deposit a thin, dark layer called Cretaceous-Tertiary boundary (**K-T boundary**) clay (**FIGURE 17-13**, p. 519). The dust and elemental-carbon soot layer from fires ignited by the impact

FIGURE 17-11 SHATTER CONES

A. Robert Hargraves, who discovered the Beaverhead impact site in Medicine Lodge Valley southwest of Dillon, Montana, points to a shatter cone cluster in an outcrop. Note that the apex of each shatter cone points upward. **B.** Note the tiny pebble at the apex of the cone (arrow) in this close view of a shatter cone at the Beaverhead impact site.

FIGURE 17-12 SPHERULES

Green glass spherules collected from the surface of the moon by Apollo 15 astronauts.

fireball 65 million years ago is generally only a centimeter or so thick and chocolate brown to almost black in color. The large amount of soot is related to worldwide fires that burned much of the vegetation on the planet.

The clay in this boundary layer contains grains of shocked quartz and other minerals, as well as spherules. The abundance of quartz in the boundary clay strongly suggests that the dinosaur-killing impact occurred in quartz-rich continental rocks, probably in an area rich in granite, gneiss, or sandstone.

The 15.1-million-year-old, 24-kilometer-diameter Ries Crater in Germany formed in limestone, shale, and sandstone over crystalline basement rocks. The impact ejected a blanket of sedimentary rock fragments that were cold when they landed 260 to 400 kilometers to the east. These fragments contain droplets of frozen melt, or *tektites*, derived from the explosive melting of underlying crystalline basement rocks. Passage through the atmosphere aerodynamically shaped the glass, but it landed cold. Shock features

FIGURE 17-13 K-T BOUNDARY

Donald Hyndman.

The Cretaceous-Tertiary (K-T) boundary layer is exposed at Bug Creek in northeastern Montana. The inconspicuous lowest dark layer (arrow) marks this important boundary.

FIGURE 17-14 MULTIPLE IMPACTS

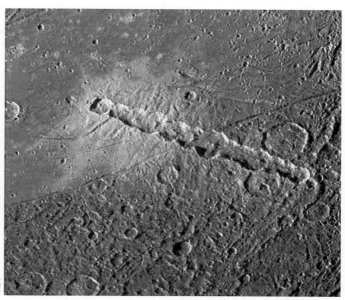

NASA.

A chain of impacts on Jupiter's moon Ganymede from fragments of the Shoemaker-Levy 9 comet that broke up on its approach to Jupiter.

are widespread. The interpreted sequence of events begins with high-speed shock waves and vapor blown out at shallow angles to Earth's surface, quickly followed by high-speed ejection of target material. The tektites formed as melted target material probably ejected above the atmosphere before falling back to Earth. The crater is circular, but the eastward fallout pattern of the tektites suggests that the impactor arrived from the west.

Also found in the boundary clay of the 65-million-year-old K-T event are anomalous amounts of iridium and other platinum-group elements, an occurrence called the **iridium anomaly**. The anomalous amounts are tiny, some 0.5 to 10 parts per billion, but those elements are essentially absent in most rocks except for meteorites and dark, dense rocks like peridotite and others derived directly from Earth's mantle. Some iridium is vaporized during the eruption of large basaltic volcanoes, such as those in Hawaii, but the amounts are too small to account for the quantities found in the K-T boundary clay. The boundary clay from the initial find contains 30 times as much as would be expected from the normal fallout of meteoric dust that rains down on Earth daily. Elsewhere, in Denmark the clay contains up to 340 times as much. Vaporization of the impacting asteroid is thought to spread the iridium worldwide, where it has now been found at roughly 100 sites. That boundary in sedimentary rocks marks the demise of the dinosaurs and 60 to 75 percent of all other species.

Multiple Impacts

An asteroid would be likely to break up in the atmosphere, so we should expect multiple impacts in a sequence. Most impacts arrive obliquely to Earth's surface, a significant proportion at 5 to 15 degrees from horizontal. Such asteroids ricochet and often break up in the atmosphere into five to ten fragments that still have about 50 percent of the original velocity.

A dramatic example of this was seen with multiple impacts on Jupiter. In 1992, the Shoemaker-Levy 9 comet broke up into 21 fragments during a close approach to Jupiter as it was pulled apart by the huge planet's intense gravitational field. Then, in July 1994, the fragments, all less than one kilometer in diameter, impacted one after another in an arc across Jupiter, over a period of six days (**FIGURE 17-14**). Comets typically travel at higher velocities than asteroids; this one was traveling at roughly 60 kilometers per second. Apparently simultaneous impacts at different sites on Earth suggest that these may be multiple fragments of larger asteroids or comets. In other cases, multiple impacts may be fragments that broke up after ricocheting off the atmosphere.

Consequences of Impacts with Earth

Impact of a modest-sized asteroid 1.5 to 2 kilometers in diameter is thought to be enough to kill perhaps one-quarter of the people on Earth and threaten civilization as we know it. The consequences of such an impact would be disastrous for life. All other hazards and disasters pale in comparison.

Immediate Impact Effects

The fireball or ejecta from the impact would ignite fires within hundreds of kilometers of the impact site. A heavy plume of smoke would linger for years in the atmosphere. Sulfate aerosols and water would be added to the atmosphere as well. A large portion of the ozone layer, which protects us from the sun's ultraviolet rays, probably would be destroyed.

If the asteroid broke into fragments, numerous smaller masses still traveling at hypervelocities would provide the heat and energy to cause widespread reaction between nitrogen and oxygen in the atmosphere. That would form nitrates that would combine with water to form nitric acid. The resulting acidic rain would damage buildings, as well as crops and natural vegetation.

At about the same time, dust blown into the stratosphere would block sunlight almost worldwide to the equivalent of an especially cloudy day. Large particles would settle out quickly, but dust particles smaller than 1/1,000 millimeter would remain in the stratosphere for months. Any temporary increase in temperature from widespread fires would quickly decrease due to less solar radiation reaching Earth's surface. The dust would be distributed worldwide because much of it would be blown out of the atmosphere before settling back into it. Agriculture would probably be wiped out for a year, and summertime freezes would threaten most agriculture after that. Many specialists view global-scale wars over food as inevitable. Such desperate conflicts would likely kill a large percentage of the world's population. Others argue that humans are more resourceful than that. If there were, say, a decade of warning before the event, sufficient food supplies could be grown and stored to outlast the period of darkness. However, that is a pretty big "if"!

Impacts as Triggers for Other Hazards

Impact of a large asteroid would trigger many other hazards. Earthquakes would occur within hundreds of kilometers of the impact site. Asteroid ocean impacts would form tsunami as water sloshed into and out of the crater. Some of these tsunami could be extremely high. The waves would inundate coastal areas for tens of kilometers inland, areas inhabited by a large proportion of the world's population. As an example, the Chicxulub collision is calculated to have produced a 200-meter-high wave. The runup likely averaged more than 150 meters in height, with a maximum of 300 meters.

An impact might also cause volcanic activity. The large, essentially circular features that cover most of the moon's surface that can be seen on a dark night are known as lunar maria ("lunar seas"). Most are filled with basalt. These are recognized as the products of major impacts, mostly during the early evolution of the moon and Earth. Many are enormous, much larger than any known impact sites on Earth. Why have no such giant sites been found on Earth, especially given that the much greater gravitational attraction of Earth should pull in many more asteroids than the moon?

Flood basalts on Earth have a curious habit of forming at about the time of major extinctions of life, so a relationship between them seems possible. The immense Deccan basalts of western India, for example, erupted at about the time of the dinosaurs' demise (along with a large proportion of other species) 65.7 million years ago. If there is a cause-and-effect relationship, what is it? Most of the other handful of major flood basalt fields on Earth erupted at times of major extinctions of animal species. One imaginative but controversial suggestion is that the impacts of giant asteroids not only caused major extinction events but also triggered mantle melting and the eruption of flood basalts through a sudden relief of pressure in the Earth's mantle. One problem with this proposal is that a layer containing shocked quartz grains has been found recently—not at the base of the 65-million-year-old Deccan flood basalts in India, but between two major periods of basalt eruption.

Mass Extinctions

The impact of an asteroid 10 to 15 kilometers in diameter 65 million years ago was associated with the demise of the dinosaurs and the death of between 40 and 70 percent of all species (see chapter introduction). If an asteroid of that size were to impact Earth today, it would annihilate virtually everyone on the planet and a large proportion of other species. Almost immediate flash incineration near ground zero would accompany strong shock waves. The 10-kilometer-diameter impactor of 65 million years ago most likely produced sufficient nitric acid rain to be the primary cause of the mass extinction.

For those species lucky enough to be far away from the impact, extinction would follow the impact from various indirect causes. Acid rain would kill vegetation and sea life around the planet. Dust, soot from fires, and nitrogen dioxide would blot out the sun, so animals that were not incinerated would freeze and starve to death. Plants would also die because of the drop in temperature and sunlight. With the impact of a 10-kilometer asteroid, land temperatures would, depending on assumptions, likely drop worldwide to freezing levels within a week to two months. Because of the large heat capacity of water, sea-surface temperatures would drop only slightly. Widespread fires would ignite from lightning strikes after much of the vegetation had died from either freezing or toxic atmospheric effects.

The vaporization of a 10-kilometer-diameter chondritic asteroid on impact with Earth would probably not only generate strong acid rain but also spread nickel concentrations of between 130 and 1,300 parts per million, even when diluted with 10 to 100 times the amount of target Earth material. A nickel-iron impactor would generate even more. This is many times the toxic level for chlorophyll production in plants. Seeds and roots would likely not recover for a long time.

Other mass extinctions befell the Earth millions of years ago, including extreme events that marked the end of Triassic time, the end of Permian time, late in Devonian time, and the end of Ordovician time (see Appendix 1 online). The end-of-Permian event, 250 million years ago, wiped out approximately 90 percent of all life species. The time of that extinction event correlates with the time of a gigantic outpouring of flood basalt lavas in Siberia, which erupted in the geologically short period of about one million years.

It has been suggested that the extinctions could have been caused by huge amounts of CO_2, methane, and other gases expelled by the basalt eruptions. Such flood basalts do not appear related to normal types of plate tectonics but rather to the rise of a deep mantle plume and hotspots, as described in Chapter 2. Another hypothesis, though controversial, is that both the extinctions and the flood basalt eruptions were caused by the impact of a massive asteroid. Although such a giant collision would almost certainly kill us all, the chances of it happening are exceedingly small.

Evaluating the Risk of Impact

Projectiles from space are not significant natural hazards on most people's horizon, although Hollywood's fictional dramatizations of such catastrophes have heightened awareness for some people. In reality, however, no other physical hazard has such a dire potential. Although the odds of a huge asteroid colliding with Earth are tiny, the consequences of such an impact would be truly catastrophic. Even without impact, grazing contact with Earth can be tremendously destructive (**Case in Point:** A Close Grazing Encounter—Tunguska, Siberia, p. 525). A large impact could wipe out civilization on Earth.

As with other hazards, small impacts are quite common, while giant events are rare (**FIGURE 17-15**). On average, a 6-meter-diameter asteroid collides with Earth every year; one 200 meters in diameter strikes the planet every 10,000 years on average. Only 1,500 or so asteroids larger than one kilometer in diameter are known to be in Earth-crossing orbits (those that pass through Earth's orbit around our sun). The largest is 41 kilometers in diameter. Most of these cross Earth's orbit only at long intervals, so the chances of a collision are fortunately extremely small. The orbits of some asteroids vary from time to time because of the gravitational pull of various planets. Innumerable smaller asteroids also cross Earth's orbit. We live in a cosmic shooting gallery with no way to predict when one will hit the bull's-eye. We can only estimate the odds.

Your Personal Chance of Being Hit by a Meteorite

Meteorites fall from the sky daily, but has anyone ever been struck? The only well-documented case occurred on November 30, 1954, in Sylacauga, Alabama, when a 3.8-kilogram meteorite crashed through the roof of a house, bounced off

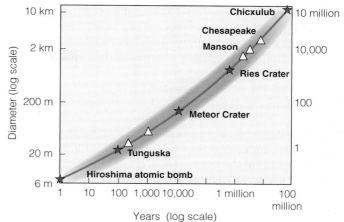

FIGURE 17-15 RISK OF ASTERIOD IMPACT

This log graph plots the approximate impact chances for asteroids of given sizes hitting Earth.

a radio, and hit Mrs. Hulitt Hodge, who was sleeping on a sofa. She was badly bruised but otherwise okay.

There have been some close calls. On the night of April 8, 1971, a 0.3-kilogram meteorite came through the roof into the living room of a house in Wethersfield, Connecticut. No one was injured. In 1982 a 2.7-kilogram meteorite struck a house just 3 kilometers away. In another unusual case, on October 9, 1992, a bright fireball seen by thousands of people from Kentucky to New Jersey came down in Peekskill, New York, where it mangled the trunk of Michelle Knapp's car. Hearing the crash, she went outside to find the 11.8-kilogram meteorite in a shallow pit under the car. It was still warm and smelled of sulfur. On June 8, 1997, a 24-kilogram meteorite crashed into a garden 90 kilometers northeast of Moscow, Russia, where it excavated a one-meter-deep crater. On the night of March 26, 2003, a 10-centimeter stony chondrite meteorite crashed through the ceiling of a house in Park Forest, near Chicago. It was one of many, most of them small, that damaged six houses and three cars over an area of about three by ten kilometers. On Saturday, June 12, 2004, at 9:30 a.m., a 1.3-kilogram stony meteorite crashed through the roof of a home in Auckland, New Zealand. The rock was 7 by 13 centimeters, gray with a black rind, and had rounded edges (see FIGURE 17-4). It was hot when it landed on the Archer family's sofa in the living room, but no one was hurt.

On November 20, 2008, a 13-kilogram meteorite fell on a farmer's land south of Lloydminster, Saskatchewan. More than 1,000 fragments of the meteor have been found, but three larger pieces tracked on eyewitness videos have yet to be located. Given their estimated sizes of 45 to 200 kilograms, they are expected to have created holes in the ground. More recently, on January 18, 2010, a fist-sized 0.3-kilogram chondrite meteorite fell through the roof of a doctor's office in Lorton, Virginia, a suburb of Washington, D.C.; it broke into several pieces on impact.

In 1860, a meteorite falling on a field near New Concord, Ohio, killed a colt, and in 1911, another killed a dog in Nakhia, Egypt. There are no records of a person having been killed—at least so far! Clearly, you need not stay awake at night worrying about the possibility of being hit by a meteorite.

Chances of a Significant Impact on Earth

Earth's gravitational field is constantly sweeping up stray asteroids, so their number should be decreasing with time. However, collisions in the asteroid belt create new asteroids that leave the belt; some of those fall into Earth-crossing orbits.

Also dangerous are comets that have orbits outside our solar system but become visible when they pass close to the sun or Earth. About ten percent of impacts on Earth and the moon are caused by comets. The periodicity of major impacts, based on the ages of impact craters and on theoretical aspects of interacting orbits and other movements within our galaxy, suggests an average interval of 33 million years. Study of fossils suggests a 26 to 31 million year periodicity of genus-extinction events, but some scientists argue against that assertion. Although the number of Earth impacts was thought to have slowed some 3.5 billion years ago, a recent study of the ages of glass-melt spherules in lunar soils indicates that after 500 million years ago, the impact rate increased again to previously high levels.

The spectacular Hale-Bopp comet, with a diameter of approximately 40 kilometers, was seen by most people on Earth between January and May 1997. It came within Earth's orbit around the sun—but on the other side of the solar system. Its closest approach to Earth on March 22 was still 320 million kilometers away (the sun is about 350 million kilometers from Earth). If it had collided with Earth, the energy expended would have been tens to hundreds of times larger than that of the dinosaur-killing asteroid of 65 million years ago. It is a long-period comet that spends most of its orbit far beyond our solar system before blowing through it again after thousands of years. Whether one of these objects annihilates us is a matter of where Earth is in this shooting gallery when one of these objects passes through.

On December 6, 1997, astronomers of the University of Arizona Spacewatch program spotted a huge chunk of space rock, an asteroid 1.5 kilometers in diameter, apparently on a near-collision course with Earth. Asteroids spotted in telescopes are tracked over time until astronomers have enough information to determine how close their trajectories will take them to Earth. In March 1998, the astronomer who spotted this particular rock, Asteroid 1997 XF11, was able to plot its path and determined it would come dangerously close to Earth; he predicted the nearest approach to Earth to be on October 26, 2028. However, in his excitement to announce the event to the press, he neglected to check earlier measurements on the same object. The corrected results indicate that the asteroid will come no closer than 2.5 times the distance to the moon.

If a mass the size of Asteroid 1997 XF11 ever did strike Earth, the energy expended would be that of two million Hiroshima-size atomic bombs. According to Jack G. Hills of the Los Alamos National Laboratory, if it struck the Atlantic Ocean, it would create a tsunami more than 100 meters high that would obliterate most of the coastal cities around that ocean. If it hit on land, the crater formed would be 30 kilometers across and darken the sky for weeks or months with dust and vapor. For comparison, the infamous asteroid that exterminated the dinosaurs and as much as 75 percent of all other species on Earth 65 million years ago was 10 to 15 kilometers in diameter.

More-recently, asteroid 2004 MN4 (Asteroid Apophis) was expected to pass very close to Earth on Friday, April 13, 2029, with collision odds initially calculated as one chance in thirty-seven. NASA's further measurements and calculations now suggest that it will miss Earth by about 30,000 kilometers (about 2.8 times Earth's diameter) and the odds of a 2029 Earth impact are less than one in 100,000.

The chance of that smaller-sized asteroid striking Earth in any one year is estimated at one in several hundred thousand (see FIGURE 17-15). This probably seems like such a minute chance as to be irrelevant. People do believe, however, in their chances of being dealt a royal flush in poker (1 chance in 649,739) or of winning the multimillion-dollar lottery jackpot (1 chance in 10 to 100 million). The chance that Earth will be struck by a civilization-ending asteroid next year is greater than either of those. As with other natural hazards, the low odds of such an event do not necessarily mean that it will be a long time before it happens. It could happen at any time.

What Could We Do about an Incoming Asteroid?

What if astronomers were to discover a very large asteroid or comet, at least as large as the one that did in the dinosaurs, and they determined it was on a collision course with Earth (see chapter-opener photo)? What would we see without a telescope, and when would we first see it? By the time it reached inside the moon's orbit, it would be three hours from impact. It would first appear like a bright star, becoming noticeably brighter every few minutes. An hour from impact, it would be as bright as Venus. Fifteen minutes from impact, it would appear as an irregular mass, rapidly growing in size. Three seconds from impact, it would enter Earth's atmosphere with a blinding flash of light, traveling at perhaps 30 kilometers per second. Then instant annihilation! After that, the impact would produce the same effects recorded in the extinction event 65 million years ago.

Clearly we could not survive the scenario outlined above. So what, if anything, can we do about it—that is, other than bury our heads in the sand and wait for the inevitable? Suggestions include blasting the asteroid into pieces with a nuclear bomb or attaching a rocket engine to it to deflect it away from Earth. For an asteroid discovered within days or months of impact, blasting it into smaller pieces might unfortunately just pepper a large part of Earth with thousands of smaller pieces—not a comforting scenario. For an asteroid discovered years or centuries before impact, blasting it into thousands or millions of pieces with a nuclear weapon would spread the fragments out over time so that most or all could miss us. Conventional explosives detonated on one side of an asteroid might deflect it rather than shatter it. Another possibility involves deflecting an asteroid by changing the amount of heat radiated from one side; for example, we could coat part of it with white paint. The effect is weak, but over a long time, it could shift its orbit enough to narrowly miss the Earth.

NASA catalogs **near-Earth objects** larger than one kilometer in diameter. The Jet Propulsion Laboratory in Pasadena, California, can detect objects as small as 10 to 20 meters or so in diameter. Smaller ones are not important; most flame out before they reach Earth's surface. However, as noted, the impact of a large one could be catastrophic. Sometimes scientists tracking asteroids do not get much warning before one comes close. On December 26, 2001, scientists spotted an object 0.3 kilometer across. Twelve days later, it came within 800,000 kilometers of Earth, roughly twice the distance to the moon, a frighteningly small distance. If it had collided with us, it would likely have destroyed an area the size of Texas or all of the northeastern states and southern Ontario. Another object 60 meters across came within 460,000 kilometers of Earth on March 12, 2002. Astronomers did not detect it until four days after it passed because it came from the direction of the sun and thus could not be seen.

Another 100-meter-diameter object was also first spotted in June 2002, three days after it missed us by only 120,000 kilometers.

The impacts of asteroids and comets with Earth only rarely affect individuals and have not killed groups of people in historic time, at least none that have been recorded. A long time without an event, and in this case an especially long time compared with human life spans, leads to the widespread belief that it will never happen, at least not to us. However, we now have enough information on objects in Earth-crossing orbits that we have a pretty good idea of the odds of a significant impact with Earth. Scientists have now discovered at least half of the potentially hazardous asteroids and we find more every year. The odds of an undiscovered one striking Earth in the next 200 years are very small. But it is quite clear that it will eventually happen—we just do not know when.

Are we prepared for that inevitable event? The short answer is, clearly not. There is no point in staying awake at night worrying about being hit by a stray space rock. Certainly, the odds of a person being hit in a human lifetime are extremely small. In addition, worrying about the possibility would do no good. We can neither predict the impact of a small but deadly rock nor see it coming before being struck. That conclusion holds for even larger impactors that could kill thousands of people. For still larger asteroids, the difficulty of spotting incoming objects heading directly for Earth creates a major predicament. For a larger doomsday object somehow spotted as it grows larger on Earth approach, we have neither decided on a formal plan of action nationally or internationally nor set up the mechanism for implementation of that action. We just do not want to think about the possibility.

If you were in a position of power or influence, what would you do?

Cases in Point

A Round Hole in the Desert
Meteor Crater, Arizona ▶

Meteor Crater is the classic open-crater impact site, 65 kilometers east of Flagstaff, that is so well known to the general public (see FIGURE 17-6). It is small as impact craters go, only 1.2 kilometers across and 180 meters deep, but it is nicely circular and has distinctly raised rims (see FIGURE 17-7). Being only 50,000 years old, it is also well preserved. The projectile was an iron meteorite with a diameter of 60 meters. It came in at 15 kilometers per second and exploded with the energy of 20 million tons of TNT, equivalent to that of the largest nuclear devices. The target rock, Coconino sandstone, shows good evidence of shock features, including shocked quartz and lechatelierite, a fused silica glass. A shepherd found a piece of iron in 1886, and a prospector found many more in 1891. One piece found its way to a mineral dealer in Philadelphia, who recognized it as an iron meteorite. The dealer visited the site and found numerous fragments of iron meteorites.

Unfortunately, no craters were then known to be formed by meteorite impacts, and even the great G. K. Gilbert, chief geologist of the U.S. Geological Survey, misinterpreted the crater as being of volcanic origin or a limestone sinkhole around which the meteorite fragments had fallen by coincidence.

Daniel Barringer, a mining engineer interested in the iron as an ore, filed claim to the site in 1903 and began intensive exploration of it with many drill holes. He found meteorites under rim debris and under boulders thrown out from deep in the crater. Barringer presented scientific papers in which he concluded that meteorite impact had formed the crater, but few scientists were convinced. A few years later, two astronomers separately visiting the crater concluded it had been caused by a meteorite, but that given its size and velocity, the meteorite would have vaporized or disintegrated completely. The scientific community remained unconvinced until Eugene Shoemaker, then a graduate student, studied the crater and its materials in detail, finding shock-melted glass containing meteoritic droplets and the extremely high-pressure minerals coesite and stishovite. Stishovite also requires temperatures greater than 750°C.

A Nickel Mine at an Impact Site
The Sudbury Complex, Ontario ▶

The Precambrian-age Sudbury intrusion is 140 kilometers in diameter. Although it is not an open crater, its widespread shatter cones lead to essentially universal acceptance that it was caused by asteroid impact. Sudbury is also the largest nickel deposit in the world. It has been suggested that the nickel originated in the impacting meteorite, but there is considerable disagreement on that point.

If the target rocks contain quartz, the extreme shock pressures produced by impact deform the quartz to produce so-called shocked quartz grains that show multiple sets of shock lamellae imposed by more than 60,000 atmospheres of pressure. These are pressures found at depths of 200 kilometers in the Earth, or deep in the mantle. In 1984, Dr. Bruce Bohor and others of the U.S. Geological Survey discovered shocked quartz in the 65-million-year-old Cretaceous-Tertiary boundary clay. Some workers suggested that shocked quartz had formed by particularly violent volcanic explosions, but volcanic grains show only a single set of deformation lamellae that can be formed at much lower deformation pressures.

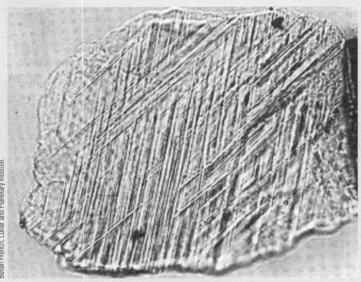

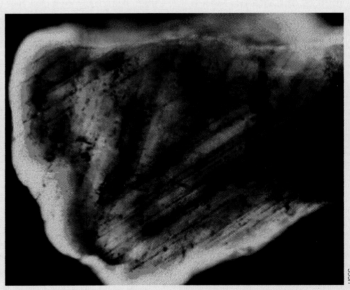

▶ **A.** *This impact-shocked quartz grain is less than one millimeter across. Multiple sets of thinly spaced deformation planes are imposed by a high-intensity impact.* **B.** *A similar shocked quartz grain as seen under polarized light. Deformation planes are visible; these colors as seen under the microscope, depend on the thickness of the grain.*

A Close Grazing Encounter
Tunguska, Siberia ▶

An asteroid estimated to be 50 meters in diameter blew down and charred some 1,000 square kilometers of forest on June 30, 1908, but failed to create a crater. It must have been a grazing encounter, approximately 8 kilometers above the ground and traveling some 15 kilometers per second before it exploded and disintegrated. Colliding with even a thin atmosphere at that altitude, its high velocity abruptly encountered severe resistance and caused instant disintegration. Its energy was equivalent to 1,000 Hiroshima-sized atomic bombs. Many people saw the huge fireball as it traveled north over remote villages in central Siberia. At the time of "impact," 670 kilometers northeast of Krasnoyarsk, people saw a bright flash and heard loud bangs. The ground shook, windows shattered, and people felt blasts of hot air. One man 58 kilometers southeast of the explosion felt searing heat and was knocked more than 6 meters off his porch by the blast. Some of the reindeer and dogs of native people near the site were killed by the explosion.

The first expedition into the remote area, in 1921, discovered that the trees were blown radially outward from the site of the explosion. Several intensive expeditions that included trenching and drilling failed to find any meteorite fragments, but microscope examination of the soil in 1957 showed tiny spheres of iron oxide meteoritic dust. It is now thought that the object was a stony meteorite.

▶ *Trees blown down by the Tunguska asteroid event in 1908.*

Chapter Review

Key Points

Projectiles from Space

- Asteroids are pieces of space rock orbiting the sun like Earth does. Comets are ice with some rock that occasionally loop through our solar system. **FIGURE 17-1**.

- When space rocks enter Earth's atmosphere, they create meteors, light-producing objects streaking through the sky. Meteorites are the pieces of rock that actually collide with Earth. **FIGURE 17-2**.

- Meteorites include those consisting of iron, chondrites composed of the dark minerals olivine and pyroxene, and achondrites that are similar to the basalt found on Earth. Proportions of the three types are similar to such rocks in the interior of the Earth. **FIGURES 17-3, 17-4**.

Evidence of Past Impacts

- The energy of a moving object is related to its mass and its velocity. The energy of an incoming asteroid doubles for every doubling of its mass and quadruples for every doubling of its velocity. **By the Numbers 17-1**.

- Impacts of projectiles from space blast a deep, round crater that compresses and melts the rock below, spraying it outward in all directions. Moderate-sized impacts have open, rimmed craters. Larger impacts collapse to form a broader, shallower basin with a central peak. **FIGURES 17-6, 17-7**.

- The asteroid that annihilated the dinosaurs and most other life forms 65 million years ago was likely 10 to 15 kilometers in diameter and struck in the Yucatán peninsula of eastern Mexico. **FIGURE 17-9**.

- Distinctive features at and around the site of a major impact include shatter cones of rock radiating downward and outward, droplets of molten glass, quartz grains showing shock features, and, for giant impacts, a layer of carbon-rich clay containing the metal iridium. **FIGURES 17-11 to 17-13**.

Consequences of Impacts with Earth

- An asteroid impact could wipe out civilization on Earth.

- Hazards from a large impact include tsunami, firestorms, soot that would block the sun and cause prolonged freezing and plant death, strong acid rain, and nickel poisoning of plants.

Evaluating the Risk of Impact

- Numerous small meteorites, far fewer large ones, and only rarely a giant one strike Earth.

- On average, a 200-meter-diameter asteroid impacts Earth once in 10,000 years. Impact of a 10-kilometer-diameter object that could wipe out most or all of civilization is expected to occur once every 100 million years on average. **FIGURE 17-15**.

What Could We Do About an Incoming Asteroid?

- NASA catalogs and tracks near-Earth objects that have the potential of colliding with Earth. Unfortunately, such objects are often undetectable.

- If an asteroid were headed for Earth, some propose that it could be blasted into smaller pieces or deflected.

Key Terms

achondrites, p. 513

asteroid belt, p. 511

asteroids, p. 511

chondrites, p. 513

comets, p. 511

iridium anomaly, p. 519

iron meteorites, p. 512

K-T boundary, p. 517

meteorite, p. 512

meteors, p. 512

near-Earth objects, p. 523

shatter cones, p. 517

stony-iron meteorites, p. 513

strewn field, p. 512

Questions for Review

1. Where do asteroids originate in space?

2. What is the path of comets around the sun?

3. How is the tail of a comet oriented? (Which way does it point?)

4. Name two general kinds of meteorites and describe their composition.

5. How fast do asteroids travel in space?

6. Why do asteroids create a more or less semicircular hole in the ground, regardless of whether they come in perpendicular to Earth's surface or with a glancing blow?

7. What is the sequence of events for the impact of a large asteroid immediately after it blows out of the crater?

8. What evidence is there that some large comets or asteroids break up on close encounter with a planet?

9. List five quite different direct physical or environmental effects of impacts of a large asteroid in addition to the excavation of a large crater.

10. Roughly what proportion of Earth's human population would likely be killed by impact of an asteroid 1.5 to 2 kilometers in diameter?

11. Sixty-five million years ago, a very large asteroid struck the Earth. Where did it apparently happen?

12. What is the relationship between the size of meteorites and the number of a given size?

13. Why would astronomers have difficulty recognizing a large incoming asteroid headed for direct impact on Earth?

Discussion Question

1. If a 15-kilometer-diameter asteroid were on collision course with Earth and you were the head of NASA, what would you do if the President gave you complete control over actions to deal with it? Why?

18 The Future: Where Do We Go From Here?

Beach-fronts are one of the most hazardous places for homes but more and more people move there in spite of the risks. This became even more obvious for those on the Texas Gulf Coast in September, 2008, when Hurricane Ike completely destroyed all of the beach-front homes near Galveston.

Hyndman.

Those who cannot remember the past are condemned to repeat it.

—George Santayana, 1905

We Are the Problem

In the introductory chapter of this book, we emphasized that a hazard exists where a natural event is likely to harm people or property. Similarly, we noted that a disaster is a hazardous event that affects humans or property. A major natural event—an earthquake, volcanic eruption, landslide, flood, hurricane, or tornado—in a remote area is merely a part of ongoing natural processes. Problems arise when people place themselves in environments where they can be impacted by such major events. It makes no sense to blame nature for natural events that have been going on for hundreds of millions of years. The problem is not the natural event, but humans.

The solution then would seem simple. Keep people from hazardous locations. Unfortunately, that is easier said than done. Nomadic cultures learned to go with the flow; they learned to live with nature rather than fight it. Impacted by major events, they merely moved to safer ground, leaving a floodplain when the water began to rise, moving away from a volcano when it began to rumble.

Part of the problem now is that there are almost seven billion of us. As civilization evolved, we moved to where we could live comfortably and where we had the resources to meet our needs for food, water, energy, shelter, and transportation. We settled along coastlines and rivers, where water was available, crops would grow, the climate was hospitable, and we could trade goods with other groups of people. We settled near volcanoes, where soil was fertile for crops. As time went on, we concentrated in groups for defensive reasons; towns grew larger but stayed essentially in the same locations that were most suitable for our basic needs.

As our cities grew, infrastructures became entrenched. Our shelters, transportation systems, and communications networks became permanent—houses, stores, factories, roads, bridges, dams, water supply systems, and power lines. When natural events impacted us, we fought back. We built levees to hold back rising rivers, not realizing we were living in the natural high-water paths of many of them. We should not have built in these places to begin with, but back then we did not know any better. We should have built our towns on higher terraces above flood levels. We could still use floodplains to grow crops but abandon them temporarily whenever water rose. Unfortunately, this is all in hindsight. Our towns grew where they did, and we have to live with the consequences.

We continue to make poor choices. As the people of industrialized countries became more affluent, and as free time and efficient transportation became widely available, we began making choices based on leisure, recreation, and aesthetics. We built homes close to streams (**FIGURE 18-1**), on the edges of sea cliffs, on offshore barrier islands, in the "shelter" and view of spectacular cliffs and majestic volcanoes. When the larger natural events come along, we complain that nature is out of control. But nature is not the problem; we are the problem. We should live in safer places.

The same feelings cloud our judgment when we decide to build or buy a house in a location that suits our fancy. That spectacular sea-cliff view of the ocean, that place on a beautiful sandy beach, that tranquil site at the edge of a scenic stream—these are all desirable places to live. These attractive spots are also prime locations for **natural hazards** that endanger our houses and perhaps our lives.

People understand that there are risks in all aspects of life and choose to accept some of them. There is little we can do about a stray meteorite unexpectedly nailing us, and we know that the odds of it happening are remote. Walking in the forest, we know that trees sometimes fall; we know the odds of one hitting us are remote, so we do not think about it. Every time we get in a car and head out on the road, we understand that some other driver may cause a collision with us by driving drunk, falling asleep, running a red light, or merely not paying attention. Yet we accept the risk because, although accidents happen, we feel one is not likely to happen to us.

Even people knowledgeable of particular natural hazards sometimes succumb to the lure of a great place. Yes, they think, catastrophes have happened to others, "but it won't happen to me." Our wanting "something special" clouds our judgment. There are those who are well aware of a particular hazard but are willing to live with the consequences, both physical and financial. In an area prone to earthquakes, people incur the expense of building earthquake-resistant homes and purchasing expensive earthquake insurance. In a landslide-prone area, they install costly drainage systems and may even pay for rock bolting the bedrock under their houses. Because landslide insurance is not generally available, they are willing to risk losing the complete value of their houses. For most of us, our home is the largest investment we will ever make; we borrow heavily, paying for it over 15 to 30 years. Unfortunately, if the house is destroyed in a landslide, we may be liable for paying off the mortgage, even if the house no longer exists.

FIGURE 18-1 FLOODING

A. Nooksack River, near Lynden, Washington, spread over its floodplain in January, 2009. **B.** Floods from 30 centimeters of rain and collapse of a county highway holding back Delton Lake in south-central Wisconsin drained the reservoir. Homes built on its muddy banks slumped when the banks eroded.

Hazard Assessment and Mitigation

The underlying problem, discussed elsewhere in this book, is that we believe we can control nature. We built levees and put in riprap to hold back an encroaching hazard. It seemed logical that we could do more of the same as the problem became worse. We did not recognize that our actions actually made the problem still worse. Governmental agencies formed to facilitate our actions came to believe that their short-term "cures" were the solution to **nature's rampages**. The public and many government officials view nature as something to be controlled or held at bay. Our overall solution to natural events is to react to them. When asked if New Orleans could survive another hurricane, a senior manager of the U.S. Army Corps of Engineers replied that it certainly could, even a Category 5, because "higher levees would solve the problem." Unfortunately, for many decades they have raised the Mississippi River levees after each major flood, yet some levees still fail and the river still floods.

Many people believe that if a river levee fails or houses and roads are washed away on a barrier island, we just did not build the levee high enough or emplace sufficient protection before the storm. We do not realize that the more we hold back the effect of a large natural event, the worse the effect will eventually be. The natural outcome is inevitable; it is just a matter of time. We need to be proactive rather than reactive. If we understand the natural processes, we can learn to live with them.

In the case of New Orleans, emergency managers have long recognized that a major event could lead to flooding of this open bowl and a catastrophe for the city. Hurricane Katrina vividly demonstrated the consequences of a major hurricane impact. Separately, a huge flood on the lower Mississippi River could overtop its big levees, which were built for 100-year flood levels and don't often fail. As noted in Chapters 11 and 12 on streams and floods, 100-year flood events can come anytime. Compounding matters, with each successive flood the channel of the Mississippi River rises because of sediment deposition from upstream. Major floods also get progressively larger over time because of continued urbanization and levee-building upstream. A levee breach upstream of New Orleans could flood low ground beyond the levees on the north, which would raise the level of Lake Pontchartrain at the northern edge of the city. If the population, thinking levees would protect them, did not evacuate, hundreds of thousands of people could be endangered after water covers their escape roads. Although warnings about New Orleans have come with some regularity from many knowledgeable individuals and organizations, no large-scale improvements have been implemented. Governmental groups and the public that they represent are seldom inclined to be proactive. Urgency only seems evident in people's reaction after a disaster.

"Soft" solutions for hazardous regions include zoning that permits building in certain restricted areas and strict building codes to minimize damage to buildings and their occupants. Over the long term, these are much less expensive than **"hard" solutions**, such as installing levees on rivers, concrete barriers and riprap along coasts, catchment basins below debris-flow channels, and drains on landslide-prone slopes.

Societal Attitudes

Even where lives are lost, collective memories of a disaster go back only a couple of years. People quickly forget that it could happen again. Homes badly damaged by some natural disaster are typically ordered abandoned or removed.

Even when such homes are condemned because they are too dangerous to inhabit, individuals and real estate groups sue for permits to rebuild and reoccupy. Often they are successful, and the problem escalates.

Homeowners commonly blame, and often sue, others for damage to property that they purchase. Who should be held responsible—the seller, the purchaser, the developer or real estate agent, the government? A good rule of thumb is the old adage "caveat emptor," or **buyer beware**. In many transactions, however, the seller is held responsible if he or she is aware of some aspect of a property that is damaged or endangered but the problem is not obvious. The issue can be a leaky roof, cracked foundation, high radon levels in a home, previous flood damage, or a house resting on a landslide. If the property is damaged but the buyer is not aware of it, the buyer is not generally held responsible. The same responsibility rests on developers or real estate agents who represent that a property is undamaged or safe from a particular hazard. If developers purchase a property and then find out that it rests on a landslide, they should be required to advise potential buyers of that fact.

Homeowners who face losing major investments such as their homes often try to pin the blame on someone else, especially someone with "deep pockets." If a developer has disappeared or declared bankruptcy, then the next in line is often a local government. Did the government provide a permit to the developer to build on a piece of property? Should a county or city be blamed for permitting building on a site that was prone to landsliding or flooding? Many areas have few restrictions on building on hazardous sites. In other areas, local ordinances require that a professional geologist or geotechnical engineer certify that a site is safe from sliding or flooding. Unfortunately, as in any profession, there are those who are less competent or less ethical than others. There are instances, in southern California, for example, where a developer has fired a local expert who provided a negative assessment that would prevent development of a piece of property and then hired another "expert" who provides a more positive view. In Seattle and many other hilly West Coast cities, real estate agents are not required to disclose the fact that landslides have been occurring in a particular area for decades.

In many areas, **zoning restrictions** concerning natural hazards are limited. In many counties in South Carolina, for example, there are few restrictions on building on active barrier islands (**FIGURE 18-2**) or standards for the quality of construction to minimize wind damage during a hurricane. In 2008, within a few months after Hurricane Ike destroyed most of the coastal houses both east and west of Galveston, Texas, numerous ads for beach-front property appeared and people began rebuilding on the same beaches even closer to the water. Although there have been dramatic advances in hurricane forecasting, coastal population increases leave roads jammed with traffic, sometimes prohibiting timely evacuation.

FIGURE 18-2 BEACHFRONT DEVELOPMENT

Nine months after Hurricane Hugo destroyed homes on the barrier bar at Pawley's Island, South Carolina, larger, more expensive houses were being built, right on the beach. The zig-zag fences help retain some blowing sand but there are no protective dunes.

After a Disaster

In many cases damage is heavily influenced by a homeowner's choices. A scenic site may be right below a steep slope that collects heavy snow in winter or nestled at the base of a spectacular mountainside capped by vertical cliffs (**FIGURE 18-3**). If a home begins breaking up on a landslide, broken water-supply pipes may leak into a slope. Did the pipes leak before movement began, thereby making the slope more prone to failure? Or did the slope fail, causing the pipes to break? Did a homeowner's roof and gutter

FIGURE 18-3 AVALANCHE

An avalanche destroyed four houses at the base of a steep slope covered with heavy snow in Cordova, Alaska, on March 1, 2000.

system, heavy watering, and septic drain field feed water into the ground to make a slope less stable? Was the slope destabilized by the cutting and filling used to create a house site on a slope?

Did building levees along a river cause flood levels to rise higher than they would have without the levees? Was a debris-flow dam not built high enough or a debris basin not built large enough to protect homes downslope? Or should the homes simply not have been built in such a hazardous site in the first place?

Should landowners be permitted to do whatever they wish with their property? **Property rights** advocates often say so. If a governmental entity permits building on land within its jurisdiction, should the taxpayers in the district shoulder the responsibility if there is a disaster? Should the government inform a buyer that a property is in a hazardous location and what the hazards are? Should a landowner be prevented from developing a piece of property that might be subjected to a disaster? If so, has the government effectively taken the landowner's anticipated value without compensation, a taking characterized in the courts as **reverse condemnation**?

We as a society don't learn very well from our mistakes. A few months after homes were lost to wildfires in southern California, people rebuilt many of them (**FIGURE 18-4**). Unfortunately, local governments control development (if it is controlled at all) but bear little responsibility for fighting fires; that responsibility (and the cost) is largely left to state and national governments. A 1999 University of California study showed that between 1970 and 1985, during a 97 percent rise in California population, the area burned rose 95 percent but the damage cost of fires rose 5,000 percent!

The availability of federal funds in the last several decades to repair and rebuild after a natural disaster has led many states, local governments, and individuals to believe that they are entitled to such help. However, since 1988, federal policy has gradually shifted to emphasize **mitigation**, but unfortunately not consistently. Federal funds are still available for rebuilding, but often only in safer ways or locations. Almost as important has been the increasing role of insurance companies. After Hurricane Andrew in 1992, Katrina in 2005, and the Mississippi River floods of 1993, many companies have either refused to renew policies or dramatically increased insurance premiums to cover anticipated losses. After the catastrophic 2004 and 2005 hurricane seasons, insurance rates increased dramatically in areas of the southeastern United States likely to be affected by large future hurricanes. These changes—coupled with land-use regulations, strictly enforced building codes, fines, and additional financial incentives—are all important in reducing losses.

Who should bear the cost of hazard mitigation—the developer or landowner who wishes to build on a parcel of land or all taxpayers in the form of local, state, or federal taxes? If one person is permitted to "protect" their property from a flood or hurricane, are they transferring that hazard to someone else downstream or farther along the coast?

Such questions are not merely hypothetical; they arise all the time. Answers are not simple. They are the subject of lawsuits, countersuits, and court cases at all levels.

Education

Given the grim assessment above, it would be easy to despair, to believe not much can be done—that the public will always pay for someone else's greed or poor judgment. Given the anti-taxation feelings of a significant proportion of the public, most people should be receptive to reducing the costs of futile efforts to control nature. Billions of tax dollars are expended to protect property in inherently high-risk areas, efforts doomed to ultimate failure. Once people realize that many major projects to protect the property of selective groups entail large costs to great numbers of people but benefit only a few, then the support for such projects should diminish. If so, the solution should be to educate the public about natural hazards and the processes that control them.

Many of these problems are political. Politicians are elected to serve their constituents. Their decisions may be made to better the lives of those they serve or for reelection purposes. People's views and decisions are commonly dictated by self interest rather than by concerns for society at large. People at all levels of government have direct control over expenditures for major publicly funded projects, so they need to be educated as to the effects and consequences of such projects. However, because officials serve at the wishes of their constituents, it is the general public who actually hold the cards.

Often an issue revolves around bringing money and jobs to a community or state, regardless of long-term

FIGURE 18-4 WILDFIRE

M. Raphael, FEMA.

Remains of a home destroyed by fire in a forested gully. San Bernardino wildfire, 2003.

advantages or disadvantages. Members of a community are often happy to have a multimillion-dollar flood-protection or shore-protection project even if the state's share of the costs is spread to others throughout the state or the country. The dollars brought in provide a significant short-term financial benefit to a broad spectrum of the community. The community may also be happy to get such a project even if its members do not clearly understand the long-term effects. Sometimes the lasting consequences, such as the loss of a beach, are detrimental to the characteristics that drew them to an area in the first place.

After Hurricane Ike in 2008, many residents, county officials, and even the state governor, argued in favor of building a protective seawall, the "Ike Dike," like the one at Galveston, for tens of kilometers along the coast in front of all of the homes destroyed or threatened by the hurricane—the cost to be borne by the state or, more likely, the federal government. They don't recognize that few people visit the beach in front of the sea wall at Galveston any longer, because increased wave-action has removed the sand that drew people to the area in the first place.

Clearly, we need to educate people at all levels, from the general public to developers and realtors, business people, banks and insurance companies, and those who hold political office. Education can be a struggle because people do not want to hear that they should not or cannot live in a location that strikes their fancy. The opportune time to effectively convey this message is within a year of a major natural catastrophe. Then the cause and effect are clearer in people's minds; we merely need to help clarify the circumstances for them and specify what individuals can do to prevent future similar occurrences. Unfortunately, even then people's memories of the trauma and costs fade quickly. They begin to believe it a rare occurrence not likely to affect them again.

If we can make people responsible for their own actions, then losses to both property and life can be minimized. The personal responsibility must be passed not only to individuals but to all involved groups and organizations. Everyone stands to gain if people can be dissuaded from living in hazardous areas. For people to accept the consequences, we must help them understand the processes of natural hazards and the consequences of how we deal with them.

Instead of addressing natural hazards on a short-term basis, both for gains and losses, we need to consider and deal with the long-term issues. We need to design for the long-term future.

Different Ground Rules for the Poor

In countries where poverty is widespread, the forces that drive many people's behavior are different than those in prosperous countries. In much of Central America and parts of southeastern Asia, for example, millions of people lack the resources to choose to live in certain places for aesthetic reasons. Food and shelter dictate where they live. In Guatemala and Nicaragua, giant corporate farms now control most of the fertile valley bottoms, leaving peasants little choice but to work for them in the fields at minimal wages and provide their own shelter on the steep landslide-prone hillsides. Others who cannot find such work migrate to urban areas but are still relegated to living in steep surrounding hills. Both groups of people clear forests to grow food, provide building materials, and gather firewood for cooking. Their choice of steep hillsides as a place to live is a hazardous one, but they have little choice in order to survive. Compounding the problem is that fertility rates and population growth in desperately poor countries are among the highest in the world, so more people are forced to live in less suitable areas. Large proportions of those populations do not minimize family sizes for religious and cultural reasons.

For such poor societies, the reduction of vulnerability to natural hazards does not depend much on strengthening the zoning restrictions against living in dangerous areas or improving warning systems, though education can help. More so, it depends on cultural, economic, and political factors. Because more affluent individuals and corporations control many of those factors, the poor are left to fend for themselves. After a major disaster, such as the 2010 Haiti earthquake, international aid poured in for rescue and rebuilding (**FIGURE 18-5**). Ultimately, however, the lack of education and grinding poverty of certain

FIGURE 18-5 BUILDINGS COLLAPSE

Joshua Kelsey, U.S. Navy; Justin Stumberg, U.S. Navy.

Rescuers search for victims following Haiti's 2010 earthquake. Because of very poor quality construction, most of the buildings in the country's largest city collapsed. Thousands died when heavy concrete floors with little reinforcing bar and weak cinderblock walls collapsed.

countries foretell poor prospects for the future of their inhabitants.

The disaster differences between poor and more-affluent countries were accentuated in early 2010 by the earthquakes in Haiti and Chile. The magnitude 7 Haiti earthquake, which struck near the capital city of Port-au-Prince, killed more than 220,000 people and left 1.3 million homeless, whereas the magnitude 8.8 earthquake in Chile killed about 800. Although the Chile earthquake was about 500 times stronger at its epicenter, it caused far fewer fatalities. The greatest difference was that in Haiti there are no building codes, and even if there were, most people have no money to follow them. The annual per capita income for Haiti is only $1,340* and three quarters of the population live on less than $2 per day. Most are uneducated and cannot even read or write. Chile, however, is the wealthiest country in Latin America with an annual per capita income of $14,299. Being along one of the world's most active subduction zones, Chile is very familiar with earthquakes; its building codes, upgraded after the 1960 earthquake (the world's largest), are among the strongest in the world.

The catastrophic December 2004 earthquake and tsunami in Sumatra and nearby areas brings into focus some of the problems faced by poor people living on low-lying coasts in much of the tropics. Large numbers of people with big families survive on fishing, working at low wages for a few large multinational companies, staffing resorts for affluent vacationers, and providing a broad range of

---

*(World Bank, 2009, adjusted for cost of living)

spin-offs from these businesses. Many of the people live virtually on beaches at the edge of normal high-tide, in homes poorly built from mud bricks, unreinforced cinderblocks, or light-weight timbers and plaster. Roofs are commonly sheets of corrugated iron. Even without a major earthquake or tsunami, any large storm will rip off roofs and parts of walls, destroying what few belongings these people have. The magnitude 9.15 earthquake crumbled many of the coastal homes in Sumatra, killing a lot of people. The giant tsunami that followed a few minutes later crushed most of the remaining homes within hundreds of meters of the beach and swept up the debris and survivors in a churning mass that only a fortunate few survived. Hundreds of kilometers away in Thailand, Sri Lanka, India, and nearby countries, thousands of people living in similar circumstances died because they were unaware of the incoming tsunami.

Pago Pago, American Samoa, was decimated on September 29, 2009, when a magnitude 8.1 earthquake on the nearby Tonga Trench triggered a tsunami. The wave amplified as it entered the long bay with the town of Pago Pago at its head. Although most people felt the earthquake, many radio stations did not interrupt their music, and there was no official local warning, so many people assumed that everything was okay. Those that quickly fled uphill survived, but dozens died (**FIGURE 18-6**). People that live near the head of a bay need to be aware that large earthquakes are often followed by tsunami, and even a small tsunami will likely be amplified as waves move into the bay.

What could have prevented the catastrophe or at least minimized its effects? Most victims lived in the coastal

---

FIGURE 18-6 TSUNAMI

Gordon Yamasaki, USGS.

Damage resulting from the September, 2009, tsunami in Pago Pago, American Samoa.

areas because of proximity to their means of livelihood or because the rugged mountainous terrain provided few other suitable building sites. The small number of buildings left standing were generally two-story reinforced concrete structures, built by the few who could afford such luxury. Clearly most of the people lacked the means to live in better locations. In hindsight, building a tsunami warning system would have helped, provided that warnings could be circulated automatically and very quickly to the populace. Since such events generally come at lengthy intervals, the public needs to be aware and reminded regularly of the possibility and what to do when a warning sounds. Education on the nature of earthquakes, their relationship to tsunami, and what to do in such events could have saved tens of thousands.

The example of Hurricane Katrina, in New Orleans in 2005, provides additional understanding of the effects of a catastrophic event on poor people (**FIGURE 18-7**). Despite clear prediction of a dangerous storm approaching, many residents had no means to leave the city and nowhere to go even if they could have left. Ten days after the storm, an estimated 10,000 people still refused to leave their homes in flooded and seriously contaminated areas. The predominantly poor people from the eastern part of the city, who were unable to evacuate before the storm's arrival, made up most of the flood survivors who crowded into the Superdome and the Convention Center during the storm. Before Katrina, 23 percent of New Orleans residents lived below the poverty line. Many survived from day to day, had no savings or working cars, and lived in rented homes or apartments. Less than half of New Orleans residents had flood insurance.

A major problem for many families—especially the poor—was separation of family members during evacuation and the storm. For example, a mother or father evacuated with children but left behind an elderly parent who couldn't be moved. In some cases, parents were separated from their children during evacuation, or one parent left to help a relative, friend, or neighbor and was unable to return. Since their home was no longer a point of reference, communication became difficult or impossible.

In the case of Hurricane Katrina, evacuees' medical problems were compounded by the storm. Patients could not find their doctors and doctors could not find their patients. Patients on prescription medications or undergoing specialized treatments often could not remember the names of their medicines or the details of their treatments. Years of medical records were lost.

Worse Problems to Come?

The prospect of climate change adds an additional dimension to these problems; it is one of the greatest challenges facing the human race. Scientists agree that global temperatures are rising. As world population grows and large numbers of people become more affluent and use larger amounts of resources, greenhouse-gas emissions increase dramatically. Our generation of greenhouse gases seems likely to cause population collapse in some parts of the world, initially—and most harshly—in poor areas most affected by natural hazards. People's living conditions will be severely disrupted. Millions will die from increased incidence of storms and coastal flooding, heat stroke, dehydration, famine, disease, and wars over water, food, heating fuel, and other resources. Population increase is susceptible to a cruel feedback mechanism that reduces populations to sustainable levels. These changes will neither be smooth nor pleasant. We can minimize adverse aspects by being proactive in the reduction of greenhouse gases. It will be difficult, but the consequences of procrastination are magnified for our children and grandchildren.

Climate change is expected to lead to more rapid erosion of coastlines, along with more extreme weather events that cause landslides, floods, hurricanes, and wildfires. Some small islands in the Indian Ocean, far from the 2004 earthquake's epicenter, were completely overwashed by tsunami waves. As sea level continues to rise, such low-lying islands will gradually submerge, even without a catastrophic event. Extensive low-lying coastal regions of major river deltas in Southeast Asia feed and are homes to millions of poor people. Deltas of the Ganges and Brahmaputra rivers in Bangladesh, the Irrawaddy River in Myanmar, and the Mekong River in Vietnam and Cambodia are subject to two-meter ocean tides more than 200 kilometers

FIGURE 18-7 FAILED TO EVACUATE

Jocelyn Augustino, FEMA.

Survivors of Hurricane Katrina wait on a roof for rescue. Why did they not evacuate when warned? Lack of transportation? Previous false alarms?

FIGURE 18-8 HOMES AT RISK

Homes along channels in the Mekong River delta in southern Vietnam are almost in the water under normal circumstances. They would be washed away in a major cyclone.

upstream (**FIGURE 18-8**). Major storms can submerge most of the deltas, including all of their rice fields and homes, under more than two meters of water, with storm waves on top of that. Sea-level rise with climate change is expected to worsen those effects, likely killing thousands in major typhoons.

The number of hurricanes has not increased significantly, but since 1990 the annual number of the most intense storms—Categories 4 and 5—nearly doubled to eighteen worldwide in 2005. Hurricane development and intensity depends on energy provided by higher sea-surface temperatures. An increase of about 0.6°C (1°F) over the last dozen years may not seem like much, but it makes a difference by fueling storms with warm, humid air. More water vapor drawn into the atmosphere in these intense storms leads to heavier accompanying rainfall—and more flooding. Whether part of the increase in hurricane strength is in response to climate change is still debated, but most of the increase is attributed to a cycle of cooling and warming over a span of 60 to 70 years called the Atlantic multidecadal oscillation. Since 1995, the Atlantic sea surface has warmed, and hurricane activity has increased significantly. The water temperatures in the Caribbean Sea and the Gulf of Mexico were extraordinarily warm in 2005. Hurricanes are already one of the deadliest natural hazards affecting the United States. Unfortunately, we don't know how much worse this problem will become (**FIGURE 18-9A**).

As would be expected, certain types of disasters impact different parts of the world. Earthquakes affect zones of tectonic activity and volcanoes depend on the locations of plate boundaries, spreading zones, and hot-spot tracks. Landslides affect many areas—not only those with steep slopes but also regions with weak soils that can be further weakened by water. Sinkholes, subsidence, and swelling clays depend heavily on the type of underlying rock, its solubility, water content, and soil composition.

Weather-related disasters are distributed according to type (**FIGURE 18-9B**). People living in areas prone to particular disaster types need to be aware of the hazards and how to avoid them. Among the most-costly disasters, hurricanes depend on warm water and are concentrated along the Gulf of Mexico coast and southeastern coast of the United States. Violent thunderstorms and tornadoes require the interaction of warm, moist, Gulf of Mexico air and cold, dry air to the northwest. Floods depend on warm, humid air and interactions with mountain ranges or colder air masses that can cause precipitation. People living on floodplains or in narrow valleys can be heavily impacted. Deadly heat waves and drought primarily affect the central to southeastern United States, not only because of high summer temperatures but also because of high humidity that amplifies the effect of heat on the body. Wildfires depend on abundant dry fuel such as trees, brush, or grass that can be ignited by lightning or careless use of fire. People with houses surrounded by abundant dry fuel are at risk, especially if they live on hillsides or narrow canyons that can channel the flames.

Finally, we emphasize that developed or prosperous countries hit by a disaster lose economic value, but in relation to their total economic viability, poorer countries sustain much greater disaster losses. In addition, underdeveloped or poor countries lose many more lives because their living conditions and surroundings are poorly able to withstand such events. After a major disaster, they spend a large proportion of their resources on disaster relief, recovery, and reconstruction rather than on improving their economy or the lives of their people. With each subsequent disaster, the scenario is repeated and those societies remain mired in poverty. Both eventualities are undesirable,

FIGURE 18-9 BILLION-DOLLAR U.S. WEATHER-RELATED DISASTERS

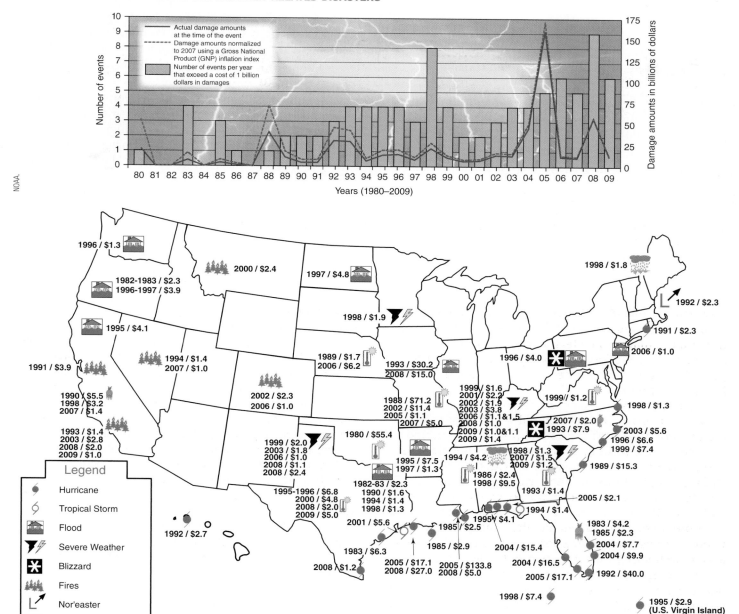

A. This graph charts the number and cost of the costliest weather-related events from 1980–2009 (constant dollars). **B.** Different types of disasters affect different parts of the U.S.

but neither seems likely to change in the near future. Both depend on societal attitudes and people's behavior. We need to begin there.

Rather than fighting nature, we need to do more to live with and accommodate its variable behavior. This would not change major events, but it would reduce the number that become hazards or disasters that affect us.

Some Persistent Problems

- We need to adapt to nature rather than trying to control it. In the long run, nature will have its way.

- As a society, we quickly forget previous disasters.

- We misinterpret the long-term variability of natural processes. We don't understand the statistical meaning of a 100-year event. That size event may happen once in a 100-year span, but it also might occur twice in a two-year span. The 100-year-average event may become a much-shorter-year event if controlling conditions change or if a larger event occurs.

- We need to focus on mitigation—reducing the likelihood of future disasters—rather than merely cleaning up and starting over by rebuilding what and where we had before.

- Business entities that deal with community growth, taxes, and tourism (developers, builders, real-estate brokers, civic leaders) often deliberately minimize or deny the existence or importance of natural hazards. Following a disaster, the same people may minimize the cost of the event for fear of negatively affecting future growth or tourism.

- Many of us resist zoning of hazardous areas because this would infringe on our individual rights to do as we wish with our property. We ignore or deny that things we do may adversely affect others. Civic leaders often cast votes based on constituents' demands rather than on what they know to be in the long-term best interests of their community or state.

- Insurance rates for natural hazards should be actuarial but typically are not; they do not reflect total real costs because state or federal governments typically pick up a significant part of real costs. Prominent examples include people living on coasts vulnerable to hurricanes and other big storms and people living in forested areas vulnerable to fire.

- Poor countries, especially those governed by repressive or dictatorial regimes—or those involved in internal conflict—are sometimes reluctant to accept international aid, especially from powerful countries like the United States or former colonial powers such as England and France. They fear that the aid may include arms or other encouragement for rebels who may undermine their authority. They also fear that external aid will emphasize the inability of their government to adequately serve the people. The May 2007 cyclone in Myanmar (Burma) is an extreme example of this type of response. Closer to home, the United States government rejected aid from some other countries after Hurricane Katrina devastated New Orleans in 2005, perhaps because accepting such aid would emphasize its own inability to quickly and effectively deal with the problem.

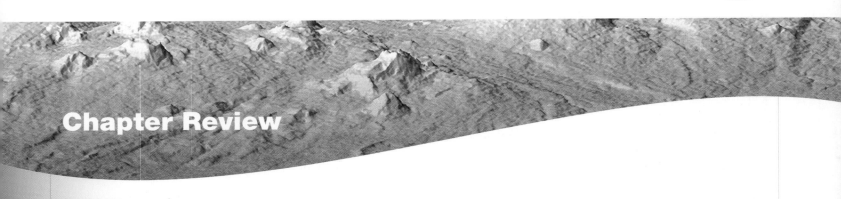

Chapter Review

Key Points

Hazard Assessment and Mitigation

- A disaster is a hazardous event that affects humans or property. We are the problem. The problem is not the natural event, but humans.

- "Soft" solutions for hazardous areas include zoning to prevent building in certain regions and strict building codes, which minimize damage and are much less expensive in the long run. "Hard" alternatives, including levees on rivers and riprap along coasts, are expensive and create other problems.

Societal Attitudes

- Members of the public largely believe that we can control nature's rampages. However, most of our solutions are short-term and create other long-term problems.

- Federal policy has gradually shifted to emphasize mitigation but decisions are heavily affected by political forces.

- In many aspects of society, the seller is held responsible if he or she is aware of some aspect of a property that is dangerous or damaged. Home owners commonly blame—and often sue—others for damages to property but would do better to remember the old adage of "buyer beware."

- In many areas, zoning restrictions concerning natural hazards may be desirable but commonly are limited.

- In some cases, damage from hazards is heavily influenced by a homeowner's own behavior.

- Who should bear the cost of hazard mitigation—the developer or landowner who wishes to build on a parcel of land or all taxpayers in the form of local, state, or federal taxes?

- We need to educate the general public; developers and realtors; business people; banks and insurance companies; and politicians at all levels. What may be in a group's short-term interest may create larger problems in the long run.

- If we can make people responsible for their own actions, then losses of both property and life can be minimized.

Different Ground Rules for the Poor

- In poor societies, the reduction of vulnerability to natural hazards does not depend on zoning restrictions or improving warning systems but on cultural, economic, and political factors.

- Affluent individuals and corporations commonly control many of those factors, so the poor are left to fend for themselves.

- Developed or prosperous countries lose economic value when subjected to natural disasters; underdeveloped or poor countries lose lives.

Worse Problems to Come?

- Climate change and long-term warming of parts of Earth potentially increases the severity of weather-related hazards.

Key Terms

buyer beware, p. 531
"hard" solutions, p. 530
mitigation, p. 532

natural hazards, p. 529
nature's rampages, p. 530
property rights, p. 532

reverse condemnation, p. 532
"soft" solutions, p. 530
zoning restrictions, p. 531

Questions for Review

1. If someone's house on a floodplain is damaged or destroyed in a flood, who is to blame? Why?

2. Why do people live in dangerous places like offshore barrier bars?

3. What is the main distinction between "soft" solutions to natural hazards and "hard" alternatives? Provide examples.

4. What is meant by hazard mitigation?

5. If I sell a house that is later damaged by landsliding, who is responsible; that is, what are the main considerations?

6. When the federal government provides funds to protect people's homes from floods or wildfires, individual homeowners benefit at the expense of taxpayers. What are the two main types of alternatives to elimination of taxpayer expense for natural hazard losses?

7. What is meant by the expression "Those who ignore the past are condemned to repeat it"? Provide an example related to natural hazards.

8. When we note that people need to take responsibility for their own actions when living in a hazardous environment, what is different about the behavior of poor people living in underdeveloped countries?

Conversion Factors

Length

1 millimeter (mm)	**= 0.03937**	**inches**
1 centimeter (cm)	**= 0.3937**	**inches**
1 meter (m)	**= 3.281**	**feet**
1 kilometer (km)	**= 0.6214**	**miles (~5/8 mile)**
1 inch (in)	*= 2.540*	*cm*
1 foot (ft)	*= 30.48*	*cm*
1 mile (mi)	*= 1609*	*m = 1.609 km*

Area

1 square centimeter (cm^2)	**= 0.1550**	**square inches (in^2)**
1 square meter (m^2)	**= 10.76**	**square feet (ft^2)**
1 square kilometer (km^2)	**= 0.3861**	**square miles**
1 hectare (510,000 m^2)	**= 2.471**	**acres**
1 in^2	*= 6.452*	*cm^2*
1 ft^2	*= 929.0*	*cm^2*
1 mi^2	*= 2.590*	*km^2*

Volume

1 cubic centimeter (cm^3)	**= 0.06102**	**cubic inches (in^3)**
1 cubic meter (m^3)	**= 35.31**	**cubic feet (ft^3) = 1.308 cubic yards (yd^3)**
1 liter	**= 61.02**	**in^3 = 1.057 qt. = 0.2642 gallons**
1 ft^3	*= 0.02832*	*m^3*
1 yd^3	*= 0.7646*	*m^3*
1 gallon U.S.	*= 3.785*	*liter = 0.1337 ft^3*

Mass

1 gram (gm)	**= 0.03527**	**oz = 0.002205 pounds (lb)**
1 kg	**= 2.205**	**lb**
1 oz	*= 28.35*	*gm*
1 pound (= 16 oz.)	*= 453.6*	*gm = 0.4536 kg*

Velocity

1 meter per second (m/sec.)	**= 3.281**	**feet per second = 3.6 km/hr**
1 mile per hour	*= 1.609*	*kilometer per hour (km/hr)*

Temperature

°C	**= 5/9 (°F–32)**
°F	*= 9/5 °C + 32*

Prefixes for SI units

Giga	$= 10^9$	$= 1{,}000{,}000{,}000$	times
Mega	$= 10^6$	$= 1{,}000{,}000$	times
Kilo	$= 10^3$	$= 1000$	times
Hecto	$= 10^2$	$= 100$	times
Centi	$= 10^{-2}$	$= 0.01$	times
Milli	$= 10^{-3}$	$= 0.001$	times
Micro	$= 10^{-6}$	$= 0.000{,}001$	times
Nano	$= 10^{-9}$	$= 0.000{,}000{,}001$	times
Pico	$= 10^{-12}$	$= 0.000{,}000{,}000{,}001$	times

Glossary

100-year flood A flood magnitude that comes along once every 100 years on average.

aa Blocky basalt with a ragged, clinkery surface.

acceleration The rate of increase in velocity. During an earthquake, the ground accelerates from being stationary to a maximum velocity before slowing and reversing its movement.

achondrite A stony meteorite that is similar to basalt in composition.

adiabatic cooling Change in volume and temperature without change in total heat content.

adiabatic lapse rate (dry and wet) The rate of change of temperature as an air mass changes elevation.

aftershocks Smaller earthquakes after a major earthquake that occur on or near the same fault. They may occur for weeks or months after the main shock.

albedo The fraction of energy reflected away from the Earth's surface.

alluvial fan A fan-shaped deposit of sand and gravel at the mouth of a mountain canyon, where the stream gradient flattens at a main valley floor.

amplitude The size of the back-and-forth motion on a seismograph.

andesite An intermediate-colored, fine-grained volcanic rock. It has a silica content of approximately 60 percent, has intermediate viscosity, and forms slow-moving lavas, fragments, and volcanic ash.

angle of repose The maximum stable slope for loose material; for dry sand it is 30 to 35 degrees.

asteroid A chunk of rock orbiting the sun in the same way that Earth orbits the sun.

asteroid belt The distribution of asteroids in a disk-shaped ring orbiting the sun.

asthenosphere Part of Earth's mantle below the lithosphere that behaves in a plastic manner. The rigid and brittle lithosphere moves over it.

Atlantic Multidecadal Oscillation (AMO) The oscillation of the sea-surface temperature of the North Atlantic Ocean by about 0.8° C over several decades, typically about 70 years.

avulsion The permanent change in a stream channel, generally during a flood when a stream breaches its levee to send most of its flow outside of its channel.

barrier island A near-shore, coast-parallel island of sand built up by waves and commonly capped by wind-blown sand dunes. Also called a *barrier bar*.

basalt A black or dark-colored, fine-grained volcanic rock. It has a silica content of approximately 50 percent and low viscosity, and it flows downslope rapidly.

base isolation A mechanism to isolate a structure from earthquake shaking in the ground; often flexible pads between a building and the ground.

base level The level below which a stream cannot erode, typically at a lake or ocean.

base surge The high energy blast of steam and ash that blows laterally during the initial stages of a volcanic eruption.

beach hardening The placement of beach "protection" structures: riprap, groins, and related features.

beach replenishment Addition of sand to a beach to replace that lost to the waves. Also called *beach nourishment*.

bedrock stream A stream that has eroded down to bedrock.

blind thrust A thrust fault that does not reach Earth's surface. It is not evident at the surface.

blizzard Blowing snow with winds greater than 56 km/hr in the United States (35 mph) or 40 km/hr in Canada. Reduced visibility is common.

braided stream A stream characterized by interlacing channels that separate and come together at different places. The stream has more sediment than it can carry, so it frequently deposits some of it.

breach Failure of part of a river's levee, leading to some flow outside the main channel.

breakwater An artificial offshore barrier to waves constructed to create calm water for a beach or for anchoring boats.

burnout A burnout fire is deliberately set to burn back toward the main fire, eliminating fuel and stopping the progress of the main fire.

buyer beware The old saying that warned a buyer to be wary that what he or she was purchasing might not be as good as it appears. The buyer pays the consequences.

caldera A large depression, generally more than 1 kilometer across, in the summit of a volcano; formed by collapse into the underlying magma chamber.

cap-and-trade A system to place limits on smokestack emissions to require industry to either cut them or trade (and pay for) the right to emit additional pollution with companies possessing more efficient plants that produce much less pollution.

carbon cycle The cyclic exchange of carbon in Earth's atmosphere, water, vegetation, and rocks.

channel scour The depth of sediment eroded during floods.

charge separation The separation distance or potential between positive and negative charges in a thundercloud. The greater the separation, the more violent the electrical discharge when positive and negative charges connect.

Chinook winds Winds that warm by adiabatic compression as they descend from high elevations of a mountain range to low elevations on the plains to the east.

chondrite A stony meteorite consisting primarily of the minerals olivine and pyroxene; the most common kind of meteorite.

cinder cone A small steep-sided volcano consisting of basaltic cinders.

climate The weather of an area averaged over a long period of time.

coastal bulge The region above a subduction zone in which the overlying continental plate flexes upward before slip on the subduction zone causes a major earthquake.

cohesion The attraction between small soil particles that is provided by the surface tension of water between the particles.

cold front The line of boundary between a large mass of cold air advancing under an adjacent large mass of warm air.

collision zone The zone of convergence between two lithospheric plates.

comet A mass of ice and some rock material traveling at high velocity in the gravitational field of the sun but traveling outside our solar system before passing through it on occasion.

continental caldera Rhyolite volcanoes characterized by their high viscosity and high volatile content but gently sloping flanks.

continental drift The gradual movement of continents as oceans spread and separate.

convective updraft The rapid rise of a heated, expanded air mass driven by its lower density. The removal of air from below draws outside air into the rising "chimney."

convergent boundary A tectonic plate boundary along which two plates come together by either subduction or continent–continent collision.

Coriolis effect Rotation of Earth from west to east under a fluid such as the atmosphere or oceans permits that fluid to lag behind Earth's rotation. Fluid flows shift to the right in the northern hemisphere and to the left in the southern hemisphere as a result of this effect. Thus a southward-moving fluid appears to curve off to the west in the northern hemisphere.

cover collapse Collapse of the roof material over an underground cavity, often a soil cavern over a limestone cavern.

cover subsidence Gradual depression of the roof material over an underground cavity.

crater A depression created as ash blasts out of a volcano.

creep (1) Surface layers of soil on a slope move downslope more rapidly than subsurface layers. (2) Slow, more-or-less continuous movement on a fault. Creeping faults either lack earthquakes or have only small ones.

crown fire A fire that burns in the treetops.

cumulonimbus cloud A thundercloud with a flat anvil-shaped top; a sign of a nearby thunderstorm.

cyclic events Events that come at evenly spaced times, such as every so many hours or years.

daylighted bed A layer inclined less steeply than the slope so that it is exposed at the surface down lower on the slope.

debris avalanche A fast-moving avalanche of loose debris that flows out a considerable distance from its source.

debris flow A slurry of rocks, sand, and water flowing down a valley. Water generally makes up less than half the flow's volume.

delta The accumulation of sediment where a river reaches the base level of a lake or ocean.

desertification The growth of new desert environments in areas where drought prevents regrowth of grass and shrubs stripped away by increased population, intensive cultivation in marginal areas, and overgrazing.

discharge Measured volume of water flowing past a cross section of a river in a given amount of time. Usually expressed in cubic meters per second or cubic feet per second.

dissolution Minerals or rocks dissolving in water.

divergent boundary A spreading plate boundary such as a mid-oceanic ridge.

downburst A localized, violent, downward-directed wind below a strong thunderstorm.

drought A prolonged dry climatic event in a particular region that dramatically lowers the available water below that normally used by humans, animals, and vegetation.

dust storm A hazard that develops when winds pick up loose, dry dust or soil.

dynamic equilibrium The condition of a system in which the inflow and outflow of materials is balanced.

elastic rebound theory The theory applied to most earthquakes in which movement on two sides of a fault leads to bending of the rocks until they slip to release the bending strain during an earthquake.

El Niño Elevated sea-surface temperatures that lead to dramatic changes in weather in some areas every few years—for example, rains in coastal Peru and western Mexico and southwestern California. *See also* La Niña.

epicenter The point on the Earth's surface directly above the focus; the initial rupture point on a fault.

equilibrium profile (1) The longitudinal profile or slope of an equilibrium stream (see graded stream); (2) the seaward slope of a beach that is adjusted so that the amount of sediment that waves bring onto the beach is adjusted to the amount of sediment that the waves carry back offshore.

eruptive rift A crack in the ground from which lava erupts.

eruptive vent The point of eruption of material from a volcano.

evapotranspiration Water returned to the atmosphere by a combination of evaporation of water from the soil and leaves, and transpiration of moisture from the leaves.

extratropical cyclone A cyclone formed outside the tropics and commonly associated with weather fronts. Nor'easters are an example.

eye The small-diameter core of a hurricane; characterized by lack of clouds and little or no wind.

far-field tsunami A tsunami generated by an earthquake far from the point of impact of the wave.

fault An Earth fracture along which rocks on one side move relative to those on the other side.

fetch The distance over which wind blows over a body of water. The longer the fetch, the larger the waves that will be developed.

firebrands Burning embers carried by the wind, potentially to ignite new fires.

firelines Paths dug down to bare mineral soil, to slow or stop a fire's spread. They can be a meter to several meters wide and dug by hand or bulldozer.

firestorm A large, extremely hot fire generating a large updraft that pulls in vast amounts of air from the sides to fan the flames.

fissure A fracture at the surface that has opened by widening. See also rift.

flash flood A short-lived flood that appears suddenly, generally in a dry climate, in response to an upstream storm.

flood crest The point where the flood reaches its peak discharge.

flood fringe The fringe or backwater areas of a flood that store nearly still water during a flood and gradually release it downstream after the flood.

floodplain Relatively flat lowland that borders a river, usually dry but subject to flooding every few years.

floodway Area of floodplain regulated by FEMA for federal flood insurance.

fluidization The process of changing a soil saturated with water to a fluid mass that flows downslope.

focus The initial rupture point on a fault. The epicenter is the point on Earth's surface directly above the focus. Also called hypocenter.

forecast The statement that a future event will occur in a certain area in a given span of time (often decades) with a particular probability.

foreshocks Smaller earthquakes that precede some large earthquakes on or near the same fault. They are inferred to mark the initial relief of stresses before final failure.

fractal Features that look basically the same regardless of size; for example, the coast of Norway looks serrated on the scale of a map of the world or on a map of only part of that coast.

frequency The number of events in a given time, such as the number of back-and-forth motions of an earthquake per second.

frictional resistance The resistance to downslope movement of a flow or landslide.

frostbite The freezing of body tissue as a result of exposure to extreme cold temperature and winds.

fuel load The amount of burnable material, such as trees and dry vegetation, available for a fire.

fuel moisture Amount of moisture in a fuel. The lower the moisture content, the more readily the fuel will burn and at a higher temperature.

Fujita scale The scale of tornado wind speed and damage devised by Dr. Tetsuya Fujita.

gaining stream A stream, typically in a climate with abundant rainfall, that lies below the water table and gains water from groundwater.

glacial-outburst flood Flooding of glacial meltwater when ice tunnels or dams collapse. Also called Jökulhlaups.

global warming Warming of Earth's surface temperatures, especially over the past 100 years. Much of the recent warming has been attributed to increases in greenhouse gases.

graded stream Stream in equilibrium with its environment; its slope is adjusted to accommodate the amount of water and sediment amounts and grain size provided to it.

gradient Slope along the channel of a streambed, typically expressed in meters per kilometer or feet per mile.

greenhouse effect Increased atmospheric temperatures caused when atmospheric gases such as carbon dioxide and methane trap heat in Earth's atmosphere.

greenhouse gases Atmospheric gases such as carbon dioxide and methane that retain heat much like a greenhouse; solar energy can get in, but the heat cannot easily escape.

groin Wall of boulders, concrete, or wood built out into the surf from the water's edge to trap sand that moves in longshore drift. The intent is to hinder loss of sand from the beach.

groundwater Water in the saturated zone below the ground surface.

groundwater mining When water pumped from aquifers for community needs exceeds the recharge from precipitation.

hailstone Single pellet of hail formed when droplets of water freeze before falling to the ground.

"hard" solutions Solutions to a problem that involve building protective structures such as riprap and levees or continuous pumping of water from a landslide-prone slope.

Hawaiian-type lava Fluid basalt lava of the type that erupts from Hawaiian volcanoes.

headland An erosionally resistant point of land jutting out into the sea (or a large lake) and often standing high.

heat-island effect Increased temperatures in urban areas due to buildings and paved areas absorbing more solar heat, while exhaust from cars and factories traps heat.

high pressure system An area characterized by descending cooler dry air and clear skies. Winds rotate clockwise around a high pressure system (in the northern hemisphere). See also right-hand rule.

hook echo A curved, hook-shaped pattern on weather service radar that is indicative of a supercell thunderstorm. The hook shape often indicates development of a tornado.

hotspot volcano An isolated volcano, typically not on a lithospheric plate boundary, but lying above a plume or hot column of rock in Earth's mantle.

hurricane A large tropical cyclone in the North Atlantic or east Pacific ocean, with winds greater than 118 kilometers per hour. Similar storms elsewhere are called typhoons or cyclones.

hurricane warning A hurricane watch is upgraded to a warning when dangerous conditions of a hurricane are likely to strike a particular area within twenty-four hours.

hurricane watch An indication that a hurricane may affect a region within thirty-six hours.

hydrograph A graph that shows changes in discharge or river stage with time.

hydrologic cycle The gradual circulation of water from ocean evaporation to the continents, where it falls to the ground as rain or snow and soaks into the ground to feed vegetation, groundwater, and streams. From there, some water evaporates and some returns to the oceans.

hydrophobic soil Soil sealed by hydrocarbon resins. These soils will not permit water to soak in.

hypothermia When a person's core body temperature drops below 35°C/95°F. Symptoms include uncontrollable shivering, drowsiness, disorientation, slurred speech, and exhaustion.

hypothesis A proposal to explain a set of data or information, which may be confirmed or disproved by further study.

ice age A period of low temperature lasting thousands of years and marked by widespread ice sheets covering much of the northern hemisphere. Two to four ice ages are recorded in North America, and as many as twenty in deep sea sediments of the last 2 million years.

ice storm Freezing rain that turns to ice that coats everything.

igneous rocks Rocks that crystallize from molten magma, either within the Earth as plutonic rocks (e.g., granite and gabbro) or at Earth's surface as volcanic rocks (e.g., rhyolite, andesite, and basalt).

indefensible location A home or building location that is especially vulnerable to wildfire. Particularly important are dry vegetation and other fuels near the building and a location high on a hill, one that fire can easily reach upslope.

insurance A service by which people pay a premium, usually annually, to protect themselves from a major financial loss they cannot afford. The insurance company that collects the premiums pays for covered losses.

iridium anomaly A thin clay-bearing layer of sediment 65 million years old that has an abnormally high content of the platinum-like metal iridium.

iron meteorite A meteorite consisting of a nickel and iron alloy.

isostacy Lower-density crust floats in Earth's higher density mantle. Also called isostatic equilibrium.

jet stream The high speed air current traveling from west to east across North America at an approximate altitude of 10 to 12 kilometers.

jetties Walls of boulders, concrete, or wood built perpendicular to the coast at the edges of a harbor, estuary, or river mouth. The intent is to prevent sediment from blocking a shipping channel.

kaolinite A non-expandable aluminum-rich clay mineral.

karst The ragged top of limestone exposed at the surface, resulting from dissolution by acidic rainfall and ground water. The ground is often marked by sinkholes and caverns.

K-T boundary The boundary between the Cretaceous and Tertiary geological time periods, from approximately 65 million years ago. The boundary is marked by a thin layer of sooty clay that contains the iridium anomaly.

Kyoto Protocol The 1997 agreement between many countries to reduce their emissions of greenhouse gases into the atmosphere.

ladder fuels Fuels of different heights that permit fire to climb progressively from burning ground material to brush, small trees, low branches, and finally to the tops of tall trees.

lahar A volcanic mudflow.

land subsidence Settling of the ground in response to extraction of water or oil in subsurface soil and sediments, drying of peat, or formation of sinkholes.

land-use planning Restriction of development according to practical and ethical considerations, including the risk of natural hazards.

La Niña The opposite of El Niño.

lake-effect snow Cold arctic air moving across a warm lake picks up moisture evaporating from the lake, rises, cools, and condenses to form tiny ice crystals that fall as snow.

latent heat The heat added to a gram of a solid to cause melting or to a liquid to cause evaporation. Also the heat produced by the opposite changes of state.

lava Magma that flows out onto the ground surface.

lava dome A large bulge or protrusion, either extremely viscous or solid on the outside, and pushed up by magma.

lava flow Molten rock flowing out of a volcano.

levee An artificial embankment of loose material placed at the edge of a channel to prevent floodwater from spreading over a floodplain.

lightning The visible electrical discharge that marks the joining of positive and negative electric charges between clouds or between a cloud and the ground.

liquefaction A process in which water-saturated sands jostled by an earthquake rearrange themselves into a closer packing arrangement. The expelled water spouts to form a sand boil. Liquefaction turns the soil to "liquid" as water surrounds the suspended grains as they settle.

lithosphere Rigid outer rind of Earth approximately 60 to 100 or so kilometers thick; it forms the lithospheric plates.

load (1) Related to landslides, the weight of material on a slope. (2) Related to flood processes, the volume of sediment a stream can carry.

longshore drift The gradual migration of sand or gravel along the shoreline resulting from waves repeatedly carrying the sediment grains obliquely up onto shore and then straight back to the water's edge.

losing stream A stream, typically in a dry climate, that lies above the water table and loses water to an aquifer.

low pressure system An area of low atmospheric pressure that is characterized by rising warmer and humid air and cloudy skies. Winds rotate counterclockwise around a low pressure cell (in the northern hemisphere). *See* also right-hand rule.

magma Molten rock. When it flows out on the ground surface, it is called *lava*.

magma chamber Large masses of molten magma that rise through Earth's crust, often erupting at the surface to build a volcano.

magnetic field The area around a magnet in which magnetism is felt.

magnitude The relative size of an earthquake, recorded as the amplitude of shaking on a seismograph. The range of amplitudes is so large that the number (open-ended but always less than 10) is recorded as a logarithm of the amplitude. Different magnitudes are based on amplitudes of different seismic waves: M_L = Richter or local magnitude; M_s = surface wave magnitude; Mb = body wave (P or S wave) magnitude; M_w = moment magnitude (most important for extremely large earthquakes).

mammatus clouds Downward rounded pouches protruding from the base of the overhanging anvil of a thundercloud.

mantle The thick layer of material below the thin crust and above Earth's core. Mostly peridotite in composition in its upper part. Its density approximates 3.2 or 3.3 g/cm^3 in the upper part and 4.5 g/cm^3 in the lower part.

meandering stream Streams that sweep from side to side in wide turns called *meanders*.

melting temperature The temperature at which a rock melts. Granite commonly melts between 700° and 900° Celsius. Basalt commonly melts near 1200 to 1400° Celsius.

meteor A piece of space rock that heats to a white hot incandescence when it streaks through Earth's atmosphere.

meteorite An asteroid that passes through the atmosphere to reach Earth.

methane hydrate Frozen methane-ice "compound", trapped in layers deeper than 1 km in continental permafrost and at shallow depths under the sea floor of many of the world's continental slopes. A potential source of greenhouse gas.

mid-oceanic ridge A high-standing rift or spreading zone in an ocean—for example, the mid-Atlantic Ridge or the East Pacific Rise.

migrating earthquakes Earthquakes that occur in a sequential manner along a fault over time.

mitigation Changes in an environment to minimize loss from a disaster.

Modified Mercalli Intensity Scale Scale measuring the severity of an earthquake in terms of the damage that it inflicts on structures and people. It is normally written as a Roman numeral on a scale of I to XII.

moment magnitude (M_w) The magnitude of an earthquake based on its seismic moment; depends on the rock strength, area of rock broken, and amount of offset across the fault.

mudflow A flow of mud, rocks, and water dominated by clay or mud-sized particles.

National Flood Insurance Program The flood insurance program guaranteed and partly funded by the U.S. government.

natural disaster A natural event that causes significant damage to life or property.

natural hazards Hazards in nature that endanger our property and physical well-being.

natural levee Natural embankment of sediment at the edge of a stream, where sediment is deposited as flood waters slow and spread over an adjacent floodplain.

nature's rampages The perception of many people that a natural catastrophe involves nature going on an abnormal rampage. Actually, nature is not doing anything that it has not done for millions of years; we humans have merely put ourselves in harm's way.

near-Earth object A space object such as an asteroid or comet that narrowly misses the Earth.

Nor'easter A strong extratropical winter storm that moves up the east coast of North America with high winds and high waves. These can be as damaging as hurricanes.

normal fault A fault (generally, steeply inclined) that has the upper block of rock moving down compared with the lower block.

North Atlantic Oscillation (NAO) The winter atmospheric pressure pattern over the North Atlantic Ocean that brings major storms every few years.

offset The distance of movement across a fault during an earthquake.

orographic effect The effect created when moisture-bearing winds rise against a mountain range: They condense and form clouds and rain.

pahoehoe Basalt lava with a ropy or smooth top.

Peléan eruption Huge eruptions characterized by giant ash columns that collapse to form incandescent pyroclastic flows.

paleoflood analysis Information on previous floods gathered from erosional and depositional features left by such a flood.

paleoseismology The study of former earthquakes from examination of offset rock layers below the ground surface.

paleovolcanology The study of former volcanic events from examination of offset rock layers below the ground surface.

Pangaea Supercontinent that began to break up to form today's continents 225 million years ago.

period The time between seismic waves or the time between the peaks recorded on a seismograph.

permafrost The condition where water in the ground remains frozen.

pillow basalt Basalt that flows into water or wet mud chills on its outside surfaces to form elongate fingers of basalt a meter or so across.

piping Water seeping through a dam or levee can begin to carry sand and cause erosion.

plate tectonics The theory that lithospheric plates that move relative to one another collide in some places, pull apart in others, and slide past one another in still others. These movements cause earthquakes and volcanic eruptions as well as build mountain ranges. The theory is supported by a wide range of data.

Plinian eruption Extremely large eruptions that involve continuous blasts of ash. Examples include Vesuvius in 79 A. D. and Mount St. Helens in 1980.

point bar Deposits of sand and gravel on the inside (concave side) of a meander bend of a stream.

precursor events Smaller events that precedes a major natural event, which may warn of the impending disaster.

prescribed burn Intentional setting of forest fires under controlled conditions to consume fuels and mitigate spread of future fires.

property rights The right to do what one wishes with one's personal property.

P wave The compressional seismic wave that shakes back and forth along the direction of wave travel.

pyroclastic flow A mixture of hot volcanic ash, coarser particles, and steam that pours at high velocity down the flank of a volcano. Also called nuée ardente or ash flow.

pyroclastic material Fragmental material blown out of a volcano—for example, ash, cinders, and bombs.

quick clay Water-saturated mud deposited in salty water tends to consist of randomly oriented flakes of clay with large open spaces between the flakes. If the salt is flushed out, the flakes are unstable and may easily collapse and flow almost like water.

recurrence interval The average number of years between an event of a certain size in a location—for example, floods, storm surges, and earthquakes. Also known as a return period.

relative humidity The percentage of moisture in air relative to the maximum amount it can hold (at saturation) under its given temperature and pressure.

retrofitting Modifying existing buildings to minimize the damage during strong earthquake motion.

reverse condemnation If the government restricts a property owner from using his or her property for some potential purpose that has effectively taken the value of the property for that purpose.

reverse fault Faults (generally, steeply inclined) that have the upper block of rock moving up compared with the lower block.

Richter Magnitude Scale The scale of earthquake magnitude invented by Charles Richter. *See also* magnitude.

rift A spreading zone on the flank of a volcano from which lavas erupt. *See also* rift zone.

rift zone An elongate spreading zone in Earth's lithosphere.

right-hand rule The rule used to remember the direction of wind rotation in a low or high pressure cell in the northern hemisphere. With your right thumb pointing in the direction of overall air movement, your fingers point in the direction of wind rotation.

rip current A short-lived surface current that flows directly off a beach and through the breaker zone. It carries water back to the sea that had piled up onto the beach.

riprap Coarse rock piled at the shoreline in an attempt to prevent wave erosion of the shore or destruction of a near-shore structure or a streambank.

risk The chance of an event multiplied by the cost of loss from such an event.

risk map A map showing the distribution and level of risk.

runout The distance of travel of a slide from its source.

run-up The height to which water at the leading edge of a wave rushes up onto shore. Also used for the height to which a tsunami wave rushes up onshore.

Saffir-Simpson Hurricane Scale The commonly used five-category scale of hurricane intensity based on wind speeds and seawater damage.

sand boil A pile of sand brought to the surface in water expelled from the ground by liquefaction at shallow depth or during flood when the water pressure under a channel forces groundwater to the surface outside a levee.

sand dune An accumulation of wind-blown sand, most commonly along the upper beach above high-tide level.

Santa Ana winds Southern California trade winds that flow southwest, bringing dry air from the continental interior.

scientific method The method used by scientists to solve problems. They analyze facts and observations, formulate hypotheses, and test the validity of the hypotheses with experiments and additional observations.

seafloor spreading Oceanic crust that spreads apart as lithospheric plates separate.

seawall Walls of boulders, concrete, or wood built along a beach front and intended to protect shore structures from wave erosion.

seismic gap A section of an active fault that has not had a recent earthquake. Earthquakes elsewhere on the fault suggest that the gap may have an earthquake in future.

seismic wave A wave sent outward through Earth in response to sudden movement on a fault. *See also* P wave, S wave, surface wave.

seismogram The record of seismic waves from an earthquake or other ground motion as recorded on a seismograph.

seismograph The instrument used to detect and record seismic waves.

sequestration The physical removal and long-term storage of CO_2.

ShakeMap Computer-generated maps of ground motion which show the distribution of maximum acceleration and maximum ground velocity during earthquakes.

shatter cone Cone-shaped features, with rough striations radiating downward and outward from the shock effect, considered diagnostic of bolide impact.

shore profile The slope of the beach.

shotcrete A fluid cement-type material that is sprayed on a slope to prevent water penetration. Also called *gunite*.

sinkhole A ground depression caused by collapse into an underground cavern.

slope angle The angle of a slope as measured down from the horizontal.

smectite Clay in which flakes have an open structure between their layers, which when filled with water cause the clay to dramatically expand.

snow avalanche Downslope movement of snow.

"soft" solutions Solutions to a problem that involve avoiding the problem through restrictive zoning and building codes that minimize damage.

soil creep Slow downslope movement of near-surface soil or rock; caused by numerous cycles of heating and cooling, freezing and thawing, burrowing animals, and trampling feet.

soluble Able to be dissolved, typically in water.

spot fire Fires ignited by firebrands which burn ahead of the main fire.

step leader Electrical charges that advance downward from a thundercloud but do not manage to reach the ground.

stony-iron meteorite Essentially a chondrite that contains some nickel-iron.

storm surge The rapid sea-level rise caused by both low atmospheric pressure of a major storm and the strong winds that accompany the storm and push water forward.

strain Change in size or shape of a body in response to an imposed stress.

streambed mining Excavation of sand and gravel from a stream bed.

stream order A way of numbering streams in a hierarchy that designates a small unbranched stream in a head-waters as first order, the stream it flows into as second order, and so on.

stress The forces on a body. These can be compressional, extensional, or shear.

strewn field Area in which fragments of large meteorites are spread out around the main impact site.

strike-slip fault A fault (generally vertical) which has relative lateral movement of the two sides.

subduction zone Convergent boundary along which lithospheric plates come together and one descends beneath the other; often ocean floor descending beneath continent.

submarine canyon A deep canyon in the continental shelf and slope offshore and extending about perpendicular to the shore.

submarine landslide Subsea level collapse of the flank of an oceanic volcano such as in Hawaii or the Canary Islands.

superoutbreak A large group of tornadoes produced along a major storm front.

surface fires Fires that spread along the ground fueled by grasses and low shrubs.

Surface runoff Water that runs off the surface of the ground.

surface rupture length The length of a fault broken during an earthquake.

surface tension The effect by which grains of sand are held together by the thin films of water between them.

surface wave The seismic wave that travels along and near Earth's surface. These waves include Rayleigh waves (which move in a vertical, elliptical motion) and Love waves (which move with horizontal perpendicular to the direction of wave travel).

surge. *See* base surge, storm surge.

S wave The seismic shear wave that shakes back and forth perpendicular to the direction of wave travel. S waves do not pass through liquids.

swelling soil A soil that expands when wet; generally, a soil that contains the swelling clay smectite.

talus Coarse, angular rock fragments that fall from a cliff to form a cone-shaped pile banked up against the slope.

theory A scientific explanation for a broad range of facts that have been confirmed through extensive tests and observations.

thrust fault Fault (generally, gently inclined) that has the upper block of rock moving up compared with the lower block.

thunderstorm A storm accompanying clouds that generate lightning, thunder, rain, and sometimes hail.

toe The outer, downslope end of a landslide.

tornado A near-vertical narrow funnel (generally only a kilometer or so in diameter) of violently spinning wind associated with a strong thunderstorm.

Tornado Alley The region of the central United States, between Texas and Kansas, that is noted for frequent tornadoes.

tornado outbreak A series of tornadoes spawned by a group of storms.

tornado warning A tornado warning is issued when Doppler radar shows strong indication of vorticity or rotation, or if a tornado is sighted.

tornado watch A tornado watch is issued when thunderstorms appear capable of producing tornadoes and telltale signs show up on the radar. At this point, storm spotters often watch for severe storms.

trade winds Regional winds that blow from northeast to southwest between latitudes 30 degrees north (or south) and the equator and centered near 15 degrees north (or south).

transform boundary A boundary marked by a transform fault.

translational slide A landslide that moves approximately parallel to the slope of the ground.

trench An elongate depression in the ocean floor at a subduction zone between two tectonic plates and most commonly at the edge of an active continental margin. Most are at the margins of the Pacific Ocean.

trimline The line along a mountainside along which tall trees in the forest upslope are bounded by distinctly shorter trees downslope.

tropical cyclone A large rotating low pressure cell that originates over warm tropical ocean water. Depending on location, they are called hurricanes, typhoons, or cyclones.

tsunami An abnormally long wavelength wave most commonly produced by sudden displacement of water in response to sudden fault movement on the seafloor. Can also form when a landslide, volcanic eruption, or asteroid impact displaces water.

tsunami warning When a significant tsunami is identified, officials order evacuation of endangered low-lying coastal areas.

tsunami watch The alert is issued when a magnitude 7 or larger earthquake is detected somewhere around the Pacific Ocean or some other ocean that may see dangerous tsunamis.

tuff A rock formed by consolidation of volcanic ash.

typhoon A large low-pressure weather system that circulates counterclockwise in the northern hemisphere and clockwise in the southern hemisphere. It is equivalent to a hurricane or a cyclone.

vog An acidic volcanic smog produced when volcanic gases react with moisture and oxygen in the air, and sun, to produce aerosols.

volcanic ash Any small volcanic particles; formally, particles less than 2 millimeters across.

volcanic weather Weather generated during a volcanic eruption. The high temperature above an erupting volcano draws in outside air that rises and cools. Moisture in the air condenses to form rain and stormy weather.

wall cloud The rotating area of cloud that sags below the main thundercloud base and from which a tornado may develop.

warm front The boundary between a large mass of cold air advancing under an adjacent large mass of warm air.

watershed That part of a landscape that drains water down to a given point on a stream.

wave energy The capacity of a wave to do work—that is, to erode the shore. Because wave energy is proportional to the square of wave height, high waves do almost all coastal erosion.

wave height The vertical distance between a wave crest and an adjacent trough.

wavelength The distance from crest to crest of a wave—for example, in an earthquake wave or a water wave.

wave refraction The bending of wave crests where one end of the wave drags bottom and slows down in shallower water.

weather The conditions—temperature, humidity, air motion, and pressure—of the atmosphere at a particular place and time.

weather fronts *See* cold front, warm front.

westerly winds Regional winds that blow from southwest to northeast and are centered 45 degrees north (or south) of the equator.

wind chill When wind evaporates skin moisture, causing heat loss.

wind shear The short-distance change in wind direction or velocity.

wing dam Segments of walls built to protrude into the current of a river to increase the depth of water in the channel to facilitate shipping.

zoning restrictions Laws that prohibit building in certain areas.

Index

Bold page numbers indicate headings and other major entries. **Bold Italic** page entries indicate Case in Point features.

INDEX   550

H

Haicheng earthquake, **68–69**
hail, **471–472**
   damages to cars, 471
   in a tornado, 483
   in thunderstorms, 468, 471
Haiti
   and hurricanes, 431, 461
   and Hurricane Gustav, 448
   and Hurricane Ike, 449
   earthquake deaths, 83
   earthquake, 2010, 3, 34–36
Halema`uma`u Crater on
      Kilauea, 149
harbor wave (tsunami), 106
hard solutions, *530*
hardening beaches, *409*
harmonic tremors, 176
   on volcano, 155
   under a volcano, 186
Hatteras Island and hurricane,
      426
Hawaiian Islands, **26**
   cinder cones, 150
   flank collapse, 109–110
   rift zone, 138
   tsunami risk, **120**
   tsunami, 109, 128
   volcanoes, 20
Hawaiian-Emperor chain, **26**
Hawaiian-type lava, **141**
Hayward Fault, California, 40, 80
Hayward-Rodgers Creek Fault, 80
hazard(s), 1
   assessment and mitigation,
      530
   associated with impact, 520
   forecasts, Mt. St. Helens, 182
   from landslides, **218–221**
   from Nor'easters, 440
   from thunderstorms, 468
   from volcanoes, 141, 163
   related to weather, **274–285**
   maps for landslides, 223
   near plate boundaries, **18**
   protection from, 10
   recognizing, 10
HAZUS, 10
headlands
   and wave energy, 389
   erode to beaches, 399
   erosion, 389
headscarp of rotational
      landslide, 211–212
health problems from volcanic
      ash, 168
heat
   buildup and storms, 268
   loss and cold, 282
   of condensation, 262
   trapped in atmosphere, 285
   waves, **281**
heating air in afternoon, 468
heat-island effect, **281**
   and tornadoes, 481
heavy objects moved by
      tornadoes, 481, 482

heavy rainfall floods, central
      Texas, 341–342
heavy rains during Hurricane
      Mitch, 458
Hebgen Lake, Montana
      earthquake, 1959, 37
height of
   barrier islands, 395
   of tsunami near shore, 111
   of tsunami wave, 107
Herculaneum
   and Mt. Vesuvius, 191–192
   destruction by Vesuvius, 178
hertz (Hz) , 50, 89
High Cascades, 22, 151, **180–185**
High Plains
   and desertification, 280
   clay deformation, **255–256**
high waves
   and beach slope, 427
   and hurricanes, 427
high winds in hurricanes, 415
higher-latitude storms, 266
high-ground refuges from
      floods, 311
high-pressure system, **264**
high-rise buildings in
      tornadoes, 480
high-risk landslide areas, 223
Hilo Bay
   1946 tsunami, 114
   1960 Chile tsunami, 115
   1975 tsunami, 121
   focuses tsunami waves,
      128–129
   tsunami hazard, 112
Himalayas, 24
   glaciers and water, 297
   and earthquakes, 100
   China earthquake, 70, 98
hoar frost, 215–216
homeowner insurance,
   floods, **365–366**
   hurricanes, **438–439**
homes, *see also house*
   and coastal storms, 427
   and stream banks, 362
   built on California beach, 400
   damage in tornadoes, 480
   flood on alluvial fan, 379
   in fire-prone areas, 497
   in forest, 497
   on high posts, 415
   on pilings, Malibu, CA, 402
   on smectite clays, 251
   raised on posts, 427
   topple in Hurricane Ike, 451
   protection from fire, 497–499
   purchase, flood insurance,
      365–366
Honduras, Hurricane Mitch,
      457–458
hook echo on radar, 477
Hope landslide, British
      Columbia, 209
hospital deaths, Hurricane
      Katrina, 445

hot ash on volcano melted by
      eruption, 169
hotels and tornadoes, 480
hotspot
   track, 26–27
   map of, 17
   volcanoes, **26–27**, **139**
house of cards clays and
      collapse, 203, 246
house, *see also homes*
   earthquake-proofing, 90
   built on debris flows, *343–344*
   deformed by swelling clay, 255
   floated, Hurricane Katrina,
      450, 444
   on alluvial fans, *343–344*
   on cliff tops, 400
   preparation for
      hurricanes, 434
   shakes off foundation in
      earthquake, 83, 86
   structure in hurricane, 424
   toppled in hurricane 415
Houston, Texas
   and Hurricane Rita, 449
   area subsidence, 242, 244
   damage from Allison, 430
   deaths, Hurricane Ike, 451
Huang Ho, China, avulsion,
      371–372
Huckleberry Ridge tuff,
      Yellowstone, 158
human impact on rivers, 352
human-caused warming, 292
hummocky ground,
      landslides, 206
hummocky landslides, 222
hurricane, *see also hurricanes*
   Agnes, flooding, 422, 430
   Andrew, 418–419, 420, 429
   area building codes, **436–437**
   arrival times, 434
   barometric pressure, 418
   beach erosion
   Bertha, Carolinas, 396
   Betsy, 422
   Bonnie, Carolinas, 396
   building damage, 430
   Camille, 418–419, 420, 425
   categories, 418
   Charley, 2004, Florida, 422,
      459–461
   classification, **417**
   costs, 421
   cross-section, 417
   damage, 417
   damages, 416, 430
   deaths, 421, **431**
   definition, 416
   evacuations, 434, 455, 460
   eye, 416–417
   flooding, 439
   Floyd, 399, 430
   formation, **416**
   Fran, 396, 418
   Frances, 2004, 419, 422,
      459–461

frequency with AMO, 273
Galveston damages, 428
generated tornadoes, 435
Gilbert, 419
Gustav, *448–449*
high tide, 425
highest winds in, 417
Hugo, 418, 420
hunter aircraft, 433
Ike, 394, 415, 418, 420, 425,
      449–451
intensity, 417
Isabel, 396, 418–419, 423,
      426, 454
Ivan, 345, 422, 432, *459–461*
Jeanne, 2004, 422, *459–461*
landfall location, deaths, 434
landfall time, 434
map view, 423
Mitch, 418, 431, *458–459*
numbers and climate
      change, 432
numbers per year, 419
path, 422, 453
planning, **433–436**
predictions, 433
preparation after "Watch," 434
rain, thunderstorms, 416
rains and floods, **430**
risk areas, **417, 419–421**
Rita, 422, 432, 449–451
rotation, 417, 442
sea temperature, 420
season, 420
shelters and deaths, 442
source areas, 417
strength and damages, 430
strengthens over warm water,
      442
surge, 418, 442
Texas Gulf Coast, 437
tornadoes, 416
tracks, worldwide, 419
traps for tornadoes, 480
uncertainty, **433–436**
updates, emergency radio, 435
warnings, **433**
watches, **433**
Wilma, 422, 432–433 437
winds, 416, 418, 427
Hurricane Katrina, 418, 420,
      439–447
   and insurance, 432
   and society, 432
   and warm water, 420
   barrier islands, 426
   costliest disaster, 453–447
hurricane-driven floods,
      344–345
hurricane-prone coasts, 434
hurricanes, *see also hurricane*
   affect coastal insurance, 438
   before-after effects, 427
   building weaknesses, 429
   Carolina Outer Banks, 454
   Central America, 431
   change the landscape, 419

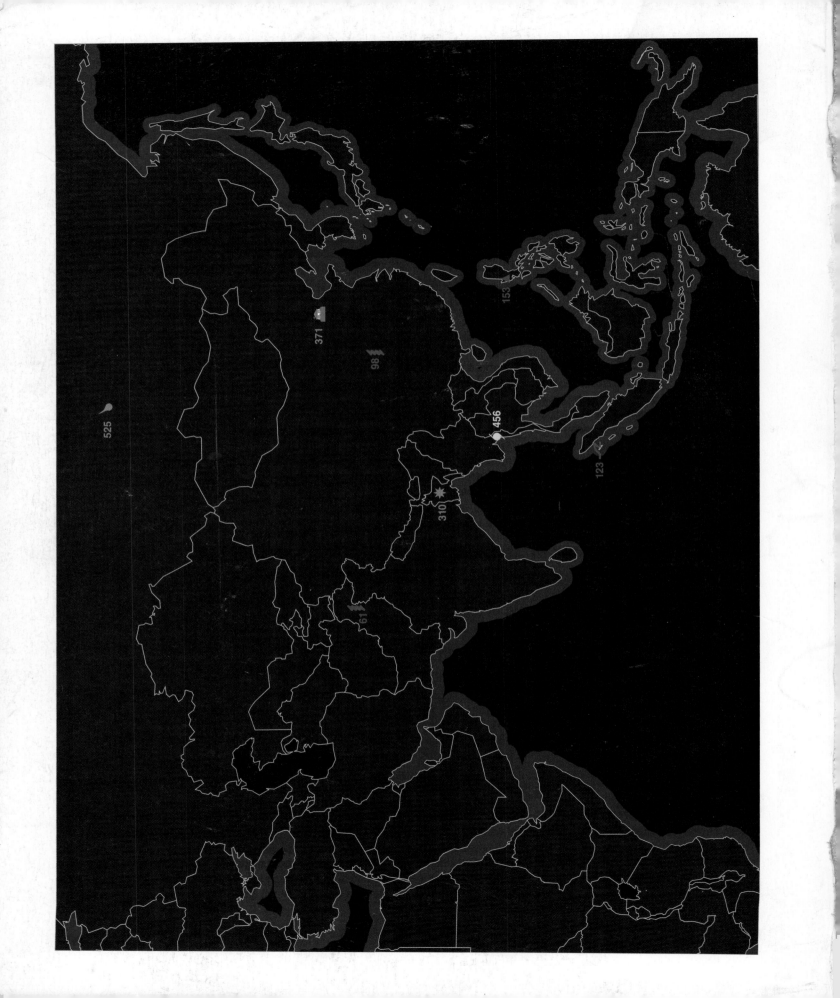